ADAPTIVE
MOBILE ROBOTICS

ADAPTIVE MOBILE ROBOTICS

Proceedings of the 15th International Conference on Climbing and Walking Robots and the Support Technologies for Mobile Machines, Baltimore, USA

23–26 July 2012

Editors

A. K. M. Azad
Northern Illinois University, USA

N. J. Cowan
Johns Hopkins University, USA

M. O. Tokhi
University of Sheffield, UK

G. S. Virk
University of Gävle, Sweden

R. D. Eastman
Loyola University Maryland, USA

World Scientific

NEW JERSEY • LONDON • SINGAPORE • BEIJING • SHANGHAI • HONG KONG • TAIPEI • CHENNAI

Published by

World Scientific Publishing Co. Pte. Ltd.
5 Toh Tuck Link, Singapore 596224
USA office: 27 Warren Street, Suite 401-402, Hackensack, NJ 07601
UK office: 57 Shelton Street, Covent Garden, London WC2H 9HE

British Library Cataloguing-in-Publication Data
A catalogue record for this book is available from the British Library.

ADAPTIVE MOBILE ROBOTICS
Proceedings of the 15th International Conference on Climbing and Walking Robots and the Support Technologies for Mobile Machines

ISBN 978-981-4415-94-1

Printed in Singapore by World Scientific Printers.

PREFACE

Robotics in one form or another has excited mankind for centuries. In the 20th century, a long list of great science fiction writers defined a grand vision of robotics, and filmmakers have brought this vision to the screen through amazing special effects. However, to date, this vision has largely escaped our technological and algorithmic abilities to fulfil except through cinematic deceits.

Recently, IBM's robotic game show contestant, Watson, demonstrated just how far artificial intelligence has progressed. This machine defeated two of the greatest Jeopardy champions of all time, moving us quickly to an era where machines can process natural language and produce sensible responses. And, like IBM's *Deep Blue* before it, the famed chess computer that defeated Kasparov, Watson pushed many to speculate anew on what lies before us. But natural language (in the case of Watson) and combinatorics (in the case of Deep Blue) are not likely the great bottlenecks of robotics. These machines certainly encapsulate important advances in machine learning and other fields that enhance robotic systems, but a common housefly can still out manoeuvre the very best robots built to date.

So, what is so different about the abstract domains of game shows and chess, as compared to the physical domains of locomotion and manipulation, and how is it that machines can compete with the human mind in the former, but insects outwit the best machines in the latter? This question remains open, but it is evident that biological mechanisms have adapted themselves remarkably well to solve the mobility problems they confront, and roboticists and engineers can take inspiration from such adaptive biological features to find new approaches of artificial creations to meet similar physical challenges.

CLAWAR 2012 is the fifteenth in a series of international conferences on Climbing and Walking Robots and the Support Technologies for Mobile Machines. The conference has brought together leading scientists and engineers focused on solving the problems of mobility in the natural world. The state-of-the-art, captured in the pages of this book in 110 technical articles by authors from 26 countries from throughout the five continents, suggests the "Watson of locomotion" may soon be within our grasp.

The editors would like to thank members of the International Program Committee, International Advisory Committee and National Organising Committee for their efforts in reviewing the submitted articles, and the authors in addressing the comments and suggestions of the reviewers in their final submissions. It is intended that the CLAWAR 2012 proceedings will be a valuable source of reference for research and development in mobile robotics.

A. K. M. Azad
Northern Illinois University, USA

N. J. Cowan
Johns Hopkins University, USA

M. O. Tokhi
Univeristy of Sheffield, UK

G. S. Virk
University of Gävle, Sweden

R. D. Eastman
Loyola University, USA

CONFERENCE ORGANISERS

Northern Illinois University
DeKalb, USA

Johns Hopkins University
Baltimore, USA

CLAWAR Association

A charitable non-profit making membership based professional organization serving the robotics community
www.clawar.org

Loyola University
Maryland, USA

University of Maryland
Maryland, USA

National Institute of Standards and Technology
Gaithersburg, USA

CONFERENCE COMMITTEES

General Chairs and Programme Chairs

of Fifteenth International Conference on Climbing and Walking Robots and the Support Technologies for Mobile Machines

A. K. M. Azad (General Co-Chair)	– Northern Illinois University, USA
N. J. Cowan (General Co-Chair)	– Johns Hopkins University, USA
M. O. Tokhi (Program Chair)	– University of Sheffield, UK
G. S. Virk (IAC Chair)	– University of Gävle, Sweden

International Advisory Committee (IAC)

of Fifteenth International Conference on Climbing and Walking Robots and the Support Technologies for Mobile Machines

H. Levent Akin	– Turkey
Manuel Armada	– Spain
Karsten Berns	– Germany
Philippe Bidaud	– France
Bryan Bridge	– UK
Anibal T de Almeida	– Portugal
Steven Dubowsky	– USA
H. Fujimoto	– Japan
Daniel Goldman	– USA
Krzysztof Kozlowski	– Poland
Giovanni Muscato	– Italy
Roger Quinn	– USA
Lakmal Seneviratne	– UK
Kenneth J Waldron	– USA
Ming Xie	– Singapore

International Programme Committee (IPC)

of fifteenth International Conference on Climbing and Walking Robots and the Support Technologies for Mobile Machines

Ahmadabadi M. N.	– Iran
Alam M. S.	– Bangladesh
Aldbrez F. M.	– UK
Arkin R.	– USA
Amar F. B.	– France
Bhattacharya B.	– India
Billingsley J.	– Australia
Bonsignorio F.	– Italy
Bouazza-Marouf K.	– UK
Cameron S.	– UK
Cheng H.	– China
Chevallerau C.	– France
Chew C-M.	– Singapore
Chugo D.	– Japan
Cruse H.	– Germany
Cubero S.	– Australia
Dai J.	– UK
Dehghani A.	– UK
de Santos P. G.	– Spain
Dillmann R.	– Germany
Dodd T. J.	– UK
Dunbabin M.	– Australia
Dutra M. S.	– Brazil
Dutta A.	– India
Fontaine J-G.	– Italy
Fu Y.	– China
Garcia E.	– Spain
Ge S. S.	– Singapore
Goher K. M.	– Oman
Gradetsky V.	– Russia
Grand C.	– France
Gross R.	– UK
Halme A.	– Finland
Heyes N.	– UK
Hossain M. A.	– UK
Howard D.	– UK
Huang L.	– NZ
Ion I.	– Romania
Jatsun S. F.	– Russia
Kadar E. E.	– UK
Kiriazov P.	– Bulgaria
Kopacek P.	– Austria
Kosuge K.	– Japan
Lewis M. A.	– USA
Lopez-Coronado J.	– Spain
Madhavan R.	– USA
Marques L.	– Portugal
Massoud R.	– Syria
Mochiyama H.	– Japan
Mohamed Z.	– Malaysia
Molfino R.	– Italy
Moon S.	– Korea
Nakamura T.	– Japan
Nefti-Meziani S.	– UK
Pratihar D. K.	– India
Prescot T. J.	– UK
Rachkov M.	– Russia
Rodriguez H. L.	– Columbia
Sano A.	– Japan
Santos V.	– Portugal
Sattar T. P.	– UK
Schilling K.	– Germany
Shaheed M. H.	– UK
Shamsudin H. M. A.	– Malaysia
Sharma S. K.	– UK
Siddique N. H.	– UK
Silva F.	– Portugal
Skibniewski M.	– USA
Steinicke L.	– Belgium
Tenreiro Machado J.	– Portugal
Vitko A.	– Slovakia
Walker R.	– UK
Watanabe K.	– Japan
Wöern H.	– Germany

Yigit A.	– Kuwait	Zielinska T.	– Poland
Zhong Z. W.	– Singapore	Zomaya A.	– Australia
Zhou C. J.	– Singapore		

National Organising Committee

of Fifteenth International Conference on Climbing and Walking Robots and the Support Technologies for Mobile Machines

Roger Eastman (Chair)	– Loyola University
Yiannis Aloimonos	– University of Maryland
Roger V. Bostelman	– NIST
Ralph Etienne-Cummings	– Johns Hopkins University
Satyandra K. Gupta	– University of Maryland

CONFERENCE SPONSORS AND CO-SPONSORS

Association for the Advancement of Medical Instrumentation
Arlington, USA

Institution of Mechanical Engineers
London, UK

CONTENTS

Section–3: Autonomous Robots

Section–4: Biologically-Inspired Systems and Solutions

Section–5: Innovative Design of CLAWAR

Section–6: Locomotion

Section–7: Miscellaneous Applications

Section–8: Modelling and Simulation of CLAWAR

Section–9: Perception and Sensor Fusion

Section–10: Planning and Control

Section–11: Service Robots

Section–12: Service Robot Standards and Standardization

SECTION–1

PLENARY PRESENTATIONS

ROBUSTNESS IN ANIMALS AS INSPIRATION FOR THE NEXT GENERATION ROBOT

ROBERT J. FULL
University of California, Berkeley, USA

Robustness is one feature that sets organisms apart from engineered devices. Robustness has been defined, in part, as persistence - the ability to withstand perturbations in structure without change in function - and often includes concepts such as modularity, redundancy, self-repair, learning and adaptation. The use of exoskeletons in polypedal locomotion by arthropods represents an excellent system to examine robustness. Insects can still locomote with the loss of legs or damage to sensors. Cockroaches maintain their speed on hard surfaces and over rough terrain even after the loss of their feet. Cockroaches transition up a wall by colliding with it head-on at over one meter or 50 body lengths per second. These small animals rely on the robustness of their exoskeleton to simplify control. The rapid design of robust exoskeletons for small robots is now possible using a process called Smart Composite Microstructures. This approach enables the construction of small, strong, lightweight structures whose ability to move comes from bending of compliant polymer hinges that connect rigid links made from carbon fiber and other lightweight composites. These structures are made as single flat pieces that are folded to form more complicated shapes and linkages. This process has resulted in legged robots such as the 10 cm long, 16 g robot, DASH (Dynamic Autonomous Sprawled Hexapod Robot) that can sustain 8-story falls without damage and run away after a 10 meter per second impact. Moreover, rapid prototyping offers the possibility of designing legged robots as physical models to test biological hypotheses. A principled understanding of robustness remains a grand challenge for biology and a potential rich source of biological inspiration for engineering.

FEEDBACK CONTROL OF PLANAR BIPEDAL LOCOMOTION

JESSY W. GRIZZLE
University of Michigan, USA

Feedback control methods for achieving very robust dynamic balance in bipedal robots will be presented. The emphasis will be on MABEL, a planar (2D) robot with series compliant actuation in the ``knee" joints, that was designed by Jonathan Hurst. MABEL has 1 m long legs and a mass of 65 kg. The robot can handle unplanned, unseen, step-down disturbances of 20 cm, step-up disturbances of 12 cm, and can run a 9 minute mile. The talk will conclude with preliminary analysis of a spatial (3D) robot ATRIAS whose construction will be finished in late summer 2012 (also built by J. Hurst, Oregon State University). Videos of MABEL and ATRIAS are available on YouTube.

USE OF COMPLIANT ACTUATORS IN PROSTHETICS AND REHABILITATION ROBOTICS

DIRK LEFEBER
Vrije Universiteit Brussel, Belgium

New robotic technologies are unable to compensate for the many drawbacks that comes along with conventional electrical actuators, like high reflective inertia, high stiffness,low force-to-weight ratio. These new robotic applications can strongly benefit from compliant actuator technology. Instead of introducing compliance on the control level through application, this approach is based on inherent adaptable compliance on a purely mechanical level. In this way intrinsic compliance is assured at all time, enhancing system safety. This is easily countered with the range of potential benefits, adding value to existing applications and also creating new applications in robotics. The lecture will provide an overview of compliant actuator technologies developed by different research groups is given. Some applications in prosthetics and rehabilitation robotics are discussed in which these compliant actuators are incorporated.

TRANSITIONS IN THE LOCOMOTION OF ROBOTS AND THE ROBOTICS OF LOCOMOTION

DANIEL KODITSCHEK
University of Pennsylvania, USA

This talk will address in the context of locomotion the interlinked problems of building experimental robots and experiments with the robots that get built. Contemporary robotics seems to be reaching a point of transition in both regards. Persistent steady state locomotion can be formally specified in the language of dynamical systems theory and has been achieved experimentally by a growing variety of legged machines over a diversity of substrates in the presence of substantial disturbances. It seems time to aim for a theory and practice of transitional locomotion behavior. At the same time, commercial legged robots can now be purchased that that arguably exceed the persistent steady state locomotion capability of any university laboratory platform past, present, and, most likely, future. It seems time to aim for a theory and practice of multi-use locomotion platform design and implementation. These two sorts of transitions - in the class of machines academics build; and in the class of behaviors they achieve - intersect in the common problem of "task specification." This talk will motivate the intersecting need and explore the prospects for a transitional task specification formalism.

ALPHADOG, THE ROUGH-TERRAIN ROBOT

MARC RAIBERT
Boston Dynamics, USA

AlphaDog is a quadruped robot being developed with funding from DARPA's Legged Squad Support System program. The goal is to provide a machine that can carry 400 lbs of payload on 20-mile, 24-hour missions. It must have significant rough-terrain mobility, follow a human leader in natural outdoor terrain, be quiet (quieter than BigDog) and pick itself up whenever it falls. We began testing AlphaDog this past January 2012 and plan extensive field tests starting in the Fall. This talk will give a status report on AlphaDog and the other dynamic robots being developed at Boston Dynamics.

SECTION–2

ASSISITIVE ROBOTS

TRACKING CONTROL SYSTEM FOR WALKING ASSIST USING INTENTION ESTIMATION

K. MATSUO*, Y. OCHI, W. ARIBOWO, R. TASAKI, and K. TERASHIMA*

Department of Mechanical Engineering, Toyohashi University of Technology, 1-1, Hibarigaoka, Tempaku-cho, Toyohashi, Aichi, 441-8580, Japan
** E-mail: k_matuo@syscon.me.tut.ac.jp, terasima@me.tut.ac.jp*
www.tut.ac.jp

T. YAMAMOTO and T. SAKAKIBARA

Department of Rehabilitation, Sawarabi-kai Fukushimura Hospital 19-14, Yamanaka, Noyori-cho, Toyohashi, Aichi, 441-8124, Japan

A tracking control system of Novel Intelligent Lift-Type Walking-Assist Mobile Robot (NILTWAMOR) is presented, where an omni-directional moving method of lift-type walking assist system is newly developed. This paper mainly proposes a method to identify the current body center of the user for estimating user's intention of walking direction. The tracking control system for walking assist is applied to not only healthy people but also patients of gait rehabilitation in order to demonstrate the effectiveness of the proposed method.

Keywords: Gait training, Localization, Tracking control system

1. Introduction

Human abilities decay slowly due to the decline of physical function with aging, and gait disorders can be generated, which narrows people's activities and increases mental and physical decline[1] . Under these circumstances, the recoveries of walking capability have been widely studied[2] .

Gait training by body weight support using treadmill-type walking machine has been recently remarked. As a general trend, body weight supporting systems pull human body upward by appropriate lifting force according to the subject's handicap level, supporting the posture of the whole body. Such systems enable handicapped people without human power to sustain their body weight using their lower limbs to stably maintain the standing posture, and therefore early rehabilitation can be realized. In this way, the recovery of the residual function can be accelerated, and it has been

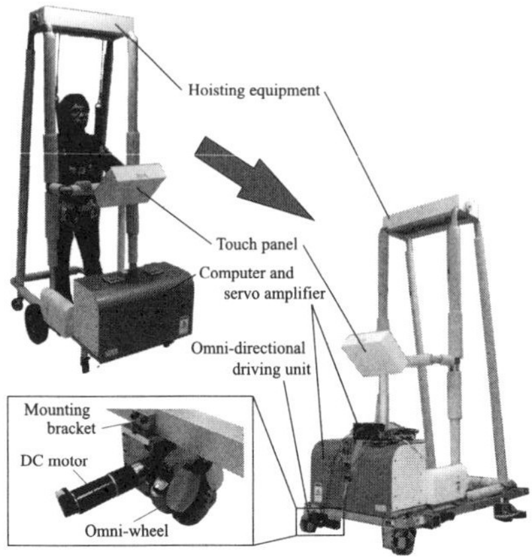

Fig. 1. Exterior photograph of NILTWAMOR

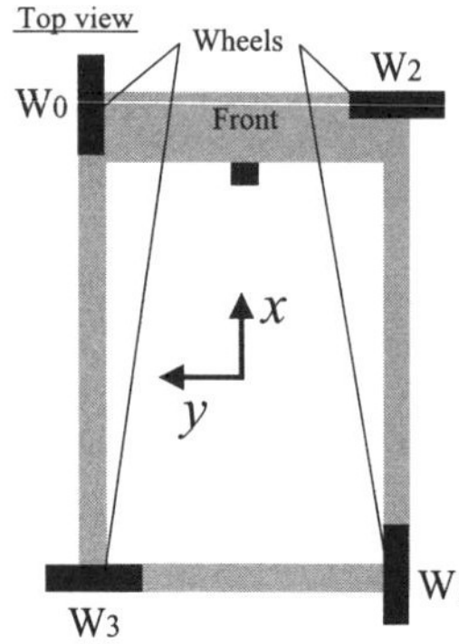

Fig. 2. Schematic illustration of omni-directional driving unit

reported that the possibility of restoring the residual function will be increased by early rehabilitation.

In recent years, we have developed a mobile walker robot, which is called NILTWAMOR, for early rehabilitation. Exterior photograph of NILTWAMOR is shown in the left part of Fig. 1. A body weight supporting system and a walking assist system such that NILTWAMOR tracks to each walking trajectory of the user have been reported[3,4] . However, these systems have been checked for only healthy people. So it has not been verified whether or not the system can be applied in a real rehabilitation training for disabled people. Furthermore, the moving direction is only straight one such as forward or backward direction. Therefore, there is few freedom to train walking, and then the straight line path restriction makes the patients easily bored and uncomfortable.

Therefore, in this paper, we newly propose a walking-training tracking control system with omni-directional movement which can be applied to not only healthy people but also patients of gait rehabilitation. The effectiveness of the proposed system is confirmed by experiments with disabled subjects.

2. Extension of NILTWAMOR to omni-directional walking-training

We have newly constbuilt omni-directional driving unit for omni-directional walking training. Schematic illustration of the omni-directional driving unit is shown in Fig. 2. Motors and Omni-wheels are assembled on four corners

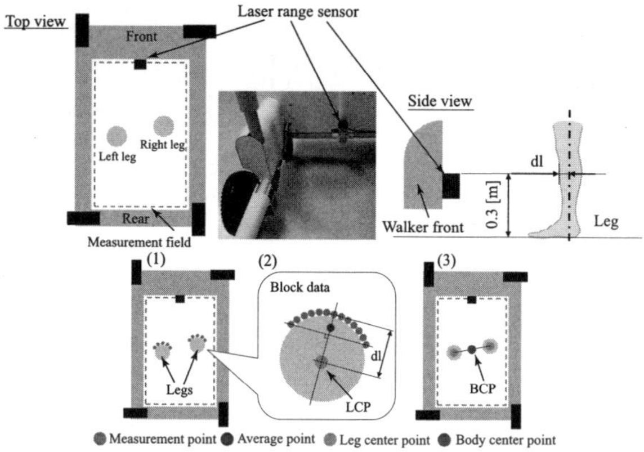

Fig. 3. Illustration of estimation method

of the walker. Wheel W_0 and W_1 drive in x-axis direction, while, wheel W_2 and W_3 drive in y-axis direction.

3. Tracking control system

When patients learn to walk by using conventional walker, they often hold handrail to support their weight, so that it is difficult to control the walker using button or joystick. Therefore, this paper presents a method to estimate patient's motion intention of the moving direction from their walking behavior, and then newly construct a walking assist system integrated with hoisting equipment.

3.1. *Estimation algorithm of body center point*

The conceptual diagram of the proposed algorithm is shown in Fig. 3. A scanning laser range sensor is attached to the front edge of the NILT-WAMOR. The sensor installation height from ground is 0.3m. Each leg position can be measured by the range sensor. By measuring the leg positions continuously and calculating the difference from the previous value, the moving velocity and the direction of each leg can be obtained. However, the transfer velocity of each leg is different from the actual gait velocity. Therefore, first we need to estimate the motion of body center point (BCP), i.e. the center point of both legs. Using the BCP value, we can then calculate the moving velocity and the direction of walker. BCP can be estimated in the following steps.

Estimation step of BCP:

(1) Measurement of the leg positions: As seen in Fig. 3, among the data measured by the sensor, only two blocks of data are extracted. Since

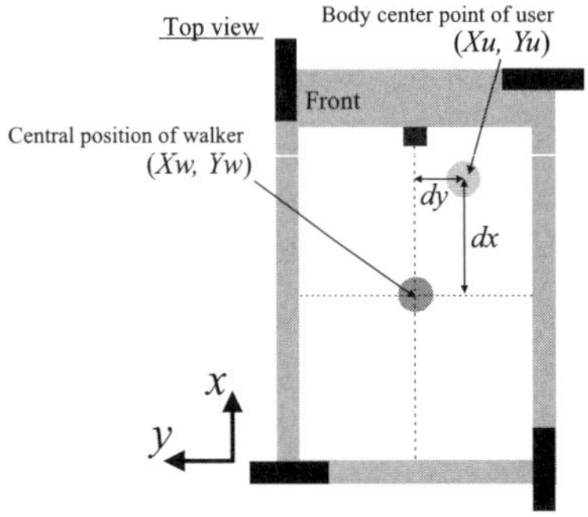

Fig. 4. Illustration of BCP moving

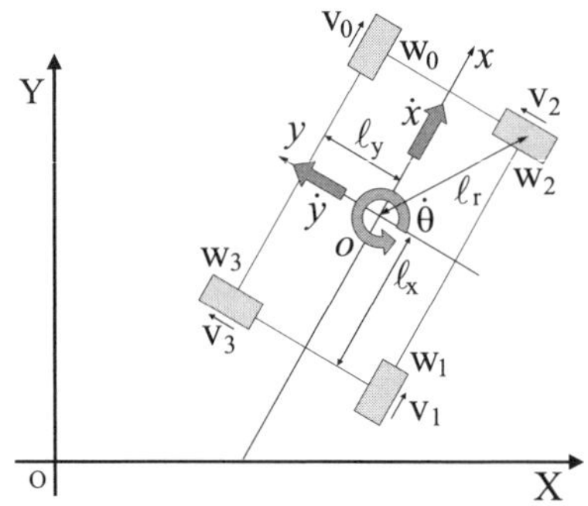

Fig. 5. Schematic model of 4 wheel driving

the sensor is scanned from right to left, the first data block is the left leg, and the second is the right leg.

(2) Estimation of the leg center point (LCP): The following operations are performed on each data block. First, the average point is calculated from the data of each block. Next, a specified line perpendicular to the straight line connecting both ends of measurement points and also passing through the average point is drawn. Then, the point located at a constant distance, dl in Fig. 3, from the measurement point on this line is defined as LCP. Here, dl is the radius of the user's leg measured at the height of 0.3m from the ground.

(3) Estimation of the body center point (BCP): The central point of the two LCPs is defined as the BCP.

3.2. *Velocity control law*

In order to realize the tracking control system, the next step is to construct velocity control law. Through this produce, the direction to BCP is throught to be such the walking direction that user wants to move. Therefore, a tracking control is required. Such as central position (X_w, Y_w) of walker is followed to (X_u, Y_u) of BCP as reference point. As shown in Fig. 4, central position $O_w(X_w, Y_w)$ of the walker, and BCP, $O_u(X_u, Y_u)$ of the patient are defined respectively. Walker moving velocity is calculated using the distance d_x and d_y between O_w and O_u. Each velocity are determined in such a way as to decrease difference d_x and d_y to 0. The walker velocities $\dot{x}$ and $\dot{y}$ are shown in the following equations, where, k_x and k_y are proportional gains.

$$\dot{x} = k_x \times d_x \tag{1}$$

$$\dot{y} = k_y \times d_y \tag{2}$$

4. Kinematic model of an omni-directional NILTWAMOR

Figure 5 shows a schematic model of the 4 wheel driving. Foregoing $\dot{x}$ and $\dot{y}$ are to be split into driving velocities V_0 ‘ V_3. Now, l_x and l_y are each axis distance between central position of walker and wheel, and, l_r is linear distance between central position of walker and wheel. The direct kinematic equation is denoted by the state vector $x = [x, y, \theta]^T$ and the input vector $u = [V_0, V_1, V_2, V_3]^T$ as

$$\dot{x} = Bu \tag{3}$$

where

$$B = \begin{bmatrix} \frac{1}{2} & \frac{1}{2} & 0 & 0 \\ 0 & 0 & \frac{1}{2} & \frac{1}{2} \\ -\frac{l_y}{4l_r^2} & \frac{l_y}{4l_r^2} & \frac{l_x}{4l_r^2} & -\frac{l_x}{4l_r^2} \end{bmatrix} \tag{4}$$

The inverse kinematic equation becomes

$$u = B^{-1}\dot{x} \tag{5}$$

where

$$B^{-1} = \begin{bmatrix} 1 & 0 & -\frac{l_r^2}{l_y} \\ 1 & 0 & \frac{l_r^2}{l_y} \\ 0 & 1 & \frac{l_r^2}{l_x} \\ 0 & 1 & -\frac{l_r^2}{l_x} \end{bmatrix} \tag{6}$$

5. Overall flow for tracking control system

Diagram of tracking control system is represented in Fig. 6. Tracking control system is composed of estimation method of body center point (BCP), velocity control and inverse kinematic model.

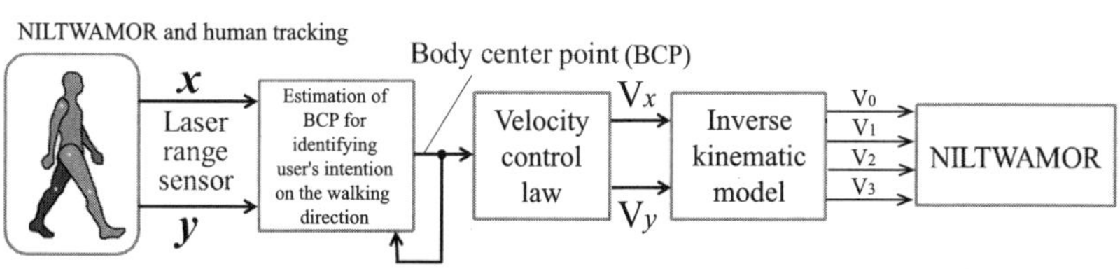

Fig. 6. Overall flow for tracking control system

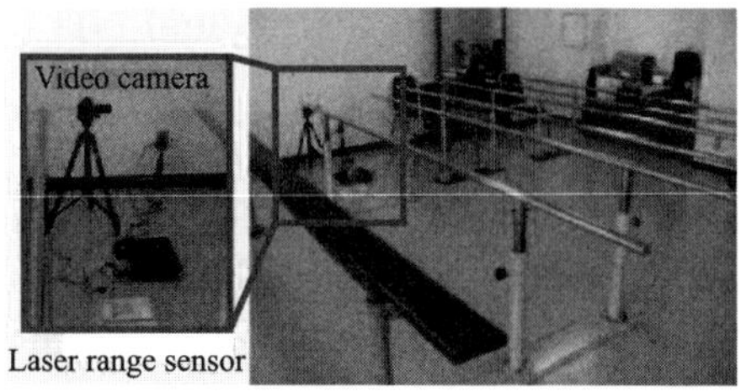

Fig. 7. Experimental setup for leg position measurement

Table 1. Symptomatic state of subjects

Subject	Age	Symptom
subject 1	80's	Parkinson disease
subject 2	70's	Parkinson disease
subject 3	80's	Left hemiplegic
subject 4	60's	Double upper limb paralysis
subject 5	40's	Ataxia
subject 6	70's	Incomplete cervical lesion
subject 7	70's	Corticobasal degeneration
subject 8	80's	Older age

6. Motion behavior measurement of gait rehabilitation

To verify the effectiveness of the proposed estimation algorithm, we have conducted several experiments by patients for gait rehabilitation. Measured data of walking behavior were recorded by using laser range sensor and video camera. The measurement environment is shown in Fig. 7. The age and symptom of each subject are shown in Table 1. The patient movement to be measured was the movement in usual conventional gait training between parallel bars. In order to walk with natural behavior, no instruction was given for gait velocity, steps and swing side for start and so on. The forward movement was measured and performed twice per person. During the measurement experiment, a physiotherapist walked alongside the patient and paid special attention to accidents such as falls.

Overview of experiment of each subject is shown in Figs. 8, 10, and 12. The estimation result of subject 1 is shown in Fig. 9. The upper figure shows the walking trajectory of LCP and BCP, and the lower figure shows the moving velocity of LCP and BCP. The result indicates that the walking behavior is very small step like shuffle. Also, the velocity of each leg can be estimated as well as healthy people. Comparing with healthy people, the velocity of each leg has changed little by little and the velocity of body

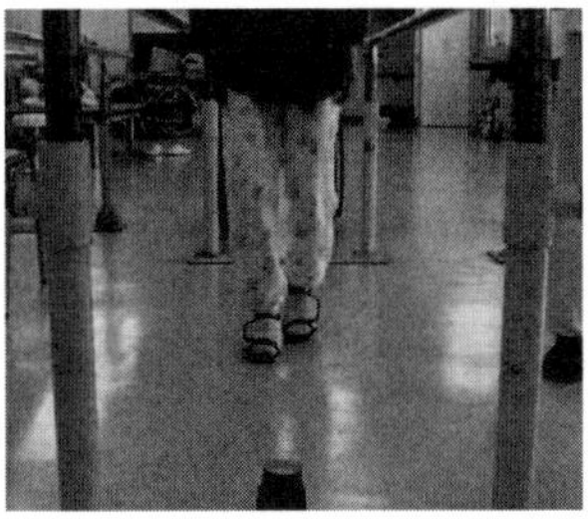

Fig. 8. Overview of experiment for subject 1

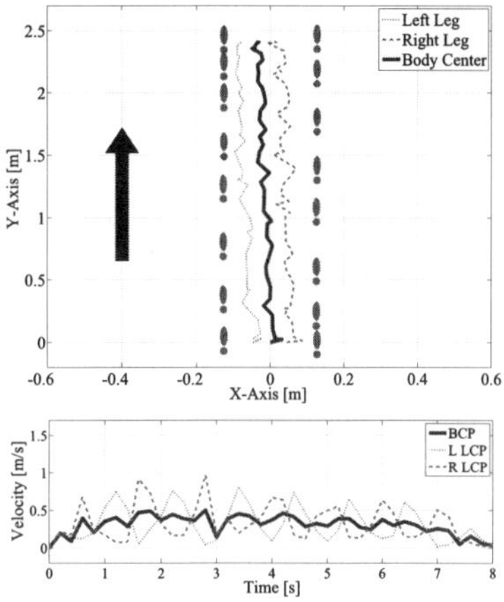

Fig. 9. Estimation results for subject 1

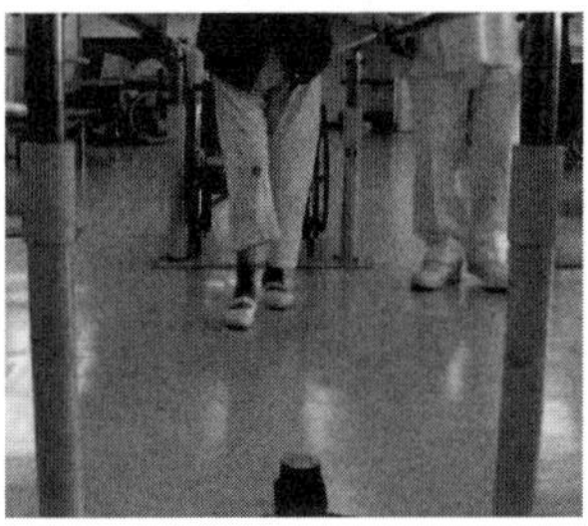

Fig. 10. Overview of experiment for subject 2

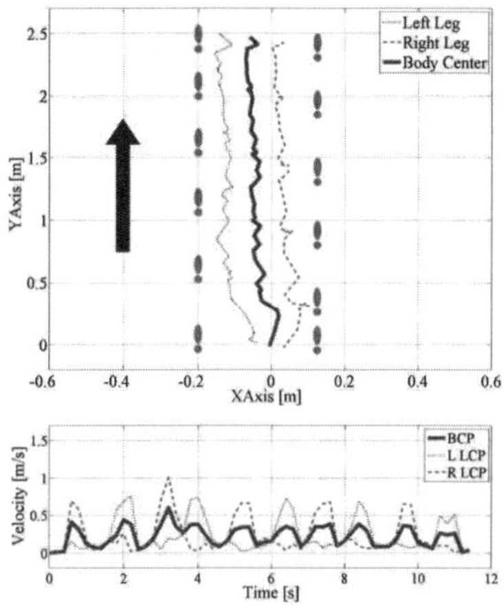

Fig. 11. Estimation results for subject 2

center point (BCP) is small. In addition, the proposed estimation algorithm was able to estimate all subjects walking behavior.

The estimation result of subject 2 is shown in Fig. 11. Although this subject has the same symptom as subject 1, it walked slowly. After touched foot on floor, its weight is shifted to the leg little by little and then the disabled people of subject 2 was walking with shifting its center of gravity carefully. However, its leg swing were unfaltering and its steps were wider than subject 1.

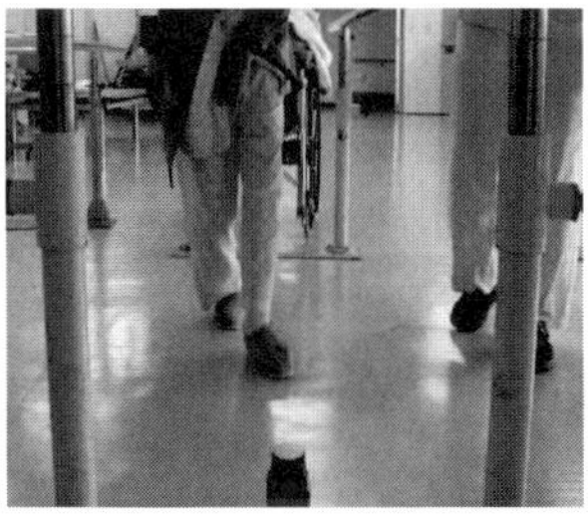

Fig. 12. Overview of experiment for subject 3

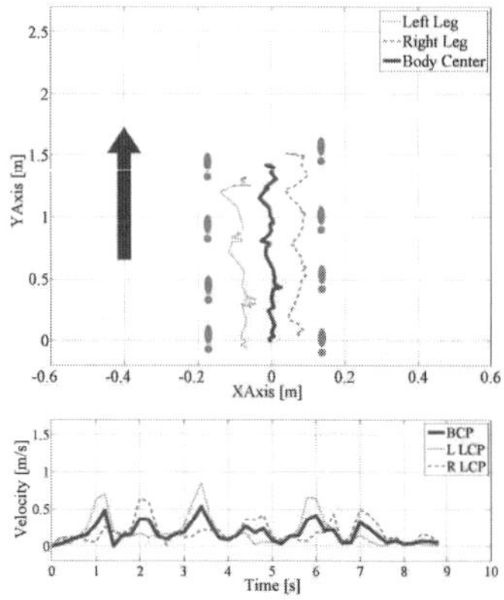

Fig. 13. Estimation results for subject 3

The estimation result of subject 3 is shown in Fig. 13. Subject 3 is left hemiplegic. It is difficult to flex its left leg so that the leg was straight and swung widely. On the other hand, when right leg was swung, its center of gravity was shifted with leaned on parallel bars. For this subject, maintaining of standing posture by body weight supporting is effective. Because, for the patient who fears to shift its weight to paralytic leg, fall prevention bring a feeling of safety, and body hoisting load on legs.

7. Conclusions

From these results, it was confirmed that the proposed estimation method is effective, even if subjects are patients of gait rehabilitation. In near future, we will examine the effectiveness of the proposed method using NILT-WAMOR system.

References

1. Akio Tsubakihara, Techological Advance in Medical RehabilitationCK-AWASAKI Journal of Medical Welfare extra No. -2(2008)
2. J. R. Franza, M. Glauserb, P. O. Rileya, U. D. Crocea, F. Newtona, P. E. Allaired, D. C. Kerrigana, Physiological modulation of gait variables by an active partial body weight support system Journal of Biomechanics 40(2007)
3. Y. Ochi, K. Terashima, K. Kubo, T. Miyoshi, K. Kakihara, T. Iwade, T. Yamamoto, T. Sakakibara, Tracking Control to Human Gait of Lift-Type Walking Assist Robot Using Range Sensor, CLAWAR2011(2011)
4. Y. Ochi, K. Terashima, K. Kakihara, Y. Nie, T. Yamamoto, T. Sakakibara, Development of a Novel Intelligent Lift-Type Walking-Assist Mobile Robot Aiming at Taylor-Made System Design for Body Weight and Walk Motion Support, CLAWAR2010(2010)

DRIVING ASSISTANCE CONTROL OF A PASSIVE-TYPE ROBOTIC WHEELCHAIR BASED ON A USER'S INTENSION

T. HIGUCHI * and D. CHUGO

Graduate School of Science and Technology, Kwansei Gakuin University, Sanda, Hyogo 6691337, Japan

Y. SAKAIDA

RIKEN-TRI Collaboration Center for Human-Interactive Robot Research, Nagoya, Aichi 4630003, Japan

S. YOKOTA

Faculty of Science and Engineering, Setsunan University, Neyagawa, Osaka 5728508, Japan

K. TAKASE

Graduate School of Information and System, The University of Electro-Communications, Chofu, Tokyo 1828585, Japan

This paper proposes a driving assistance control with estimation of a user's intension for our passive-type robotic wheelchair. In our previous research, we propose a novel intelligent wheelchair based on the passive robotics. Our proposed assistive wheelchair is simple and low cost mechanism, which consists of a frame, casters, wheels and servo brakes. Controlling a torque of the servo brake, our wheelchair changes its dynamics and realizes the reference tracks. For realizing a driving assistance control for our developing wheelchair system to fit the trajectory which the user wants, we developed two topics in this paper. One is the estimation scheme of the direction which the user wants to move. For realizing this estimation, we use the minimum jerk trajectory model, which expresses the typical human movement. The other key topic is novel driving assistance algorithm, which controls the servo brake in cooperation with the user's operation. We test our proposed assistance system by the experiments with our prototype and verify its effectiveness.

1. Introduction

Wheelchairs are widely used by many people who are mobility impaired in daily activities. However, many accidents often occur to wheelchair users and its risk is serious. More than 80[%] of wheelchair accidents are caused by environmental problems [1]. Especially, the inclination of sidewalk has high risk for wheelchair user. In Japan, it is permitted to incline the sidewalk up to 5[deg] (Figure.1) [2]. This inclination leads the wheelchair user to run out from the sideway to the roadway and it causes the accident between the wheelchair and the car. Therefore, the driving assistance system for wheelchair is important.

In previous works, a lot of assistive technologies for wheelchairs are developed. In general, many handicapped people traditionally use power wheelchairs [3] and previous researchers have tried to realizing assistance functions by adding wheels with actuators and controlling them based on the robot technology such as motion control technology [4]. These intelligent wheelchairs provide many functions, such as suitable motion, obstacle avoidance and navigation; thus, they provide a maneuverable system. However, many wheelchair users have the upper body strength and dexterity to operate a manual wheelchair. For these wheel chair users, its cost is too expensive and is not acceptable. On the other hand, a manual wheelchair without servomotors, which consists of a frame, wheels, casters and hand brakes, is commercially available and widely used. Its mechanism is simple and low cost. There is no risk by malfunction of servomotors and patients can use it intuitively because they use these wheelchairs passively with their own intentional force.

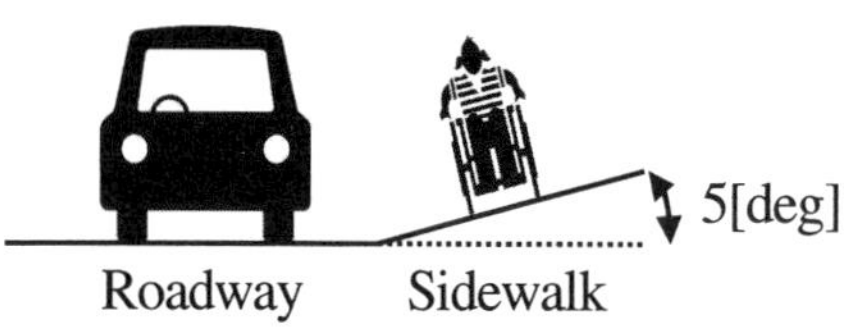

(a) Inclination of the sidewalk

(b) The example of typical sidewalk

Figure 1. Risks of the sidewalk

In our previous research, we are developing a passive-type robotic wheelchair with servo brakes [5]. For developing it, we use the concept of passive robotics [6]. This system moves passively based on external force without actuators. The system uses servo brakes, which can change brake torque, and realizes the desired motion changing the physical parameters using servo brakes according to the applied force and reference tracks. The passive robotics system requires no actuator and its mechanism is simple, therefore, the system will be low cost. In the research area of assistive robotics, the passive robotics concept has been used for the walker and its performance is useful [7-8]. Thus, we adapt this concept for the wheelchair assistance system.

In this paper, we propose the novel driving assistance controller for our wheelchair system. We use passive robotics system and develop the driving assistance control scheme which is based on the velocity control. Our proposed system estimates the direction which the user wants to move. Our estimation scheme measures a beginning part of row motion by the user and compares a measuring trajectory and a minimum jerk trajectory model which expresses the typical human motion [9-10]. After estimation the user's intention, our system sets the reference trajectory based on this result and controls the servo brake for realizing it.

2. System-Configuration

Figure.2 (a) shows our prototype. Our proposed wheelchair utilizes a powder brake, which is one of a servo brake. Comparing the walker, the wheelchair requires larger brake torque and we choose the powder brake as figure.2(b) and (c) (ZKG-YN50, Mitsubishi Electric Corp.), which can generate enough brake torque for stopping 4[km/h] moving wheelchair with a 100[kgf] user by 1[sec]. Our system has compatibility with a general wheelchair, which fulfills Japan Industrial Standard (JIS T9201). This means the user can build our system into their wheelchair without special construction.

Figure.3 shows its controller. Our wheelchair has two powder brakes with the tension controller on each wheel. Our system can measure the rotational velocity using encoders on each wheel.

(a) Overview (b) Installed brake (c) Powder brake

Figure 2. Our prototype

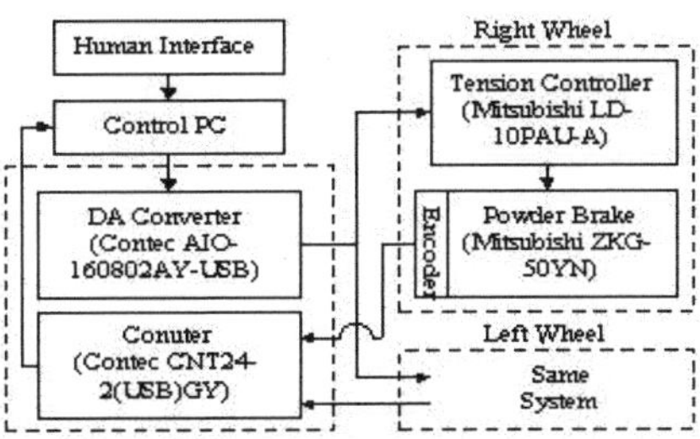

Figure .3 Control system

3. Driving Assistance Controller

3.1 *Minimum Jerk Trajectory Model*

In the field of neurophysiology, Flash *et al.*, analyzed voluntary human arm movements and demonstrated that they can be closely approximated by minimum jerk trajectory model with characteristic velocity profile [9-10]. In the minimum jerk trajectory model, human arm movements can be formulated as the trajectories minimizing the following equation.

$$C_j = \int_0^{t_f} \left(\frac{d^3 x(t)}{dt^3} \right)^2 dt \tag{1}$$

Where t_f is the final time of the movement and $d^3 x(t)/dt^3$ is the differential of acceleration, called jerk. C_j assumes an extremum when $x(t)$ is the solution of Euler-Poisson equation. The resulting equation is the following:

$$\frac{d^6 x(t)}{dt^6} = 0 \tag{2}$$

This condition shows that $x(t)$ is the fifth order polynomial:

$$x(t) = a_5 t^5 + a_4 t^4 + a_3 t^3 + a_2 t^2 + a_1 t + a_0 \tag{3}$$

If applying the following boundary conditions as (4):

$$\begin{aligned} x(0) &= x_0, \quad \dot{x}(0) = 0, \quad \ddot{x}(0) = 0 \\ x(t_f) &= x_f, \quad \dot{x}(t_f) = 0, \quad \ddot{x}(t_f) = 0 \end{aligned} \tag{4}$$

$x(t)$ is expressed as the following:

$$x(\tau) = x_0 + (x_f - x_0)(6\tau^5 - 15\tau^4 + 10\tau^3) \tag{5}$$

$$\tau = \frac{t}{t_f} \tag{6}$$

Using minimum jerk trajectory model, we define the human arm behavior characteristic on row movement on the wheelchair. Figure.4 shows the velocity of each wheel and minimum jerk trajectory model calculated from the measuring values. In this case, the subject rows the wheelchair to the right direction. t_f is 3 [sec]. x_f in left wheel (x_f^{left}) is 0.8[m] and in right wheel (x_f^{right}) is 0.5[m]. From these results, we can verify that the minimum jerk trajectory model expresses the row motion on the manual wheelchair.

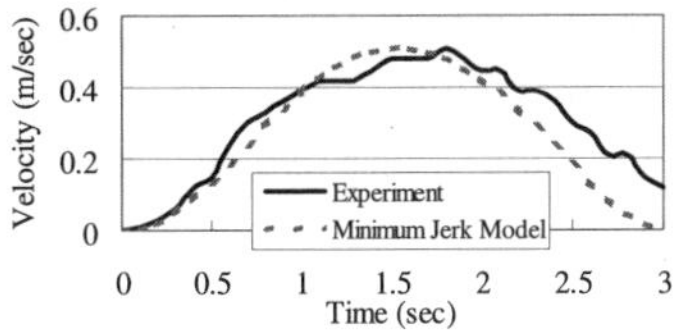

(a) Left Wheel

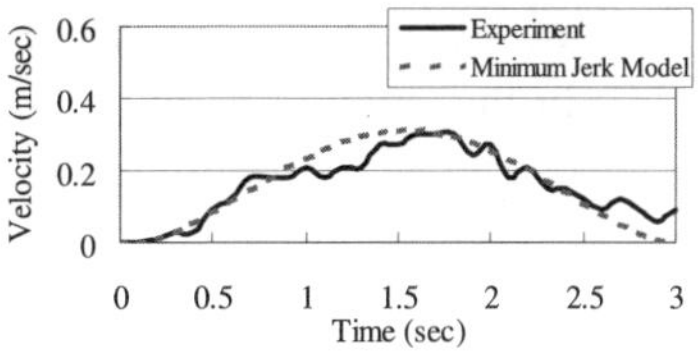

(b) Right Wheel

Figure 4. The velocity of each wheel

3.2 *Estimation of the User's Intension*

Using the minimum jerk trajectory model, we estimate the direction which the user wants to move. For realizing it, we propose the following method.

- Our system measures the rotation velocity of each wheel. If the rotation velocity of right or left wheel increases Δt continuously, our system judges the user rows the wheelchair and starts to estimate the user's motion using the minimum jerk trajectory model. In this study, we set $\Delta t = 0.2$ [sec] experimentally and in our future work, we will discuss on this coefficient value.
- Our system sets the time $t = t_0$ at the moment when the rotation velocity increases. At the same time, our system measures the position (θ_0), velocity ($\dot{\theta}_0$) and acceleration ($\ddot{\theta}_0$). We assume that the boundary condition of the minimum jerk trajectory model is as follows:

$$\begin{aligned} &\theta(t_0) = \theta_0, \quad \dot{\theta}(t_0) = \dot{\theta}_0, \quad \ddot{\theta}(t_0) = \ddot{\theta}_0 \\ &\theta(t_0 + t_f) = \theta_f, \quad \dot{\theta}(t_0 + t_f) = \dot{\theta}_0, \quad \ddot{\theta}(t_0 + t_f) = -\ddot{\theta}_0 \end{aligned} \tag{7}$$

- Using this assumption, the minimum jerk trajectory model is derived as (8) and (9).

$$\begin{aligned} \theta(t) = \theta_0 &+ (\theta_f - \theta_0)(6\tau^5 - 15\tau^4 + 10\tau^3) \\ &+ \dot{\theta}_0 (t - t_0)(-3\tau^4 + 8\tau^3 - 6\tau^2 + 1) \\ &+ \frac{\ddot{\theta}_0}{2}(t - t_0)^2(-\tau^3 + 3\tau^2 - 3\tau + 1) \end{aligned} \tag{8}$$

$$\tau = \frac{t - t_0}{t_f} \tag{9}$$

- In (8) and (9), unknown values are θ_f and t_f. Therefore, our system uses a pattern matching method as (10) and derives θ_f and t_f which minimize $c(\theta_f, t_f)$. Please note $\theta_{real}(t)$ is measuring value. If the

$c(\theta_f, t_f) > c_0$, our system judges the user does not row the wheelchair. In this study, we set $c_0 = 3.0$ experimentally. Furthermore, we set the candidate of these values as (11) and these values are derived experimentally.

$$c(\theta_f, t_f) = \int_{t_0}^{t_0+2\Delta t} \{\theta(\theta_f, t_f)(t) - \theta_{real}(t)\} dt \qquad (10)$$

$$\theta_f = \{0.05, 0.10, 0.15, \cdots, 1.50\} \; t_f = \{1.0, 1.1, 1.2, \cdots, 5.0\} \qquad (11)$$

- Our system excuses this matching process on each wheel. Therefore, our system estimates θ_f in right wheel (θ_f^{right}) and in left wheel (θ_f^{left}), and estimates t_f in right wheel (t_f^{right}) and in left wheel (t_f^{left}). Our system uses θ_f^{right} and θ_f^{left} for direction estimation which the user wants to go as Table.1. In this time, our system estimates only three directions. In our future work, we will improve our system which can estimate various directions.

Table 1. Estimation Pattern

	Estimated Direction
$\theta_f^{right}/t_f^{right} > \theta_f^{left}/t_f^{left}$ or $(\theta_f^{right}/t_f^{right} > 0 \;\; and \;\; \theta_f^{left} = NA)$	Left
$\theta_f^{right}/t_f^{right} < \theta_f^{left}/t_f^{left}$ or $(\theta_f^{right} = NA \;\; and \;\; \theta_f^{left}/t_f^{left} > 0)$	Right
$\theta_f^{right}/t_f^{right} = \theta_f^{left}/t_f^{left}$	Straight

3.3 *Wheel Controller*

Based on the estimated result, our system controls its wheels by PID controller. For preventing the risk of the slope on the sidewalk as Fig.1, the first, we implement the straight assistance control to our system. Our proposed controller works as follows:

- If the system estimates the straight trajectory, our system starts the PID controller. In other cases, our system does not control its wheel and it is our future work.
- The control error integrates from $t = t_0$ in I controller and this means our system maintains the direction at this time.
- If the rotation velocity of right or left wheel increases Δt continuously, our system stops the wheel controller.

4. Experiments

4.1 *Estimation Direction Test*

In this experiment, we test our estimation system in two cases. In first case, the subject rows the wheelchair twice to the straight direction on the flat floor. In second case, the subject rows the wheelchair to the right direction on the same floor. As the result, in first case, our system estimates the user's motion as Figure.5.At the first motion, our system derives $\theta_f^{right} / t_f^{right} = \theta_f^{left} / t_f^{left} = 0.22$ and our system estimates the straight direction. At the second motion, our system derives $\theta_f^{right} = \theta_f^{left} = 0.17$ and estimates the straight direction, too. In second case, our system estimates the user's motion as Figure.6. Our system derives $\theta_f^{left} = 0.14$ and $\theta_f^{right} = NA$, and estimates the right direction.

From these results, our scheme succeeds to estimate the user's intension.

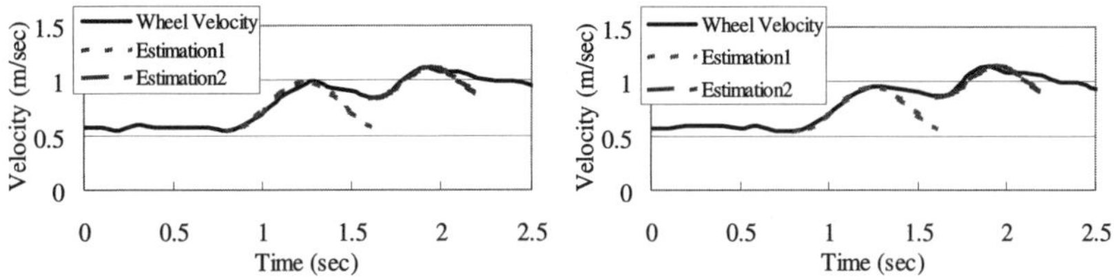

Figure 5. The velocity of each wheel (Straight direction)

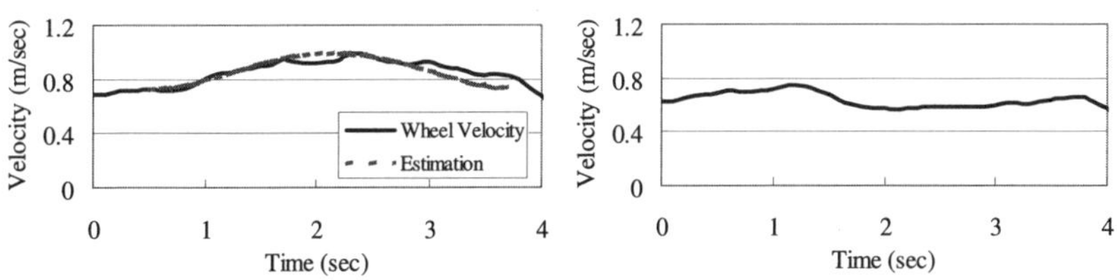

(a) Left Wheel (b) Right Wheel

Figure 6. The velocity of each wheel (Right direction)

4.2 *Running Test on the Slope*

In this experiment, the user goes straight with our wheelchair on the test load with inclination as Figure.9. As the result, the user can go straight with our system as Figure.9. On the other hand, without our assistance system, it is difficult to go straight by the inclination as Figure.8. Figure.7 shows the velocity of each wheel during the experiment. In Figure.7 (b), the velocities are almost same and this means that the wheelchair goes straight comparing with Figure.7 (a) which shows the case without proposed assistance.

From these results, we can verify the effectiveness of our proposed passive-type assistance system for a wheelchair.

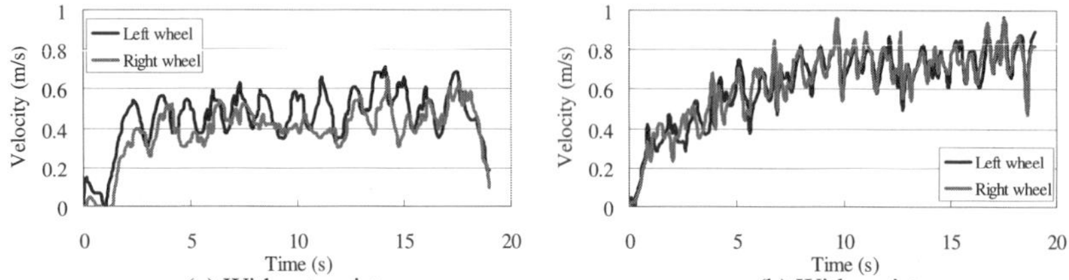

(a) Without assist (b) With assist

Figure 7. The velocity of each wheel

Figure 8. Test run without our assistance system

Figure 9. Test run with our assistance system

5. Conclusions

In this paper, we develop a novel assistance control for a passive-type wheelchair for healthy users who have enough upper body strength. For realizing the assistive wheel control, we develop the estimation scheme for the user's intension and our system controls its wheels based on the estimated results.

In our final paper, we will discuss the direction estimation scheme more closely (Now, our system estimates only three directions) and path navigation scheme including the curve based on the estimated results.

Acknowledgments

This work was supported in part by Kawanishi Memorial ShinMaywa Education Foundation.

References

1. National Consumer Affairs Center of Japan, "Considering the safety of self-propelled manual wheelchairs," NCAC News, Vol.14, No.3, pp.4 (2002).
2. "A Sidewalk guideline for smooth mobility, 2nd edition," Japan Institute of Construction Engineering, pp.3-2-19, 2008, ISBN: 4802893922. (in Japanese)
3. "JW II," Yamaha Motor Co., Ltd., http://www.yamaha-motor.jp/wheelchair/unit/jw2/ (Available at April, 2012)
4. D. P. Miller, M. G. Slack, "Design and Testing of a Low-Cost Robotic Wheelchair Prototype," Autonomous Robots, Vol.2, pp.77-88, (1995).
5. D. Chugo, T. Higuchi, Y. Sakaida, S. Yokota and K. Takase, "Development of a Passive-type Robotic Wheelchair", Proc. of 14th Int. Conf. on Climbing and Walking Robots, pp.27-34, (2011).
6. A. Goswani, M. A. Peshkin and J. Colgate, "Passive robotics: An exploration of mechanical computation," Proc. of IEEE Int. Conf. on Robotics and Automation, pp.279-284, (1990).
7. A. J. Rentschler, R. A. Cooper, B. Blaschm and M. L. Boninger, "Intelligent walkers for the elderly: Performance and safety testing of VA-PAMAID robotic walker," J. of Rehabilitation Research and Development, Vol.40, No.3, pp.423-432, (2003).
8. Y. Hirata, A. Hara and K. Kosuge, "Motion Control of Passive Intelligent Walker Using Servo Brakes," IEEE Trans. on Robotics, Vol.23, No.5, pp.981-990, (2007).
9. T. Flash and N. Hogan, "The coordination of arm movements: an experimentally confirmed mathematical model, " J. of Neurosicence, Vol.5, No.7, pp.1688-1703, (1985).
10. H. Seki and S. Tadakura, "Minimum Jerk Control of Power Assisting Robot based on Human Arm Behavior Characteristics," Proc. of IEEE Int. Conf. on Systems, Man and Cybernetics, pp.722-727, (2004).

EXPERIMENTAL ANALYSIS OF INTERACTIVE BEHAVIORS FOR A PERSONAL MOBILE ROBOT

C. GRANATA*, P. BIDAUD*, C. BECK ** and P. MAYER**

** Institut des Systèmes Intelligent et de Robotique*
Université Pierre et Marie Curie (CNRS - UMR7222), Paris
E-mail: granata@isir.upmc.fr; philippe.bidaud@isir.upmc.fr
Université Pierre et Marie Curie (CNRS -UMRS 7210), Paris
*** Vienna University of Technology, Vienna*
E-mail: domeo-for@fortec.tuwien.ac.at

For social robots, the synthesis of interactive behaviors that fit in with different situations of everyday life is one of the most important problem to consider. In this paper we propose a framework for the synthesis of interactive behaviors for social mobile robots. We show the framework flexibility through the interactive person following behavior. The system performances are evaluated by a set of experiences conducted with a mobile platform in different true life situations.

1. Introduction

The person following for human-robot interaction (HRI) is a central function, even essential, because it carries out the robot ability to move autonomously in order to bring all its functionalities at the service of the person. Different factors have to be taken into account to define the person following function appropriately: the task to achieve (e.g. provision of an interface, remote control of the person activity), the characteristics of the environment (e.g. obstacles, person occultations) and above all, the person's safety and comfort. This interactive following process may well not simply be based on the control of the distance[1] and/or of the relative speed[2] with respect to the person. It must exploit appropriated reasoning mechanisms. Gockley et al.[3] present an interactive approach of person following discussing two methods, direction-following and path-following the robot automatically selecting which method to use. A similar system is described by Yuan:[4] a combination of direction-following, path-following and parallel-following behaviors depending on the environment and on the relative position of the followed subject. Another approach of person following is based on the visibility of the person: the robot speeds are computed considering the movements of the pixels related to the person in the image.[5] Zender et al.[6] propose an approach combining person tracking and SLAM (Simultaneous Localization and Mapping): they use a semantic map in order to adjust the robot speeds according to its position in the environment. In this paper, we propose a framework for robot behaviors design composed by: 1. A person detector based on the onboard laser sensor. 2. An Extended Kalman Filter (EKF) for person's state estimation and prediction. 3. A Decision Making System (DMS) which uses the fuzzy logic mechanisms exploiting the SpirOps AI software. The framework has been implemented on the Kompaï[7] robot (Robosoft) and its performances are been evaluated with different scenarios of assisted living home (work supported by the DOMEO project AAL 2008-2008-1-159).

2. Person detection

The location of the person uses a detection process based on a low-cost laser sensor. This approach allows to define the performance level that can be achieved using a single sensor. It is clear that today this central issue in social robotics is the subject of numerous investigations with the aim of obtaining a high level of robustness in particular by fusing multiple sources ([8,9]). The laser sensor is used to detect the legs of the persons assuming that they are in upright position or seated (the laser on the robot is 0.3 m above the ground). First we have modeled the legs size and posture (distance between 2 legs) by the means and the standard deviations (in m.) given in the equation 1. They are obtained experimentally from data collected from a sample of 100 scan capturing several persons static or moving close and far.

$$\begin{cases} (\mu, \sigma)_{radius} = (0.13, 0.03) \\ (\mu, \sigma)_{posture} = (0.23, 0.04) \end{cases} \tag{1}$$

Using these models, the proposed approach is composed by the 3 following steps. 1. We first look interest points by detecting "blobs" in the laser scan. A blob is a set of consecutive laser reflection points supposed appertain to the same object. Two points belong to the same blob if their distances are in the range of 0.1 m. 2. For the recognition of a blob as leg, we apply the defined Gaussian model $(\mu, \sigma)_{radius}$ (the used threshold is $= 3\sigma_{radius}$). 3. Two legs are considered as a pair of legs if their distance is in the Gaussian model $(\mu, \sigma)_{posture}$ (the used threshold is $= 3\sigma_{posture}$). The detector shows experimentally a correct detection rate $> 30\%$ within $3m$ (this means an average of 4 detections per second).

3. Person state estimation

The state of the person (position and velocity) cannot be determined by direct observation this for obvious reasons related to false or lack of detection and noise in the measurement. To determine the person movements and to predict its evolution, the information provided by the person detector are processed by an Extended Kalman Filter (EKF). The choice of this filter is motivated by the short computational time required (compared e.g. to particle filters[10,11]). Assuming that humans move with constant velocity in a certain interval of time and that the system and measurement noises have Gaussian distribution, we can describe the human motion in polar coordinates by the following discrete state variable equations.

$$x_{k+1|k} = F_k x_{k|k} + w_k$$
$$y_k = C_k x_k + v_k$$

$$x_k = \begin{bmatrix} \rho_k \\ \theta_k \\ \dot{\rho_k} \\ \dot{\theta_k} \end{bmatrix} ; K_k = \begin{bmatrix} 1 & 0 & \Delta t_k & 0 \\ 0 & 1 & 0 & \Delta t_k \\ 0 & 0 & 1 & 0 \\ 0 & 0 & 0 & 1 \end{bmatrix} ; C_k = \begin{bmatrix} 1 & 0 & 0 & 0 \\ 0 & 1 & 0 & 0 \end{bmatrix}$$

where $x_{k+1|k}$ is the state equation, y_k is the measurements equation (y_k is the person position provided by the detector described in Section 2), ρ_k and θ_k are the polar coordinates at the sampling time k; $\dot{\rho_k}$ and $\dot{\theta_k}$ are the velocities of ρ_k and θ_k at k; F_k is the state transition matrix; C_k is the observation matrix and w_k and v_k are respectively the system noise due to mis-modeling and the observation noise due to the sensor and to the detector errors. At each sampling time k, the velocities $\dot{\rho_k}$ and $\dot{\theta_k}$ are updated in the state equation computing the person velocities from its motion during the 10 last sampling times. Note that the use of the human state predictor has two advantages. First, the predicted trajectories are smoother than the detected trajectories (aberrant detections and noises are filtered). Second, it solves the problem of lack of detection in a lot of situations. Therefore, we investigated the use of a human locomotion oriented prediction by relying on Hitcheur's studies[12] who have shown that humans tend to keep a constant speed when walking straight and to reduce their speed when turning. The walking speed is linked to the turning radius by an exponential low and this exponent depends on the path shape. In Figure 1 the predictor performances taking into account and not the speed reduction are shown.

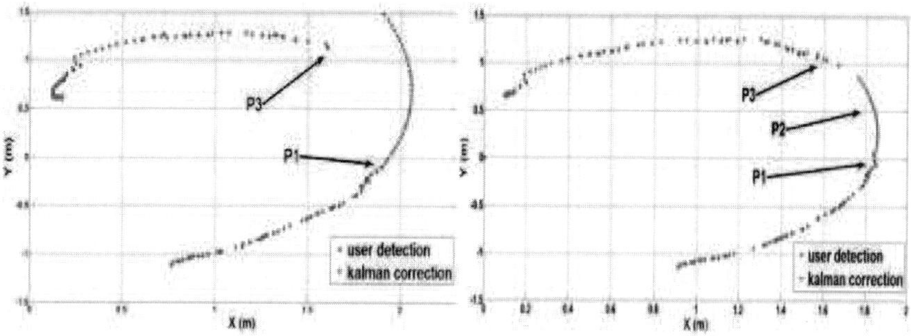

Figure 1. Example of person state prediction in a case of occultation. Left: The predictor keeps constant speed. Right: The predictor considers the speed reduction. P1= loss of detection; P2= reduction of speed considered by the filter; P3= person detected again.

4. Decision making system

The deterministic approaches for the decision making processes are appropriate to the problems in which the available information are precise. Their use in situations where the information are scarce, vague, contradictory, missing or incomplete, can lead, in many cases, to inappropriate decisions (behaviors). The decision making problem under uncertain for confused or vague goals (often in conflict), can be treated by using the fuzzy logic mechanisms. Fuzzy logic has already been used for robot navigation in cluttered environments[13,14] and for person following application.[15] We propose a Decision Making System (DMS) which uses the fuzzy logic mechanisms exploiting the SpirOps AI software. See[16] for more details about the use of SpirOps. In order to define an appropriate global behavior, the robot uses different strategies that may be selected dynamically according to the available information about the state of the person, the context and the environment. In this paper we will restrict to the use of 4 strategies (designed to realize the person following function): *GoToUser* (the robot has to approach the person), *FollowUser* (the robot has to follow the person), *GlobalSearch* (the robot has to patrol the environment locking for the person) and *LocalSearch* (the robot has lost the detection and it has to search for the person around the location of the last detection). A strategy corresponds to a basic action and includes different parameters (i.e. point to reach and speed). The decision about the strategy to apply is made through fuzzy rules. These rules assign a degree of interest to each strategy. The degree of interest is computed as the product of different factors (e.g. the person detection confidence, the time elapsed from the last visit to a particular

place etc.) and it can take any value between 0 and 1. In the developed DMS, the computation of the interests is left to SpirOps.
For the robot movement generation we exploit the collision-free paths provided by the SLAM integrated on the robot and we apply a trajectory following algorithm by using the control law proposed by Morin et al.[17]

5. Experiments

The robustness of the person state estimation and of the following behavior are been tested on the premises of the Vienna University of Technology (TUW) (the layout of the test room is showed in Figure 2 left) and of the Institut des Systèmes Intelligent et de Robotique (ISIR) (Figure 2 right) . The geometric characteristics of the environment are known by the robot through a preliminary exploration step (functionality provided by the SLAM). In this section, 4 experiments are described and the results are analyzed. The first 2 experiments focus the person localization in order to evaluate the robustness to the perturbations. The third experiment demonstrates the ability of the proposed framework to select the robot strategy by analyzing the context and the specific situation. The last experiment shows the anticipating skill of the robot.

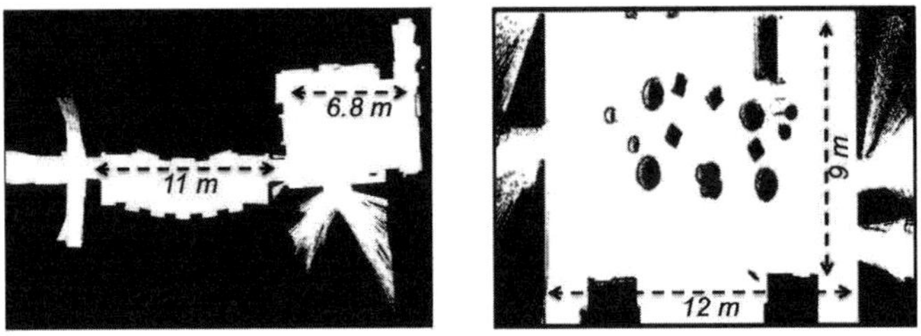

Figure 2. Layouts test environment. Left: TUW premises. Right: ISIR premises.

5.1. *Experiment 1*

The first experiment was carried out in the environment shown in Figure 3 (left) and its main purpose is to test the robustness of the proposed person localization method. The robot stands at starting position and is turned on. At the start of the experiment, no person is in the room, thus the robot starts doing a patrol of the environment. Then, a person appears (from several entry points in the room) and starts to walk around the room (purple circle). The robot should detect the person (*target person*) and follow him/her. From time to time another person (*perturbation person*) walks through the room passing between the robot and the followed person (the green and blue lines show the path of other people traversing the scene) in order to introduce perturbations. Robot must not be distracted from following the *target person.*

Two *target persons* were followed by the robot for a total time of 2 hours each. During each hour several *perturbation persons* walked through the room 12 times (some of them unintentional). The speed of the *target person* was about 0.6 m/s. If the *target person* walked on a circular path and no disturbance occurs, the robot followed him/her in 10 out of 10 cases without a problem. If a *perturbation person* crossed and distracted the robot walking more quickly than the *target person*, the robot continued to follow the *target* 8 out of 10 times. If the *perturbation person* crossed the robot walking at about the same speed of the *target person*, the robot was distracted and switched the *target* 9 out of 10 times. That happened because when the *perturbation person* walked enough slowly between the robot and the *target*, the robot was forced to slow down and it lost the *target* detection for too long. Since in the tested configuration the trajectories of the *target* and the *perturbation person* were enough close, the EKF accepted the new detection (the *perturbation person*) as *target*. This is a compromise choice between the flexibility and the precision of the EKF.

Figure 3. Pictures of test environments. Left: First test environment. Right: Second test environment.

5.2. *Experiment 2*

The second test session was done in a hallway (Figure 3 right). The aim is to prove yet the robustness of the person localization to different perturbations. The test was done by first walking up and down the hallway (pink path). Then a disturbing walking by people was added (green path).

The procedure was repeated by 2 different *target persons*, 10 times each. At each time, 1 or 2 *perturbation person* disturbed along the green path walking at different speeds. The robot was never distracted by the people on the green path. This prove the prediction system robustness in the situations where the disturbances occur close to the *target person*, but not between the *target persons* and the robot.

5.3. *Experiment 3*

The third experiment demonstrates the ability of the proposed framework to select the robot strategy by analyzing the context and the specific situation (environmental constraints, person detection confidence, person and robot state). At the beginning of the experiment, the person was not moving (phase A, Figure 4) and the robot applied the GoToUser strategy. Then, the person started walking at natural speed and passed between two obstacles (phase B); the robot started following him/her. During C the robot lost the person and the LocalSearch strategy was selected. The detection was found again in phase D, thus the robot applied UserFollow strategy and it decided to reach the point P_1. Soon it lost again the person (phase E) and it applied the LocalSearch strategy selecting the point provided by the EKF (point P_2). At this moment it decided to turn around the obstacle (collision-free path). Note that the EKF prediction was not correct (because the person had stopped in F). While the robot was going to reach P_1, it finded the person again.

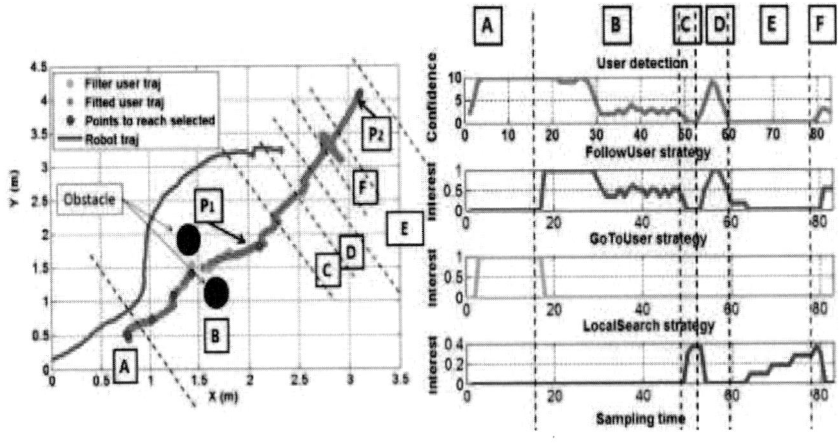

Figure 4. Left: Robot and person trajectories. Right: Strategy selection and person detection confidence during the experience.

5.4. *Experiment 4*

The last experiment presented in this paper shows the anticipating skill of the robot. In this case the FollowUser strategy was deselected, thus the robot task was to approach to the person and not to follow him/her. The robot started moving toward the person until the loss of detection, when the person was occulted by an obstacle. Thus, the LocalSearch strategy was activated and the robot moved to reach the predicted person position. In this case the collision-free generated path was a shortcut (Figure 5). This robot behavior only occurs under particular conditions: the robot as to be at a certain distance from the person (not too close) and the path of the shortcut has to be shorter than the path which should be covered skirting the obstacle.

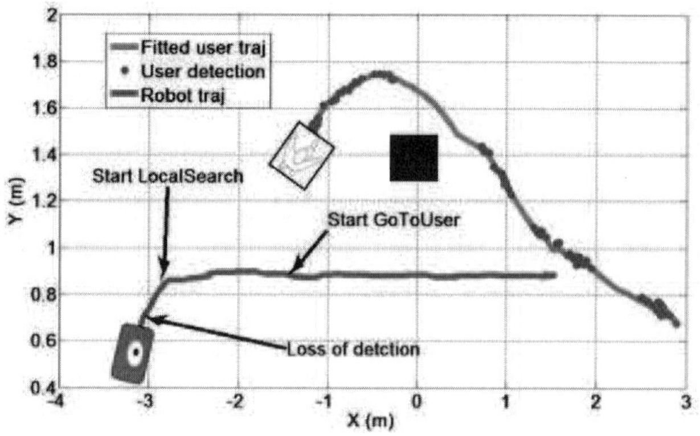

Figure 5. The robot takes a shortcut to reach the user.

Bibliography

1. R. C. Luo, Y. J. Chen, C. T. Liao and A. C. Tsai. Mobile robot based human detection and tracking using range and intensity data fusion. In Proceedings of the IEEE ARSO 2007 workshop on advanced robotics and its social impacts, Hsinchu, Taïwan, 2007.
2. J. Satake and J. Miura, Robust stereo-based person detection and tracking for a person following robot. In Proceedings of the IEEE ICRA 2009 workshop on people detection and tracking, Kobe, Japan, 2009.
3. R. Gockley, J. Forlizzi and R. Simmons. Natural person-following behavior for social robots. In Proceedings of the IEEE ACM International conference on Human-Robot Interaction , Washington DC, USA, 2007.
4. F. Yuan, M. Hanheide and G. Sagerer. Spatial context-aware person-following

for a domestic robot. In Proceedings of the International Workshop on Cognition for Technical Systems, Munich, Germany, 2008.
5. C. Hu, X. Ma,X. Dai. A Robust person tracking and following approach for mobile robot.In Proceedings of the ICMA/IEEE International Conference on Mechatronics and Automation, ICMA 2007.
6. H. Zender, P. Jensfelt and G.J.M. Kruijff. Human-and situation-aware people following. In Proceedings of the 16th IEEE RO-MAN 2007 International Symposium on Robot and Human Interactive Communication, Jeju Island, Korea, pages 1131-1136, 2007.
7. http://www.doc-center.robosoft.com/Kompaï_home_page
8. X. Wang, X. Clady and C. Granata. A Human Detection System for Proxemics Interaction. In Proceedings of the ACM/IEEE international conference on Human-Robot Interaction (Late Breaking Results). Lausanne, 2011.
9. O.M. Mozos, R. Kurazume and T. Hasegawa. Multi-layer people detection using 2d range data. In Proceedings of the IEEE ICRA 2009 workshop on people detection and tracking, Kobe, Japan, 2009.
10. C. Yardim, P. Gerstoft and W. S. Hodgkiss. Tracking Refractivity from Clutter Using Kalman and Particle Filters. In Proceedings of IEEE Transactions on Antennas and Propagation, vol. 56, issue 4, 2008, 1058-1070.
11. J. Woo, Y. J. Kim, J. Lee and M. T. Lim. Localization of Mobile Robot using Particle Filter. In Proceedings of the SICE-ICASE International Joint Conference 2006, Bexco, Busan, Korea, 2006.
12. H. Hicheur, S. Vieilledent, M. J. E. Richardson, T. Flash and A. Berthoz. Velocity and Curvature in Human Locomotion along Complex Curved Paths : a Comparison with Hand Movements. Experimental Brain Research, vol. 162, issue 2, 2005, 145-154.
13. W. Li. Fuzzy logic-basedperception-action behavior control of a mobile robot in uncertain environments. In Proceedings of the Third IEEE Conference on Fuzzy Systems, IEEE World Congress on Computational Intelligence, pages 1626-1631, 1994.
14. P. Rusu, E.M. Petriu, T.E. Whalen, A. Cornell and H.J.W. Spoelder. Behaviorbased neuro-fuzzy controller for mobile robot navigation. Instrumentation and Measurement, IEEE Transactions on, 52(4) :1335-1340, 2003.
15. C.H. Hu, X.D. Ma and X.Z. Dai. Reliable person following approach for mobile robot in indoor environment. In IEEE International Conference on Machine Learning and Cybernetics, volume 3, pages 1815-1821, 2009.
16. C. Granata and P. Bidaud. Interactive person following for social robots. In Proceedings of CLAWAR 2011, 11th International Conference on Climbing and Walking Robots and the Support Technologies for Mobile Machines, Paris, 2011.
17. P. Morin and C. Samson. Motion control of wheeled mobile robots. Springer Handbook of Robotics (B. Siciliano and O. Khatib, eds.), Springer Berlin Heidelberg, 799-826, 2008.

STRATEGIES TO REDUCE ENERGY EXPENDITURE IN THE MOTION CONTROL OF LOWER-LIMB ORTHOSES

D. Sanz-Merodio*, M. Cestari, J.C. Arevalo and E.Garcia

Centre for Automation and Robotics, CSIC-UPM
La Poveda, 28500 Madrid, Spain
**E-mail: daniel.sanz@csic.es*
www.iai.csic.es/dca

By analyzing the dynamic principles of the human gait, an economic gait control strategy is performed and passive elements are included to increase energy efficiency in the motion control of active orthoses. Traditional orthoses use position patterns from clinical gait analysis (CGA) from healthy people, denormalized and adjusted to each user. These orthoses maintain a very rigid gait and their energetic cost is very high, reducing its autonomy. In this work, first, to take advantage of the inherent dynamics of the legs, a state machine pattern with different gains in each state is applied, to reduce actuator consumption. Next, different possibilities of using passive elements like springs or brakes in the joints to reduce power are analyzed. After an off-line parameter optimization and a heuristic improvement with genetic algorithms a reduction in power consumption of 16.8% is obtained applying state machine pattern control and a reduction of 18.9% using passive elements. Finally, combining both strategies a more natural gait is obtained and a reduction of power expenditure of 24.6% is achieved with respect to pure following CGA pattern.

Keywords: Lower-limb active orthoses, energy efficiency, bipedal walking.

1. Introduction

The term exoskeleton is usually used to describe a wearable device that augments the user's strength and endurance, whereas the term orthosis is typically used to describe a device that is used to assist a person with a limb pathology.[1] In designing a lower-limb active orthosis two fundamental issues must be taken into account: Stability and energy consumption. The energy required for these devices is excessive for a portable battery pack and this is currently the biggest obstacle to shooting their practical realization. A research device like Vanderbilt Orthosis[2] provides autonomy of one hour, while commercial lower-limb orthoses like active orthosis (Argo

Medical Technologies[3]) and eLEGS (Berkeley Bionics[4]) provide eight and six hours of autonomy, respectively. In this paper, two main approaches to be implemented in an active orthosis in order to increase its energy efficiency are presented. The proposal for energy-saving is centered on two key approaches:

(1) Design of a gait pattern, taking advantage of the inherent dynamics of the legs.
(2) Introduction of passive elements attached to the limbs, either as springs or rubber bands guided by pulleys placed at the joints to accumulate and release energy.

The traditional motion control of active orthoses is based on the tracking of reference patterns taken from CGA, which implies high power consumption. This work is focused on the sagittal plane motion of the hip, knee and ankle joints, due to the fact that the most energy consumption occurs in this type of movement.[5] In Section 2 the configuration of the biped robot used in simulation is presented. Lately, in Section 3, the methodology used is presented. In Section 4, the results obtained when a passive gait is implemented are presented and in Section 5 the results applying passive elements in the joints are shown. In Section 6 the viability of using both previous techniques together is analyzed and the results obtained on energy consumption are shown. Finally, in Section 7 the main ideas and the future lines of work are remarked.

2. Simulation set up

The ultimate goal of the ATLAS project[6] is to design an active orthosis that allows walking to a quadriplegic girl. So, masses, lengths, center of masses and moments of inertia have been adapted to the patient in the simulation with the features that show Table 1, following the estimations and adjustments given in[7] for the body inertial segment parameters (BISPs).

Table 1. Physical parameter of the simulated biped robot.

	Torso and head	Thick	Shank	Foot	Total
Mass (kg)	11.96	2.92	0.9	0.2	20
Length (m)	0.571	0.285	0.292	0.124	1.15

In order to estimate power consumption, a biped robot has been designed using a robot dynamics simulation package (*Yobotics SCS*[8]).

Through optimization by minimization of the mechanical power used, different techniques are going to be used to reduce consumption and its implementation in active orthoses is going to be analyzed.

3. Methodology of work

As it has been discussed earlier, traditional orthoses base their movement on the tracking of CGA patterns, usually using a PD control strategy. So, some preestablished and normalized angles and velocities, adapted to the height and weight of each user, will be used as reference angles and speeds in each actuator, performing a torque given by:

$$\tau_{joint} = K_p(\theta_{real} - \theta_{CGA}) - K_d(\dot{\theta}_{real} - \dot{\theta}_{CGA}) \tag{1}$$

where θ_{real} and $\dot{\theta}_{real}$ are actual angle and speed in the joint and θ_{CGA} and $\dot{\theta}_{CGA}$ are the desired position and speed values, and K_p and K_d are the gains of the PD control.

Therefore, the mechanical power consumed in a time T is given by:

$$P = \frac{\int_0^T \sum_{i=1}^{2}\sum_{j=1}^{3}(\omega_{ij}(t)\tau_{ij}(t))dt}{T} \tag{2}$$

where $\omega_{ij}(t)$ is the angular speed and $\tau_{ij}(t)$ is the torque at the joint j of leg i.

Mean velocity could differ in each experiment, for this reason, a non-dimensional parameter is considered for comparison purposes, which is the specific resistance:

$$\varepsilon = \frac{E}{WL} = \frac{P}{Wv} \tag{3}$$

where E is the energy required to travel a distance L, by a vehicle with weight W. P is the power consumed and v is the speed of the vehicle. The specific resistance considers the power consumed per unit of mass and unit of speed.

This gait tracking of a CGA pattern has been implemented in simulation, adjusting control gains and the power consumption has been measured and the specific resistance has been calculated. This measure will be taken as a reference to the rest of experiments.

4. Energy-efficient gait pattern

From the pendulum model for the gait,[9] the inherent swing properties of the leg must be taken into account to reduce energy expenditure while the legs follow the desired trajectory. Once the swing phase starts, the leg can continue its trajectory almost without energy intervention. In the single support phase of the gait, the biped behaves like two coupled pendulums formed by the stance and swing legs. The swing phase is divided into three sub-phases. In the early swing constant torque will be applied to both, the hip and the knee joint until each limb reaches the necessary speed, using a ballistic approach.[10] In the middle of the swing phase hip and knee will move freely then almost no torque will be needed. At the end of the swing phase, PD controller is used at the joints so that the desired posture at the moment of touch down can be realized.

Each leg will have a simple state machine, and in each state different control strategies will be used (see Figure 1 (a)). Table 2 shows energy consumption applying this control strategy and the reduction achieved over the pure CGA pattern PD control.

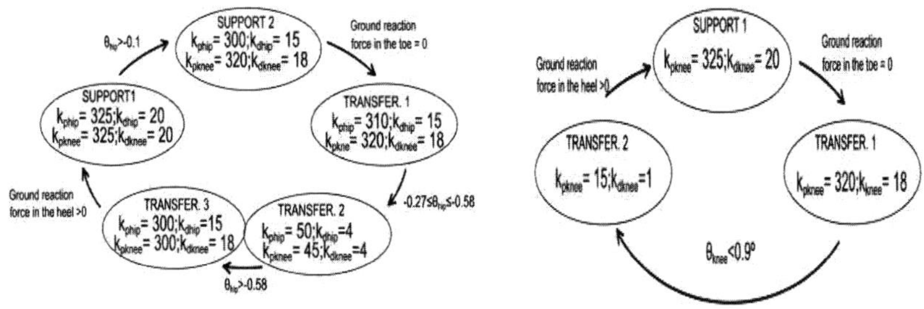

(a) (b)

Fig. 1. State machine controller. In each state it shows PD constants and the transition condition. (a) For hip and knee ; (b) For the knee

The movement of the thigh and the shank are independent and each of them reaches the larger kinetic energy in a different point of the gait cycle (see Figure 2). So, we have achieved a more efficient control algorithm implementing two state machines, one for the hip, shown in Figure 1 (a), and the other one for the knee, shown in Figure 1 (b).

Table 2 shows that with this new strategy a reduction in mechanical energy expenditure of 16.8% is achieved.

This section has been focused on the active control of hip and knee in order to reduce power. Also, ankle joint can be controlled by passive elements that provide ground clearance. This will be analyzed in the next section.

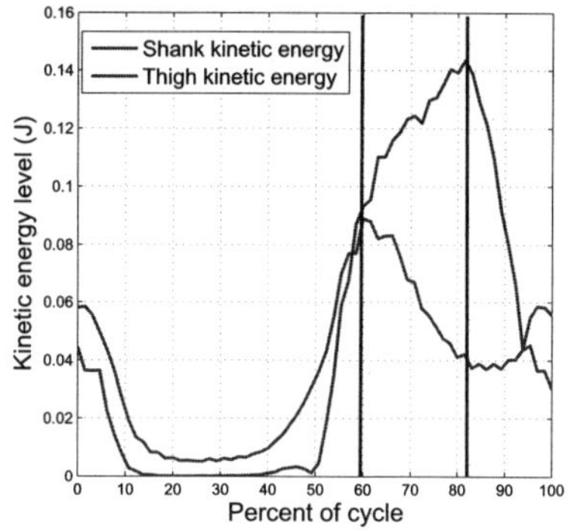

Fig. 2. Gap in the maximum kinetic energy in thigh and shank.

Table 2. Result applying a passive control strategy.

Control strategy	Specific resistance	Energy saving %
PD Control of CGA references	1.13	
A state machine for hip and knee together	0.97	14.9
Two state machine one for hip and another for knee	0.94	16.8

5. Including passive elements

Human tendons are passive elastic elements that store energy and release it without metabolic cost. In the ankle joint more than 60% of energy is provided by tendons.[11]

Analyzing biomechanical features of clinical gate analysis (CGA) of a healthy individual, some important conclusion about how to reduce power consumption can be reach with passive elastic elements. From walking gait data at low speeds, it shows that the hip and the ankle joints have periods of negative power followed by a period of positive power, so a spring could store energy during the negative power period and release it during the positive power period. Moreover, power at the knee is predominantly negative that means that it basically dissipates power. So, it could be a good deal to incorporate in the knee brace some braking or damping elements.

5.1. *Incorporating hip and ankle springs*

The use of a spring in the ankle and another at the hip has been analyzed in simulation. From the torques τ and the angles θ in these joints obtained from the CGA, the mechanical power computed from equation (2) has been minimized taking into account that a torsion spring exerts torque if the joint ankle exceeds a threshold angle θ_{thres}. This torque is given by:

$$\tau_{spring} = \begin{cases} k_{spring}(\theta - \theta_{thres}) & if\ \theta > \theta_{thres} \\ 0 & if\ \theta < \theta_{thres} \end{cases} \tag{4}$$

being k_{spring} the spring constant and θ the joint angle.

Making use of the MATLAB function *fminsearch* with these CGA values, statically optimal values of spring constants and threshold angle, shown in table 3, are obtained.

Table 3. Values for spring constants and threshold angle in hip and ankle joints, with a static optimization.

	From CGA optimization		From genetic algorithm adjustment	
Joint	$k_{spring}(Nm/rad)$	$\theta_{thres}(rad)$	$k_{spring}(Nm/rad)$	$\theta_{thres}(rad)$
Hip	120	-0.28	200	0.05
Ankle	230	-0.05	80	0.05

In this static optimization, it has been assumed that the kinematics of the gait does not change when including passive elements. But when these springs are introduced in the simulated biped, it is observed that although the power consumption is reduced in the joint in which the spring has been added, the rest of the joints are forced and the overall power expenditure increases. Then, from the static optimized values, a heuristic adjustment based on genetic algorithms has been built-in and the values for spring constants and threshold angle shown in Table 3 have been obtained. Table 4 shows energy consumption reduction including one spring at the hip, one at the ankle and both together.

Figure 3 shows in red the torque exerted with pure CGA control, in green the torque when a spring is included and in black the torque exerted by the spring, just at the hip and at the ankle. Remark the important reduction in energy expenditure achieved when two springs in hip and ankle are working together, proving the synergy between these joints.

Usually these elastic elements do not incorporate electronics and then the recycled energy is released too soon. Some devices[12] incorporate some mechanisms that force to release energy in the exact moment of the ankle push-off and it consumes a 10% less than the spring alone.

5.2. *Damping at the knee*

As mentioned before, the knee dissipates energy most of the cycle time. The use of dissipative device could help save the energy lost during knee braking. Torque exerted by a damper is, without taking into account hysteresis:

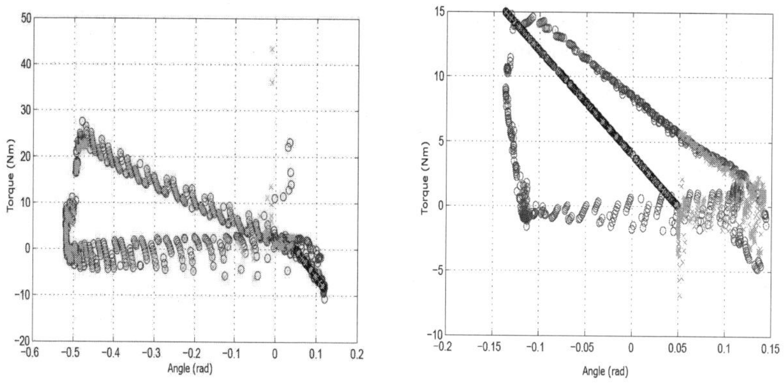

(a) (b)

Fig. 3. Spring torque. (a) Hip ; (b) Ankle.

$$\tau_{damper} = \begin{cases} B\dot{\theta} & if \;\; damper \; is \; enable \\ 0 & if \;\; damper \; is \; disable \end{cases} \tag{5}$$

where B is the damping constant and $\dot{\theta}$ is the joint angular speed.

Just like in the previous subsection, an optimization has been performed with MATLAB based on the CGA angles. With this static optimization a damping constant $B = 3.8\ Nms/rad$ has been applied to the range of the gait cycle from 55% to 64%. But this part of the gait corresponds to toe-off phase. If the damping coefficient is increased during the toe-off phase, then the foot will not be allowed to provide ground clearance and the consequence will be the biped stumbling against the ground. So, the static theoretical results obtained are deceitful, and therefore the damping constant and the range of application in the gait cycle have been obtained using genetic algorithms. Table 4 shows that the energy reduction with a constant $B = 4\,Nms/rad$ from 15% to 35% of the gait cycle is 1.2%.

Table 4. Results applying passive elements.

Control strategy	Specific resistance	Energy reduction %
Control PD CGA	1.13	
Hip spring	1.11	1.7
Ankle spring	1.04	7.2
Knee damping	1.11	1.2
Locking the knee	1.07	4.8
Hip and ankle springs	0.96	14.48
Hip and ankle springs and locking the knee	0.91	18.9

Some previous works[13] that use magneto-rheological damping at the knee, have not achieved their goal of reducing energy and have obtained an increment of the metabolic rate consumption. The causes for this increment are the added weight, caused by the introduction of passive devices and a non zero friction introduced by the magneto-rheological damper in off-state.

5.3. *Introducing a lock at the knee*

By examining the gait, it is observed that, in same phases, the knee angle is almost constant. So a lock has been considered for blocking the knee when the swing leg moves forward and when the support leg is in the single support phase. Adjusting the range of activation of the lock again by genetic algorithms, it is obtained that the lock from 30.3% to 40.2% and from 67.1% to 85.2% of the gait cycle. This lock of the knee results in a power consumption reduction of 4.8%.

After analyzing the benefits of each passive element (springs, damping and brakes), an interesting analysis is performed to consider the combined case of all these mechanism and their benefits in power saving. Using all the passive elements, springs and lock, together in the dynamic simulation an important reduction in energy consumption of 18.9% is achieved.

6. Passive walking and passive elements all together

In an active orthosis, all the power must be provided by the actuators, then if the goal is to reduce energy consumption the next logical step is to join together all the previous ideas: passive control pattern and passive elements.

The control pattern described in section 4 will be used based on finite states machines, but now the response in each sub-phase should be changed.

In phase SUPPORT 2 (see Figure 1) most of the work that the actuator needs to do, is made by the spring at the hip, so no actuator torque is needed. More over in the middle and last phase of swing (phases TRANSFER. 2 and 3 in Figure 1), a passive control scheme is applied to take advantage of the thigh inertia.

At the same time, if a damper that brakes the knee in the early stance is included, then a state machine could be implemented for the knee to exploit the potential energy stored in mid swing and to reduce the actuator consumption.

Figure 4 shows the finite state machines employed for hip and knee incorporating passive elements.

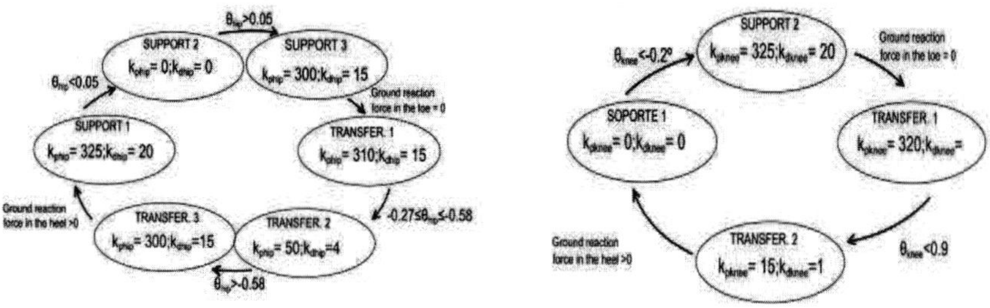

(a) (b)

Fig. 4. Finite state machine with passive elements at the hip. (a) At the hip ; (b) At the knee.

Putting together passive elements and a strategy of passive control a reduction of 24.6% in energy expenditure is achieved.

7. Conclusions and future works

This paper presents a deep analysis of the biped gait and multiple options to reduce power consumption, looking for implementing these strategies on an active orthosis. First, on dynamics simulation of the different passive gait control schemes, a reduction in power consumption of 16.8% has been obtained. Next, considering passive elements like springs, dampers and lockers, a reduction of 18.9% has been achieved. Joining together both ideas a reduction of 24.6% in power consumption is achieved.

The focus of this work is purely theoretical, but the results presented in this paper will be applied to the experimental platform ATLAS.

8. Acknowledgments

This work has been partially funded by the Spanish National Plan for Research, Development and Innovation through grant DPI2010-18702 which has funded personnel costs and by AECID through grant PCI-Iberoamerica A1/039883/11 which has funded equipment and consumable costs.

References

1. H. Herr, Exoskeletons and orthoses: classification, design challenges and future directions, *Journal of NeuroEngineering and Rehabilitation* **6**, p. 21 (2009).
2. R. Farris, H. Quintero and M. Goldfarb, Preliminary evaluation of a powered lower limb orthosis to aid walking in paraplegic individuals, *Neural Systems and Rehabilitation Engineering, IEEE Transactions on* , 102 (2011).

3. A. M. Technologies, *www.argomedtec.com*, (2012).
4. B. Bionics, *www.eksobionics.com*, (2012).
5. B. Siciliano and O. Khatib, *Springer Handbook of Robotics* (Springer, London, 2008).
6. E. Garcia, D. Sanz-Merodio, F. Sanchez, J. Arevalo and G. de Santos P., Development of the atlas lower-limb active orthosis, in *14th Int. Conf. Climbing and Walking Robots and the Support Technologies for Mobile Machines, Paris, France*, 2011.
7. R. Dumas, L. Cheze and J. Verriest, Adjustments to McConville et al. and young et al. body segment inertial parameters, *Journal of biomechanics* **40**, p. 543553 (2007).
8. Yobotics Inc., Boston, MA., *Yobotics! Simulation Construction Set: Users Guide*, (2002). Available: `http://www.yobotics.com/`.
9. G. A. Cavagna, N. C. Heglund and C. R. Taylor, Mechanical work in terrestrial locomotion: two basic mechanisms for minimizing energy expenditure, *American Journal of Physiology-Regulatory, Integrative and Comparative Physiology* **233**, p. R243 (1977).
10. S. Mochon and T. A. McMahon, Ballistic walking, *Journal of Biomechanics* **13**, p. 4957 (1980).
11. E. N. Brown, H. M. Herr and P. Krishnaswamy, Human leg model predicts ankle Muscle-Tendon morphology, state, roles and energetics in walking, *Computational Biology* **7**, p. 16.
12. S. H. Collins and A. D. Kuo, Recycling energy to restore impaired ankle function during human walking, *Plos One* **5**, p. e9307 (2010).
13. A. Valiente, *Design of a quasi-passive parallel leg exoskeleton to augment load carrying for walking*, tech. rep., DTIC Document (2005).

STUDY ON THE CONTROL BASED ON VIRTUAL SPRING MODEL FOR LOWER EXTREMITY EXOSKELETON

JONG-WON LEE
Department of Intelligent Robot Engineering, University of Science and Technology (UST), Ansan, Gyeonggi-do, 426-791, Korea
E-mail: jongwonia@ust.ac.kr

HYOGON KIM, JAEHO JANG, SANGDEOK PARK
Field Robot R&D Group, Korea Institute of Industrial Technology (KITECH), Ansan, Gyeonggi-do, 426-791, Korea
E-mail: {kimhg, jaeho, sdpark}@kitech.re.kr

In recent years, many research groups have developed the lower extremity exoskeleton composed to fully or partially passive components to reduce energy consumption. However, it is difficult to apply the above exoskeleton to real life because of the complex mechanism and bulky design. In this paper, we propose the control method based on virtual spring model for lower extremity exoskeleton and the method for determination of stiffness of virtual spring and desired position. The preliminary simulation result shows that the control method is applicable to the lower extremity exoskeleton.

1. Introduction

Human locomotion is the complex process including phases of the generation and dissipation of mechanical energy (i.e. the region of positive and negative power) [1]. In these phases, human muscles are activated to product or absorb energy by concentric (shortening) and eccentric (lengthening) contractions. In the field of robotics, most of the passive components are used to store energy in the region of the negative power and release the energy in the region of the positive power. In recent years, many research groups have developed the lower extremity exoskeleton composed to fully or partially passive components to reduce energy consumption [2-6]. Although the exoskeletons mentioned above have advantage of low energy consumption, it is difficult to apply the exoskeleton to real life because of the complex mechanism and bulky design. For example, the characteristic of the passive components must be changed according to the change of wearer or wearer's motion (e.g. from walking to running). However, the exoskeletons mentioned above are difficult to change

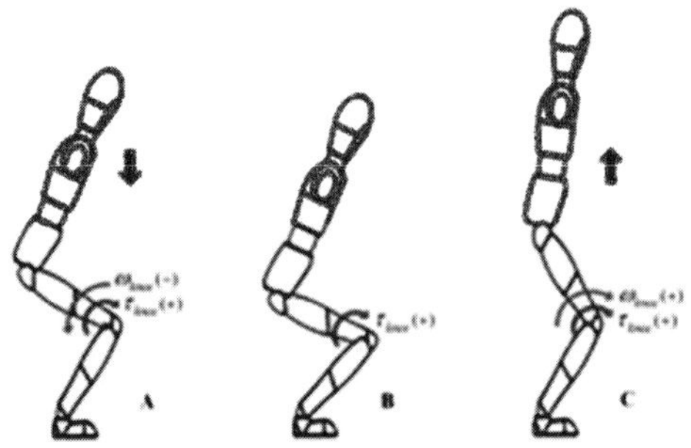

Figure 1. Description of the sit-down motion (A), semi-squat position (B), and stand up motion (C) in sagittal plane.

passive components. In this paper, to solve the above problem, we propose the control method based on virtual spring model. Section 2 contains the brief background on the biomechanics in sit-to-stand motion for clear description of our control concept. Section 3 describes the modeling and control method based on virtual spring model. Section 4 includes the simulation method for verification of proposed control method and its results. Finally, the summary of this research and future works are described in section 5.

2. Motivation

Before getting into the explanation about our control concept, we describe the results of human motion analysis. It is very useful for reader to better understand our concept. In this section, we focus only on knee joint for brief description.

2.1. *Biomechanics of Sit-to-Stand Motion*

Sit-to-stand motion of human is a good example to show our concept (see Figure 1). Fig. 2 show the behavior of the knee joint during sit-to-stand motion through one cycle defined from the beginning of the sit-down motion to the ending of the stand-up motion. A normal and healthy subject is participated in the measurements. Kinematics was measured by camera based optoelectronic system. An inverse dynamics approach was applied to calculate joint moment. The joint power was calculated by multiplying the joint moment and velocity data. During sit-down motion (at ≈ 0-30 % cycle time), the knee joint muscles dissipates energy to decelerate the bodyweight. On the other hand, during stand-up motion (at ≈ 60-100 % cycle time), the knee joint power is predominantly positive, i.e. the energy is generated to lift the bodyweight. At semi-squat position (at ≈ 30-60 % cycle time), the knee joint power was relatively small but the moment of the knee joint is required to maintain the position. Note that the

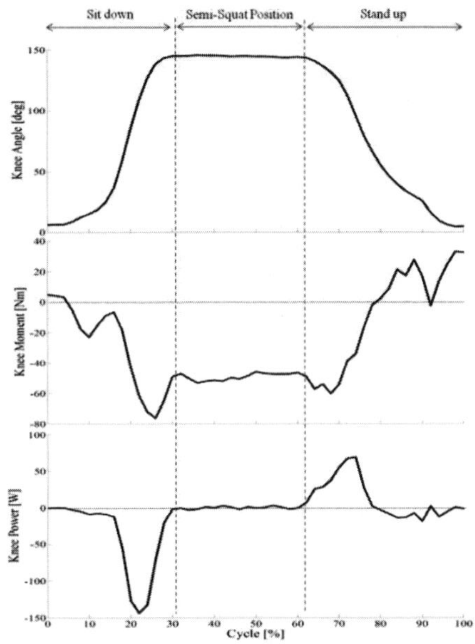

Figure 2. Knee joint angle, moment, and power during sit-to-stand motion. The cycle starts with sit-down motion and finishes with the end of the stand-up motion. The values are scaled for an 80 kg subject.

behavior of generation and dissipation of the energy is not limited to sit-to-stand motion, can be found a variety of human locomotion such as level walking and ascent and descent ramp and stair walking [1].

2.2. *Control Concept*

Based on the Figure 2, even if the exoskeleton transfers payloads to the ground throughout their own legs, human feels fatigued easily in uncomfortable poses such as semi-squat position. Therefore the exoskeleton can support not only payloads but wearer's bodyweight also. Figure 3.(A) illustrates our control concept. The exoskeleton has two degrees of freedom and three links per leg. Let's assume that the tension spring is utilized between the end-effecter of the exoskeleton and arbitrary point on the ceiling. During sit-down motion, the spring is extended and provides forces to support the bodyweight. Thus, at the semi-squat position, human can feels reduced the bodyweight. During the stand-up motion, the spring is released and adds the force to lift the wearer's body. As mentioned above, if this concept is realized by mechanical devices, the system will be not only complex but bulky devices also. In the following section, we describe the control method based on virtual spring model.

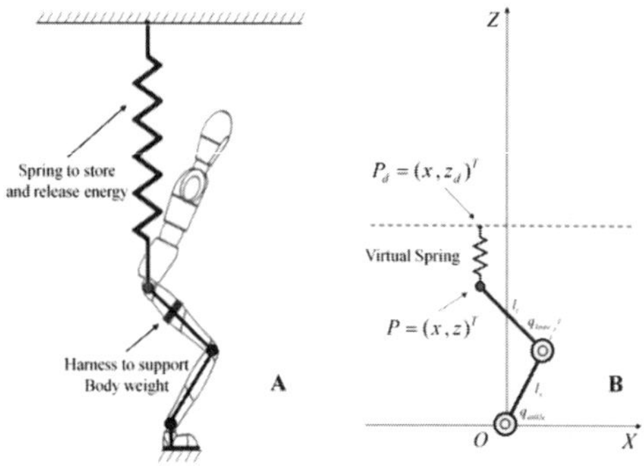

Figure 3. The concept of the control method (A) and kinematic model of the lower extremity exoskeleton (B).

3. Control

The human locomotion can be classified into the stance phase and swing phase. In stance phase, the important role of human joints is to support and move forward the weight of human and payloads. On the other hand, in swing phase, the role of human joints is to swing the legs. Thus, in most of human locomotion, the magnitude of required moment in the swing phase is relatively small compared to stance phase. If the exoskeleton has lightweight legs, the impedance by inertia of legs will be small. In this paper, based on the above assumption, we only considered on the control method in stance phase.

3.1. *Kinematic Model*

In stance phase, it can be assumed that the exoskeleton is a manipulator in two-dimensional space under gravity. As illustrated in Figure 3.(B), the exoskeleton has two joints in total and the base of the exoskeleton is fixed at contact point, *O*. *P* denotes the actual position of the end-effector in task space and P_d is the desired position that human wants to move. The virtual spring is modeled between P_d and *P*.

3.2. *Control input based on Virtual Spring Model*

The control input of the exoskeleton is as follows.

$$u = -J_p(q)^T F_{vs} + g(q) \tag{1}$$

$$F_{vs} = k(P_d - P) \tag{2}$$

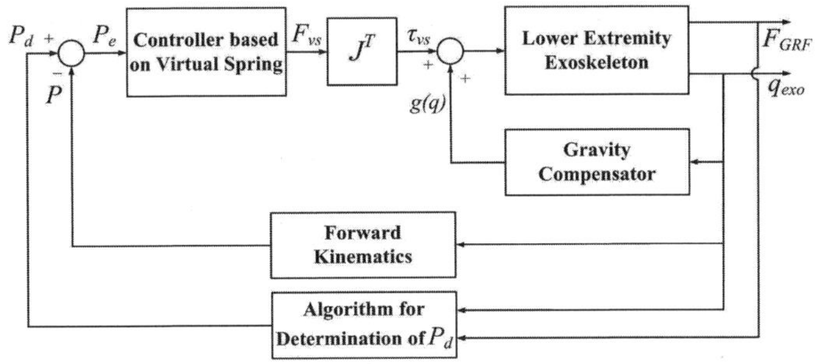

Figure 4. Control method based on virtual spring model with the algorithm for determination of desired position.

In Eq. (1), the control input is composed of two terms, $-J_p(q)^T F_{vs}$ is the joint torque term generated by virtual spring model, and $g(q)$ is a term for gravity compensation [7]. In other word, the joint torque to support payloads is generated by second term and the joint torque to support the wearer's bodyweight is generated by third term. The block diagram of the control method for the exoskeleton is shown in Figure 4.

3.3. *Determination of Desired Position, P_d*

In this section, how to decide the desired position is discussed. In industrial robots, the desired position is decided by object tasks. However, under human robot interaction, the desired position should be determined in the light of user's safety. We got an idea from the behavior of human. We know from experience that human tends to change their poses in a more comfortable and stable positions. The comfortable and stable position means the position at which the energy consumption is minimized. The desired position at single and double support phases are illustrated in Figure 5.

3.4. *Determination of Spring Coefficient, k*

The stiffness of the spring, *k*, is important variable. If the value of *k* is very large, it can interfere with the movement of human. Conversely, if the *k* has very small value, it can deficient to support human locomotion. To determine the stiffness of the spring, in this paper, we modeled simple mass-spring system equivalent to human model (see Figure 5). The stiffness of the spring can be determined by using Hooke's law.

$$k = \frac{M_{eq} g}{\Delta z} \tag{3}$$

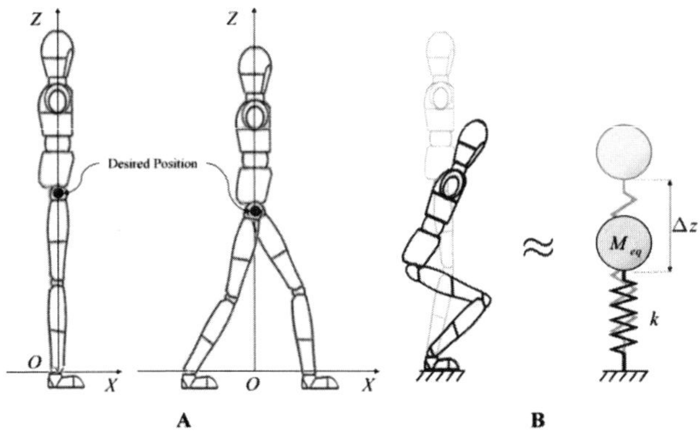

Figure 5. Desired position in the case of single and double support phase (A), and the simple 1 DOF mass-spring system equivalent to human model (B).

Where M_{eq} denotes a mass equivalent to wearer's bodyweight. g is the gravity acceleration and Δz is the maximum displacement in sit-to-stand motion. Note that M_{eq} can be determined by the literatures about the mass distribution ratio of the human body and Δz is also determined by the data of motion analysis. In this paper, the stiffness of the spring is approximately 1000 N/m for simulation (Δz is 0.52 m and M_{eq} is about 53 kg).

4. Simulation and Results

In order to verify the control method, we conducted the simulation. Figure 6 illustrates the simulation method. The human model makes a sit-to-stand motion with and without control. At that time, we captured and compared the moment and power at the knee joint of human model.

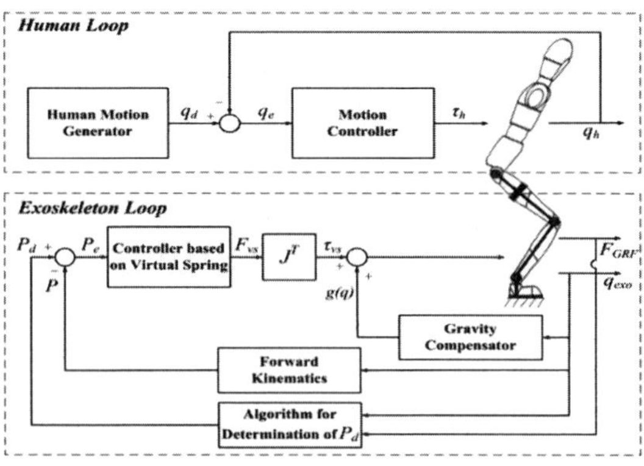

Figure 6. The block diagram of the simulation method to verify the performance of the control method based on virtual spring model.

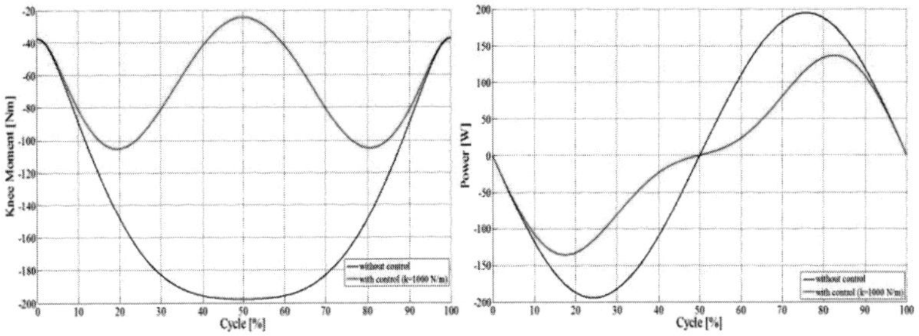

Figure 8. Simulation results: black line – without control; red line – with control (k = 1000 N/m)

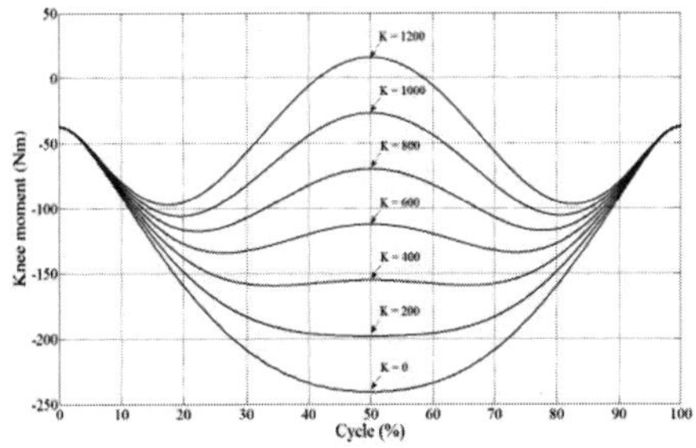

Figure 7. The knee moment variation according to spring stiffness.

Figure 7 shows the data of knee moment and power during a cycle of sit-to-stand motion. This result shows that the knee moment decreased significantly with the controlled exoskeleton compared to the case without controlled exoskeleton. However, few changes of the knee moment occurred at the beginning and ending of the cycle (at 0-5 % and 95-100 % cycle time). The reason is that the configuration of the exoskeleton is near singular. The effect of the stiffness of virtual spring on the knee joint is shown in Figure 8. As the stiffness of virtual spring increases, the knee moment to support and lift the bodyweight is decreased. When the value of the stiffness is over 1000 N/m, the knee moment is rather increased.

5. Conclusion

In this paper, the control method based on virtual spring model was proposed. To apply the control method to the lower extremity exoskeleton, we also introduced the methods of the determination of the spring coefficient and desired position. The simulation results show the validity of the proposed control method and applicability to lower extremity exoskeleton. Future works will focus on the simulation on a variety of human motions and implementation to actual lower extremity exoskeleton.

References

1. P. DeVita, J. Helseth, and T. Hortobagyi, "Muscle do more positive than negative work in human locomotion," *in journal of Experimental Biology*, vol. 210, pp. 3361-3373 (2007).
2. Collins SH and Kuo AD, "Recycling energy to restore impaired ankle function during human walking," *in PLoS ONE*, vol. 5 (2010).
3. Y. Hirata, T. Isoda and K. Kosuge, "Development of passive wearable walking support system based on brake control," *in proceeding of the IEEE International Conference on Mechatronics and Automation*, Kagawa, Japan, pp. 363-368 (2007).
4. C. J. Walsh, K. Pasch, and H. Herr, "An autonomous, underactuated exoskeleton for load-carrying augmentation," *in proceeding of the IEEE/RSJ International Conference on Intelligent Robots and Systems*, Beijing, China, pp. 1410-1415 (2006).
5. A. M. Dollar and H. Herr, "Design of a quasi-passive knee exoskeleton to assist running," *in proceeding of the IEEE/RSJ International Conference on Intelligent Robots and Systems*, Nice, France, pp. 747-754 (2008).
6. S. Krut, M. Benoit, E. Dombre, and F. Pierrot, "MoonWalker, a lower limb exoskeleton able to sustain bodyweight using a passive force balancer," *in proceeding of the IEEE/RSJ International Conference on Robotics and Automations*, Alaska, USA, pp. 2215-2220 (2010).
7. J. K. Salisbury, "Active stiffness control of a manipulator in Cartesian coordinates," *in proceeding of the 19th IEEE International Conference on Decision and Control*, Albuquerque, New Mexico, pp. 95-100 (1980).

STATE SPACE MODELLING AND CONTROL OF SQU-TWO-WHEELED MOBILITY VEHICLE (SQU-TWMV): AN ENERGY ANALYSIS APPROACH

A Al-Harrasi[1]*, K M Goher*[1]*, S Al-Abdali*[1]*, A Al-Siyabi*[1]*, J Al-Abri*[1]*,*
M O Tokhi[2] *and A M Almeshal*[2]
[1]Department of Mechanical and Industrial Engineering
Sultan Qaboos University, Oman, e-mail: kgoher@squ.edu.om
[2]Department of Automatic Control and Systems Engineering
The University of Sheffield, United Kingdom

This paper addresses the control of on a two-wheeled single-seat vehicle project built at Sultan Qabbos University, Oman. The controller design is based on the Linear Quadratic Regulator (LQR) and the pole placement techniques. The state space model of the SQU-Two-Wheeled Mobility Vehicle (SQU-TWMV) that mimics the inverted pendulum scenario has been obtained from the linearized equations of the mathematical model. Estimation of the vehicle energy requirements is of great importance in terms of the cost requirements, the design of appropriate actuators and the development of suitable controllers. This work specifically carries out an energy analysis for the SQU-TWMV. Further investigation of the robustness of the developed LQR and the pole placement techniques is emphasised through various experiments using an applied impact load on the vehicle.

1. Introduction

The research on two-wheeled balancing robot has gained momentum over the last decade due to the nonlinear and unstable dynamics system. Various control strategies had been proposed by numerous researchers to control the two-wheeled balancing robot such that the robot able to balance itself. Fiacchini et al. (2006), designed a personal pendulum vehicle. The vehicle is simply a two-wheeled inverted pendulum with steering ability. They have presented various controller types for both linear and non-linear control mechanisms such as partial linearization, energy shaping, velocity stabilization and non-linear controller tuning. Experiments and simulations have been carried out to compare the behaviour of the Linear Quadratic Regulator (LQR) and energy-shaping controllers. The results show a very similar behaviour around the equilibrium point of the angle. Xiong et al, (2010) presented a new method to find the optimal LQR control parameters for a double inverted pendulum system. The authors compared the system response with optimized parameters and heuristic parameters. Optimal selection of the LQR controller parameters was superior and tended to equilibrium point faster than heuristic selection of the parameters. Bogdanov (2004) reported and applied different optimal controllers on a double inverted pendulum system. The author compared LQR, state-dependent riccati equation (SDRE) and optimal neural network controllers. The author proved via simulations that the SDRE controller coupled with optimal neural network controller has optimal control over the double inverted pendulum system. Evolutionary algorithms have been used for optimizing existing controllers of the inverted pendulum systems. Wongsathan and Sirima (2008) have demonstrated the application of genetic algorithm to design an LQR controller for the inverted pendulum system. Authors have provided a brief comparison between the experimental LQR and the genetically optimized LQR controller, where the latter has achieved a better system response. Naser *et al,* (2010) presents investigations into the performance comparison of Linear Quadratic Regulator (LQR) and PID-PID controllers for a highly nonlinear 2–wheels balancing robot. The performances of the LQR and PID-PID controllers are examined in terms of input tracking and disturbances rejection capability. In the work presented by Nawawi et al (2006), a robust controller based on Sliding Mode Control is proposed to perform the robust stabilization and disturbance rejection of the system.

In this paper an investigation of the system performance is carried out, based on the linear state space model of the vehicle, using Linear Quadratic Regulator and pole placement techniques. The mathematical model of the vehicle is derived based on the Lagrangian Euler (LE) formulation.

2. SYSTEM MODELLING AND DESCRIPTION

In this project, the dynamic model of the whole system is represented by the stabilization mode of the vehicle at stage 3 as shown in Figures 1 and 2.

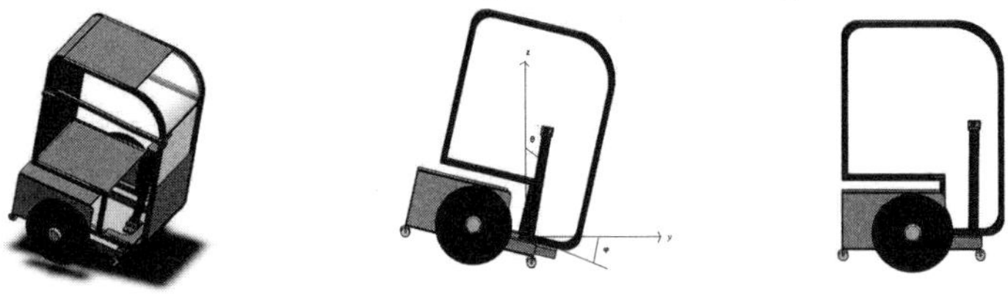

(a) (b) (c)

Fig. 1 SOLIDWORKS design of the SQU-TWMV

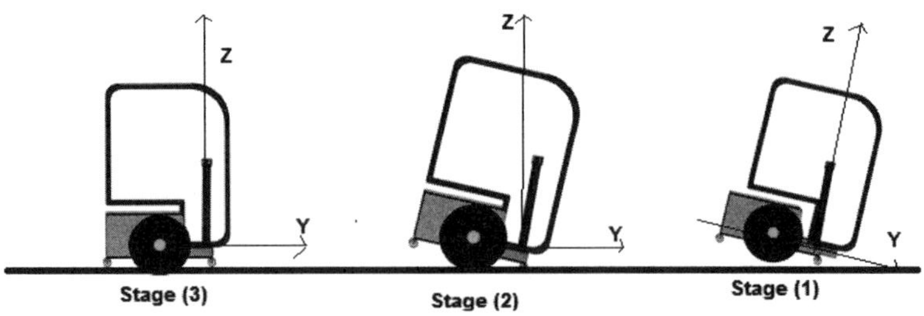

Fig. 2 Schematic diagram for the sliding and stabilization modes

The vehicle is considered of three motion coordinates as shown in Figure 3; linear motion of the system in the Y direction, angular motion of the body about the rod on the Z direction (tilt angle), and angular motion of the body about the cart (base) on the Y direction (inclination angle). The system inputs are the force driving the wheels and a disturbance force acting on the vehicle.

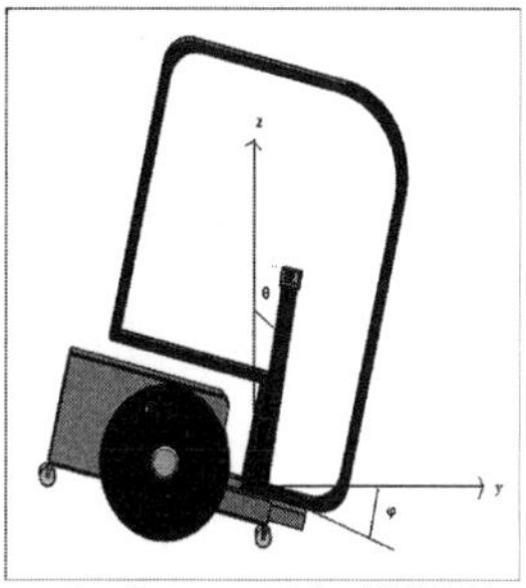

Fig. 3 Schematic diagram of the TWMV with the rod

Applying LE dynamic formulation on the vehicle yields the following equations:

$$\frac{\partial}{\partial t}\left(\frac{\partial L}{\partial \dot{Y}}\right)-\frac{\partial L}{\partial Y}=F_1 \tag{1}$$

$$\frac{\partial}{\partial t}\left(\frac{\partial L}{\partial \dot{\theta}}\right)-\frac{\partial L}{\partial \theta}=F_2 \tag{2}$$

Where, Y is the linear displacement of the vehicle, θ is the tilt angle between the rod and Z axis, M_c is the mass of the cart including the mass of the motors and batteries, m_r is the mass of the rod including the mass of the rider, L_c is the distance from the centre of gravity to the centre of mass, L_r is the distance of the rod, F_1 is the driving force of the wheel, and F_2 is an external disturbance force.

Using equations (1) and (2), the systems equation of motion can be described as follows:

$$(M_c+m_r)\ddot{Y}-M_cL_c(\cos(\theta)+\sin(\theta))\left(\ddot{\theta}+\dot{\theta}^2\right)=F_1 \tag{3}$$

$$\begin{aligned}&(m_rL_r^2+M_cL_c^2)\ddot{\theta}+(m_rL_r\cos(\theta)-M_cL_c\sin(\theta))\ddot{Y}\\&+2\left(M_cL_c^2-M_rL_r^2\right)g\cos(\theta)\sin(\theta)=F_2\end{aligned} \tag{4}$$

Linearization of the above non-linear 2nd order differential equations yields a linear model of the system in the state-space form.

$$\dot{X}=AX+Bu \tag{5}$$

$$Y=CX+Du \tag{6}$$

Equations (5) and (6) are two equations of general state-space model for a dynamic system, which are called state and output equations respectively.

Where

A is the state matrix,
B is the input matrix ,
C is the output matrix ,
D is the direct transmission matrix ,

Six state space variables were chosen for the dynamic system of two wheeled vehicle where the state vector of dynamic system (X) can be formulated as follows:

$$X = \begin{bmatrix} y & \theta & \varphi & \dot{y} & \dot{\theta} & \dot{\varphi} \end{bmatrix} \tag{7}$$

From the system equations, the state space model is derived using MATLAB code as shown below:

$$\begin{bmatrix} \dot{x}_1 \\ \dot{x}_2 \\ \dot{x}_3 \\ \dot{x}_4 \\ \dot{x}_5 \\ \dot{x}_6 \end{bmatrix} = \begin{bmatrix} 0 & 0 & 0 & 1 & 0 & 0 \\ 0 & 0 & 0 & 0 & 1 & 0 \\ 0 & 0 & 0 & 0 & 0 & 1 \\ 0 & A_1 & A_4 & 0 & A_7 & A_{10} \\ 0 & A_2 & A_5 & 0 & A_8 & A_{11} \\ 0 & A_3 & A_6 & 0 & A_9 & A_{12} \end{bmatrix} \begin{bmatrix} x_1 \\ x_2 \\ x_3 \\ x_4 \\ x_5 \\ x_6 \end{bmatrix} + \begin{bmatrix} 0 & 0 \\ 0 & 0 \\ 0 & 0 \\ B_1 & 0 \\ B_2 & 1 \\ B_2 & 0 \end{bmatrix} \begin{bmatrix} F_1 \\ F_2 \end{bmatrix} \tag{8}$$

Where $A_1, A_2, \ldots\ldots\ldots, A_{12}$ *and* $B_1, \ldots\ldots, B_4$ are the matrix parameters coeffecients identified as follows:

$$\begin{aligned} A_1 = & \left(F_1 \sin(x_2)/L_r - 2L_r m_r g \cos(x_2^3) + 4L_r m_r g \sin(x_2^2)\cos(x_2) + L_r m_r \cos(x_2) x_5^2\right) \\ & /\left(1 - m_r \cos(x_2^2) - M_c \sin(x_3^2)\right)/\left(m_r + M_c\right) - 2\left(F_1\left(1 - \cos(x_2)/L_r + \sin(x_3)/L_c\right)\right. \\ & \left. - 2L_r m_r g \sin(x_2)\cos(x_2^2) + M_c L_c \cos(x_3) x_6^2 - 2M_c L_c g \cos(x_3)\sin(x_3^2) + L_r m_r \sin(x^2) x_5^2\right) \\ & /\left(1 - m_r \cos(x_2^2) - M_c \sin(x_3^2)\right)^2 /\left(m_r + M_c\right) m_r \cos(x_2)\sin(x_2) \end{aligned} \tag{9}$$

$$\begin{aligned} A_2 = & -2L_r^2 \sin(x_2^2) m_r g + 2L_r^2 \cos(x_2^2) m_r g + 1/L_r \sin(x_2)\sin(x_2)(F_1(1 - \cos(x_2)/L_r + \sin(x_3)/L_c) \\ & 2L_r m_r g \sin(x_3)\cos(x_3^2) + M_c L_c \cos(x_3) x_6^2 - 2M_c L_c g \cos(x_3)\sin(x_3^2) \\ & + L_r m_r \sin(x_2) x_5^2)/(1 - m_r \cos(x_2^2) - M_c \sin(x_3^2))/(m_r + M_c) - 1/L_r \cos(x_2)(F_1 \sin(x_2)/L_r \\ & + L_r m_r \cos(x_2) x_5^2)/(1 - m_r \cos(x_2^2) - M_c \sin(x_3^2)/(M_c + m_r) \\ & + 2/L_r \cos(x_2^2)(F_1(1 - \cos(x_2)/L_r + \sin(x_3)/L_c)) - 2L_r m_r g \sin(x_3)\cos(x_3^2) + M_c L_c \cos(x_6^2) \\ & - 2M_c L_c g \cos(x_3)\sin(x_3^2) + L_r m_r \sin(x_2) x_5^2)/(1 - m_r \cos(x_2^2) - M_c \sin(x_3^2))^2 /(M_c + m_r) m_r \sin(x_2) \end{aligned} \tag{10}$$

$$\begin{aligned}A_3 &= 1/L_c\sin(x_3)(F_1\sin(x_2)/L_r - 2L_rm_rg\cos(x_2^3) + 4L_rm_rg\sin(x_2^2)\cos(x_2)\\&+ L_rm_r\cos(x_2)x_5^2)/(1-m_r\cos(x_2^2) - M_c\sin(x_3^2))/(m_r+M_c) - 2/L_c\sin(x_3)(F_1(1-\cos(x_2)/L_r\\&+\sin(x_3)/L_c) - 2L_rm_r\sin(x_2)\cos(x_2^2) + L_cM_c\cos(x_3)x_6^2)\\&-2M_cL_cg\cos(x_3)\sin(x_3^2) + L_rM_r\sin(x_2)x_5^2)/(1-m_r\cos(x_2^2)\\&-M_c\sin(x_3^2))^2/(m_r+M_c)m_r\cos(x_2)\sin(x_2)\end{aligned} \tag{11}$$

$$\begin{aligned}A_4 &= (F_1\cos(x_3)/L_c - M_cL_c\sin(x_3)x_6^2 + 2M_cL_rg\sin(x_3^3) - 4M_cL_rg\sin(x_3)\cos(x_3^2))\\&/(1-m_r\cos(x_2^2)) - M_c\sin(x_3^2)/(m_r-M_c) + 2(F_1(1-\cos(x_2)/L_r) + \sin(x_3)/L_c)\\&-2/L_c\sin(x_3)(F_1(1-\cos(x_2)/L_r + \sin(x_3)/L_c) - 2L_rm_r\sin(x_2)\cos(x_2^2) + L_cM_c\cos(x_3)x_6^2\\&-2M_cL_cg\cos(x_3)\sin(x_3^2) + L_rM_r\sin(x_2)x_5^2)/(1-m_r\cos(x_2^2)\\&-M_c\sin(x_3^2))^2/(m_r+M_c)M_c\cos(x_3)\sin(x_3)\end{aligned} \tag{12}$$

$$\begin{aligned}A_5 &= -1/L_r\cos(x_2)(F_1\cos(x_3)/L_c - 2L_rm_rg\cos(x_3^3) + 4L_rm_rg\sin(x_3^2)\cos(x_3)\\&-M_cL_c\sin(x_3)x_6^2 + 2L_cM_cg\sin(x_3^3) + 4M_cL_cg\cos(x_3^2)\sin(x_3))/(1-m_r\cos(x_2^2)\\&-M_c\sin(x_3^2))/(m_r+M_c) - 2/L_r\cos(x_2)(F_1(1-\cos(x_2)/L_r + \sin(x_3)/L_c)\\&-2L_rm_rg\sin(x_3)\cos(x_3^2) + M_cL_c\cos(x_3)x_6^2 - 2M_cL_cg\cos(x_3)\sin(x_3^2)\\&+L_rm_r\sin(x_2)x_5^2)/(1-m_r\cos(x_2^2) - M_c\sin(x_3^2))^2/(M_c+m_r)M_c\sin(x_3)\cos(x_3)\end{aligned} \tag{13}$$

$$\begin{aligned}A_6 &= -2L_c^2\cos(x_3^2)M_cg + 2L_c^2\sin(x_3^2)M_cg + 1/L_c\cos(x_2)\sin(x_2)(F_1(1-\cos(x_2)/L_r + \sin(x_3)/L_c)\\&-2L_rm_rg\sin(x_2)\cos(x_2^2) + M_cL_c\cos(x_3)x_6^2 - 2M_cL_cg\cos(x_3)\sin(x_3^2)\\&+L_rm_r\sin(x_2)x_5^2)/(1-m_r\cos(x_2^2) - M_c\sin(x_3^2))/(m_r+M_c)\\&-1/L_c\sin(x_2)(F_1\cos(x_2)/L_c - L_cM_c\sin(x_3)x_6^2) + 2M_cL_cg\sin(x_3^3)\\&+4L_cM_cg\cos(x_3^2)\sin(x_3))/(1-m_r\cos(x_2^2) - M_c\sin(x_3^2)/(m_r+M_c)\\&+2/L_c\sin(x_3^2)(F_1(1-\cos(x_2)/L_r + \sin(x_3)/L_c - 2M_rL_rg\sin(x_2)\cos(x_2^2) + M_cL_c\cos(x_3)x_6^2\\&-2L_cM_cg\cos(x_3)\sin x_3^2) + L_rm_r\sin(x_2)x_5^2)/(1-m_r\cos(x_2^2) - M_c\sin(x_3^2)^2/(m_r+M_c)M_c\cos(x_3)\end{aligned} \tag{14}$$

$$A_7 = 2L_rm_r\sin(x_2)x_5/(1-m_r\cos(x_2^2) - M_c\sin(x_3^2)/(m_r+M_c) \tag{15}$$

$$A_8 = -2m_r\cos(x_2)\sin(x_5)/(1-m_r\cos(x_2^2) - M_c\sin(x_3^2)/(m_r+M_c) \tag{16}$$

$$A_9 = 2/L_cL_rm_r\sin(x_2)\sin(x_3)x_5/(1-m_r\cos(x_2^2) - M_c\sin(x_3^2)/(m_r+M_c) \tag{17}$$

$$A_{10} = 2L_cM_c\cos(x_3)x_6/(1-m_r\cos(x_2^2) - M_c\sin(x_3^2)/(m_r+M_c) \tag{18}$$

$$A_{11} = -2/L_rL_cM_c\cos(x_3)x_6/(1-m_r\cos(x_2^2) - M_c\sin(x_3^2)/(m_r+M_c) \tag{19}$$

$$A_{12} = 2M_c\sin(x_3)\cos(x_3)x_6/(1-m_r\cos(x_2^2) - M_c\sin(x_3^2)/(m_r+M_c) \tag{20}$$

$$B_1 = (1-\cos(x_2)/L_r + \sin(x_3)/L_c)/(1-m_r\cos(x_2^2) - M_c\sin(x_3^2)/(m_r+M_c) \tag{21}$$

$$B_2 = -1/L_r\cos(x_2)(1-\cos(x_2)/L_r + \sin(x_3)/L_c)/(1-m_r\cos(x_2^2) - M_c\sin(x_3^2)/(m_r+M_c) \tag{22}$$

$$B_3 = 1/L_c\sin(x_3)(1-\cos(x_2)/L_r + \sin(x_3)/L_c)/(1-m_r\cos(x_2^2) - M_c\sin(x_3^2)/(m_r+M_c) \tag{23}$$

3. Control Strategy

LQR is a method in modern control theory that uses state space approach to analyze such a system. Using state-space methods it is relatively simple to work with a multi-output system. The system can be stabilized using full state feedback. The schematic of this type of control system is shown in Figure 4.

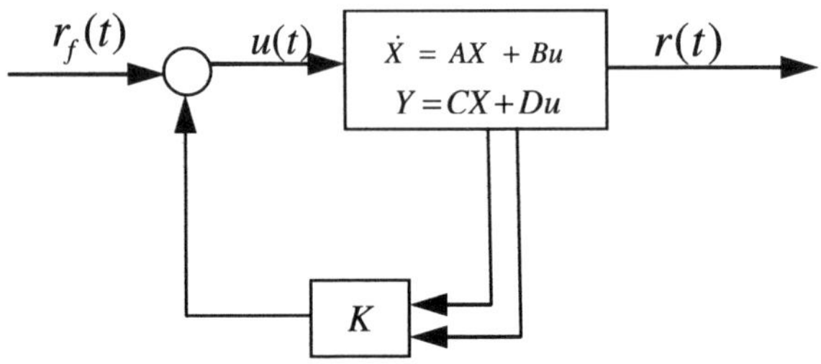

Fig. 4 Block diagram of the LQR and poles placement control

4. Simulation Results

Results of the simulation for both the pole placement and LQR techniques are shown in Figures 5-7 and Figures 8-10 respectively. For the pole placement technique, the system achieved the desired values; both the linear displacement and the tilt angle, in less than 1.5 seconds. The required control effort shown in Figure 6, measured in volts, is limited to 30 volts at the start while decay in respond to the developed control strategy. As can be noticed from the comparison of Figures 7 and 10 that using the pole placement technique causes the overshoot in the tilt angle to be decreased to half its value if compared with the one obtained using LQR.

A sudden impact force has applied at 2 seconds intervals on the vehicle body. The developed LQR has succeeded to a satisfactory level to absorb the impact disturbance while been able to bring back the vehicle to the desired vertical upright position. However, an occurrence of accumulated errors has been noticed in both the tilt angle and the linear displacement of the entire system. Increasing the time interval between each time the disturbance applies is recommended to avoid such occurrence.

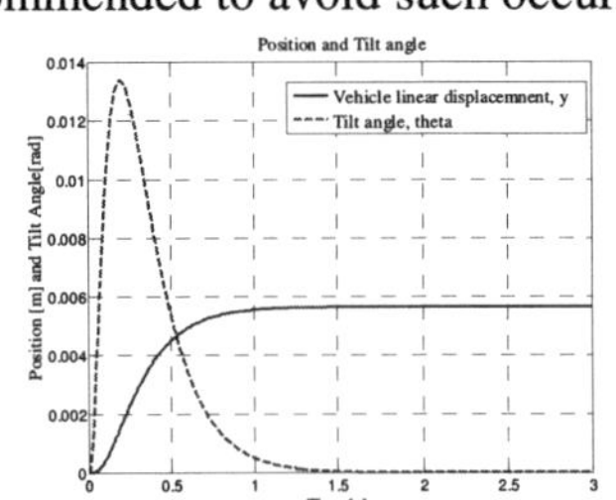

Fig. 5 Tilt angle and linear displacement, no disturbance, Pole placement technique

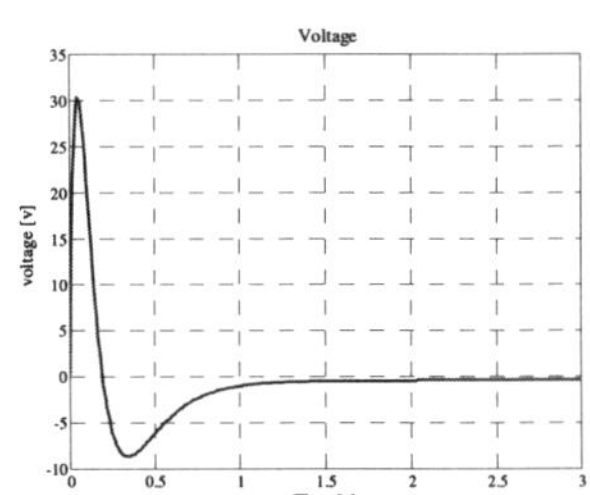

Fig. 6 Control Output no disturbance Pole placement technique

5. Conclusion

Mathmetical modelling of two-wheeled vehicle is presented based on Lagrangian Euler formulation. The linearized state space model has been derived along with a design and implementaion of LQR as wel as pole placement techniques. The robustness of the LQR control scheme has been investigated based on the system response where there is an impact disturbance on the vehicle.

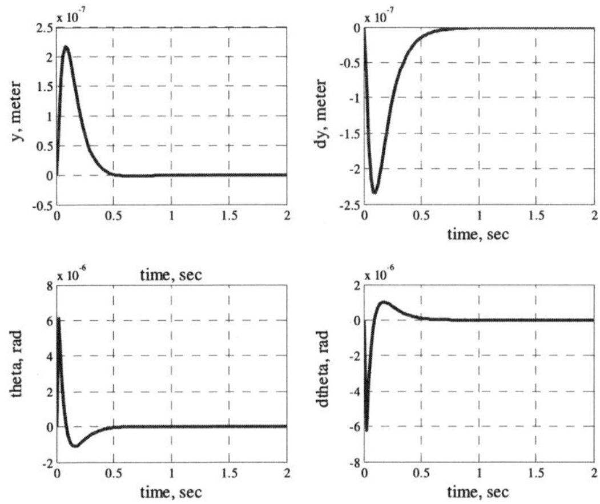

Fig. 7 System performance with no disturbance
Pole placement technique

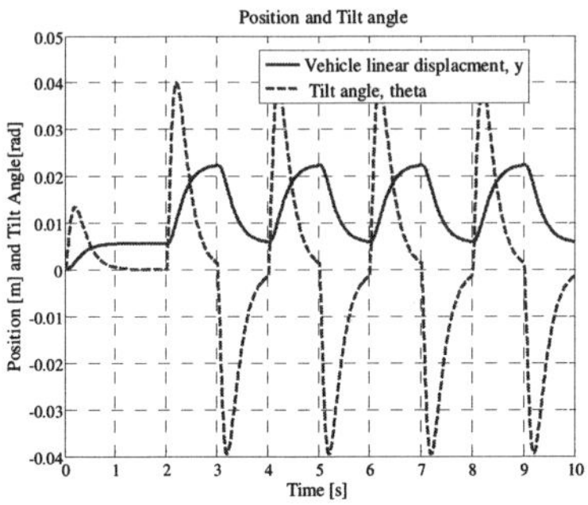

Fig. 8 System performance, with disturbance
Linear Quadratic Regulator

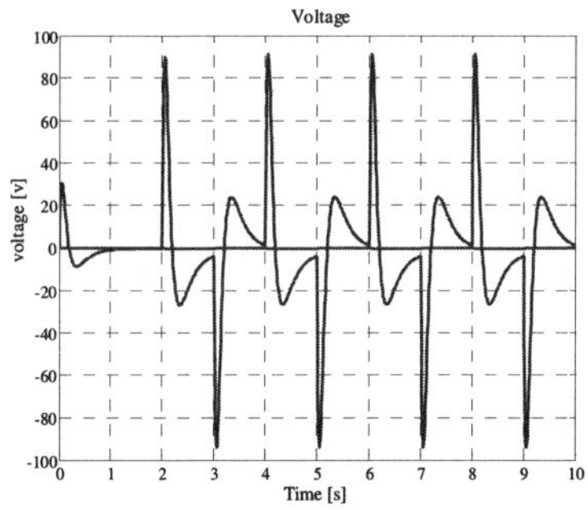

Fig. 9 Control Output, volt with disturbance
Linear Quadratic Regulator

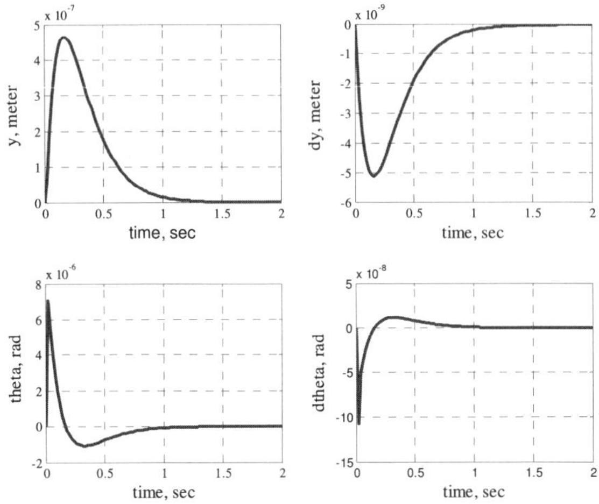

Fig. 10 System performance with no disturbance Linear Quadratic Regulator

References

1. Bogdanov A, Optimal Control of a Double Inverted Pendulum on a Cart, *Oregon Health and Science University Technical Report CSE*, 2004.
2. Fiacchini M, Viguria A, Cano R, Prieto A, Rubio F R, Aracil J and Canudas-de-Wit C, (2006), Design and Experimentation of a Personal Pendulum Vehicle, *The 7th Portuguese Conference on Automatic Control (CONTROLO 06)*, pp. 11-13.
3. Goher K. M. and Tokhi M. O., (2008), Balance control of a TWRM with a static payload, in the Proceedings of the 11th International Conference on Climbing and Walking Robots and the Support Technologies for Mobile Machines (CLAWAR 2008), Coimbra, Portugal, September 8-10.
4. Nasir A N K, Ahmad M A and Raja Ismail R M T, (2010), The Control of a Highly Nonlinear Two-wheels Balancing Robot: A Comparative Assessment between LQR and PID-PID Control Schemes, Proceedings of World Academy of Science, Engineering and Technology 70, pp. 227 – 232.
5. Nawawi S W, Ahmad M N, Osman J H S, (2006), Control of Two-wheels Inverted Pendulum Mobile Robot Using Full Order Sliding Mode Control, Proceedings of International Conference on Man-Machine Systems 2006, September 15-16, Langkawi, Malaysia.
6. Scokaert P O M and Rawlings J B, (1998), Constrained linear quadratic regulation," IEEE Trans. Automatic Control, 43(8):1163-1169.
7. Wongsathan, C, Sirima, C, (2009), Application of GA to design LQR controller for an Inverted pendulum System. *IEEE International Conference on Robotics and Biomimetics (ROBIO)*, 22-25 Feb., pp. 951-954.
8. Xiong, X, Wan Z, (2010), The simulation of double inverted pendulum control based on particle swarm optimization LQR algorithm, *IEEE International Conference on Software Engineering and Service Sciences (ICSESS)*, 16-18 July, pp. 253-256.

TAKING ADVANTAGE OF THE JOINT SYNERGY FOR THE ACTUATION IN A LOWER-LIMB ACTIVE ORTHOSIS

D. Sanz-Merodio*, M. Cestari, J.C. Arevalo and E.Garcia

Centre for Automation and Robotics, CSIC-UPM
La Poveda, 28500 Madrid, Spain
**E-mail: daniel.sanz@car.upm-csic.es*
www.car.upm-csic.es/control

Lower-limb exoskeletons and powered orthoses are external devices that assist patients with locomotive disorders to achieve correct movement of their limbs. These powered devices have energy requirements for long periods that current batteries cannot cope with, and this has become a major challenge in the development of these portable robots. Conversely, legged locomotion in animals and humans is very efficient; to emulate their behavior, biomimetic actuation designs try to incorporate elements that resemble biological elements such as tendons and muscles in the mechanical systems. In this paper we present a mechanism that resembles a human tendon to take advantage of synergic actuation of leg joints and achieve efficient locomotion.

Keywords: Active orthoses, synergic actuation, bipedal gait, efficient locomotion.

1. Introduction

Current estimations inform about 265.000 spinal cord injured (SCI) individuals in the United States.[1] The spinal cord injury derives in paraplegia or quadriplegia resulting in the loss of mobility. In an effort to restore mobility of the legs of these patients, lower limb active orthoses have been developed.[2] Some research devices like Vanderbilt Orthosis[3] and also some commercial devices like the Rewalk active orthosis (Argo Medical Technologies) and eLEGS (Berkeley Bionics) have been developed for this purpose. All the referred devices are not actuated on the ankle, some of them only include a passive element to provide ground clearance. In order to achieve a more efficient and natural gait is necessary to actuate the ankle. The addition of an actuator or passive element increases weight of the limbs and consequently the moment of inertia of the leg, resulting in larger power consumption in hip and knee joints. In the theoretical work by Van der Bogert,[4] an energetic optimization is performed, joining hip and ankle by an exotendon and achieving a 47% reduction in joint power consumption. In this paper we present a mechanism that resembles a human tendon to achieve and take advantage of synergic actuation of leg joints. At the ankle, a steel cable has been attached, acting like a synergic biarticular linkage and transferring motion from hip and knee joints. In this way, we prevent high inertia in the more distal limbs and at the same time provide power to actuate the ankle joint. This mechanism has been integrated in a lower-limb active orthosis, actuated at hip and knee joints. Taking advantage of these two joint rotations and thanks to the attached cable, a movement in the ankle joint is induced, that has to be mapped to a CGA ankle joint behavior to achieve natural motion. We focus our experiment in the toe-off, swing and heel-strike phases, which are considered the most determinants of gait energetics.[5] The fitted ankle rotation helps the leg motion during the toe-off phase, provides ground clearance by preventing foot dragging during the swing phase, and accommodates for a proper heel-strike, resulting in a more natural gait. Section 2 presents a brief background on the ankle motion. Section 3 describes the implementation details of the ankle motion strategy while Section 4 shows the results after apply these ideas in the real ATLAS[6] orthosis. Finally Section 5 presents the main conclusions.

2. Ankle motion along the gait cycle.

Figure 1 shows the angle follow by every joint by a normal and healthy person, that is call Clinical Gait Analysis (CGA) pattern. To follow the behavior of the CGA pattern, the gait cycle has been divided into three phases. The first stage corresponds to the support phase, in the beginning of this phase no cable will act. In the second period, that correspond to the final part of support, the trailing leg gives impulse before the leading leg begins heel strike. From an energetic point of view this is the determinant phase of the gait.[5]

Once the foot has left the ground, phase 3 begins, that correspond to the swing phase. In this phase, the ankle movement has to provide ground clearance.

2.1. *Providing impulse at the end of support phase*

In the mentioned gait second phase, to provide the required impulse in the ankle, a cable has been attached, joining the hip to the foot through the

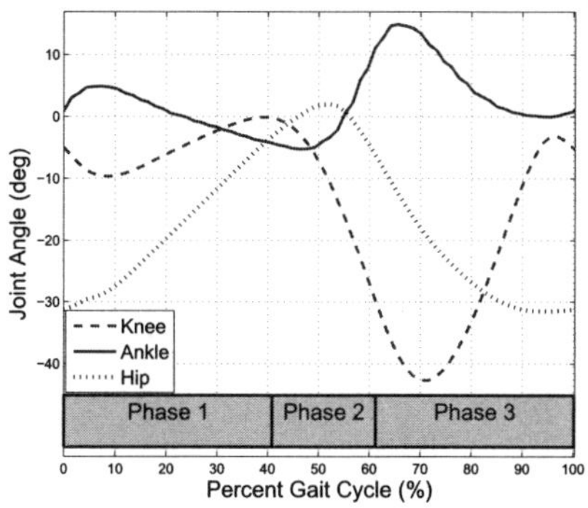

Fig. 1. Gait cycle has been divided into three phases.

knee. Analyzing Figure 2 it follows that the ankle angle is related to:

$$\Delta(\theta_{ankle}) \propto R_{ankle}^{-1}[R_{hip}\Delta(\theta_{hip}) + R_{knee}(\theta_{knee})\Delta(\theta_{knee})] \qquad (1)$$

where R_{ankle}, $R_{knee}(\theta_{knee})$ and R_{hip} are the active ratios, calculated to achieve an adequate movement in the ankle. It can be observed that the ratio R_{hip} is larger than the ratio R_{ankle} because a small variation in θ_{hip} has to provide an important variation in θ_{ankle}. An eccentricity has been

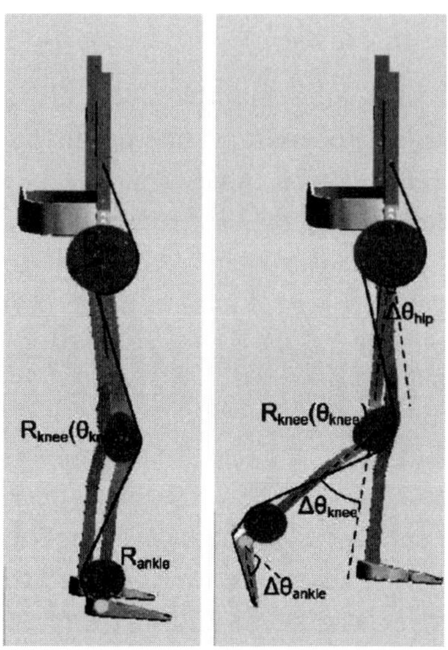

Fig. 2. Ankle angle depends on the motion in the hip and the knee.

included on the knee so the active ratio $R_{knee}(\theta_{knee})$ is not constant and it depends on the knee angle. Thus, the impulse of the trailing leg is given at the proper time. Several simulation adjustments have been made in the active ratios to give the impulse in the right moment in the cycle, see Figure 3 (a).

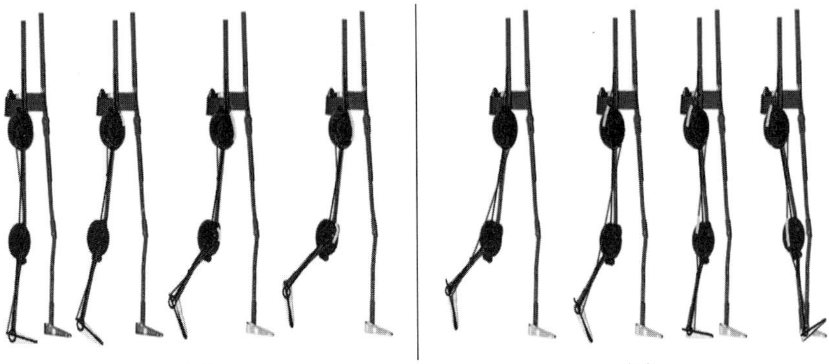

(a) (b)

Fig. 3. Cables attached to move the ankle (a) A cable provides the impulse in the ankle. (b) Another cable provides ground clearance in the swing.

2.2. *Providing ground clearance in the swing phase*

Based on the previous idea, another cable takes advantage of the motion of the hip to move the ankle, to avoid stumbling in the third gait phase, the swing. The attached cable causes that when θ_{hip} increases, advancing the leg, θ_{ankle} also increases, lifting the foot to clear the ground and preparing for heel strike. Again, active ratios have been adjusted to achieve the correct ankle motion, see Figure 3 (b). In this case, no eccentricity is needed in the knee.

3. Implementation details.

An active orthosis prototype ATLAS will be used in this experimental work. Two cables have been connected through the joints of the powered orthosis, acting like exotendons, and a goniometer has been attached to the ankle for measuring the angular variation. The adjustments in active ratios that have been done in simulation for every joint, have been incorporated in the ATLAS prototype orthosis, see Figures 4 (a) and (b). Notice that both cables in the mechanisms cannot act simultaneously. The ankle angle variation has been evaluated, fitting it to adjust the correct behavior of the ankle based on CGA data. Figure 5 shows snapshots of a sequence in real experiments.

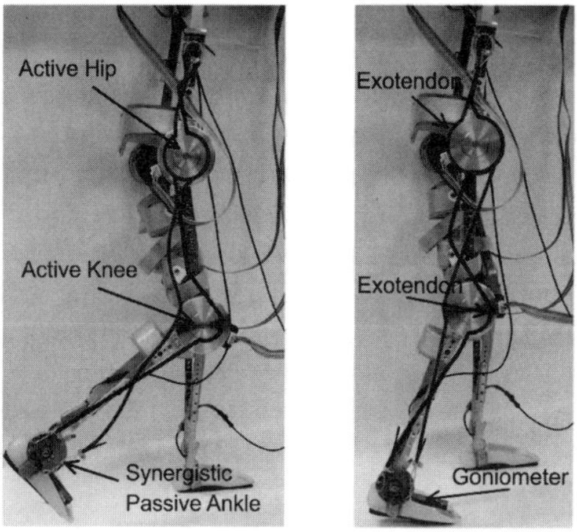

(a) (b)

Fig. 4. Cable attached in real ATLAS orthosis: (a) First cable giving impulse; (b) Second cable providing ground clearance. The cable paths have been drawn by lines in the pictures to clarify the illustration.

The passive synergic ankle motion provides impulse in the toe-off, increases support force in stance and helps providing ground clearance.

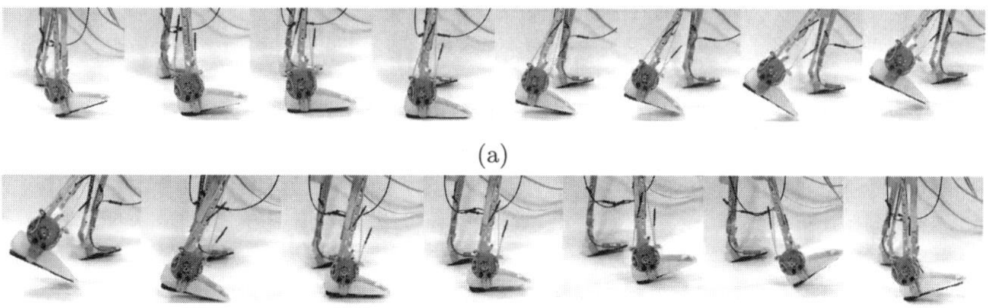

(a)

(b)

Fig. 5. Passive ankle movement: (a) Right leg support phase and toe-off; (b) Right leg swing phase.

4. Experimental evaluation

Instead of following fixed walking pattern, in ATLAS orthosis, parametric trajectories have been implemented in each supporting or swinging leg. In the supporting leg, a direct pendulum trajectory has been parametrized to achieved a specific step length and in the swinging leg an inverted pendulum trajectory has been parametrized to ensure an adequate step height and strike angle. Therefore, these parametric trajectories provide a safer, configurable and more versatile walking pattern. Then the hip and knee angle vary with respect to the CGA. Figure 6 shows the angles implemented for the actuator in hip and knee, and the angle measured by the goniometer in the ankle. The experiments have been done with the orthosis walking on the ground and hold only to keep balance. Comparing with the angles in Figure 1, measured in CGA of normal and healthy persons, it is observed that in phase 2 and 3, correct behavior is achieved. In phase 1, the behavior is nearly identical because the foot is supported on the ground and moves passively.

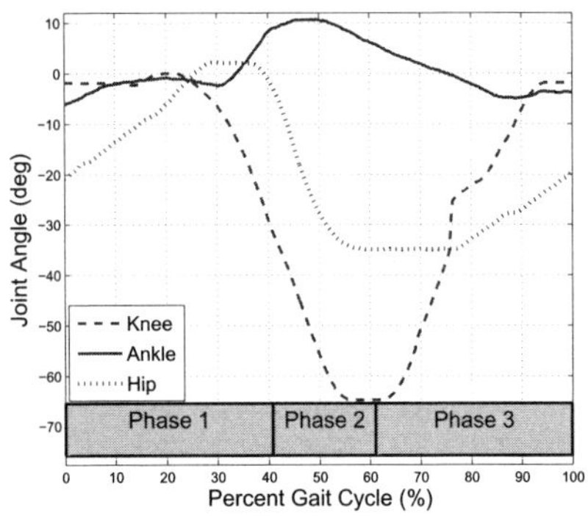

Fig. 6. Angles at the exoskeleton's right leg joints in real experiment.

5. Conclusions.

In order to achieve the correct and efficient motion in the ankle of an active orthosis, two steel cables have been attached in the ATLAS orthosis. These cables act like a synergic biarticular linkage and transferring motion from hip and knee joints. Synergic ankle motion provides impulse in the toe-off, increases support force and provides ground clearance. These goals are achieved with a low energy expenditure due to synergical actuation and preventing high inertia in the more distal limb.

6. Acknowledgments

This work has been partially funded by the Spanish National Plan for Research, Development and Innovation through grant DPI2010-18702 which has funded personnel costs and by AECID through grant PCI-Iberoamerica A1/039883/11 which has funded equipment and consumable costs.

References

1. N. S. C. I. S. C. (NSCISC), *Spinal Cord Injury Facts and Figures at a Glance, Available on: https://www.nscisc.uab.edu*, (Febrary 2011).
2. A. M. Dollar and H. Herr, Lower extremity exoskeletons and active orthoses: Challenges and state-of-the-art, *Robotics, IEEE Transactions on* **24**, p. 144158 (2008).
3. R. Farris, H. Quintero and M. Goldfarb, Preliminary evaluation of a powered lower limb orthosis to aid walking in paraplegic individuals, *Neural Systems and Rehabilitation Engineering, IEEE Transactions on* , 102 (2011).
4. A. J. Van Den Bogert, Exotendons for assistance of human locomotion, *Biomedical engineering online* **2** (2003).
5. J. M. Donelan, R. Kram and A. D. Kuo, Simultaneous positive and negative external mechanical work in human walking, *Journal of Biomechanics* **35**, 117 (2002).
6. E. Garcia, D. Sanz-Merodio, F. Sanchez, J. Arevalo and G. de Santos P., Development of the atlas lower-limb active orthosis, in *14th Int. Conf. Climbing and Walking Robots and the Support Technologies for Mobile Machines, Paris, France*, 2011.

SECTION–3
AUTONOMOUS ROBOTS

REAL TIME PID CONTROL OF A TWO-WHEELED ROBOT

A. R. SHORT[1], O. K. SAYIDMARIE[1], S. A. AGOURI[1], M. O. TOKHI[1], K. M. GOHER[2], AND A. M. ALMESHAL[1]

[1]*Department of Automatic Control and System Engineering, The University of Sheffield, Sheffield, UK,*
o.sayidmarie@sheffield.ac.uk
[2]*Department of Mechanical and Industrial Engineering, College of Engineering, Sultan Qaboos University, Oman*

In this paper, a two-wheeled inverted pendulum robot prototype is used. PID control is designed for balancing and manoeuvring of the system considering the tilt angle and wheels speed as feedback. Tilt angle sensing error is reduced by adopting a sensor fusion technique. The control structure is extended allowing self-erecting of the robot from a resting position. The controller is implemented using an AVR microcontroller. The experimental results show that the robot has the ability to balance during its movement.

1. Introduction

Two-wheeled robots, which operate on the inverted pendulum principle, are in nature unstable, non-linear, and under actuated. Since the introduction of the two-wheeled robot JOE [1], this kind of robots has induced great interest as they provide opportunities for application of control design, signal processing, distributed control systems and consideration of real-time constraints during implementation. Moreover, they offer advantages over other ground robotic vehicles such as zero-turn radius, simple structural design, and small footprint for navigation in tight spaces. However, owing to the inherent instability, the system needs to be dynamically controlled, and sensors need to be installed on the system for manoeuvring. Because of these intriguing requirements, two-wheeled robots are widely used in research institutions to study control techniques with relevant sensors [2-6].

Commercial products, based on two-wheeled robot, have also been developed, such as the Segway human transporter, PUMA (Personal Urban Mobility and Accessibility), and the wheel chair iBot [7-9].

The aim of this paper is to implement simple control techniques for a two-wheeled robot prototype, developed at the University of Sheffield, for upright balance and smooth manoeuvring.

The control system includes two cascaded PID controllers which are implemented using an AVR microcontroller. The first is responsible for balancing the robot at upright position, while the second PID controls the robot's horizontal movement. Accurate tilt angle measurement is obtained by implementing a sensor fusion technique. The robot is remotely guided by client PC software through Bluetooth link, which is used for debugging, tuning the PID gains and recording the experimental data for offline analysis.

2. Robot Hardware

The developed prototype robot is shown in Figure 1. The robot chassis is constructed from aluminium frame (20x20x31 cm) and consists of three layers. Two DC motors with the driving board, which drive the robots motion, are attached to the bottom layer. Each motor has a magnetic encoder of 360 pulses per revolution, and its shaft is attached to wheels of 5 cm radius. The MD-25 board, a 2.8 dual H-Bridge DC motor controller, is used. This board has two modes of operation, common and individual wheel control, and receives speed and turn commands over Serial or I2C Interface.

The middle layer houses two main unites. The first is the ATmega32 microcontroller development board. The CPU of this board is a low-power 8-bit AVR® Microcontroller with eight channels 10-bit ADC. The second unit is The IMU (Inertial Measurement Unit), which is used to measure the posture of the robot. It consists of the ADXL203, 2-axis ±1.7g accelerometer and the ADXRS614, 1-axis ±30°/sec gyroscope.

The upper layer houses the Bluetooth model, which is used to communicate with the client PC. In addition, the batteries pack is placed on this layer to shift the centre of gravity up.

Figure 1: Two-wheeled robot prototype

3. Sensor Fusion

Accurate and fast sensing of the tilt angle of the robot is very important to maintain robot balance; such accurate measurement could be done using expensive IMU. However, replacing this unit with a low cost tilt sensor causes sufficient error in the measurement. For instance, inclinometers have slow time response, and could not follow the robot dynamics. A gyroscope has fast response, and can follow the robot dynamics, but suffers from time drift error which represents the accumulated error by integration to obtain angle information. An accelerometer could be used to measure the tilt angle, it provides accurate steady-state measurement, but the measurement is highly corrupted by noise due to movements of the robot. Figure 2 shows the tilt angle measurements with a gyroscope and accelerometer. Thus, a sensor fusion technique is realized by implementing a complementary filter, which combines the gyroscope and the accelerometer sensor measurements to obtain accurate angle information as shown in Figure 3.

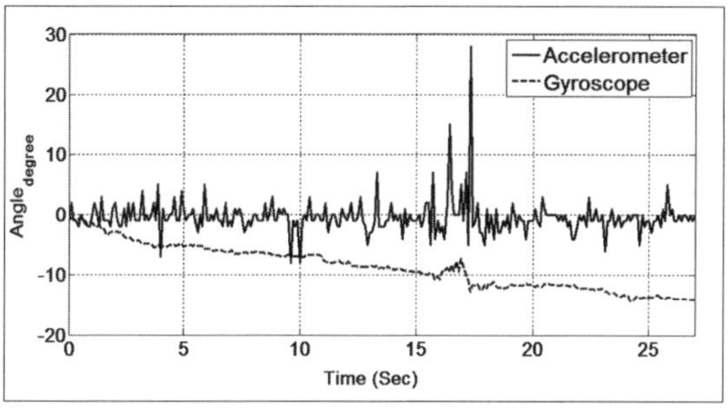

Figure 2: Angle measurements

The digital implementation of the complementary filter can be expressed as

$$y(k) = a\left(y(k-1) + x_{Gy}(k) * \Delta t\right) + (1-a)x_{Acc}(k) \tag{1}$$

where $x_{Gy}(k)$, $x_{Acc}(k)$ are the gyroscope and the accelerometer signals respectively, Δt is the sampling time, and

$$a = \frac{\tau}{\tau + \Delta t} \tag{2}$$

Accordingly, the important factor for the performance is the time constant τ, and the corresponding proper cut-off frequency, which mainly depend on the sensors specification, is found by investigating the characteristics of the sensors experimentally, and τ = 0.5Sec was founded for the system. Figure 4 shows angle measurement with the complementary filter. As noted it is relatively free of measurement noise and drift.

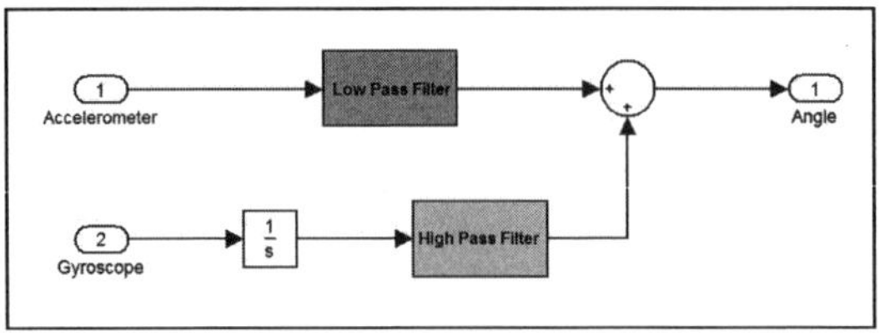

Figure 3: Complementary filter structure

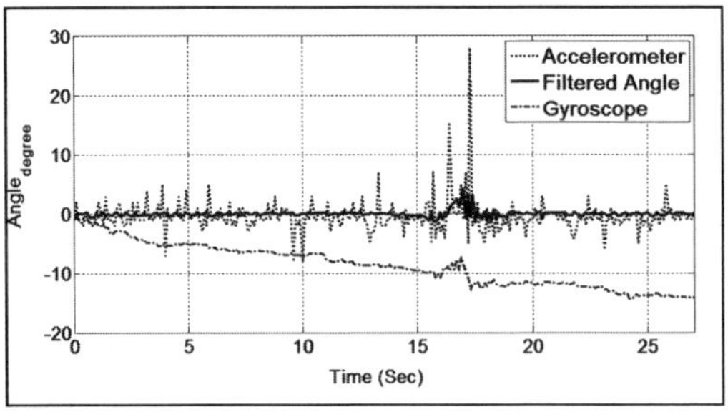

Figure 4: Complementary filter angle measurements

4. PID Control

PID control is simple to implement, it generally provides adequate results, and thus is the most common controller type used. The controller consists of three tuning parameters (Kp, Ki and Kd) which represent the gains of the error, integral of the error and change in the error respectively. Combination of these terms can benefit from individual advantages. Methodologies for tuning these parameters are available, such as Cohen-Coon and Ziegler-Nichols as well as some software tools.

In order to balance the robot in the vertical position, the control system should keep the tilt angle close to zero. The tilt angle measurement is filtered and used as feedback signal for the controller, while the controller output is applied to the motors of the wheels, adjusting the speed and direction of movement accordingly; to reduce the error in the tilt angle, see Figure 5.

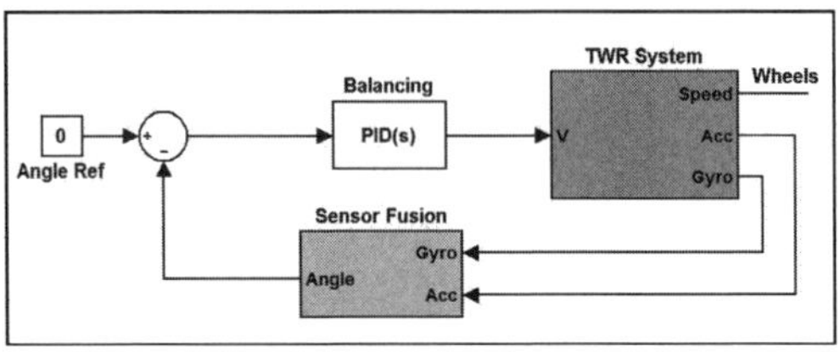

Figure 5: Balancing PID controller

The controller parameters are manually tuned using a heuristic approach. Figure 6 shows experimental results of tilt angle and wheels speed; as noted the robot moved effectively keeping the tilt angle between ±1°.

As the balancing controller maintains the tilt angle close to zero, the horizontal movement could be achieved by choosing a non-zero set point for the balancing controller. A positive or negative set-point will result in a forward or backward movement respectively. However, if a constant non-zero set point value is assigned, the robot will keep increasing the wheels speed until it violates physical constraints and become unstable; therefore, a second controller is implemented to adjust the balancing set point according to the desired robot travelling speed as shown in Figure 7.

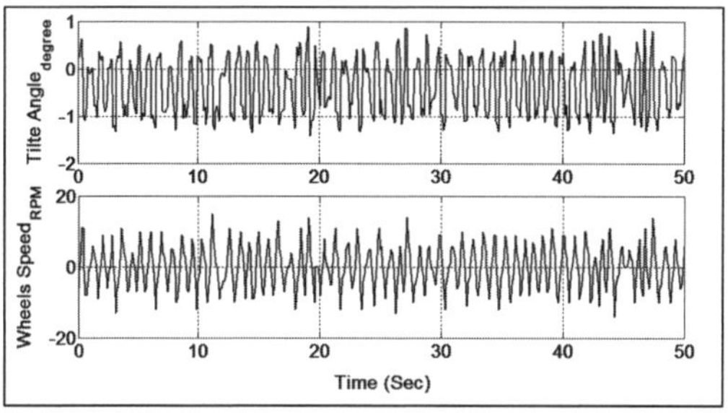

Figure 6: Experimental results of the balancing controller

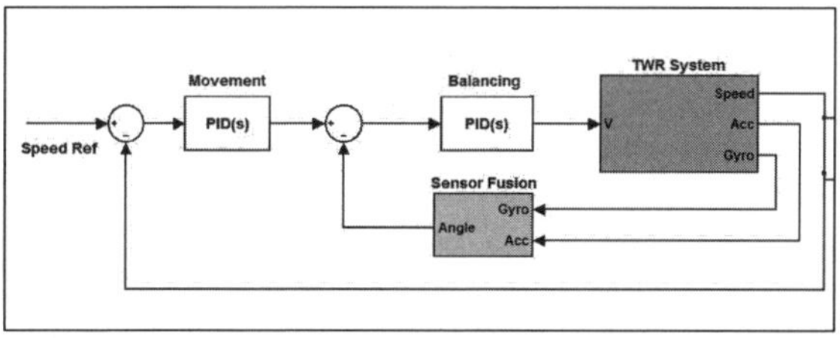

Figure 7: Movement controller

The feedback of the second controller is the wheels speed; hence, it tries to reduce the speed that is required for balancing. This effect was reduced by making the controller slower with respect to the balancing controller. Figures 8 shows the experimental results for wheels speed, distance travelled and tilt angle while the robot performs a two-meter forward.

The travelled distance is measured by integrating the wheels speed. Ideally, the wheels speed would be constant for a smooth linear movement, but due to the fact that the controller is also maintaining robot balance, this causes a non-linear travelling distance curve.

The steering of the robot is achieved by implementing a differential steering mechanism, where a turn speed value is added to and subtracted from the applied control signal of the left and right wheels respectively, see Figures 9 and 10 for the associated experimental results. This mechanism maintains the average speed of both wheels, and the balancing mechanism would not be affected. Table 1 defines the robot's motion behaviour for different values of travel and turn speeds.

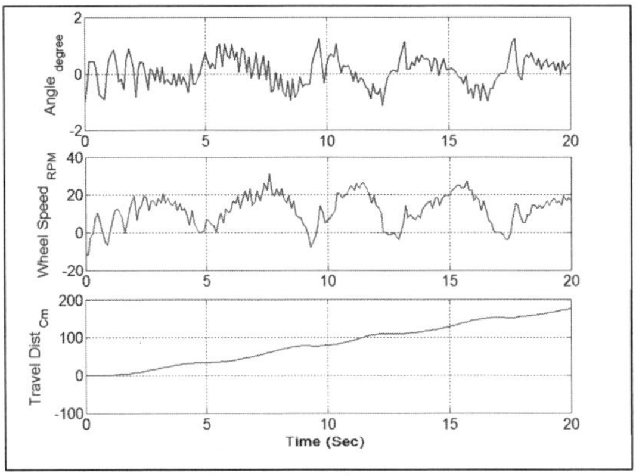

Figure 8: Forward movement

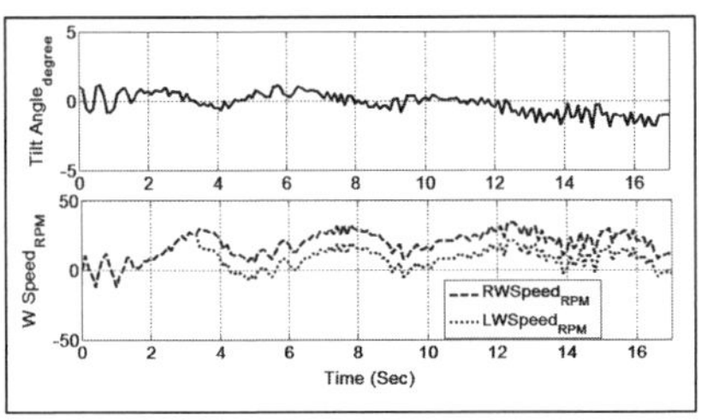

Figure 9: CW arc turn around corner

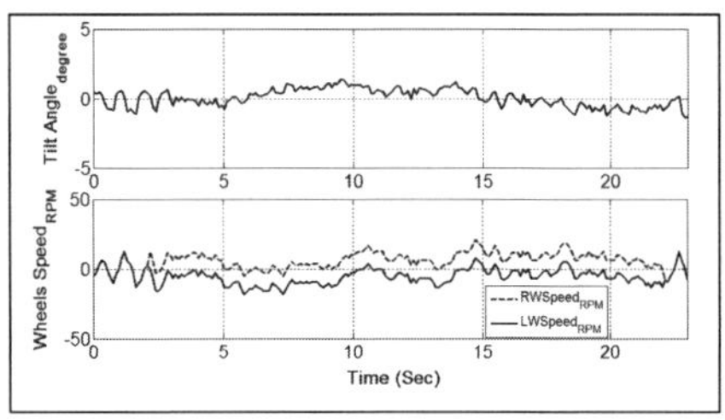

Figure 10: 360° CCW rotate around the vertical axis

Table 1: Robot's motion behaviour

Travel speed ref	Turn speed ref	Robot's motion behaviour
0	0	Balancing
+Ve	0	Forward
-Ne	0	Backward
0	+Ve / -Ne	CW/CCW rotate around its vertical axis
+Ve / -Ne	+Ve / -Ne	CW/CCW arc turn movement

5. Self erecting

This term refers to the ability of the robot to stand up from the start position and balance without human intervention. The challenge is when the robot tries to balance from a starting point so far from the set-point; it results in large gains and unacceptable overshoot. Consequently, leading to large and oscillatory wheel speeds and eventually driving the robot outside of the safety boundaries. Therefore, a dual-mode approach is considered. The first mode is responsible for reducing the tilt angle to a manageable value, while the second mode enables normal balancing and keeps the error in tilt angle close to zero.

The self erecting control structure is shown in Figure 11. The idea is to use the same controller which is responsible for balancing the robot, except altering its characteristics. In fact, only a change in the proportional gain is found to be enough. Reducing this gain effectively reduces overshoot and drives the robot into a safe region, even with the reduced proportional gain; the robot is able to maintain stability. However, the performance is degraded and therefore the second mode restores the original parameters and resumes the normal balancing behaviour. Figures 12 and 13 demonstrate the system performance with positive 6° and negative 7.5^{o} initial tilt angles.

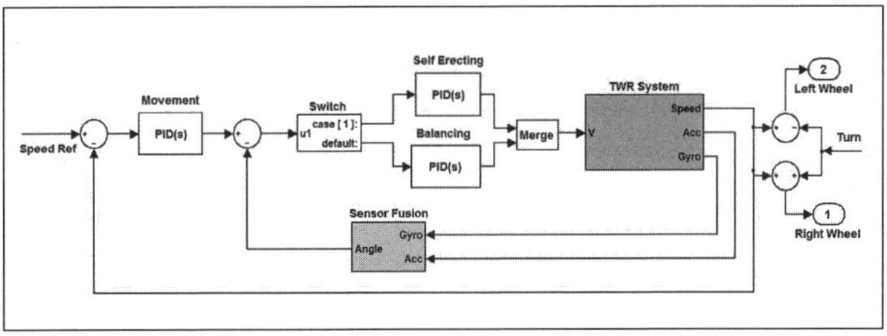

Figure 11: Self erecting control structure

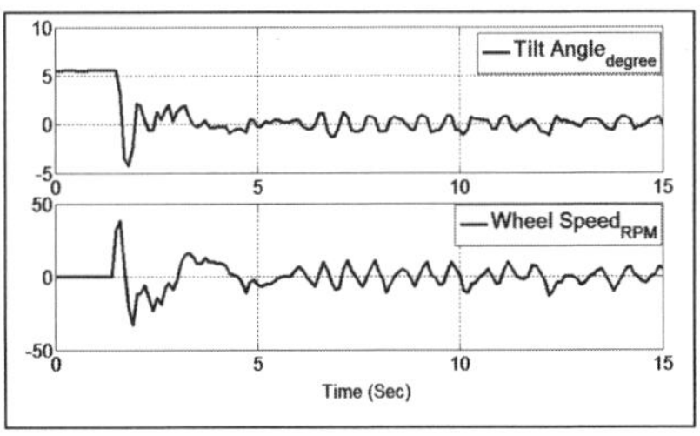

Figure 12: Self erecting +Ve initial angle

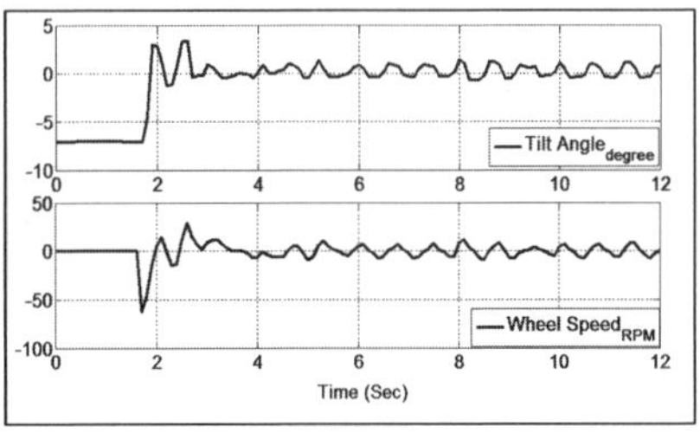

Figure 13: Self erecting -Ne initial angle

6. Conclusion

The development of a PID based control mechanism has been presented for balancing a two-wheeled robot. The controller has been implemented on a microcontroller and tested in real-time with a two-wheeled robot prototype. To ensure accurate measurements of tilt angle are obtained, a sensor fusion mechanism based on filtering of measurements of accelerometer and gyroscope has been realised. Furthermore, the control structure has been extended allowing self-erecting of the robot from a resting position. The experimental results have shown the ability of the robot to maintain balance during manoeuvres, and the designed control system has achieved the requirements imposed by speed, low cost, flexibility and stability especially during self-erecting.

References

[1] F. Grasser, A. D'Arrigo, S. Colombi, and A. C. Rufer, "JOE: a mobile, inverted pendulum," *Industrial Electronics, IEEE Transactions on INDUSTRIAL ELECTRONICS,* vol. 49, pp. 107-114, 2002.

[2] D. P. Anderson. (2002). *nBot balancing robot.* Available: http://www.geology.smu.edu/~dpa-www/robo/nbot/, accessed: 13-Jan-2012.

[3] S. W. Nawawi, M. N. Ahmad, and J. H. S. Osman, "Development of a two-wheeled inverted pendulum mobile robot," *2007 5th Student Conference on Research and Development,* pp. 235-239, 2007.

[4] J. Leavitt, A. Sideris, and J. E. Bobrow, "High bandwidth tilt measurement using low-cost sensors," *Mechatronics, IEEE/ASME Transactions on,* vol. 11, pp. 320-327, 2006.

[5] R. Imamura, T. Takei, and S. Yuta, "Sensor drift compensation and control of a wheeled inverted pendulum mobile robot," in *Advanced Motion Control, 2008. AMC '08. 10th IEEE International Workshop on*, 2008, pp. 137-142.

[6] K. T. Goher, MO "Modelling, simulation and balance control of a two-wheeled robotic machine with static variation in load position," in *Proceedings 22nd European Conference on Modelling and Simulation ©ECMS*, Nicosia, Cyprus, 2008.

[7] D. Kamen. (2001). *The Segway® Personal Transporter (PT), the first self-balancing, zero emissions personal transportation vehicle.* Available: http://www.segway.com/, accessed: 13-Jan-2012.

[8] I. Technology©. (2001). *iBOT® Mobility System.* Available: http://www.ibotnow.com/index.htm, accessed: 13-Jan-2012.

[9] (2009). *P.U.M.A. (Personal Urban Mobility & Accessibility).* Available: http://www.segway.com/puma/, accessed: 13-Jan-2012.

A STEREO VISION MODEL FOR A SMALL FORM FACTOR AUTONOMOUS MOBILE ROBOT "KCLBOT"

EVANGELOS GEORGIOU

Centre for Robotics, King's College London-
London, WC2R 2LS, United Kingdom

PROFFESOR JIAN S DAI

Centre for Robotics, King's College London-
London, WC2R 2LS, United Kingdom

PROFFESOR MICHAEL LUCK

Division of Informatics, King's College London-
London, WC2R 2LS, United Kingdom

In the field of mobile robot research, one of the design challenges is the size constraint of the robot. When designing a small form factor mobile robot, due to the size constraint, critical aspects are sacrificed such as the computing required to do high level operation. At King's College London, a mobile robot, nicknamed the 'KCLBOT', has been designed as a small form factor mobile robot with computation power to carry out high level functions locally. The motivation for this paper is to present the result of introducing stereo vision to a small form factor mobile robot.

1. Introduction

In mobile robot research, the key to a mobile robot system being autonomous is effective self-location, path planning and obstacle avoidance. To ensure the success of these key elements, the system requires accurate sensor data. Typically on small mobile robot platforms, to counter the limitation of processing power and memory, infrared or ultrasound range sensors are attached to a microcontroller. This type of configuration gives the mobile robot limited information about the environment and an incomplete map of the navigation space. The ideal solution would incorporate a computer vision system that could provide detailed data of the navigation space that can be used for self-location, path planning, and obstacle avoidance. Computer vision systems, however, require significant processing power and memory. Fortunately, due to the innovations of the Pico-ITX framework [1], it is possible to add this processing power and memory to a small form factor mobile robot.

Hamzah and Salim [2] present a similar approach, which uses Tsai's method for un-calibrated cameras and a block matching algorithm for image correspondence. With respect to miniature mobile robot platforms, Haverinen et al [3] present a mobile robot that is capable of stereo vision, which was designed for swarm robotics. For vision system, Huber and Kortenkamp [4] present PRISM-3 which is a stereo vision system used for tracking and pursuing moving targets.

While the systems mentioned above offer solutions to stereo vision, they do not offer an autonomous solution at the form factor of the KCLBOT [5]. This paper presents a novel mobile robot configuration using a Pico-ITX system to manage a stereo vision system that can be used for self-localization, path planning and obstacle avoidance.

2. The Experimental Platform

2.1. *The KCLBOT: An Autonomous Non-holonomic Mobile Robot*

The KCLBOT [5] is a mobile robot platform, developed at King's College London, that is equipped with some of the state of the art equipment that enables high level computing and stereo vision, locally and autonomously.

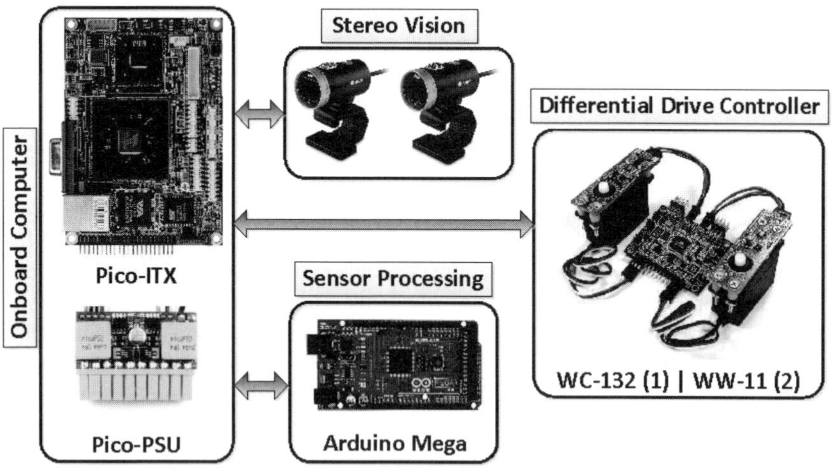

Figure 1. KCLBOT System Critical Hardware Overview

The KCLBOT is packed with some of the latest cutting edge robotic hardware. Figure 1 provides an overview of this. The mobile robot's hardware is managed by VIA's Pico-ITX [1] computer system with a non-mechanical solid state disk. The mobile robot's drive system is managed by Nu-botics [6] WC-132 Wheel-commander and two WW-11 Wheel-watchers. The sensor data is processed using the Arduino Mega 2560 [7], which manages the shaft encoder, digital compass, ultra-sonic rangers, and 3D accelerometer data. The stereo vision system is implemented using two Microsoft LifeCam Cinema [8] high definition cameras.

Figure 2. The KCLBOT: An Autonomous Non-Holonomic Mobile Robot

With the Pico-ITX onboard computer and the dual camera configuration, depicted in Figure 2, the mobile robot platform is capable of implementing stereo vision.

2.2. *The Kinematics and System Dynamics*

The KCLBOT was designed as a maneuverable classified [9] wheeled mobile robot. The mobile robot platform is constrained to rotate about its central axis before the platform can translate to a desired position.

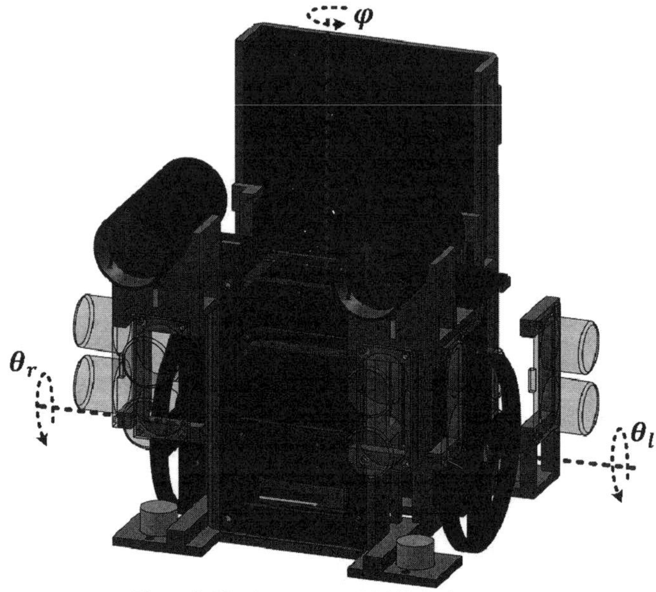

Figure 3. The Autonomous Mobile Robot Model

From Figure 3, $\theta_{l,r}$ describes the rotation angle of the left and right wheel, φ describes the mobile robot's orientation angle, and $F_{l,r}$ describes the left and right camera frame reference. As the mobile robot platform, depicted in Figure 3, is driven using a dual wheel configuration and because of the slip generated by the rotating wheels and the limited number of controllable degrees of freedom, the system is considered to be non-holonomic [10].

The non-holonomic mobile robot is modeled using Lagrange-d'Alembert principle [11]. Using this principle, the mobile robot is represented in state space form as follows:

$$M(q).\ddot{q}+V(q,\dot{q})=E(q).u+A^{T}(q).\lambda_{n} \tag{1}$$

where $\ddot{q}$, $\dot{q}$, and q describe the system's acceleration, velocity and position, respectively. $M(q)$ matrix is used to describe the systems inertia. $V(q,\ddot{q})$ is a column vector describing the system's velocity dependent forces. $E(q)$ is a column vector describing the system's actuator torques that are generated by the rotating wheels. $A(q)$ describes the system's holonomic and non-holonomic constraints and λ_n are the Lagrange multipliers used to optimize the control of the mobile robot.

For the purpose of this paper, the full decomposition of the model [5] is not required, but it is important to understand that stereo vision will input into equation (1). Stereo vision will enable the mobile robot to self-localize and this will provide the dynamic model with mobile robots position, velocity and acceleration.

3. Stereo Vision

The key to the successful implementation of an autonomous navigation mobile robot is for the autonomous platform to be able to self-localize, path plan and avoid obstacles. Having a stereo vision configuration, the mobile robot will be able to build 3D environment maps for self-localization, path planning, and using the depth information, the mobile robot can avoid obstacles.

3.1. *3D Recovery from 2 Images*

Using two cameras attached to the mobile robot platform, it is possible to correlate the images from both cameras and produce a 3D co-ordinate structure [2]. To recover the 3D position of a point from the two cameras, two projection matrices need to be set up, as follows:

$$P_{1,2} = \begin{bmatrix} p_{11} & p_{12} & p_{13} & p_{14} \\ p_{21} & p_{22} & p_{23} & p_{24} \\ p_{31} & p_{32} & p_{33} & p_{34} \end{bmatrix} \qquad (2)$$

The projection matrix holds the information about the cameras' 3D position, the pixel pitch information, the angle of view of the camera and the effective focal length of the camera. Using the projection matrix defined in equation (2), a 3D world to 2D object equation can set up as follows:

$$\begin{bmatrix} a_{1,2} \\ b_{1,2} \\ w_{1,2} \end{bmatrix} = P_{1,2} \begin{bmatrix} X \\ Y \\ Z \\ 1 \end{bmatrix} \qquad (3)$$

where the image object point co-ordinates can be described as $x_{1,2} = a_{1,2}/w_{1,2}$ and $y_{1,2} = b_{1,2}/w_{1,2}$, and where $w_{1,2}$ is not assumed to be 1.

$$\begin{bmatrix} x_1 p_{31}^1 - p_{11}^1 & x_1 p_{32}^1 - p_{12}^1 & x_1 p_{33}^1 - p_{13}^1 \\ y_1 p_{31}^1 - p_{21}^1 & y_1 p_{32}^1 - p_{22}^1 & y_1 p_{33}^1 - p_{23}^1 \\ x_2 p_{31}^2 - p_{11}^2 & x_2 p_{32}^2 - p_{12}^2 & x_2 p_{33}^2 - p_{13}^2 \\ y_2 p_{31}^2 - p_{21}^2 & y_2 p_{32}^2 - p_{22}^2 & y_2 p_{33}^2 - p_{23}^2 \end{bmatrix} \begin{bmatrix} X \\ Y \\ Z \end{bmatrix} = \begin{bmatrix} p_{14}^1 - x_1 p_{34}^1 \\ p_{24}^1 - y_1 p_{34}^1 \\ p_{14}^2 - x_2 p_{34}^2 \\ p_{24}^2 - y_2 p_{34}^2 \end{bmatrix} \quad (4)$$

Having expanded and manipulated equation (3) to get equation (4), to solve the 3D world co-ordinates, we simplify the equation to the form $AX = B$. This is solved by using Singular Value Decomposition SVD [12].

$$X = (A^T A)^{-1} A^T B \quad (5)$$

With equation (5) providing the solution to solve for the 3D world co-ordinates, all that is required to enable the solution is a matching set of images from the camera and the projection matrix (2) with constants.

3.2. *Camera Calibration: Deducing the Projection Matrix*

It is not possible to recover the 3D position (5) from the two camera images until the projection matrix (2) for each camera is deduced. To achieve this, equation (4) is restructured as follows:

$$Xxp_{31} + Yxp_{32} + Zxp_{33} + xp_{34} = Xp_{11} + Yp_{12} + Zp_{13} + p_{14} \quad (6)$$

$$Xyp_{31} + Yyp_{32} + Zyp_{33} + yp_{34} = Xp_{21} + Yp_{22} + Zp_{23} + p_{24} \quad (7)$$

Using equations (6) and (7), the projection values are isolated as follows:

$$\begin{bmatrix} X & Y & Z & 1 & 0 & 0 & 0 & 0 & -Xx & -Yx & -Zx & -x \\ 0 & 0 & 0 & 0 & X & Y & Z & 1 & -Xy & -Yy & -Zy & -y \end{bmatrix} p = 0 \quad (8)$$

$$p = [p_{11} \ \ p_{12} \ \ p_{13} \ \ p_{14} \ \ p_{21} \ \ p_{22} \ \ p_{23} \ \ p_{24} \ \ p_{31} \ \ p_{32} \ \ p_{33} \ \ p_{34}]^T \quad (9)$$

where equation (9) holds the constants stacked in a vector, which are the values in the projection matrix (2). To find the constant values in equation (9), the *nullspace* of matrix multiplied by p needs to be found. This can be achieved by using SVD [12]. The next step is to populate the matrix in equation (8) with 2D co-ordinate values from the matching images.

Left Camera Image　　　　　Right Camera Image

Figure 4. Sample Images captured from the mobile robots left and right cameras

The next step after having the two coinciding camera images is to find multiple matching points in the image to correlate against. This is achieved by using a chessboard image, as depicted in Figure 4. The corners of the 9 x 6 chessboard were used as matching reference points. To identify the pixel co-ordinates in the images, the Harris feature detector [13] algorithm was used.

3.3. *Experimental Results*

Using a sample of 9 matching images with a variety of different poses of the chessboard, depicted in Figure 4, the matrix in equation (8) is filled with values and SVD [12] to solve the projection matrix for the left and right camera.

$$P_1 = \begin{bmatrix} 537.31 & 0 & 326.76 \\ 0 & 537.16 & 250.61 \\ 0 & 0 & 1 \end{bmatrix} \tag{10}$$

$$P_2 = \begin{bmatrix} 533.72 & 0 & 341.76 \\ 0 & 533.85 & 236.10 \\ 0 & 0 & 1 \end{bmatrix} \tag{11}$$

Using the project matrices (10) and (11), it is possible to solve equation (5) with two matching images and return 3D co-ordinates for those matching images.

4. Conclusion

The paper outlines a novel mobile robot configuration for stereo vision system that can be used for self-localization, path planning and obstacle avoidance, at a small form factor for an autonomous mobile robot, depicted in Figure 1. The paper also provides a model for recovering 3D world co-ordinates from two images from a stereo camera configuration. It presents the initial experimental results of the solution to projection matrices needed for the recovery processes.

References

1. Via Technologies Inc. "Pico-ITX," http://www.via.com.tw
2. R. Hamzah, Salim, S., "Software Calibration for Stereo Camera on Stereo Vision Mobile Robot using Tsais Method," *International Journal of Computer Theory and Engineering,* vol. 2, no. 3, pp. 390-394, 2010.
3. J. Haverinen, Parpala, M., Roning, J., "A Miniature Mobile Robot With a Color Stereo Camera System for Swarm Robotics Research." pp. 2483-2486.
4. E. K. Huber, D., "Using stereo vision to pursue moving agents with a mobile robot." pp. 2340 - 2346.
5. E. Georgiou, Dai, J., Luck, M., "The KCLBOT: A Framework of the Nonholonomic Mobile Robot Platform Using Double Compass Self-Localisation " *Mobile Robots - Current Trends*, Intech Open, 2011.
6. "WheelCommander Motion Controller," http://www.nubotics.com
7. Arduino. "Arduino Mega 2560," http://arduino.cc.
8. Microsoft. "Microsoft LifeCam Cinema," http://www.microsoft.com
9. G. Campion, Bastin, G., and D'Andrea-Novel, B. , "Structural Properties and Classification of Kinematic and Dynamic Models of Wheeled Mobile Robots," *IEEE Transactions on Robotics and Automation,* vol. 12, no. 1, pp. 47–62, 1996.
10. I. A. M. Kolmanovsky, N., "Developments in nonholonomic control problems," *IEEE Control Systems Magazine*, pp. 20-36, 1995.
11. P. Johnson, *Classical Mechanics with Applications*, pp. 79-96: World Scientific, 2010.
12. W. Press, Teukolsky, S., Vetterling, W., Flannery, B. , *Singular Value Decomposition*, p.^pp. 59-70: Cambridge University Press, 1992.
13. C. Harris, Stephens, M., "A Combined Corner and Edge Detection." pp. 147-151.

PRECISE SELF-LOCALIZATION OF A WALKING ROBOT ON ROUGH TERRAIN USING PTAM*

Dominik Belter[‡] and Piotr Skrzypczyński[‡]
Institute of Control and Information Engineering, Poznan University of Technology, 60-965 Poznan
ul. Piotrowo 3A,
[‡]*E-mail: {Dominik.Belter, Piotr.Skrzypczynski}@put.poznan.pl*

This paper presents the concept and selected results of the localization system for a robot walking on rough terrain. The Parallel Tracking and Mapping (PTAM) algorithm and the Inertial Measurement Unit (IMU) are used to determine the 6-DOF pose. The system operates on-line on the real robot. The inclination of the robot's platform is determined by using IMU. The localization system is used together with the RRT-based motion planner which allows to walk autonomously on rough, previously unknown terrain. Efficiency and precision of the proposed solution are demonstrated by experiments.

1. Introduction

Autonomous locomotion of a robot in unknown environment is a fundamental problem in the field of mobile robotics. Robots which can explore previously unseen areas have wide application. For instance such robots can be used for Urban Search and Rescue (USAR) missions. These include search for survivors in collapsed buildings, search for persons who are lost in disaster areas, or localization of chemical and biological sources of pollution.

During a real mission, a robot has to map the environment while traversing the terrain and then it has to use the environment model to plan further movements. Important issues are raised by leaving the assumption that the robot knows the terrain in advance. This assumption often holds even in recent research concerning motion planning for walking robots[3]. The LittleDog robot used in Ref. 3 knows the shape of the terrain, and also the position of the robot itself is precisely determined by using a motion capture system.

*This work was supported by NCN grant no. 2011/01/N/ST7/02080

Recently, we presented a terrain map building system for a walking robot equipped with an active 2D laser range sensor[1,2]. The system updates the elevation map while the robot is walking on rough terrain. The data from the laser rangefinder and current position of the robot are necessary to update the map. We use legged odometry to determine the 2D position of the robot (x_R,y_R,z_R). It means that the robot counts steps and assumes that the executed motion is the same as the planned motion. The IMU is used to measure inclination of the robot's platform. Such system returns erroneous values in presence of slippages or even when the drives in leg's joints work imprecisely. In real mission an on-board, real-time localization system based on exteroceptive sensing is required.

2. Related work

To deal with GPS-denied rubble areas the robot requires a high precision localization system. The Messor robot, used in this research, employs an elevation grid map of the terrain as the basis for motion planning. The size of the elevation map cell is 1.5×1.5 cm. The position error should be below these values to correctly update the elevation map and to plan the motion of the robot. Large scale self-localization[11] is less important than the precise pose estimation in small workspaces. Moreover, the localization system should operate in real time. A visual, real-time, monocular SLAM can be used on the walking robot[10]. Unfortunately the precision is not sufficient for our purposes. The drift of the pose estimate can be corrected by detecting the loop closures[7]. However, during an USAR mission the robot often tries to explore new areas and avoids returning to previously visited places.

The pose of the robot can be precisely estimated by using scan matching[8]. However, in our case (rubble area) finding a transition that maximizes the overlap of consecutive scans is not possible. The laser scanner, mounted at the front of the robot, is tilted down so the consecutive scans contain measurements about different areas of the environment.

In opposite to SLAM the Parallel Tracking and Mapping algorithm was proposed[4]. In PTAM features tracking and mapping are separated into two parallel tasks. It enables the real-time implementation and tracking of thousands of landmarks. The accuracy and robustness is better than in other SLAM-based systems. The PTAM algorithm uses single camera. It requires many computing resources (dual core processor is preferred). However, PTAM was also implemented on the smartphone[5]. Only a simple stereo initialization is required[4] before normal operation. The efficiency of the PTAM system was also presented on the mobile robot equipped with a laser range finder[9].

3. On-board localization

3.1. *Control strategy*

The Messor robot used in this research uses a deliberative approach while walking on rough terrain. It measures the terrain profile by using the laser rangefinder and then precisely plans its motion. The core of the motion planning method is the RRT-Connect (Rapidly-exploring Random Trees)[6] algorithm.

Our implementation of the motion planner uses an adaptation of the RRT-Connect algorithm to find the 2D position of the robot on the map. To determine the full state of the robot (in 24D space) we use supporting software modules. The foothold selection module[1] allows to determine the position of the feet. Then the posture of the robot is determined to assure the maximal values of the kinematic and stability margin of the robot. The feet motion planner determines the secure path of the feet above the obstacles. Each posture of the robot is verified by the collision detection module, static stability and kinematic margin checking procedures. The motion planning procedure allows to walk autonomously over rocky terrain. The robot avoids obstacles and decides how to climb terrain irregularities by using the same RRT-based algorithm.

During execution of the planned motion the robot acquires new knowledge about the environment. Precise information about position of the robot is necessary to update the elevation map. In case of imprecise estimation of the position the mapping algorithm updates wrong cells of the elevation map. As a result the map contains phantom obstacles and the motion planning algorithm works improperly trying to find the path over the phantom obstacles.

3.2. *Localization for a walking robot*

To determine the requirements for the localization system and to verify the influence of the localization error on motion execution a series of simulations was conducted. The simulations were performed on the map of the rocky terrain mockup (Fig. 1). During execution, a random error within a defined range was added to the estimated value of the position of the robot. The results (success ratio) are shown in Tab. 1.

Even when the robot knew its position exactly or the position error

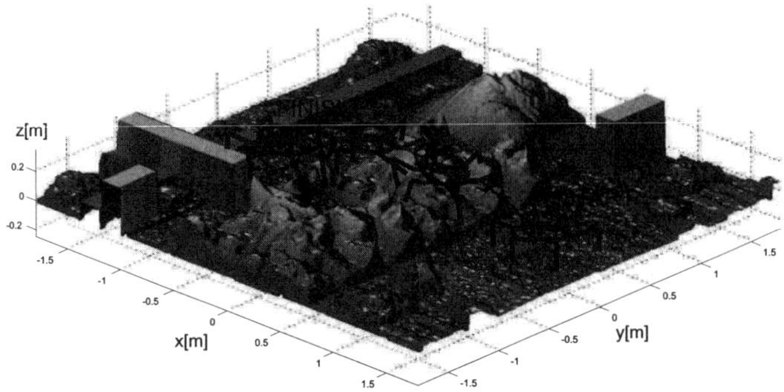

Fig. 1. Simulation environment for the experiment which verifies influence of the localization error during motion planning (note that the robot is planning the motion only through the rocky terrain in the middle of the map)

Table 1. Results of simulation – efficiency of motion execution in presence of localization errors.

Localization error [cm]	0.0	1.0	2.0	3.0	4.0
Success ratio [%]	90	90	60	50	10

was small (within the range of ± 1 cm), the execution failed in one of 10 trials. The failure is connected to the parameters of the motion planner. The robot failed because it didn't have the kinematic margin to execute the motion. The settings of the motion planning algorithm allow to choose a posture which has a small value of the kinematic and stability margin. This greedy approach allows to find faster a path to the goal position. The robot can find a way to deal with demanding obstacles. On the other hand, the postures chosen by the algorithm might be risky and the execution of the motion might fail.

The results presented in Tab. 1 show that the efficiency of execution of the motion decreases when the localization error is bigger than ±2 cm. Such a small value is difficult to obtain when SLAM is used. In practice, when the terrain is not as demanding as in this example, the acceptable position error can be higher but only the PTAM system can provide the estimated position of the robot with given precision.

4. Experiments

The PTAM algorithm returns the (x_P,y_P,z_P) values defined in the O_P coordinate system (located in the initial position of the camera). To express the position in the global coordinate system O_R and to recover the scale each experiment starts with a calibration procedure. The camera is translated along each axis by the known distance. The values returned by PTAM are used to recover the transformation from the O_R to O_P coordinate system and then to determine the position of the robot in the global coordinate system.

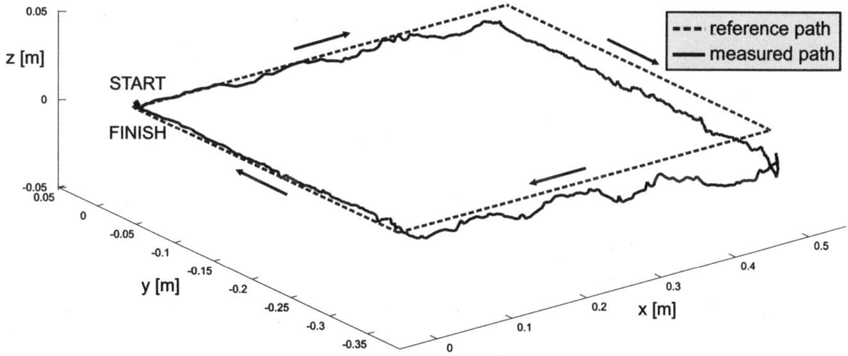

Fig. 2. The difference between the reference path and the path obtained by using PTAM

To verify the efficiency of the PTAM algorithm we performed an experiment on the rough terrain mockup. During the experiment the camera is located on a cart to provide the ground truth and the perspective view like on the real robot. The observed scene should contain enough features for tracking and for pose estimation. The operating environment for the walking robot (rocky terrain, rubble) seems perfect according to this requirements. The efficiency of the PTAM algorithm was confirmed during the experiment. The reference and measured path of the camera are presented in Fig. 2.

Table 2. Percentage distribution of localization error δ. Each column represents the percentage share of points with localization error bigger than threshold.

	δ>2.0 cm	δ>1.5 cm	δ>1.0 cm	δ>0.5 cm
δ_x [%]	0.0	10.8	28.5	59.8
δ_y [%]	0.0	0.0	0.0	5.1
δ_z [%]	4.8	19.9	40.5	58.9

The position error at the end of the path is 3.6 mm. The results are obtained without loop closures detection and optimization. Table 2 presents the percentage share of estimated camera positions for which the localization error δ is bigger than the given value. The localization error along z axis was bigger than 2 cm for only 4.8% of measured positions. For x and y axis the error was always smaller than 2 cm. The localization accuracy makes the PTAM algorithm feasible for the robot walking on rough terrain.

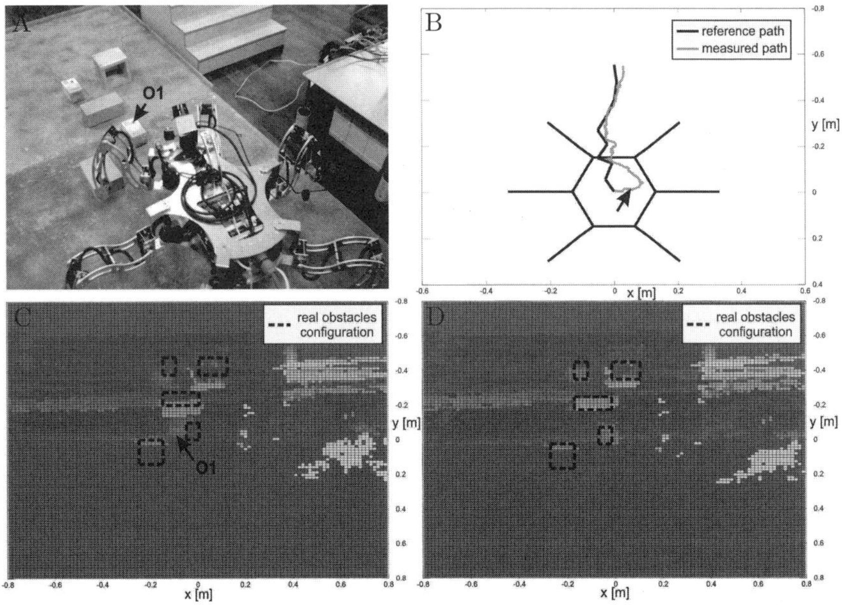

Fig. 3. Experiment on the real robot: experimental setup (A), planned and executed path (B), the elevation maps obtained by using legged odometry (C) and PTAM (D)

Next, the algorithm was verified on the real robot. Do determine the orientation of the robot's platform IMU is used. The IMU returns more reliable data independently of the observed scene. We don't have the motion capture system to measure the position of the robot so the ground truth for the robot's path is unknown. Instead we can compare the known configuration of the obstacles on the scene (Fig. 3A) and the configuration of the obstacles on the obtained elevation map. The planned and measured path of the robot are shown in Fig. 3B. At the beginning of the motion the robot placed its foot on the edge of the obstacle and felt down. As a result the real path differs significantly from the planned path (an error indicated by the arrow in the Fig. 3B). Without the localization system the robot can't compensate any of the unexpected motions. As a result the obtained map does not represent properly the real configuration of the obstacles (especially the obstacle O1 in Fig. 3C). The obtained map can't be used for further motion planning because of imprecise environment representation. On the other hand when the data about the position of the robot obtained from the PTAM are used the elevation map represents properly the environment configuration (Fig. 3D).

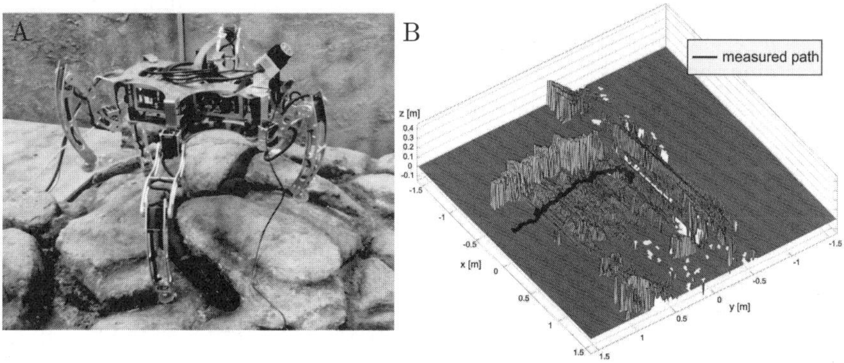

Fig. 4. Messor robot during experiment on the rough terrain mockup (A); the obtained path and elevation map (B)

Finally, an experiment on the rough terrain mockup was conducted (Fig. 4A). During the experiment the robot was using RRT-based planner to plan the motion (select footholds, determine the posture, avoid collisions and preserve static stability). The robot was using data from the PTAM to update the elevation map. The reliable localization system gives the possibility to correctly map the obstacles and successfully finish the task. Despite of some small slippages the robot was able to continue the task. The path measured by PTAM is shown in Fig. 4B. It represents properly behavior of the robot during the experiment – first the robot climbed up the rocks and then it climbed down the mockup[a].

5. Conclusions and future work

The estimation of the robot's pose plays an important role in the deliberative systems where the robot measures the state of the environment and plans its motion. The PTAM-based localization system of the robot enables robust walking on rough terrain and mapping. The system operates on-line

[a]A video is available on http://lrm.cie.put.poznan.pl/clawar2012DBPS.avi

and can be located on the robot's platform. The results of the experiments show that the position error is small enough to allow proper mapping using a laser range finder. Despite of the slippages and falls the robot can precisely estimate its position and successfully recover from errors.

In the future we are going to replace the current IMU with more reliable Attitude and Heading Reference System to estimate orientation of the robot's platform. Now, the localization system is limited to small workspaces. It is possible to create hybrid system which uses PTAM to determine the position of the robot when the walking distance is small and SLAM to determine the position of the robot in large areas.

References

1. Belter D, Skrzypczyński P (2011), Rough terrain mapping and classification for foothold selection in a walking robot, Journal of Field Robotics, Vol. 28(4), 497–528
2. Belter D, Skrzypczyński P (2011), Integrated motion planning for a hexapod robot walking on rough terrain, 18th IFAC World Congress, Milan, Italy, 6918–6923
3. Kalakrishnan M, Buchli J, Pastor P, Mistry M, Schaal S (2010) Fast, robust quadruped locomotion over challenging terrain, Proc. IEEE Int. Conf. on Robotics and Automation, Anchorage, USA, 2665–2670
4. Klein G, Murray D (2007) Parallel tracking and mapping for small AR workspaces, Proc. Int. Symp. on Mixed and Augmented Reality, Nara, Japan, 225–234
5. Klein G, Murray D (2009) Parallel tracking and mapping on a camera phone, IEEE Int. Symposium on Mixed and Augmented Reality, Orlando, USA, 83–86
6. Kuffner J J, LaValle S M (2000) RRT-Connect: An efficient approach to single-query path planning, Proc. IEEE Int. Conf. on Robotics and Automation, San Francisco, USA, 995–1001
7. McDonald J, Kaess M, Cadena C, Neira J, Leonard J J (2011) 6-DOF multisession visual SLAM using anchor nodes, Proc. European Conference on Mobile Robots, 69–76
8. Skrzypczyński P (2009) Simultaneous localization and mapping: A feature-based probabilistic approach, Int. J. of Applied Mathematics and Comp. Sc., vol. 19(4), 575–588
9. Jinbo S, Tano S, Songmin J (2011) Mobile robot localization and map building based on laser ranging and PTAM, Proc. IEEE Int. Conf. on Mechatronics and Automation, Beijing, China, 1015–1020
10. Stasse O, Davison A, Sellaouti R, Yokoi K (2006) Real-time 3D SLAM for humanoid robot considering pattern generator information, Proc. IEEE Int. Conf. on Intelligent Robots and Systems, Beijing, China, 348–355
11. Strasdat H, Montiel J M, Davison A (2010) Scale drift-aware large scale monocular SLAM, Proc. Robotics: Science and Systems, Zaragoza, Spain

MOBILE ROBOT NAVIGATION VIA OLFACTION

S.M.ARMAN ULLAH

South Westphalia Univ of Applied Sciences Campus Soest, Germany

GURVINDER S. VIRK

University of Gävle, Faculty of Eng and Sustainable Det, SE-801 76 Gävle, Sweden

Mobile robot navigation via olfaction has been receiving considerable interest over recent years to assist in the development of advanced cognitive robots which are able to operate in natural environments. With the related advances in electronic noses, it is now possible to detect and recognize and locate a range of different odours for a variety of applications. This paper involves designing and building a chemical sensing mobile robot that is able to navigate using these latest olfactory sensors. A smart robot equipped with olfactory sensors is designed for performing various search tasks in chemical fields; these search tasks include locating point sources, following chemical trails and plume following and investigating how the search efficiencies can be enhanced.

1. Introduction

Animals use a variety of senses, namely, sight, smell, taste, hearing, tactile feeling, etc. to navigate in nature. Chemical sensing is important in smelling and tasting and it plays a significant role in the life of animals since it is useful for finding food, avoiding threats and coordinating key behaviours in social communities. Although playing a major role in nature, chemical sensing has been a largely disregarded sense in robotics [1] until recently and the main dependence for robot navigation has been on vision (which is a very compute intensive operation). Interest in chemical sensing has grown since the 90s and it is proving to be valuable, but it is still being actively developed because it is not still fully understood due to the complexities and instabilities of the operational environments, the sensing methodologies and the variety of uncertainties of the actual decision making processes that exist. Some of the recent issues and developments in olfaction based robotics are presented in [2]-[4] and much of the inspiration for producing solutions in this area has been based on nature since there are some creatures with impressive olfactory capabilities that can work over large ranges. Some examples include the following:

- seabirds can pinpoint food-rich habitats over hundreds of kilometres based on plankton-derived olfactory cues;
- some pigeons (Columba livia) seem to use olfactory landscapes for homing from long distances;

- male silk moths are able to detect and follow pheromones released by female moths over 1km away;
- ants leaving their nest are able to follow long scent trails to find food;
- sharks can smell tiny amounts of blood in the water to locate prey from over a mile away; and
- catfish can smell certain amino acids and other substances from long distances to find their prey.

This paper presents research to investigate some of the key issues in chemical sensing for realising effective olfactory based navigation of mobile robots. The paper is set out as follows: Section 2 presents an overview of the current chemical sensing methods; Section 3 describes the key search tasks that arise in chemical searching; Section 4 presents details of the mobile smelling robot and some ideas for the planned work items; Section 5 presents the experimental setup and some experimental results; and Section 6 presents the conclusions.

2. Chemical sensing

Chemical sensors produce signals as the result of the chemical or biochemical interaction with their target compound as shown in Figure 1(a); other chemical sensing methods devices also exist and include those where alternative forms of chemical analysis is performed, eg, molecular or atomic spectroscopy and "lab-on-a-chip" sensor concepts as shown in Figures 1(b) and (c), respectively.

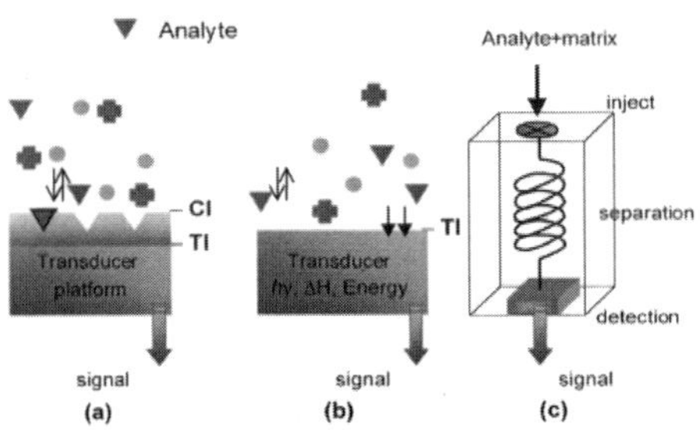

Figure 1: Sensors for chemical analysis [5]

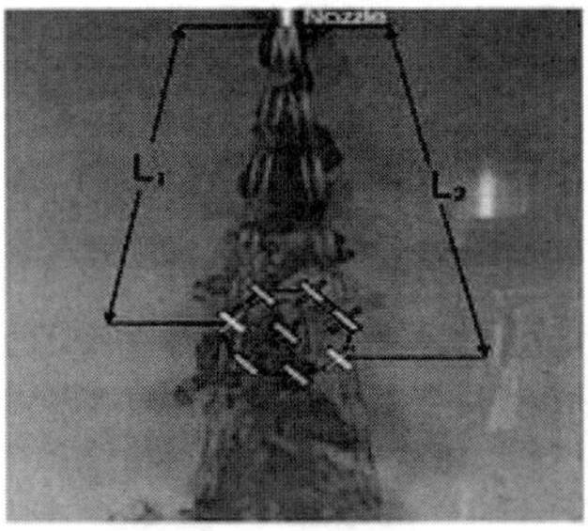

Figure 2: Visualizing chemical plumes [7]

There exist many sensory systems for detecting and analyzing volatile chemicals; they range from simple metal oxide sensors that can selectively detect gases such as hydrogen, water vapour, methane and carbon monoxide to mass spectrometers and gas chromatographs that can analyze complex chemical mixtures. Chemical sensors for robot applications should have desirable properties like high sensitivity, low power consumptions, rapid response times, be compact, low cost and be widely commercially available. Examples of chemical sensors that meet these requirements and which have been employed in olfactory robotic research are metal oxide gas sensors, quartz microbalance sensors and conducting polymer sensors [6].

3. Search tasks and animal strategies

Odour source localisation is regarded as the base of many tasks using olfactory mechanisms. So far, the main problem studied in olfactory navigation has been the tracking and localization of static odour sources. This can be broken into the three following tasks: 1) finding traces of the chemical of interest; 2) tracking the respective odour plume; and 3) localizing the odour source. Such odour plume tracking has been the problem that has been studied the most using heuristic and bio-inspired algorithms; the odour tracking is sometimes complemented by other sensing modalities, like vision [1].

Understanding the mechanism of chemotaxis practiced by some animals is useful for the construction of chemical plume tracking robots (in air and in water). In a steady-state turbulent plume, at any point located distance L from the source, the instantaneous concentration $C(t)$ can be expressed as the sum of the mean value $\langle C \rangle$ and the fluctuating concentration $c(t)$, ie, $C(t) = \langle C \rangle + c(t)$, [7]. Figure 2 shows a photograph of red dye isokinetically injected into a fully developed turbulent boundary layer with an average flow velocity of 5 cm/s to help in the visualisation of a chemical plume. The position of a hypothetical eight-sensor array in the chemical plume is shown by the X-marks and the pair of sensors used in the correlation analysis is highlighted and they are L_1 and L_2 cm away from the source. The overlaid stripes represent two channels realized in the bench top virtual plume [7].

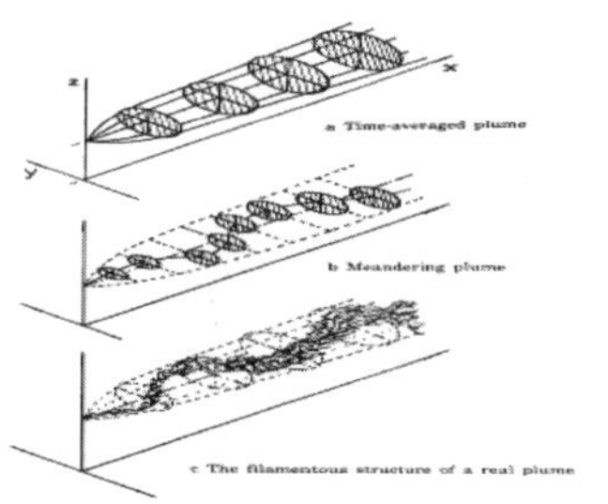

Figure 3: Models of pheromone diffusion from a point source[8]

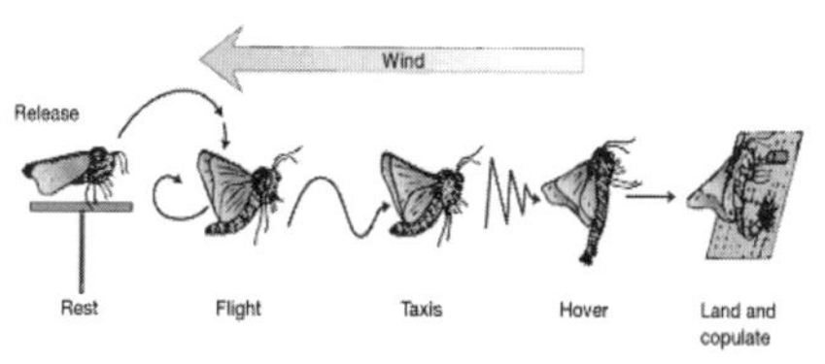

Figure 4: Male moth's response to sex pheromone released by the female[8]

The odour from a point source defuses in a cone-shaped plume that expands downwind and the shape of the plume depends on airspeed. The resulting plume is neither straight nor homogeneous over long distances but is influenced by the turbulence of the air stream that forms pockets of higher concentration or absence of chemical vapour. This can be modelled and Figure 3 shows the time averaged aspects where the symmetrical expansion of the plume from the point of emission can be seen together with the meandering aspects with the concentrations of discs distributed around the meandering centre line, and the filamentous structure of real plumes [5]. An insect downwind of such a plume would detect the plume as odour bursts rather than as a constant stream; Figure 4 shows a typical response from a male noctuid to sex pheromone released by female moths [8].

Other chemical sensing tasks in nature involve trail following. Ants leaving the nest are able to follow scent trails to find food and they are incredibly sensitive to trail marking chemicals; it has been estimated that one milligram of substance could lead a trail of ants three times round the world. While following a trail, ants detect the scent marks on the ground with their two antennae which sense the volatile chemicals by contact or at close proximity and thus provide accurate information about the trail's location [9].

4. Smelling robot design and key issues in chemical searching

The designed robot, SOMR-1 (Soest Olfactory Mobile Robot version 1), is a round with diameter ≈20 cm. It consists of two wheels, two castors, four IR sensors, four olfaction sensors and a WIFI system to communicate with a PC user console. Both wheels are controlled by motor controllers and have independent control. The differential steering mechanism is used to control the motion of the robot. The sensors (IR and smelling) are attached to the robot at 90° spacing. Figaro TGS 2620 olfactory sensors are used for detecting the chemicals in the plume and tracking the direction of the search path whereas, IR sensors are used for obstacle avoidance. The microcontroller used is an AXON II (Atmega 640) and the programming language used is C++. Figures 5 and 6 show the designed mobile robot.

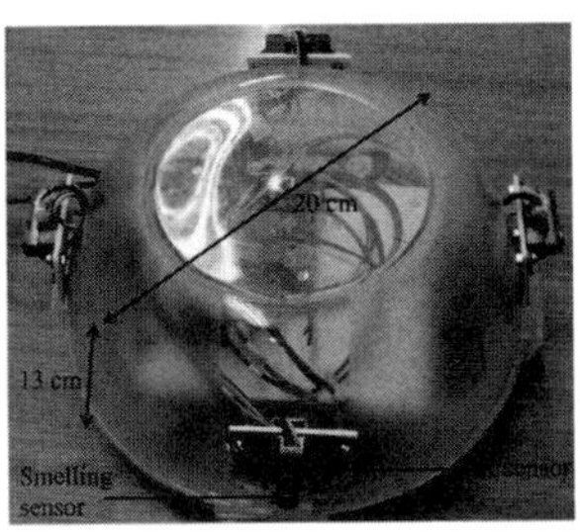

Figure 5: Robot with IR and smell sensors

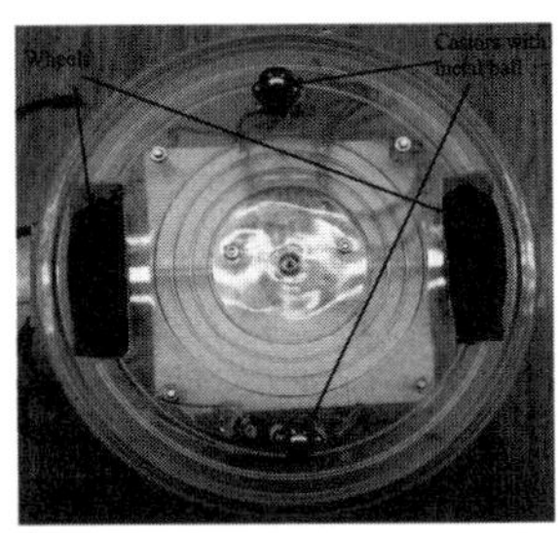

Figure 6: Wheels and castors of the robot

The further plan for the robot is to increase the number of sensors so that they are at 45° spacing to enhance the search performance during implementation of the various navigation strategies. Figure 7 shows a typical search task in a cluttered environment that is using the sensors and Figure 8 shows the top view of robot. The robot communicates via WIFI for logging and studying the search tasks as shown in Figure 9. Figure 10 shows the test arena built for performing the search trials; it's size is 1.5m×1.5m×0.3m and it is made of plexiglass covering an aluminium frame.

Chemical searching is often based on chemotaxis which requires two sensors which provide the minimum directional information needed to orientate the robot towards the increasing chemical gradient (ie, towards the chemical source). This is because chemosensors are not directionally sensitive and can only provide point readings (in contrast to visual sensors) so that a single chemical sensor reading should not be able to provide sufficient information on the direction to move to locate the chemical source [4]. However, most bisensory creatures are able to rely on unisensory strategies if needed and this leads to uncertainty as to what are the fundamental issues in chemical sensor based navigation, ie, how a unisensory based strategy can work in chemical searching is not yet fully understood.

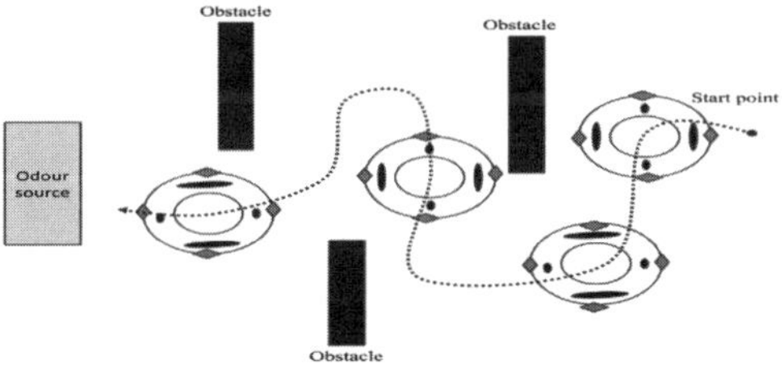

Figure 7: Chemical searching in cluttered environments

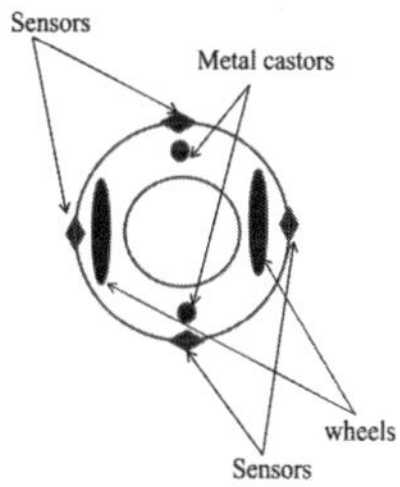

Figure 8:Top view of the olfactory robot

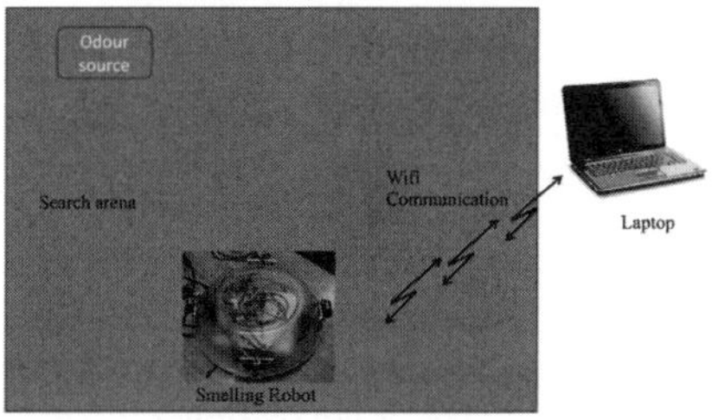

Figure 9: Test arena for trials with robot comms via WIFI

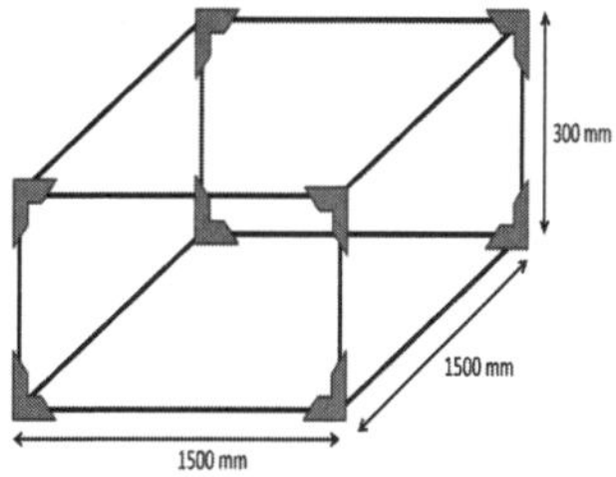

Figure 10: Closed arena for search trials

The project intends to investigate this issue as well as other key issues via implementation and testing on the mobile robot system developed. Some of these issues include the following:

1. Matching of search tasks with different search strategies: the work investigates chemical sensing search tasks to locate point sources. Plume and non-plume situations are explored.
2. Do the number of smell sensors matter? If so how? Biological studies suggest both unisensory and bisensory strategies have a role in chemical searching but details still need to be demonstrated.
3. How can chemical sensing search efficiencies be enhanced? Biological search patterns in olfactory navigation are typically slow and animals seem to be 'aware' of this problem and often try to find ways to enhance their performances. There are several possibilities for this, namely

- o Some species have developed strategies to benefit from detecting the direction of the wind (eg, moths) or the water-currents (eg, salmon, crabs and lobsters) and the creatures simply move upstream as long as they keep detecting the attractant chemicals carried by the medium; such plumes can be set up and explored as a part of the project to assess chemical search efficiencies
- o Sniffing in air is another smart strategy used by numerous species, such as dogs and cats to improve the performance. Such sniffing can be explored in robots by incorporating suction pumps to draw in air to enhance the smelling and assess if this improves the searching.

For performing the chemical search trials in the search arena, SOMR-1 is controlled by a remote PC and the odour source tracked using gas smelling sensors (such as SnO_2). Various methods are being explored to develop simple yet effective navigational strategies; these include using odometry to maintain a reference of the travel path from the known start position.¨

5. Experimental setup and results

The smelling robot used for the experiments initially will be the basic plume following strategy. In this experiment, the odour source used is Ethyl alcohol and the robot containing four smelling sensors is about 1.5 m away from it. Also a buzzer is installed for alarm generation to indicate the target has been reached when a threshold level is exceeded. A small fan is used to create the plume in the test arena. The robot uses differential steering mechanism to drive wheels for the movement .The environment used for the testing is a space where natural airflow is interrupted. The strategy used here to detect the odour source:

1. Sensors detect the odour and compare the smell data with other sensors.
2. The highest readings are used to determine the heading direction for the motion.
3. The robot moves to the chosen direction until it gets a new higher value in another direction.
4. If the robot senses an obstacle, it moves back, waits, turns right or left (by 45°) and then follows the sensor readings if the path is clear.

The robot reaches the odour source when one of the sensor reaches a defined threshold value. In that case, the robot stops and generates alarm to indicate a successful search.

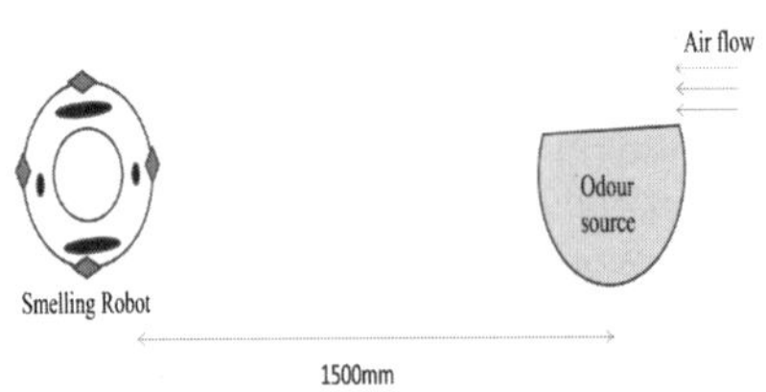

Figure 11: Setup for odour plume tracking

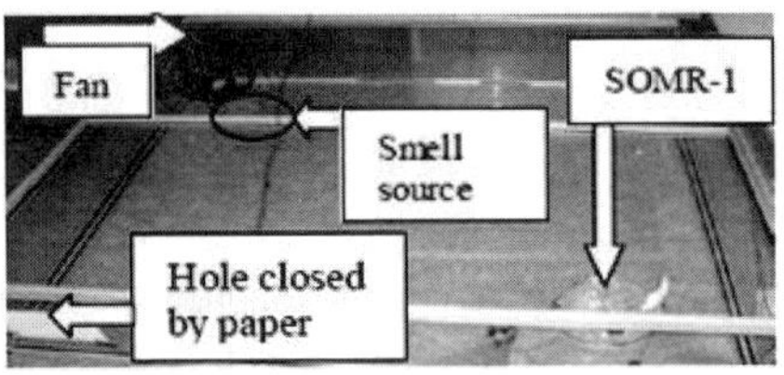

Figure 12: Setup in test environment

Figures 11 and 12 show the set up for the odour plume tracking where the plume experiments show that the sensors can detect the chemical smell, compare and follow the highest reading to reach the odour source. In this way it can navigate to follow the gradient in the chemical field to find source. Table 1shows the results of search trials to find the source.

Table 1. Results of plume following search trials

Trials no.	Distance travelled, mm	Time to reach source, sec	Successful/Bad Search
1	1,670	53	Successful Search
2	1,800	57	Successful Search
3	2,590	140	Unsuccessful Search
4	1,590	51	Successful Search
5	1,740	55	Successful Search
6	3,360	251	Unsuccessful Search
7	1,700	54	Successful Search
8	2,470	131	Unsuccessful Search
9	1,935	59	Successful Search
10	1,630	52.50	Successful Search

The average travel time to find the source for the majority of the successful searches led to selecting 60 seconds as the limiting time between successful and unsuccessful searches. The characteristics of the chemical field are extremely important in designing the navigational strategy and it experiments to study it have been designed. The chemical field has been visualised by use of fog generating methods to ts key characteristics but the input and exhaust capabilities could not be tuned sufficiently with the equipment available and more specialised hardware is being organised. Figure 13 below shows the visualization of the smoky fog which was covered by 20 consecutive snaps taken at a sampling rate of 1.2 sec.

6. Conclusions and future plans

Biological strategies for performing search tasks in chemical fields have been discussed and how these solutions can be implemented on smell-based navigating mobile robots presented. The success rate to find the odour source is almost 70-80% using this strategy. The robot shows good behaviours in detecting the chemical field strength for determining the concentration gradient and is able to travel correctly by following the chemical gradient. Sometimes the slow dynamic response of the chemical sensors lead to incorrect motions if the robot moves too fast; this happens if the front sensor reaches the highest value to drive the motion of the robot in changing direction. The time to reach the threshold and to go below the threshold is a major problem of the sensor for performing these search tasks. Increasing the number of sensors can enhance the search performance and lead to improved navigation efficiency. Future plans to investigate the key issues in olfactory navigating robots include exploring the fundamental issues and how the basic principles may be enhanced by increasing the number of sensors, using sniffing strategies and cooperating to perform a variety of search tasks such as locating multiple sources and trail following. These issues are expected to be inter-related and application specific.

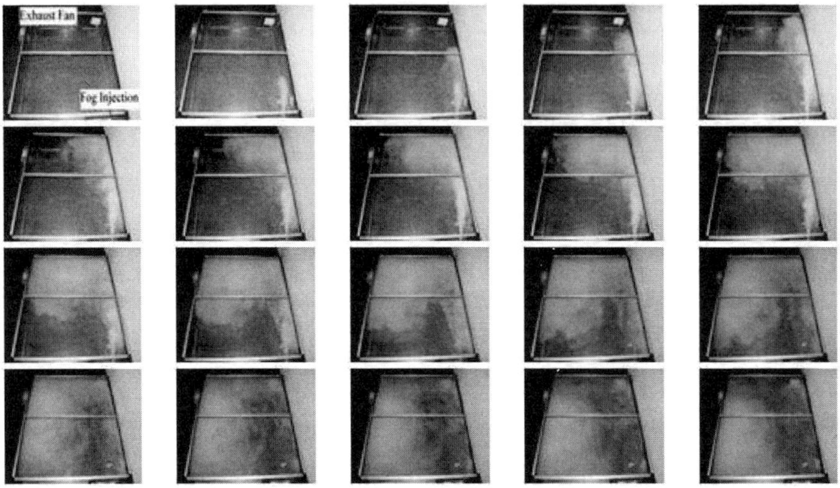

Figure 13: Visualisation of plume: photos to realise the characteristics of the chemical field; the fog is generated and injected at the bottom right; on the diagonally opposite corner, there is an exhaust fan. The pictures 1-20 at a sampling rate of 1.2 second and the fog was stopped at photo 13.

References

1. Marjovi A, Nunes J, Sousa P, Faria R, Marques L, An olfactory-based robot swarm navigation method, IEEE Int Conf on Robotics & Automation, Anchorage, Alaska, USA, May 3-8, 2010.
2. Lytridis C, Virk GS, Rebour Y, Kadar EE, Odour-based navigational strategies for mobile agents, adaptive behaviour, 9(3-4), pp171-187, 2001.
3. CLytridis L, Kadar EE, Virk GS, Systematic approach to odour source localization, Autonomous Robot, Vol20, pp261-276, 2006.
4. Kadar EE, Virk GS, Lytridis C, In search principles of odour source localisation, CLAWAR'07, pp623-631, 2006.
5. Stetter JR, Penrose WR and Yao S, Sensors, chemical sensors, electrochemical sensors and ECS, Journal of the Electrochemical Society, 2003.
6. Tzanos PC, Information theoretic motion algorithms for biochemical source localization, PhD Thesis, Univ of Illinois at Chicago, 2008.
7. Kikas T, Ishida H, Webster DR and Janata J, Chemical plume tracking 1-Chemical information encoding, Analytical Chemistry, Amer Chem. Society, Vol. 73, No. 15, Aug 2001.
8. Schowalter TD, Insect ecology: an ecosystem approach, Elsevier, UK, p99-101, 2006.
9. Russell RA, Odour detection by mobile robots, World Sci. Pub., London, UK, p22-31, 1999.

TURN AROUND BEHAVIOR GENERATION AND EXECUTION FOR UNMANNED GROUND VEHICLES OPERATING IN ROUGH TERRAIN

M. M. DABBEERU AND P. SVEC

Department of Mechanical Engineering, University of Maryland, College Park, Maryland 20742, USA, E-mail: {mmd,petrsvec}@umd.edu

J. D. LANGSFELD

Energetics Technology Center, St. Charles, Maryland 20603, USA, E-mail: jdlangs@umd.edu

S. K. GUPTA

Department of Mechanical Engineering and the Institute for Systems Research, University of Maryland, College Park, Maryland 20742, USA, E-mail: skgupta@umd.edu

We present an approach for turn around behavior generation and execution for an unmanned ground vehicle (UGV) operating in a rugged terrain. The generated behavior allows the vehicle to rapidly execute N-point turn maneuver in a narrow space by exploiting the vehicle's dynamics as well as the properties of the terrain. The developed approach incorporates model-predictive planning and optimization into the behavior execution. It combines offline synthesis of an atomic maneuver model for a given vehicle and terrain model, fast online computation of motion goals determining the execution sequence of atomic maneuvers and their extent in the state space, and atomic maneuver selection and adaptation to the terrain shape. We have successfully evaluated the turn around behavior using an augmented reality setup by having the vehicle execute 5-point turn maneuver on an inclined platform.

Keywords: N-point turn; Turn-around behavior; Unmanned Ground Vehicle.

1. Introduction

In order to operate on a rugged terrain, UGVs must have the capability to rapidly turn around. A terrain is considered rugged if it contains many features that are challenging for the vehicle to negotiate, such as steep slopes that may or may not be climbable based on the vehicle's approach velocity, varied ground material, and randomly distributed obstacles including ditches, rocks, and debris. The need to turn around arises because of limitations of perception accuracy and limited situational awareness. In many situations, executing turn around requires a large number of turning maneuvers in a tight space. The main challenge when planning complex maneuvers autonomously is in determining the effective coupling of velocity and steering angle profiles that allow the vehicle to reach its desired state safely and in the minimum time, given its dynamics and intricacies of the terrain (e.g., different inclination, bumpiness, friction, etc.).

Dynamically feasible trajectory planners for ground vehicles operating in a rough terrain have been previously studied in different contexts.[1–3] Specifically, planning and control was developed for an Ackerman steered vehicle executing N-point turn maneuvers in an orchard environment.[4] In addition, receding horizon model-predictive control and trajectory planning techniques were recently combined to allow following intricate paths including turning maneuvers.[5] In our approach, in order to achieve high computational speed and motion precision, we automatically generate a set of atomic maneuvers of various complexity for different initial and final states of the vehicle on a rough terrain. We then use the generated maneuver profiles as seeds for fast online adaptation to a specific terrain shape. In addition, we develop a cost function policy model that biases the local search for the maneuver along the most promising direction towards the goal state. We evaluate the developed approach on a physical platform.

2. Approach

The overall approach is divided into three components: (1) offline atomic maneuver model synthesis, (2) online motion goal computation, and (3) online atomic maneuver selection, adaptation, and execution (see Fig. 1 for an overview of the entire approach). We use the genetic algorithm[6] (GA) for automated generation of a set of atomic maneuvers for different initial and final poses of the vehicle and terrain inclination using a high-fidelity physics-based simulation (see Fig. 2(b)). During the online maneuver planning and execution, appropriate motion goals are determined based on the terrain constraints and vehicle kinematics. The atomic maneuvers are used as reference maneuvers for adaptation to a specific terrain shape using on-board physics simulations. The adapted atomic maneuvers are then executed in a sequence by the vehicle to realize the entire N-point maneuver.

An atomic maneuver is represented as two B-splines that correspond to steering angle and velocity profiles (see Fig. 2(c) and (d)). This allows the

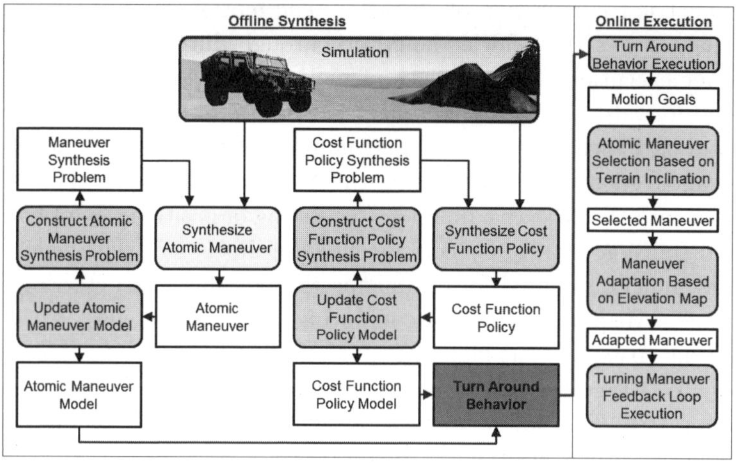

Fig. 1. Overview of the approach.

vehicle to synchronously control its steering angle ϕ and velocity v to be able to safely and efficiently reach a desired state $\mathbf{x_T} = [x_T, y_T, \theta_T, v_T]^T$ from its current state $\mathbf{x_I}$, where x_T and y_T are the position coordinates, θ is the orientation, and v_T is the velocity of the vehicle, while satisfying a set of dynamic constraints. During the maneuver synthesis and adaptation, we use the control points of both B-spline profiles and their execution time as the variables for optimization.

The profiles are synthesized over a five dimensional space of position and orientation differences $\triangle x$, $\triangle y$, $\triangle \theta$ between x_I and x_T, and terrain inclination angles $\triangle \alpha$ w.r.t x axis and $\triangle \beta$ w.r.t. y axis. Using the atomic maneuvers, we constructed a complete atomic maneuver model represented as a look-up table. We validated the maneuver model by conducting sensitivity analysis at the boundaries of the five dimensional parameter space. The optimization problem is defined to minimize the objective function $F = \omega_1 f_1 + \omega_2 f_2$ as the weighted sum of (a) the difference f_1 in position and orientation between $\mathbf{x_I}$ and $\mathbf{x_T}$, and (b) the execution time f_2 of the maneuver. The initial population of solutions was seeded with profiles obtained by manually driving the vehicle in the workspace from $\mathbf{x_I}$ to $\mathbf{x_T}$.

As a first step in the turn around behavior execution, the vehicle explores the space around it and computes a sequence of reachable motion goals $X_G = \{\mathbf{x_{G,1}}, \mathbf{x_{G,2}}, \ldots, \mathbf{x_{G,n}}\}, \mathbf{x_{G,i}} = [x_G, y_G, \theta_G, v_G]^T$ to approach its target state $\mathbf{x_T}$ based on its initial state $\mathbf{x_I}$, kinematic constraints, and the geometric configuration of obstacles. The motion goals specify the intermediate states the vehicle should go through during the N-point turn maneuver execution. The sequence of motion goals is computed by planning and forward simulating a sequence of back and forth maneuvers that are optimized such that the required number of turns is minimized. The local maneuvers are computed using the Newton-Rhapson method[7] in accordance with the objective function (consisting of position C_P, orientation C_Θ, and collision penalty C_T components; see Eq. 1) on a flat terrain. The objective function is used for optimizing the parameterized steering angle and velocity B-spline profiles for computing the motion goals.

In order to explore the planning space effectively, the navigation system utilizes a cost function policy model in which each policy is represented as a decision tree. The policy model is designed iteratively (see Fig. 1) and each policy returns a position ω_P, orientation ω_Θ, and terrain ω_T weights as the components of the objective function. For computation of each motion goal, the position weight ω_P is set to zero until a user-specified threshold μ of the orientation error falls below this threshold. Once the orientation is within the desired limits, the final state is computed by assigning equal weights to both the position and orientation parameters. The optimization problem for computing a motion goal can be formulated as

$$\begin{aligned} &\text{Minimize:} \quad C = \omega_P\, C_P + \omega_\Theta\, C_\Theta + \omega_T\, C_T \\ &C_P = \frac{\|[x_T, y_T]^T - [x_I, y_I]^T\|}{D_{max}} \\ &C_\Theta = \frac{-dot(\mathbf{u}.\mathbf{v})}{2} + \frac{1}{2} \\ &C_T = \begin{cases} 1 & \text{if collision} \\ 0 & \text{otherwise} \end{cases} \end{aligned} \tag{1}$$

where $\mathbf{u}$ is the current normalized heading of the vehicle, $\mathbf{v}$ is the desired normalized heading of the vehicle, and D_{max} is the maximum dimension of the workspace. The decision tree thus represents a cost function policy model (see Fig. 1) that is used for determination of suitable motion goals in the environment for efficient and safe execution of the N-point turn based on the context of the environment.

Having generated the motion goals, the vehicle can query the corresponding maneuvers (i.e., velocity and steering angle profiles) from the atomic maneuver model. The maneuvers are further adapted using the conjugate gradient method[8] based on the elevation map of the environment and concatenated into the entire N-point turn maneuver. In this way, the vehicle can minimize the turn around maneuver execution time by exploiting the terrain as well as vehicle properties. Finally, the maneuvers are executed using a PID controller.[9]

3. Results and Discussion

The steering and velocity angle profiles were represented using two B-splines with five control points. We considered a chromosome of length 11 that included y coordinates of 10 control points $\{Q_1, \ldots, Q_{10}\}$ as the design variables (i.e., five for each profile) along with the execution time t for GA optimization (see Fig. 2(c) and (d)). We generated 76800 atomic maneuvers to construct the maneuver model. The initial population was seeded with velocity and steering angle profiles obtained by manually driving the vehicle in the workspace between two states. The workspace was segmented over a 5 dimensional space of $\triangle x$ (-4.0 m, 4.0 m), $\triangle y$ (-4.0 m, 4.0 m), $\triangle\theta$ (-80, 80°), $\triangle\alpha$ (0, 16°)(angle w.r.t x-axis), $\triangle\beta$ (w.r.t y-axis) (0, 12°) in a high fidelity simulation environment. The GA parameters for synthesis of atomic maneuvers were as follows: population size $N_s = 100$, crossover probability $P_c = 0.5$, mutation probability $P_m = 0.4$, reproduction probability $P_r = 0.1$, and the number of generations $N_g = 80$. The computation time on Intel Core i7 processor per generation was 9.5 s.

In the designed experiments, the motion goals were computed on a flat terrain by optimizing the steering angle profile of the vehicle. A simple car kinematic model of the vehicle[10] was used with the maximum steering angle of 30° and the length of 0.4 m. The average time it took to compute a single motion goal on Intel Core i7 processor computer was 0.3 s.

To be able to reach the computed motion goals efficiently, the vehicle needs to exploit its dynamics as well as terrain conditions through the use of precomputed atomic maneuvers. The atomic maneuvers are obtained from the generated atomic maneuver model. We represented the maneuver model as a lookup table with key values in a 5-tuple $(\triangle x, \triangle y, \triangle\theta, \triangle\alpha, \triangle\beta)$. For a given state of the vehicle and terrain inclination, the nearest neighbor interpolation was used to estimate the parameters of the selected atomic maneuver. For instance, for the 5-tuple $(0.98, 1.23, 40, 4, 8)$, the interpolation method outputs an atomic maneuver that produces an error of $x_{err} = 0.2$ m, $y_{err} = 0.064$ m, and $\theta_{err} = 5.1°$.

The conjugate gradient optimization technique[8] is a powerful line search method when compared to the steepest descent method[7] and is often applied to problems involving a large number of variables. We implemented the adaptive line search method, where the adaptive step size is determined

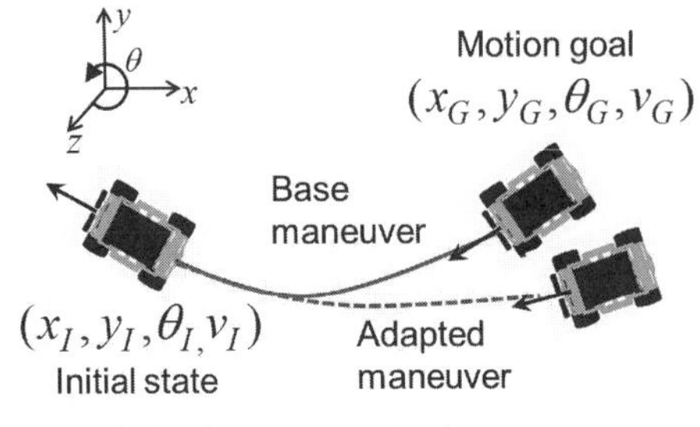

(a) Reverse atomic maneuver.

(b) Physics-based Vortex® simulation.

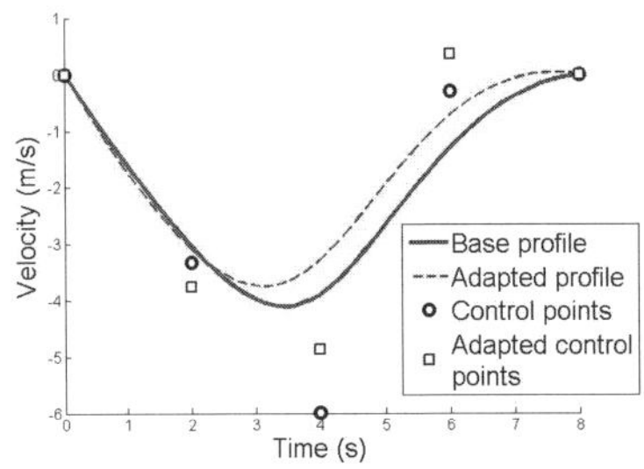

(c) Velocity profile.

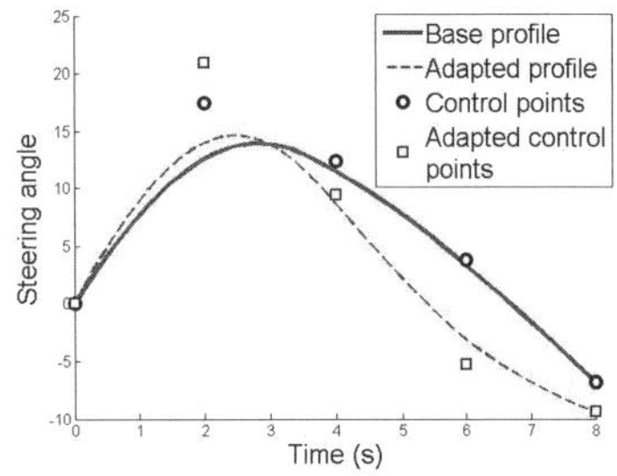

(d) Steering angle profile.

Fig. 2. An atomic maneuver allows the vehicle to synchronously control its steering angle ϕ and velocity v to be able to reach a desired motion goal $\mathbf{x_G} = [x_G, y_G, \theta_G, v_G]^T$. The steering angle and velocity are represented as two B-splines.

by using the Fletcher-Reeves formula. With the steepest descent method, the adaptation time was more than 16 s, whereas with our method the adaptation time was reduced to 4 s for a flat ground maneuver with errors of $x_{err} = 0.004$ m, $y_{err} = 0.02$ m, and $\theta_{err} = 1.3°$. Fig. 2 shows the maneuver profiles with and without adaptation.

Our ability to synthesize and adapt maneuvers is greatly aided by the Vortex® physics engine (see Fig. 2(b)). We are able to simulate the key features of the vehicle, providing a model with very similar behavior to the physical vehicle used in our experiments. We are able to take the geometric features of the vehicle directly from a CAD model, defining the collision behavior. The simulated dynamic components include the electric motor, the powertrain differentials, and the independent suspension on each wheel. The parameters of maximum engine speed, engine torque, gear ratio, suspension stiffness, and coefficient of friction of the tires were measured on the physical vehicle and used in the simulation. Even with these multiple

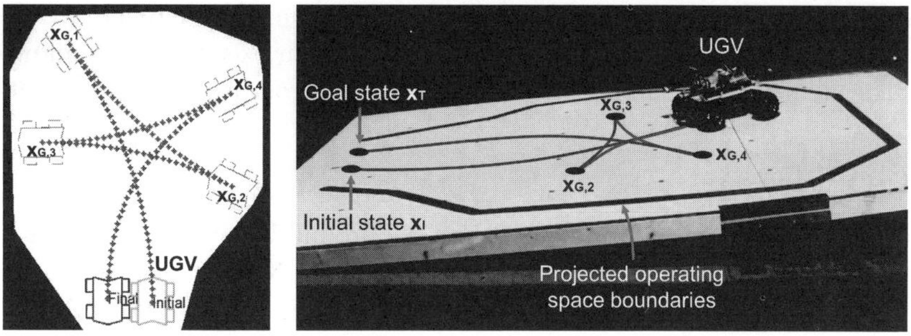

(a) Maneuver planning. (b) Maneuver execution on an inclined platform.

Fig. 3. A ground vehicle executing 5-point turn maneuver. A video demonstrating the maneuver is available on the Maryland Robotics Center YouTube channel (http://youtu.be/16JR-ebxL7E).

dynamic components being simulated, simulation performance is still quite high. On an Intel Core i7, execution time of the simulation can be sped up to 12 times with no observed loss in fidelity.

We have developed an augmented reality setup (see Fig. 2(b)) for evaluation of the turn around behavior on a physical platform. The behavior was tested using a ground vehicle on a flat, inclined platform with dimensions 5×3 m. The user interface of the evaluation setup allows the user to define the boundary of the operating terrain, location of obstacles, and the initial and final poses of the vehicle. The planning region of this interface is shown in Fig. 3(a). The user-specified terrain boundary is projected onto the platform using a ceiling-mounted projector, where the blue border lines shown in Fig. 3(b) divide the space into obstacle and obstacle-free regions. We use a fish eye lens camera to locate the vehicle's position and orientation. The camera is calibrated for the operating space and is able to communicate its boundaries to the vehicle with an error of less than 3 cm.

In this setup, after launching the vehicle from its initial position and orientation, a sequence of motion goals is computed first by the vehicle to reach its goal position and orientation. After computing the motion goals, the vehicle queries the atomic maneuver model for a sequence of reference maneuvers connecting the motion goals. However, the vehicle is not able to reach the motion goals with high precision by solely using the reference maneuvers due to the necessary discretization. In addition, the pose error gradually accumulates when attempting to reach motion goals in a sequence. Hence, by the use of the adaptation algorithm, the vehicle refines the atomic maneuvers in order to reduce the execution error. Nevertheless, some residual error may still remain due to the unavoidable differences between the simulation and the physical setup when executing the adapted maneuvers. This remaining error is handled by the on-board PID controller.

4. Conclusions

We have presented an approach for a turn around behavior generation for an UGV to be able to efficiently plan and execute N-point turn maneuvers in a highly constrained region of a rough terrain. We have developed an augmented reality setup for evaluation of the behavior. The behavior was successfully tested using a physical ground vehicle executing the maneuver on an inclined platform.

Acknowledgments

This research has been supported by Office of Naval Research N00014-10-1-0438. Opinions expressed in this paper are those of the authors and do not necessarily reflect opinions of the sponsors.

References

1. T. Howard and A. Kelly, *The International Journal of Robotics Research* **26**, 141 (2007).
2. M. Likhachev and D. Ferguson, *The International Journal of Robotics Research* **28**, 933 (2009).
3. M. Pivtoraiko and A. Kelly, Kinodynamic motion planning with state lattice motion primitives, in *IEEE/RSJ International Conference on Intelligent Robots and Systems (IROS)*, 2011.
4. D. Ferguson, Path generation and control for end of row turning in an orchard environment, Master's thesis, Carnegie Mellon University Robotics Institute (2010).
5. T. Howard, C. Green and A. Kelly, Receding horizon model-predictive control for mobile robot navigation of intricate paths, in *Field and Service Robotics*, 2010 pp. 69–78.
6. S. Sivanandam and S. Deepa, *Introduction to Genetic Algorithms* (Springer Publishing Company, Incorporated, 2010).
7. S. Russell and P. Norvig, *Artificial intelligence: a modern approach* (Prentice Hall, 2009).
8. J. C. Gilbert and J. Nocedal, *SIAM Journal on Optimization* **2**, 21 (1992).
9. A. Visioli, *Practical PID control* (Springer Verlag, 2006).
10. F. Fahimi, *Autonomous robots: modeling, path planning, and control* (Springer Verlag, 2008).

IMPACT OF PAYLOAD INERTIA ON THE SYSTEM DAMPING CHARACTERISTICS OF A TWO-WHEELED ROBOTIC MACHINE

K M Goher[1], M O Tokhi[2], A M Almeshal[2], O Sayidmarie[2] and S A Agouri[2]
[1]Department of Mechanical and Industrial Engineering
Sultan Qaboos University, Oman, e-mail: kgoher@squ.edu.om
[2]Department of Automatic Control and Systems Engineering
The University of Sheffield, United Kingdom

In real life applications, two-wheeled vehicles are considered to carry a payload which may vary in the sense of the size, position or the speed of motion along the intermediate body. Studying the impact of those parameters should be considered in detail in order to accommodate as many as possible parameters which could affect the vehicle stability and the control system behaviour. This paper investigates the impact of changing the payload size and associated inertia on the system damping characteristics.

1. Introduction

The nature of two-wheeled vehicles poses several interesting control questions. Industrial applications of such vehicles will arise to a great extent in the coming few years; for instance, material handling in narrow paths to different heights etc. (Goher and Tokhi 2008).

In the work carried out by Shiroma *et al.* (1996), they showed the interaction of forces between objects and the robot by taking into account the stability effects due to these forces. In his research Ooi (2003), an indirect Kalman filter configuration combining a piezo rate gyroscope sensor and an inclinometer was implemented to obtain an accurate estimate of the tilt angle and its derivative. Therefore the suitability and examine the performance of linear control systems like the Linear Quadratic Regulator and Pole-placement controller in stabilising the system was investigated. B. Browning and others in (2004) presented a new domain, called Segway Soccer, for investigating the coordination of dynamically formed, mixed human-robot teams within the realm of a team task that requires real-time decision making and response.

Salerno and Angles (2003, 2004, 2007), presented a multivariable control by differentiable state feedback of semi-autonomous self-balancing two-wheeled quasiholonomic robots. Pathak *et al.* (2005) analyzed the dynamic model of a wheeled inverted pendulum (e.g., Segway, Quasimoro, and Joe) from a controllability and feedback linearizability point of view. Kim *et al.* (2005) discussed the enhancement of floor-driving ability of a two wheeled inverted-pendulum-type autonomous vehicle based on the dynamics of this robotic mechanism by using Kane's dynamic modelling. Randal (2005) applied the idea of non-linear control strategy.

One of the most important contributions discussing the dynamics of the two-wheeled inverted pendulum robot is the work carried by Kim and co-workers (Kim and Byung, 2005, 2007; Kim et al., 2005, 2006). The exact dynamics have been investigated in order to provide the necessary information for further design and production. Tsai *et al.* (2006), developed techniques for system design, modelling and adaptive control of a personal two-wheeled transporter driven by two DC servomotors. Two adaptive two-degree-of-freedom (DOF) controllers were designed to achieve self-balancing and rotation control. Tsai and Hu (2006) investigated the robust motion pilot control of an auto-balancing two-wheeled cart where the feedback control system is solely based on the angle of inclination without using any velocity information.

Choi and Oh (2008) developed a wheeled inverted pendulum and investigated the interaction between human and the robot through applying a pulling or pushing force. Imamura *et al.* (2008) discussed the configuration of a baggage transportation system by an inverted pendulum robot and realized a navigation system. Jeong and Takahashi (2007) discussed the mobile control and the standing and sitting motions of an inverted pendulum type assistant robot aiming at the coexistence of safety and work capability. Nawawi *et al.* (2006, 2007) designed and developed a two wheeled inverted mobile robot for use as flexible platform, comprising an embedded unstable linear plant intended for research and teaching purposes. Zheng *et al.* (2007) discussed the mechanical design and control system of a miniature autonomous surveillance robot ("BMS-1") for indoor reconnaissance tasks.

One of the areas not covered in literature is the investigation of certain parameters on the stability of the system of two-wheeled machines. The impact of a set of varying parameters on the capability of the control strategies and the system behaviour have been considered in the work carried out by Goher and Tokhi (2008) including mass, position and speed of the payload, amplitude, position and duration of an external impact force acting on the system.

The paper is organized as follows: Section II describes the system dynamic modeling process and effect on changing the size of the payload on the system dynamic equations. Section III describe the control strategy used based on PD control technique. A numerical simulation is carried out in Section IV.

2. System modelling

This section describes the system dynamics in case there is a load attached to the upper end of the rod as shown in Figure 1. Adding a payload will affect not only the form of the wheels dynamic equations derived by Goher and Tokhi (2008), but will also affect the values of the reaction forces between the rod and the axle at the connection.

2.1 *System governing equations*

In this section, the dynamics of the system are derived with due consideration of behaviour of a virtual payload with variable position.

Consider Figures 1 and 2 which describe the vehicle with the effect of an attached payload M at a variable distance Q from the centre of the wheels O. An external disturbance force F is applied at a distance S from the same position.

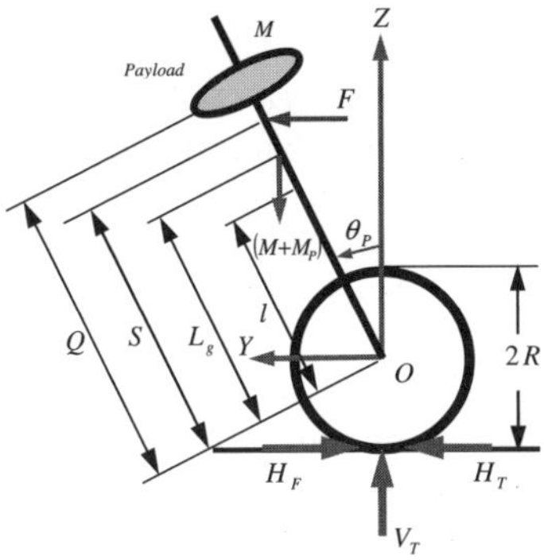

Fig. 1 Schematic diagram of the vehicle

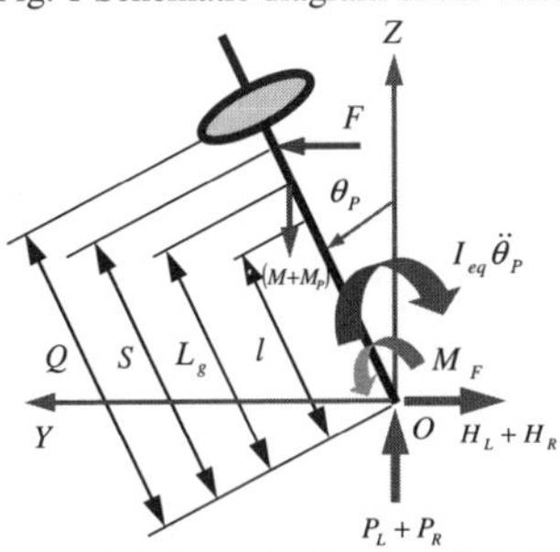

Fig. 2 Schematic diagram of the IB

Applying Newton Euler dynamic formulation yields the following dynamic equations describing the system with the effect of an external disturbance and an attached payload; detailed steps of the mathematical model can be found in the work done by Goher and Tokhi (2008).

$$\ddot{\theta}_P = \frac{1}{\left(I_{eq} + (M + M_p)L_g^2\right)}\left(-\frac{k_m k_e}{RR_w^2}(\dot{y}_L + \dot{y}_R) + \frac{k_m}{RR_w}(V_L + V_R) + M_F \right. \tag{1}$$

$$\left. -(M + M_P)L_g g \sin\theta_p + (M + M_P)L_g \ddot{Y}\cos\theta_p + F\cos\theta_P\left(S - 2L_g\right)\right)$$

and

$$\ddot{Y} = \frac{1}{\left(2\left(M_w + \frac{J_{wy}}{R_w^2}\right) + (M + M_p)\right)}\left(-\frac{k_m k_e}{RR_w^2}\dot{Y} + \frac{k_m}{RR_w}(V_L + V_R) \right. \tag{2}$$

$$\left. -(M + M_P)L_g \ddot{\theta}_p \cos\theta_p + (M + M_P)L_g \dot{\theta}_p^2 \sin\theta_p - F\right)$$

Careful look at Equations (1) and (2) will yield to the following observation; increasing the inertia of the various system components tends to add more damping to the system and decrease possible oscillations. However, this behaviour of the system will require a lot of control effort to manoeuvre the vehicle and hence more power consumption.

3. Control Strategy

An external disturbance force F is applied on the intermediate body at a distance S from the wheels origin O. The disturbance force is applied after 10 seconds at the start of the simulation and then after each 90 second interval. This interval is determined as the safe limit for repeating such impact forces on the rod. Below such limit, it was found that the developed controller would not completely keep the IB at a stable position.

The control inputs are the error and the derivative of error for the two measured variables of the system and whereas the control outputs are the terminal voltage of the DC motors. As the type of motion considered in this study is linear, the two control signals from both the feedback loops are identical and hence the motor terminal voltage considered as the system input is a summation of both the voltage to the left and right wheels as shown in Figure 3.

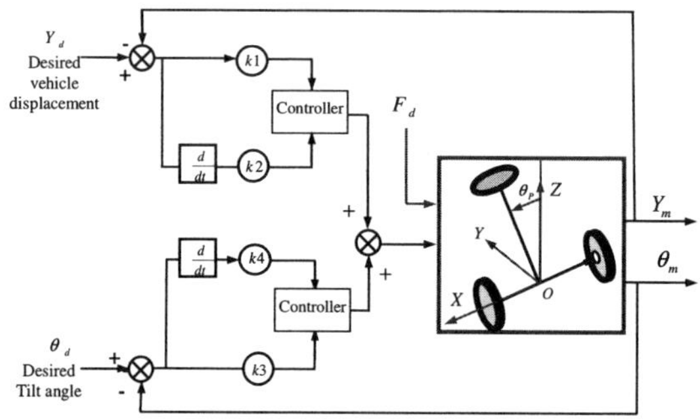

Fig. 3 Implementation of a PID control strategy

The inputs to the system, in this case, are the driving voltage of the DC motor and the external disturbance force F while the outputs are the overall linear displacement of the vehicle and the tilt angle of the intermediate body.

4. Numerical Simulation

An investigation on the impact of the payload size is examined in this section where three different payload sizes are used; 5 kg, 20 kg and 50 kg respectively as shown in Figure 4. A disturbance effect is applied on the system every 20 seconds intervals for the entire simulation course.

Increasing the size of the payload tends to increase the time required for the intermediate body to stabilize at the upright position. The system rise time is also affected by changing the payload size; larger payload size increases the rise time of the system. Increasing the payload size tends to increase the damping of the system and in turn decrease the amount of oscillation associated with the intermediate body. Details of the corresponding effects on the rise time, the settling time and the percentage overshoot are presented in Table 1.

Table: System performance characteristics (different payload size)

PD control										
M, kg		5			20			50		
y, m		0	1	2	0	1	2	0	1	2
θ	t_r, s	7	7	7	10	10	10	14	14	14
	t_s, s	24	24	24	30	30	38	40	40	57
	%*Os*	12	8	7	8	3	2.5	5	1	1

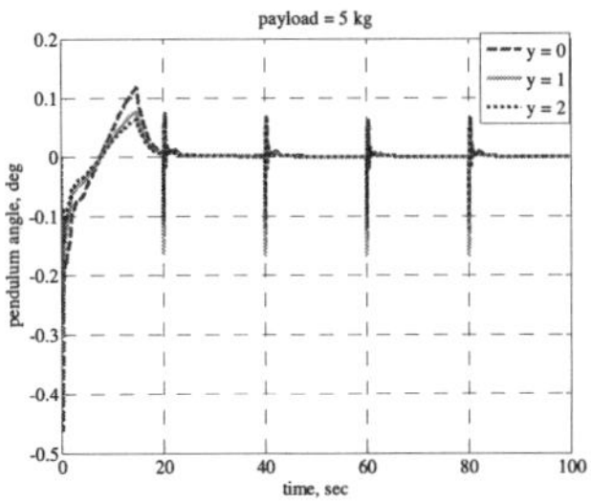

(a) 5*kg* payload

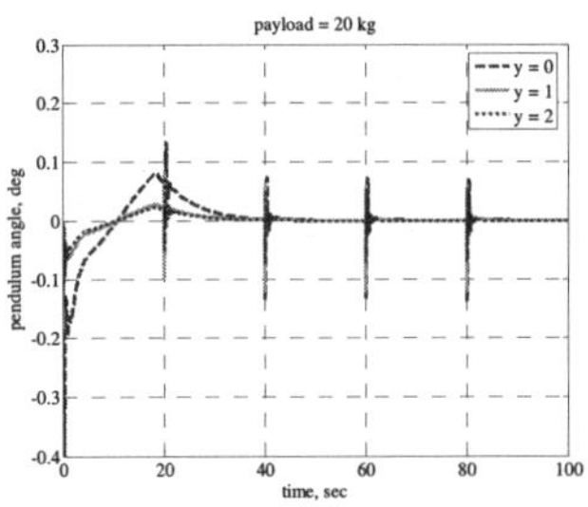

(b) 20*kg* payload

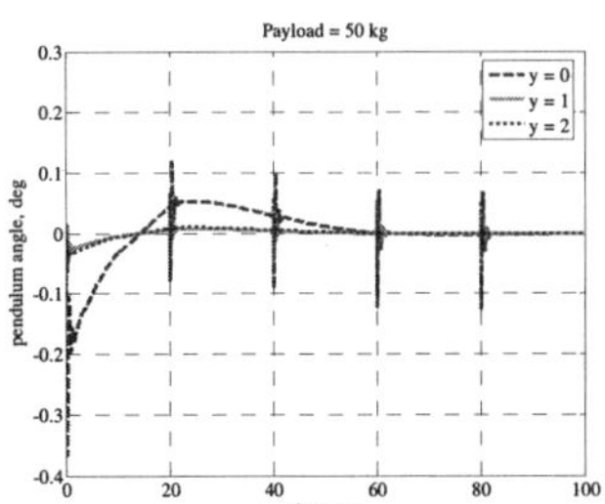

(c) 50*kg* payload

Fig. 4 Tilt angle with different payloads

4.1 *Impact of payload inertia on the system damping characteristics*

It has been observed from the previously derived dynamic equations of the system that the oscillations of the system decrease by increasing the attached payload. Two different cases are examined by testing the system performance without any attached payload and with a payload of 70 kg. The system performance is measured in terms of the intermediate body tilt angle and the vehicle linear displacement. In order to detect the frequency contents of the measured values, a power spectral density analysis is calculated for both the measurement variables as noted in Figures 5 and 6.

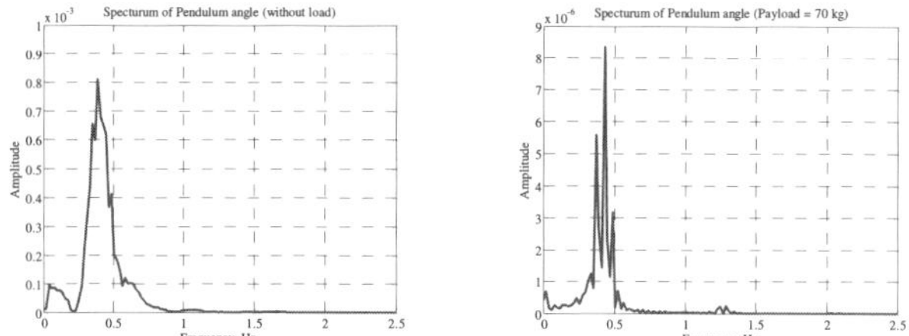

(a) No payload (b) 70kg payload

Fig. 5 Spectral density of the tilt angle

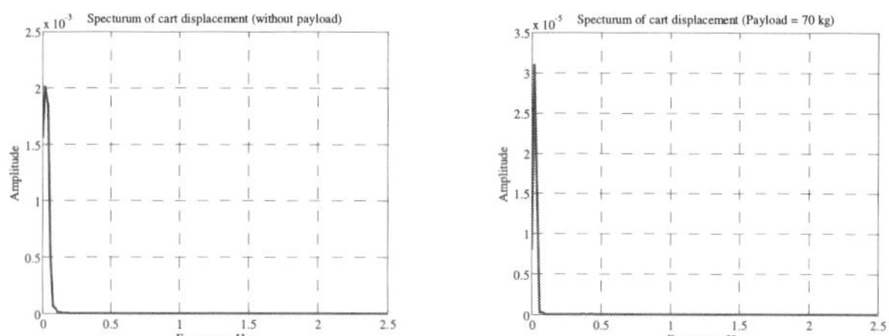

(a) No payload (b) 70kg payload

Fig. 6 Spectral density of the vehicle displacement

5. Conclusions

Based on the mathematical model derived in earlier work, an investigation of the system balance has been carried out. PD controller has been used to achieve the system target. The system balance condition has been investigated based on the effect of an external disturbance. In order to simulate a real life sudden impact, a real disturbance force has been applied on the intermediate body. Simulations have been carried out for different cases of changing the size of the attached payload. The system dynamic behaviour has been presented while giving further consideration of the impact of changing the size of the payload on the system damping characteristics. Direct relation has been detected between the change in the payload size and the system convergence towards to the upright position and the required control effort to achieve such balance condition.

References

[1] Browning B., Rybski P. E., Searock J., and Veoso M. M. (2004), Development of a soccer – playing dynamically – balancing mobile robot, Proceedings of the International Conference on Robotics & Automation, New Orleans, LA, 1752 – 1757.

[2] Choi D. and Oh J. H. (2008), Human-friendly motion control of a wheeled inverted pendulum by reduced-order disturbance observer, Proceedings of the IEEE International Conference on Robotics and Automation, Pasadena, CA, USA, 2521 – 2526.

[3] Goher K. M. and Tokhi M. O., (2008), Balancing of a two-wheeled robotic machine with payload using PD/FL controllers, in the Proceedings of the 27th IASTED International Conference on Modelling, Identification and Control, Innsbruck, Austria, Feb. 11-13, pp. 226-231.

[4] Goher K. M. and Tokhi M. O., (2008), Modelling, simulation and balance control of a two-wheeled robotic machine with static variation in load position, in the Proceedings of the 22nd European Conference on Modelling and Simulation, Nicosia, Cyprus, June 3-6.

[5] Goher K. M. and Tokhi M. O., (2008), Robust control of a two-wheeled robotic machine, in the Proceedings of the 8th Portuguese Conference on Automatic Control (CONTROLO 2008), Vila Real, Portugal, July 21-23.

[6] Goher K. M. and Tokhi M. O., (2008), Balance control of a TWRM with a static payload, in the Proceedings of the 11th International Conference on Climbing and Walking Robots and the Support Technologies for Mobile Machines (CLAWAR 2008), Coimbra, Portugal, September 8-10.

[7] Goher K. M. and Tokhi M. O., (2008), Balance control of a TWRM with a dynamic payload, in the Proceedings of the 11th International Conference on Climbing and Walking Robots and the support Technologies for Mobile Machines (CLAWAR 2008), Coimbra, Portugal, September 8-10.

[8] Imamura R, Takei T, Yuta S, (2008), Sensor drift compensation and control of a wheeled inverted pendulum mobile robot, IEEE Workshop on Advanced Motion Control, 137 – 142.

[9] Jeong S. H. and Takahashi T. (2007), Wheeled inverted pendulum type assistant robot: inverted mobile, standing, and sitting motions, Proceedings of the IEEE/RSJ International Conference on Intelligent Robots and Systems, San Diego, CA, USA, 1932 – 1937.

[10] Kim Y., Kim H. K., and Kwak Y. K. (2005), Dynamic analysis of a nonholonomic two-wheeled inverted pendulum robot, Journal of Inteligent and Robotic Systems, 44(1), 25-44.

[11] Kim Y., Kim H. K., and Kwak Y. (2006). Improving driving ability for a two-wheeled inverted pendulum-type autonomous vehicle, Proceedings of the Institution of Mechanical Engineers, Part D: Journal of Automobile Engineering, 220(2), 165-175.

[12] Kim C. H. and Byung K. K. (2005), Energy – saving 3 – step velocity control algorithm for battery – powered wheeled mobile robots, Proceedings of the IEEE International Conference on Robotics and Automation, Barcelona, Spain.

[13] Kim C. H. and Byung K. K. (2007), Minimum-energy translational trajectory generation for differential-driven wheeled mobile robots, Journal of Intelligent and Robotic Systems, 49, 367–383.

[14] Nawawi S. W., Ahmad M. N., Osman J. H. S. (2006), Control of two-wheels inverted pendulum mobile robot using full order sliding mode control, Proceedings of International Conference on Man-Machine Systems, September 15-16, Langkawi, Malaysia

[15] Nawawi S. W., Ahmad M. N., Osman J. H. S. (2006), Variable structure control of two-wheeled inverted pendulum mobile robot, Proceedings of the Regional Postgraduate Conference on Engineering and Science, Johore, Malaysia, 197 – 201.

[16] Nawawi S. W., Ahmad M. N. and Osman J. H. S. (2007), Development of a two-wheeled inverted pendulum mobile robot, The 5th Student Conference on Research and Development, 11-12 December, Malaysia.

[17] Ooi R. C. (2003), Balancing a two-wheeled autonomous robot, Final Year Thesis, The University of Western Australia, School of Mechanical Engineering.

[18] Pathak K., Franch J. and Agrwal S. K. (2005), Velocity and position Control of a wheeled inverted pendulum by partial feedback linearization, IEEE Transactions on Robotics, 21(3), 505-513.

[19] Randal H. K. C. (2005), Balancing wheeled robot, Research Project Toward the Degree of the Bachelor of Engineering, Faculty of Engineering and Surveying, University of Southern Queensland.

[20] Salerno A. and Angles J. (2003), On the nonlinear controllability of a quasiholonomic mobile robot, Proceedings of the IEEE Conference on Robotics and Automation, Taipie, Taiwan, 3379 – 3384.

[21] Salerno A. and Angles J. (2004), The control of semi-autonomous self balancing two-wheeled quasiholonomic mobile robot, Proceedings of the 15th CISM-IFToMM Symposium on Robot Design, Montreal , Canda.

[22] Salerno A. and Angles J. (2004), The control semi-autonomous two-wheeled robots undergoing large payload-variations, Proceedings of the IEEE Conference on Robotics and Automation, New Orleans, LA, 1740 – 1745.

[23] Salerno A. and Angles J. (2007), A new family of two – wheeled robots: modeling and controllability, IEEE Transactions on Robotics, 23 (1)., 169 – 173.

[24] Shiroma N., Matsumoto O., Kajita S. and Tani K. (1996), Cooperative behaviour of a wheeled inverted pendulum for object transportation, Proceedings of the IEEE/RSJ International Conference on Intelligent Robots and Systems (IROS 96), Osaka, Japan, 396 – 401.

[25] Tsai M. C. and Hu J. S. (2006), Robust control of auto-balancing two-wheeled cart motion pilot, Proceedings of the Automatic Control Conference, Tamsui, Taiwan.

[26] Tsai C. C., Lin S. C. and Luo W. L. K. (2006), Adabtive steering of a self-balancing two-wheeled transporter, Proceedings of the Automatic Control Conference, , November 10-11, Tamsui, Taiwan.

[27] Zheng C., Zheng C., Huang Q., Sang W., Liu Y., Wang L. and Huang Y. (2007), Mechanical design and control system of a miniature autonomous surveillance robot, Proceedings of the IEEE International Conference on Mechatronics and Automation, Harbin, China, 1752 – 1757.

AUTONOMOUS NAVIGATION OF A 5 GRAM CRAWLING MILLIROBOT IN A COMPLEX ENVIRONMENT

Allison Mathis, Jared Russell, Tom Moore,
Jason Cohen, Brian Satterfield

Lockheed Martin Advanced Technology Laboratories
Cherry Hill, NJ, 08002, USA

NicHolas Kohut, Xiao-Yu Fu, ronald S. Fearing

University of California, Berkeley
Berkeley, CA, 94720, USA

We present the Medic robot: a novel 5-gram autonomous crawling and climbing millirobot designed to navigate through dark, cluttered, confined and electrified environments. Due to payload constraints, all sensing is performed by a single low-resolution camera and processing is split between an onboard unit and a base station. We present an integrated object detector, localizer, and planner that have been successfully demonstrated to enable the millirobot to determine its position within the environment, plan and navigate the most efficient path between user-specified start and goal positions, and recover from stuck conditions.

1.1. Introduction

Millirobotics offers a unique solution to the problem of diagnosis and repair of large electronic systems where human inspection is often costly, impractical, or error-prone. In many cases, simply gaining access to electronics installed in satellites or aircraft is enormously expensive; a resident millirobot equipped with basic circuitry diagnosis capabilities is an attractive low-cost alternative. Our research investigated whether a resident millirobot could successfully autonomously navigate through a powered electronic environment to specified components of interest. We envision that future work will develop actual diagnosis and intervention capabilities. Circuit boards are generally cluttered with surface-mount components that pose significant navigational challenges for a robot of this size (5 grams, 55mm x 35mm x 18mm). Our solution uses an onboard video sensor to perform visual landmark-based localization and to estimate egomotion using optical flow techniques.

Successful navigation at this scale poses several unique challenges. Operation on circuit boards requires a high degree of precision in order to avoid getting the vehicle stuck on surface-mount components and pins. Due to size and weight constraints, our crawling millirobot is mechanically compliant, leading to difficult-to-predict interactions with varying surfaces and obstacle configurations. Moreover, the inherent payload constraints limit the sensors that may be carried. Consequently only a single CMOS camera was employed during the course of this research.

To address these challenges, we designed a novel four legged platform based on previously described methods of manufacturing [1] that could carry a modest payload and mount small obstacles with minimal force requirements [2]. Image-capture and settings, as well as low-level actuator control are processed onboard, but image processing and high-level planning are performed offboard via an IEEE 802.15.4 link to a PC. Our approach to detecting landmarks was adapted from a method developed by the leaders of the PASCAL Visual Object Classification challenge [3]. Due to observed variability in the steps taken by the robot and the uncertainty inherent in estimating position from visual landmarks, we use a probabilistic approach to localization using particle filters. Path planning was performed via a hybrid planner that combined A* and Field D* methods. Additionally, methods to detect and recover from "stuck" conditions were implemented.

1.2. Hardware Overview

The details of the construction and actuation were originally presented in [2], however the following is a high level summary to aid understanding of the overall system. The Medic robot (Figure 1) measures 55 mm long, 35 mm wide at the legs, 25 mm wide at the hips, and is 18 mm high. Its total mass is 5 grams, including the onboard battery. It is constructed primarily of S2-glass [1] with integrated circuitry for mobility control. Mobility is enabled by four legs (each a member of a connected pair on either side of the chassis) that are actuated by coiled shape memory alloy (SMA) springs. Forward, reverse, and turn-in-place maneuvers are enabled by sequencing the actuation of the SMAs. It has been designed and built to traverse horizontally mounted computer motherboards and is capable of mounting obstacles up to 5.5 mm high. For further information on the physical parameters and construction method used for rapid fabrication see [2].

A consequence of the robot's lightweight design is that significant uncertainty was observed in the motion achieved by each type of step. This is due to two factors: the minimal mass of the Medic robot results in compliance in structural members when the actuators are exerting force, and the numerous unsensed obstacles (such as individual pins and traces) on PCBs result in unpredictable interactions between the robot's legs and the environment. We characterized the motion achieved by each step over a variety of surface conditions with normal distributions (Table 1).

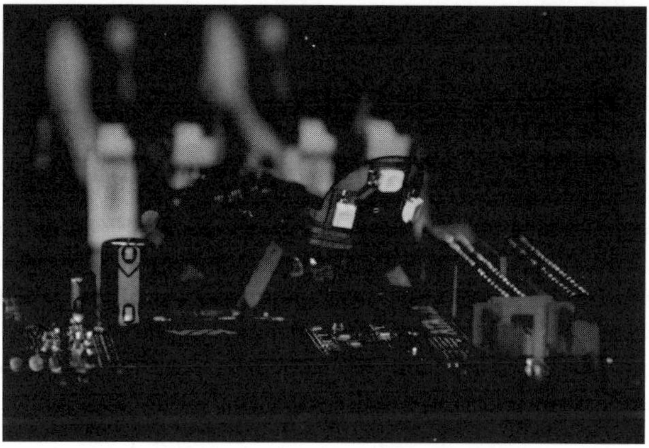

Figure 1. MEDIC robot in an electronic environment

Table 1. Step motion characterization.

Step Type	Trans. Mean (mm)	Trans. Var. (mm^2)	Turn Mean (rad)	Turn Variance (rad^2)
Fwd	6.0	0.25	0	0.048
Rev	-4.0	0.25	0	0.048
Turn	2.0	0.25	+/- 0.15	0.048

1.3. Localization

The Medic navigation system combines a labeled overhead map of the circuit board (with only major components labeled, e.g., capacitors and slots but not individual pins or traces) with a library of visual appearance features of specific electronic component categories. Localizing the millirobot within its environment using only the 160x100 pixel onboard grayscale camera and limited a priori maps requires the fusion of multiple pieces of evidence about robot pose. A sequential importance resampling (SIR) particle filter approach was chosen, as particle filters can easily represent non-Gaussian, multi-modal probability distributions and can be used for fusion simply by applying multiple successive "observe" functions to the distribution sequentially [4]. All processing was performed off-board via a ZigBee link to a laptop PC.

The millirobot's component detection algorithm was derived from successful techniques used for higher-resolution classification [6, 7]. Invariant features formed the lowest level of the detector [8]. Feature clustering via a Pyramid Match Kernel with spatial pooling provided compact feature representations that were used to perform convex optimization [9]. A support vector machine was used to solve the optimization problem and find decision boundaries between component types based on hand-annotated ground truth imagery.

Detected components were integrated into the particle filter by comparing what was found with what each particle expected to see based on the annotated overhead circuit map. Ray tracing was conducted from each component to each particle, removing candidate components that were out of the particle's field of view or occluded. An exponential reward function was used to re-weight the particles according to the amount of overlap between the set of detected and expected components and their relative locations within the camera frame. This approach minimized the effect of spurious detections, as particles that confirmed the existence of multiple components were more heavily weighted than those that confirmed only a few or a single component. False negatives were not penalized due to frequent occlusions and poor image quality, allowing the detectors to be tuned for specificity rather than sensitivity. Table 2 summarizes the performance of the component detection algorithm for each component type.

Table 2. Component detection and classification performance over a set of test imagery captured in an electronics environment (64 images, each subdivided into 25 regions).

Component Type	# of Regions	Detection Rate	Avg. False Alarms Per Image
No Component	1108	98.1%	0.76
Buzzer	3	100.0%	0
Capacitor	54	88.9%	0.03
Fan Header	10	90.0%	0
Heat Sink	63	92.1%	0.05
IDE Connector	17	52.9%	0
Inductor	14	92.9%	0
Molex Power	9	88.9%	0.02
PCI	272	94.1%	0.52
PCI-Express	58	82.8%	0

1.4. Planning

The density of the obstacles in the robot's environment make traditional grid-based planners ill-suited for path generation. Such planners typically discretize the world into a 2D grid whose cells represent some small region of the environment. Planners based on the ubiquitous A* algorithm [10] find optimal paths given a map and some cost function. However, when A* plans across a 2D grid, the generated paths consist of turns that are multiples of π / 4 radians. The Medic robot has highly constrained battery life, so paths that consist of superfluous turns severely limit the operational range. Furthermore, the robot typically exhibits 2 mm of translation while turning, making turn-in-place maneuvers impossible and all turns subject to more uncertainty than pure forward/backward translation.

A solution to this problem is to use a planner that allows for a continuous action space. A planner based on the Field D* [11] algorithm was implemented. Field D* generates a continuous cost field by using discrete cost maps and bilinear interpolation between nodes. The cost field can be traversed using gradient descent until the goal selection of steps available to the Medic robot necessitated using a discretized search algorithm over this continuous cost map location is reached. Our solution was to apply a hybrid planning approach wherein a Field D* cost field would be generated, and a modified version of A* descended the gradient while considering the robot's repertoire of steps.

When the Field D* cost field had been generated, A* was run in the opposing direction, i.e., toward the goal. Instead of using a 2D grid and a canonical eight-connected A* search graph, the expansion of nodes was carried out with respect to the robot's action space. From a given A* node (which represented a real-valued location), the neighboring nodes would be the predicted destination locations of each of the robot's actions. The resulting paths were optimal with respect to the cost field given the robot's mobility constraints.

An experiment was carried out in simulation in which 64 random start/goal location pairs were generated. Both a standard A* planner and our hybrid planner were used to generate paths for each pair, the first averaged a total path length of 1.52e+04 nodes compromising 70,331 steps while the second averaged 1.63e+04 nodes and only 5,066 steps. The 8-connected A* planner generated paths that were slightly shorter, but at the expense of many sharp turns within the paths, is evidenced by the order-of-magnitude difference in the number of steps required by the A* planner.

1.5. Detection and Recovery from "Stuck" Conditions

Even with optimal plans and accurate localization, the robot would often become stuck, typically due to one of the legs getting caught on an unsensed and unmodeled surface-mount component or pin during SMA actuation. Errors of a few millimeters in localization often made the difference between completely clearing an obstacle and "stubbing a toe," resulting in undesired vehicle behavior such as yawing around the obstacle rather than translating forward as commanded. The compliance of the legs and body of the Medic robot made accurately predicting these situations and necessitates a new approach.

In order to identify and overcome these conditions, we implemented an optical flow algorithm to compare camera frames captured before and after each step. The SURF descriptor [5] was computed on each frame, and feature points were correlated between successive frames using a RANSAC algorithm to estimate the robot's egomotion. We developed a model of expected observed motion (direction and magnitude) for each type of step, and compared the results against these thresholds to determine whether nominal movement was observed. This heuristic successfully identified situations where a leg had become ensnared with an obstacle 100% of the time during testing. Extensions to the extrication algorithms that will attempt to reason about the likely cause of entanglement through observation of motion are planned for future work.

1.6. Results

The Medic autonomous navigation system has been demonstrated successfully during "A-to-B" type tests on a powered, enclosed PC motherboard. Typical runs consisted of 40-60 steps (with an image taken before each), and lasted 10-20 minutes. Due to the entanglement hazards mentioned previously, many trials (approximately 2/3) ended with the robot irreparably entangled on an obstacle. However, when entanglement was completely averted (approximately 1/6 of trials) or correctly remedied via our heuristic process (approximately 1/6 of trials), the Medic robot did achieve its goal location within 1 cm of the commanded goal position.

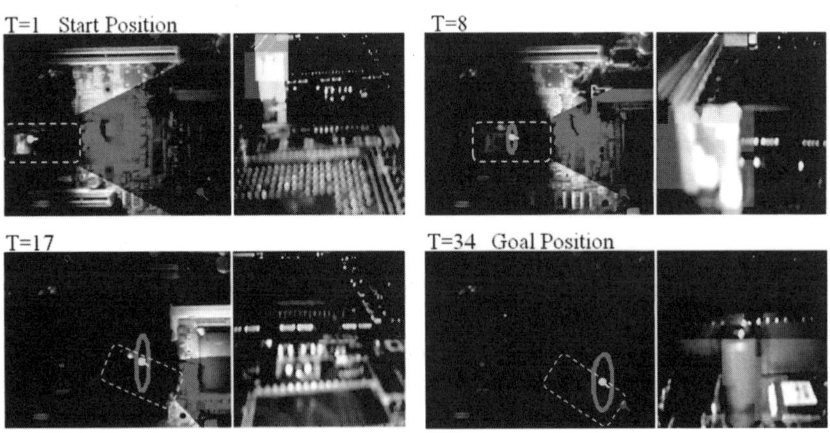

Figure 2. The robot (yellow dashes), its estimated position (yellow dot), the camera data and object detection results at four instances over the course of a run. The objects detected are shown superimposed over the images collected at that moment. The robot's FOV is the shaded triangle, the goal location is represented by the red dot. Capacitors have a red mask, power connectors have green, and PCI slots are shown in blue.

The series of images in Figure 2 represents a typical successful trial, shown as captured by an overhead camera over a series of steps through the test environment (these are best viewed in color). In each instance, the image on the left shows the actual millirobot in the test environment, while the right is the view from the onboard camera. The light blue circle encompasses the possible locations within one standard deviation of the most likely estimate, which is represented by the yellow dot in the center. The tail coming off both the maximum likelihood estimate (MLE) and the goal represents the orientation of the millirobot.

1.7. Conclusions and Future Work

This paper has presented the perception and autonomous navigation design for the millirobot Medic. The use of landmark-based visual navigation for millirobots in the 5-gram class is novel and useful due to the constraints for size, weight and power. The Medic robot is capable of autonomous navigation in rough terrain and of detecting and sometimes removing itself from stuck conditions. The performance of the perception system is comparable to the state-of-the-art in higher resolution object detection despite challenges in resolution, bit depth and illumination. The autonomous system reliably guided the millirobot to within 1 cm of the goal during live testing. We believe this represents the first visually guided millirobot of less than 6 grams capable of autonomous navigation through rough terrain.

The current system could be improved to increase the performance and robustness of navigation and perception. These improvements would include an improved localization method that performed simultaneous global and local state estimation, and stuck-detection reasoning that would determine how the platform was stuck and employ the correct action for extraction to increase robustness. Replacing the current camera with a newer model would significantly increase the performance of the perception system by providing higher-resolution and improved light sensitivity. Additionally, alternate sensing modalities, such as tactile or infrared imaging, are worth investigation.

References

1. R.J. Wood, S. Avadhanula, R. Sahai, E. Steltz, and R.S. Fearing, "Microrobot design using fiber reinforced composites," Journal of Mechanical Design, vol. 130, no. 5, p. 052304, 2008. [Online]. Available: http://link.aip.org/link/?JMD/130/052304/1

2. N. Kohut, A. Hoover, K. Ma, S. Baek, and R.S. Fearing, "MEDIC: A Legged Millirobot Utilizing Novel Obstacle Traversal," IEEE Int. Conf. on Robotics and Automation, Shanghai May 2011.
3. PASCAL Challenge: http://pascallin.ecs.soton.ac.uk/challenges/ VOC/.
4. N. Gordon, D. Salmond, and A. Smith, "Novel approach to nonlinear-non-Gaussian Bayesian state estimation," IEEE Proceeding-F, Vol.140 No.2 April 1993.
5. H. Bay, A. Ess, T. Tuytelaars, and L. Gool, "SURF: Speeded Up Robust Features," Computer Vision and Image Understanding (CVIU), Vol. 110, No. 3, pp. 346-359, 2008.
6. K. Yu, T. Zhang, and Y. Gong, "Nonlinear Learning using Local Coordinate Coding," in Neural Information Processing Systems, 2009. NIPS 2009.
7. X. Zhou, K. Yu, T. Zhang, and T. Huang, "Image Classification using Super-Vector Coding of Local Image Descriptors," European Conference on Computer Vision, 2010. ECCV 2010.
8. D.G. Lowe, SIFT flow: dense correspondence across difference scenes. Proc. ICCV 1999, Kerkyra, Greece, pp. 1150-1157.
9. K. Grauman and T. Darrell. "The Pyramid Match Kernel: Discriminative Classification with Sets of Image Features," IEEE International Conference on Computer Vision (ICCV), Beijing, China, October 2005.
10. P.E. Hart, N.J. Nilsson, and B. Raphael, "A formal basis for the heuristic determination of minimum cost paths," Systems Science and Cybernetics, IEEE Transactions on, vol. 4, no. 2, pp. 100-107, 1968.
11. Nabbe and M. Hebert, "Extending the Path-Planning Horizon," International Journal of Robotics Research, pp. 997-1024.

HOLONIC MOTION CONTROL DESIGN FOR A CLASS OF MODULAR OUTDOOR WALKING ROBOTS

THEODOR BORANGIU[1], ION ION[2], SILVIA ANTON[1], FLORIN ANTON[1]

[1]*Department of Automation and Applied Informatics,* [2]*Department of Technology of Manufacturing, University Politehnica of Bucharest, 313, Spl. Independentei, sector 6, 060042 Bucharest, Romania*

The paper describes the outdoor walking robot MERO3 used for heavy load tranportation on irregular surfaces with obstacles, designed as a modular mechatronic system the mechanical structure of which is reconfigurable with 2-legged modules. Each leg of the structure has three degrees of freedom and a tactile sensor for detecting the contact with the ground. A coordinate robot motion control system is developed as a *holonic motion execution system* (HMES), based on the original concept of *holonic motion control* (HMC) derived from the holonic paradigm developed for social organisations and living organisms. The novelty of the control system design consists into proposing a reference architecture for holonic motion control addressed to mechatronic devices with multi-body moving structures, for which modularity, decentralization, autonomy, scalability, and re-use of software components are provided.

1. Introduction

The creation of autonomous walking robots integrated in exploration, search and rescue, transportation and service tasks is an objective of many research groups. Such robots must have capabilities of autonomous navigation, obstacle detection and reasoning for avoidance, heavy load transportation and handling, perception and learning, reasoning and knowledge udating.

More and more applications requiring services in a natural, unstructured terrain, determine that feet-movement (walking) solutions become more and more attractive. In agriculture and sylviculture, due to the specific conditions created by the vegetation and state of the ground, and in order to protect the environment, caterpillar or wheeled machines used nowadays for performing the tasks have limited mobility and usually destroy and/or damage the environment, vegetation and trees in their movements from one place to another [1], [2].

The successful fulfillment of the mission of such a robot depends both on its initial knowledge relative to fixed, time and / or space references and pre-configured behavior (e.g. for obstacle avoidance), and on information accessed, processed in real-time through its control system during its outdoor movement.

The new modular walking robot MERO2 is designed as a multi-functional mechatronic system able to perform planned movements aimed at accomplishing several off line or on-line scheduled targets. The walking robot carries out and completes tasks by permanently interacting with the environment in which there are known or unknown physical objects and obstacles. The robot environmental interactions may be technological (caused by mechanical effort or contact) or contextual ones (route identification, obstacle avoidance, orientation etc).

2. Mechanical Design of MERO3 and Walking Strategy

The walking robot MERO3 is made up of the following parts:

a) the mechanical system made up of, two or three modules articulated and shaped according to the requirements of the movement on an uneven ground, the robot's shift system is thus built that it may accomplish many toes' trajectories, which can alter by each step.
b) the actuating system of legs have a hydraulic drive;
c) the distribution system is controlled by 12 or 18 servo-valves, according to the robot's configuration;
d) the energy feeding system;
e) the system of data acquiring on the shift the system's configuration and the environment;
f) the control panel processing signals received from the driving and the acquiring systems.

The MERO walking robot's main technical characteristics are the following:

Table 1. MERO robot technical characteristics.

Robot type	four-legged robot	hexapod
robot length	920 mm	1,440 mm
width	750 mm	750 mm
height	520 mm	520 mm
drive	hydraulic	hydraulic
weight	84.8 kg	124 kg
payloads	110kg	160 Kg (predicted)

The walking robot MERO3 is a variant of the robot MERO1 where the anthropomorphic leg is replaced with the leg structure RRP, this version being under construction.

The legs are symmetrically arranged relative to the platform axis. The three legs on the right with respect to the movement direction are superscript marked with $2i$, $i = 1, 6$ whereas the three legs on the left with $2i{-}1$. Each platform of the rear modules is connected to the platform of first module by a 3-DOF kinematic chain with two links and three rotational pairs. The axes of these pairs are concurrent and orthogonal. The joints of each leg are numbered consecutively from A - pair with link 1 to C - pair with link 3. The robot platform is designed as link no. 0, followed by links 1, 2, 3 for each RRP leg (Fig. 1).

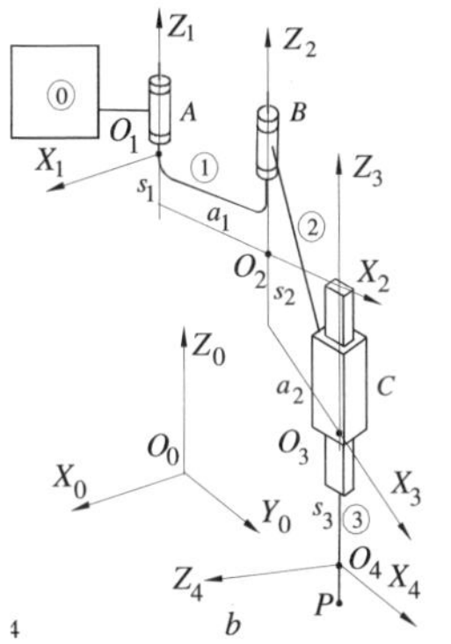

Figure 1. The kinematic representation of a RRP MERO 3 leg.

The Denavit-Hartenberg formalism is used to derive the Direct Kinematics model of the leg: the $\mathbf{A}_3^j$ matrix performs the coordinate transformation of a point belonging to link 3 from the $O_4^j X_4^j Y_4^j Z_4^j$ frame to the $O_3^j X_3^j Y_3^j Z_3^j$ frame attached to link 2. Similarly, the coordinates of the lower end point belonging to link 3 is mapped from $O_4^j X_4^j Y_4^j Z_4^j$ to $O_1^j X_1^j Y_1^j Z_1^j$ attached to the platform (link 0) as follows:

$$\left\| \begin{matrix} 1 \\ X_{1P}^j \\ Y_{1P}^j \\ Z_{1P}^j \end{matrix} \right\| = \prod_{k=1}^{3} A_k^j \left\| \begin{matrix} 1 \\ X_{4P}^j \\ Y_{4P}^j \\ Z_{4P}^j \end{matrix} \right\|, j = 1,..,6. \qquad (1)$$

This matrix equation describes the geometrical model of legs 1-6 of the walking robot. Thus, the location of the end point P can be expressed in terms of the pair variables θ_i^j, $1 \le i \le 3$, $1 \le j \le 6$. For the Inverse Kinematics analysis, (2) is solved for θ_i^j, considering known the position of point P and of the platform relative to the frame $OXYZ$ attached to the ground.

The graphical representation in Fig. 2 was used to define the gait parameters of the MERO3 modular robot and to derive the gait strategies implementing the various types of motion of the 6-legged platform: walking, climbing / descending and obstacle avoidance (W, C/D, OA).

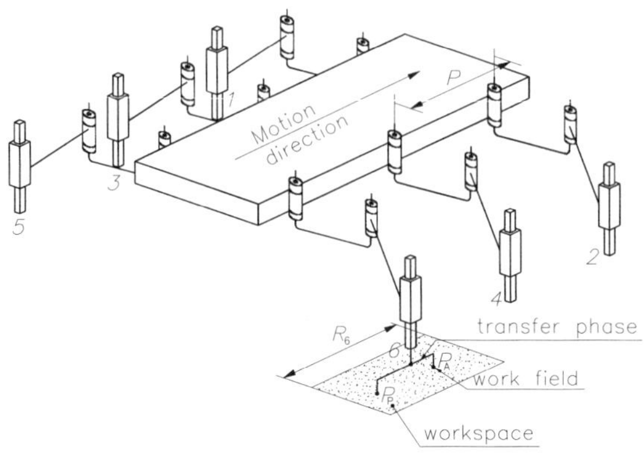

Figure 2. Graphic representation of the MERO3 platform and LMS for gait strategy definition.

The walking robot leans successively on some of its feet in a stable posture, moving the rest of its feet in a new position. The step of the walking robot was defined as the distance along which its gravity center displaces in a motion cycle. The gait of the robot was defined, for each type of motion (W, C/D, OA), by the leg motion sequences realizing particular displacement modes: moving forward, turning, stepping over, a.o.

Each leg can be at each moment of time in one of the following two phases: (i) *Transfer phase*: the time period τ when the leg' is not in contact with the terrain; (ii) *Lean phase*: the time period ξ during which the leg is in contact with the terrain. The duration of a leg motion cycle is $T = \tau + \xi$. In Fig. 2, *PA* is the theoretical contact point of the leg with the terrain when the leg is placed on soil, and *PP* is the theoretical contact point of the leg with the terrain when the leg is raised from the soil. $R_i, 1 \leq i \leq 6$ is the distance on which the leg moves during the lean phase, relative to the robot platform.

For the MERO3 robot, a *periodic* gait was chosen for straight line walking: similar phases of the same leg, during successive stepping, occur at constant time interval for all six legs. Further, a *wave-like* walking style was imposed for the implementation of the periodic gait, because it provides good global stability. For obstacle avoidance, a "traversing" motion style was defined, based on real-time estimation of the obstacle's dimensions. Function of the obstacle's size and shape, collision can be avoided either by a "detour" sequence or by a "step over" strategy. In this latter case, the gait type switches from periodic to *non periodic* motion of the type "Follow The Leader", when the legs are placed in the allowed areas around the obstacle and rear legs occupy successively the collision-free positions of the front legs [3].

The selection of the strategy implementing an outdoor robot displacement mode took into consideration operational aspects such as: terrain irregularities and inclination, soil consistency, gait stability, maximal payload, and speed.

3. Motion Control of the Walking Robot

The robot motion control will be implemented on two layers:

- *low level layer* which individually controls the motion of the robot legs' joints and synchronizes data transfers between leg's axes, robot legs and the low- and high level motion control layers, and
- *high level layer* which performs: navigation task, detection, feature-based description and locating of obstacles, the obstacle avoiding strategy and the coordinated control of the legs' motion according to the robot kinematic model and the constraints imposed by the terrain profile (inclination, etc.).

The low level layer is distributed in a number of microcontrollers, each of them acting as a motion controller for one robot leg by implementing a PWM-type closed-loop drive control of the RRP leg's joint axes. These six dedicated

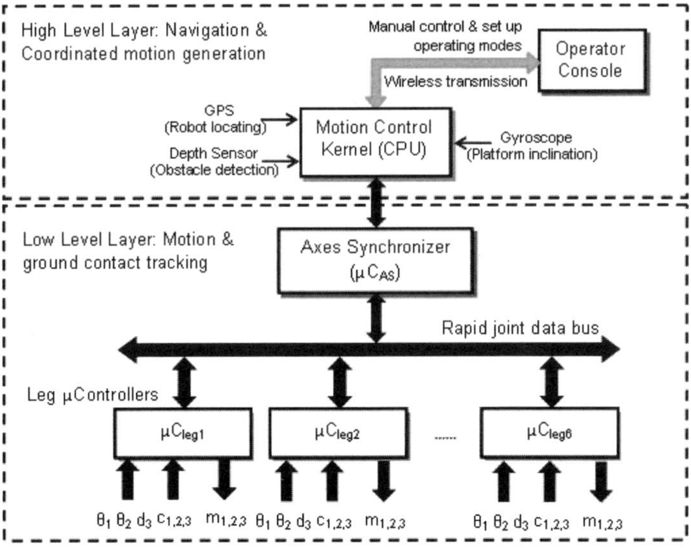

Figure 3. Two-layer hierarchical modified robot motion control system.

leg motion controllers are fed back with joint displacement data measured from axis encoders; they will also receive signals from the effort sensors mounted on the ground contact plates of a robot leg. This allows monitoring the completion of successive motion programs prescribed for robot leg steps with respect to the status of multiple robot legs contacts with the ground (*simultaneity of motion completion and ground contact*; number of legs in contact with the ground; rest of leg displacements to be cancelled / transferred to a next step; a.o.). One additional microcontroller is used for coordinating and synchronizing the movements of legs within one step, also acting as a communication bridge for data and information transfers between the high level layer motion coordination kernel and the low level layer (distributed leg motion controllers), see Fig. 3.

The high level layer communicates with the low level layer in order to receive movement and leg position information (current position and motion status for each leg axis (in progress / completed / aborted) as well as ground contact information $c_{1,2,3}$ from the pressure sensors, and also to send movement commands $m_{1,2,3}$ and displacement references θ_1 θ_2 d_3 to each leg axis involved in a current robot step.

3.1. *Hierarchical Modified Architecture for Holonic Motion Control*

The multi-legged robot motion process is a multivariable one, distributed on 18 atomic axis motions which must be coordinated in order to perform it according to one gait strategy selected from the set of predefined platform displacement modes: straight forward / backward or left / right; turn; climb / descend, a.o.

Because of the distributed nature of the robot motion process, a *hierarchical modified control architecture* is considered; the components of this architecture are grouped on two layers and interconnected both top-down and horizontally.

A system of this type serves many simple motion processes - the individual joint displacements of the legs - performed on the lower level; coordination of these simple processes is performed centralized on the higher level in two stages:

1. *Leg coordination to implement the walking strategy* currently activated to obtain a certain displacement mode (turning, climbing, stepping) according to predefined tasks, leg associations, motion increments and velocity specs.
2. *Axis coordination and tracking for each active leg* to implement its individual motion step, according to the kinematic model of the leg.

From the point of view of leg coordination, the control structure behaves *strictly hierarchically*, delivering displacement references to the leg controllers. From the point of axis tracking, the control structure behaves as a *modified hierarchical* (*semi-heterarchic*) one by implementing the controlled axis motion services as independent agents without a superior, but communicating between them. This means that on one hand timing of motion execution is synchronized (by an axis synchronizing processor) for all active axes, and on the other hand there is a direct motion conditioning between the axes of active legs, triggered by the axis' temporal relationship "**completion of motion program - contact with ground**". As long as this simultaneity relationship is satisfied (with a certain tolerance for particular combinations of legs), the low level leg processors continue to operate in the synchronized mode. When the relationship is altered for some axes in certain leg combinations, the control switches from the *heterarchic collaborative mode* at axis level to the *hierarchical top-down* mode which updates leg coordination by closing at high level the position feedback loop for all leg axes [4].

The software system for this type of semi-heterarchical structure is currently developed as *holonic motion execution system* (HMES), based on the innovative concept of *holonic motion control* (HMC).

The basic building blocks of the holarchy for coordinated holonic multi-body motion control are:

- A **resource holon** contains a physical part, i.e. a motion axis (leg joint, link, drive and encoder) of the robot structure, and an information processing part that controls the resource.
- A **motion holon** holds the process and displacement knowledge to assure the correct motion execution (path following, collision avoidance, etc).
- An **order holon** represents a task in the motion control system.
- **Staff holons** are included in the HMES architecture to assist the basic holons for: coordination, synchronization, re planning walking at terrain climbing, re planning trajectories for obstacle avoidance, orientation in the landscape, etc and provide them with sufficient information such that these can take the correct decision to solve the problem. [5].

3.2. *Navigation and Collision Avoidance*

The robot is designed to operate in an unstructured outdoor environment. In order to navigate in such environment, where there are not spatial reference points, a GPS or AGPS system is integrated and used to determine the absolute robot position or, during its displacement, the position relative to its current destination. The GPS is connected to the higher level of the robot motion control system providing a feedback about its current location (distance and orientation) on its path towards the programmed destination, but also proposing a trajectory for avoidance of large objects or areas (buildings, lakes, rivers, etc.) helping thus the command system to set up the direction of travel.

In order to avoid small obstacles (objects, pits, excavations), the robot control system will be equipped with a depth sensor which will offer information about the obstacle's shape and dimension and the structure of the road in front of the robot. This information is provided for obstacles which are within a spatial sector of 3 meters length and ± 90° angle relative to the sensor's location; the depth sensor will be mounted on the mobile robot's platform, aligned with its current displacement axis.

4. Conclusion

The walking robot MERO3 is a multifunctional mechatronic system designed to carry out planned transportation tasks in outdoor unstructured environment. The walking robot is able to tightly interact with the environment for orientation and obstacle avoidance. The coordinate motion control system is developed as a *holonic motion execution system* (HMES), based on the original concept of *holonic motion control* (HMC) derived from the holonic paradigm developed for social organisations and living organisms. A set of simulations were performed in order to validate the motion strategies for gait implementing. The software application were built using the Visual Studio .NET integrated development environment, and integrated in the DirectX display SDK package supporting graphical representations of the modular robot configuration.

References

1. S.M. Song and J.K. Waldron, An Analytical Approach for Gait Study and its Application on Wave Gait, *Int. J. of Robotics Research*, **6**, *2*, 87 (1987)
2. I. Ion, I. Simionescu and A. Curaj, MERO Modular Walking Robot Support of Technological Equipments, *8th Int. Conf. on Climbing and Walking Robots*, London (2005)
3. R.B. McGhee and G.I. Iswandi, Adaptive Locomotion of a Multi-legged Robot Over Unarranged Terrain, *IEEE Trans. on Systems, Man. and Cybernetics*, Vol SMC-9, **4**, 176 (1979)
4. Th. Borangiu, P. Gilbert and A. Rosu, An Implementing Framework for Holonic Manufacturing Control with Multiple Robot-Vision Stations, *Int. J. EAAI, Elsevier*, 505, doi: 10.1016/j.engappai2009.03.001 (2009)
5. T. Berger, et al., Personal Rapid Transit in an Open-Control Framework, *Comp. & Ind. Engineering*, **61**, 300, doi: 10.1016/j.cie2010.12.010 (2010)

MATHEMATICAL MODELLING AND PID CONTROL OF SQU-TWO-WHEELED MOBILITY VEHICLE (SQU-TWMV)

A Al-Siyabi[1], K M Goher[1], S Al-Abdali[1], A Al-Harrasi[1], J Al-Abri[1], M O Tokhi[2], A M Almeshal[2], O Sayidmarie[2] and S A Agouri[2]

[1]Department of Mechanical and Industrial Engineering
Sultan Qaboos University, Oman, e-mail: kgoher@squ.edu.om
[2]Department of Automatic Control and Systems Engineering
The University of Sheffield, United Kingdom

This paper presents the design, mathematical modelling and control of a two-wheeled single-seat vehicle built at Sultan Qabbos University, Oman. The design of the vehicle is similar to the Personal Urban Mobility and Accessibility (PUMA) vehicle developed in 2010 by General Motors in collaboration with Segway. The body of the vehicle is designed to have two parts. The lower part comprising driving motors, wheels and lower base contains all accessories and wiring. The upper part of the vehicle comprises the driver seat, driving mechanism and the outer frame. There are mainly three motors driving the vehicle; a linear Direct Current (DC) motor causing the upper part to slide over the upper part starting from the rest position and two brushless DC motors activating the vehicle while moving forward/backward and/or manoeuvring. The model of the vehicle is divided into two models based on the Lagrange Euler (LE) formulation. Two stages PID control scheme are designed and implemented on the system models.

1. Introduction

GM and Segway have developed in 2009 an electrically powered, two-seat prototype vehicle that has only two wheels. The Personal Urban Mobility and Accessibility (PUMA), is a self balancing two-wheeled vehicle which can travel up to a speed of 35 mph for 8-10 hours for each single charge of the battery. The weight PUMA is 300 lb (136 kg) without passengers [6]. PUMA is equipped by a lithium battery as well as vehicle-to-vehicle communications system for reducing the risk of accidents and regulating the flow of traffic. This vehicle represents a solution to the world's urban transportation problems while affording the advantages of being a fast, safe, inexpensive and a clean alternative compared to traditional cars and trucks for cities across the world. It has been equipped by high technology in controlling, electric components, autonomous driving and parking and vehicle-to-vehicle communications.

Several control approaches have been developed to control two wheeled machines using the conventional PID control strategy; among recent approaches are the following:

Tinkir *et al.*, (2010) have presented a comparison of a PID control and Interval type-2 fuzzy logic control (IT2FL) of a double inverted pendulum system. The authors have aimed to control the swing up position of the double inverted pendulum system. The modelling of the system has been carried out using MatLab/SimMechanics toolbox and using pre-defined system parameters. The comparison results show an unstable response with the PID controller while a fast convergence stable response has been achieved with the IT2FL controller. The ITF2L controller has stabilized the system within 3.5 seconds approximately. Tatikonda *et al.* (2010) applied an ANFIS controller on an inverted pendulum system. A comparison between the conventional PID controller and ANFIS were presented. The authors presented responses of the system through simulations and tested the system with different masses of the pendulum. The ANFIS controller proved to be more robust and bring the pendulum to balance in the upright position in about 6 seconds. Jingato *et al.* (2008) conducted control experiments on a wheel-driven mobile inverted pendulum using a PID controller combined with neural networks. The authors have shown experiments to maintain the balance of the inverted pendulum system and its ability to track a desired trajectory. The neural controller controls both the balance and the position and shows robustness even with disturbances. Han *et al.* (2010) developed a self-adaptive fuzzy PID with fuzzy controller for the inverted pendulum system. The controller consists of three double-input single output fuzzy controllers to adjust the PID controller gains based on the error of change of error. The controller was implemented and tested on a Googol GLIP2003 IP system. The IP system reaches stability in two seconds. However, the response shows some minor oscillations that need improvements.

2. System modelling and description

The SQU-TWMV vehicle, with a similar design to PUMA, will be designed depending on using electric rechargeable batteries. Also, the vehicle will be designed to be compact in order to allow easy maneuvering as well as to avoid the parking restrictions. In addition, that type of small vehicles allows the disabled people to drive and hence to have the feel of self independency. The design of the vehicle has been guided by a set of design constraints including: less cost, small size, less weight and cheap and light materials. Details of the realistic constraints which have been taken in consideration when designing the vehicle can be formulated as follows:

Dimension constraints: The dimension of SQU-TWMV has been selected to be comfortable for the rider. It can move in tight places and can park at any place. Its height allows it to enter the places that have low roofs.

Wight constraints: This constraint is the most important issue in the balancing. When the vehicle is light, this mean less load can be moved. It also means less torque needed and less energy consumption. An aluminum parts are chosen as the primary construction material. Furthermore, plastic will be used at the upper parts in order to provide less weight.

Velocity constraints: TWMV is designed to travel up to a speed 60 Km/h for each single charge of the battery.

Safety and health constraints: Four cluster wheels are added to the vehicle to avoid any problems due to possible failure of the control, electric connections or while stopping the vehicle. An aluminum and plastic are selected for safety reasons because they will not crash if any sudden accidents happened. The door will give the vehicle more safety. The vehicle has some open areas in its upper body to help on exchanging the air inside and outside the vehicle.

The paper is organized as follows: Section II describes the system dynamic modelling process. Section III describe the control strategy used based on PID control technique. A numerical simulation is carried out in Section IV.

In this project, the dynamic model of the whole system consists of three separate sub-models; linear model of the brushless DC motor (BLDC) activating the vehicle wheels, modeling of sliding mode which make transfer between is stage 1 and stage 2 and the stabilization mode from stage 2 to stage 3 as shown in Figures 1 and 2.

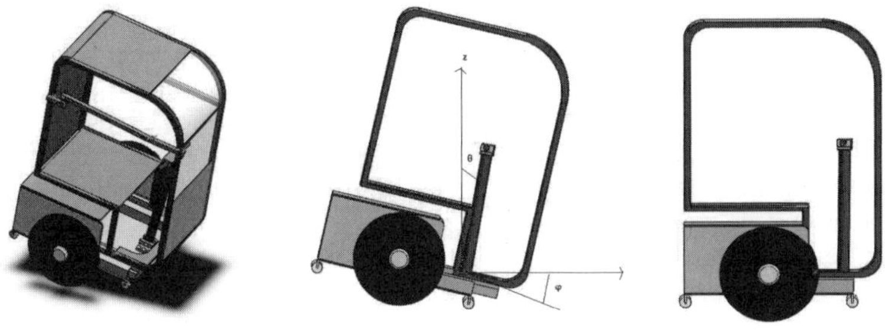

(a) (b) (c)

Fig. 1 SOLIDWORKS design of the SQU-TWMV

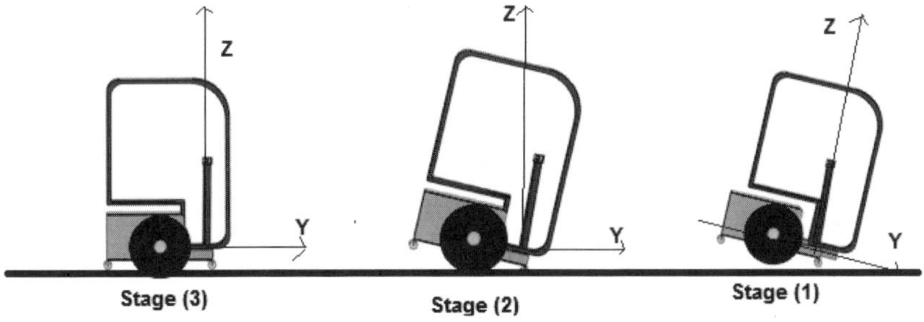

Fig. 2 Schematic diagram for the sliding and stabilization modes

2.1. *Sliding Mode Modeling*

In this model, the equations of the sliding motion of the upper part of the vehicle are derived based on the Lagrangian dynamic approach. The upper part is simplified by single mass sliding up to an inclined plane with an inclination angle φ as shown in Figure 3. The system input is the linear force developed by the linear DC motor.

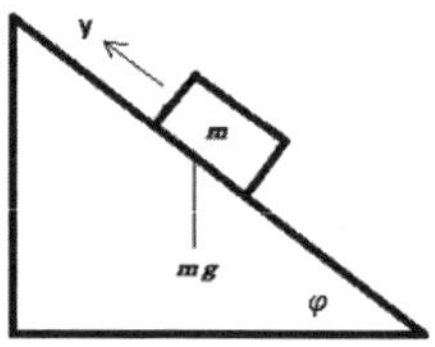

Fig. 3 Schematic diagram of the vehicle sliding

Appling Lagrangian dynamic formulation on the vehicle yields the following equations:

$$\frac{d}{dt}\left(\frac{\partial L}{\partial \dot{y}_i}\right) - \frac{\partial L}{\partial y_i} = F_L \tag{1}$$

where y is the linear displacement of the vehicle in m, φ is the inclination angle in degree, m is the mass of the entire vehicle in Kg, F_L is the force of the linear motor in N, where, the mass of the entire vehicle in Kg include the mass of the motors, batteries and rider.

Using equation (1), the Lagrange function of the system can be expressed as follow:

$$L = \frac{m\dot{y}^2}{2} - mgy \sin \varphi \tag{2}$$

This yields the following equation of motion representing the vehicle dynamics during the sliding mode:

$$F_L = m\ddot{y} + mg \sin \varphi \tag{3}$$

where $\ddot{y}$ represents the linear acceleration of the upper part of the vehicle in response to the force developed by the linear actuator F_L.

2.2. *Stabilization Mode of the Vehicle*

The vehicle can be described schematically by three motion coordinates as shown in Figure 4; linear motion of the system in the Y direction, angular motion of the body about the rod on the Z direction (tilt angle), and angular motion of the body about the cart (base) on the Y direction (inclination angle). The system inputs are the force driving the wheels and a disturbance force acting on the vehicle.

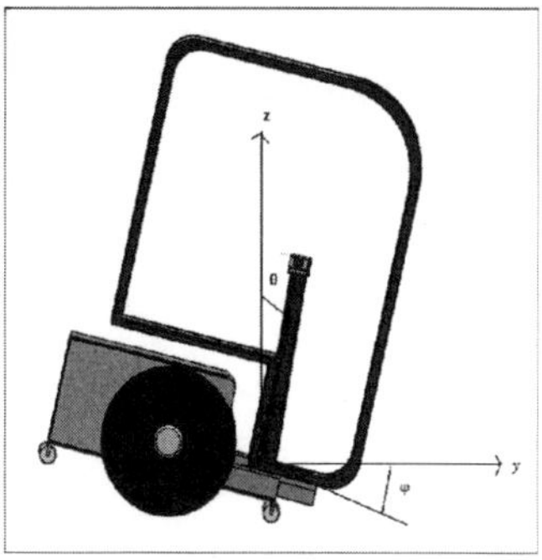

Fig. 4 Schematic diagram of the TWMV with the rod

Appling LE dynamic formulation on the vehicle yields the following equations:

$$\frac{\partial}{\partial t}\left(\frac{\partial L}{\partial \dot{Y}}\right) - \frac{\partial L}{\partial Y} = F_1 \tag{4}$$

$$\frac{\partial}{\partial t}\left(\frac{\partial L}{\partial \dot{\theta}}\right) - \frac{\partial L}{\partial \theta} = F_2 \tag{5}$$

where, Y is the linear displacement of the vehicle, θ is the tilt angle between the rod and Z axis, M_c is the mass of the cart including the mass of the motors and batteries, m_r is the mass of the rod including the mass of the rider, L_c is the distance from the centre of gravity to the centre of mass, L_r is the distance of the rod, F_1 is the driving force of the wheel, and F_2 is an external disturbance force.

Using equations (4) and (5), the systems equation of motion can be described as follows:

$$(M_c + m_r)\ddot{Y} - M_c L_c (\cos(\theta) + \sin(\theta))(\ddot{\theta} + \dot{\theta}^2) = F_1 \tag{6}$$

$$\begin{aligned}&(m_r L_r^2 + M_c L_c^2)\ddot{\theta} + (m_r L_r \cos(\theta) - M_c L_c \sin(\theta))\ddot{Y} \\ &+ 2(M_c L_c^2 - M_r L_r^2)\, g \cos(\theta)\sin(\theta) = F_2\end{aligned} \tag{7}$$

3. Control strategy System Modelling

A schematic diagram of PID controller is displayed in Figure 5 which shows the implementation of the developed control scheme on the stabilization mode (stage 3) described in Figure 2. Three PID controllers are considered where the parameters are adjusted heustrically. The system response will be analysed in terms of the sliding displacement of the vehicle upper part, the tilt angle and the linear motion of the entire vehicle in addition to the control effort, measured in voltage, required for the linear motor as well as the two DC motors required to achieve the stabilisation mode of the vehicle in the upright position. Results of the aforementioned results are represented in Figures 6-11.

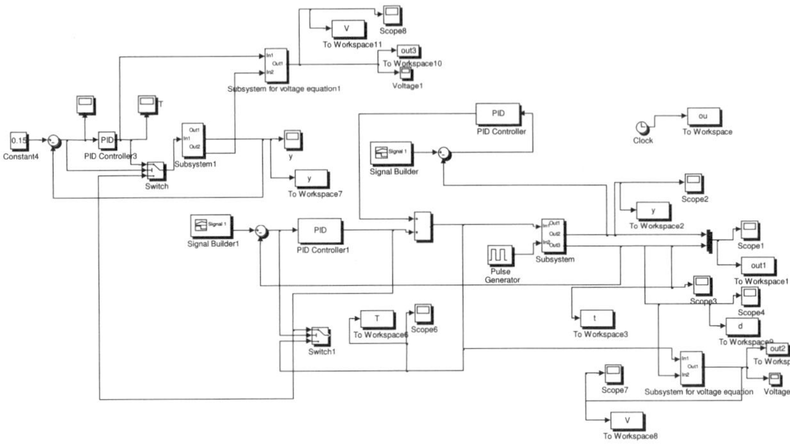

Fig. 5 Implementation of a PID control strategy

The control inputs are the error and the derivative of error for the three measured variables of the system whereas the control outputs are the terminal voltage of the linear motor and the two DC motors activating the vehicle wheels. As the type of motion considered in this study is linear, the two control signals from both the feedback loops are identical and hence the motor terminal voltage considered as the system input is a summation of both the voltage to the left and right wheels.

4. Numerical simulation

Controlling the upper part of the vehicle while sliding; transfer between stage 1 and stage 2, described in Figures 6, 7 and 8 is achieved utilising a PID control approach to control the sliding distance of the upper part. According to the vehicle design, the upper part has to slide, with constant angle of 18 degree, a distance of 15 cm before stopping at stage 3.

The system response during the stabilization mode is described in Figures 9-11. The linear displacement of the vehicle is described in Figure 10. The vehicle has to move from initial reference position a distance of 1.5 meters while keep balancing in the upright position. The controller achieved its target within around 7 seconds. The inclination angle is shown in Figure 9 were it took the same time as the inclination angle to achieve the target of the controller. The control effort required by the DC motor torque required to achieve the control target, described in Figure 11, tends to be high at the beginning of the motion and then decay until the system reach the stabilization within 10 sec.

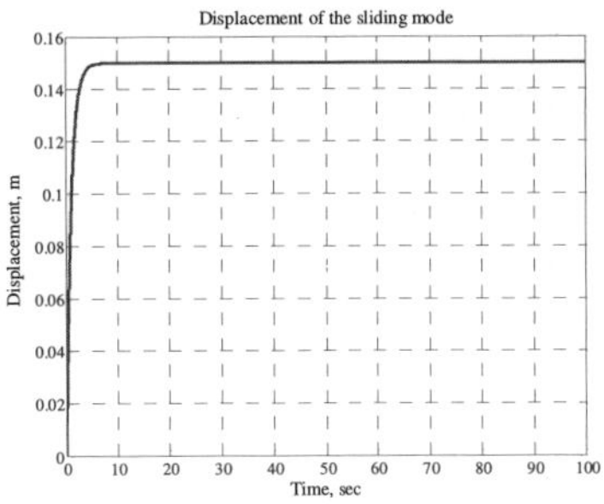

Fig. 6 Sliding displacement of the vehicle moving part

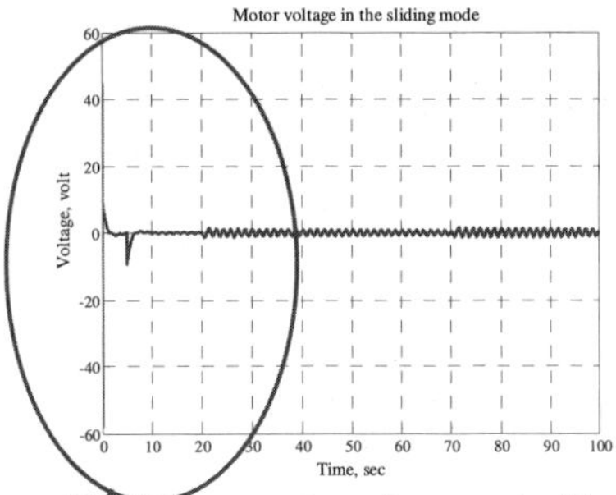

Fig. 7 Linear motor voltage required for Sliding

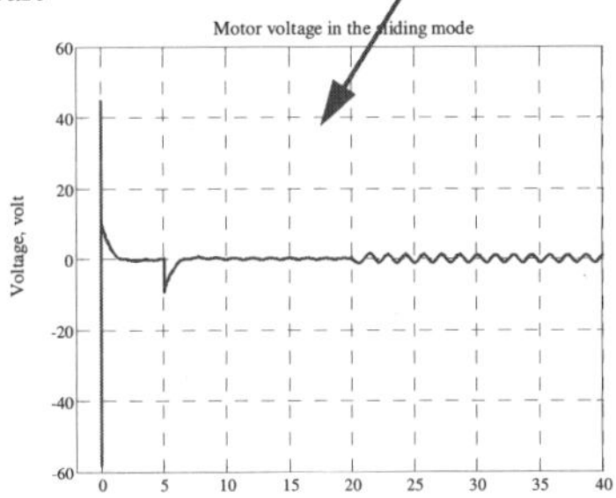

Fig. 8 Linear motor voltage required for Sliding

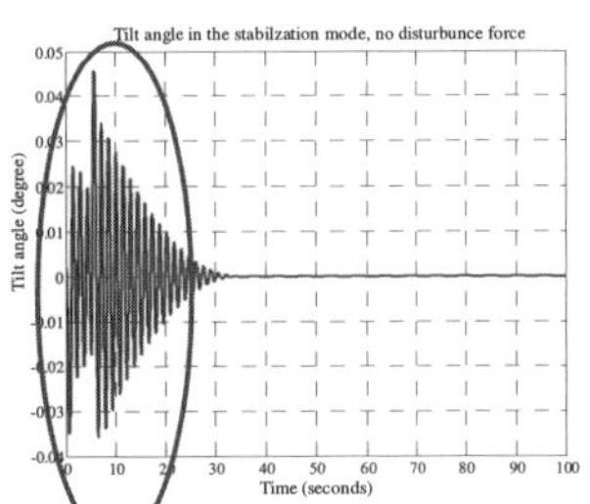

Fig. 9 Tilt angle, no disturbamce force

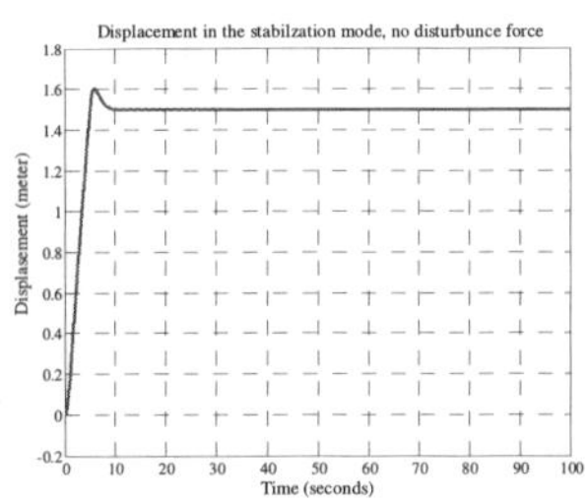

Fig. 10 Linear disp., no disturbance force

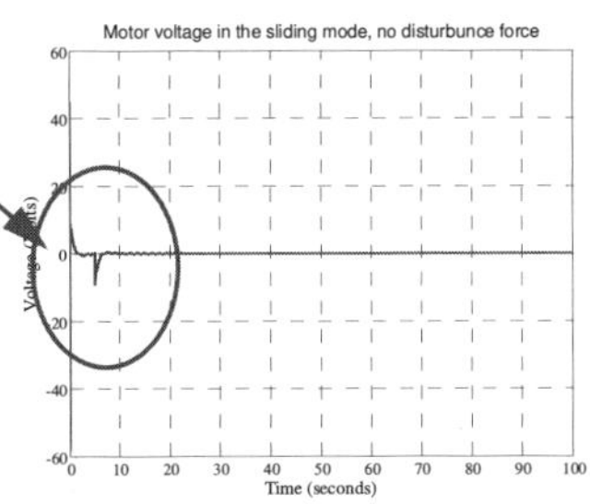

Fig. 11 Stablization motor voltage

5. Conclusion

Mathmetical modelling of a new design of two-wheeled veicle is presented along with a design and implementaion of a simple robust PID control strategy. The response of the system has approved to be satisfactory. Furthermore, developing an entire control strategy for the different stage of the vehicle; sliding mode (stage 1 – stage 2), lifting (stage 2 to stage 3) and stabilization (stage 3) will be provided in the finbal version with necessary switching mechanisms between each stage. Mesurement of the control effort, measured in voltage, for the entire journey of the vehicle (from stage 1 to stage 3) has been provided based on the developed PID control strategy.

References

1. Goher K. M. and Tokhi M. O., (2008), Balance control of a TWRM with a static payload, in the Proceedings of the 11th International Conference on Climbing and Walking Robots and the Support Technologies for Mobile Machines (CLAWAR 2008), Coimbra, Portugal, September 8-10.
2. Goher K. M., (2010), Dynamic Analysis and Control of a Two Wheeled Robotic Machine with an Extended Rod, Department of Automatic Control and Systems Engineering, The University of Sheffield, UK.
3. Jingtao L, Xueshan G, Qiang H and Matsumoto, O, (2008), Controller Design of a Two-Wheeled Inverted Pendulum Mobile Robot, *IEEE International Conference on Mechatronics and Automation (ICMA 2008),* pp.7-12.
4. Tinkir M, Onen U, Kalyoncu M and Botsali F M, (2010), PID and Interval Type-2 Fuzzy Logic Control of Double Inverted Pendulum System, *The 2nd International Conference on Computer and Automation Engineering (ICCAE),* pp.117-121.
5. Tatikonda R C, Battula V P and Kumar V, (2010), Control of Inverted Pendulum Using Adaptive Neuro Fuzzy Inference Structure (ANFIS), Proceedings of IEEE International Symposium on Circuits and Systems (ISCAS), pp.1348-1351.
6. http://www.segway.com/puma/ (last updated: 16/04/2012)

ADAPTIVE MOBILE ROBOT PATH PLANNING BASED ALGORITHM FOR DYNAMIC ENVIRONMENTS

HUSSEIN HAMDY SHEHATA*, JOSEF SCHLATTMANN

Systems Technology and Design Methodology, Hamburg University of Technology 21073 Hamburg, Germany

It is intractable to deal with dynamic environments, which are totally or partially unknown or even dynamically changing. Several algorithms have been proposed and among such algorithms, potential fields are the core of a class of effective navigation schemes for autonomous robots. This paper proposes an adaptive algorithm for a mobile robot path planning in a dynamic environment where the target and obstacles are moving. This algorithm relies on the integration between a dynamic virtual obstacle concept and a robot size factor, which represents a threat factor to any possible collisions and enables the robot to plan its motion not only with right positions, but also with reasonable velocities. The convergence of this proposal is discussed to guarantee its reliability.

1. Introduction

The motion planning problem in dynamic environments is to plan the robot motion to follow a moving target in a desired behavior while avoiding moving obstacles [1]. In general, there are two main categories in solving this problem, which are Artificial Intelligence (AI) algorithms and Artificial Potential Field (APF) algorithms. APF approaches are much convenient than AI approaches due to its elegant mathematical analysis and simplicity [2]. However, APF has some limitations, such as Local Minimum Problem (LMP) [2–5]. In an attempt to solve the problem of path planning in a dynamic environment, Munasinghe *et al.* [6] presented the velocity dipole field and its integration with the conventional APF to form a new real-time obstacle avoidance algorithm. Prassler *et al.* [7] developed a velocity obstacle method for obstacles moving on arbitrary known trajectories. Sugiyama *et al.* [8] proposed a hydrodynamic APF to guide the mobile robot towards the target while avoiding moving obstacles. Grundel [9] presented a methodology for tracking multiple moving targets.

Despite the fact that significant results on the path planning problems have been acquired in the above literature, the problem of path planning in uncertain dynamic environments has not been fully investigated.

* Corresponding author: Tel.: +49(0)40-42878 4430; E-mail: shehata@tu-harburg.de

2. Motivations and Objectives

In most real applications, the environment is dynamic. This means not only the obstacles are moving, so does the target. In such situations, the most common methods ignore the trend of moving target and obstacles. When the obstacle is moving away from the robot, the robot's path should not be affected by the obstacle as shown by the solid line in Figure 1. Furthermore, while we are in total agreement to think about the relative velocity of the robot with respect to the obstacles, we disagree with the incorporation of the absolute velocity and acceleration of both the target and the obstacles as in [1, 10], which resulting in high computational complexity and limit the real-time implementations. To deal with all these unrealistic assumptions, we present an adaptive algorithm entitled as **A**daptive **V**irtual **O**bstacle **A**lgorithm (AVOA).

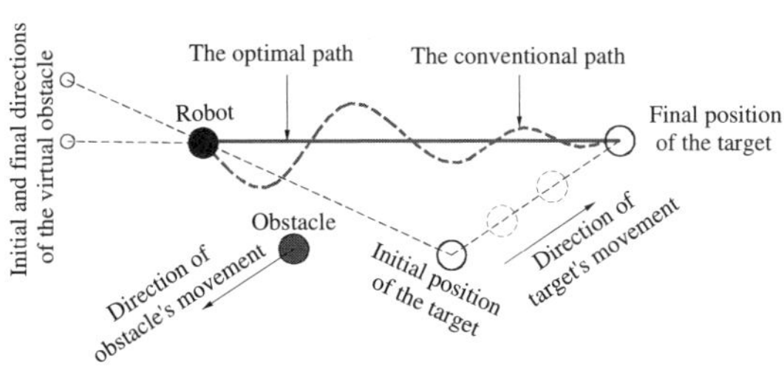

Figure 1. The oscillations of a mobile robot in a dynamic environment

3. Adaptive Virtual Obstacle Algorithm (AVOA)

Generally, there are no constraints on the velocity of either the target or the obstacles except that the maximum robot's velocity should be higher than the velocity of both the target and the obstacle. In order to clarify the analysis, this proposal assumes two assumptions as follows: a) the obstacles are convex polygons whose positions can be accurately measured *on-line*, and b) the mobile robot can move in any direction smoothly.

3.1 *The Proposed Model*

The proposed AVOA basically relies on a dynamic virtual obstacle, which has the role of adjusting not only the speed of the robot but also the steering angle. In our algorithm, we imagine that there is an additional force associated with an additional virtual obstacle in the workspace. The positions of the robot and the target are denoted by $q_r = [x_r \; y_r]^T$and $q_t = [x_t \; y_t]^T$, respectively. While the positions of the obstacles and the virtual obstacle are denoted by $q_{o_i} =$

$[x_{o_i}\ y_{o_i}]^T$ and $q_{vo} = [x_{vo}\ y_{vo}]^T$, respectively, where $i = 1, 2, \ldots n$, and n is number of obstacles. The position of the virtual obstacle is located along the extension line from the target to the robot and in the opposite side of the target. This virtual obstacle exerts a virtual repulsive force, $f_{vo}(q_r)$, toward the robot. The reasonable virtual repulsive force should be configured as follows: a) it should increase as the robot gets closer to the virtual obstacle, b) it should increase quickly in some range of distance so that the robot can avoid collision successfully, c) it should increase slowly outside this range, and d) it should decrease as the robot gets closer to the target in order to allow a soft landing on the target. Accordingly, the need for the virtual obstacle must be a function of the nature and status of the obstacles and the target in the workspace. Consequently, the magnitude of $f_{vo}(q_r)$ is inversely proportional to $\rho(q_r, q_{vo})$, where $\rho(q_r, q_{vo})$ is the minimal distance from the robot to the virtual obstacle, and should be initiated if and only if $\rho(q_r, q_{vo}) \leq \rho_o$, where ρ_o is a positive constant denoting the distance of influence of the obstacle. The relative distance between the robot and the nearest obstacle, and the relative distance to the target, in each step, efficiently determine whether the AVOA should be implemented or not. The placement of the virtual obstacle is chosen as follows:

$$\rho(q_r, q_{vo}) = \frac{\rho(q_r, q_t) + \min\left(\rho\left(q_r, q_{o_i}\right)\right)}{2} \tag{1}$$

where $\rho(q_r, q_t)$ and $\min\left(\rho\left(q_r, q_{o_i}\right)\right)$ are the minimum distances from the robot to the target, and to the nearest obstacle, respectively.

When $q_r \neq q_t$, the repulsive force of the ith obstacle, $f_{rep_i}(q_r)$, and $f_{vo}(q_r)$ are presented as follows:

$$f_{rep_i}(q_r) = \begin{cases} \eta\left(\frac{1}{\rho\left(q_r, q_{o_i}\right)} - \frac{1}{\rho_o}\right) \\ \left(\frac{1}{\rho^2\left(q_r, q_{o_i}\right)}\right)\nabla\rho\left(q_r, q_{o_i}\right), & \text{if } \rho\left(q_r, q_{o_i}\right) \leq \rho_o \\ 0, & \text{otherwise} \end{cases} \tag{2}$$

$$f_{vo}(q_r) = \begin{cases} \eta\left(\frac{1}{\rho(q_r, q_{vo})} - \frac{1}{\rho_o}\right) \\ \left(\frac{1}{\rho^2(q_r, q_{vo})}\right)\nabla\rho(q_r, q_{vo}), & \text{if } \rho(q_r, q_{vo}) \leq \rho_o \\ 0, & \text{otherwise} \end{cases} \tag{3}$$

where η is a positive scaling number, $\rho(q_r, q_{o_i})$ is the minimum distance from the robot to the ith obstacle, and $\nabla\rho\left(q_r, q_{o_i}\right)$ and $\nabla\rho(q_r, q_{vo})$ are two unit vectors pointing from the ith obstacle to the robot and from the virtual obstacle to the robot, respectively.

Consider the case where the robot and the obstacle are moving toward each other, it is noted that $f_{vo}(q_r) > f_{rep_i}(q_r)$ in some range of distances, while $f_{vo}(q_r) < f_{rep_i}(q_r)$ if the robot is close enough to the obstacle. This may lead to some kind of oscillations or collision with the obstacle's boundaries. To avoid this phenomenon, we should dampen the oscillatory motion and to keep a proper safety margin between the robot and the obstacle.

3.2 *The Significance of the Robot Size Factor*

A particularly novel feature of this work is the incorporation of a robot size factor, λ. The value of λ is linearly proportional to the size of the robot, while the magnitude of $f_{vo}(q_r)$ is inversely proportional to λ. It is defined as a positive real number belongs to $\lambda \leq 1$. When the obstacle is far enough from the robot, so there is no threat on the robot, λ is set to minimum but not less than the actual size of the robot, i.e., λ = 0.1. If the obstacle is close enough to the robot, then λ = 1. This means; the proper choice of λ will produce not only a proper virtual force but also a reasonable speed. We define a safety margin between the robot and the obstacle, which in turn depends on the difference (λ − maximum normalized robot size), which represents a safety index (*SI*). If $\rho(q_r, q_{o_i})/SI > 1$, the robot will continue moving toward the target without deviation. If $\rho(q_r, q_{o_i})/SI \leq 1$, the robot will slightly deviate away from the obstacle. Therefore, the robot size factor is considered a threat factor to any possible collisions. Through different complex scenarios, which were extensively tested, we constructed a calibration curve between λ and the robot's diameter. The calibration curve is shown in Figure 2, while Figure 3 shows the influence of the robot size factor on the safety margin. The modified virtual force, $f_{vo}(q_r)_{mod}$, is presented in (4).

$$f_{vo}(q_r)_{mod} = \begin{cases} \eta\left(\frac{1}{\lambda}\cdot\frac{1}{\rho(q_r,q_{vo})} - \frac{1}{\rho_o}\right) \\ \left(\frac{1}{\lambda^2}\cdot\frac{1}{\rho^2(q_r,q_{vo})}\right)\nabla\rho(q_r,q_{vo}), & \text{if } \rho(q_r,q_{vo}) \leq \rho_o \\ 0, & \text{otherwise} \end{cases} \tag{4}$$

Apparently, the introduction of λ has a tri-effect. It defines not only a safety margin of possible physical collision, but also can be used to accommodate any possible measurement uncertainty. Sometimes the applications require a soft landing on the target, other times a hard landing is required. All these requirements can be acquired by controlling the robot size factor.

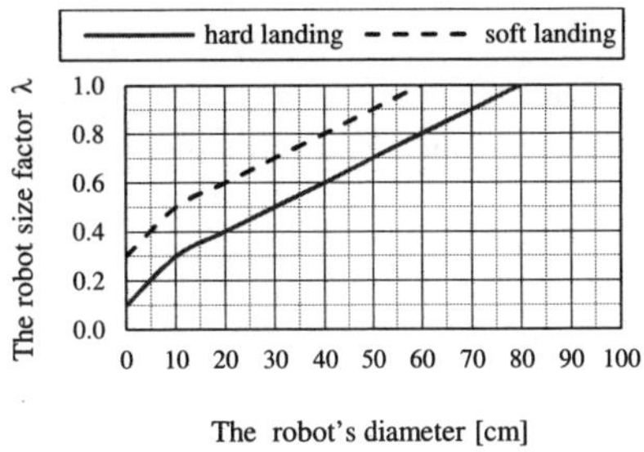

Figure 2. The calibration curve

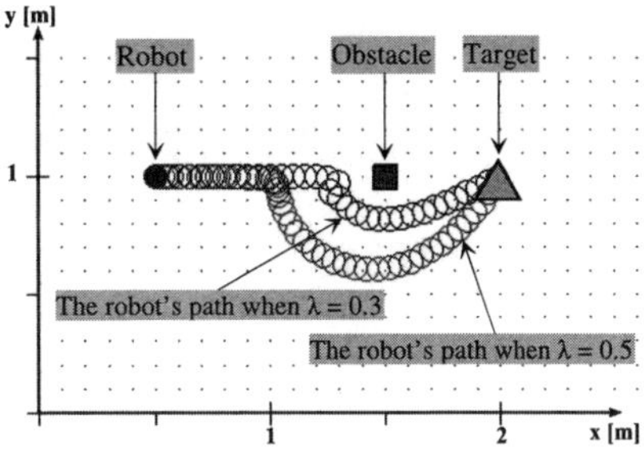

Figure 3. The influence of the robot size factor

Also, AVOA can be used to adjust the robot's velocity with respect to the moving target. The attractive force, $F_{att}(q_r)$, is presented as follows:

$$F_{att}(q_r) = \zeta(q_t - q_r) + f_{vo}(q_r)_{mod} \tag{5}$$

where ζ is a positive scaling factor.

The total force $F_{total}(q_r)$ is given in (6).

$$F_{total}(q_r) = F_{att}(q_r) + \sum_{i=1}^{n} f_{rep_i}(q_r) \tag{6}$$

It is apparent that the relative distances are interpreted as forces. As a result, it is a reasonable idea to develop these forces to generate a direct input signal to the controller. Consequently, the speed of the robot is proportional to the magnitude of $F_{total}(q_r)$. Furthermore, the conjunction between $f_{vo}(q_r)_{mod}$ and $f_{rep_i}(q_r)$ represents a damping effect, which in turn depends on the robot size factor. It is clear that $\rho(q_r, q_{o_i})$ and $\rho(q_r, q_{vo})$ influence the magnitude of $F_{total}(q_r)$, hence they influence the robot's velocity. Therefore, the speed mechanism should be implemented as follows:

If $\rho(q_r, q_{vo}) < \rho_o$ and $\rho(q_r, q_{o_i}) \geq \rho_o$, the robot will rush with high speed toward the target directly. If $\rho(q_r, q_{vo}) \leq \rho(q_r, q_{o_i}) < \rho_o$, the robot will move with medium speed to a direction near to the target. If $\rho(q_r, q_{vo}) > \rho(q_r, q_{o_i})$ and $\rho(q_r, q_{o_i}) > 2R_r$, the robot will deviate with medium speed away from the obstacle. If $\rho(q_r, q_{vo}) > \rho(q_r, q_{o_i})$ and $\rho(q_r, q_{o_i}) \leq 2R_r$, the robot will move with high speed to the opposite direction of the obstacle, where R_r is the robot's radius. Generally, the maximum velocity of the robot is subjected to physical constraints, however, it is important to restrict the maximum travel per instant no more than the robot's radius in order to acquire a smooth path. To guarantee this, the following condition must be satisfied:

$$T\,\mathbb{V}_{max} \leq R_r \tag{7}$$

where T and $\mathbb{V}_{max}$ denote the sampling period, and the maximum speed of the robot, respectively. $\mathbb{V}_{max}$ should be defined at the beginning of the run.

At each sampling instant, the velocity, v_{robot}, and the angular velocity, ω_{robot}, of the robot can be calculated as follows:

$$v_{robot} = \begin{cases} \mathcal{K}_v F_{total}(q_r), & \text{if } |\mathcal{K}_v F_{total}(q_r)| \leq \mathbb{V}_{max} \\ \mathbb{V}_{max} \dfrac{F_{total}(q_r)}{|F_{total}(q_r)|}, & \text{otherwise} \end{cases} \tag{8}$$

$$\omega_{robot} = \begin{cases} \mathcal{K}_\omega(\alpha - \theta), & \text{if } \omega \leq \omega_{max} \\ \omega_{max}, & \text{otherwise} \end{cases} \tag{9}$$

where $\mathcal{K}_v$ and $\mathcal{K}_\omega$ are the velocity and angular velocity gain coefficient, respectively. α and θ denote the steering angle and the direction angle of the current movement, respectively. ω_{max} is the maximum allowable steering rate.

Assuming there is no slipping, the linear and angular velocity at the center of the differential drive mobile robot, as shown in Figure 4, are presented in (10) and (11), respectively. The enhancement of the movement toward the target is shown in Figure 5. The thick dashed line in Figure 5 represents the direction of movement when the AVOA is not applied.

$$v_{robot} = \frac{R_w}{4G_r}\left(\omega_r(t) + \omega_l(t)\right) \tag{10}$$

$$\dot{\theta}(t) = \frac{2}{L} v_{robot} - \frac{R_w}{LG_r}\omega_l(t) \tag{11}$$

where $\omega_r(t)$ and $\omega_l(t)$ are the angular velocity outputs of the right and left wheel, respectively. R_w, G_r, and L are the diameter of the drive wheels, the gear ratio, and the distance between the driven wheels, respectively.

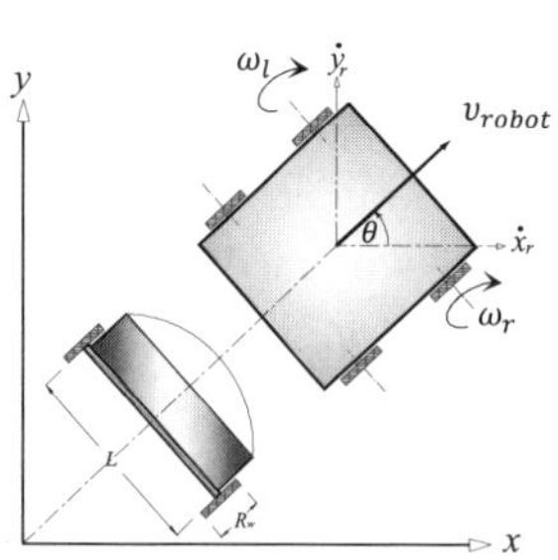

Figure 4. Schematic drawing of the differential drive mobile robot

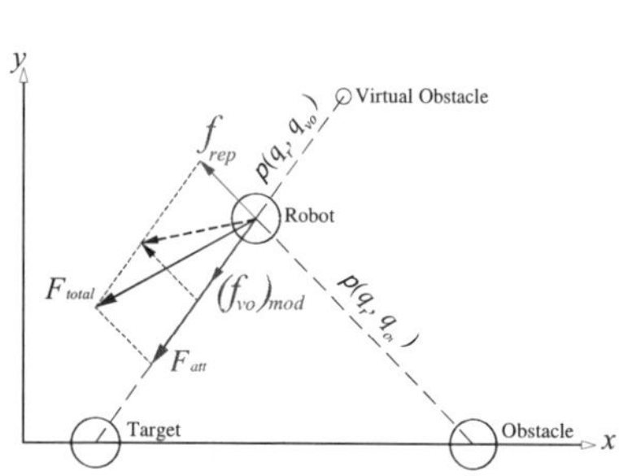

Figure 5. The l ayout of the virtual obstacle

4. Results and Discussion

Extensive simulation studies are carried out under realistic assumptions to ensure the validity of AVOA. In the operational space, the initial positions of the robot, the target, and the obstacle are represented by a black circle, a green triangle, and a red square, respectively. Simulations were conducted using our own software (ROBINAV-V2) developed by the authors. The values of all parameters are set as $\zeta = \eta = 1, \rho_o = 2$, and $\mathcal{K}_v = \mathcal{K}_\omega = 1$. The simulations are based on a sampling period $T = 0.1$ second.

4.1. *Dynamic Environment with Moving Obstacles*

We investigate the validity of the AVOA in a dynamic environment where both the target and the obstacles are moving. In this case, the positions of the target and the obstacles concurrently influence the motion behavior. The target is moving linearly at a constant velocity $[0.013\ -0.015]^T$ starting from point $[3\ 7]^T$. There are two moving obstacles whose initial positions and velocities are $q_{o_1} = [2.5\ 1.6]^T, q_{o_2} = [7\ 7]^T, v_{o_1} = [0\ 0.03]^T$, and $v_{o_1} = [-0.008\ -0.03]^T$, respectively. The robot starts from its initial position $[0.7\ 0.55]^T$ with zero initial velocity. The blue and black trajectories in Figure 6a show the motion behaviors when λ is adjusted to 0.4 and 0.6, respectively. While the solid and dashed lines in Figure 6b represent the velocity histogram for the same situations. This simulation is based on 20 cm robot's diameter and $\mathbb{V}_{max} = 1$ m/s. Consider the case when $\lambda = 0.4$. At first, the robot is far from the obstacles and is influenced only by the target. At $t = 0.71$ s, the robot enters the influence range of the first obstacle. Hence, the repulsive force increases gradually and the robot slows down its velocity. At $t = 10.88$ s, the safety margin reaches to its minimum value, which is equal to the robot's diameter. Therefore, the robot decelerates its velocity very quickly and keeps its direction nearly parallel to the obstacle's movement in order to allow the obstacle passes first. At $t = 15.56$ s, the robot succeeded to detour behind the obstacle and avoids collision, however the robot still inside the influence range of the obstacle. Hence, the robot starts to accelerate again. At $t = 19.78$ s, the obstacle moves away from the robot, and its threat decreases, therefore the robot decelerates its velocity to allow soft landing on the target. At $t = 24.17$ s, the obstacle is no longer a threat but the relative distance between the robot and the target increases; therefore, the robot starts to accelerate again. At $t = 26.75$ s, the robot caught the target.

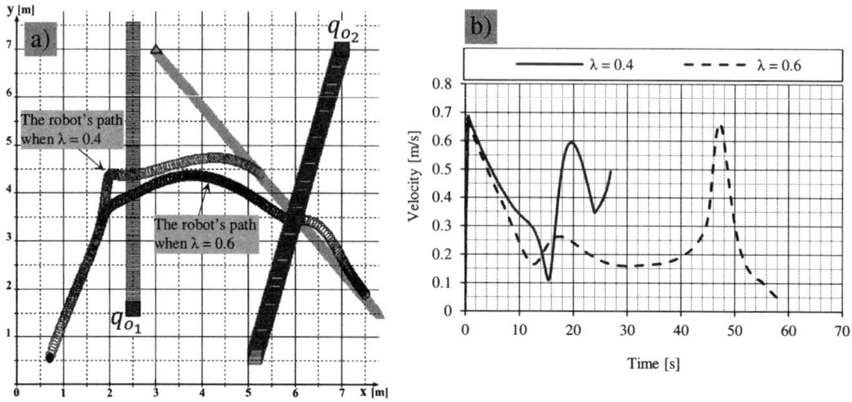

Figure 6. a) Trajectories of the robot, target and obstacles; b) The velocity histogram

4.2 *Dynamic Environment with LMP*

The likelihood of the local minima does exist in the dynamic environment. For simplicity, let us consider the case when the robot, the obstacle, and the target are collinear and the obstacle is in between. They all move in the same direction along the same line. Assuming that; the target's velocity is greater than or equal the obstacle's velocity. The simplest way to solve this problem is to wait for the obstacles or the target to change their motion. However, we will make slight deviations by considering the distance to one edge of the obstacle. Accordingly, the direction of the total force will slightly deviate upward or downward and the scenario will be changed. Figure 7 shows the simulation results for 10 cm robot's diameter and $\mathbb{V}_{max}$ = 0.5 m/s.

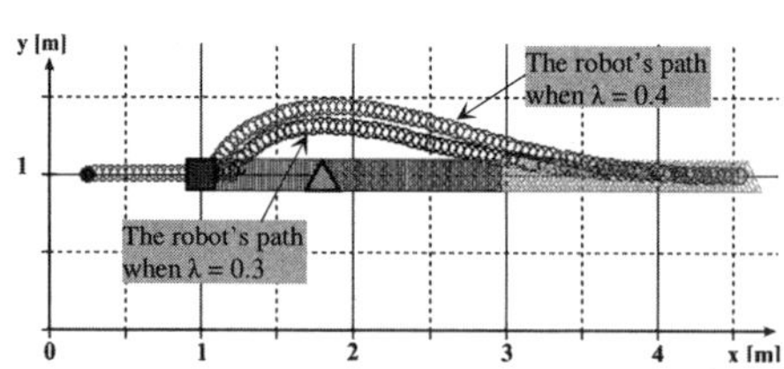

Figure 7. The simulation results of LMP

5. Conclusions

This paper illustrates an adaptive algorithm to solve the problem of path planning in a dynamic environment. This algorithm relies on the integration between the dynamic virtual obstacle and the robot size factor, which represents a threat factor to any possible collisions. This integration has the role of adjusting not only the speed of the robot but also the steering angle. Furthermore, this algorithm affords high capabilities to deal with local minima. The convergence of this proposal is discussed to guarantee its reliability and validity in dynamic environments.

References

1. S. S. Ge and Y. J. Cui, Dynamic Motion Planning for Mobile Robots using Potential Field Method, *International Journal of Autonomous Robots*, **13(3)**, 207 (2002).
2. H. H. Shehata and J. Schlattmann, Mobile Robot Path Planning and Obstacle Avoidance Based on a Virtual Obstacle Concept, in *Proceedings of the 21st International Conference on Flexible Automation and Intelligent Manufacturing*, Vol. 2, No. 1, pp. 905–914, Taichung, Taiwan (2011).
3. Y. Koren and J. Borenstein, Potential Field Methods and their Inherent Limitations for Mobile Robot Navigation, in *Proceedings of the IEEE Conference on Robotics and Automation*, Vol.2, pp.1398–1404, Sacramento, California (1991).
4. J. Ren, K. A. Mcisaac and R. V. Patel, Modified Newton's Method Applied to Potential Field-Based Navigation for Mobile Robots, *IEEE Transactions on Robotics*, **22(2)**, 384 (2006).
5. E. Rimon and D. E. Koditschek, Exact Robot Navigation using Artificial Potential Functions, *IEEE transactions on Robotics and Automation*, **8(5)**, 501 (1992).
6. S. R. Munasinghe, C. Oh, J. -J. Lee and O. Khatib, Obstacle Avoidance using Velocity Dipole Field Method, in *Proceedings of the International Conference on Control, Automation, and Systems,* pp. 1657–1661, Kintex, Korea (2005).
7. E. Prassler, J. Scholz and P. Fiorini, A Robotic Wheelchair for Crowded Public Environments, *Robotics and Automation Magazine*, **8(1)**, 38 (2001).
8. S. Sugiyama, J. Yamada and T. Yoshikawa, Path Planning of a Mobile Robot for Avoiding Moving Obstacles with Improved Velocity Control by using the Hydrodynamic Potential, in *Proceedings of the IEEE/RSJ International Conference on Intelligent Robots and Systems*, pp. 1421–1426, Taipei, Taiwan (2010).
9. D. A. Grundel, Searching for a Moving Target: Optimal Path Planning, in *Proceedings of the IEEE International Conference on Networking, Sensing and Control*, Vol. 2, pp. 867–872, Arizona, USA (2005).
10. C. Qixin, H. Yanwen and Z. Jingliang, An Evolutionary Artificial Potential Field Algorithm for Dynamic Path Planning of Mobile Robot, in *Proceedings of the IEEE/RSJ international Conference on Intelligent Robots and Systems*, pp. 3331–3336, Beijing, China (2006).

SECTION–4

BIOLOGICALLY-INSPIRED SYSTEMS AND SOLUTIONS

EFFECT OF INERTIAL TAIL ON YAW RATE OF 45 GRAM LEGGED ROBOT*

N.J. KOHUT, D. W. HALDANE
Department of Mechanical Engineering, University of California, Berkeley
Berkeley, CA 94709, USA

D. ZARROUK, R.S. FEARING
Department of Electrical Engineering, University of California, Berkeley
Berkeley, CA 94709, USA

1. Introduction

Small legged robots have unique potential for widespread application in search and rescue, hazardous exploration, battlefield reconnaissance, and almost any remote, inaccessible or dangerous situation not reachable by humans. Dynamic running robots have used a variety of steering means, including differential velocity drive [1][2] and actively changing leg kinematics [3][4]. Previous robots have also used tails for various functions, such as turning in an aquatic environment [5], active pitch control [6][7], or stabilization during climbing [8]. The robot discussed here uses a tail to produce a turn on rough or flat ground. This paper examines the way in which this is done, through the use of angular momentum exchange and internal impacts that generate a turning impulse.

2. Description of Robot and Model

2.1. *The TAYLRoACH Robot*

The TAYLRoACH (Tail Actuated Yaw Locomotion RoACH) is a 45 gram robot that features three 7 mm brushed DC motors, two which independently control a set of 3 legs on one side of the robot. It is an evolution of the OctoRoACH platform [2], with a similar configuration. TAYLRoACH is also equipped with a 4 gram, 11.5 cm tail, driven by a custom gearbox and the third DC motor. This gives the tail a moment of inertia of 5.3×10^{-5} kg-m^2 about its

---

* This work was supported by the Army Research Laboratory under the Micro Autonomous Systema and Technology Collaborative Technology Alliance.

base, while the TAYLRoACH has a moment of inertia of 2.3×10^{-4} kg-m^2 about its center of mass. The center of mass of the body is located 6 cm away from the base of the tail.

Figure 1: The TAYLRoACH robot. Electronics and battery have been removed to show the body.

The gearbox for the tail features an 85.3:1 ratio for high acceleration. It is also equipped with two stops that limit the rotation of the tail to approximately 260° statically. These stops may decelerate the tail quickly, producing forces from impact.

Figure 2: The tail gearbox, with stops.

2.2. *Model*

Below, we see an abstract representation of the tail-body system.

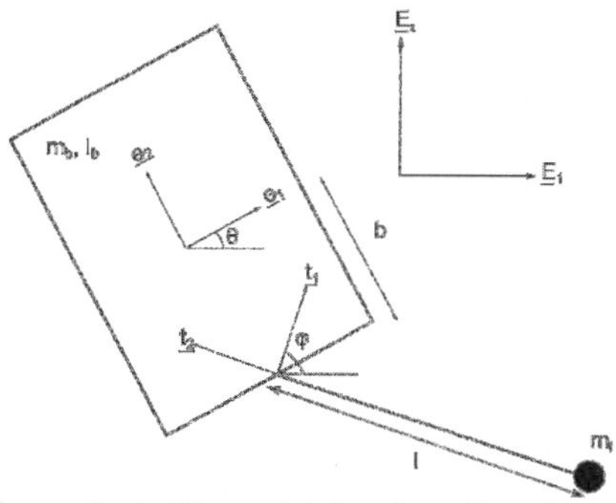

Figure 3: A 2D model for the tail and body.

The body has mass m_b and inertia I_b about the body's center of mass, while the tail consists of a massless rod and a point mass m_t. The body's angle with respect to the world frame is θ, and the tail's angle with respect to the world frame is φ. The distance from the body's center of mass to the center of the tail's rotation is b. From here, we can express the angular momentum about the center of mass of the body-tail system as

$$\boldsymbol{H}_{COM} = \boldsymbol{I_b \omega_b} + \boldsymbol{r}_{b,COM} \times m_b \boldsymbol{v}_{b,COM} + \boldsymbol{r}_t \times m_t \boldsymbol{v}_{t,COM}$$

If no friction or other external torques were present, angular momentum would be conserved, and the rotation of the body would be a function of the rotation of the tail and the total angular momentum [6]. However, friction is present and changes things significantly. If the tail moves slowly, no motion of the body may occur at all, due to friction holding the body in place. If the tail moves quickly, the friction can be overcome, turning the body. Additionally, if the tail hits the hard stop at the end of its travel, a moment may be produced by the impact to turn the robot. This turn will be opposite in direction to the turn caused by tail movement. The friction is a result of six moving, flexible legs with varying contact. A detailed discussion of the friction will not be undertaken here, but it warrants further study.

3. Experiment and Results

TAYLRoACH was driven from a stop on a tile surface at a fixed throttle for the legs for two seconds. The tail is initially set to its full right position, and is free, with no torque applied to it. Velocity control is implemented on the leg speeds, but neither position control of the legs nor steering control of the body is used. This means that the phasing of the left and right legs is not controlled, and there is no feedback on the robot heading to ensure straight ahead motion. One second into the experiment, the tail is actuated with maximum torque for 300 milliseconds moving the tail from full right to full left position. Yaw data for the tail and body were recorded using an OptiTrack [9] motion capture system.

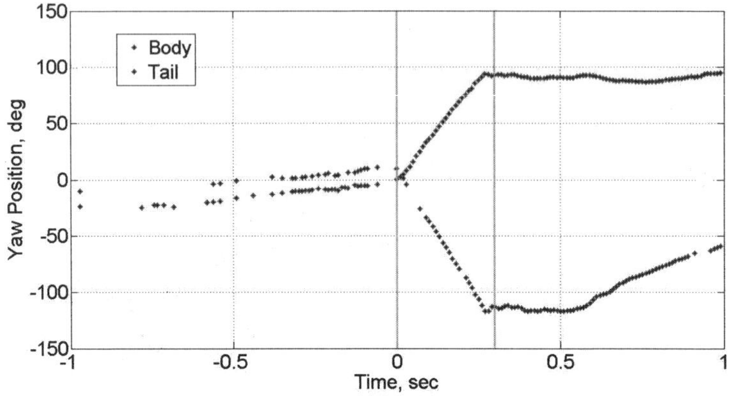

Figure 4: Data from an example trial showing the yaw displacement of the body and tail with respect to time. At the beginning of the trial the robot moves forward without any steering input. At the first vertical line, the tail is actuated. This actuation ends at the second vertical line while the robot continues running.

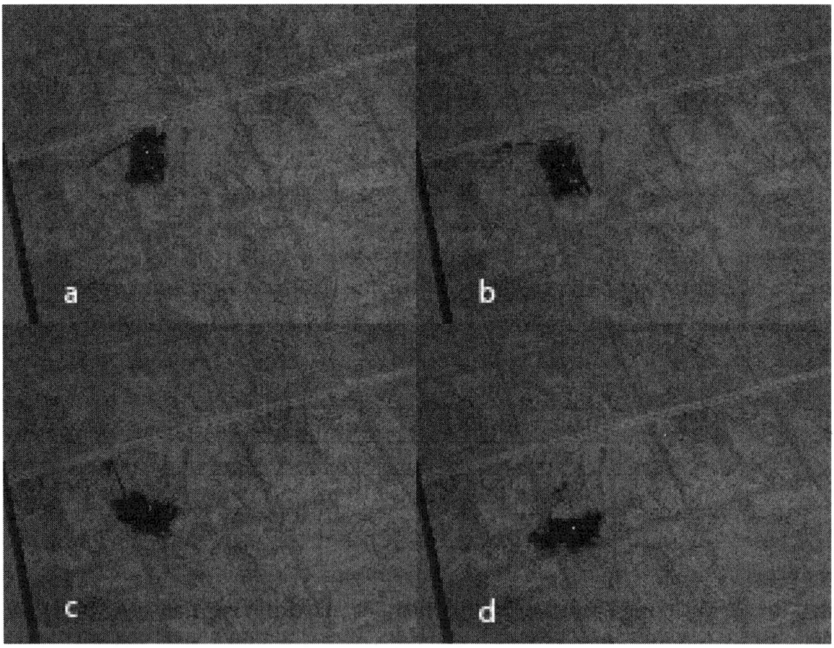

Figure 5: Stills from a video showing the robot turn approximately 90°. a) t = 0s, b) t = 0.10s, c) t = 0.23s, d) t = 0.37s.

As seen in figure 4, the robot drifts before tail actuation, has a high yaw rate during the tail actuation, and then changes direction after the tail actuation, possibly due to an internal impact. TAYLRoACH is a small, minimally actuated robot, and many variables, such as ground contact, roll, pitch, and leg position are not controlled during this experiment. This leads to variation among trials, shown in figure 5.

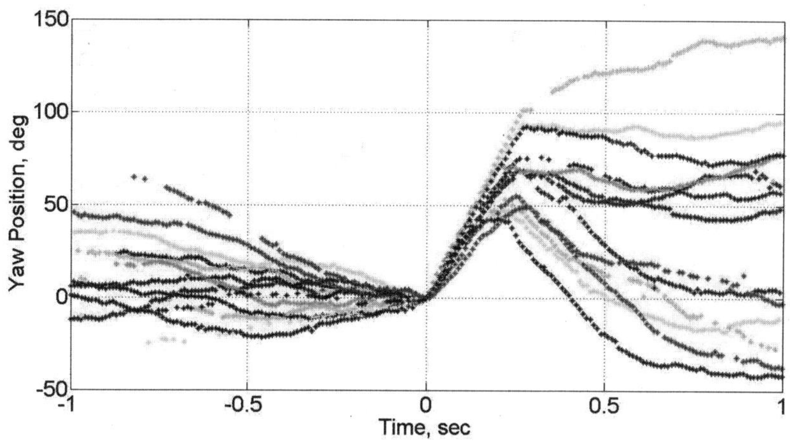

Figure 6: Body yaw vs. time. Fourteen trials were used to measure the variation in behavior. This plot is normalized so that at the point of tail actuation, time and body yaw are zero.

We break these data into three events: before tail actuation, during tail actuation, and after tail actuation. We find that each event has distinct characteristics.

As shown in figure 7, yaw rate is very high during tail actuation (mean 348° sec^{-1}). Before tail actuation, mean yaw rate is basically zero (-7° sec^{-1}), suggesting the robot is going on average straight, though it is still subject to variation. After tail actuation, average yaw rate is -41° sec^{-1}. The difference in yaw rate before and after actuation (both periods in which the tail is unactuated and free) may be accounted for by the impact of the tail. This impact reduces the total turn angle of the robot and is undesired. Since these yaw rates are measured over different time periods, it is useful to look at the total yaw as well.

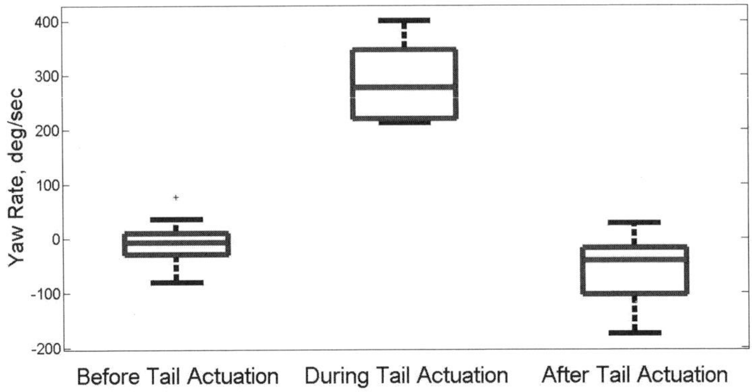

Figure 7: Body yaw rate, before, during, and after tail actuation. Red line marks mean, blue box marks one standard deviation, and black bars mark minimum and maximum values.

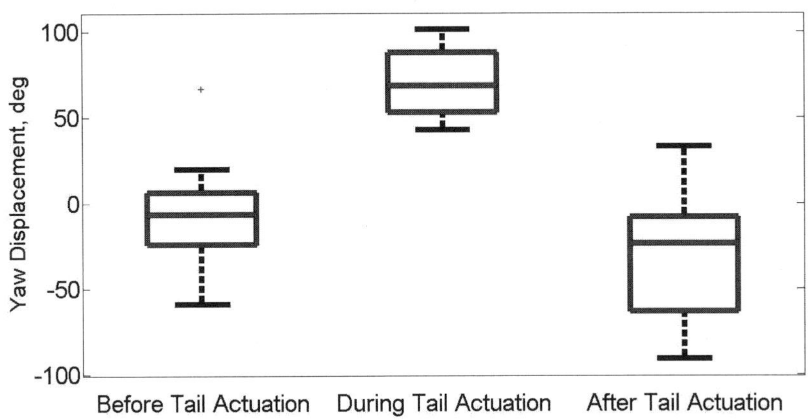

Figure 8: Mean, standard deviation, min, and max of yaw displacement, before, during, and after tail actuation.

In figure 8 we see that the yaw displacement before actuation is small (7°), while it is large during actuation (68°) and after actuation (-24°). The impact of the tail decreases the total turn of the robot, on average, by 35%.

4. Discussion and Conclusion

Using a tail to turn a legged robot running over ground is a complex phenomenon. Several factors, such as leg contact, normal load, surface variation, and others, can determine the amount of friction and constraint forces imposed on a robot during a turn. Additionally, the dynamics and mechanical constraints of the tail show high performance during tail actuation, but this can be mitigated by impact forces which reduce the overall turn angle and increase the settling time of the robot.

Overall performance can be very high, producing net turns of 90° at 300° sec^{-1}. However, the variation shown needs to be accounted for and controlled for consistent high performance. Controlling leg position may be crucial to this, as currently the left and right sides of the robot are driven independently without phase locking. In addition, developing a running gait with a predictable airborne phase may allow a tail actuation during a period where no friction is present, giving more repeatable results. Finally, position sensing on the tail will allow more intelligent control.

Acknowledgments

We would like to thank the Army Research Laboratory and the Micro Autonomous Systems and Technology Collaborative Technology Alliance for supporting this work. We would also like to thank Kevin Peterson, Fernando Garcia Bermudez, and Andrew Pullin for help with figure preparation, OptiTrack operation, and Python support, respectively.

References

1. U. Saranli, M. Buehler, and D.E. Koditschek, "RHex: A simple and highly mobile hexapod robot," *The Int. J. of Robotics Research*, vol. 20, no. 7, pp. 616-631, 2001.
2. A.O. Pullin, N.J. Kohut, D. Zarrouk, R.S. Fearing, "Dynamic turning of 13 cm robot comparing tail and differential drive," to appear in *IEEE International Conference on Robotics and Automation*, May 2012.
3. J.G. Cham, S.A. Bailey, J.E. Clark, R.J. Full, and M.R. Cutkosky, "Fast and Robust: Hexapedal robots via shape deposition and manufacturing," *The Int. J. of Robotics Research,* vol. 21, no. 10-11, pp. 869-882, 2002.
4. A. McClung, "Techniques for dynamic maneuvering of hexapedal legged robots," Ph. D. dissertation, Stanford University, 2006.
5. K. Hirata, T. Takimoto, K. Tamura, "Study on turning performance of a fish robot," in *First International Symposium on Aqua Bio-Mechhanisms*, August 2000.
6. T. Libby, T. Y. Moore, E. Chang-Siu, D. Li, D. J. Cohen, A. Jusufi, R. J. Full, "Tail-assisted pitch control in lizards, robots, and dinosaurs." *Nature*, vol. 481, pp. 181-184, 2012.
7. E. Chang-Siu, T. Libby, M Tomizuka, R. Full, "A Lizard-Inspired Active Tail Enables Rapid Maneuvers and Dynamic Stabilization in a Terrestrial Robot," IEEE International Conference on Intelligent Robots and Systems, September 2011.

8. W. Provancher, S. Jensen-Segal, M. Fehlberg, "ROCR: An Energy-Efficient Dynamic Wall Climbing Robot," *IEEE Transactions on Mechatronics,* October 2011, vol. 16 no. 5, pp. 897-906.
9. http://www.naturalpoint.com/optitrack/

APPLICATION OF NEW BRAIDED SOFT ACTUATOR DESIGN IN BIOMIMETIC ROBOT LOCOMOTION

M.F. A. 'ATHIF, M.N. ILI NAJAA AIMI, R. NUR FATIHA, M.R. M. RUSYDI
Department of Mechatronics and Robotics Engineering, Faculty of Electrical Engineering, Universiti Teknologi Malaysia,
Johor Bahru, Johor 81310, Malaysia

D.HIROOKA, S. WAKIMOTO, K. SUZUMORI
Graduate School of Natural Science and Technology, Okayama University,
3-1-1 Tsushima-Naka, Okayama city, Okayama 700-8530, Japan

This paper presents applications of novel soft actuator design; a combination of different braided angles of artificial muscle actuator in biomimetic locomotion. The actuator construction is based on the integration of artificial muscle contraction and extension theory. The actuator comprises of fiber-reinforced inside silicone rubber and is capable to create extension, contraction and one-sided bending motion. Based on the experimental results, the developed soft actuator can realize leech, swimming-frog and inchworm-like locomotion behavior after standard air pressure driving experiment is executed.

1. Introduction

Actuators have been vigorously utilized in many applications in modern society these days. The need of soft-surfaced actuator that can produce soft movement is increasing, as researchers are focusing on potential research where soft actuator could be useful in producing higher flexibility in its movement in various applications of robotics and automation. Existing developed soft actuator have shown promises in producing soft movements to objects and human [1-4].

Soft actuator actuation principle of extension and contraction can be achieved by applying different actuator design as developed by the previous researchers which includes chambers [1], fiber reinforced [1-3], and shape-type [4]. For the fiber reinforced design, the fiber may be braided by having certain degrees of angles or just in circular direction. This type of actuators is used by researchers in many fields of studies. A device utilizing soft actuator with braided angle technology has been developed to help doctors in performing colonoscopy insertion by changing its stiffness. The technology is used to reduce collision and prevent injuries during insertion process. The soft actuator will only change its stiffness and limit its function from extend, contract or bend when the air is supplied [2].

The braided angle type actuator also has been used in development of combining the contraction and extension principle in realizing the one-sided bending motion [3].

Recently, animals locomotion actively being imitate in building a robot which referred as biomimetic. The exploration in the biomimetic robotics field has produced a soft starfish robot constructed using pneumatic balloon channels. The actuator will bend due to inflation of the balloons thus mimics the starfish locomotion [4]. There is also a swimming robot that mimicked the swimming locomotion of manta stingrays. The manta swimming robot used a circular braided actuator with two internal chambers that enable it to bend in both sides [1].

In this paper, we developed a new braided angle soft actuator design in mimicking leech, swimming frog-like and inchworm locomotion. The purpose of this study is to broaden the application for the braided angle type actuator in the biomimetic locomotion.

2. Braided Angles and Its Application on the Developed Actuators

Study conducted by Iwata *et al.* [3] has proven that a single chamber hydraulic actuator knitted with fiber of angle 23.5° around it will lead to contraction and angle of 66.5° will result in extension of the actuator length. The attachment of both hydraulic actuators; contracted and extended actuators, as shown in Fig. 1 has made both actuators bended to the side of contracted actuator.

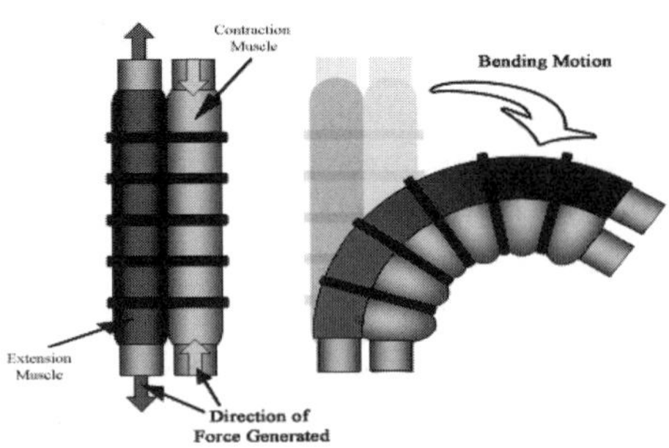

Figure 1. Generation of bending motion to the contracted side when actuators with extension muscle and contraction muscle are banded together [3].

Our pneumatic soft actuator design was inspired by his finding but we enforced different material of fibers with two angles knitted differently at each half of a single chamber actuator where (θ_1) on the top side and (θ_2) on the bottom side; as demonstrated in Fig. 2. We fixed the inside and outside diameter, thickness and length of the soft actuator for all models to be 8mm, 14mm, 3mm and 148mm respectively.

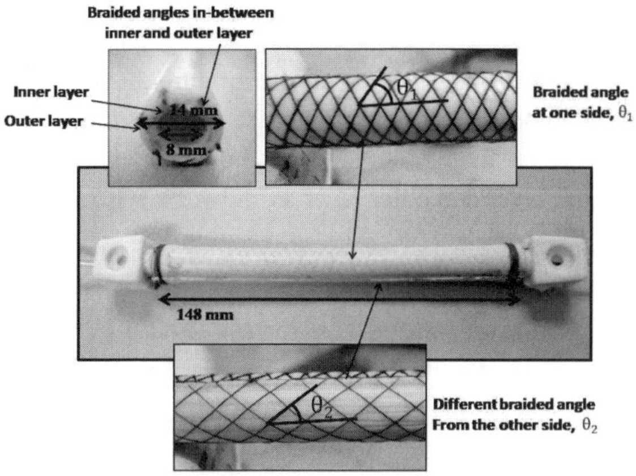

Figure 2. A single chamber actuator, braided with two different angles.

3. Fabrication Process

The developed pneumatic soft actuator was designed using 3D-CAD software INVENTOR™ from Autodesk. The molds designed are used to construct and give the shape of the soft actuator. The molds will then be fabricated by CNC machine based on the 3D model designed.

Figure 3 and 4 shows the head and tail part of actuator with vacuum suction cup to control the locomotion of the actuator. The head has two insertion tubes, one for supplying air pressure into the inner chamber and the other one is for vacuum suction while the tail has only one tube connected to it which is the vacuum suction.

We have fabricated 6 models of actuators namely *Model A, B, C1, C2, D* and *E*. All actuator models have the same basic construction consist of head, tail and body parts. The construction of the actuator body part comprises of three layers. The first inner layer was fabricated beforehand by using silicone rubber material mixed with curing agent.

Vacuum tube

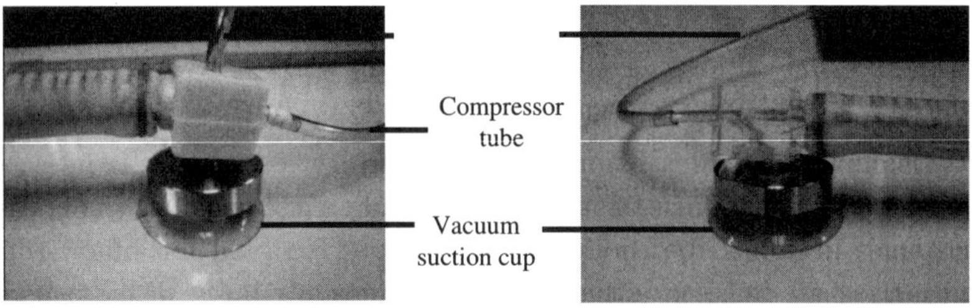

Figure 3. Head part.

Figure 4. Tail part.

A rod as in Fig. 5 is placed on top of the mixture that has been poured into the mold before the upper and bottom molds are attached together. This rod will be placed in the center of molds to create a chamber to actuate the actuator. We used 2 molds of inner layer and outer layer as shown in Fig. 6.

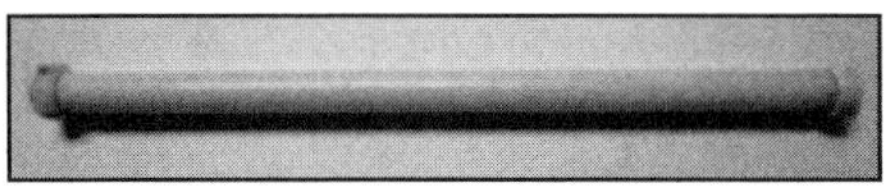

Figure 5. Rod placed in the center of molds

The following process is to knit the fabricated inner layer with reinforced fiber. Our models differ by these knitting angles. Next, the outer rubber layer will cover the inner layer with the knitted fiber. For *Model C2*, we used other types of rubber material mixture to compare actuator from which material could produce better bending ability.

Figure 6. Upper and bottom parts of molds for inner layer.

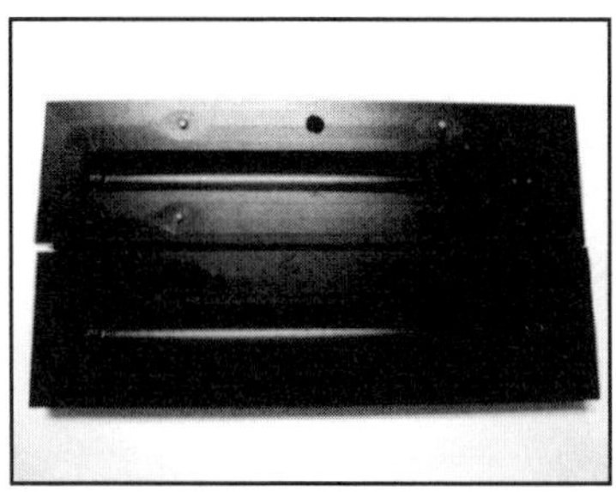

Figure 7. Upper and bottom parts of molds for outer layer.

The mixture is heated in oven at temperature of 80°C for about 40 minutes. The heating process will fasten the curing process. It could take 24 hours for the mixture to cure without the heating process.

3.1. *Experimental Testing*

Experimental setup as in Fig. 8 was prepared. Two valves are used; one to actuate the motion form air compressor and the other to control the movement using vacuum pump. The experiment was conducted to validate whether the developed actuators could locomote as predicted or not. The efficiency of movement based on the maximum displacement they can achieve at every cycle of locomotion can also be evaluated. Based on the result, it is proven that the developed actuators are able to locomote and the movement appears to be similar like a leech, a swimming frog and an inchworm.

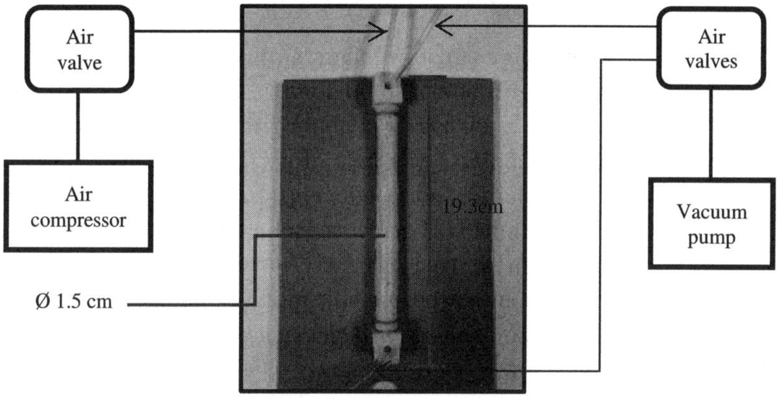

Figure 8. Experimental setup.

3.2. *Actuator Locomotion*

Fig. 9 to 14 shows the motion of soft actuator before and after pressurized with air in one cycle of locomotion. The air pressure and vacuum pressure applied to all models was 0.15MPa and -0.4MPa respectively.

Model A is able to elongate its body due to extension of fiber-reinforced in axial direction when body chamber is pressurized while *Model B* can shorten its body because of contraction of whole fiber-reinforced in the body. For *Model C1* and *C2*, the side with knitted θ_1° is contracted while the other side θ_2° is extended; thus producing bending motion to the contracted side.

Model C1 and *C2* use the same angles with different mixing material. Silicon rubber used in *Model C1* is from Dow Corning Corporation, using Silastic P-1 type (as base) and curing agent while *Model C2* is made from Shin Etsu Silicones Corporation using room temperature vulcanizing type (KE-1603-A-B). In the experiment conducted, the actuators designed for *Model C1* and *C2* produced bending angle 32° and 31.5° respectively. We found that both rubber materials give not much different in the deformation.

Some optimization in the bend-type actuator used in mimicking swimming frog has been made, where the actuator can bend up to 105°. We tested the optimized version of bend type actuator in mimicking swimming frog, as in Fig. 13. Finally we tested the same actuator for inchworm locomotion as in Fig. 14.

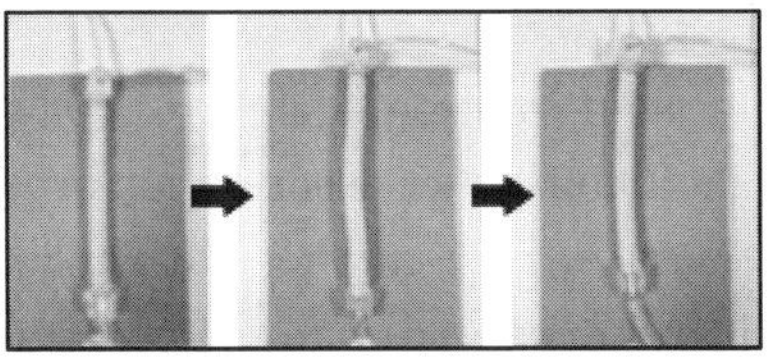

Figure 9. *Model A*, leech-like locomotion (extension).

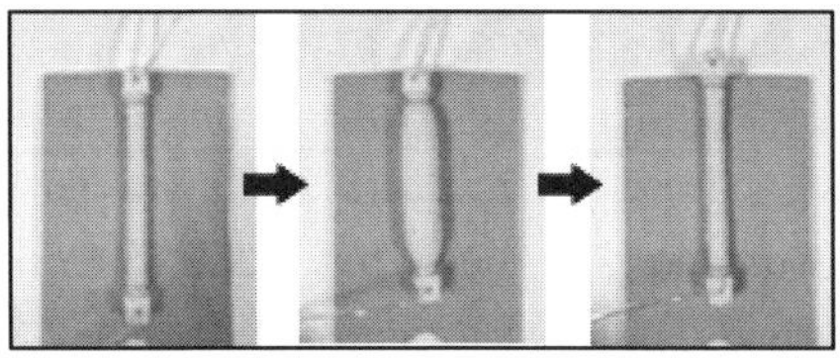

Figure 10. *Model B*, leech-like locomotion (contraction).

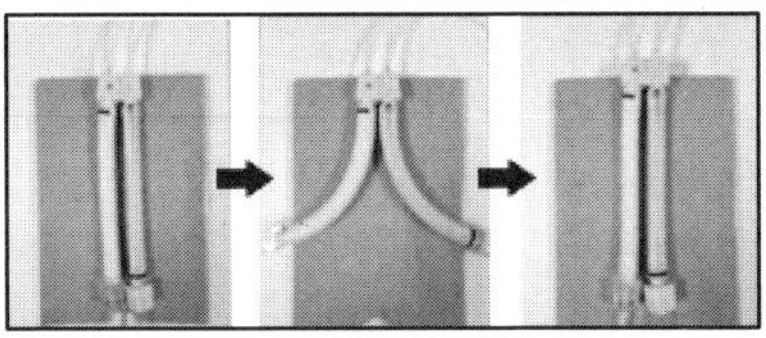

Figure 11. *Model C1*, swimming frog-like locomotion (extension and contraction).

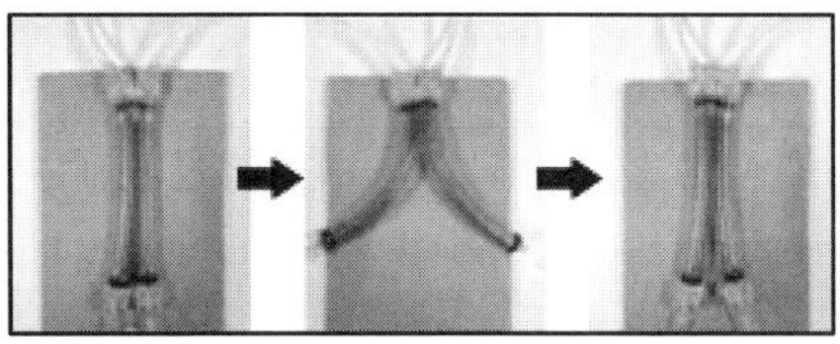

Figure 12. *Model C2*, swimming frog-like locomotion (extension and contraction).

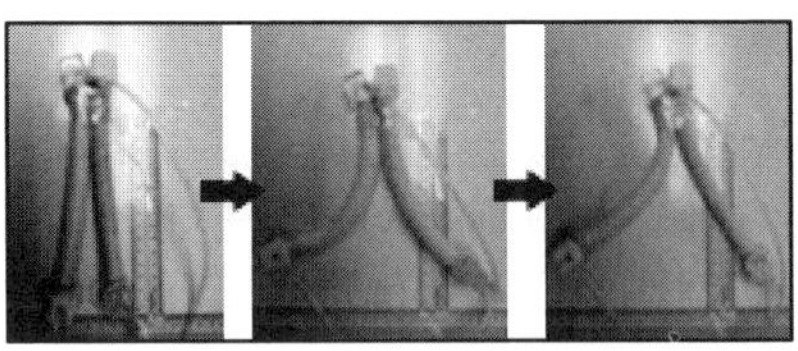

Figure 13. *Model D*, swimming frog-like locomotion (extension and contraction).

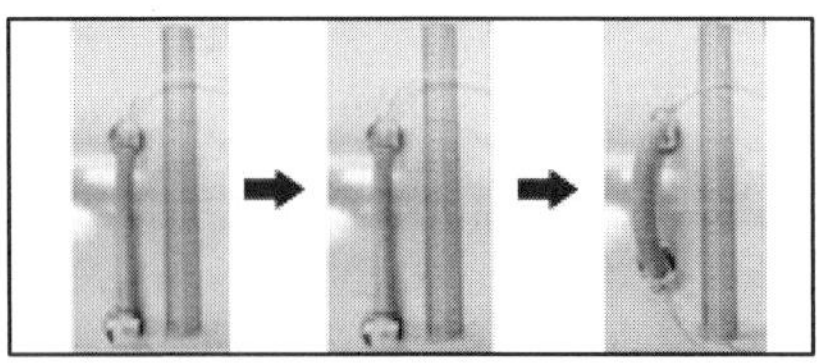

Figure 14. *Model E*, inchworm locomotion (extension and contraction).

All robots start to move by sucking air under its head to make sure the position of the head is fixed. Then air pressure is applied inside the robots body producing extension, contraction or bending motion. Next, the tail will suck air under its tail to fix the position of the tail. After that, the head and the pressure inside the body will released its suctions and pressure to produce movement.

Table 1 shows the maximum forward movement (displacement) of 6 robots models at every cycle of locomotion. The maximum displacement in 1 cycle of locomotion is achieved by *Model E*, the inchworm robot followed by *Model A, D, B, C1* and *C2*.

Table 1. Maximum forward displacement of 6 actuator models at every cycle of locomotion.

ACTUATOR MODEL	θ_1(°)	θ_2(°)	MOVEMENT GENERATED	DISPLACEMENT (cm/cycle)
A	78	78	Elongate (leech)	0.9
B	14	14	Shorten (leech)	0.5
C1	73	52	Bend	0.3
C2	73	52	Bend	0.2
D (optimized model from *C*)	84	52	Bend (swimming frog)	0.68
E	84	52	Bend (inchworm)	6.2

Selected robot of *Model A, B, D* and *E* were tested to know the fastest walking robot. Speed test was conducted at walking distance of 30 cm and the result is illustrated in Table 2. The fastest robot is achieved by *Model E*, inchworm robot.

Table 2. Speed test result at a fixed distance, 30cm.

ACTUATOR MODEL	DISTANCE (cm)	TIME (s)	SPEED (cm/s)
E	30	18	1.67
A	30	39	0.77
D	30	137	0.22
B	30	142	0.21

4. Discussion

Based on our experiment, we observed that *Model E* produced highest speed followed by *Model A, D* and *B*. *Model A* and *B* showing slower performance because of its structure produces less elongation and contraction compare to others and this could lead to slower performance.

Model D and *E* use the same braided angle. Even though the actuators of both models are same, *Model E* of inchworm robot can generate higher speed than *Model D* of swimming frog robot. This might due to high friction resulting from the robot body to the floor. *Model D* has larger area that is directly in contact with the floor compared to *Model E* where this could lead to slower performance.

In the future, the bending angle of the bend-type actuator can be further improve to generate greater displacement than the contracted and extended type actuator so that the robot will move faster like real animal. We also found that the actuator made from rubber material that have similar characteristic with silicon mixture yields better deformation.

5. Conclusion

In this research, we realized 6 models of braided type soft actuator which are able to generate movement based on the principle of extension and contraction. The technique of knitting fiber in actuator give benefit to the actuator's bending motion. With this ability, the actuator is found to be suitable in biomimetic robot locomotion application.

Acknowledgments

The authors would like to give a special thanks to Universiti Teknologi Malaysia (UTM), Ministry of Higher Education (MOHE) Malaysia, Professor Koichi Suzumori, Associate Professor Shuichi Wakimoto, Daisuke Hirooka from Okayama University for their help and support.

References

1. Koichi Suzumori, Satoshi Endo, Takefumi Kanda, Naomi Kato and Hiroshi Suzuki, "A Bending Pneumatic Rubber Actuator Realizing Soft-bodied Manta Swimming Robot", *IEEE International Conference on Robotics and Automation, 2007.*
2. Shuichi Wakimoto, Issei Kumagai, and Koichi Suzumori, "Development of large Intestine Endoscope Changing Its Stiffness", *IEEE International Conference on Robotics and Biomimetics (ROBIO), 2009.*
3. K. Iwata, K. Suzumori and S. Wakimoto, "Development of Contraction and Extension Artificial Muscles with Different Braid Angles and Their Application to Stiffness Changeable Bending Rubber Mechanism by Their Combination", *Journal of Robotics and Mechatronics Vol. 23, No. 4, 2011.*
4. Filip Ilievski, Aaron D. Mazzeo, Robert F. Shepherd, Xin Chen and George M. Whitesides, "Soft Robotics for Chemists", *Journal of Angewandte Chemie, Vol. 123, Issue 8, pages 1930–1935, 2011.*

INCLINATION DETECTION AND ADAPTIVE WALKING FOR LOW-COST SIX-LEGGED WALKING ROBOTS USING ORGANIC COMPUTING PRINCIPLES

A. Al-Homsy*, J. Hartmann and E. Maehle

Institute of Computer Engineering, University of Lübeck,
23562 Lübeck, Germany
** E-mail: alhomsy@iti.uni-luebeck.de*
www.iti.uni-luebeck.de

For a legged robot that walks on rough terrain it is very important to avoid falling down while executing its mission. For this purpose, maintaining the six legged robot statically stable is considered an important issue and challenging task. This paper will shed light on the applied approach for detecting the direction of ground inclination, obstacle avoidance as well as the effect of amputation on statically stable up- and downhill walking based on evaluating the local current consumption and angular position of each leg's joint. The body posture will be adjusted in relation to the direction of inclination. Our approach is based on an organic computing architecture and was tested on a low-cost version of the OSCAR robot.

Keywords: Organic Computing, Walking Robots, Adaptive Uphill and Downhill Walking, Reflexes, Robot Leg Amputation

1. Introduction

Locomotion behaviour of insects in their environment is remarkable and fascinating. At first sight, it looks very easy and as if done without any effort, but in fact is very complex. Insects swerve, climb, turn, walk, stop or run without losing their balance. To maintain a statically stable walking behaviour, for six legged insects, at least three legs have to touch the ground.

Walking is executed by simple coordination rules: each leg performs a swing and a stance phase independently and neighbour legs are not allowed to be in the swing phase at the same time.[5] Besides the leg coordination, the walker has to control its body posture to maintain a statically stable walking in rough terrain.[6] The control of body position of insects has been studied intensively by Cruse et al.,[4,6,7] showing that each leg acts as a height controller to the distance between the insect's body and substrate, independent of the other legs. The applied force of the legs then affects the insect's thoracic segments and contributes to controlling the body height. In addition, load information that is produced from each leg plays an important role in achieving stable walking, because stable walking requires effective load distribution among all stance legs.[1,8] Several types of legged robots have been designed to demonstrate biologic walker locomotion for statically stable locomotion in rough terrain e.g. LAURON[9] or the six-legged robot AMOS-WD06.[12] AMOS-WD06 depends on a neural circuit as a central pattern generator (CPG). On a tilted plane, the motor current consumption is compared with the stored default current when the robot walks on flat terrain. The error signal is used to enable neural learning while the CPG selects appropriate behaviours biased on the synaptic output. When AMOS-WD06 encounters a slope again, the inclination sensor will immediately be triggered which leads to the same output and to directly selecting the slow wave gait without further learning.

In this work, somatosensory feedback is used to detect the ground inclination direction and adjust the body posture to maintain statically stable walking and also avoid obstacles like small objects. The approach applied to detect the inclination solely and avoiding the obstacles depends on the evaluation of the consumed current in the leg's joints. This work has been implemented in the Organic Robot Control Architecture (ORCA)[3] and tested on the latest version of the hexapod robot OSCAR (Organic Self Configuring and Adaptive Robot),[11] based on the Bioloid robot kit which uses low-cost off-the shelf servos.

2. Test Framework

The developed approach has been tested using the Organic Robot Control Architecture (ORCA) and the hexapod robot OSCAR. ORCA is a modular, hierarchically organized framework which enables the design of organically inspired systems.[3] The architecture consists of two main units: the BCU (Basic Control Unit) and the OCU (Organic Control Unit). BCUs form the backbone of a system and provide sufficient functionality for the system under normal conditions. OCUs are a supplemented unit to the BCUs. They monitor and observe the BCU and other OCU signals and have the authority to change the BCUs parameters as soon as the BCU's signal shows anomalies, ensuring the system's function in case of faults.

Fig. 1. The test platform: hexapod robot OSCAR. The general test setup is shown left, the robot geometry right.

ORCA has been previously applied on several versions of the hexapod robot OSCAR.[11] In this work, experiments will be conducted on a new version of OSCAR, which is based on the robot kit Bioloid[a], showing that complex tasks like adaptive walking can be accomplished using low-cost commodity hardware. The new OSCAR robot (see Fig. 1) has six roughly symmetrically distributed legs which are divided into front, middle and rear legs. Each leg has three joints which will, as in the previous versions, be called α, β and γ in the remainder of this paper, with α being closest to and γ furthest from the body. Each joint is presented by a Dynamixel AX-12 servo motor which offers the properties to read, among other parameters, consumed current, the actual servo position and speed. Additionally, each leg is equipped with a binary switch for detecting ground contact. All robots' components use two batteries (NiMH3900) as power supply and the robot is controlled from an external PC via USB.

Each leg is now controlled by a set of BCUs that set the target position and read the actual position as well as the consumed current of each servo. The OCU unit for each leg BCU monitors the consumed current and reacts on detected anomalies, e.g. high current consumption in case of ground inclination, by changing the leg's BCU parameters, e.g. the anterior (AEP) and posterior (PEP) leg position as well as the step height.

3. Inclination Detection

A common method to detect the inclination and its direction is to use an inclinometer or acceleration sensors. Those increase the cost of the robot and its fault probability and make the control system more complicated. Therefore, our work will discuss how to detect the inclination and its direction

[a]Bioloid Robot Kit, `http://www.robotis.com/xe/bioloid_en`

and adapt the robot body posture accordingly depending on somatosensory sensing by measuring the consumed current in each γ servo motor.

On inclined ground, the consumed current in the γ servo motor is higher than for a robot walking on flat ground. This is because when the robot walks uphill or downhill, the weight of the robot is not equally distributed on the legs, e.g. for uphill walking the rear legs carry more load than the front legs. Therefore, the consumed current in the affected joints will increase significantly during the stance phase.

Now, if the current consumption in the γ servo motor of the considered legs exceeds a predefined threshold during the stance phase for either both front legs or both rear legs, downward or upward inclination will be detected respectively. Denoting the consumed current for the γ servo of leg n in the stance phase at time t as $I_n(t)$ and the threshold as T_1 for downward inclination and T_2 for upward inclination, the inclination test reads:

$$\text{If } I_1(t) > T_1 \;\; \& \;\; I_0(t) > T_1 \Rightarrow \text{downhill walking} \tag{1}$$

$$\text{If } I_3(t) > T_2 \;\; \& \;\; I_4(t) > T_2 \Rightarrow \text{uphill walking} \tag{2}$$

Having successfully detected an inclination, the next section will discuss how to react to the inclination.

4. Reaction to Inclination

After the inclination detection the robot has to adjust its posture to maintain statically stable walking, moving the center of gravity inside the vertical projection of the support feet to avoid an overturning of the robot. If an uphill or downhill inclination is detected, the reaction step will change the front and back leg parameters. The legs, which detect the inclination, will update their position to expand more and lift up the downward facing part of the robot's body. The legs, which are on the opposite side, will similarly flex more. The decision function DI now updates the position of the β and γ servos for the legs BG which detect the inclination and the opposite legs $\overline{BG}$ as follows

$$DI = \begin{cases} \begin{cases} BG(t+1) = BG(t) + step \\ \overline{BG}(t+1) = \overline{BG}(t) - step \end{cases} & \text{if inclination detected} \\ \begin{cases} BG(t+1) = BG(t) \\ \overline{BG}(t+1) = \overline{BG}(t) \end{cases} & \text{else} \end{cases} \tag{3}$$

The decision process is run cyclically. In case the condition becomes false due to a dropping of the consumed current under the threshold, the leg keeps the last position and the robot continues its mission with its new body posture.

4.1. *Reflexes and Amputation*

In this work, we will further analyze the influence of disturbances, e.g. triggered reflexes and leg amputation, on the proposed inclination detection algorithm and vice versa. The elevator reflex serves as an example in the experiments. Here, the step height of a single leg will be increased when encountering an obstacle.[11] An obstacle is detected by evaluating the α servo current consumption. Obviously, the elevator reflex will be limited by a steep slope. Similarly, a triggered elevator reflex may affect the inclination detection.

Adaptive walking with leg amputation has been studied and tested on the OSCAR[2] and DLR[10] six-legged robots. Yet, it was focused on the walking on flat ground. In this paper, leg amputation will be simulated in software, i.e. a leg will be stiffened and moved on top of the body to minimize its effect on the remaining legs. In our experiment, as soon as the robot loses one of its legs, the servomotor speed will be decreased and the length of leg step will be shortened to ensure stable walking. This behaviour is also observed by walking insects.[10] Further, the robot reconfigures its legs depending on the inclination direction to maintain statically stable walking. Results of the influence of leg amputation on the inclination detection will be discussed at the end of the next section.

5. Experimental Results

The setup of the test ground consists of a board 240x50 cm indoor area with a 15° inclination. The board is covered with rough carpet. The robot is equipped with an inclination sensor as a reference to our adjustment of the body posture. The following experiments focus on the most common critical situations which maybe encountered by the robot during its locomotion in an unstructured environment. They comprise the uphill and downhill inclination detection as well as inclination detection with small obstacle avoidance by reflexes and leg failures, i.e. leg amputation.

5.1. *Uphill-Downhill Walking with Robot Body Adaptation*

The first experiment in Figs. 2 and 3 and shows how the robot detects inclination and direction and adjusts its posture according to the inclination direction. For uphill walking, the γ servo current consumption of the

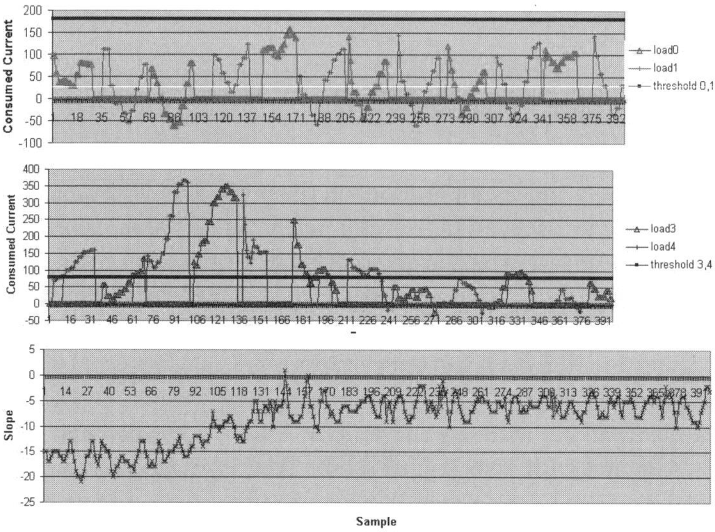

Fig. 2. Correction of an uphill inclination. Top: consumed current in front legs (0 and 1) during the stance phase. Center: consumed current in rear legs (3 and 4) during the stance phase. Bottom: reference inclination of the robot obtained from a tilt sensor.

rear legs exceed the threshold at times 11 and 61 respectively and is thus successfully detecting the upward inclination during the first stance phase. The body inclination is reduced from -15° to -5° at time 240, after which the current consumption of both rear legs drops below the threshold.

For downhill walking, the γ servo current consumption of the front legs exceeds the threshold at times 22 and 50 respectively and is thus again detecting the downward inclination during the first stance face. The body inclination is reduced from 15° to 6° at time 150. The current consumption of leg 0 drops below the threshold at time 135, after which the body posture is not further adjusted.

5.2. *Elevator Reflex*

Fig. 4 shows experimental results of the inclination detection algorithms with obstacles on the path with the robot walking downhill. The current consumption in α joint is monitored during the swing phase. As soon as the current consumption exceeds the predefined threshold an elevator reflex will be triggered to correct the leg trajectory. The left front leg will be disturbed by putting an obstacle in the way during its swing phase. The obstacle is detected at time 1184 when the current consumption in the α-joint exceeds

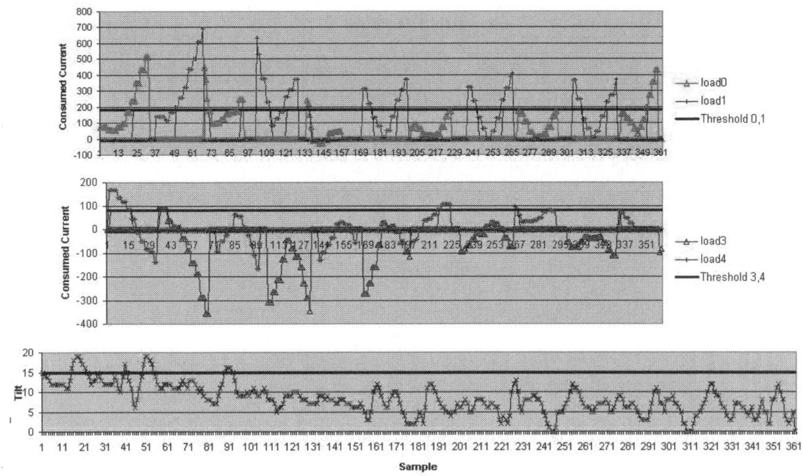

Fig. 3. Correction of an downhill inclination. Top: consumed current in front legs (0 and 1) during the stance phase. Center: consumed current in rear legs (3 and 4) during the stance phase. Bottom: reference inclination of the robot obtained from a tilt sensor.

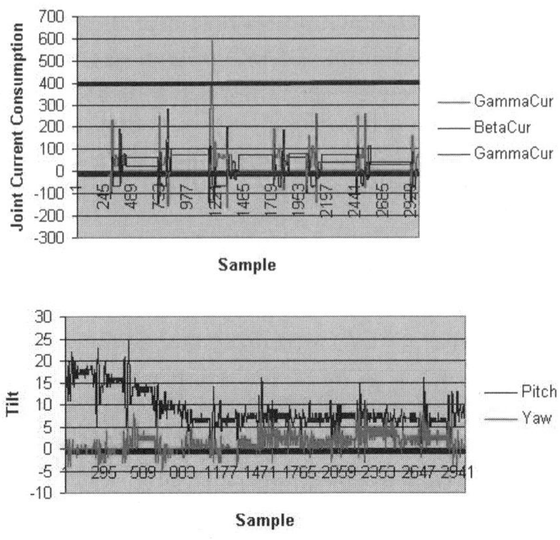

Fig. 4. Left: α-, β- and γ-joint current consumption for the leg on which the elevator reflex is triggered during downhill walking. Right: corresponding pitch and yaw readings from the reference sensor.

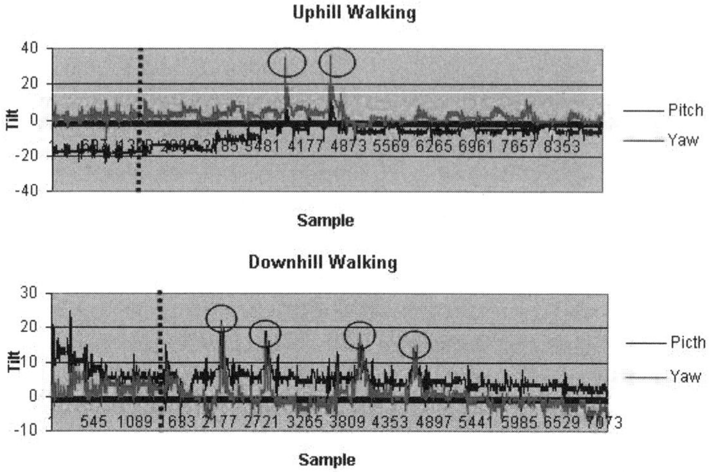

Fig. 5. Correction of an uphill (left) and downhill (right) inclination with amputated leg. Leg amputation is at data time 1497 for uphill walking and at data time 1449 for downhill walking.

the threshold of 400 triggering the elevator reflex. As shown by readings of the reference tilt sensor, the inclination detection is not disturbed by the reflex. The initial body inclination of 18° is reduced to 7° at time 900 and does not increase when the elevator reflex is performed.

5.3. *Leg Amputation*

Fig. 5 shows experimental results on the effect of amputation of the left middle leg. In case of leg amputation on a slope, the robot reconfigures its legs depending on the direction of the inclination. The left front and left rear legs shift their PEP and AEP toward the amputated legs to balance the robot's weight during walking and the right rear leg shifts its PEP and AEP in clockwise direction to enable the robot to walk stable during uphill walking and reducing the robot's tilting. Similarly, the remaining legs' AEP and PEP can be adjusted in downhill walking. In Fig. 5, the left middle leg is amputated at data time 1449 during downhill walking und at data time 1497 during uphill walking. The pitch signal presents the inclination in the walking direction (x) and the yaw signal presents the inclination in the lateral direction (y). Generally, also in case of leg amptutation, the ground inclination is successfully detected and the body inclination reduced to well below 10° as indicated by the pitch value of the reference sensor.

The remaining problem is therefore a tilting due to the loss of one middle leg. The robot tilts twice shortly after the amputation shown by the circles. Nevertheless, the robot maintains a stable walking afterwards. In case of downhill walking, the robot tilts slightly four times after the amputation and then maintains a stable walking.

6. Conclusion

This paper has been focused on improving the locomotion behaviour of a low-cost version of the hexapod robot OSCAR which is based on the Bioloid robot kit, through body posture adaptation on uneven terrain. Ground inclination is detected through monitoring the current consumption in the distal joint (γ-joint) in the front and rear legs during the stance phase. The test cases for adaptation to detected inclination have been focused on the two critical situations of uphill and downhill walking. The presented approach enables the robot to adapt its position and continue its mission without using a tilt sensor.

Further experiments have been conducted with disturbances along the robot path, i.e. small obstacles and leg failure. An elevator reflex is triggered by hitting an obstacle during the swing phase. It is detected by monitoring the current consumption in the α-joint for the front and middle legs and α- or γ-joint for the rear legs. Additionally, the amputation of the middle leg has been discussed and presented. The first results have demonstrated that the proposed inclination detection and body posture adaptation approach also work under such disturbances.

In the future we will focus on the extension of this approach to contain more cases of the robot walking on different terrain slopes and terrain types, i.e. sand or gravel.

7. Acknowledgments

This work was funded in part by the German Research Foundation (DFG) within priority programme 1183 under grant reference MA 1412/8-2 and the ALBAS University, Syria.

References

1. T. Akay, B. Ch. Ludwar, M. L. Göritz, J. Schmitz, and A. Büschges. Segment specificity of load signal processing depends on walking direction in the stick insect leg muscle control system. *The Journal of Neuroscience*, pages 3285–3294, 2007.

2. A. El Sayed Auf, F. Mösch, and M. Litza. How the six-legged walking machine oscar handles leg amputations. *From Animals to Animats*, 9, 2006.
3. W. Brockmann, E. Maehle, K. E. Grosspietsch, N. Rosemann, and B. Jakimovski. *ORCA: An Organic Robot Control Architecture*, pages 385–398. 2011.
4. H. Cruse. The control of the body position in the stick insect (carausis morosus), when walking over surfaces. *Biol. Cybernetics*, 24:25–33, 1976.
5. H. Cruse, J. Dean, V. Dürr, T. Kindermann, J. Schmitz, and M. Schumm. Control of hexapod walking - a decentralized solution based on biological data. *RTO MP-052*, 2000.
6. H. Cruse, D. Riemenschneider, and W. Stammer. Control of body position of a stick insect standing on uneven surfaces. *Biol. Cybernetics*, 61:71–77, 1989.
7. H. Cruse, J. Schmitz, D. U. Braun, and A. Schweins. Control of the body height in a stick insect walking on a tread wheel. *J. exp. Biol*, 181:141–155, 1993.
8. J. Duysens, F. Clarac, and H. Cruse. Load-regulating mechanisms in gait and posture: comparative aspects. *Physiology Rev*, 80:83–133, 2000.
9. B. Gaßmann, K.-U. Scholl, and K. Berns. Locomotion of LAURON III in rough terrain. In *Proceedings IEEE International Conference on Advanced Intelligent Mechatronics (ASME)*, pages 8–12, 2001.
10. M. Görner and G. Hirzinger. Analysis and evaluation of the stability of a biologically inspired, leg loss tolerant gait for six- and eight-legged walking robots. In *Proceedings IEEE International Conference on Robotics and Automation (ICRA)*, pages 1525–1531, 2010.
11. E. Maehle, W. Brockmann, K. E. Grosspietsch, A. El Sayed Auf, B. Jakimovski, S. Krannich, M. Litza, R. Maas, and A. Al-Homsy. *Application of the Organic Robot Control Architecture ORCA to the Six-legged Walking Robot OSCAR*, pages 385–398. 2011.
12. S. Steingrube, M. Timme, F. Woergoetter, and P. Manoonpong. Self-organized adaptation of simple neural circuits enables complex robot behavior. *Nature Physics*, 6:224–230, 2010.

COORDINATES MATTER: VIRTUAL CHASSIS AND MINIMUM PERTURBATION COORDINATE REPRESENTATIONS

D. ROLLINSON*, R. L. HATTON AND H. CHOSET

The Robotics Institute, Carnegie Mellon University,
Pittsburgh, PA 15213, USA
**E-mail: drollins@cs.cmu.edu*

Although any set of generalized coordinates can be used to represent the motion of a system, the choice of coordinates can greatly effect the ease of its analysis. Notably, the choice of body frame can affect a system's representational clarity as well as the computational accuracy of any linear approximations. In this paper we examine how departing from the typical convention of rigidly fixing a body frame to a system and instead carefully choosing the body frame can provide these benefits to a number of articulated locomoting systems, namely modular snake robots and planar swimming systems.

1. Introduction

In principle, any set of generalized coordinates can represent the motion of a system. In practice, the choice of coordinates can greatly effect the tractability and the ease of analysis of that system. For example, if the body frame of a car were fixed to one of its wheels, instead of its chassis, the mathematical expressions of the car's motion through the world would be significantly more complex and more difficult to understand. Similarly, if we consider a satellite floating in space, assigning a body frame to the center of mass simplifies the analysis because, by conservation of momentum, the origin cannot translate as a result of internal shape changes.

Recently we have explored this topic in two contexts. The first is with theoretically-driven planar locomoting systems using a framework that generalizes the center-of-mass analysis of inertially-isolated systems into a broader notion of *minimum-perturbation* coordinates.[1] The second is with real-world systems, such as snake robots,[2–5] that do not have a stable point on the body analogous to a wheeled vehicle's chassis. In this case, we calculate an averaged body frame to serve as a *virtual chassis*.[6]

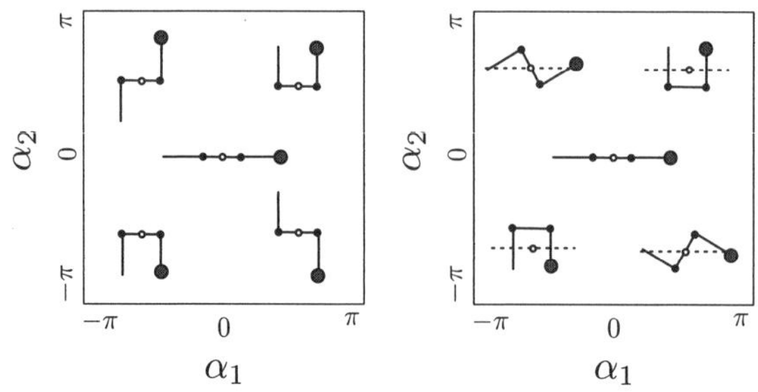

Fig. 1. Diagrams of the three-link swimming snake shown at zero orientation across corresponding shapes in a body frame fixed to the middle link (left) and a body frame defined by mean orientation (right).

Fig. 2. A montage showing a snake robot at 3 positions in a gait used for climbing poles (right) viewed in a frame fixed to the head module (top row) and a frame aligned with the virtual chassis (bottom row).

In this paper we unify the results from these two contexts, show that same underlying philosophies are at work, and show that our theoretically grounded results with minimum perturbation coordinates can help us to better understand the benefits of using the virtual chassis on a real system. Our philosophy is built on two subsequent ideas. The first is that the motion of a locomoting system should be studied primarily from the perspective of that system's body frame, as opposed to an external world frame. The second is that the choice of body frame has a significant effect on the mathematics and techniques that can be brought to bear in studying that system.

2. Coordinate Choice

The roots of our investigation of coordinate choice lie in the observation that the overall motion of an articulated system is often simpler and more intuitive than the motion any single part of that system. Based on this observation, we propose that a "good" body coordinate frame is one that represents a system's overall external motion independently from its internal shape changes. For articulated locomoting systems, this means departing from the typical convention of rigidly fixing a body frame to the system and instead defining it with respect to the system's constraints or overall shape.

A key result from our prior geometric work is that good choices of body frame are those for which the pose of the system fluctuates the least in response to internal changes in shape. For planar locomoting systems, this can be formalized by optimizing our frame such that differential changes in position induced by differential changes in shape as small as possible at all the points in our shape space. Specifically, this is accomplished using the *Hodge-Helmholtz Decomposition*,[7] which separates the vector fields that represent the system's motion into a conservative component and a divergence-free remainder. We term such frames *minimum perturbation* body frames.

For real-world systems, calculating the minimum perturbation coordinates is rarely feasible, since we lack a model for how the robot moves in response to its shape changes. However, experience and intuition have shown that the minimum perturbation coordinates are often close to those defined by the center of mass and mean orientation of the system. Therefore, for our lab's snake robots, we calculate the *virtual chassis* by continually aligning the body frame with the robot's principle moments of inertia. Specifically, this is done by taking the singular value decomposition (SVD) of the positions of all of the robot's links with respect to the robot's center of mass at discrete time steps as the robot executes the gait.

Unlike the minimum perturbation coordinates, the virtual chassis does not explicitly take into account the robot's interaction with the world. However, in practice it is a good approximation to how the robot's orientation changes in response to internal shape changes and we see benefits similar to those provided by the minimum perturbation coordinates. These benefits can be characterized as providing either better representational clarity or improved computational accuracy.

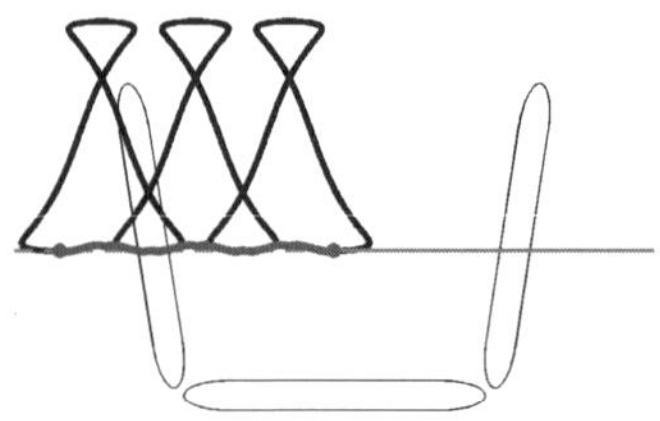

Fig. 3. The path traced by a three-link swimmer, represented in a body frame fixed to the middle link (black line) and a minimum perturbation body frame (red line). Note that the black path has been shifted vertically so that the start and stop locations are aligned.

3. Representational Clarity

The virtual chassis and minimum perturbation frameworks provide representational benefits that allow us to describe the motion of articulated systems in an intuitive and natural way.

An example of this improved representation involves the three-link planar swimming snake. Figure 1 shows a swimmer at zero orientation 5 different internal shape configurations. Unlike the middle-link body frame, the minimum perturbation representation takes into account the constraints of the swimmer, and as such it has the property that moving from the center shape to any of the 4 outer shapes would leave the system in a configuration whose pose is almost the same. Figure 1 shows the configuration of our lab's modular snake robot in 3 frames of the pole climbing gait. The robot's motion in the virtual chassis frame stays oriented with the pole, rather than the head when using a fixed frame. These effects come from some key traits of this representation:

(1) As an average point on the system, the body frame moves in a straight line with approximately constant velocity as the system executes gaits.
(2) Similarly, the body frame rotates very little, meaning that as the robot executes a gait, the direction it translates per cycle is independent of the starting phase. Together with the previous benefit, this means that executing a partial cycle of a gait moves the system a proportional fraction of the motion over a full gait cycle.
(3) The swept volume of the robot's body over the range of shapes corresponding to a given pose is minimized.

The first and second traits are highlighted by the example of the three-link swimmer in Fig. 3, which shows the paths of the exact same motion

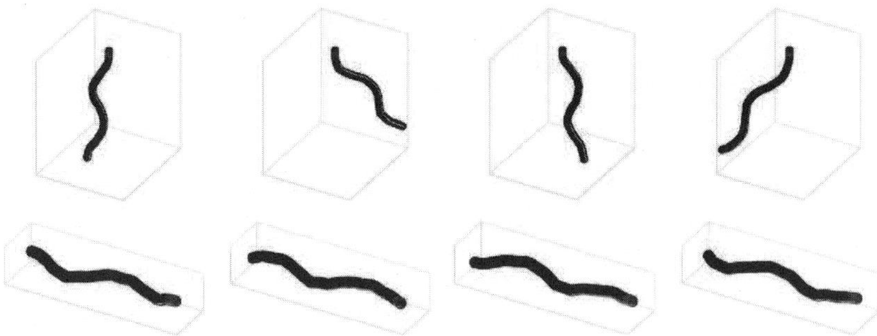

Fig. 4. A montage of the robot in sidewinding, shown in different positions from one gait cycle. The top row of images show the pose of the robot in a body frame fixed to the head link. The bottom row shows the pose corresponding to the virtual chassis body fame at the same points in the gait cycle.

viewed in different body frames. The pose represented by the minimum perturbation body frame changes monotonically towards its final value, without the large "wind up" and "power stroke" seen when a body frame is fixed to the middle link.

For real-world snake robots, we benefit primarily from the third trait. By applying the virtual chassis to the wide range of gaits we have developed, we have found that the orientation of the robot viewed in the virtual chassis corresponds closely to the way the robot interacts with the environment. Figure 3 shows the motion of the snake robot executing the sidewinding gait in a frame fixed to the head module and a frame aligned with the virtual chassis.

4. Computational Accuracy

Working in minimum perturbation coordinates, or a practical implementation of them, also offers benefits with regard to the accuracy of computations that rely on linearizing of the system dynamics. Almost all interesting locomoting systems are non-linear, but the degree of non-linearity depends heavily on the system's coordinate representation. Because of this, the same traits from the previous section that provide representational benefits also provide computational benefits.

In simple planar systems, using minimum perturbation coordinates allows us to easily identify effective motions by identifying the net displacement produced by different gaits with area integrals of the system dynamics over regions of the shape space.[1] These area integrals require linearization of the dynamics, and so were previously only applicable to small-scale mo-

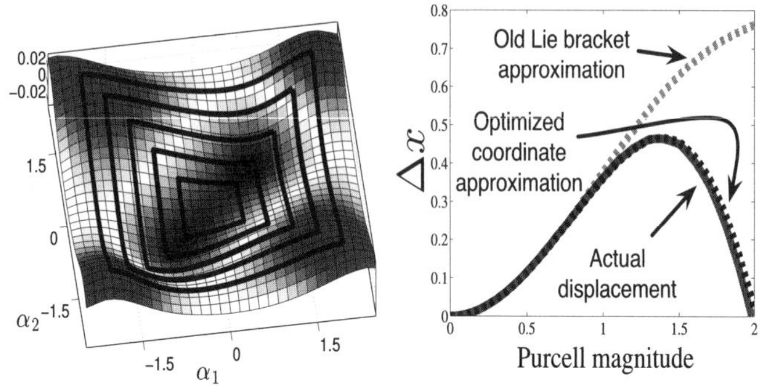

Fig. 5. Using the minimum perturbation coordinates greatly improves the accuracy of approximating the motion of a box stroke using the body velocity integral. Left, strokes of increasing magnitude in the shape space of the swimmer. Right, the displacement of the stroke shown in red, along with the approximation using a fixed frame (gray dashed) and minimum perturbation frame (black dashed).

tions, limiting their use in identifying maximally effective motions. Figure 4 shows the improved accuracy gained by using the minimum perturbation coordinates to suppress the nonlinearities of the three-link swimmer executing a box stroke. When using a minimum perturbation coordinate body frame, the body velocity integral remains an accurate approximation of the swimmer's true displacement, even for large magnitude strokes.

For snake robots, the virtual chassis allows us to fuse the redundant proprioceptive data distributed throughout the robot into a state estimate for pose using an extended Kalman filter (EKF).[8] One of the challenges of this technique is determining a process model for how the robot's pose changes over time. Models that are highly non-linear are problematic for an EKF because the linearization that the filter performs at each timestep becomes a poor approximation of the true system.[9] Working in the virtual chassis body frame not only improves our linearization, it allows a much simpler process model to be used in the first place. Because a true model of how the robot moves in the world is unknown, a constant velocity model is used as the prediction of body frame angular velocity between filter updates.

Normally, this model would be a poor predictor of the robot's motion, especially when the robot stops, starts, or changes its motions. Figure 4 shows the body frame angular velocities as the robot begins at rest, starts rolling rapidly, and then stops. This example shows how the dramatically smoother motion of the robot in the virtual chassis better matches the model's constant-velocity assumption, enabling the EKF to accurately esti-

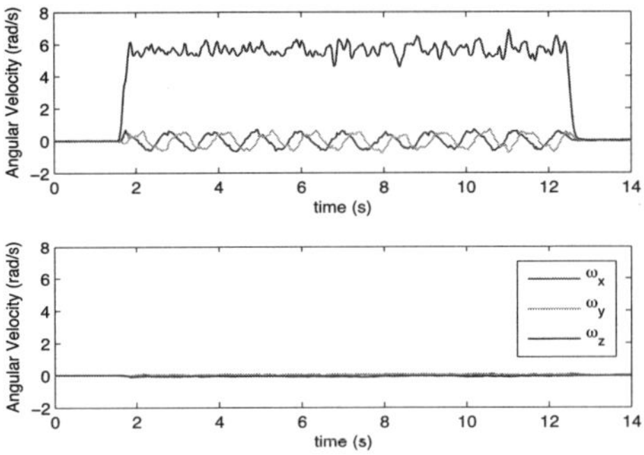

Fig. 6. A comparison of estimated angular velocities for the rolling gait, observed in a body frame fixed to the middle link (upper plot) and a body frame aligned with the virtual chassis (lower plot). The red, green and blue lines are respectively the x, y and z body frame angular velocities.

mate the robot's pose even during highly dynamic motions. This is because the virtual chassis separates the robot's internal shape changes (which are highly dynamic, but accurately measured with the robot's joint encoders) from it's external motions (which are less dynamic, and approximated with the constant velocity process model).

When using the virtual chassis the state estimator is also more robust to sensor errors, and far less sensitive to the filter's tuning parameters. Furthermore, the generality of the constant velocity model has allowed it to be successfully applied to a range of different gaits, including slithering, sidewinding, pipe crawling, and pole climbing.

5. Discussion

We have demonstrated the importance of coordinate choice in two domains of our lab's research, and shown that in each area the same underlying philosophies are at work. In particular, we have shown that when studying a locomoting system there are benefits from using a body frame that captures the system's overall motion, either by taking into account the system's dynamics or by simply averaging the system's kinematic configuration.

Using the virtual chassis lets us begin to treat articulated mobile robots more like wheeled robots, where models of the robot's motion in the world are much simpler and intuitive. For our lab's modular snake robots, we are currently working to build generic motion models using only knowledge of the robot's internal shape changes.[10] This is accomplished by averaging the movement of modules on the bottom of the robot, as defined by the flattest principle direction of the robot's shape in the virtual chassis. As a tool for designing new motions for the robot, the virtual chassis allows us to easily observe the macroscopic motion of the robot, without having to iteratively test the motions on a real robot or in physical simulator that models ground contact.

We also believe the virtual chassis is applicable to many systems beyond snake robots and planar locomoting systems. In general, we feel that any system that lacks a fixed point that is representative the system's overall motion can benefit from using the virtual chassis.

References

1. R. L. Hatton and H. Choset, "Geometric motion planning: The local connection, Stokes's theorem, and the importance of coordinate choice," *The International Journal of Robotics Research*, June 2011.
2. S. Hirose, *Biologically Inspired Robots.* Oxford University Press, 1993.
3. G. Chirikjian and J. Burdick, "The kinematics of hyper-redundant robot locomotion," *IEEE Transactions on Robotics and Automation*, vol. 11, no. 6, pp. 781–793, 1995.
4. A. Transeth, R. Leine, C. Glocker, and K. Pettersen, "3-D Snake Robot Motion: Nonsmooth Modeling, Simulations, and Experiments," *IEEE Transactions on Robotics*, vol. 24, pp. 361–376, Apr. 2008.
5. C. Wright, A. Buchan, B. Brown, J. Geist, M. Schwerin, D. Rollinson, M. Tesch, and H. Choset, "Design and Architecture of the Unified Modular Snake Robot," in *IEEE International Conference on Robotics and Automation (accepted)*, 2012.
6. D. Rollinson and H. Choset, "Virtual Chassis for Snake Robots," *IEEE International Conference on Intelligent Robots and Systems*, 2011.
7. G. B. Arfken, H. J. Weber, and F. Harris, *Mathematical Methods for Physicists, Sixth Edition: A Comprehensive Guide.* Academic Press, 6 ed., July 2005.
8. D. Rollinson, A. Buchan, and H. Choset, "State Estimation for Snake Robots," *IEEE International Conference on Intelligent Robots and Systems*, 2011.
9. S. Thrun, W. Burgard, and D. Fox, *Probabilistic Robotics.* MIT Press, 2005.
10. F. Enner, D. Rollinson, and H. Choset, "Simplified Motion Modeling for Snake Robots," in *IEEE International Conference on Robotics and Automation (accepted)*, 2012.

USING PNEUMATIC ARTIFICIAL MUSCLE IN DEVELOPMENT OF IN-PIPE INSPECTION ROBOT FOR NARROW PIPES BASED ON AN EARTHWORM'S PERISTALTIC CRAWLING MOTION

MEGUMI IKEUCHI, SHOTA HORII, DAI MATSUBARA and TARO NAKAMURA

Faculty of Science and Engineering, Department of Precision Mechanics
Chuo University
1-13-27 Kasuga, Bunkyo-ku, Tokyo 112-8551, Japan

In this study, we developed a robot that mimics the peristaltic crawling motion of earthworms as a locomotion mechanism. This robot consists of five units, each of which has an artificial muscle that is driven by pneumatics. In addition, we performed two experiments in a 16 mm diameter acrylic pipe. These experiments were designed to examine the relationship between locomotion speed and motion patterns. The other is to examine the relationship between traction force and motion patterns.

1. Introduction

Recently, many pipe related accidents have been reported. Therefore, in-pipe inspection is important to prevent accidents caused by the bursting of pipes or corrosion. Currently, fiberscopes are used for the inspection of pipes. However, they cannot be inserted into pipes that are longer than 15 m or complex pipes such as elbow pipes because of friction. Therefore, in-pipe inspection robots are required to be self-propelled inside the pipes. Consequently, in-pipe inspection robots were developed, such as the one that uses an active scope camera driven by a ciliary vibration mechanism[1] and another whose motion is similar to that of a snake[2]. However, these robots have some drawbacks, in that the former cannot go backward and the latter needs a space that is wider than itself. Wheel-type robots[3-5] have been developed, but their bodies are not easily miniaturized because of their complex structures. In addition, robots that can move in narrow pipes such as the 15A, which has an inner diameter of approximately 16 mm, have not yet been reported. On the other hand, the peristaltic crawling robot exhibits these features. It has a simple mechanism for movement and can easily move backward and forward. There is a space inside it into which an endoscope can be mounted. In addition, it can continue to propel itself even if the pipe's diameter changes.

Therefore, we have developed several peristaltic crawling robots[6-10]. However, it is difficult to downsize the body of the robot using servo motors because of the nature of its mechanism. We are currently developing a robot that is capable of propelling itself through various pipes such as long pipes and elbow pipes, specifically, a peristaltic crawling robot using a pneumatic artificial muscle for in-pipe inspection.

In this study, we developed a peristaltic crawling robot for the in-pipe inspection of the pipe 15A. We derived theoretical formulas for the robot's locomotion speed. In addition, we performed experiments with the robot to examine the locomotion speed, and compared the experimental results and theoretical values. Moreover, we also performed experiments to examine the traction force.

2. Peristaltic Crawling Pattern of an Earthworm

An earthworm moves by peristaltic crawling, in which extension and contraction waves are propagated along the anteroposterior direction by varying the thickness and length of the segments. Figure 1 shows the peristaltic crawling pattern of an earthworm. First, the earthworm contracts its anterior segments, thereby increasing friction between the surface and the segments. This is because the thicker segments are in contact with the surface during locomotion, while contraction propagates toward the rear. This movement pulls the rear segments in the direction of the movement. The anterior segments of the earthworm are extended in the axial direction after contraction.

Figure 2 shows an analysis of the peristaltic crawling motion of an actual earthworm whose total length was 137 mm. The earthworm's motion was analyzed using movement analysis software (MOVIAS Pro: NAC Image Technology, Inc.). Marks were made on the earthworm's body and these were captured on video. The figure shows that contraction waves propagate through the entire body from the head to the tail.

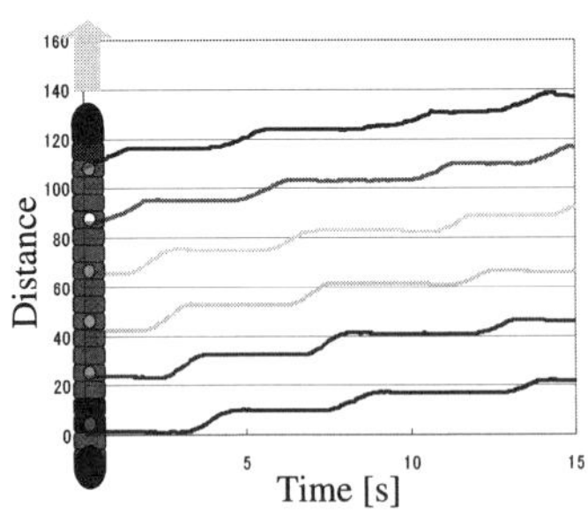

Figure 2. Experimental result of the analysis of an actual earthworm's movement

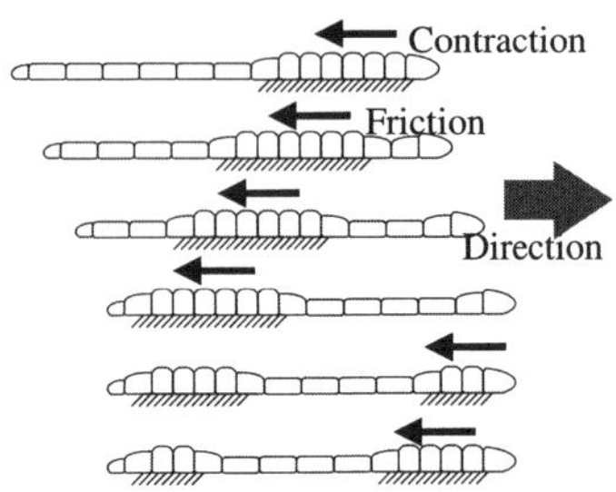

Figure 1. Peristaltic crawling pattern of an earthworm

3. In-Pipe Inspection Robot

3.1. *Overview of the Robot*

Figure 3 shows an overview of the pipe inspection robot. This robot was developed by implementing an earthworm's actual motion. This robot comprises five units. One unit of the robot is equal to a segment of an actual earthworm. The robot weighs 9.8 g and each of length 190.5 mm when fully extended. The structure of the unit is shown in Figure 4 and its specifications are given in Table 1. This unit is composed of a spring-rolled aluminum sheet, two flanges, and artificial muscles. When air is inhaled into the chamber, the artificial muscle expands in the radial direction and contracts in the axial direction. When the air is discharged from the chamber, the artificial muscle extends in the axial direction. In this way, a motion similar to that of an earthworm can be realized by repeated the extension and contraction motion.

Figure 3. Developed peristaltic crawling robot

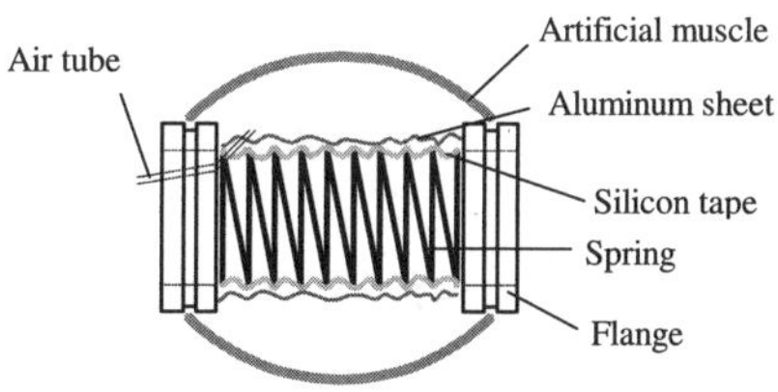

Figure 4. The structure of the unit

Table 1. Specification of the unit

Weight [g]		1.8
Length [mm]	Extension	38.1
	Contraction	33.9
Width [mm]	Extension	11.0
	Contraction	21.1

3.2. *Pneumatic Artificial Muscle*

Figure 5 shows an overview of the pneumatic artificial muscle. The artificial muscle is composed of micro-carbon fiber and low ammonia natural rubber latex. Because carbon fiber was placed in the axial direction, the rubber expands but the fibers do not extend when air pressure is supplied to the artificial muscle. As a result, the artificial muscle contracts in the axial direction and expands in the radial direction. The advantages of an artificial muscle are that it is lightweight and inexpensive.

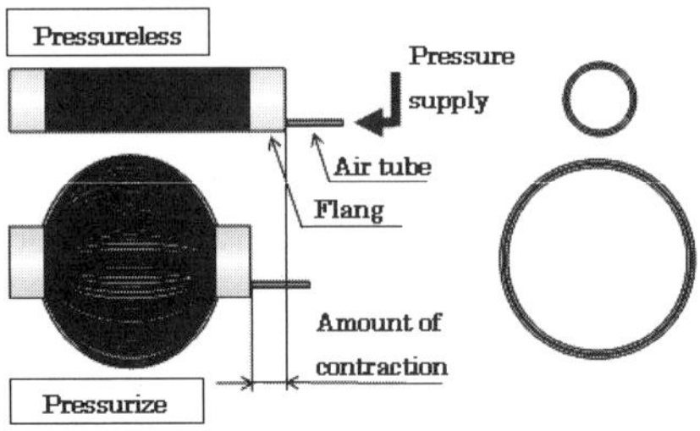

Figure 5. Schematic diagram of an artificial rubber muscle

3.3. *Method of Operation*

The robot is controlled using a microcomputer (H8-3052F). The digital signal that is the output from the H8 microcomputer is converted into an analog signal by a D / A converter. The analog signal is inputted into the proportional solenoid valve. The air pressure that is supplied to the robot through the manifold from the compressor is proportional to the input voltage. Therefore, the units will be able to contract and extend independently.

4. Motion Patterns

By altering the extension or contraction of each unit can achieve several motion patterns. Motion patterns are determined by the wavelength, propagation speed, and number of waves. Here the wavelength is the number of units that are extended in the axial direction. The propagation speed is the number of units that are propagated at rear. For example, Motion3-1-1 (wavelength-propagation speed-number of the waves) is shown in Figure 6.

In addition, each parameter needs to satisfy the following equations:

$$(l+s)\times n < N \tag{1}$$

$$N-l > s \tag{2}$$

$$l \geq s \tag{3}$$

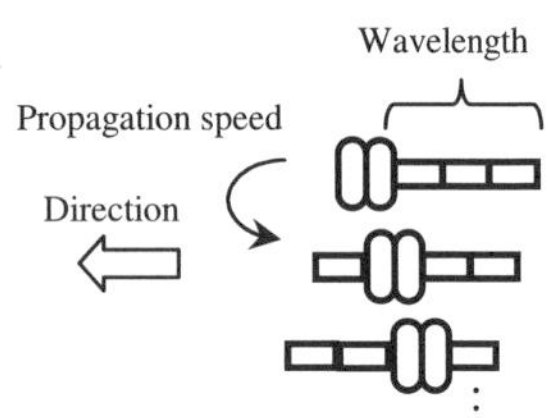

Figure 6. Motion pattern 3-1-1

Here l, s, n, and N are the wavelength, propagation speed, number of waves, and number of units, respectively.

5. Theoretical Model

First, we derive the theoretical locomotion speed of several basic motion patterns that were considered. In each basic motion pattern, the distance progressed by the robot and the time required for a wave to propagate from the head to the tail are d_B and T_B, respectively. We express d_B and T_B as

$$d_B = rnl, \tag{4}$$

$$T_B = \frac{Nt}{s}. \tag{5}$$

Here r is the amount of contraction and t is the time taken by each unit. From (4) and (5), we express the theoretical locomotion speed of the basic motion pattern v_B as

$$v_B = \frac{d_B}{T_B} = \frac{rnsl}{Nt} \tag{6}$$

The theoretical values of the locomotion speed obtained from (6) are shown in Figure 7. We can see that the locomotion speed increases with increasing wavelength, propagation speed, or number of waves. The maximum locomotion speed of the robot is the value of motion, 2-2-1 with a time interval of 0.1 s. In addition, from (6), the amount of contraction depends on the locomotion speed. Moreover, the units can be arranged roughly because the distance between the units does not depend on the locomotion speed. In addition, if one of the units breaks down, the broken unit is considered to be the distance between the units, and the robot can still move.

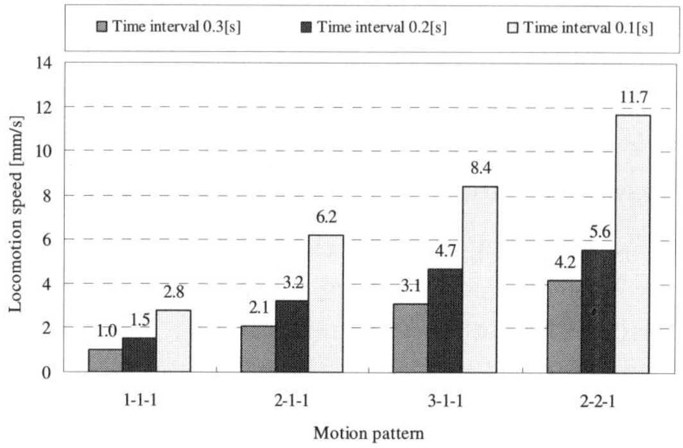

Figure 7. Theoretical locomotion speed of the robot

6. Experimental Results and Discussion

6.1. *Locomotion Speed of the Robot*

We tested the developed robot and measured the locomotion speed for each motion pattern. These experiments were performed using an acrylic pipe that was 16 mm in diameter. In addition, an air pressure of 0.09 MPa was applied to the unit. The relationship between the locomotion speed and motion patterns is shown in Figure 8. From this figure, we observe that the maximum speed of the robot is 7.8 mm/s with a time interval of 0.1 s and Motion3-1-1. We also observe that the locomotion speed increases with increasing wavelength when the propagation speed is 1.

However, the experimental values are less than the theoretical values for the Motion 2-2-1 with a time interval of 0.1 s. This is explained as follows. When the propagation speed is 2, the time taken for the units to expand is short. Because the supply of air to the unit cannot be maintained, sufficient expansion of the unit cannot be realized. The robot will then slip to the rear. This is the reason that the experimental value of the locomotion speed is lesser than the theoretical value.

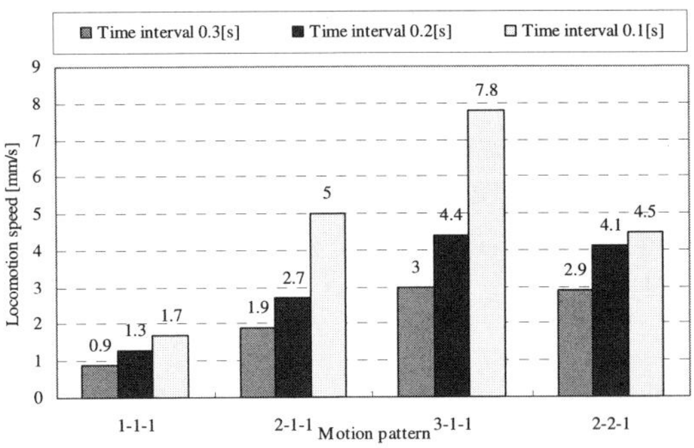

Figure 8. Relationship between the locomotion speed and motion patterns

6.2. *Traction Force of the Robot*

The traction force is important because the robot pulls its own air tube as it moves. The traction force of the developed robot was measured for each motion pattern using a load cell in the experimental setup shown in Figure 9. The relationship between traction force and motion patterns is shown in Figure 10. From this figure, we observe that the maximum traction force is 12.16 N in a time interval of 0.3 s and Motion1-1-1, and that the traction force decreases with increasing wavelength.

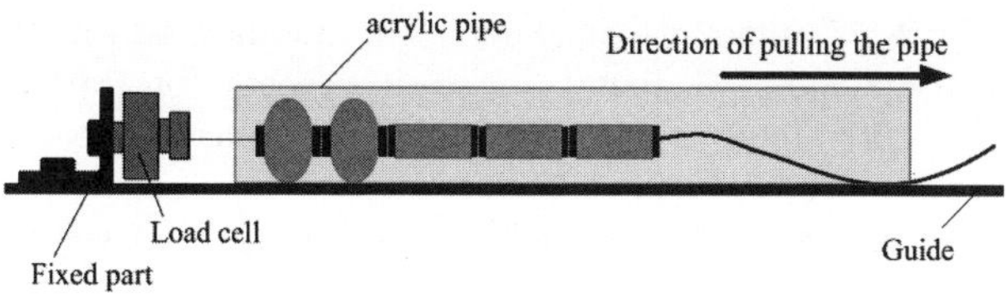

Figure 9. Experimental setup for measuring the traction force

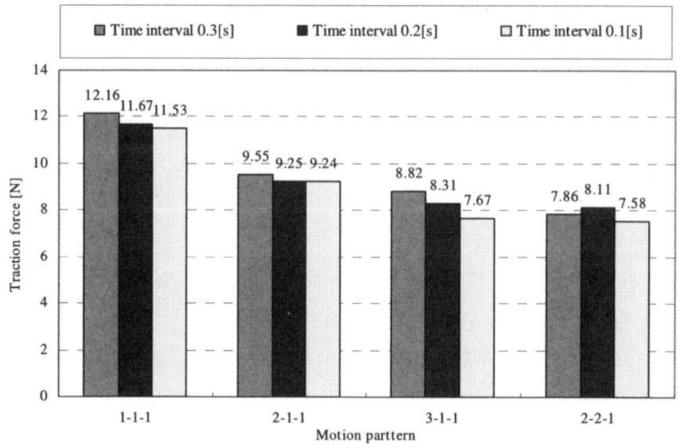

Figure 10. Relationship between traction force and motion patterns

7. Conclusion

For an industrial endoscope we developed an in-pipe inspection robot based on an earthworm's peristaltic crawling using pneumatic artificial muscles. The conclusions are as follows.

1. The robot can pass through a pipe having an inner diameter of 16 mm.
2. We derived theoretical formulas describing the robot's locomotion speed.
3. We examined the relationship between the locomotion speed and motion pattern and found that the maximum locomotion speed of the robot is 7.8 mm/s for motion 3-1-1 with a time interval of 0.1 s.
4. We examined the traction force of the robot and observed that the maximum traction force is 12.16 N in a time interval of 0.3 s for Motion1-1-1

In future, the robot will be improved to increase the maximum locomotion speed.

References

1. M. Konyo, K. Hatazaki, K. Isaki, and S. Tadokoro (Tohoku University) "Development of an Active Scopecamera Driven by Ciliary Vibration Mechanism," Proc. of the 12th ROBOTICS symposia, pp.460-465, 2007.

2. A. Kuwada, K. Tsujino, K. Suzumori, T. Kawada (Okayama University) "Intelligent Actuators Realizing Snake–like Robot for Pipe Inspection," Proc. IEEE 2006 International Symposium on Micro-nano Mechatronics and Human Science, 2006-11.
3. Peng Li, Shugen Ma, Bin Li, and Yuechao Wang, "Development of an Adaptive Mobile Robot for In-pipe Inspection Task," Proc. IEEE International Conference on Mechatronics and Automation, pp. 3622-3627, 2007
4. Tokuji Okada and Tsuyoshi Sanemori, "MOGER: A Vehicle Study and Realization for In-pipe Inspection Tasks", IEEE Journal of Robotics and Automation, Vol. RA-3, No. 6, December 1987
5. Amir H. Heidari, M. Mehrandezh, R. Paranjape and H. Najjaran, "Dynamic Analysis and Human Analogous Control of a Pipe Crawling Robot," Proc. IEEE/RSJ International Conference on Intelligent Robots and Systems, pp. 733-740, 2009
6. Taro Nakamura, Takashi Kato, Tomohide Iwanaga and Yoichi Muranaka, "Development of a Peristaltic Crawling Robot Using Servo Motors Based on the Locomotion Mechanism of Earthworms", Proceedings of IEEE International Conference on Robotics and Automation（ICRA 2006）, pp.4342-4344, 2006
7. Taro Nakamura and Tomohide Iwanaga, "Locomotion Strategy for a Peristaltic Crawling Robot in a 2-Dimensional Space", Proc. IEEE International Conference on Robotics and Automation, 2008
8. Taro Nakamura, Takashi Kato, Tomohide Iwanaga and Yoichi Muranaka, "Development of a Peristaltic Crawling Robot Using Servo Motors Based on the Locomotion Mechanism of Earthworms", Proceedings of IEEE International Conference on Robotics and Automation（ICRA 2006）, pp.4342-4344, 2006
9. Hayato Omori, Takeshi Hayakawa and Taro Nakamura, "Locomotion and Turning Patterns of a Peristaltic crawling Earthworm Robot Composed of Flexible Units", Proceedings of IEEE/RSJ 2008 International Conference on Intelligent Robots and Systems, pp.1630-1635, 2008
10. S. Horii, T. Nakamura, and K. Aoki "Development of an in-pipe mobile robot for use as an industrial endoscope based on an earthworm's peristaltic crawling" Proc. the 13th International Conference on Climbing and Walking Robots and the Support Technologies for Mobile Machines, pp. 141-149, 2010.

PATH PLANNING WITH ROBOT'S TRANSFORMATION FOR TRAVELING-WAVE-TYPE MOBILE ROBOT (TORO-IV)

TATSUYA OSAWA, TERUYOSHI OGAWA, and TARO NAKAMURA

Faculty of Science and Engineering, Department of Precision Mechanics
Chuo University
1-13-27 Kasuga, Bunkyo-ku, Tokyo 112-8551, Japan

An omnidirectional movement mechanism that can move a robot in a narrow complicated passage is needed. However, existing mechanisms cannot achieve stable motions. We noted that a snail uses traveling waves for moving and can achieve stable motion because of a large contact area. Therefore, we developed a traveling-wave-type omnidirectional mobile robot (TORo-IV) using a snail's locomotion mechanism. In this study, we propose a path planning method that allows the robots to be transformed. Furthermore, we confirmed the usefulness of this proposed method through experiments.

1. Introduction

Recently, many robots that can operate in narrow, indoor spaces such as those found in factories and hospitals have been developed. They require an omnidirectional movement mechanism that does not require a large space to turn. The conventional approach is to use an omniwheel to achieve this omnidirectional movement[1]. However, an omniwheel cannot achieve sufficient frictional forces with the ground using a mechanism that slides in the axial direction and contacts the ground at a point. On the other hand, a snail's locomotion mechanism uses traveling waves[2, 3]. A snail moves forward by propagating traveling waves from its tail to its head. An advantage of this movement mechanism is that it enables stable motion because of a large contact area. Therefore, we previously developed the traveling-wave-type omnidirectional mobile robot (TORo)[4-7], which uses the locomotion mechanism of a snail. Moreover, this robot can move in all directions and can change its form from being circular to straight. In the circular form, this robot can move in all directions and when in the straight form, it can move in narrow spaces. We believe that the robot can move more efficiently when combining movements both in the circular and straight forms.

In this study, we propose a path planning method that is based on a potential method[8, 9] and allows the robot to be transformed. Furthermore, we confirmed the usefulness of this proposed method by performing experiments.

2. The locomotion mechanism of the snail

The common belief is that the snail moves by a traveling wave locomotion mechanism. This produces a traveling wave in the gastropod's foot that propagates in the direction of locomotion[3]. The state of the traveling wave when moving forward is shown in Figure 1. When the traveling wave occurs, the part that is not forming the traveling wave in the gastropod's foot remains in contact with the ground. The ground generates a frictional force with the grounded foot of the gastropod.

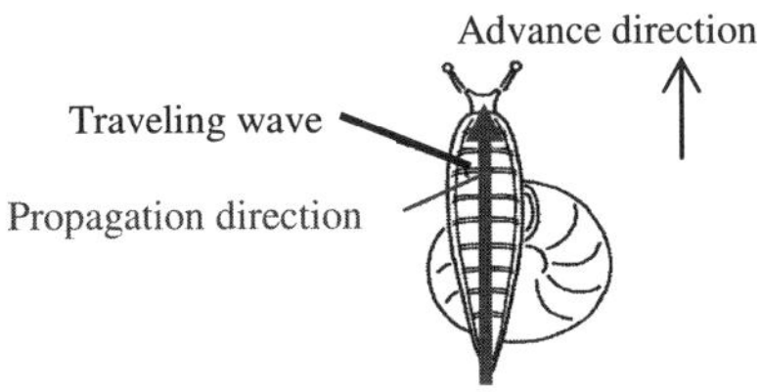

Figure 1. Propagation of traveling wave[3]

3. Traveling-wave-type omnidirectional mobile robot (TORo-IV)

3.1. *Overview*

TORo-IV is shown in Figure 2 and its specifications are shown in Table 1. As shown in Figure 1, the structure of the robot consists of eight units.

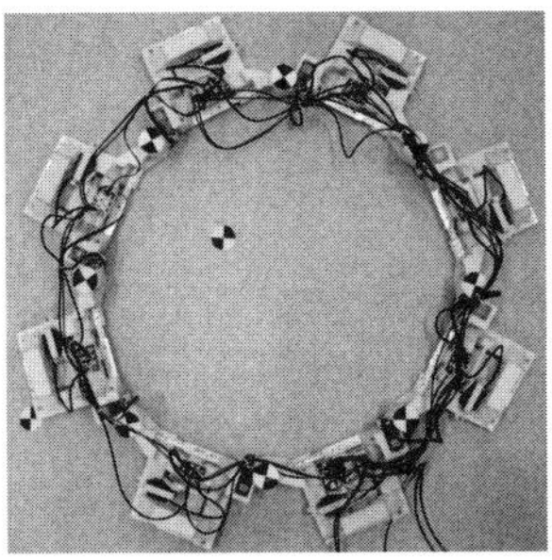

Figure 2. TORo-IV

Table 1 Specifications of TORo-IV

Maximum diameter [mm]	794
Weight [kg]	3.9
Number of unit	8

3.2. *Mechanism of the unit*

Figure 3 shows the design of the units. The unit has two servo motors, denoted as servomotor I and servomotor II. Servomotor I can control the contraction and extension of the unit. In addition, a friction plate moves up and down as the unit extends and contracts, respectively. During extension, the friction plate makes contact with the ground and the unit is fixed to the ground. During contraction, the friction plate leaves the ground and the unit can be easily moved. Servomotor II can control the angle between the units.

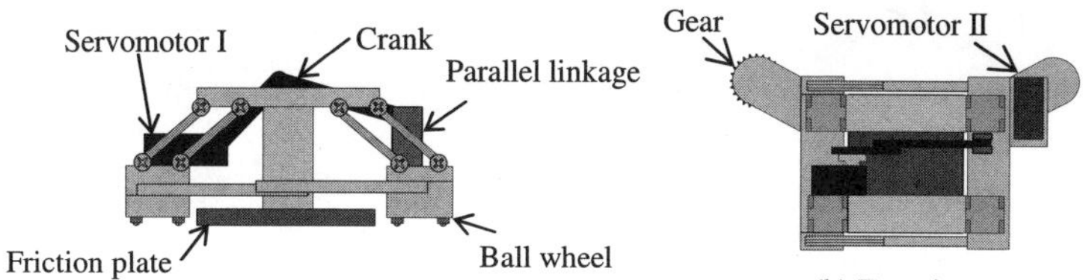

(a) Side view (b) Top view

Figure 3. Mechanism of the unit

3.3. *Modeling the shape of the robot*

We modeled the shape of the robot as shown in Figure 4. The robot can be modeled in an octagonal shape consisting of contraction links and joints. The shape of the robot can be expressed by solving the coordinates of the joints by kinematics, as Eq. (1):

$$\mathbf{P}_i = \begin{bmatrix} x_i \\ y_i \end{bmatrix} = \begin{bmatrix} x_{i-1} + L_{i-1}\cos(\theta_{def} + \sum_{j=0}^{i-1}\theta_j) \\ y_{i-1} + L_{i-1}\sin(\theta_{def} + \sum_{j=0}^{i-1}\theta_j) \end{bmatrix} \quad (1)$$

$$\left(\begin{array}{l} \mathbf{P}_0 = \begin{bmatrix} x_0 \\ y_0 \end{bmatrix}, \ \mathbf{P}_1 = \begin{bmatrix} x_1 \\ y_1 \end{bmatrix} \\ \theta_{def} = \mathrm{arctan2}\{(y_1 - y_0), (x_1 - x_0)\} \\ i = 2,3,\cdots, N-1 \end{array} \right)$$

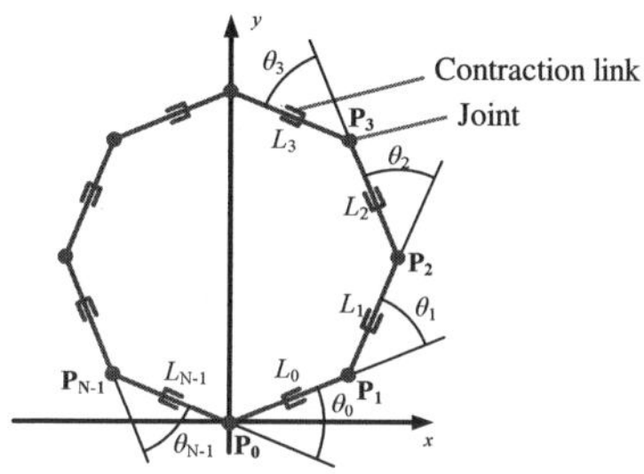

Figure 4. Modeled shape of TORo-IV

Here, θ is the angle between the units, L is the length of each unit, $\mathbf{P}(x,y)$ is coordinate of a joint, and N is the number of units in the robot.

3.4. *Robot's motion*

An imaginary wave propagation line is used to realize the movement of this robot. If the wave propagation line and unit overlap, then the unit moves in the direction of the wave propagation line, as shown in Figure 5. By changing the direction of the wave propagation, TORo-IV can move in arbitrary directions[7].

In addition, this robot can change from a circular form, which enables omnidirectional movement, to a straight form, which allows movement in narrow spaces, as shown in Figure 6.

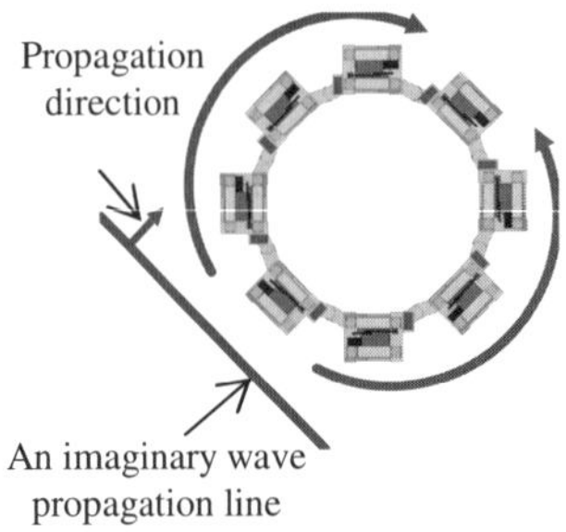

Figure 5. Motion of the robot

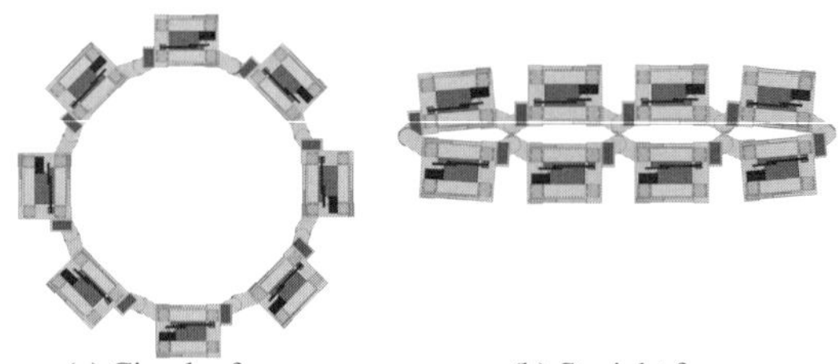

(a) Circular form (b) Straight form

Figure 6. Form of the robot

4. Path planning

4.1. *Path planning based on potential method*

We generated a path that is based on the potential method and that allows the robot to move while avoiding obstacles. In this study, obstacles are treated as points. First, we generated a virtual potential field. The potential field consists of two potential elements, namely attractive potential and repulsive potential.

The attractive potential is the potential for the robot to be moved to the destination. The attractive potential U_a is expressed as Eq. (2).

$$U_a = K_1 d \tag{2}$$

Here U_a is the attractive potential, K_1 is the positive coefficient, and d is the distance to the destination. The repulsive potential is the potential that causes the robot to avoid the obstacles. The repulsive potential U_r is expressed as Eq. (3).

$$U_r = \sum_{j=1}^{N_0} \frac{K_2 D}{\left| P_{obj} + P_g - 2P_c \right|} \tag{3}$$

Here P_{ob} is the position of the obstacle, P_g is the position of the gauge point of the robot, P_c is the position of the center of the robot, D is the width of the robot, N_0 is the number of obstacles, and K_2 is the positive coefficient. From the potential field, we define the desired path of the robot. The desired path is determined by the gradient of the potential field. The discretized desired path T is expressed as Eq. (4).

$$\boldsymbol{T}_{di} = T_{di-1} + \begin{bmatrix} \dfrac{\partial (U_{\mathrm{a}} + U_{\mathrm{r}})}{\partial x} \\ \dfrac{\partial (U_{\mathrm{a}} + U_{\mathrm{r}})}{\partial y} \end{bmatrix} \tag{4}$$

The discretized path T is connected along the range of the robot's motion. Then the connected path is adjusted to be aligned with the direction of the robot's motion, and its size becomes the robot's migration length.

4.2. *Path planning with robot's transformation*

We generated a path with robot's transformation. An example is shown in Figure 7. First, we determine the basic path using the potential method and the width when in the straight form. Next, we identify the portion for which the distance traveled with an obstacle is smaller than the width when in the circular form. The robot cannot traverse this portion while in the circular form. In addition, we search for the point in this portion that has the shortest distance with an obstacle, and consider this as the point with the most neighbors. Two points are used to determine a position that does not interfere with obstacles in a direction that is perpendicular to the line connecting the obstacle to the point having the most neighbors. This point is considered as a subgoal. The space between the two subgoals is made into a straight line, and the robot progresses in the straight form. Between the starting point and the first subgoal, and between the second subgoal and the destination, the robot progresses in the circular form and the potential method is used with the width being that of the circle form. In addition, the robot undergoes a transformation at the subgoal. Then, we determine the path for which the robot transforms into the straight form in a narrow passage.

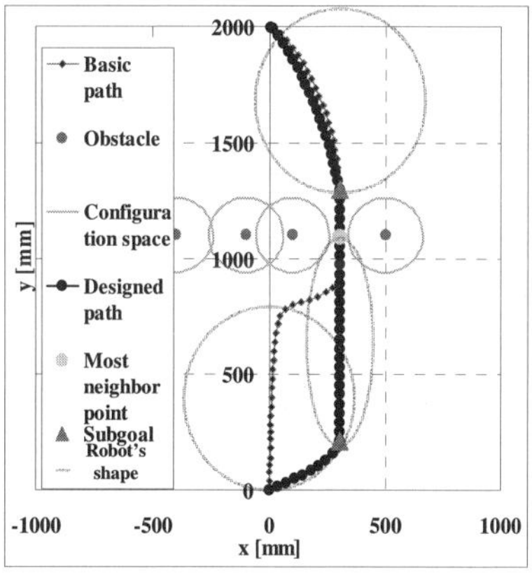

Figure 7. Designed path

5. Locomotion experiment

5.1. *Path planning experiment*

We planned a path using the proposed method. Our path planning experiments were performed only for the circular form and in the case involving transformation within the same environment. The environment is shown in Figure 8 and Table 2. We set obstacles along a line between the start and the destination. The path planning results when only in the circular form and when the transformation is included are shown in Figures 9 and10, respectively.

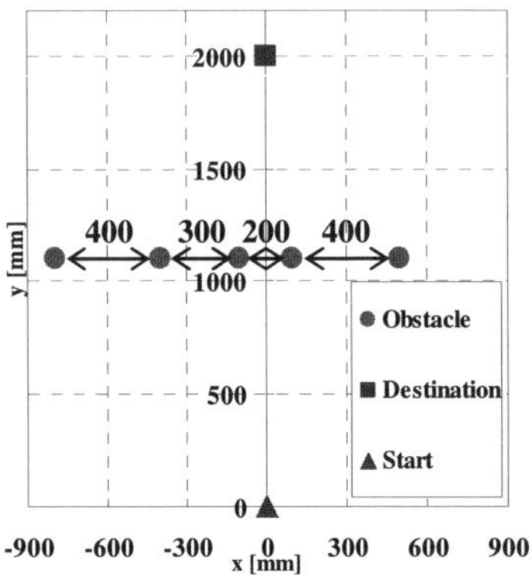

Figure 8. Environment

Table 2 Environment

	x	y
Start	0	0
Destination	0	2000
Obstacle	-800	1100
Obstacle	-400	1100
Obstacle	-100	1100
Obstacle	100	1100
Obstacle	500	1100

Table 3 Robot's width

Circular form [mm]	794
Straight form [mm]	330

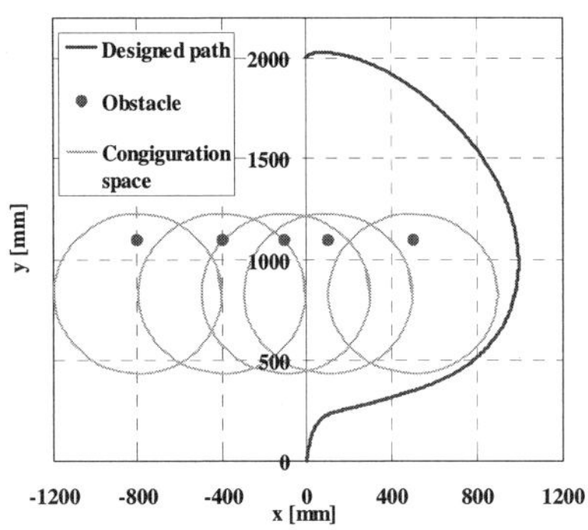

Figure 9. Path of only circle form

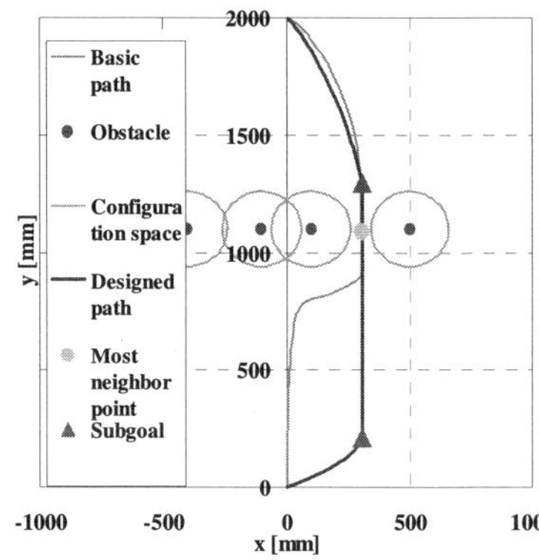

Figure 10. Path with transformation

Figure 9 shows that when the robot has only the circular form, it chooses the path through which the circular form can pass. Using the potential method, the robot having only the circular form can efficiently find and complete the path planning process in the environment. Figure 10 shows that when the robot has both the circular and straight forms, the robot chooses the path through which it can pass while in the straight form. By transforming to the straight form, less space is required and the robot can more easily pass between obstacles. For the robot to pass through narrow spaces, it is automatically transformed from the circular form to the straight form. After comparing Figures 9 and 10, we find that the robot with transformation capabilities can adapt to its environment using the proposed method. The path length has also been effectively reduced.

5.2. *Locomotion experiment*

We experimentally verified the ability of the robot to follow a specific path and also verified the usefulness of the transformation. We conducted locomotion experiments in planned paths shown in Figures 8 and 9. The locomotion tracks were analyzed using movie analysis software (MOVIAS Pro: Nac Image Technology Inc.). The locomotion experimental results are shown in Figures 11 and 12.

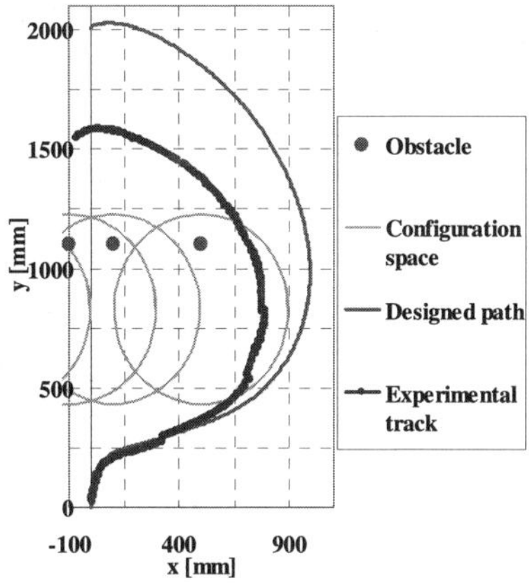

Figure 11. Track of only circle form

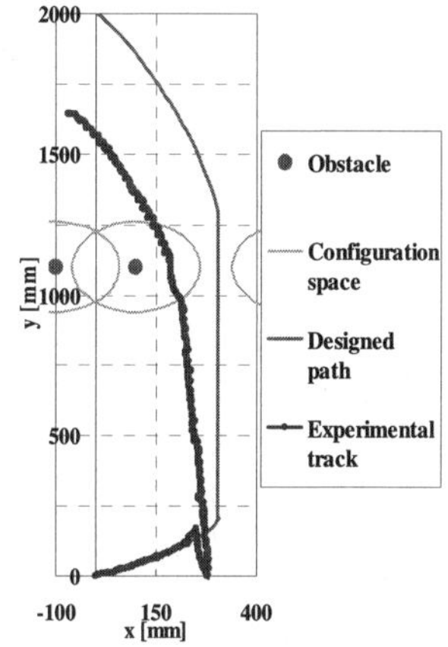

Figure 12. Track with transformation

Figures 11 and 12 show the deviation between the desired path and the experimental track. We believe that this is because of slippage. Since we are controlling the robot using only the feedforward technique, we cannot consider the sliding motion of the robot and the effect of the inclination at the time of transformation on the movement of the robot. Because the shape of the trajectory is similar to the shape of the designed path, the path planning method can be adapted for the robot.

When only in the circular form, it took 245 [s] to reach the destination, whereas when the transformation was included, it took 187 [s]. The transformation enabled the robot to arrive at the destination 58 [s] earlier than in the case when only in the circular form, which is a time reduction of about 25%. Therefore, we believe that planning a path that considers the robot's transformation allows for more efficient movement.

6. Conclusion

In this study, we proposed a path planning method that allows the robot to be transformed. Furthermore, we confirmed the usefulness of this proposed method by performing experiments. In future, we will control the robot with feedback that uses internal sensors in the servomotor. In addition, we will apply an external sensor to the robot to enable the autonomous detection of obstacles.

Reference

1. J. Urbano, K. Terashima, T. Miyoshi and H. Kitagawa, "Velocity Control of an Omni-directional Wheelchair Considerring Use's Comfort by Suppresing Vibration", IEEE/RSJ 2005 International Conference on Intelligent Robots and Systems, pp. 3169-3174, 2005
2. Brian Chan, N.J.Balmforth, and A.E.Hosoi, "Building a better snail: Lubrication and adhesive locomotion", Physics of Fluids, pp. 113101:1-10, 2005
3. Ryousen Fujiwara, Hirohisa Morikawa, Yuya Hukuya, Hiroshi Sakai, and Shunichi Kobayashi, "Pedal-Like Locomotion Mechanism Modeled on Pedal Crawling of Terrestrial Gastropod," The Japan Society of Mechanical Engineers, JAPAN, Vol. 70, No. 695, pp. 215-221, 2004
4. Taro Nakamura and Kuniaki Sato, "Locomotion Strategy for an Omni-directional Mobile Robot Using Traveling Waves Propagation" IEEE International Conference on Robotics and Automation, pp.3769-3774, 2010
5. Taro Nakamura and Kuniaki Sato, "Development of an omni-directional mobile robot using traveling waves based on snail locomotion", Journal of Industrial Robot, Vol. 35, No. 3, pp. 206-210, 2008
6. Kuniaki Satoh and Taro Nakamura, "Development of an omni-directional mobile robot based on snail locomotion", Proceedings of 7th International Conference on Climbing and Walking Robots and the Support Technologies for Mobile Machines, pp. 144-152, 2007
7. Teruyoshi Ogawa, Masayuki Yamamoto, Taro Nakamura, "Proposed omnidirectional locomotion strategy for traveling-wave-type mobile robot (TORoIII)" , Proceeding of 14th International Conference on Climbing and Walking Robots and the Support Technologies for Mobile Machines, pp127-134, 2011
8. Robert B. Tilove, "Local Obstacle Avoidance for Mobile Robots Based on the Method of Artificial Potentials", IEEE International Conference on Robotics and Automation, pp. 566-571, 1990
9. Naemeh Nejatbakhsh and Kazuhiro Kosuge "User-Enviroment Based Navigation Algorithim for an Omnidirectional Passive Walking Aid System", IEEE International Conference on Rehabilitation Robotics, pp. 178-181, 2005

PROPOSE OF V-SHAPED CLOSED ELASTICA AND ITS APPLICATION TO FISH-LIKE ROBOT

MIKIRO NAGASAWA, TATSUHIKO SEKIYA, NAOYUKI TAKESUE*
and HIROMI MOCHIYAMA

Tokyo Metropolitan University,
6-6 Asahigaoka, Hino, Tokyo, 191-0065, Japan
**E-mail: ntakesue@sd.tmu.ac.jp*
University of Tsukuba,
1-1-1, Tsukuba, Ibaraki, 305-8573, Japan

The closed elastica has been proposed as a functional mechanical structure which can generate impulsive motions repeatedly. In this paper, we propose a new type of V-shaped closed elastica (VSCE) and develop a fish-like robot using it. The VSCE consists of two elastic strips and two links. When the links are driven, the strips behave like a fish motion waving a tail fin. We examine the swimming performance of the robot by changing the structural parameters.

Keywords: Fish-like robot, Closed elastica, Snap-through buckling

1. Introduction

Underwater robots are one of the effective solutions for the underwater survey. Although a propeller screw is a popular propulsive method for underwater vehicles, fish's swimming motion can realize more effective and efficient mobility than a propeller screw propulsion.

The aquatic creatures such as carp or dolphin swim freely under water where they live in.[1] A lot of researchers study to realize such smooth and quick motion of creatures by robots. However, it is not easy to develop such robots, and in fact there are still few in number.

On the other hand, the structure called "Closed Elastica"[2] can easily produce impulsive motion and the application of the mechanism to robot has been studied.[3–5] This structure generates snap-through buckling with bending and distortion, and the motion can be repeated. Therefore, it is expected to use the mechanism as a propulsive method of the fish robot.

2. V-Shaped Closed Elastica (VSCE)

The closed elastica generates impulsive motion using the characteristic of elastic material. The simplest type of closed elastica consists of two rotating joints (one is passive and the other is active) and a elastic strip between them. When the active joint is driven, the elastic strip changes rapidly the shape through S-shape in two-dimensional plane (Fig. 1).

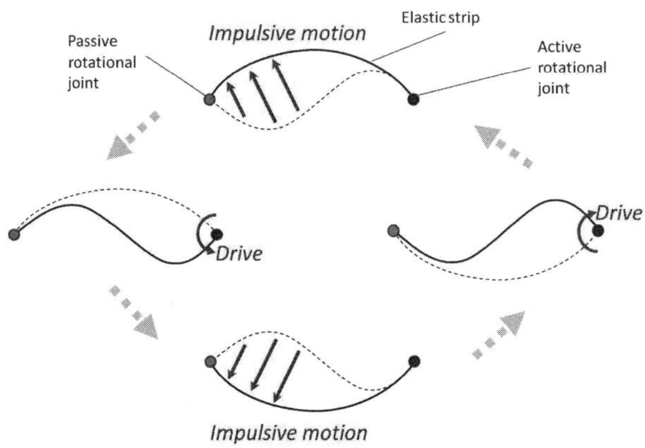

Fig. 1. Shape transition of closed elastica (basic type).

Mochiyama attached a fin to the passive joint and developed a swimming robot. Furthermore, by attaching a fin to the elastic strip directly, a high acceleration swimming robot was realized.[4] However, the shape and motion of robot are quite different from real fish.

In this paper, we propose a novel spatial V-shaped closed elastica (VSCE) as shown in Fig. 2. It consists of two elastic strips and two rigid links. The ends of strips are connected each other at a passive free joint. The other ends are fixed at the links in a certain angle θ. The links are connected at an actuated joint.

At the initial state, the strips are bending and the active joint between the links has a certain angle as shown in Fig. 2. When the active joint is rotated, the strips are greatly deformed and snapped beyond a certain posture as shown in Fig. 3. As a result, the tip of the strips (the free joint) moves and swings rapidly. Since the motion seems to be that of fish swimming, we apply the mechanism to the fish-like robot in the next section. There are some structural parameters such as the fixed angle θ, the length

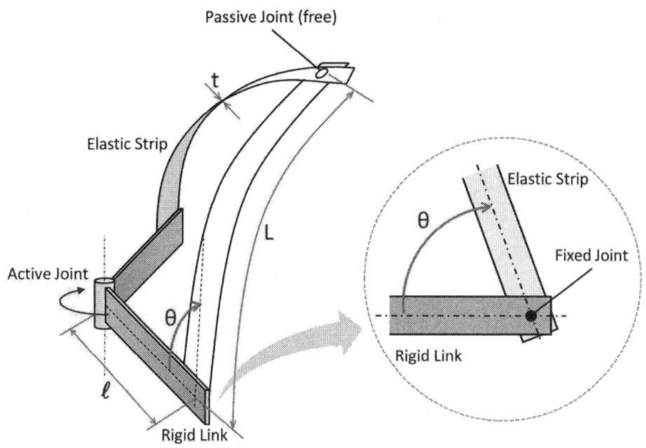

Fig. 2. V-Shaped Closed Elastica (VSCE).

of links, the size of strips (length, width, thickness) and material properties. In this paper, some experiments are examined to investigate the influences of the structural parameters to the swimming performances of the robot.

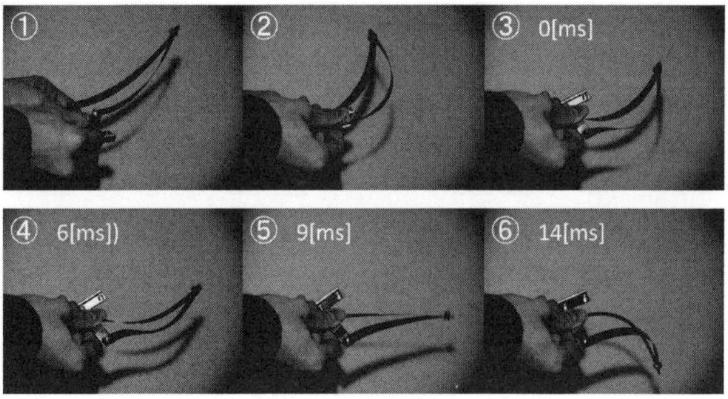

Fig. 3. Shape transition of VSCE : The values in these figures describe the time of snap through buckling motion.

3. Fish-Like Robot with VSCE

Figure 4 shows the developed fish-like robot using the VSCE. It consists of a body and a tail. In the body, the electric devices such as micro-computer, battery and servo motor are arranged with water-proof. In the tail, VSCE is used to drive (swing) the tail fin at the tip of strips. The servo motor drives one of the rigid links and the other link is fixed on the body. In the preliminary experiment, it was confirmed that the robot swam smoothly like a real fish as shown in Fig. 5.

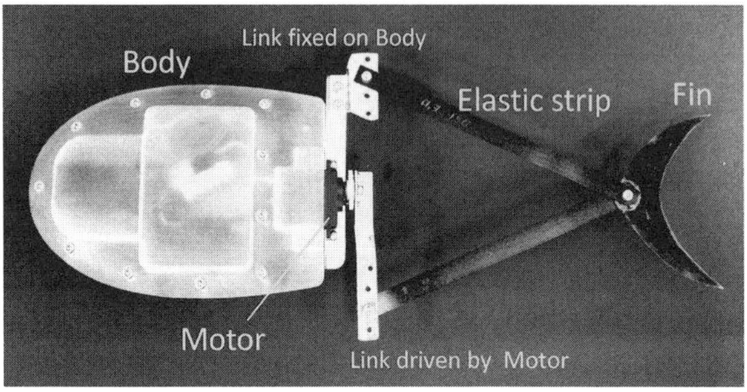

Fig. 4. The developed fish-like robot.

Table 1. Spec of swimming robot

parameter	value	[unit]
Length of body	150	[mm]
Height of body	107	[mm]
Thickness of body	36	[mm]
Full length	about 300	[mm]
Full weight	0.4	[kg]
Input voltage	5.0	[V]

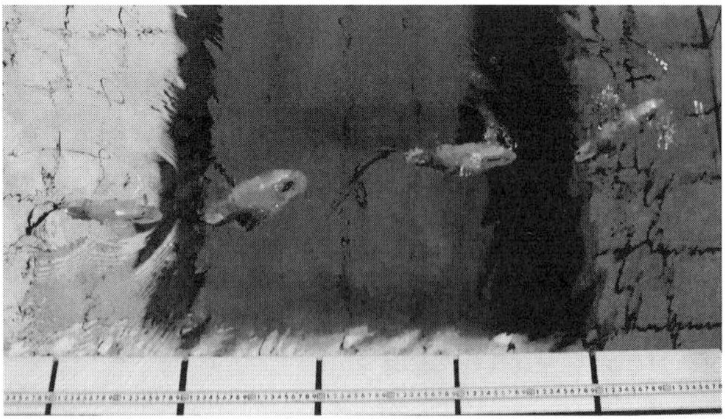

Fig. 5. Photo of swimming robot in pool.

4. Experiments

We conducted experiments to investigate the swimming velocity of the robot by changing the structural parameters such as the length of links, the length and thickness of strips, and the fixed angle between them.

4.1. *Influence of link length / fixed angle*

First, the link length and the fixed angle were changed and the experiments were conducted. Table 2 shows experimental parameters of structure. The experimental results are shown in Fig. 6.

Swimming velocity tends to go fast when the fixed angle is increased to 90° because the bending deformation of the strips becomes larger. However, the large bending deformation makes it difficult to rotate the motor. On the other hand, if the fixed angle is too small, the robot swims slowly. In other words, there is the appropriate range in θ corresponding to the link length.

The longer the link length is, the smaller the maximum of the fixed angle becomes. In this experiment, the robot recorded about 0.25[m/s] as the maximum swimming velocity under 70[mm] of link length, 80[deg] of fixed angle and 1.35[Hz] of frequency.

Table 2. Experimental Parameters 1

Elastic strip	0.3 10 150 [mm]
Link length l	40, 50, 60, 70 [mm]
Fixed angle θ	70 ‘ 90 [deg]
Driving frequancy f	1.00, 1.35 [Hz]

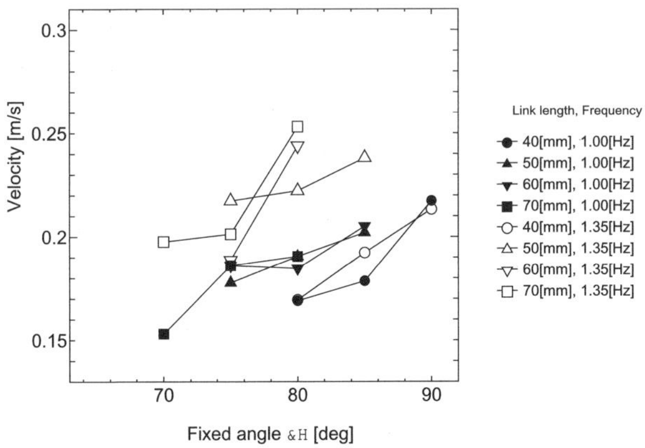

Fig. 6. Swimming velocity in case of changing the link length and the fixed angle.

4.2. *Influence of elastic strip / fixed angle*

As the next experiment, we changed the size of elastic strips (the link length is constant to 50[mm]). We prepared six kinds of elastic strips and compared swimming velocity. Table 3 shows experimental parameters of structure. The experimental results are shown in Fig. 7 and Fig. 8.

The thicker the thickness of elastic strips, t, is, the faster the robot swims because the bending motion is stronger. Conversely, when we used thin elastic strips, although the swimming velocity was slow, the robot swam smoothly and the range of fixed angle was expanded. The longer the length of elastic strips, L, is, the faster the robot swims. In this experiment, the robot recorded the maximum swimming velocity 0.29[m/s] which is higher than the last record (t=0.3[mm], L=200[mm], θ=90[deg], f=1.35[Hz]).

Table 3. Experimental Parameters 2

Thickness of elastic strips t	0.2,0.3 [mm]
Length of elastic strips L	100,150,200 [mm]
Width of elastic strips	10 [mm]
Link length(fixled) l	50 [mm]
Fixed angle θ	65 ‘ 90 [deg]
Driving frequancy f	1.00, 1.35 [Hz]

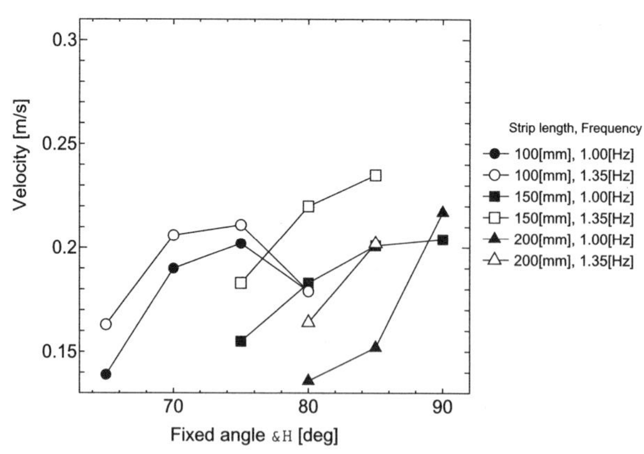

Fig. 7. Swimming velocity (t=0.2[mm]).

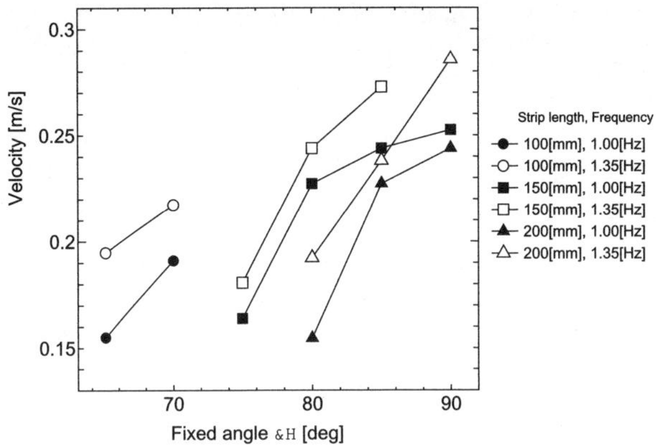

Fig. 8. Swimming velocity (t=0.3[mm]).

4.3. *Turning performance*

As seen in Fig. 5, the fish robot swims forward when the driven link is oscillated between the left and right as shown in Fig. 9(b).

When the link is oscillated between the center and the left/right as shown in Fig. 9(a) and (c), the tip of elastic strips is shaken without causing snap-though buckling motion. Since the tip of elastic strips waves on only the left/right in this case, the robot swam turning as shown in Fig. 10. In this experiment, the robot turned in the water tank at about 160[mm] of radius. Therefore we found that this robot can turn in the water using only one actuator.

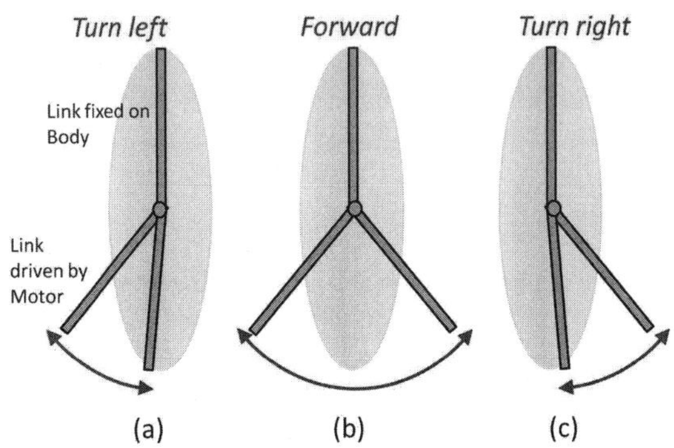

Fig. 9. The way to oscillate the link. (back view of the robot).

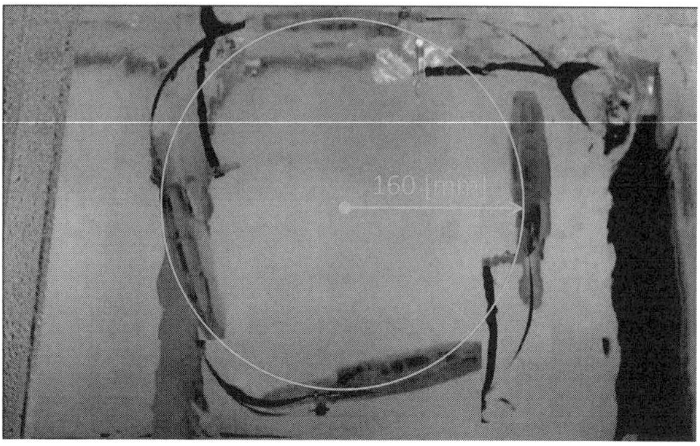

Fig. 10. Turning in the water tank.

5. Conclusion

In this paper, we proposed a new type of spatial V-Shaped Closed Elastica and applied it to a fish-like robot. In order to investigate the swimming performance of the robot, experiments were conducted and the influences of the structural parameters were clarified.

In addition, this robot can swim turning when the robot waves the tail fin on one-side from the center to the left/right. Therefore we found that VSCE is effective method for the swimming robot.

As future works, we will design optimal parameters including the shape of fin, and make the robot swim faster.

References

1. M. Nagai: "Thinking Fluid Dynamics with Dolphins", ISBN 4-274-90492-X(Ohmsha), ISBN 1 58603 231 3(IOS Press), 2002.
2. H. Mochiyama, M. Watari, H. Fujimoto: A robotic catapult based on the closed elastica and Its application to robotic tasks, Proc. of IROS 2007, pp.1508-1513, 2007.
3. H. Mochiyama, A. Yamada and H. Fujimoto: "Impulse Force Generator based on Closed Elastica", J. of RSJ, Vol. 29, No. 6, pp.492-495, 2011. (in Japanese)
4. H. Mochiyama, H. Esaki, M. Watari and H. Fujimoto: "Impulsive Swimming Robot by Robotic Catapults based on Closed Elastica", Proc. JSME Robotics and Mechatronics 2007, 2P1-A09, 2007. (in Japanese)
5. A. Yamada, H. Maeda, Y. Sugimoto, H. Mochiyama and H. Fujimoto: "An Impulsive Turning Mechanism for Swimming Robots Based on Repeated Snap-through Buckling", CLAWAR 2010, pp.253-260, 2010.

SNAKE-LIKE, TRACKED, MOBILE ROBOT WITH ACTIVE FLIPPERS FOR URBAN SEARCH-AND-RESCUE TASKS*

MARC NEUMANN†, THOMAS PREDKI, LEIF HECKES, PATRICK LABENDA

Chair of Engineering Design, Ruhr-University Bochum
Universitätsstr. 150, Bochum, 44801, Germany

After a building collapse people buried alive have to be localized and rescued. This requires the damage site's inspection and surveillance. These tasks are dangerous and challenging due to the area's hard-to-reach and hazardous environment. The damage site cannot be actively entered but must be inspected from a safe distance. In this context, mobile robots gain in importance. These can be operated semi-autonomously or remote-controlled without exposing the first responders to the risk. The paper at hand introduces a novel robot which stands out due to its high locomotion and mobility capabilities. It has a snake-like build-up, uses tracks and active flippers for locomotion and negotiates completely structured as well as extremely unstructured and rough terrain. The system's slender, segmented and modular structure is actively articulated by the use of overall 30 degrees-of-freedom which allow the robot's flexible adaptation to a given terrain. System-terrain-interaction is detected by the use of an innovative, RFID-based sensory integrated in the system's tracks. The paper at hand presents the mobile robot's basic features as well as first experimental results for semi-autonomy and tele-operation.

1. Introduction

Urban search-and-rescue (USAR) tasks endanger the lives of the first responders at place especially if a damage site has to be actively entered for inspection and surveillance. Consequently, there is a great need for mobile robots for USAR tasks which can be operated semi-autonomously or remote-controlled [1].

But mobile robots for USAR tasks must possess enhanced rough terrain mobility and outstanding locomotion capabilities [2]. They must be able to constantly produce enough tractive power, must have the ability to maneuver through tight and narrow passes as well as to negotiate diverse forms of discrete obstacles such as high steps and wide gaps.

---

* This work has been supported by the German Federal Ministry of Education and Research (BMBF). Many thanks for financing the project "Intelligent securing localization system for search and rescue of buried people (Intelligentes sicherndes Lokalisierungssystem für die Rettung und Bergung von Verschütteten – I-LOV)" within which this contribution has been worked out. The project has been part of Germany's high-tech strategy "Research for Civil Security".

† For correspondence: neumann@lmk.rub.de.

Mobile robots with corresponding, significant potentials are especially snake-like, actively articulated vehicles. These have a highly modular and flexible structure which in combination with their given numerous degrees-of-freedom can be used for an extensive terrain adaptation as well as traction optimization [3, 4]. Further on, the systems' structure and kinematic redundancy can lead to an impressive terrainability with respect to the negotiation of obstacles. Finally, the systems have a slender build-up with a small cross-section and are therefore able to navigate through tight and narrow passes as well as to avoid insurmountable obstacles [5]. The described capabilities and potentials are illustrated in Figure 1.

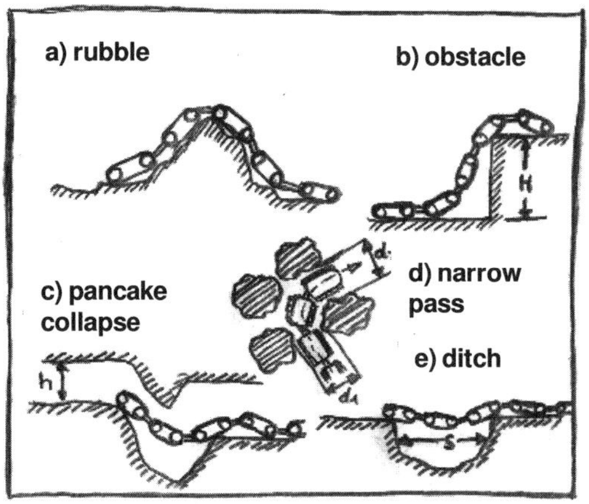

Figure 1. Mobility capabilities of snake-like, actively articulated vehicles.

These capabilities can be further improved by the use of tracks for propulsion as well as the integration of active flippers e.g. for additional terrain adaptation.

2. Related work

Snake-like, actively articulated robots are under development since the 1970s [6]. By now, numerous systems have been elaborated [7, 8]. This also applies to robots using tracks as propulsive elements [9-11]. Until now, the given systems do not combine a snake-like, actively articulated structure with active flippers. Anyhow, the given systems have already impressively demonstrated their mobility capabilities. But most of the systems can only be either completely manually tele-operated or rely on control strategies depending on exhaustive environmental models which cannot be adequately generated for an extremely unstructured and rough terrain [12, 13]. The reason for this is that no reactive or hybrid control strategies can be used due to missing external, tactile sensors for a direct detection and determination of physical system-environment-interaction.

3. Objectives

We develop snake-like robots and investigate their locomotion and mobility capabilities. The systems' development is challenging, on the one hand, due to an adequate realization of kinematic and actuator functionalities and the distribution of power to all propulsive elements. On the other hand, challenges result from the realization of sensor functionalities especially with respect to the systems' actual, physical interaction with the environment. Finally, the systems' locomotion and mobility capabilities also come along with an adequate incorporation of strategies for tele-operation and semi-autonomy.

4. Robot Moebhiu²s

Our robot Moebhiu²s is composed of four approximately identical modules and three active joint units. The system is illustrated in Figure 2. Also major characteristics are given.

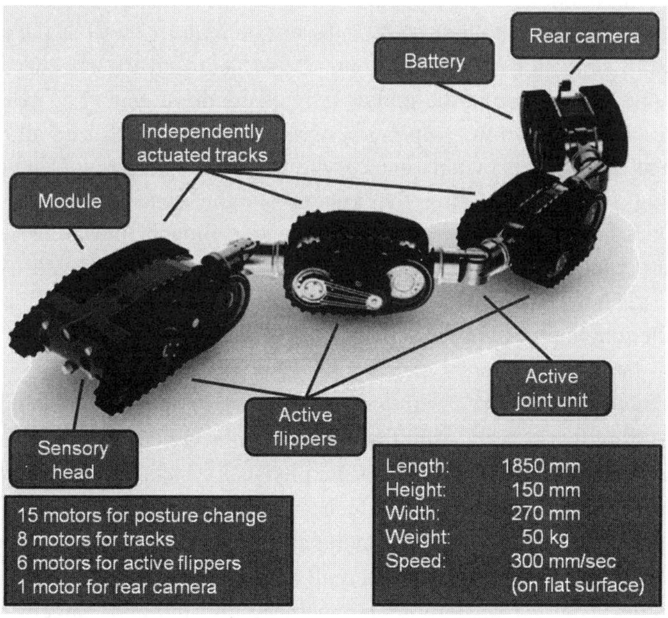

Figure 2. Snake-like, tracked, mobile robot with active flippers Moebhiu²s.

Each module is equipped with two independently actuated tracks at its left and right side which are used not only for propulsion but also for steering and maneuvering. The system has a symmetric structure and is therefore able to operate upside-down. Further on, the first three modules are equipped with two active flippers each which, too, feature tracks and again are actuated independently. The robot's tail module does not possess active flippers due to space-saving circumstances and the necessity of integration of on-board energy supply devices within.

The individual modules are connected with each other by powerful, active joint units with five degrees-of-freedom each which are basically responsible for the system's abilities to negotiate remarkable obstacles as well as to adapt its posture and configuration, respectively, for traction optimization. Each joint unit has a serial structure with alternating degrees-of-freedom. Here the individual degrees-of-freedom are pitch, roll, yaw, roll, pitch. Thus the joints alternate by 90° with respect to each other.

For the actuation of the joint units as well as for the main tracks and active flippers a modular drive unit was designed that can be exchanged easily in case of a damage. Each drive unit uses a powerful brushless torque motor by Robodrive with three hall sensors for commutation and a Harmonic Drive gear transmission. Each joint is equipped with an incremental encoder for joint position detection. The hollow shaft of the torque motor as well as the geometry of the wave generator of the Harmonic Drive gear and the encoder enable a simple cable guide through the middle axis of the drive unit. For maintenance and decomposition a screw coupling between the joints is realized allowing the detachment of the joints by only one easy to loosen coupling ring (see Figure 3).

Taking into consideration the system's eight motors for independently actuated tracks, six actuators for integrated active flippers as well as 15 motors for posture change it can be stated that the robot is equipped with overall 29 actuators when dealing with its locomotion and mobility degrees-of-freedom.

Further on, the robot is equipped with diverse sensors. Each track motor and each joint has an incremental encoder for velocity and joint position detection. The first module is equipped with an inertial measurement unit. The tail module carries a tilting rear camera. Additionally, all tracks are equipped with multiple, external, tactile sensors. These detect the physical system-terrain-interaction by mechanical bumpers using RFID technology for signal transmission [14]. Finally, the first module carries a modular and exchangeable sensory head which can be equipped with diverse external sensors (see Figure 4). Our basic configuration features a stereoscopic camera, an infrared camera, distance sensors, a CO_2 sensor and acoustic and light emitting devices.

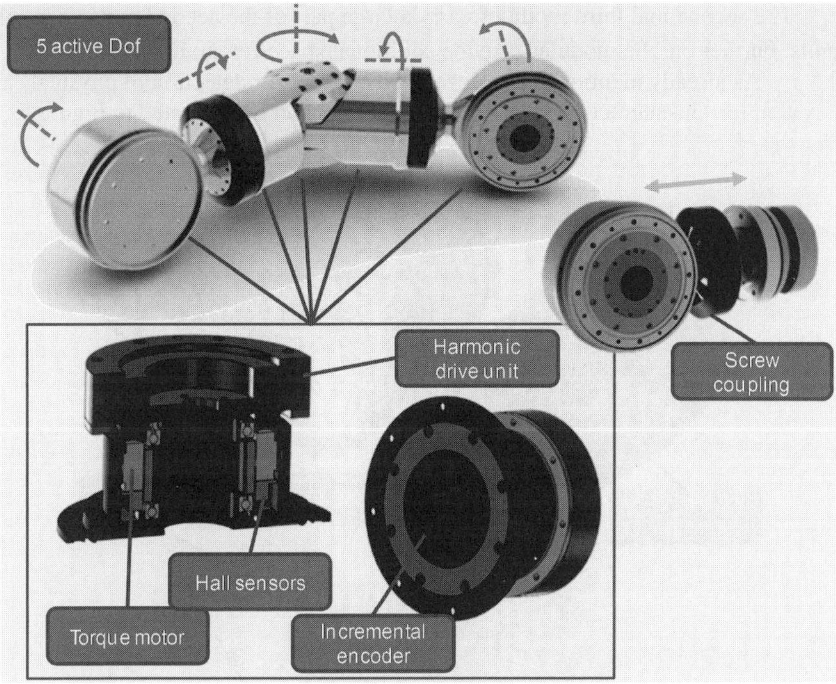

Figure 3. Joint unit with five active degrees-of-freedom.

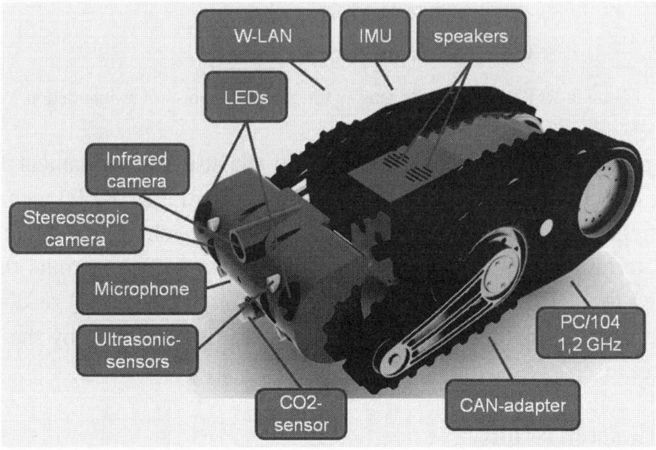

Figure 4. First module and integrated components.

Additionally, the first module is equipped with a W-LAN device and the robot's control unit.

The second and third modules carry a large part of the needed motor control units. Further on, the modules provide additional space for auxiliary goods.

As already mentioned our developed concept for detection of physical system-terrain-interaction uses RFID technology and is illustrated in Figure 5.

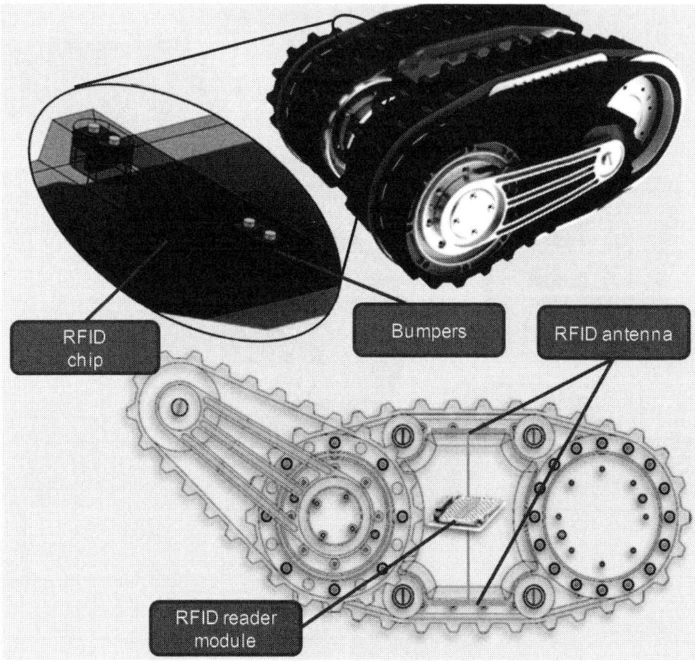

Figure 5. RFID-based tactile sensors for physical robot-terrain-interaction.

The physical interaction is detected by the use of mechanical bumpers. These are directly integrated into the given, rotating tracks. RFID technology is used for signal transmission and energy supply. If a mechanical contact between track and terrain has been detected the RFID transponder transmits this signal radio-based to a RFID antenna and finally to an integrated RFID reader module within the specific module. Afterwards the signal is processed by the system's control unit.

5. Experimental results

First experimental results have been elaborated with respect to, both, the robot's locomotion on structured as well as unstructured terrain. Locomotion on structured, flat terrain includes a semi-autonomous control enabling the system's maneuverability in a tunnel and sewer system [15]. The human operator only controls the configuration of the head module. All other degrees-of-freedom are controlled by the system's automation. This automation uses a follow-the-leader strategy imitating the behavior of a passive truck-trailer system and is described in detail in [16].

Locomotion on unstructured terrain is actually investigated by the use of a completely manual control involving two tele-operators. But semi-autonomous control strategies including virtual tracks, innovative user interfaces, the developed RFID-based sensors for physical robot-terrain-interaction and impedance control for the given active flippers are currently under development.

Figure 6 shows experiments dealing with the robot's locomotion and mobility on structured and unstructured terrain.

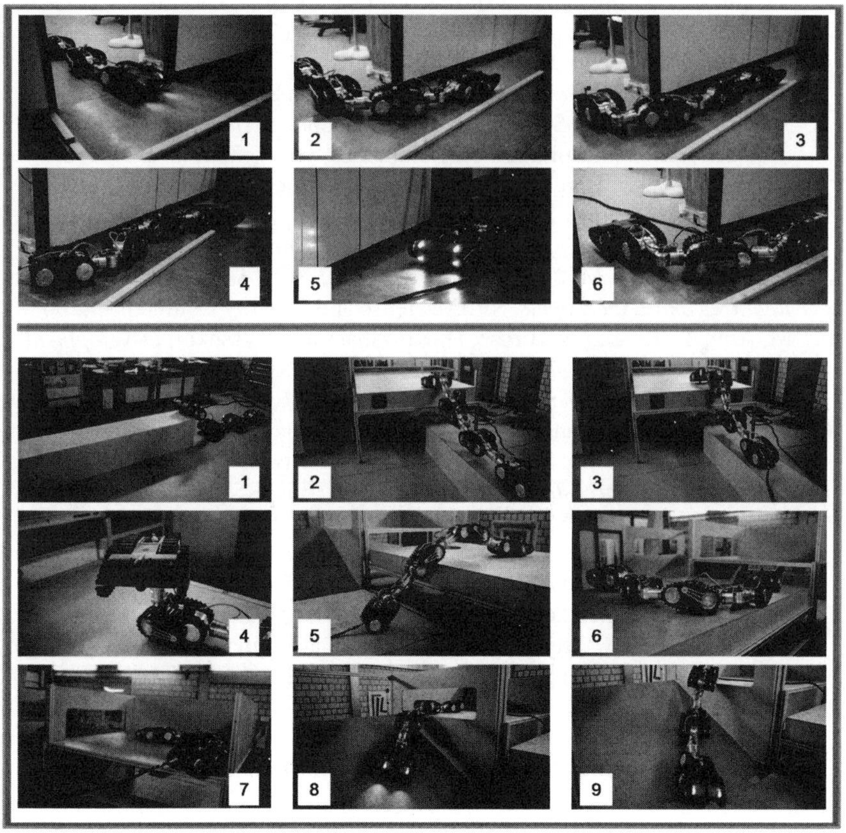

Figure 6. Extracts from practical experiments; locomotion on structured terrain (top); locomotion on unstructured terrain (bottom).

References

1. A. Davids, *Urban search and rescue robots: from tragedy to technology.* Intelligent Systems, vol. 17 (2002).
2. R. R. Murphy, S. Tadokoro, D. Nardi, A. Jacoff, P. Fiorini, H. Choset and A. M. Erkmen, *Search and Rescue Robotics*, In: B. Siciliano and O. Khatib (ed.), *Springer Handbook of Robotics*, Berlin, New York, Springer (2008).
3. P. Labenda and T. Sadek, *Design for trafficability of kinematically redundant locomotion systems.* Int. Design Conference – DESIGN (2010).
4. P. Labenda, T. Sadek and L. Heckes, *Considerations for enhanced rough terrain mobility of kinematically redundant locomotion systems.* IFAC Symposium on Mechatronic Systems (2010).
5. P. Labenda, T. Sadek and T. Predki, *Controlled Maneuverability of an Articulated Tracked Mobile Robot.* ASME IDETC & CIE (2010).
6. S. Hirose, *Biologically Inspired Robots: Snake-Like Locomotors and Manipulators.* Oxford, Oxford University Press (1993).
7. G. Granosik, *Hypermobile Robots*, In: H. Aschemann (ed.), *New Approaches in Automation and Robotics.* I-Tech Edu. and Pub. (2008).
8. J. K. Hopkins, B. W. Spranklin and S. K. Gupta, *A survey of snake-inspired robot designs.* Bioinspiration &Biomimetics, vol. 4 (2009).
9. G. Granosik, J. Borenstein and M. G. Hansen, *Serpentine Robots for Industrial Inspection and Surveillance*, In: L. K. Huat (ed.), *Industrial Robotics: Programming, Simulation and Application* (2006).
10. J. Borenstein and M. G. Hansen, *OmniTread OT-4 Serpentine Robot – new Features and Experiments.* SPIE Defense and Security Conference (2007).
11. M. Arai, Y. Tanaka, S. Hirose, H. Kuwahara and S. Tsukui, *Development of "Souryu-IV" and "Souryu-V": Serially Connected Crawler Vehicles for in-Rubble Searching Operations.* Journal of Field Robotics, vol. 25 (2008).
12. Y. Okada, K. Nagatani, K. Yoshida, T. Yoshida and E. Koyanagi, *Shared Autonomy System for Tracked Vehicles to Traverse Rough Terrain Based on Continuous Three-Dim. Terrain Sensing.* IEEE/RSJ IROS (2008).
13. K. Kurose, S. Saga, S. Okamoto, K. Ohno and S. Tadokoro, *Designing of online simulation environment for development control algorithms for robots operating in rough terrain.* IEEE/RSJ IROS (2008).
14. L. Heckes, P. Labenda and T. Sadek, *A Wireless Contact Point Sensor for Tracked Robots.* 4th Annual IEEE International Conference on Technologies for Practical Robot Applications (TePRA) (2012). *accepted for publication.*
15. Q. Hamp, O. Gorgis, P. Labenda, M. Neumann, T. Predki, L. Heckes, A. Kleiner and L. Reindl: *Study of efficiency of USAR operations with assistive technologies.* Journal Advanced Robotics (2012). *submitted for review.*
16. P. Labenda: *Shared Control System to Safeguard Mobility of Wheeled, Actively Articulated Truck-Trailer Vehicles on Structured and Rough Terrain.* Intelligent Transportation Systems Conference (ITSC) (2012). *submitted for review.*

BARRIER NAVIGATING AND SHELTER SEEKING ROBOT INSPIRED BY COCKROACH BEHAVIOR*

VICTORIA A. WEBSTER, NICHOLAS S. SZCZECINSKI, BRIAN R. TIETZ, KATHRYN A. DALTORIO and ROGER D. QUINN

Biologically Inspired Robotics Laboratory, Mechanical and Aerospace Engineering Dept., Case Western Reserve University, 10900 Euclid Ave. Cleveland, OH, 44106, USA

DOUGLAS PORR, JOHN RICHARDS and ROY E. RITZMANN

Department of Biology, Case Western Reserve University, 10900 Euclid Ave. Cleveland, OH, 44106,USA

The robotic controller in this study uses dynamically weighted behavior control architecture to obtain emergent behavior inspired by the cockroach *Blaberus discoidalis*. The effect of sensori-motor subsystems is weighted based on the current sensory state of the robot. Neither history nor prediction was considered, leading to a flexible navigation algorithm with minimal computation. Cockroach behavior was recorded to compare with robotic results.

1. Introduction

Biological organisms often provide useful sample systems for robotics. With inspiration from animals, engineers can design robots that are potentially more versatile and robust than traditional designs [1]. Cockroaches are being studied to investigate decision making, since they are capable of performing tasks that are currently difficult for robots and are conducive to neurobiological manipulation. By studying how cockroaches manage multiple tasks (such as following a wall while searching for a dark shelter), one can achieve a better understanding of how emergent behavior (i.e. the ability to avoid obstacles in order to reach a shelter) arises.

* This work is supported by in part by an NSF graduate student research fellowship, an NDSEG fellowship, SPUR grant from HHMI, and a NSTRF fellowship (Grant number #NNX11AN15H). It was also supported by a grant from DARPA (Grant W91CRB-11-1-0003 M3 Program) and AFOSR grant FA 9550-10-1-0054

Experiments run in this study with cockroaches suggested that the observed behavior was not caused by any single motivation, but was instead the product of a dynamic balance between visual and tactile inputs. Under this proposed mechanism, a cockroach would adjust the relative importance of each local goal based on its current state to generate the overall behavior.

The robotic controller developed in this study has no explicit overarching behavioral goal, but obtains the desired behavior by weighting sub-system goals based purely on the current state of the robot's sensory systems. This controller architecture may be considered subsumption architecture [2] in so far as it is a decomposition of action into modular behavior based controllers. However, it depends on a behavioral weighting mechanism that controls the effect of each module on the final changes to the actuators.

Other researchers have had success with the potential energy field method to weight various behaviors [3]. A robot uses its inputs to create a smooth potential field for each sensory system. Each input is then weighted by the gradient at that point, leading to smooth transitions between behaviors. Other algorithms exist that rely primarily on sensory subsystems, such as the BUG Algorithm [4]. The controller toggles between behavioral modes such as goal seeking and obstacle avoidance. In the algorithm presented in this paper behaviors are not switched on and off.

The effectiveness of our robotic controller was compared to success in seeking a shelter of cockroaches in an identical (but scaled) arena. Previous and concurrent studies were used to validate the robotic model as well as provide additional insight into the decision making processes of the cockroaches [5].

2. Background

The goal of this project was to develop a controller for a small, mobile robot capable of navigating an unfamiliar environment, negotiating a barrier and reaching a shelter using sensory inputs analogous to those in *Blaberus discoidalis*. Previous studies of these cockroaches by Daltorio *et al.* showed that they were motivated to find a dark shelter in a well-lit arena with little or no obstructions [5]. Additionally, Daltorio *et al.*'s work showed that cockroaches will locate a darkened space over 90% of the time when not challenged by transparent partial walls, and 86% of the time when confronted with them in 1 minute trials in a 90 cm square arena.

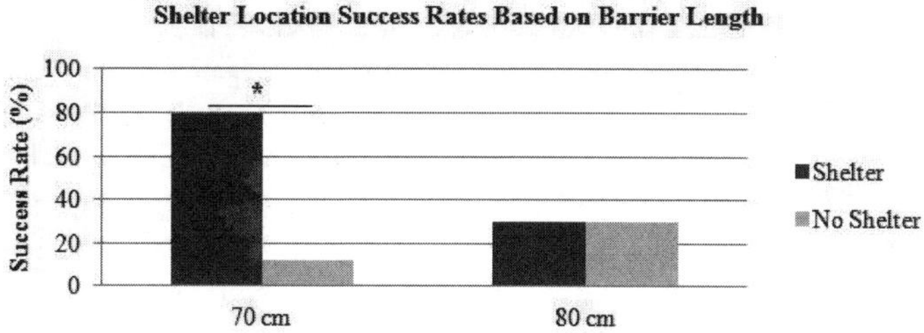

Figure 1. Cockroach success rate in reaching shelter area based on barrier length. The dark area represents trials where a shelter is present and the light area where a shelter was not present.

Previous studies have shown that cockroaches use their antennae to avoid obstacles and run along walls [6] [7]. Just like shelter location, wall following contributes to a cockroach's total behavior when in an environment that includes walls, but is not likely the solitary factor that causes the cockroach to act in any particular way. The total behavior of the cockroach is produced by the complex interaction of numerous goals, but this study focused on a balance between the tactile antennae system and the visual system.

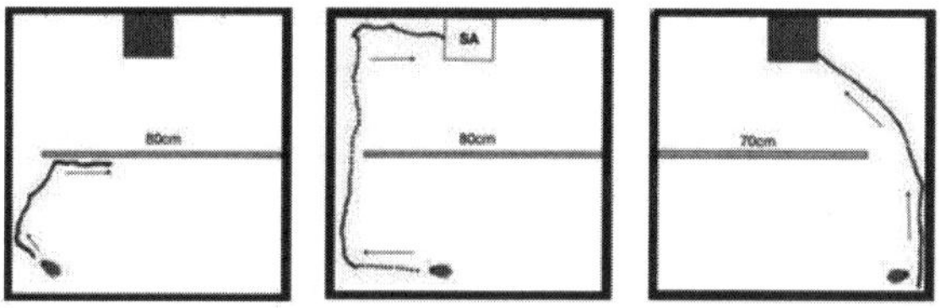

Figure 2. Traces of paths taken by cockroaches in different arena conditions. The square at the top of each image represents the shelter area, darkened if the shelter is present, and light if the shelter is not present. The light bar in the center represents the transparent barrier.

Research on cockroach barrier navigation and shelter location indicates that cockroaches reach the shelter area more frequently with a 70cm barrier than with an 80cm barrier as well as reaching the area more frequently with a shelter present when a 70 cm barrier is present [8](Figure 1). Cockroach path traces during shelter seeking around both 70 and 80 cm barriers were used to determine possible causes of the higher success rate in a shelter-less arena when faced with an 80 cm barrier (Figure 2). This observation in the biological system led to the development of the dynamically weighted behavior control architecture described here.

3. Design

The robotic platform developed in this study used a modified Surveyor SRV-1 Blackfin with an integrated camera (Figure 3).

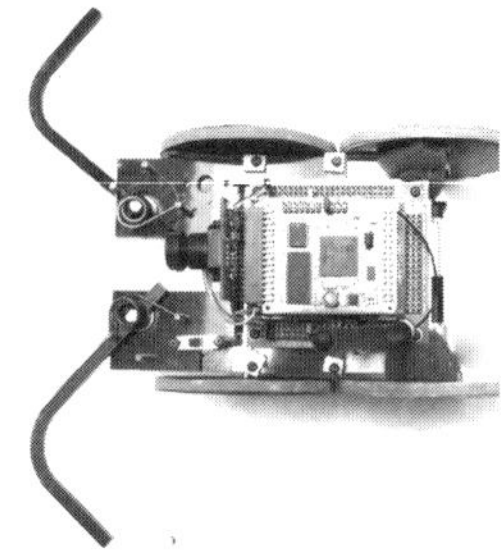

Figure 3. Small mobile robot with visual and tactile sensors.

3.1. *Hardware*

Antennae were designed based on cockroach hair-plates. When deflected, the antennae rotate a potentiometer to produce a voltage difference. A torsional spring returns them to the "home" position. The potentiometers provide Coulomb damping. When cockroaches follow walls, they drag an antenna on the wall behind them, maintaining contact with the wall and measuring the deflection of the antenna along its length and at the hair-plate[9]. Mimicking this required the addition of an analog-to-digital converter between the antennae and the processor.

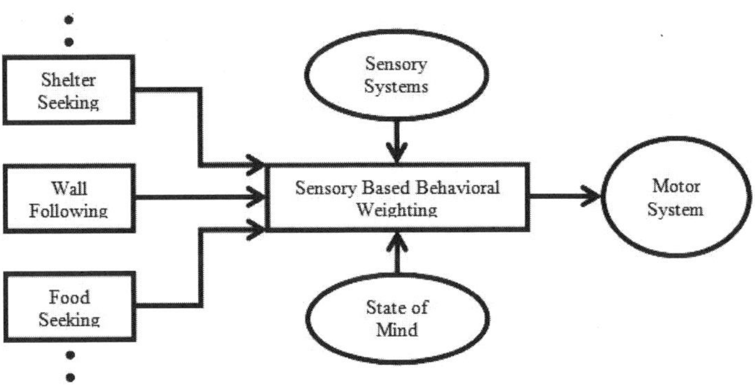

Figure 4. Dynamically weighted behavior control architecture applied to a cockroach inspired robot. On the left, each module in the behavioral module level is entirely independent from the others. These modules suggest independent motor changes which are weighted based on the current sensory data in order to produce the final actuator changes

3.2. *Software*

The control system in this study does not provide the robot with a global task. Instead, the controller is comprised of a series of modular finite state machines which each control a specific behavior. Rather than possessing a hierarchy, this controller uses parallel modular systems that suggest outputs to the behavioral weighting module. These behavioral outputs are then dynamically weighted based on the current state of the sensory input and controller "State of Mind"

(Figure 4). This allows the importance of each module to vary dynamically, rather than remain in a static hierarchy.

Three modules were developed. The first two, Wall Following and Shelter Seeking, were used in trials to compare with data collected from cockroach experiments. The third, Food Seeking, was tested to examine the ability of the controller to dynamically vary behavior importance among additional behaviors.

3.2.1. *Wall following*

In order to quickly converge on a wall and accurately follow it, the antennae were programmed with a proportional-integral-derivative (PID) feedback loop. The state variable of interest is the yaw angle of the robot. As suggested but not implemented in previous wall following research [9] [10], an integral term was found to be useful in stabilizing the robot's path along the wall.

The PID controller on board this robot differed from traditional PID controllers in that, when either antenna is deflected too far, the robot exits the PID loop and simply backs away from the wall with a turning bias. Although this behavior is not always seen in cockroaches, backing away from a barrier is a response seen during a head on collision with a barrier [7]. Additionally, an antenna's input has zero weight if not engaged in following a wall.

3.2.2. *Shelter seeking and food seeking*

Both the modules for shelter seeking and food seeking behavior contained the same controller. Due to the lack of odor sensors on the robot, "food" was denoted by a target of a specific color and tracked with the visual system much like the shelter. Location of visual goals was achieved using a PD controller where the variable of interest was yaw.

3.2.3. *Behavioral weighting*

As mentioned earlier, the output of each module is weighted based on the current state of the sensory system and the "State of Mind" of the system. Even though the system possesses no global goals, global behavior emerges as a result of this weighting. For the simple system in this study, the state of the sensory system was defined by whether an antenna was in contact with an obstacle, whether food or shelter was visible and on which side of the robot each was located. For arena trials to be compared to cockroach data, only Wall Following and Shelter Seeking were used. The weight needed for each was determined experimentally in order create behavior qualitatively similar to that of cockroaches. The "State of Mind" of the system refers to secondary weighting, which may cause one behavior to be more important than another. For example, hunger would be a "State of Mind". When hunger is high, the importance of finding food may be higher than that of finding a shelter. For the trials conducted in this study hunger was provided as a counter which incremented with time. The weight applied to the food module was then the total hunger divided by the maximum possible hunger.

4. Methods

Cockroach trials to which the robotic model was compared were performed in a 90 cm square acrylic arena that a cockroach was allowed to explore for 1 minute. The cockroach was presented with 70 and 80 cm clear barriers. Robotic trials were conducted in a 6.1 by 6.1 m racquet ball court. Here it faced 4.57 and 5.48 m barriers that permitted it to see the shelter from anywhere in the arena.

Robotic trials were extended to two minutes due to scaling between the robot and cockroach. The robot ran autonomously in the arena during each trial.

5. Results

Robotic trials using a controller with Wall Following and Shelter Seeking behavior modules were performed in an open arena, both with and without a dark shelter, in order to confirm that the robot is motivated to find the shelter. Additional trials were run with barriers of lengths 4.57 and 5.48 m both with and without the shelter present. These trials were compared to the cockroach results with 70 and 80 cm barriers, respectively. These results are presented in Figure 5. For each case of barrier length and shelter condition the difference in results of the robotic trials and the cockroach trials is not statistically significant (Table 1).

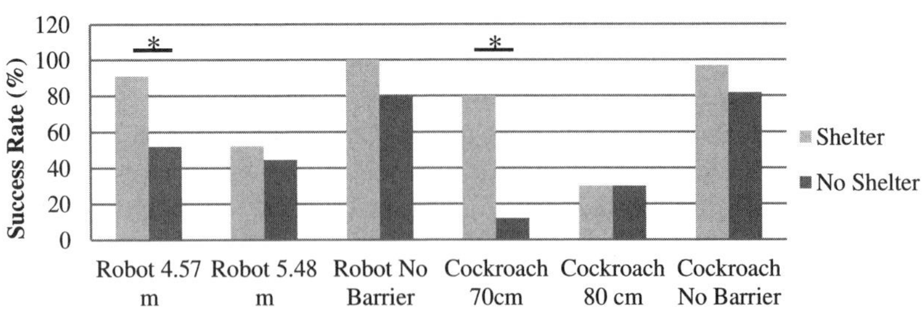

Figure 5. Comparison of task success rates based on arena conditions for both cockroaches and the robot

Table 1. p-values calculated via Fisher's Test to determine whether the difference in success rates was statistically significant.

System 1	*System 2*	*Barrier*	*p*
Cockroach No Shelter	Robot No Shelter	No	>.99
Cockroach With Shelter	Robot With Shelter	No	>.99
Cockroach No Shelter	Robot No Shelter	Short	0.010
Cockroach With Shelter	Robot With Shelter	Short	0.405
Cockroach No Shelter	Robot No Shelter	Long	0.675
Cockroach With Shelter	Robot With Shelter	Long	0.216
Robot No Shelter	Robot With Shelter	No	>0.99
Robot No Shelter	Robot With Shelter	Short	0.001
Robot No Shelter	Robot With Shelter	Long	0.450

Additionally, the total amount of time the robot spent following the walls and in corners over the length of each trial was found to vary. Average values for each trial are shown below in Figure 6.

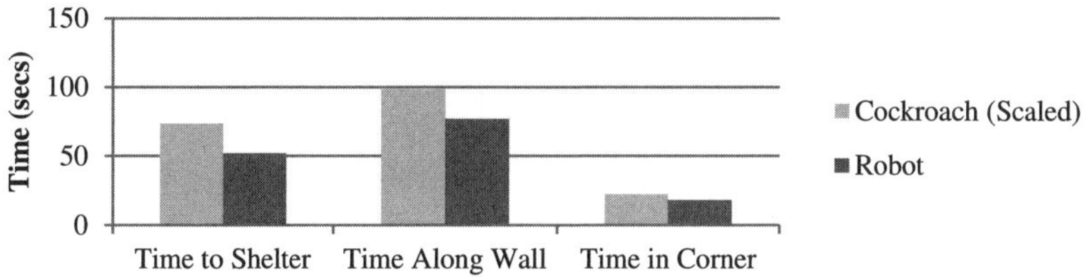

Figure 6. Comparison of cockroach and robot data for the amount of time it took to reach the shelter area, time along the wall, and time spent in corners

6. Discussion

This study sought to create a robotic system inspired by cockroach behavior that pursued dynamically changing goals in an unknown environment. While the system did not directly mimic the behavior of the cockroach, it did succeed in completing the desired task with no environmental map. This is an important process in autonomous navigation.

The Food Seeking mechanism was tested on a stationary test stand where strips of the appropriate color for both the shelter and the food source could be moved in front of the robot. When 'hunger' values were low the robot would ignore the food source if a shelter was present. Likewise if hunger values were high the robot would ignore the shelter if a food source was visible. This is much like behavior seen in cockroach trials where the cockroach would stop running in order to drink if it found water in the arena.

With respect to the biological model, the robot was more successful than the cockroach in reaching the shelter area when a shelter was present. The cockroach is a complex organism with motivations that are often unknown. The robotic model lacks this and therefore behaves more consistently. This randomness has been quantified and a new controller, RAMBLER, which takes a statistical approach to the same problem, and more accurately mimics the movement and behavior of the cockroach[5]. Both controllers essentially solve the same problem, but the one discussed in this paper is perhaps more predictable and presents an easily adaptable control architecture.

7. References

[1] R. E. Ritzmann, R. D. Quinn, J. T. Watson, and S. N. Zill, "Insect walking and biorobotics: A relationship with mutual benefits," *BioScience*, vol. 50, no. 1, pp. 23-33, 2000.

[2] R. Brooks, "A robust layered control system for a mobile robot," *Robotics and Automation, IEEE Journal of*, vol. 2, no. 1, pp. 14–23, 1986.

[3] T. L. Anderson and M. Donath, "AUTONOMOUS ROBOTS AND EMERGENT BEHAVIOR : A Set of Primitive Behaviors for Mobile Robot Control," *Autonomous Robots*, no. 1, pp. 723-730.

[4] J. Ng and T. Bräunl, "Performance Comparison of Bug Navigation Algorithms," *Journal of Intelligent and Robotic Systems*, vol. 50, no. 1, pp. 73-84, Apr. 2007.

[5] K. A. Daltorio et al., "A Stochastic Algorithm for Explorative Goal Seeking Extracted from Cockroach Walking Data," *ICRA 2012*, 2011.

[6] N. J. Cowan, "Task-level control of rapid wall following in the American cockroach," *Journal of Experimental Biology*, vol. 209, no. 15, pp. 3043-3043, Aug. 2006.

[7] Y. Baba, A. Tsukada, and C. M. Comer, "Collision avoidance by running insects: antennal guidance in cockroaches.," *The Journal of experimental biology*, vol. 213, no. 13, pp. 2294-302, Jul. 2010.

[8] B. R. Tietz, "Models of Cockroach Shelter Seeking Implemented on a Robotic Test Platform," 2012.

[9] N. J. Cowan, E. J. Ma, M. Cutkosky, and R. J. Full, "A Biologically Inspired Passive Antenna for Steering Control of a Running Robot," *Star*, pp. 541- 550, 2005.

[10] S. N. Sponberg, O. Y. Loh, a. G. Lamperski, R. J. Full, and N. J. Cowan, "Templates and Anchors for Antenna-Based Wall Following in Cockroaches and Robots," *IEEE Transactions on Robotics*, vol. 24, no. 1, pp. 130-143, Feb. 2008.

MODELING HALF-CIRCULAR COMPLIANT LEG BY USING SIMMECHANICS

M. O. OGUCU*, A. SARANLI§, and Y. YAZICIOGLU‡

**Dept. of Electrical and Electronics Engineering,*
§Dept. of Electrical and Electronics Engineering, ‡Dept. of Mechanical Engineering,
Middle East Technical University, Ankara 06800, Turkey
*E-mail: *orkunogucu@gmail.com, §afsars@metu.edu.tr, ‡yigit@metu.edu.tr*

Despite the well-understood effects and benefits, half-circular compliant legs are difficult in terms of modeling and analyzing. In many studies, simplified models have been used, which have a number of assumptions and whose validity need to be verified either against the actual physical system or against a more accurate simulation model. Leaving the experimental validation issues to the future works, the aim of this study is to come up with a more accurate simulation model of the half-circular compliant leg with the minimum number of assumptions, approximations and simplifications. For this purpose, Simulink/SimMechanics modeling and simulation environment is considered. We believe that our simulation model to be a better approximation of the system behavior because, it allows for a more general leg shape during locomotion. Also, slippage conditions can be carried out. The simulation is more general overall, allowing a distributed leg mass, non-planar leg behavior as well as allowing wider contact area with the surface.

Keywords: RHex, Half-Circular Compliant Leg, SimMechanics

1. Introduction

In mobile robotic field, the variety of tasks to be fulfilled, requirements, conditions, needs and other factors, have led to the emergence of different approaches, including legged, wheeled and tracked robot platform applications. Although each approach has more powerful aspects, legged robots stand out from others by the capability to move on any kind of surface and ability to use different kinds of gaits for achieving various motion modes, like walking, running, climbing, galloping, bounding and flipping.

RHex, six-legged dynamically stable robot platform, is not a slavish imitation, but is a successful interpretation of the nature with superior acting ability and mechanical simplicity.[1] The latest version of it, including our experimental platform SensoRHex shown in Fig. 1, have half-circular compliant legs, which contributes to the whole hexapedal system in favor of energy efficiency and locomotion stability.[2]

Fig. 1. Simulation model and experimental platform of SensoRHex, six-legged dynamically stable Sensor Riched Hexapedal robot

Despite the well-understood effects and benefits, half-circular compliant legs are difficult in terms of modeling and analyzing. Thence, in many studies, simplified leg models have been used, which makes rigid rod[3] and linear spring[4] assumptions and disregards circular geometry. Whereas, due to the circularity and resulted rolling motion, the leg-ground contact point moves, the effective leg length changes and the stiffness of the leg varies during locomotion. In[5] , the circularity of the leg has been considered, however, compliance has been neglected. A more realistic model has been presented for the half-circular compliant leg which derives the relationship between the ground reaction force and the amount of deflection.[6] However, it is assumed that the leg preserves its circular morphology during deflection and there is no slippage occurrence throughout the locomotion.

These studies have a number of assumptions whose validity need to be verified either against the actual physical system or against a more accurate simulation of the system. We have a physical robot, but no ability to obtain necessary measurements related with the flexible leg structure. In the present problem, we also have leg's dynamic behavior with the robot body which is not measurable with our current ability. Thence, leaving the experimental validation issues to the future works, the aim of this study is to come up with a more accurate simulation model of the half-circular compliant leg with the minimum number of assumptions, approximations and simplifications. For this purpose, Simulink/SimMechanics environment is considered, which is used for modeling and simulation of complex dynamical systems[7] and flexible bodies with great success as understood from related studies. We believe SimMechanics model to be a better approximation of the system behavior because of the following:

(1) The SimMechanics model allows for a more general leg shape during deflection while the analytical model assumes circularity,
(2) The SimMechanics model allows for friction and slippage while the analytical model assumes no slippage,

The simulation is more general overall, allowing a distributed leg mass, non-planar leg behavior as well as allowing wider contact area with the surface. In fact, by means of our SimMechanics model, we were able to quantify how much the circularity assumption remained valid.

2. SimMechanics Model of Mobile Robot Platform and Half-Circular Compliant Leg

SimMechanics modeling and simulation environment is best suited for structures which form kinematic chains, due to the necessity of connection between an object and a stationary point by using a joint. However, mobile robot platforms can be still modeled by settling a joint between body and global coordinate frames, which has degrees of freedom (DOF) suitable for the application. The dynamic model of the compliant leg is constructed by adding a revolute joint at every segment connection. Each such joint in its default form, contains a torsional spring element as well as an optional torsional damper to allow the leg to compress in the leg circle plane. In fact, the model allows for each joint to have zero or more DOFs. Zero DOF implies a rigid leg and 2-DOF implies the leg can flex out of plane. In this study, we particularly focus on 1-DOF (in plane) motion of the leg. The implemented simulation model allows experimentation with different

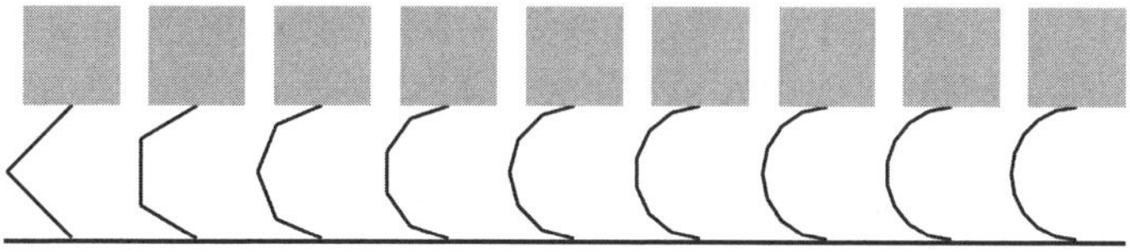

Fig. 2. The configuration of our multi-segmented leg approximations (2-, 3-, ..., 10-segmented structures)

number of leg segments as illustrated in Fig. 2, corresponding to different precisions of simulating our physical legs which are continuous half-circular composite springs. Segments are placed over a circle of given diameter with the constraint that segment connections are necessarily on the periphery. The only two independent parameters in our leg shape are the number of leg segments and the leg diameter. The uncompressed angles between each pair of leg segments are all equal and can be computed from the circular geometry.

3. Ground Reaction Force Model

In simulating the locomotion of the leg, one needs to model the interaction of the leg with the ground. SimMechanics environment does not provide a standard ground reaction model or a collision detection mechanism. The ground reaction force model is implemented by considering a kinematic collision event as well as a soft-contact model[8] to determine the overall leg-ground interaction. The model is validated by considering a standardized object-ground interaction test from the sport of Bowling. The model parameters are optimized to result in the expected bouncing behavior of a standard rigid bowling ball as shown in Fig. 3. We claim that this results in a "realistic" ground model that can be used for the remainder of the study.

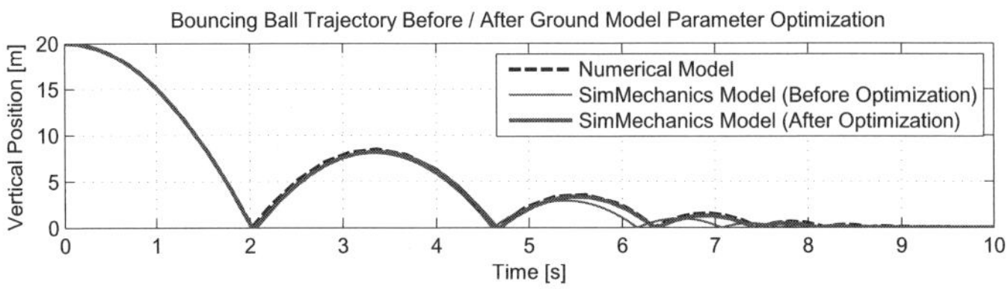

Fig. 3. Bouncing rigid ball (0.65 for the restitution coefficient and 6 [kg] for the mass property as an USBC Equipment Specification Standard) trajectory before/after ground parameter optimization

4. Simulation Results

Although we are able to simulate the dynamic behavior of the biped and hexapod, for simplicity one-legged hopper is considered. The results are examined under three headings. The first subsection is devoted to the validation works of the implemented SimMechanics leg models by using analytical counterparts. At the second subsection, the circularity assumption of the analytical models is tested by the capability of SimMechanics leg model which is deflected in a more general shape. Finally, at the third subsection, various slippage scenarios are simulated by changing tangential friction coefficient of the ground model.

4.1. *SimMechanics and Analytical Leg Models*

We first consider the SimMechanics and analytical SLIP models whose proper behavior is well known under the effect of gravity. Actually, these models dissociate between each others by the means of the "ground effect".

SimMechanics model provides leg-ground interaction by using ground reaction force relationships. However, analytical model sets the velocity and acceleration quantities of the leg to zero until the next liftoff event, whenever a touchdown event is detected, therefore, does not lead to energy dissipation.

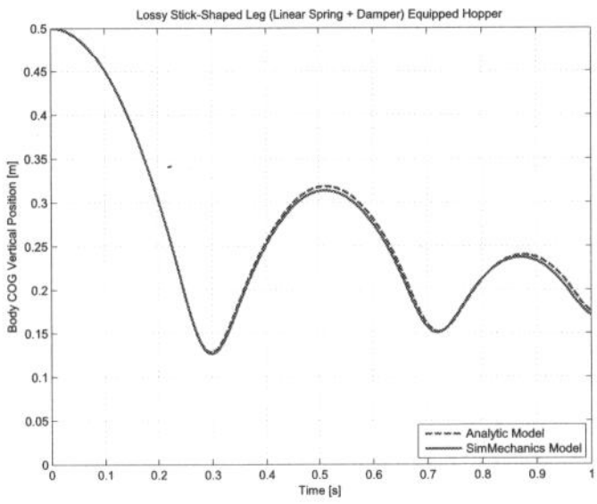

(a) Lossy Stick-Shaped Leg

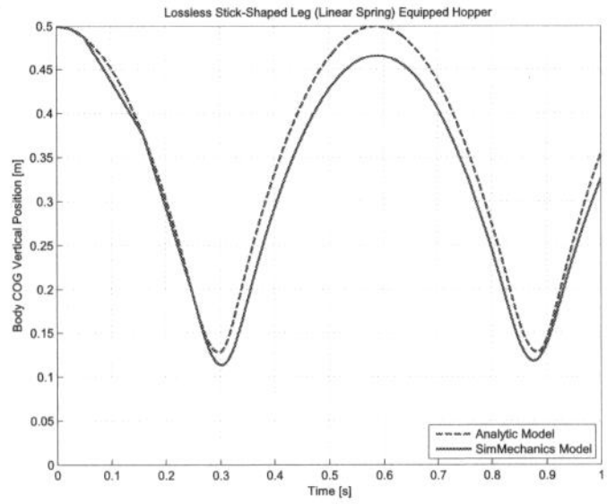

(b) Lossless Stick-Shaped Leg

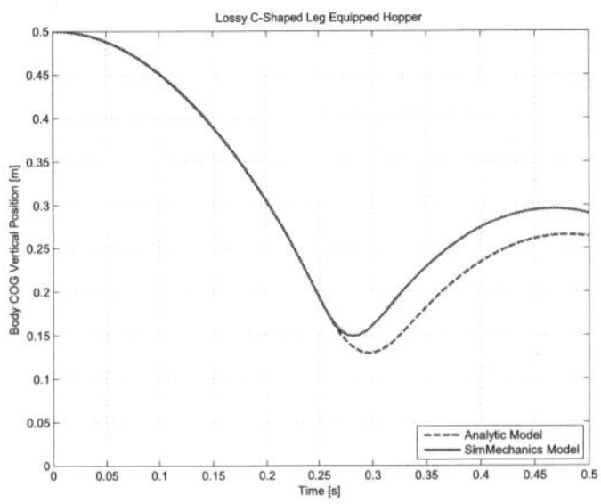

(c) Lossy C-Shaped Leg

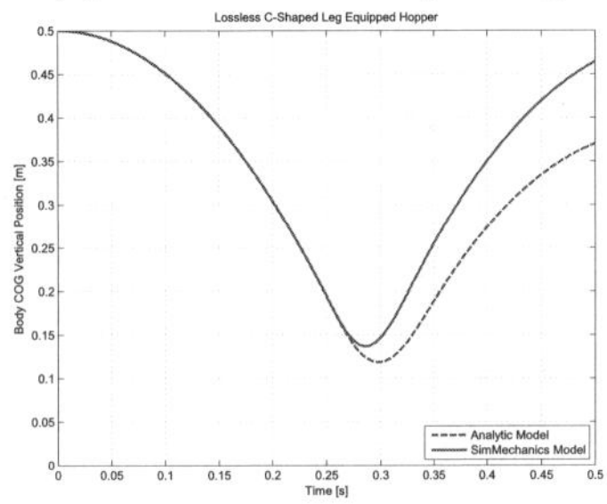

(d) Lossless C-Shaped Leg

Fig. 4. The vertical position of the hip point of one-legged hopper which is attached to the center-of-gravity of the body frame

In Fig. 4(a) and Fig. 4(b) the vertical position of the one-legged hopper is given for the lossy (spring-damper) and lossless (spring) leg models, respectively. We observe that for a reasonable damping value, the SimMechanics SLIP model closely match the known dynamics of the analytical SLIP model. However, in lossless leg simulations, the results do not overlap due to energy absorption caused by the damping of the ground model in SimMechanics, which is irrelevant for the analytical model. In this way, we have an idea about what the impact of the ground model is. In Fig. 4(c) and Fig. 4(d), the results are given for approximate 36-segmented SimMechanics and analytical[9] half-circular compliant leg models at lossy and lossless conditions, respectively. In lossy leg simulations, the dynamical behaviors seem to be fairly close to each other with an offset value, which may be originated from dissimilarity between damping coefficients of these models.

In lossless leg simulations, the dynamical behaviors seem quite different. It is already expected an energy absorption for the SimMechanics model due to the "ground effect" as mentioned above. However, the energy dissipation of the analytical model is an unforeseen situation caused by the unpredictable nonlinearities.

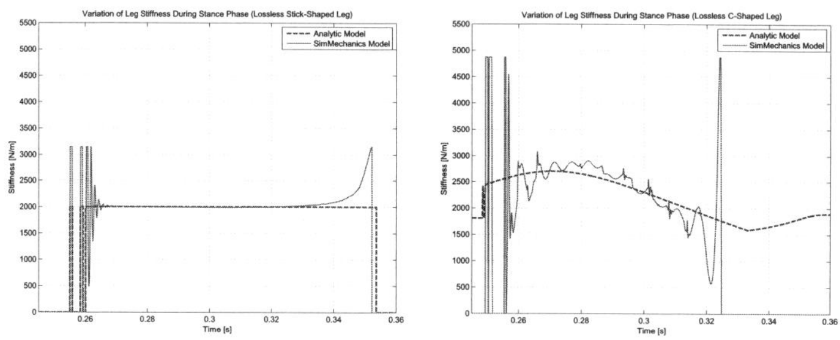

(a) Lossless Stick-Shaped Leg (b) Lossless C-Shaped Leg

Fig. 5. Leg stiffness calculated by using ground reaction forces (for SimMechanics model) and calculated directly from leg model (for analytical model)

Also, the stiffness values of the SimMechanics and analytical leg models are compared. Note that, for the half-circular analytical leg model, the spring stiffness value is already calculated as a part of the simulation model. However, in segmented half-circular SimMechanics leg model, there are torsional springs between each consecutive leg segments, whose values are meaningless in respect of whole leg stiffness. Still we are consider that, an equivalent leg stiffness "induced" between hip and ground contact points can be calculated by using ground reaction forces generated by our ground model. Although we know its actual value (2000 [N/m]), to test whether this approach is valid or not, we calculate leg stiffness of the SimMechanics SLIP model as shown in Fig. 5(a) by using generated ground reaction forces and resulted spring displacement value. It is observed that, there are some peaks and oscillatory behavior on calculated stiffness value until the leg becomes stationary after it hits to the ground first-time. This result shows that, once transient responses die out, the leg stiffness can be calculated accurately by using the ground reaction forces. Similarly, we obtain leg stiffness for the segmented half-circular leg model and compare with the results taken from analytical half-circular compliant leg model[9] as shown in Fig. 5(b). Clearly, the main trend of the responses match well, however, due to the impacts occurred at touchdown and lift instants, peaks and oscillatory behavior is observed in SimMechanics model.

4.2. *A More General Leg Shape During Deflection*

Analytical models assume that the leg preserves its circular morphology during deflection. By means of our SimMechanics leg model, which allows a more general shape during deflection, we were able to quantify how much the circularity assumption remained valid. Fig. 6(a) shows deflected leg shape and best circle fits at specific instants of the stance phase, which captures the error arising from the circularity assumption visually. It is clear that, the circularity assumption does not violated roughly, however, it remains an open question how this mismatch would effect the performance of any control algorithm. In Fig. 6(b), the normalized error computed by the disparity between leg shape and half circle is shown. Also the variation of leg radius is given in Fig. 6(c).

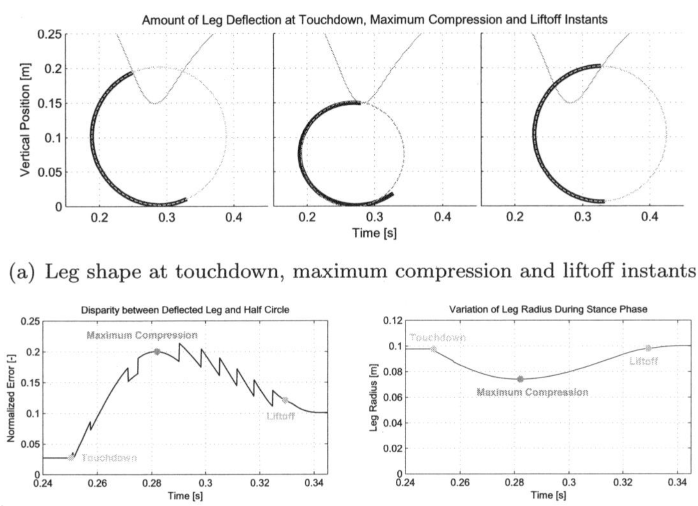

(a) Leg shape at touchdown, maximum compression and liftoff instants

(b) Leg/half circle disparity

(c) Leg radius

Fig. 6. Deflected leg shapes at specific instants, as well as the variation of normalized error and leg radius during stance phase

4.3. *Slippage Scenarios on Different Surface Properties*

One of the subject we focus on is the dynamical behavior of legged locomotion on different surfaces. Therefore, in some conditions, slippage may be too important to be neglected due to the dependence on surface properties. As shown in Fig. 7, various slippage scenarios are simulated by changing the tangential friction coefficient (μ) of the ground model.

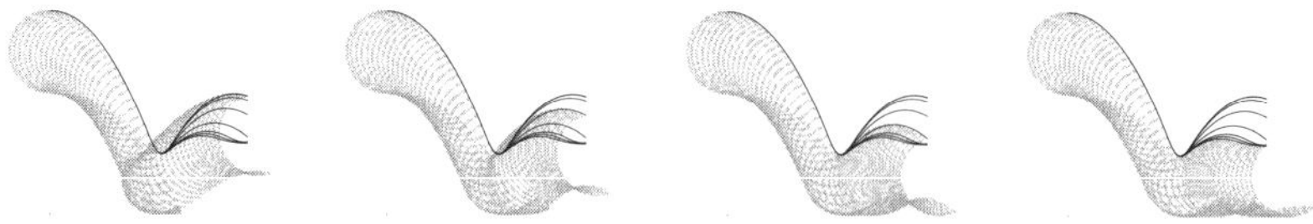

(a) $\mu = 1$ (b) $\mu = 0.01$ (c) $\mu = 0.005$ (d) $\mu = 1e - 6$

Fig. 7. Slippage scenarios on different surfaces (μ: tangential friction coefficient)

5. Conclusion

In this paper, we implemented a SimMechanics simulation model which represents the half-circular compliant leg structure in a better way. By using this model, we got rid of circularity and no slippage assumptions which are the cases for studies in related literature. Moreover, we afforded more realistic results, because our model provides non-planar leg behavior with distributed mass and wider contact area with the surface. We quantified that how much the circularity assumption remained valid. Also, we simulated various slippage scenarios for different friction coefficients.

References

1. U. Saranli, M. Buehler and D. Koditschek, Design, modeling and preliminary control of a compliant hexapod robot, in *IEEE International Conference on Robotics and Automation*, 2000.
2. E. Moore, D. Campbell, F. Grimminger and M. Buehler, Reliable stair climbing in the simple hexapod 'rhex', in *Robotics and Automation, 2002. Proceedings. ICRA '02. IEEE International Conference on*, 2002.
3. U. Saranli and D. E. Koditschek, Back flips with hexapedal robot, in *Proceedings of the IEEE International Conference on Robotics and Automation*, 2002.
4. U. Saranli, M. Buehler and D. E. Koditschek, Rhex: A simple and highly mobile hexapod robot, *International Journal of Robotics Research* (2001).
5. E. Sayginer, T. Akbey, Y. Yazicioglu and A. Saranli, Task oriented kinematic analysis for a legged robot with half-circular leg morphology, in *Robotics and Automation, 2009. ICRA '09. IEEE International Conference on*, may 2009.
6. Y. O. Aydin, K. C. Galloway, Y. Yazicioglu and D. E. Koditschek, Modeling the compliance of a variable stiffness C-shaped leg using Castigliano's theorem, in *ASME Conference Proceedings*, (44106)2010.
7. W. J. Evers, I. Besselink, H. Nijmeijer and A. Van der Knaap, Development and validation of a modular simulation model for commercial vehicles, *International Journal of Heavy Vehicle Systems* **16**, 132 (2009).
8. P. T. Bibalan and R. Featherstone, A study of soft contact models in simulink, in *Australasian Conference on Robotics and Automation*, 2009.
9. M. M. Ankarali, E. Sayginer, Y. Yazicioglu, A. Saranli and U. Saranli, A dynamic model of running with a half-circular compliant leg, in *International Conference on Cimbing and Walking Robots, CLAWAR (In Press)*, 2012.

INERTIAL REDIRECTION OF THRUST FORCES FOR FLIGHT STABILIZATION

A. DEMIR[†], M. MERT ANKARALI[†], J. P. DYHR[‡], K. A. MORGANSEN[*], T. L. DANIEL[‡], and N. J. COWAN[†]

[†] *Dept. of Mechanical Engineering, Johns Hopkins University, Baltimore, MD 21218 USA*

[‡] *Dept. of Biology,* [*] *Dept. of Aeronautics & Astronautics, University of Washington, Seattle, WA 98195 USA*

Insects are highly maneuverable fliers. Naturally, engineers have focused much of their efforts on understanding the role of insect wing design and actuation for maneuvering and control of bio-inspired micro air vehicles. However, many insects exhibit strong visually mediated abdominal reflexes. The hawkmoth, *Manduca sexta*, has a particularly large abdomen, and recent evidence suggests that these visuo-abdominal reflexes are used to inertially redirect thrust forces for control. In a biologically inspired control framework, we show that the stability of a quadrotor can be categorically improved by redirecting aerodynamic forces using appendage inertia.

Keywords: insect flight control, flight stabilization, pitch control, inertia, dynamics, flexible airframe

1. Introduction

Absent aerodynamic forces, terrestrial animals and robots have little affordance over their net angular momentum during flight, e.g. after a jump. Nevertheless, adjustments to internal configuration degrees of freedom (*i.e.* motions in the "shape-space" of the locomotor system[11]) can capitalize on momentum conservation to orient the body during free flight maneuvers. Based on this idea, Libby *et al.*[9] recently discovered that the tail on certain animals such as the gecko—and possibly dinosaurs—helps keep the animal upright after it jumps. The mechanism involved is remarkably simple: any net angular momentum imparted to the animal during lift-off can be counteracted by counter-rotating the tail at the appropriate speed. This principle was demonstrated on a robot where, following a jump, the robot simply controlled its tail using feedback from an internal sensor to maintain its body angle while in midair.

How might this principle—intersegmental reorientation via inertial forces—be used in flight control? To sharpen this broad question, we focus here on insect flight control. An insect has three primary body segments: the head, thorax, and abdomen with actively controlled joints between each segment. For example, head motions are critical components of the visual tracking control system in insect flight.[13] The motion of the abdomen relative to the middle thoracic segment has been implicated in adjustments of the center of pressure.[3] But, given that the abdomen constitutes at least 50% of the mass of the flying animal suggests that inertial effects could be substantial.[7] This is particularly true in large flying insects such as the hawkmoth *Manduca sexta*. Indeed, these animals produce sensory mediated abdominal responses,[2,8,10] but the role they play in flight control has been unclear. Recently Dyhr *et al.*[4,6] reported evidence that the hawkmoth might use its abdomen to reorient its thorax, much like a gecko uses its tail to reorient its body as described above.

In the case of the hawkmoth, however, there is an interesting twist. As described by Dyhr *et al.*,[4,6] as the abdomen and thorax are reoriented in space, the flight apparatus—which is attached to the thorax—is also reoriented in space, and therefore subsequent thrust forces are redirected. Dyhr *et al.* modeled the biomechanical flight control "plant," and performed system identification of the moth sensorimotor controller. They determined that the visuo-abdominal reflex[2] is consistent with the hypothesis that the moth uses its abdomen to help stabilize flight.

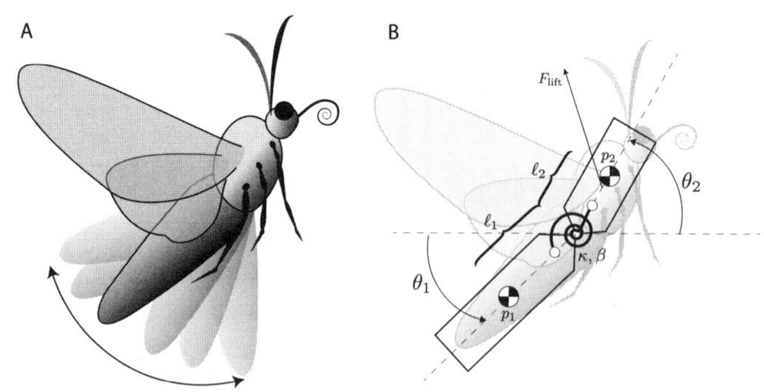

Fig. 1. **(A)** The abdomen of a moth comprises 50% of the mass of the animal. It was recently discovered that strong visually mediated abdominal reflexes of the moth may serve to redirect flight forces for control.[4,6] **(B)** A two-link rigid-body model of sagittal plane dynamics can be used to describe how rapid adjustments of the thoracic abdominal angle, namely $\theta_2 - \theta_1$, can be used to redirect wing forces for control.

Can flexible frames such as those observed in the moth enhance robotic flight control[5]? We devised a mechanism that enables us to test the ideas proposed by Dyhr *et al.*[4–6] in a robot. The mechanism is similar to that presented by Bouabdallah *et al.*[1] for a coaxial helicopter, for which the battery pack was moved to adjust the center of mass position relative to the center of lift. In that work, strong gyroscopic effects from large counter-rotating propellers[12] may limit the connection to insect flight. Here, we focus on building on the ideas above to enhance stability of a flying robot via actuation of an abdomen-like appendage.

2. Methods

2.1. *Aircraft: The X-3D-BL ResearchPilot*

Our tests were performed with an *X-3D-BL ResearchPilot* quadrotor by *Ascending Technologies, Inc.* The aircraft, composed of a magnesium chassis with carbon-fiber sandwiched balsa arms, is approximately 50 cm × 50 cm and weighs approximately 450 g with the battery. It is capable of carrying a payload weighing approximately 400 g for about 10 minutes.

The X-3D-BL has onboard roll and pitch stabilization facilitated by three piezo gyroscopes and a triaxial accelerometer. Data from these sensors are fused onboard and filtered to give accurate absolute roll, ρ, and pitch, ϕ, angles and their rates, $\dot{\rho}$ and $\dot{\phi}$, respectively. These computed measurements are then utilized by separate onboard PD controllers whose gains can be set prior to each flight test.

The quadrotor is piloted via a radio frequency (RF) remote controller (RC) unit (DX7se by *Spektrum*), which gives roll, pitch and yaw control in absolute angles and a dimensionless thrust magnitude control with adjustable resolution. Piloting functions of the RC were partially transferred to our ground computer. This separate communication channel also carried all sensor data from the quadrotor. In addition, we used this channel to remotely set internal stability gains and other flight parameters as well as to conduct sensor calibration and to map remote control functions. The manufacturer provided the specifications of the serial protocol API necessary for our in-house designed flight control software.

2.2. *Inertial Appendage: Servo-Actuated Battery*

We attached the quadrotor's own battery below its chassis in a manner similar to the mechanism proposed by Bouabdallah *et al.*[1] The battery of the X-3D-BL is a rectangular prism weighing about 150 g, *i.e.* one third of the entire quadrotor mass (including battery).

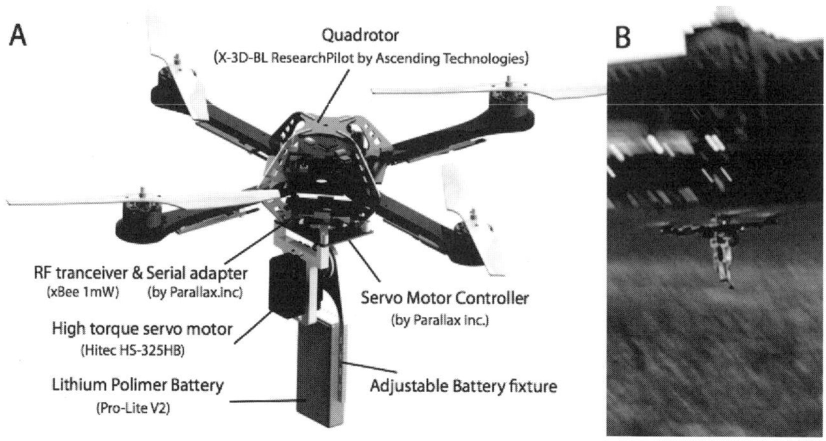

Fig. 2. **(A)** Quadrotor abdomen design and integration.**(B)** Stable flight of the abdomen integrated quadrotor.

The full quadrotor assembly and relevant components are depicted in Figure 2A. The battery angle, θ, with respect to the quadrotor chassis's z-axis (the yaw axis) is set by an available high-torque (3.7 kg/cm) servo motor that has a range of about $-90°$ to $+90°$ and a maximum angular rate of 400 deg/sec at no load.

The battery pivots about the servo shaft at an adjustable distance via a laser-cut balsa fixture. A separate 1 mW 100 m range RF module (XBee 802.15.4 by Digi) is used for data transmission between the ground station and the "abdomen" assembly. The servo motor, servo controller, and RF module are powered by the quadrotor battery through a voltage converter.

Attaching this assembly on the bottom of the X-3D-BL increases the total mass by 110 g and lowers the center of mass by 6 cm. The inertia about the pitch axis is also increased by about 70% (from 0.26 to 0.44 kg m^2).

2.3. *Simple Abdominal Controller*

In our experiments, the abdominal angle, θ_{ctrl}, was controlled via PD feedback from the pitch angle and pitch rate:

$$\theta_{\text{ctrl}}(t) = K_P \cdot \phi(t) + K_D \cdot \dot{\phi}(t). \tag{1}$$

The gains K_P and K_D were hand tuned such that the controller output led to significant stabilization but at same time stayed within the bandwidth limitations of our servo motor. The gains for the final data collection were 0.25 for K_P and 0.07 for K_D.

2.4. *Gain Settings for Onboard Controller*

We used two distinct sets of onboard pitch-control gains for the propellers, separate from the abdomen pitch gains in (1). From the point of view of abdominal pitch control, the plant includes the propeller gains, and we use two different settings for these gains. The first set of propeller gains were set so that the quadrotor was stable but highly oscillatory, which we call the "Stable Plant" gains. The second set of propeller gains were insufficient to stabilize pitch, which we call the "Unstable Plant" gains.

2.5. *Perturbation Experiments*

During our experiments, the human pilot provided thrust control through the RC. The pilot did not adjust pitch, roll or yaw during the experiment. The quadrotor abdomen assembly was tethered from top and below to define the indoor "no-fly zone" for safety reasons. A soft stick was used to perturb the aircraft. The perturbations were provided directly beneath the front or back propeller, creating a large sagittal-plane pitching moment. The perturbations were spaced approximately 10 seconds apart in two consecutive 120 second long trials. Before each trial, the quadrotor sensors were recalibrated. We compared perturbation responses between the trials where the abdominal control was on (closed-loop) and off (open-loop).

The pitch responses to each perturbations were normalized to 15° at their first peak, and then the responses were averaged. The resulting decaying oscillatory signal was fit assuming a second-order response, *i.e.* $\ddot{\phi} + 2\zeta\omega_n\dot{\phi} + \omega_n^2\phi = 0$. The damping ratio, ζ, was calculated via logarithmic decrement, and the period was estimated to recover the the natural frequency $\omega_n = \frac{2\pi}{T\sqrt{1-\zeta^2}}$. This yielded an almost perfect fit (depicted as dashed black line of Fig 3) to the averaged curve.

3. Results

As depicted in Figure 3, the performance of the quadrotor can be enhanced with the complementary inertial redirection of aerodynamic forces provided by the PD-controlled abdomen. In these experiments, the system was open-loop stable (that is, it was stable in the absence of abdominal feedback), but the performance improved significantly in the presence of abdominal feedback. The first two columns of Table 1 show the change in poles for this experiment.

Our second result shows that a pitch-wise unstable quadrotor (X-3D-BL pitch derivative gain is set to 0) can be stabilized with the same feedback

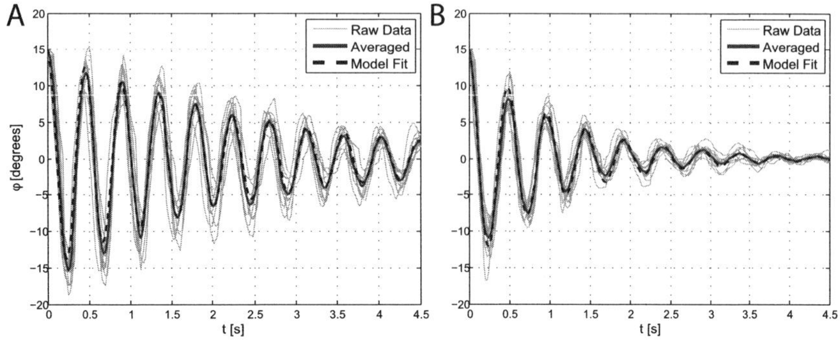

Fig. 3. Pitch angle versus time after pitch perturbations, where the quadrotor propeller gains were "de-tuned" so that pitch was stable but highly oscillatory. **(A)** The open loop (no abdominal control—the abdominal servo-angle is held fixed). **(B)** The servo is controlled in closed-loop via the simple PD scheme described controller (1).

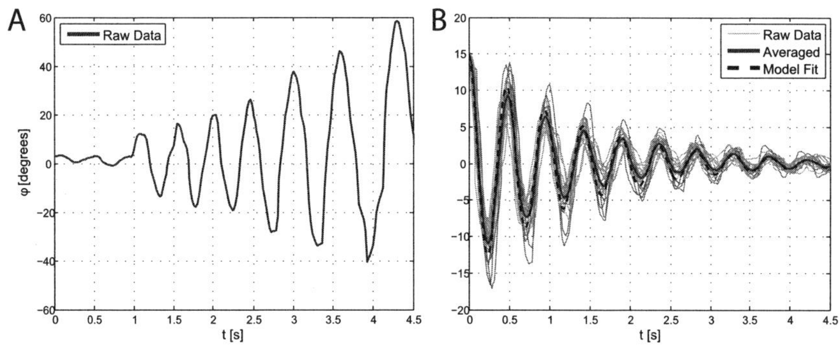

Fig. 4. Pitch angle versus time after pitch perturbations, where the quadrotor propeller gains were "de-tuned" so that pitch was unstable. **(A)** Response with no abdominal control. **(B)** Response with abdominal control.

controller (same PD gains). Fig 4A shows a sample unstable response curve of the pitch angle upon the initial 15 degree perturbation. Fig 4B shows of the same configuration with the active abdominal control. The parameters of the second order model and the associated eigenvalues (poles) are presented in the third column of Table 1.

4. Discussion

Dyhr *et al.*[4] hypothesized that active feedback control of an inertial appendage could be used to enhance the stability of a moth during flight and

Table: Characterization of stability improvement with abdominal control. **(A) Stable Plant:** *Open-loop* refers to the abdominal control being turned off, and *Closed-loop* refers to the abdominal control being turned on. As can be seen, the damping increases significantly when the abdominal control is turned on. **(B) Unstable Plant:** Stability is recovered in *Closed-Loop* (with abdominal feedback).

	(A) Stable Plant		**(B) Unstable Plant**
	Open-loop (abdomen off)	*Closed-loop* (abdomen on)	*Closed-loop* (abdomen on)
$\lambda_{1,2}$	$-0.40 \pm 14.02j$	$-0.90 \pm 13.03j$	$-0.76 \pm 13.38j$
ω_n	14.03 rad/s	13.061 rad/s	13.40 rad/s
ω_d	14.02 rad/s	13.03 rad/s	13.38 rad/s
ζ	0.028	0.069	0.057

suggested this effect could be used for artificial machines.[5] Here, we instantiated their biological hypothesis as an engineering design concept: flexible airframes for active inertial redirection of aerodynamic forces. In our experiments, lift forces generated by propellers were redirected by changes in the pitch angle of the main chassis (analogous to the thorax of the hawkmoth). The pitch angle of the main chassis of the quadrotor was itself mediated by PD-controlled abdominal articulation in the sagittal plane.

There are obvious limitations of our quadrotor aircraft as a physical model of insect flight: the robot and animal operate at dramatically different physical scales and employ radically different propulsion mechanisms. Yet, the combination of mathematical analysis and biological modeling presented by Dyhr *et al.*[4] and physical experiments put forth in this paper, lay a foundation for future work in flexible frames for flight control.

There remain many engineering improvements before us. First, implementing absolute position control would enable tetherless testing of abdomen-mediated maneuvers involving flips and rolls, which is a worthwhile future direction. Most importantly, the controllers we have implemented are ad hoc, and more systematic controller design should greatly enhance the performance of the system.

Acknowledgments

JD, KM, and TD acknowledge support of an ONR MURI grant to KM. This material is based upon work supported by the NSF under Grant No. 0845749 to NC, which was used to support the work of AD, MA and NC.

This paper was included as part of a special session at the 2012 Climbing and Walking Robotics (CLAWAR) conference called "Using Appendage Inertia," organized by Aaron M. Johnson and Daniel E. Koditschek.

References

1. S. Bouabdallah, R. Siegwart, and G. Caprari. Design and control of an indoor coaxial helicopter. In *Proc. IEEE Int. Conf. Robot. Autom.*, pages 2930–2935, Oct. 2006.
2. J. Camhi. Sensory control of abdomen posture in flying locusts. *J. Exp. Biol.*, 52(3):533, 1970.
3. J. Camhi. Yaw-correcting postural changes in locusts. *J. Exp. Biol.*, 52(3):519–531, 1970.
4. J. Dyhr, K. A. Morgansen, T. Daniel, and N. Cowan. Flexible strategies for flight control: an active role for the abdomen. *J. Exp. Biol.*, in prep.
5. J. P. Dyhr, N. J. Cowan, D. J. Colmenares, K. A. Morgansen, and T. L. Daniel. Autostabilizing airframe articulation: Animal inspired air vehicle control. In *Proc. IEEE Int. Conf. on Decision Control*, 2012. Submitted.
6. J. P. Dyhr, N. J. Cowan, A. J. Hinterwirth, K. A. Morgansen, and T. L. Daniel. Flexible frames for flight. In *Soc. Int. and Comp. Biol.*, 2012.
7. T. L. Hedrick and T. L. Daniel. Flight control in the hawkmoth manduca sexta: the inverse problem of hovering. *J. Exp. Biol.*, 209(16):3114–3130, Aug 2006.
8. A. J. Hinterwirth and T. L. Daniel. Antennae in the hawkmoth Manduca sexta (Lepidoptera, Sphingidae) mediate abdominal flexion in response to mechanical stimuli. *J. Comp. Physiol. A*, 196(12):947–956, Dec 2010.
9. T. Libby, T. Y. Moore, E. Chang-Siu, D. Li, D. J. Cohen, A. Jusufi, and R. J. Full. Tail-assisted pitch control in lizards, robots and dinosaurs. *Nature*, 481:181–184, Jan. 2012.
10. T. Luu, A. Cheung, D. Ball, and M. V. Srinivasan. Honeybee flight: a novel streamlining response. *J. Exp. Biol.*, 214(13):22152225, July 2011.
11. J. Ostrowski. Computing reduced equations for robotic systems with constraints and symmetries. *IEEE Trans. Robot. Automat.*, 15(1):111–123, 1999.
12. C. Pradalier. pers. comm., 2012.
13. G. Taylor and H. Krapp. Sensory systems and flight stability: what do insects measure and why? *Advances in Insect Physiology*, 34:231–316, 2007.

DIRECT CONTROL DESIGN WITH BACTERIAL FORAGING ALGORITHM

S. M. HASSAN, H. SUPRIYONO[†] and M. O. TOKHI

Department of Automatic Control and Systems Engineering,
The University of Sheffield, UK

This paper presents the control of hub-angular displacement of a single-link flexible manipulator using proportional-derivative joint-based collocated control optimised by bacterial foraging algorithm. A finite difference model representing the dynamic behaviour of the flexible manipulator is used. Two objective functions, namely mean squared error and the integral squared error, are used in the optimisation process. Analysis of the time-domain results show that the optimisation process using integral squared error results in better performance in terms of response overshoot, undershoot, settling-time and steady-state error as compared to using mean squared error.

1. Introduction

It can be noted from the literature that previous works on hub-angular displacement control of a single-link flexible manipulator have been proposed by researchers [1]. A proportional-derivative (PD)-type joint-based collocated (JBC) control, referred to as JBC PD control, has been used [2-6]. There are only two parameters to tune in JBC PD, i.e. proportional gain in the feed-forward path of hub-angle reference input and derivative gain in hub-velocity feedback path.

The root-locus approach has been previously used to design the JBC PD controller. In the root-locus approach, the two parameters of JBC PD are obtained by finding the poles of the closed-loop transfer function of the plant graphically [7]. Thus, in order to obtain an accurate controller an accurate model of the plant under study is needed. Another technique which can be utilised to find optimal parameters of JBC PD controller is to use an optimisation algorithm.

2. Basic bacterial foraging algorithm

Bacterial foraging algorithm (BFA) [8] is an optimisation method developed based on the foraging strategy of Escherichia Coli (E. Coli) bacteria, which live in the human intestine. Foraging strategy is a method that animals locate, handle

---

[†] Corresponding author, email: herusupriyono@gmail.com.

and ingest their food. In the BFA, the following parameters should be pre-determined: p is dimension of the search space (number of parameters to optimise), S is the number of bacteria in the population (for simplicity, S is chosen for even number), N_c is the number of chemotactic steps per bacterium lifetime between reproduction steps, N_s is maximum number of swim of bacteria in the same direction, N_{re} is the number of reproduction steps, N_{ed} is the number of elimination and dispersal events, p_{ed} is the probability that each bacteria will be eliminated/dispersed. The computation step of BFA [8] is given in the flowchart below:

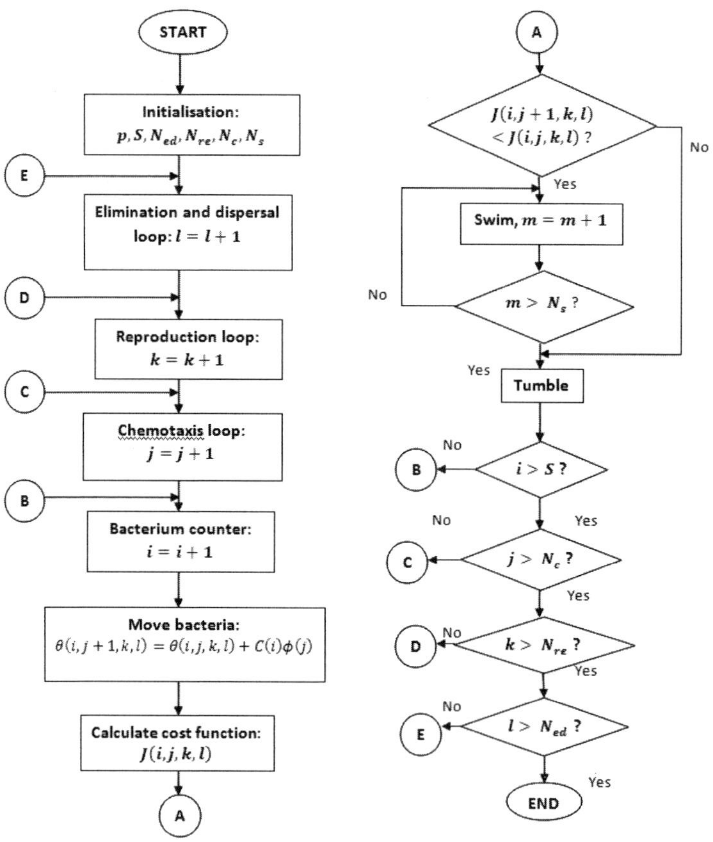

Figure 1. Flowchart of original BFA [8]

In the current work, BFA is used to optimise the JBC PD controller for set-point tracking of single-link flexible manipulator.

3. Flexible manipulator model

A schematic representation of the flexible manipulator considered in this work is depicted in Figure 2 [2], where the stationary and moving coordinates are represented by POQ and $P'OQ'$ respectively, θ is the hub-angular displacement, τ is the applied torque at the hub by a drive motor, M_p represents a payload mass, I_h is the hub inertia, I_p is inertia associated with the payload.

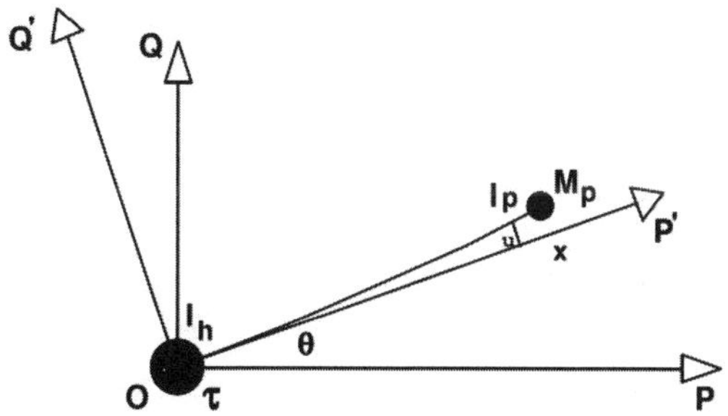

Figure 2. Schematic representation of the flexible manipulator.

A finite difference representing a single-link flexible manipulator under study developed by [2] and then enhanced by [4] is used in the simulations in this work.

4. Control approach

A block diagram of JBC PD control with BFA optimisation for the flexible manipulator is shown in Figure 3; where $u(t)$ is the control command, K_p is proportional gain, K_v is the derivative gain, $R(t)$ is the reference angular displacement and $\theta(t)$ is the actual angular displacement. The nutrient media which will be optimised by BFA is the cost function which is formulated based on the hub-angle error.

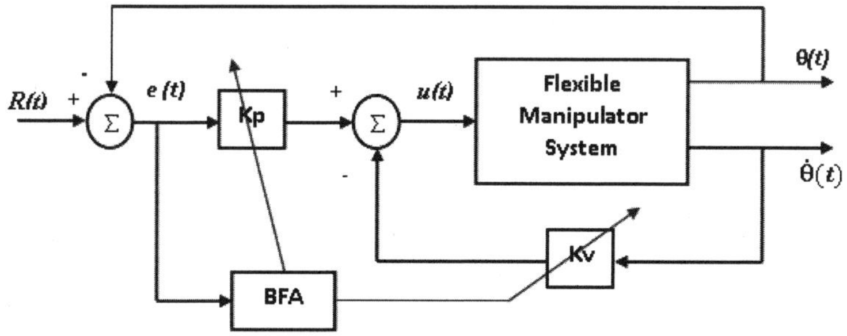

Figure 3. JBC PD-BFA block diagram.

The control signal for the JBC PD-BFA control can be formulated as:

$$u(t) = K_p e(t) - K_v \frac{d\theta(t)}{dt} \quad (1)$$

where *e*(*t*) is the error signal given as:

$$e(t) = R(t) - \theta(t) \quad (2)$$

The mean squared error (MSE) and the integral of squared error (ISE) are considered for optimisation in this work. The mean squared error was chosen based on its features of being mathematically convenient and easy to formulate [9]. The mean squared error is formulated as:

$$J_1 = \frac{1}{N} \sum_{k=1}^{N} (e(k))^2) \quad (3)$$

where $e(k) = R - \theta(k)$ is the angular displacement error, N is the number of data points considered. Similar to MSE, the ISE was chosen based on its feature of being mathematically convenient for computational and analytical purposes [9]. The ISE is formulated as:

$$J_2 = \int_0^T (e(t))^2 dt \quad (4)$$

where *e(t)* is as defined in Eq. (2) and T is chosen arbitrarily and is should at least be equal to the settling time.

5. Results and discussions

In this work, following a set of trials the parameters of the bacterial foraging algorithm were selected as: $p = 2$, $S = 4$, $N_c = 15$, $N_{re} = 4$, $N_s = 4$, $N_{ed} = 3$, $p_{ed} = 0.25$, $C(i) = 0.01$ and $S_r = S/2$. Bacteria were initially placed randomly across the nutrient media.

The performance of the JBC PD control is evaluated based on time-domain characteristics of the hub-angular displacement based on the convergence of the two cost functions. For the convergence analysis of the JBC PD-BFA control with cost functions J_1 and J_2, the convergence will be given for the lowest cost function value achieved by bacteria in every iteration step against the total steps taken by the BFA, given by $N_c \times N_{re} \times N_{ed}$, and for every step. JBC PD control based on root-locus design is also considered for a comparative assessment of performance of the JBC PD-BFA control.

5.1. Reference input

A staircase reference signal with $\pm 60^\circ$ peak to peak amplitude, as shown in Figures 4, is considered in the simulations in this work. The staircase reference signal thus constructed corresponds to a set of bang-bang signals commonly used as input to flexible manipulators performing repetitive manoeuvring tasks.

5.2. *Performance without payload*

Time-domain hub-angle responses of the flexible manipulator to the staircase reference without payload for with the controls are shown in Figure 5 while their numerical results are presented in Table 1. It is noted that the performance of the JBC PD-BFA with J_2 was, by far better than the other two controls, in terms of both response overshoot and undershoot. Comparing the convergence of the optimized controllers in Figures 6 and 7 it can be seen that JBC-BFA with J_2 had a faster convergence (needed smaller number of steps) than JBC PD-BFA with J_1.

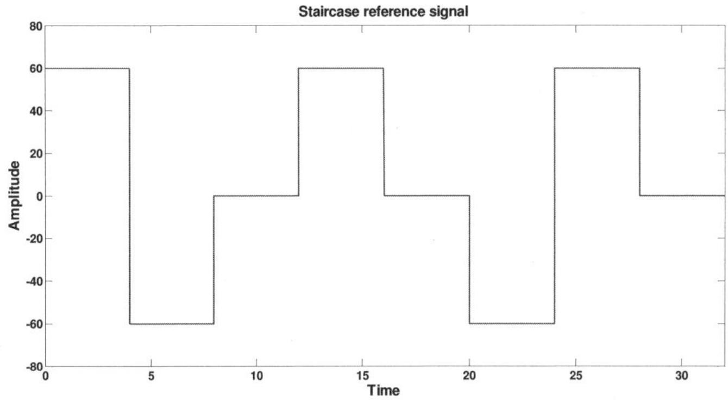

Figure 4. Staircase reference signal used throughout the work.

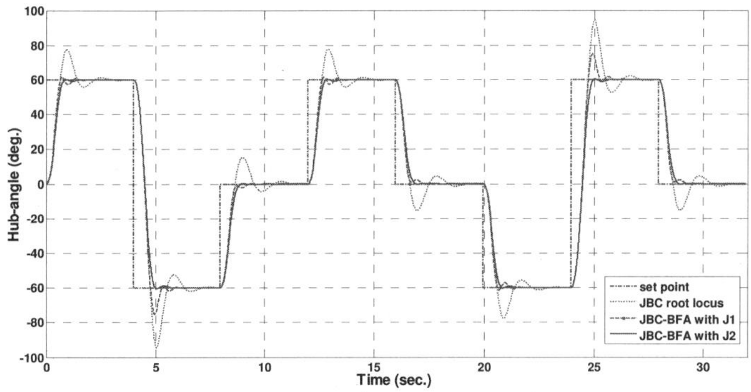

Figure 5. Hub-angle time-domain responses of flexible manipulator to staircase input with various controls without payload.

Table 1. Performances of the controls for staircase input without payload.

Control	K_p	K_v	$\sigma_o(\%)$	$\sigma_u(\%)$	Optimum J
Root locus	1.000	0.300	29.33	57.18	-
BFA with J_1	2.5	0.803	2.40	25.17	479.965
BFA with J_2	1.322	0.543	0.396	0.7649	$9.072e^5$

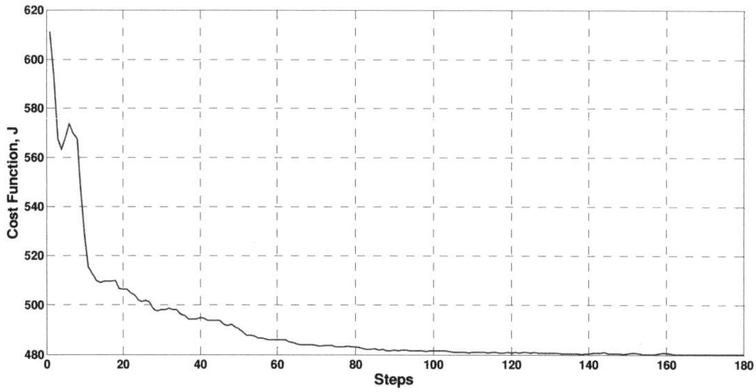

Figure 6. Convergence of JBC PD-BFA for staircase reference input without payload with J_1.

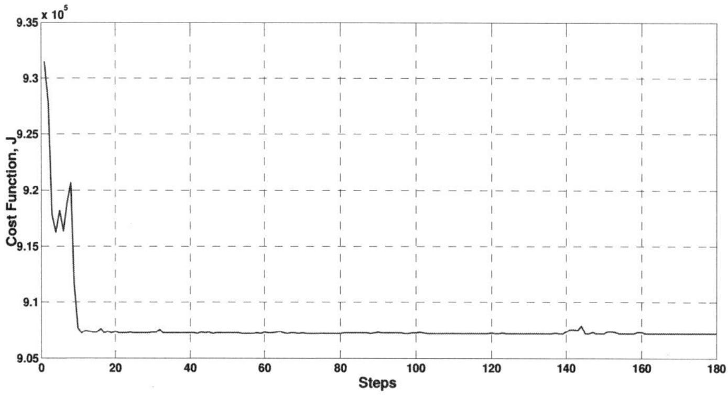

Figure 7. Convergence of JBC PD-BFA for staircase reference input without payload with J_2.

5.3. Performance with 20g payload

Hub-angle responses of the system to the staircase input with 20g payload and different controllers are shown in Figure 8, and their corresponding numerical results are presented in Table 2. It can clearly be seen from both the figure and the table, that the performance of the JBC PD-BFA with J_2 was better than the other two controllers in terms of response undershoot, while that of JBC PD-BFA with J_1 was better than the other two in terms of response overshoot. However, due to the introduction of payload, an increase in the overshoot and undershoot for results of both JBC PD-BFA with J_2 and JBC PD-BFA with J_1 was observed. The convergence plots of JBC PD-BFA controls using J_2 and J_1 are presented in Figures 9 and 10 respectively.

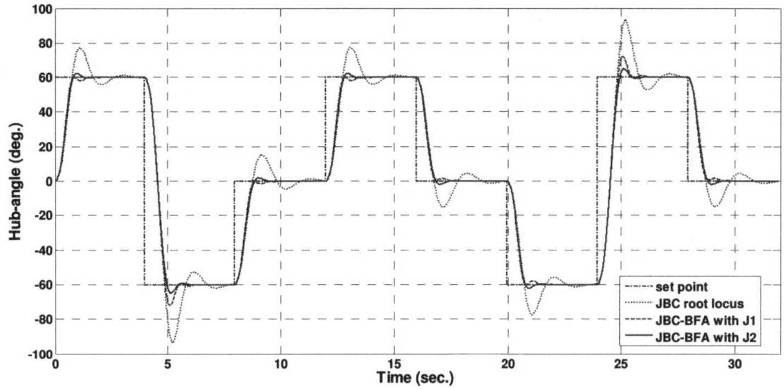

Figure 8. Hub-angle time-domain response of the system to staircase input with 20g payload.

Table 2. Performances of all controls for staircase input with 20g payload.

Controller	K_p	K_v	$\sigma_o(\%)$	$\sigma_u(\%)$	Optimum J
Rootlocus	1.000	0.340	28.97	55.40	-
With J_1	2.591	0.952	1	19.93	540.822
With J_2	1.609	0.667	3.72	7.87	$9.0747e^5$

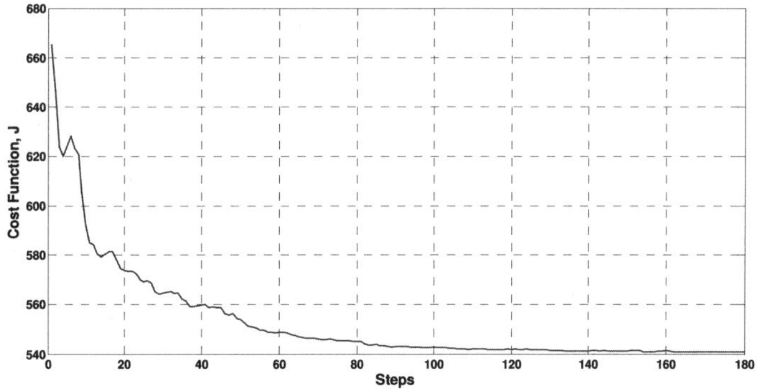

Figure 9. Convergence of JBC-BFA for staircase reference input with 20g payload using J_1.

6. Conclusion

A JBC PD-BFA control strategy has been developed for hub angular displacement of a single-link flexible manipulator using two different objective functions: MSE (J_1) and ISE (J_2). From the time-domain results obtained, it was found that the JBC PD-BFA using J_2 result in better performance, in terms of response overshoot, settling-time and undershoot (where applicable). The results also showed that JBC PD-BFA using J_1 achieved faster response than that using

J_2. In terms of convergence speed, it has been found that the JBC PD-BFA with J_2 converged faster than with J_1.

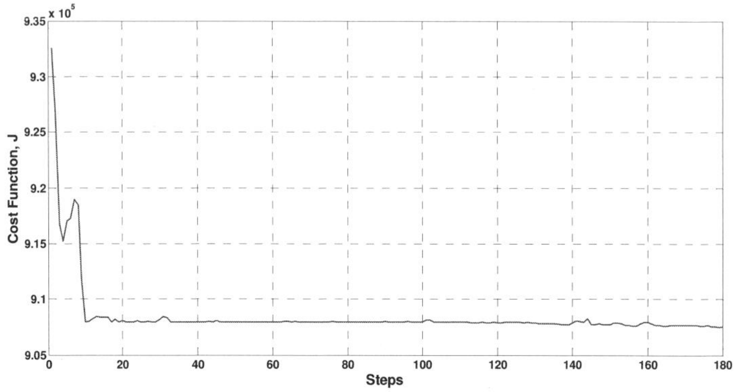

Figure 10. Convergence of JBC-BFA for staircase reference input with 20g payload using J_2.

References

1. Tokhi, M. O. and Azad, A. K. M. (2008). Flexible manipulators: modelling, simulation and control. *Institution of Engineering and Technology*. UK.
2. Azad, A. K. M. (1994). *Analysis and design of control mechanisms for flexible manipulator systems*. PhD thesis, Department of Automatic Control and Systems Engineering, The University of Sheffield, UK.
3. Md Zain, M. Z. Tokhi, M. O. and Alam, M. S. (2005). Robustness of hybrid learning acceleration feedback control scheme in flexible manipulators. *World Academy of Science, Engineering and Technology*, 6, pp. 143-146.
4. Poerwanto, H. (1998). *Dynamic simulation and control of flexible manipulator systems*. PhD thesis, Department of Automatic Control and Systems Engineering, The University of Sheffield, UK.
5. Tokhi, M. O. and Md Zain, M. Z. (2006). Hybrid learning control schemes with acceleration feedback of a flexible manipulator system. *Proceedings of IMechE Vol. 220 Part I: Journal of Systems and Control Engineering*, 220, pp. 257-267.
6. Tokhi, M. O. Md zain, M. Z., Alam, M. S. and Mohamed, Z. (2004). An iterative learning control approach for vibration suppression of flexible manipulator systems. *Proceedings of Control 2004, University of Bath*, UK, September 2004, ID-091, pp. 1-5.
7. Ogata, K. (2002). *Modern control engineering fourth edition*. Prentice-Hall Inc. Upper Saddle River, New Jersey.
8. Passino, K. M. (2002). Biomimicry of bacterial foraging for distributed optimization and control. *IEEE Control Systems Magazine*, June 2002, pp. 52-67.
9. Dorf, R. C. and Bishop, Bishop, R. H. (2008). *Modern control systems, eleventh edition*. Pearson Education International, New Jersey.

INTEGRATING CAPTURE POINT INTO BIOLOGICALLY MOTIVATED CONTROLLED BIPED FOR MAINTAINING STABLE STATE

JIE ZHAO† and KARSTEN BERNS

Robotics Research Lab, University of Kaiserslautern,
Kaiserslautern, 67655, Germany
†E-mail: zhao@informatik.uni-kl.de

In nature, humans can always find a way to maintain the balance when there is an external push in a large magnitude. Most of present bipedal robots walk properly on the even terrain without any external disturbances. However, they cannot deal with the unexpected disturbance very well, such as, uneven terrain or abrupt strong push. The Capture Point[1,2] has been presented to compute the region on the ground where the bipedal robots must land so as to complete a safe stop. Except for the model-based control approaches, model free control methods have been recently researched, such as, a biologically motivated control for a biped has been researched by.[3,4] The integrated behavior-based control framework[5] has been as well included in this simulated biped which simplifies the complexity of control system in a large scale and enhances its adaptability to different environments. This paper will describe, firstly, how to establish the model of a linear double link inverted pendulum with a flywheel. Afterwards, the calculated torques will be integrated into the motor patterns of biologically motivated control to enhance the capability of the biped in more complex environments. Finally the simulation results will be demonstrated and analyzed.

Keywords: Capture Point; Biologically motivated control; iB2C; Biped robots; Inverted pendulum

Introduction

Maintaining body balance is very necessary for bipedal robots performing in complex human environments. Due to the unexpected push during walking or standing still on an uneven terrain, it is critical for bipedal robot to maintain the stable state and re-stabilize its posture. Well-known stability theory such as Zero-Moment-Point theory[6] can be only applied to predefined trajectories. However, it cannot be suitable for maintaining stability since influences are always unknown. Therefore the trajectories and posture adjustments cannot be calculated in advance. There is a requirement of a theory that prevents fall and can be easily implemented. A Capture Point[1] is a point on the ground on which the bipedal robot can stride so as to make itself to a complete stop. It considers the states and actions that let itself take a step to maintain balance.

The biologically motivated control is based on the observation of human motions. The control of human motion is hierarchically organized(shown in Fig. 1), in which *Locomotion Modes*, *spinal Pattern Generators*, *Motion Phases*, *Reflex Controller* and *Motor Patterns* have been placed from the most top layer to the lowest layer. This approach utilizes feedback control (reflex controllers) as well as feed-forward control (motor patterns) that allow us to easily and rhythmically control the biped walking and other motion behaviors.

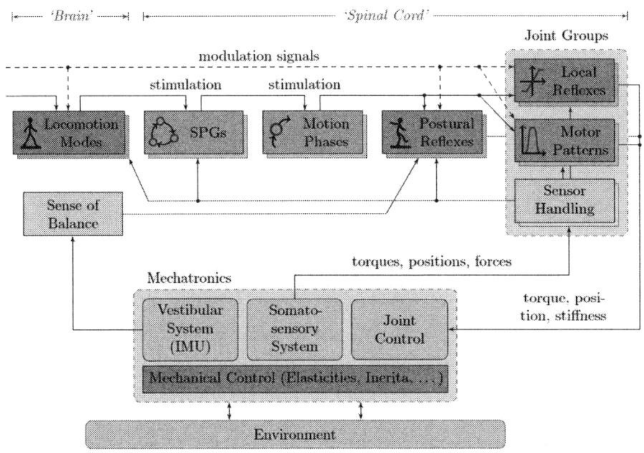

Fig. 1. Hierarchical organization of control units and the control systems, the system mechatronics and the environment.

Capture Point and Capture Region

The Capture Point and Capture Region allows the bipedal robot to take a step to a full stop at the proper timing and place. The location of a Capture Point is dependent on the trajectory of swinging leg and other behaviors in the leg swing phase and is therefore not unique.[1] Hence, a collection of such *Capture Points* is called a *Capture Region* . The Fig. 2 shows the Capture Region on which the bipedal robot takes step to maintain its stability. Since the motion in sagittal and frontal plane can be treated separately, we can consider the biped system as a planar inverted pendulum with a flywheel in

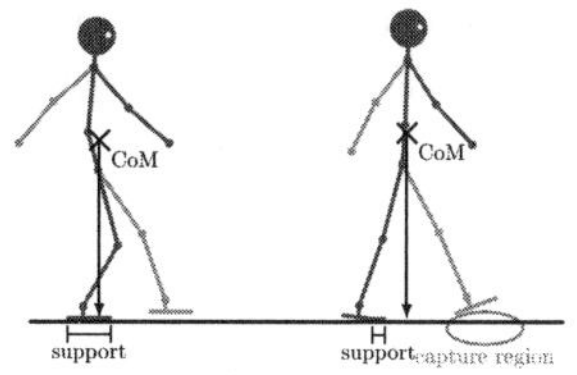

Fig. 2. Capture region for a recovery step.

sagittal plane as shown in Fig. 3. This system has three actuators mounted at the hip, knee and ankle joint in which the hip and upper body can be intuitively dealt as a flywheel. Because the actuator at ankle normally works for the *Toe Off* and the *Gesture Adjustment* of foot during leg swing phase,[3] we can ignore its influence in determining *Capture Point*. The actuator at the knee normally operates very slightly during leg swing phase, since the passive dynamics has been taken into account.[4] Consequently we mainly consider the torque generated at the hip joint in locating the *Capture Point*. The dynamics of the system can then be written as:

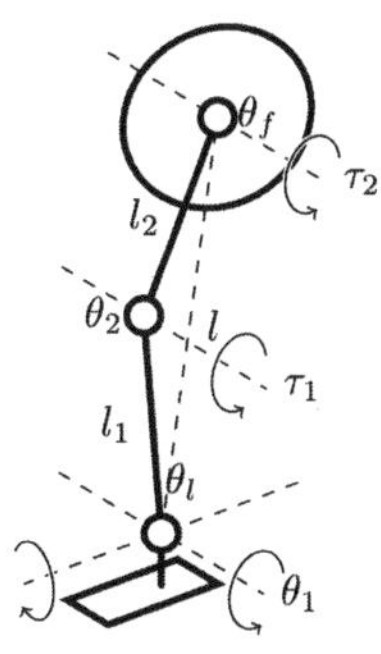

Fig. 3. Simplified model of a biped in the heel contact phase with a flywheel body and legs. Three actuators in the sagittal plane are mounted at the hip, knee and ankle respectively.

$$\begin{aligned} m_a\ddot{x} &= F\sin\theta_l - \frac{\tau_2}{l}\cos\theta_l \\ m_a\ddot{z} &= F\cos\theta_l + \frac{\tau_2}{l}\sin\theta_l - m_a g \\ J\ddot{\theta}_f &= \tau_2 \\ l &= \sqrt{l_1^2 + l_2^2 + 2\cos(\theta_1 - \theta_2)l_1 l_2} \\ \theta_l &= \arcsin\left(\frac{l_1\sin\theta_1 + l_2\sin\theta_2}{l}\right), \end{aligned} \tag{1}$$

where m_a and J are the mass and inertial moment of the flywheel, g is the gravitational acceleration constant, $\ddot{x}$ and $\ddot{z}$ determine the acceleration of center of mass in sagittal and vertical plane. τ_2 is the motor torque on the flywheel, and F is the force along the leg. The l denotes the distance from ankle joint to the center of flywheel, whereas l_1 and l_2 indicating the length of shank and upper leg. θ_1, θ_2 and θ_l are the angles of ankle, knee and approximated leg with respect to the vertical axis respectively. The θ_f is the angle of the center of the flywheel depending on the vertical axis. For bringing the biped to a stop at the maximum flywheel angle, the torque can be solved as:[1]

$$\tau_2(t) = \tau_{max}u(t) - 2\tau_{max}u(t - T_1) + \tau_{max}u(t - T_2) \tag{2}$$

where τ_{max} is the maximum torque that the hip joint can provide, and $u(t)$ is the unit step function. T_1 and T_2 are the time at which the deceleration begins and then it comes to a stop respectively. Where

$$\begin{aligned} T_2 &= \sqrt{\frac{4J}{\tau_{max}}(\theta_{f_{max}} - \theta_{f_0})} \\ T_1 &= \frac{1}{2}T_2 \end{aligned} \tag{3}$$

θ_{f_0} and $\theta_{f_{max}}$ denote respectively the initial and the maximum angle of flywheel allowed with respect to the vertical axis. The detailed solution for determining T_1 and T_2 and a Capture Region can be found in .[1] In this paper, we will discuss how to generate the required *Motor Patterns* and related *Reflex Controllers* in the biologically motivated control system.

Integrating Capture Point into Framework of Biologically Motivated Control

Due to lacking of model of the system in biologically motivated control, figuring out an integration of *Capture Point* is hence emerging. As mentioned above, there exists state machines that are responsible for switching walking and standing phases. Here we would like to add the *Capture Point* as a block into this state machine. The Fig. 4 illustrates the different states and their triggering events for stable standing.

Motor Patterns for a Capture Point

As mentioned, the motor pattern is a basic control unit in the framework of biologically motivated control. It works in a feed-forward way that depends on the certain event or timing during the walking. Its main advantage is

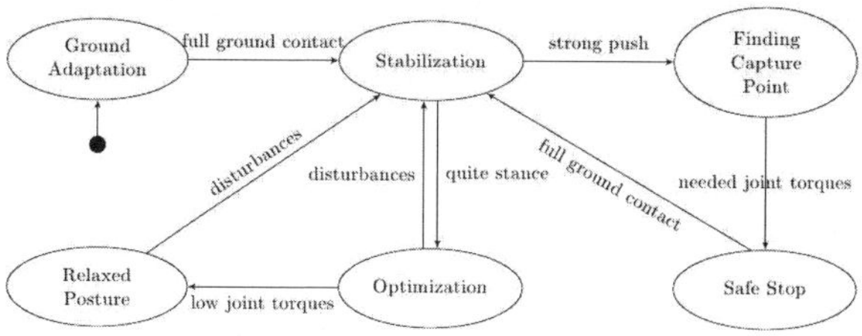

Fig. 4. State machine for standing and push recovery.

that it is working rhythmically and the parameters, such as, rising and falling time of the pattern can be easily adjusted. The motor patterns have normally the following parameters, rising time t_1, duration time t_2 and falling time t_3 and the magnitude of the maximum torque which mitigates the complexity of calculating motor patterns in a large scale.

Eq. 4 shows the function that is used as a general form for motor patterns,

$$\tau(t) = \tau_{max} \cdot A \begin{cases} \frac{1}{2} + \frac{1}{2}\sin(\ \pi(\frac{t}{t_1} - \frac{1}{2})\ ) & 0 \leq t < t_1 \\ 1 & t_1 \leq t < t_2 \\ \frac{1}{2} - \frac{1}{2}\sin(\ \pi(\frac{t-t_2}{t_3-t_2} - \frac{1}{2})\ ) & T_2 \leq t \leq t_3 \end{cases} \quad (4)$$

Three parameters t_1 , t_2 , and t_3 are responsible for adapting the length of the smooth rise of the function, the duration of the plateau at maximum peak, and the time until the smooth fall of the function arrives zero.[7] A denotes the adjustable parameter that is normalized in $[-1, 1]$. Thus the output torque τ is limited in the range of $[-\tau_{max}, \tau_{max}]$. Fig. 5 illustrates how those time parameters are defined and sign and amplitude can be varied by the factor A. Let us consider the Eq. 2 again, we will have to define the

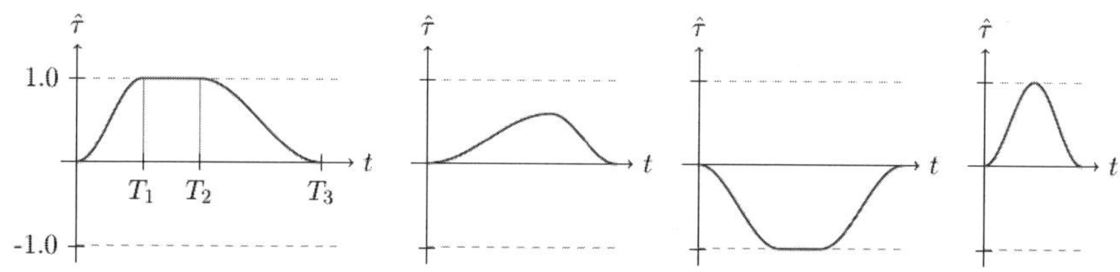

Fig. 5. General form for motor patterns. t_1 is the end of rise time, whereas t_2 and t_3 indicating the start and end of fall. A is the adjustable parameter for maximum torque output.[4]

time T_1 for decelerating the upper body. Comparing it to Eq. 4, we have to only add a negative motor pattern to realize the behavior of deceleration. Afterwards, a positive motor pattern which starts at T_2 is then generated. The torque form can be found in Fig. 6.

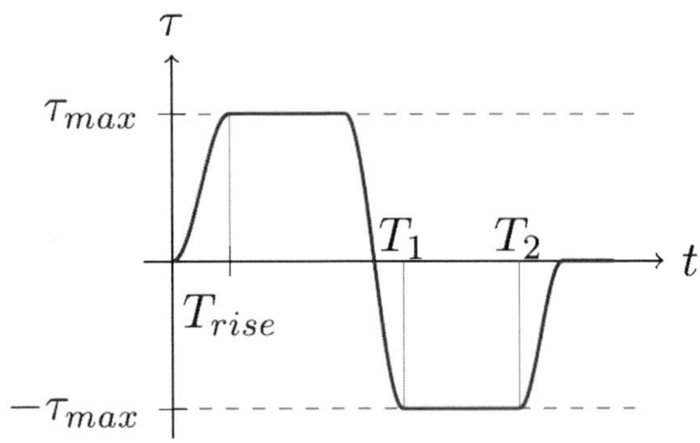

Fig. 6. The calculated torques for determining a Capture Point.

Related Reflex Controllers

We have just mentioned only one motor pattern that manages the hip swing, other reflex controllers and motor patterns must be included in addition to realize the whole process. The walking process can be separated into 5 different phases,[4] thus we will analyze the *Leg Swing* phase and *Weight Acceptance* phase of the same leg. From Fig. 7, we can see that the *Hip*

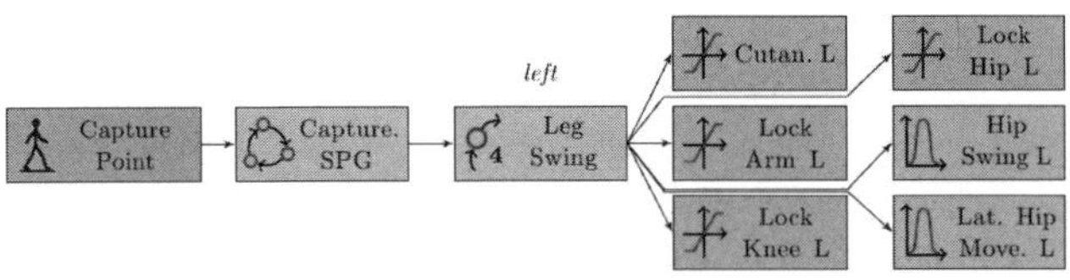

Fig. 7. Control Units stimulated during *Leg Swing* phase

Swing is the torque generator for finding a capture point while the reflex controller *Lock Hip* is responsible for keeping hip joint swinging in the allowable range. It will stimulate the stiffness controller[4] when hip joint is approaching its maximum angle.

Considering the *Weight Acceptance* phase in Fig. 8, the *Hip Brake* motor pattern will be deployed to decelerate the upper body mentioned in Fig. 6.

Another important reflex controller that should be seriously considered is the *Ankle Brake.* It works as a torque generator and a stiffness controller, which provides decelerating torques as a motor pattern in case the ankle is not yet close to its neutral angle. When the angle is close enough to the neutral angle, the ankle would become stiffer to prevent the over-swinging of leg.

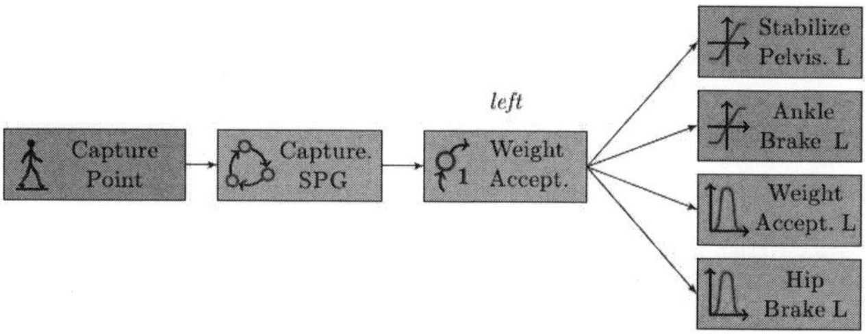

Fig. 8. Control Units stimulated during *Capture Point* phase 1

Simulation Results

The suggested method is tested in the framework of MCA2[a] with the physics engine *Newton Game Dynamics*[b]. The biped has in total $m = 80kg$

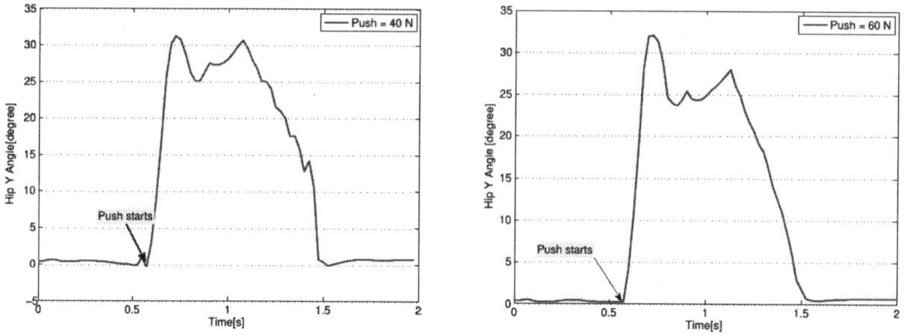

Fig. 9. (a) Left Hip Y Angle when Push = 40 N. (b) Left Hip Y Angle when Push = 60 N

in weight and $h = 180cm$ in height. In which the weight of upper body is $50kg$ with the inertia of $J = 2.5kgm^2$. The maximum flywheel angle is $\theta_{f_{max}} = 0.2\pi$ and the maximum hip torque is $\tau_{max} = 50Nm$.

[a]http://rrlib.cs.uni-kl.de
[b]http://newtondynamics.com/forum/newton.php

Fig. 9(a) and 9(b) show the course of left hip angle in sagittal plane after suffering a strong push at respectively $40N$ and $60N$ for locating the *Capture Point* at the maximum speed $0.8m/s$. we can see that when the push is stronger, the angle changes also quicker because of the shorter existence of the positive torque. It decreases sharper as well than that of in slighter push due to the longer emergence of negative torque. The video of finding *Capture Point* can be found on `http://agrosy.informatik.uni-kl.de/en/robot-gallery/biped/`.

Conclusion and Future Works

Although the model established for an inverted pendulum is too simplified for a real biped, the behaviour-based control can easily integrate this method and thus implement the experiment. However, we have just made the implementation while it is standing, the push during walking is still missing. Next step would be the realization of push recovery in both direction during walking. Furthermore, N- step capturability could be tested if the disturbances are higher. Online learning controllers could be also used to determine the place and timing to step automatically after suffering disturbances.

References

1. J. Pratt, J. Carff, S. Drakunov and A. Goswami, Capture point: A step toward humanoid push recovery, in *Humanoid Robots, 2006 6th IEEE-RAS International Conference on*, dec. 2006.
2. T. Koolen, T. de Boer, J. Rebula, A. Goswami and J. Pratt, Capturability-based analysis and control of legged locomotion(March 2011).
3. J. Zhao, T. Luksch and K. Berns, Controlling human-like locomotion of a biped by a biologically motivated approach, in *14th International Conference on Climbing and Walking Robots (CLAWAR)*, September 6-8 2011.
4. T. Luksch and K. Berns, Control of bipedal walking exploiting postural reflexes and passive dynamics, in *IEEE International Conference on Applied Bionics and Biomechanics (ICABB)*, (Venice, Italy, 2010).
5. M. Proetzsch, T. Luksch and K. Berns, Development of complex robotic systems using the behavior-based control architecture iB2C, *Robotics and Autonomous Systems* **58**, 46(January 2010).
6. M.Vukobratovic and B. Borovac, Zero-moment point –thirty five years of its life, *International Journal of Humanoid Robotics* **1**, 157(January 2004).
7. T. Luksch and K. Berns, Controlling dynamic motions of biped robots with reflexes and motor patterns, in *Fourth International Symposium on Adaptive Motion of Animals and Machines (AMAM)*, (Cleveland, USA, 2008).

SECTION–5

INNOVATIVE DESIGN OF CLAWAR

A NEW APPROACH TO THE MODELLING OF VORTEX-BASED SUCTION CUPS FOR CLIMBING ROBOT ADHESION

FILIPPO BONACCORSO, GIOVANNI MUSCATO
Dipartimento di Ingegneria Elettrica Elettronica e Informatica,
Università degli Studi di Catania, Viale A. Doria 6, 95125 Catania Italy
e-mail: filippo.bonaccorso@diees.unict.it, gmuscato@diees.unict.it

ARTURO PAGANO, ALBERTO FICHERA
Dipartimento di Ingegneria Industriale e Meccanica,
Università degli Studi di Catania, Viale A. Doria 6, 95125 Catania Italy
e-mail: apagano@diim.unict.it, afichera@diim.unict.it

In this work an innovative approach to the modeling of vortex-based suction cups is presented. This new kind of active suction cup provides a strong adhesion force. The cup is of non-contacting type, reducing friction between cup and wall, allowing thus to be used as a conventional sliding suction cup with a very low friction, since no sealing brushes or gaskets are required. Several sets of measurements of the cup internal pressure and associated payload have been acquired. In order to better understand and describe the cup behavior, a model has been developed allowing to properly identify the relation between the adhesion force and the fan angular velocity, using information on internal pressure. This relation has been validated by a comparison with experimental data.

1. Introduction

Actual research at DIEEI Robotic Laboratory comprises new gripping methods for climbing robot application and their modeling and control [1]: magnetic adhesion [2], [3], standard suction cups [4], [5], vortex-based suction cups [6], [7] and electrostatic adhesion [8]. A stronger consideration was given to high payload systems in order to build more efficient and useful climbing robots. In this work a sliding suction cup, based on the idea of a vortex attractor, has been addressed. This new kind of suction cups was originally developed by Duke University and adopted on some robot prototypes [9], and it was also recently used by ETH researchers [10]. In the next sections a short description of the hardware will be shown while more attention will be given to new results on the identification of the static relation between the adhesion force, generated by the vortex-based suction cup, and the fan angular velocity.

2. Vortex-based suction cup structure

The adopted active suction cup, shown in Figure 1, uses a special centrifugal fan that, by means of its rotation, produces a vortex inside the cup. The pressure distribution related to the vortex generates a normal force that allows the robot to adhere to the surface. The cup and the fan were built so that the air is recycled inside the cup. Being the cup of non-contacting type it can be used as a sliding cup.

This system has been successfully adopted to built the Alicia-VTX robot [6], [7], shown in Figure 1.

Figure 1. A section of the active suction cup and a picture of the Alicia-VTX climbing robot.

3. Experimental test bed

The experimental test-bed, already shown in [7], allows to simultaneously acquire both the pressure inside the cup and the adhesion force.

Pressure profile stands for the pressure trend inside the cup along a radius. Being the system characterized by a central symmetry, as reported in [6], pressure profiles have been measured along a radius of the cup using five evenly spaced sensors while a set of three load cells, mounted between the cup and the sliding surface, have been used to measure the adhesion force.

Figure 2 shows the experimental test bed. An initial set of measurements have been performed at room pressure in order to compensate different sensor offsets and the value of the environmental normal pressure. The average for each sensor of these measurements represents the mean offset of that sensor at current room condition and is indicated with P_{0a}.

Figure 3 reports data acquired by the five sensors, i.e. the instantaneous pressure difference, ΔP, between room pressure and the pressure inside the cup at each position. It is calculated as $\Delta P_i = P_i - P_{0a}$.

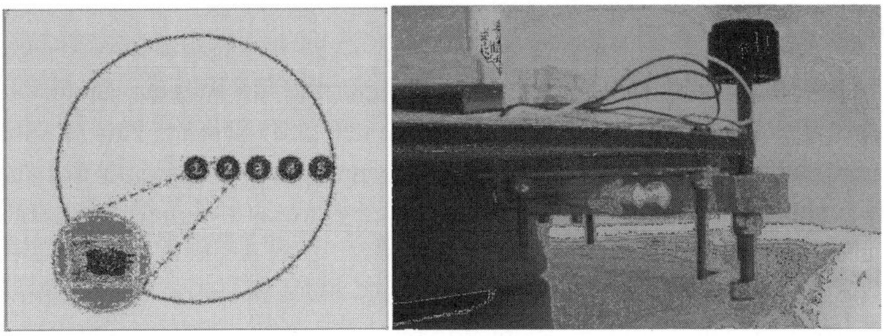

Figure 2. The experimental test bed.

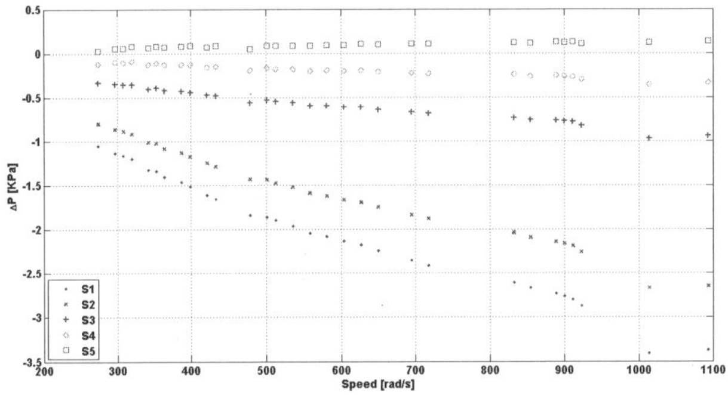

Figure 3. Data Acquired by pressure sensors.

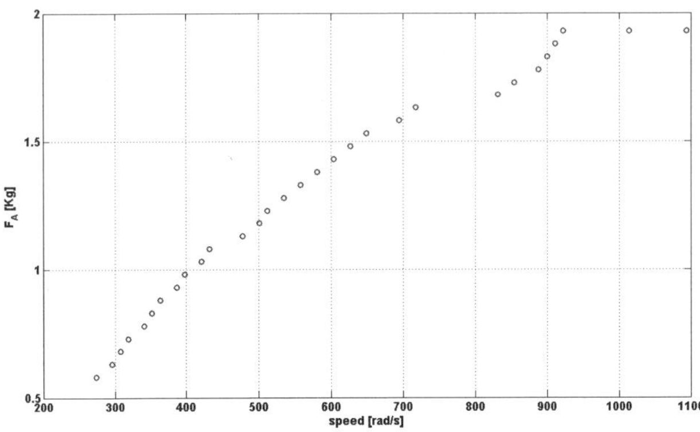

Figure 4. Adhesion force F_A in Kg.

Figure 4 shows the adhesion force measured at the values of the fan angular velocity corresponding to the pressure measurements shown in Figure 3.

4. Model identification

The system that has to be modelled has as input the fan angular velocity ω and as output the adhesion force, F_A, between the cup and the wall. This relationship is represented by the function *Γ()* and has been already discussed in [6] and [7]. Figure 5 shows the model structure where S_I are pressure signals while *H* is the gap between the cup and wall.

The subsystem *Φ()* of *Γ()* has been considered separately. The function *Φ()* used in [6] is a static law that relates the pressure inside the cup to the force F_A and depends from the pressure profile and from the gap between the cup and the wall. During measurements, the fan angular velocity was held at constant value, to allow the reading of the F_A value. At the same time, the pressure time series measured by the five sensors were recorded.

Figure 5. The proposed model.

In [6] a neuro-fuzzy network, trained with ANFIS algorithm, has been used to identify *Φ()*. Being this a static law, the dynamic model of the adhesion system has to be identified too. Having modeled the relationship between the pressure signals and the attractive force F_A, the model of the whole system has been identified using as output the attractive force F_A as estimated by *Φ()* and as input the fan angular velocity. Since the system is strongly non-linear, a NARX model has been considered, as reported in [6] and [7].

Even if it is simple to realize and use vortex-based suction cup, as already presented in [6] [7] and [10], there are few results concerning the modeling of the system behavior.

A new approach to evaluate the static part of *Γ()* will be presented in the next section.

5. A new approach to F_A evaluation

Figure 6 reports the mean values of the pressure measured at each of the five positions indicated in Figure 3 for the experimental condition performed at maximum angular velocity and *H=0* mm; it is worth noting that the first two pressure measurements only slightly differ from one another, so that the pressure can be assumed approximately constant in the fan inlet region ($r<r_2$), corresponding to the pressure measured by the first sensor. Moreover, the whole set of experimental conditions have been observed to show the same trend of distribution of the mean values of the measured pressure.

The approach herein proposed aims at determining a static law $F_A(\omega)$ between the fan angular velocity, ω, and the adhesion force F_A. This can be done on the basis of two successive estimations.

The first consists, for a given value of the fan angular velocity, in the determination of the function that fits the experimental mean values of the pressure and, therefore, describes the radial distribution of the pressure inside the cavity, $p(r)$.

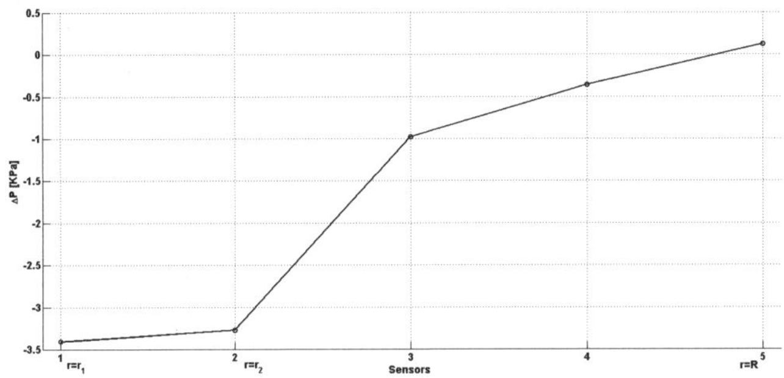

Figure 6. Pressure measured by the five sensors with maximum fan angular velocity and H=0.

In particular, a piece-wise function made up of two terms can be adopted to describe the pressure distribution inside the cavity. The first piece of the function can be set constant and equal to the pressure p_1, measured by the first sensor at $r=r_1$; therefore $p(r)=p_1$ has been assumed to describe the pressure distribution for $r<r_2$. The second piece of the function is obtained through a simple polynomial fitting of the mean pressure values measured at position from 2 to 5 of Figure 2. In order to estimate this part of the fitting function, both a second order and a third order polynomials in the variable r have been determined, as shown in Eq. (1) and (2). Therefore, for $r>r_2$, the function $p(r)$ can be approximated using one of the following two equations:

$$p(r) = ar^2 + br + c \quad (1)$$

$$p(r) = \alpha r^3 + \beta r^2 + \gamma r + \delta \quad (2)$$

It is now important to observe that the coefficients of both polynomials, i.e. *[a, b, c]* and *[α, β, γ, δ]*, are constant terms only for a fixed value of the fan angular velocity. For this reason, different arrays *[a(ω), b(ω), c(ω)]* and *[α(ω), β(ω), γ(ω), δ(ω)]* of coefficients have been calculated for each of the experimental values of the fan angular velocity.

Therefore, the second estimation simply consists in the determination of a new set of fitting functions expressing the dependence of each of the above coefficients on the angular velocity ω. Clearly, a similar observation hold for the first piece of the pressure fitting function, i.e. $p(r)=p_1$, considering that the measured value of p_1 varies with ω, and can therefore be expressed by determining the fitting function $p_1(\omega)$. At the end of the procedure, the pressure distribution inside the cavity is identified and can be expressed in either of the following two forms corresponding to (1) and (2):

$$p(r,\omega) = \begin{cases} p_1(\omega) & 0 \le r < r_2 \\ a(\omega)r^2 + b(\omega)r + c(\omega) & r_2 \le r \le R \end{cases} \quad (3)$$

$$p(r,\omega) = \begin{cases} p_1(\omega) & 0 \le r < r_2 \\ \alpha(\omega)r^3 + \beta(\omega)r^2 + \gamma(\omega)r + \delta(\omega) & r_2 \le r \le R \end{cases} \quad (4)$$

A linear dependence on the angular velocity ω, has been found to perform satisfactorily for all of the coefficients appearing in previous equations. Therefore, it is straightforward to proceed to the integration of the fitting function of the pressure distribution over the entire adhesion surface, with the aim of determining the adhesion force.

For the case of the second order polynomial fitting, the integration can be expressed according to:

$$F_A(\omega) = \int_0^{r_2} p_1(\omega) 2\pi r\, dr + \int_{r_2}^{R} [a(\omega)r^2 + b(\omega)r + c(\omega)] 2\pi r\, dr \quad (5)$$

while, for the case of the third order polynomial fitting, it becomes:

$$F_A(\omega) = \int_0^{r_2} p_1(\omega) 2\pi r\, dr + \int_{r_2}^{R} [\alpha(\omega)r^3 + \beta(\omega)r^2 + \gamma(\omega)r + \delta(\omega)] 2\pi r\, dr \quad (6)$$

It has to be observed that in both cases the integration is very simple, since all the coefficients depend on ω but are independent from *r*. Figure 7 reports both the experimental and the calculated adhesion force for both the second and the third order fittings.

From the analysis of Figure 7 it is possible to observe that the predictions of the experimental adhesion force can be considered suitable:

- for $\omega < 700$ for the second order fitting estimation, Eq (5);
- for $\omega > 700$ for the third order fitting estimation, Eq(6).

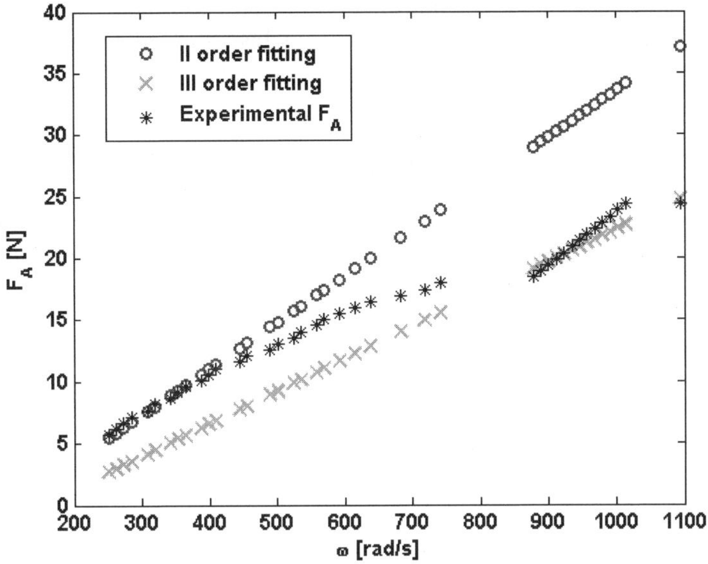

Figure 7. Estimated vs experimental adhesion force.

6. Conclusion

A novel approach to the modeling of vortex-based suction cups has been presented. The approach presented in previous section formally introduces a new way to identify the static part of the model of the suction cup.

Although the proposed approach is simple no one has yet modeled the suction-cup behavior.

Pressure measurements inside the cup have been used to identify a function correlating the adhesion force of the cup to the fan angular velocity in static conditions. The relation found has shown a non-linear behavior which is still under investigation.

This relation has been validated by comparison with experimental data as shown in Figure 7.

References

1. D. Longo, G. Muscato, "Adhesion techniques for climbing robots: State of the art and experimental considerations", *11th International Conference on Climbing and Walking Robots and the Support Technologies for Mobile Machines*, CLAWAR'08, 08-10 September 2008, Coimbra, Portugal.

2. L. Fortuna, A. Gallo, G. Giudice, G. Muscato, "ROBINSPEC: "A Mobile Walking Robot for the Semi-Autonomous Inspection of Industrial Plants", *Robotics and Manufacturing: recent trends in research and applications*, Vol. 6, ASME PRESS New York (USA), pp. 223-228, May 1996.
3. D. Longo, G. Muscato, "A Small Low-Cost Low-Weight Inspection Robot with Passive-Type Locomotion", *Integrated Computer-Aided Engineering*, Vol. 11 N. 4, pp. 339-348, 2004.
4. G. La Rosa, M. Messina, G. Muscato, R. Sinatra, "A low cost lightweight climbing robot for the inspection of vertical surfaces", *Mechatronics* vol. 12, 2002, pp. 71-96, Pergamon.
5. D. Longo, G. Muscato, "The Alicia[3] climbing robot for automatic wall inspection", IEEE *Robotics And Automation Magazine*. vol. 13 pp. 42-50 ISSN: 1070-9932.
6. F. Bonaccorso, C. Bruno, D. Longo G. Muscato, "Structure And Model Identification Of A Vortex-based Suction Cup", *Proceedings of the11th International Conference on Climbing and Walking Robots and the Support Technologies for Mobile Machines* CLAWAR'08, 8-10 September 2008, Coimbra, Portugal.
7. F. Bonaccorso, D. Longo, G. Muscato, "Modeling Of An Innovative Actuator For Climbing Robot Adhesion"; *Proceedings of the 12th International Conference on Climbing and Walking Robots and the Support Technologies for Mobile Machines* CLAWAR'09, Istanbul, Turkey, 09-11 September 2009, pp.891-900.
8. D. Longo, G. Muscato, G. Tarantello, "Performance evaluation of electrostatic adhesion for climbing robots", *13th International Conference on Climbing and Walking Robots and the Support Technologies for Mobile Machines* CLAWAR'10, 31 August - 03 September 2010, Nagoya, Japan.
9. http://www.clarifyingtech.com/public/robots/robots_public.html
10. L. Geissmann et al., " PARASWIFT – A Hybrid Climbing And Base Jumping Robot For Entertainment", *Proceedings of the 14th International Conference on Climbing and Walking Robots and the Support Technologies for Mobile Machines* CLAWAR 2011, Paris, France, 6 – 8 September 2011, pp 317 - 324.

DESIGN AND SCALING OF VERSATILE QUADRUPED ROBOTS

C. SEMINI*, H. KHAN, M. FRIGERIO, T. BOAVENTURA, M. FOCCHI,
J. BUCHLI and D. G. CALDWELL

Department of Advanced Robotics, Istituto Italiano di Tecnologia
Via Morego 30, 16163 Genova, Italy
**E-mail: claudio@semini.ch*
http://www.iit.it/en/advanced-robotics.html

The rough terrain mobility of legged robots is expected to exceed the performance of their wheeled or tracked counterparts. To fully take advantage of the legs, such robots need to be versatile by achieving highly dynamic motions at the same time as careful navigation over rough terrain. Highly dynamic robots need to be designed to be fast and strong enough to run and jump. A dynamic robot needs to be light and powerful at the same time. Two requirements that are conflicting and therefore have to be traded off. In this work we present a tool that helps quadruped robot designer to better select and size joint actuators for various robot sizes. We use the squat jump as characteristic motion of a highly dynamic robot and estimate required joint torque and velocity in relation to maximum jump height, body mass and leg segment length.

Keywords: Quadruped Robot Design; Scaling; Actuator Sizing; Jump.

1. Introduction

Legged machines promise higher mobility in difficult terrain in comparison to more traditional vehicles with wheels or tracks. Nature shows that legged animals including humans are able to navigate through challenging grounds by putting their limbs on suitable footholds to support their body. Depending on the difficulty of the terrain, this may not only include climbing but also highly dynamic maneuvers such as jumping where static equilibrium is temporarily lost.

Today's legged robots lack the ***versatility*** to perform both highly dynamic motions **and** rough terrain walking. Only few examples exist, e.g. *Boston Dynamic's BigDog.*[1] Whether such machines will ever reach the above-mentioned higher mobility, and therefore open up a vast range of applications in dangerous and difficult terrain, largely depends on the robot designers' abilities to construct truly versatile machines of various sizes.

A crucial part of the design process of versatile legged robots is the careful sizing and selection of actuators. A highly dynamic robot needs to be light-weight and powerful at the same time. The higher the required robot performance, the higher the required joint torque and speed, and the higher the actuator weight. A heavier robot requires higher joint torques and thus heavier actuators. To come up with a good compromise, most robot designers go through several iterations before finding a suitable actuator.

The aim of this work is to create a useful tool for designers of versatile legged robots that allows them to make good decisions about the size and type of joint actuators. For this study we selected the *squat jump* as characteristic motion for highly dynamic robots to obtain peak values of joint speed and torque in relation to robot mass, leg segment lengths and desired jump height. We use the leg segment length as a measure of the robot's size, as the leg length usually scales in accordance to overall size. The tool should be straight-forward and intuitive to provide a rough estimate with the least amount of parameters that need tuning. Experiments with a highly dynamic quadruped robot performing squat jumps confirm the validity of our simulation. As a result, we present plots that show required joint torque and velocity for various robot weights, leg segment lengths and jump heights.

Related Work Similar studies were conducted for the electric DC motor sizing of a bounding robot,[2] hopping monopod robot[3] and a vertical climbing robot.[4] A genetic algorithm was used by Smith *et al.*[5] to obtain design parameters including joint torque for statically stable walking robots. Alexander[6] summarized and compared the jumping capabilities of animals ranging from bushbabies to frogs and fleas. In an earlier study he compared the maximum forces exerted by animals in relation to their body weight.[7] Similar animal studies were conducted by Fedak *et al.*[8] and Heglund *et al.*[9]

2. Experimental Platform: HyQ Robot

The experimental platform used in this study is the versatile, quadruped robot HyQ,[10,11] Fig. 1(a), a hydraulically and electrically actuated machine that weighs $70kg$, is $1m$ long and has a leg segment length of $0.35m$. The robot's legs have three degrees of freedom each, two hydraulic joints in the sagittal plane (hip and knee flexion/extension) and one electric joint for hip adduction/abduction. Each joint has 120° range of motion and is controllable in torque and position. The maximum joint torque is 145Nm for the hydraulic and 152Nm for the electric joints. Semini *et al.*[11] describe HyQ's design and specifications in detail.

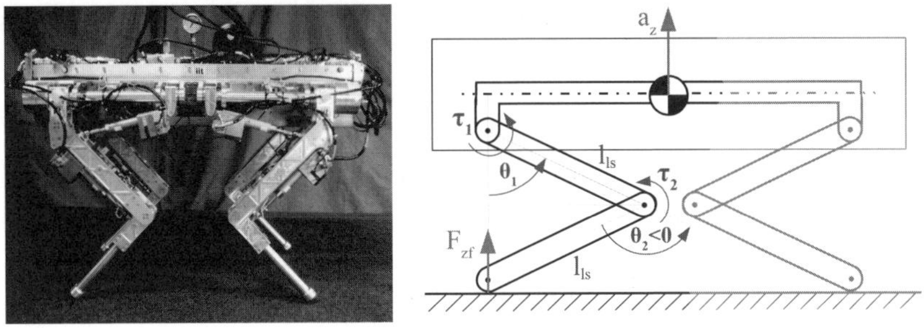

(a) (b)

Fig. 1. HyQ: Hydraulic Quadruped robot. **(a)** picture and **(b)** sketch of side view of robot in squat posture, defining the centre of mass (COM) acceleration vector a_z; joint angles $\theta_{1,2}$ and torques $\tau_{1,2}$ of the hip and knee joint, respectively; leg segment lengths l_{ls} and ground reaction force vector at the foot F_{zf}.

3. Squat Jump Simulation

As mentioned above we selected the squat jump as a characteristic motion for a highly dynamic robot. Such a *squat jump* is composed of several phases: first, a vertical **acceleration phase** from a squatting posture until **lift-off** (when the feet loose contact with the ground); then, a parabolic **flight phase** with the legs moving to a suitable landing posture.

To reduce the amount of tunable parameters, we made several assumptions: the legs are massless, all weight is equally distributed in the robot's torso, and the vertical acceleration is constant during the whole acceleration phase. Furthermore, the lengths of the upper and lower leg segment (distance between the hip and knee axis (upper); knee axis and foot (lower)) are equal and the hip joint axes are always vertically above the contact point of their corresponding feet, as shown in Fig. 1(b). This leads to the following relationship between the hip and knee joint angles θ_1 and θ_2:

$$\theta_1 = -\frac{\theta_2}{2} \tag{1}$$

Furthermore, these assumptions lead to ground reaction force vectors F_{zf} that intersect (or *point through*) the hip axes during the whole acceleration phase, resulting in zero required hip torque. Section 5 will address the consequences of these assumptions and discuss the limitations of this approach.

The jump height is the crucial input to our simulation and is measured as the vertical distance that the COM travels from the time the body lifts off the ground to the end of the upward motion. The maximum jump height

h_{max} of any object undergoing a parabolic flight phase is directly related to its lift-off velocity v_{lo}:

$$v_{lo} = \sqrt{2gh_{max}} \tag{2}$$

and can be obtained by equating the kinetic energy E_{kin} at lift-off with the potential energy E_{pot} at the maximum jump height:

$$E_{kin} = \frac{1}{2}mv_{lo}^2 = mgh_{max} = E_{pot} \tag{3}$$

where g is the gravity constant and m_{robot} the robot mass. Note that v_{lo} is independent from m_{robot}.

The next step is to calculate the constant vertical acceleration a_z necessary to reach the velocity v_{lo}. To further reduce the number of input parameters, we defined the distance z_{ap} of vertical travel of the COM before lift-off to be equal to the length of the leg segments l_{ls}.

$$a_z = \frac{1}{2}\frac{v_{lo}^2}{z_{ap}} = \frac{1}{2}\frac{v_{lo}^2}{l_{ls}} \tag{4}$$

This results in the required vertical force F_{ap} during the acceleration phase:

$$F_{ap} = (a_z + g)m_{robot} \tag{5}$$

This force should be equally spread over the four legs and therefore results in a vertical ground reaction force of $F_{zf} = \frac{1}{4}F_{ap}$ at each foot. The required torque in the knee joint depends on the momentary joint angles during the motion and is obtained as follows:

$$\tau_2 = \frac{1}{4}F_{ap}l_{ls}\sin(-\theta_1(t)) = \frac{1}{4}F_{ap}l_{ls}\sin(\frac{\theta_2(t)}{2}) \tag{6}$$

with the joint angles and torques as defined in Fig. 1(b). The joint angle trajectories are obtained through the foot trajectory in Cartesian coordinates and the leg Jacobian.[11] Using (2), (4)-(6) we get the following result

$$\tau_2 = \frac{1}{4}gm_{robot}l_{ls}\sin(\frac{\theta_2(t)}{2})(\frac{h_{max}}{l_{ls}} + 1) \tag{7}$$

4. Experimental Results

We performed several squat jump experiments with HyQ. Figure 2(a) and 3(a) show the results of a jump with $0.2m$ jump height that can be seen in a *youtube* video[12] at minute 1:06.

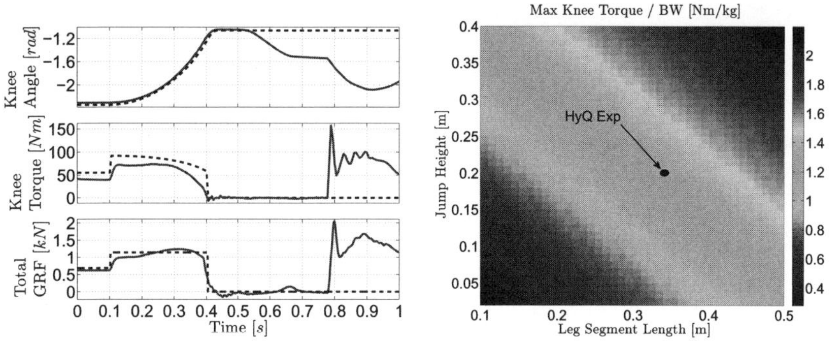

(a) (b)

Fig. 2. **(a)** Plot of experimental data (red solid) and simulated data (black dashed) for a squat jump motion of 0.2m jump height. *Top:* knee joint angle θ_2; *middle:* knee torque τ_2; *bottom:* total ground reaction force $4F_{zf}$. **(b)** plot of maximum knee joint torque scaled by the robot's body weight (BW) for different leg segment lengths and jump heights. The arrow marked with *HyQ Exp* shows the experimental results obtained with HyQ as an example.

The three subplots of Fig. 2(a) show the data of the experiment (red solid line) and of the simulation (black dashed line) for the knee joint angle (top), knee joint torque (middle) and vertical ground reaction force (bottom). The acceleration phase starts at $0.1s$ and lasts till $0.4s$ when the torques go to zero. The robot touches down again at $0.78s$. The simulation calculates values only during the acceleration phase.

Similarly, the three subplots of Fig. 3(a) report again the data for the knee joint angle on the top (to facilitate the comparison of the plots), and the knee and hip joint velocities in the middle and bottom plot, respectively.

The plots show that the simulation results match well for joint position, joint velocities and ground reaction forces. The simulated knee torques are slightly higher due to the above mentioned assumptions. See Section 5 for a discussion about these assumptions and their effect on the results.

Based on the simplified squat jump simulation presented in Section 3 we created three 3D plots illustrating estimations of knee joint torque, velocity and power for a selection of leg segment lengths between $0.1m$ and $0.5m$ and jump heights ranging from $0.02m$ to $0.4m$. Figure 2(b) shows a plot of the maximum knee joint torques scaled in relation to body weight (BW). Figure 3(b) shows a similar plot for the maximum knee joint velocities. As the jump height mentioned in (2) is independent of body weight and only related to the lift-off velocity, this plot does not need to be scaled by

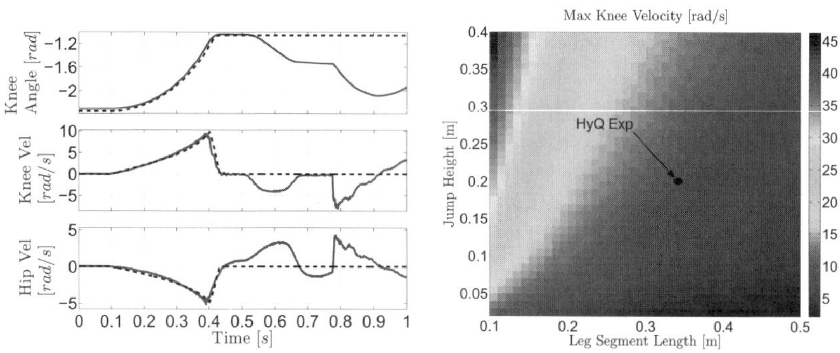

(a) (b)

Fig. 3. **(a)** Plot of experimental data (red solid) and simulated data (black dashed) for a squat jump motion of $0.2m$ jump height. *Top:* knee joint angle θ_2; *middle:* knee joint velocity $\dot{\theta}_2$; *bottom:* hip joint velocity $\dot{\theta}_1$. **(b)** plot of maximum knee joint velocity for different leg segment lengths and jump heights. The arrow marked with *HyQ Exp* shows the experimental results obtained with HyQ as an example.

body weight. Finally, Fig. 4 shows the product of the knee joint torque and velocity plots resulting in the maximum required knee joint power scaled in relation to body weight.

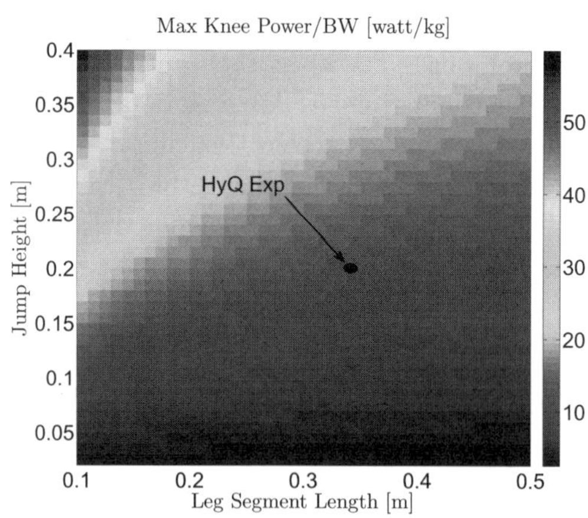

Fig. 4. Plot of maximum knee joint power scaled by body weight BW for different leg segment lengths and jump heights. The arrow marked with *HyQ Exp* shows the experimental results obtained with HyQ as an example.

As a reference we added the results of the squat jump experiment with HyQ presented in Section 4 to each of the plots with an arrow marked

HyQ Exp. A jump height of $0.2m$ and leg segment length of $0.35m$ results in a value of $1.3Nm/kg$ for the maximum knee torque. Multiplied by HyQ's weight of $70kg$, this results in $91Nm$, matching the peak value of the simulated torque shown in Fig.2(a). In terms of joint velocities we obtain $10rad/s$. Figure 4 shows an estimated maximum knee joint power of 910W for HyQ $(13W/kg * 70kg)$.

5. Discussion

In this section we discuss the limitations of our method and evaluate the consequences of the assumptions mentioned in Section 3. *Massless legs:* If the mass of the legs is small with respect to the torso's weight, this first assumption does not significantly influence the results. However, Alexander[6] showed that increasing leg mass reduces the jump height, especially if the additional mass is located in the lower leg segment. *Equally distributed body weight:* If the body weight is not equally distributed, the joint torques in the four legs are not equal, however the average will match our simulation. A wise rule for the design of versatile, quadruped robots is an equal distribution of body mass, since it simplifies balancing during locomotion. *Constant vertical acceleration:* In reality the acceleration and thus the pushing force does not follow the step input as shown in Fig. 2(a), this leads to a smaller jump height. *Hip joint axes vertically above feet:* In reality, it is difficult to achieve perfectly synchronized motions of the hip and knee joints. Thus, the ground reaction forces will not always be only vertical, but non-zero tangential components result in internal forces that might lead to foot slippage if the foot-ground friction coefficient is low. For this reason, the hip torque during the experiment is not zero. The torque plot of Fig. 2(a) shows that our simulation results in an overestimation of required joint torques.

A major limitation of our approach is that it currently only considers the squat jump as characteristic motion of a versatile robot. It does not provide any information about leg workspace and thus joint range of motion, neither does it take running at different speeds into consideration.

6. Conclusion and Future Works

This paper presented a simple method to estimate maximum joint torques and velocities for the design of highly dynamic quadruped robots with articulated legs. A jumping motion out of a squat posture was used as characteristic motion of dynamic robots. The maximum jump height, the body mass and the leg segment length were the input parameter for the simulation. Maximum joint torques, velocities and power were the output. Our method should provide robot designers with a useful tool helping them in correctly sizing and selecting joint actuators.

Future Works The presented method only looks at one aspect of versatile robots. We will therefore extend it by a similar estimation of maximum joint torques and velocities for running with forward speed, body mass and leg segment length as input parameters. We are currently working on the design of a smaller version of HyQ, whose actuators will be sized and selected based on these estimations.

References

1. M. Raibert, K. Blankespoor, G. Nelson, R. Playter and the BigDog Team, Bigdog, the rough-terrain quadruped robot, in *IFAC World Congress*, 2008.
2. P. Chatzakos and E. Papadopoulos, The influence of dc electric drives on sizing quadruped robots, in *IEEE ICRA Conference*, 2008.
3. P. Gregorio, M. Ahmadi and M. Buehler, Design, control, and energetics of an electrically actuated legged robot *IEEE Transactions on Systems Man and Cybernetics, Part B: Cybernetics, Vol 27,* 1997.
4. A. De, G. Lynch, A. Johnson and D. Koditschek, Motor sizing for legged robots using dynamic task specification, in *IEEE TePRA Conference*, 2011.
5. B. Smith, C. Saaj and E. Allouis, Evolving legged robots using biologically inspired optimization strategies, in *IEEE ROBIO Conference*, Dec. 2010.
6. R. M. Alexander, *Principles of Animal Locomotion* (Princeton University Press, 2002).
7. R. M. Alexander, The maximum forces exerted by animals *Journal of Experimental Biology, Vol 115,* 1985.
8. M. A. Fedak, N. C. Heglund and C. R. Taylor, Energetics and mechanics of terrestrial locomotion. kinetic energy changes of the limbs and body as a function of speed and body size in birds and mammals. *Journal of Experimental Biology, Vol 79,* 1982.
9. N. Heglund and C. Taylor, Speed, stride frequency and energy cost per stride: how do they change with body size and gait? *Journal of Experimental Biology, Vol 138,* 1988.
10. C. Semini, HyQ - design and development of a hydraulically actuated quadruped robot, PhD thesis, Italian Institute of Technology and University of Genoa, 2010.
11. C. Semini, N. G. Tsagarakis, E. Guglielmino, M. Focchi, F. Cannella and D. G. Caldwell, Design of HyQ - a hydraulically and electrically actuated quadruped robot *IMechE Part I: Journal of Systems and Control Engineering, Vol 225,* 2011.
12. *http://www.youtube.com/watch?v=wPxXwYGZmd8*

ENVIRONMENTALLY ADAPTIVE CONTROL POLICIES FOR EXPENSIVE SYSTEMS

Matthew Tesch[‡], Jeff Schneider, and Howie Choset
Robotics Institute, Carnegie Mellon University, Pittsburgh, PA 15213
[‡] *mtesch@cs.cmu.edu*

Running experiments on physical robotic systems typically requires significant resources (personnel, time, and money), restricting the number that can be run during training. Additionally, these systems typically operate over a range of external conditions and can be highly non-linear. We focus on the problem of learning a globally optimal policy to adapt controllers for such systems based on the value of these external conditions in order to maximize expected performance. We propose two novel myopic search methods for this problem, and present results comparing these algorithms with various other approaches. Finally, we use these methods to train both simulated and physical snake robots to automatically adapt to changing terrain.

Many robotic systems are effectively a "black-box" — there is no known analytic process model or expression for performance, and it is infeasible to sample the control space densely enough to obtain such a model. In particular, we focus on systems for which evaluation of a single control parameter may take significant effort. Careful *experiment selection* is therefore necessary in order to minimize the number of evaluations required for selection of globally optimal control parameters.

Furthermore, during typical operation of such systems, the environment can change significantly. For example, a locomoting snake robot may move over gently up-sloped terrain, traverse a slightly bumpy horizontal area, and move downhill through many large obstacles. Consider a parameterized controller which can be tuned for each of these environments; an example of such a controller would be a set of gait parameters which close the loop on actuator position at a low level. Obviously, the same set of gait parameters would not be optimal method in each environment; rather one would expect the parameters for best locomotive performance to vary based on the environment. In this work, we seek to develop experiment selection

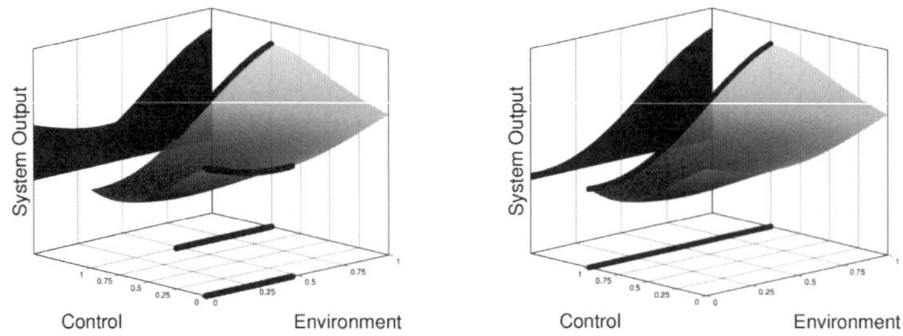

(a) (b)

Fig. 1: We are interested in problems for which the optimal control parameter changes significantly depending on the environmental conditions. Here we show a system objective function for a toy problem with 1-D environment and control spaces; the policy is shown below the function and the policy score is projected to the left. An optimal adaptive policy is illustrated in (a), presenting the best control parameter for every environment parameter. The constant policy shown in (b) results in a significantly lower overall score.

methods which enable intelligent generation of *control policies* which adapt to changes in the environment by selecting the best controller parameters given a particular environment (Fig. 1). This control policy then acts to close the loop at a high level (typically a relatively slow timescale), changing the parameters of the low-level (typically fast timescale) controller as significant changes in the environment are detected.

Related Work

A key idea in this work is the use of a *surrogate* to represent a function which is expensive to evaluate, and basing search methods on this cheap model. These ideas have been extensively explored in the global optimization community to minimize the number of evaluations required to globally maximize expensive functions.[1] Often these methods rely on stochastic processes to create a surrogate function;[2–4] following this lead we use Gaussian processes[5] as a function approximation method that generates an estimate of expected experiment output along with a measure of confidence in that estimate.

In these cases where function evaluations are costly (hours to days), careful sample choice is extremely important in order to best use the experimental budget. A number of heuristics and statistical methods have been derived to use information from the surrogate function to best choose sample locations (Jones[1] provides a survey of many existing methods). These metrics include the upper confidence bound of the predicted function,[6,7] the probability of improvement,[8,9] and the expected improvement.[10,11] In particular, expected improvement has been shown to effectively trade off exploration and exploitation without requiring algorithm parameters to be carefully tuned. However, existing expensive global optimization methods are not directly applicable to the addition of environment parameters.

Perhaps the most closely related work is that of contextual bandits,[12] an extension of the multi-armed bandit literature. The choice of which arm to pull (analogously, which experiment to run) in the contextual setting is informed by a context (the environment parameter in our work). The contextual bandit work presents a framework for learning an adaptive arm-selection policy, but assumes an online setting where the context is given to the algorithm; the algorithm must then minimize continuous regret over a series of experiments. In our work, we are interested in efficient training of a controller in an offline setting, where we can set the context (environment parameters) as well as the arm (control parameters); here we wish to minimize absolute expected regret over all contexts. Note that this learned controller could then be used later in an online setting.

Problem Definition

This work aims to find a control policy that optimally adapts to external variables, while minimizing the number of experiments done to perform the optimization. In order to formalize this problem, we define several terms and then more precisely state the goal of this work.

Definition 1 (Control Parameter). *The control parameter space X_c is a compact subset of $\mathbb{R}^{m_c}$. Each $x_c \in X_c$ represents a particular value of the system of interest that can be fully specified during normal operation. Some examples of control parameters include the value of a set of gains in a PID controller, the relative concentration of two reactants in an industrial process, or the prescribed dosage of a drug during drug development.*

Definition 2 (Environment Parameter). *The environment parameter space X_e is a compact subset of $\mathbb{R}^{m_e}$. This space contrasts with the control space in that values $x_e \in X_e$ cannot be controlled under normal real-world operation, but can be specified in laboratory trials. Furthermore, the value of x_e can be measured during normal operation. Therefore, these parameters represent continuous valued external factors of the system, such as terrain steepness for a locomoting system, wind strength and direction for a UAV, particulate size in an industrial process, or disease strain during drug development.*

Definition 3 (System Output). *The system output, denoted as* $f\colon X_e \times X_c \to \mathbb{R}$*, is a continuous, real-valued function of the environment and control parameters. This function represents the performance of the system, given some environmental conditions and some specified control parameter. Example system outputs include the speed of a locomoting system over a terrain, the efficiency of a mechanical process, or the turbulance of a wing design calculated from a wind tunnel or computational fluid dynamics experiment. For the methods we propose here, we assume that sampling this system output is time intensive or computationally expensive, and therefore there is a limit to the number of times this function can be evaluated.*

Definition 4 (Control Policy). *The control policy defined in this work is a mapping* $\gamma\colon X_e \to X_c$ *(not necessarily surjective), such that* $\gamma(x_e)$ *represents the control parameter set by* γ *in reaction to sensing environment parameter* x_e*.*

The policy score,

$$\mathcal{S}(\gamma) = \int_{X_e} \omega(x_e) f(x_e, \gamma(x_e))\, dx_e, \tag{1}$$

represents the expected performance of a policy for a given environment distribtion $\omega\colon X_e \to \mathbb{R}$*.*

The optimal policy γ^* is defined as $\operatorname{argmax} \mathcal{S}(\gamma)$. In Fig. 1, γ^* is shown projected onto the control-environment plane, and $\mathcal{S}(\gamma^*)$ is visualized on the system output-environment plane. Note that γ^* is independent of ω, because $\hat{\gamma}^*(x_e)$ can be independently determined for each $x_e \in X_e$. Although the weighting function can make a significant difference during experiment selection and policy comparison, the optimal policy is unaffected by the relative importance of different environments.

The goal is to find the highest scoring policy γ after a number of system output evaluations. In other words, the quantity of interest is not a single point, but a mapping from environment to control parameters. We break this problem into the subproblems of *policy generation* and *experiment selection*:

(1) **Policy Generation:** Given the results of n system output evaluations, choose the best estimate for γ^*.
(2) **Experiment Selection:** Choose the sequence of points $X = \{x^1, x^2, \ldots x^n\}$, $x^i \in X_e \times X_c$, where the choice of x^{k+1} is informed by $\{f(x^i) \mid i \le k\}$, which maximizes the score of the policy produced by the chosen policy generation algorithm.

Proposed Methods

One of the difficulties in solving these problems lies in the fact that the true objective function we wish to maximize is $\mathcal{S}$, but as this involves an integral over the expensive f it is impossible to calculate for any γ, let alone optimize via standard optimization techniques. By using Gaussian process regression to find a surrogate $\hat{f}$ for f, we obtain the probability density function $p_x(y)$ of the predictive distribution at each $x \in X_e \times X_c$. These distributions allow us to calculate statistically meaningful quantities for a potential sample, such as the approximate expected improvement over the current predicted optimal policy's score if that experiment were run.

Policy Generation

For the policy generation problem, we can use $\hat{f}$ to select the policy which maximizes our best prediction of the score. In low dimensions, $\hat{\gamma}^*$ can be estimated via a dense sampling of the (relatively) cheap surrogate $\hat{f}$; in higher dimensions one might perform global optimization (e.g., with a simulated annealing or genetic algorithm) on a chosen policy parameterization, using $\hat{f}$ to cheaply evaluate policies. As this work mainly focuses on experiment selection methods, we assume a policy generation method based on maximization of (1) via dense sampling (replacing f with $\hat{f}$).

Experiment Selection

Standard local and global optimization methods directly applied to optimizing system output perform poorly. This is because these methods are not optimizing the true objective; rather they focus on improving knowledge in environments with high system output results, and ignore environments with low system output results. This causes the resulting policy to be very weak in "difficult" environments (ones with low system output results), lowering the overall score. However, we have included results from the popular global search algorithm EGO[11] for comparison. At each iteration, this algorithm selects the parameters for the next experiment as those which maximize the expected improvement over the best previous experiment.

Our first proposed algorithm for experiment selection adapts the idea of expected improvement to the explicit separation of X_e and X_c. Instead of measuring improvement over the best evaluation of f so far, we consider the improvement over the maximum predicted value of f for the *same environment* as the test point. This gives the *unbiased expected improvement* (UEI) of a point $x^t = (x_e^t, x_c^t)$:

$$\mathrm{UEI}(x^t) = \omega(x_e^t) \int_{y^*_{x_e^t}}^{\infty} (y - y^*_{x_e^t}) p_x^t(y)\, dy, \text{ where } y^*_{x_e^t} = \max_{x_c \in X_c} (\hat{f}(x_c, x_e^t)) \quad (2)$$

UEI reduces the bias of standard expected improvement towards good environments. However, UEI only considers improvement of the policy score at one environment. To measure true expected improvement of the policy score from sampling a point x^t, the expectation must be computed over the predictive distribution $p_x^t(y)$, where each potential y involves regression of a new surrogate $\hat{f}_y$ conditioned on the addition of this potential sampled value. This second proposed method is termed the *expected policy score improvement*, or EPSI:

$$\mathrm{EPSI}(x^t) = \int_{-\infty}^{\infty} p_x^t(y) \int_{X_e} \omega(x_e) \max\Big(\hat{f}_y(x_e, \gamma_y^*(x_e)) - \hat{f}(x_e, \gamma^*(x_e)), 0\Big)\, dx_e\, dy \quad (3)$$

Using this method generates a more complete estimate of the effect of sampling a point on the policy score. Of course, computing a large numeric integral can also take significantly more time, and the quality of the solution can vary based on the resolution of the integral.

Experimental Results and Concluding Remarks

We evaluated the performance of the two proposed algorithms on analytic test functions as well as on simulated and physical snake robots. The three test functions of varying complexity were created specifically so that a static control policy would not be effective over the entire space.

To measure the score of a particular algorithm, we non-deterministically generate an initial set of k points through guaranteed-coverage sampling method, such as a latin hypercube,[13] and then sequentially select $n - k$ more points, evaluating the system output after each choice. The predicted $\hat{f}$ is used after each evaluation to generate a policy, which is scored via a numeric integral on system output. This entire process was repeated 10 times for each algorithm, the results averaged, and standard errors calculated to provide a rigorous comparison of methods. The algorithms were written in MATLAB, and used the open-source Gaussian Processes for Machine Learning package provided by Rasmussen and Williams.[14]

A summary of these results is shown in Fig. 2; more thorough comparisons[15] are omitted for brevity. Standard global optimization methods

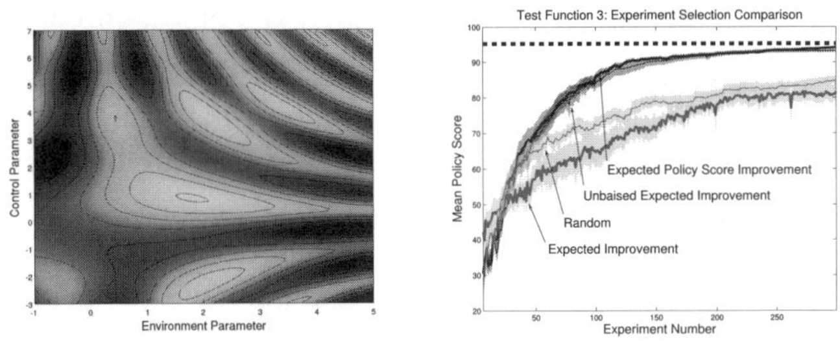

Fig. 2: **(Left):** One of the three analytic test functions used for algorithm comparison. **(Right):** Scores of policies learned for this test function (average over 10 trials); the dotted black line represents the best possible policy score for that function. Policies were generated and scored after each new experiment (x-axis represents experiment number).

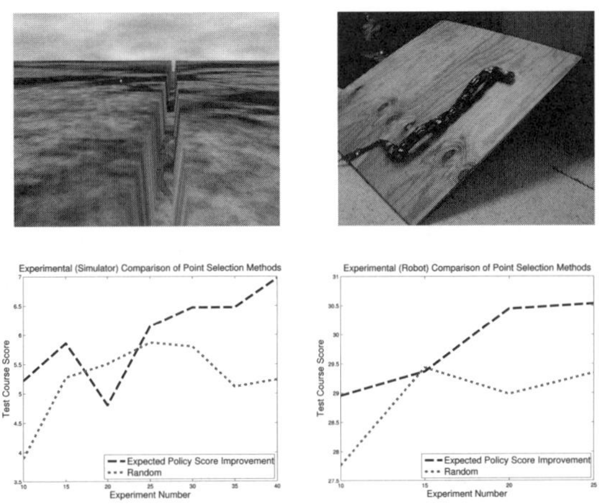

Fig. 3: **(Top):** Experimental setup; system performance in a measure of locomotive energy efficiency for a snake robot crawling through a crevice (simulated robottests) and climbing up an incline (physical robot tests). **(Bottom):** Performance of policies generated from points selected randomly versus using EPSI for the robot tests.

initially have good performance, but the policy score tends to stagnate quickly because these methods are optimizing f rather than $\mathcal{S}$. Random point selection also performs suboptimally, showing that it is important to carefully select experiments. UEI and EPSI both perform a better global search; the latter outperforms the former only slightly, indicating that UEI provides a simpler and quicker method which produces similar results.

As such a complete analysis could not be run on physical systems due to the expensive nature of system output evaluation and the inability to compute a true policy score, we instead set up a range of environmental conditions in a "test course", and then used the above algorithms to generate policies which were scored on this test course. These policies map environment parameters (slope and crevise width) into a 2-D gait parameter control space.[16] The EPSI algorithm was compared to random point selection; the results are shown in Fig. 3.

Although demonstrated here on snake robots, this framework is applicable to a rich set of problems both within and outside the field of robotics. We have described two approaches for experiment selection, proposed a simple policy generation method, and demonstrated the efficacy of these algorithms on analytic test functions and a physical snake robot. Future work involves application to other systems, restricted environment selection during training, and derivation of theoretical performance bounds.

References

1. D. R. Jones, *Journal of Global Optimization* **21**, 345 (2001).
2. R. G. Regis and C. A. Shoemaker, *INFORMS Journal on Computing* **19** (2007).
3. H.-M. Gutmann, *Journal of Global Optimization* **19** (1999).
4. K. Holmström, *Journal of Global Optimization* **41** (2008).
5. C. E. Rasmussen and C. K. I. Williams, *Gaussian Processes for Machine Learning* (The MIT Press, 2006).
6. A. W. Moore and J. Schneider, *Advances in Neural Information Processing Systems* , 1066 (1996).
7. D. Cox and S. John, A statistical method for global optimization, in *1992 IEEE International Conference on Systems, Man, and Cybernetics*, (Ieee, 1992).
8. H. J. Kushner, *Journal of Basic Engineering* **86**, 97 (1964).
9. A. Žilinskas, *Journal of Global Optimization* **2**, 145(June 1992).
10. J. Mockus, V. Tiesis and A. Zilinskas, *Towards Global Optimization* **2**, 117 (1978).
11. D. R. Jones, M. Schonlau and W. J. Welch, *Journal of Global Optimization* **13** (1998).
12. T. Lu, D. Pál and M. Pál, *13th International Conference on Artificial Intelligence and Statistics* (2010).
13. M. D. McKay, R. J. Beckman and W. J. Conover, *Technometrics* **21**, 239 (1979).
14. C. E. Rasmussen and C. K. I. Williams, Gaussian Processes for Machine Learning.
15. M. Tesch, J. Schneider and H. Choset, Adapting Control Policies for Expensive Systems to Changing Environments, in *International Conference on Intelligent Robots and Systems*, (IEEE/RSJ, San Francisco, 2011).
16. M. Tesch, K. Lipkin, I. Brown, R. Hatton, A. Peck, J. Rembisz and H. Choset, *Advanced Robotics* **23**, 1131(June 2009).

IN-PIPE INSPECTION ROBOT*

BASEM F. YOUSEF†
Department of Mechanical Engineering, United Arab Emirates University
Al-Ain PO Box.17555, UAE

NABIL BASTAKI, NIKOLAOS MAVRIDIS, SHAMMAH ANTALI
JAMEELA MENHALI, KLAITHEM AL-NUAIMI, NOURA AL DHAHERI
College of Engineering, United Arab Emirates University, UAE

GEN ENDO, SHIGEO HIROSE, HIROKI SEKINE, DAIGO SETO
YUKI SASAKI
Aerospace Mechanical Engineering, Tokyo Institute of Technology, Japan

HUGUES DUPONT
Schlumberger –SKK, Japan

A novel in-pipe inspection robot is designed to overcome the difficulties and challenges that face other existing in-pipe inspection robots. The sophisticated design and configuration of the robot enables it to navigate autonomously through a pipeline. It consists of two clamping modules and one module for steering and locomotion. The clamping modules utilize flexible spring sheets to enable the robot to dynamically adjust its radial dimension and clamping force to attach to pipes of different inner diameters and overcome the obstacles associated with sudden or gradual changes in pipe diameters. A sophisticated middle module enables the robot to not only propel through the pipeline, but also to steer into the desired direction when approaching a junction or a bend. In addition, the robot is supported by a hole-finding capability through a set of smart sensors mounted on the robot nose, so the robot can find its travel trajectory along turns and T-junctions.

1. Introduction

Pipelines are extensively used to carry natural gas and oil to destinations throughout the world. They range from high-pressure transmission lines to low-pressure distribution lines. In order to prevent a failure/leakage in a pipeline due to corrosion and/or high pressure, the interiors of the pipes need to be routinely

* This work was supported by Research Affairs at the United Arab Emirates University under Research Grant # 1605-07-01-10 (PI: Basem F. Yousef). Features of the design are reported and claimed in US provisional patent file No. 61577020.

† basem_yousef@uaeu.ac.ae

monitored and inspected to evaluate the need to maintain or repair the pipeline, and to decide the most effective means of doing so. Although cutting one or more samples of the pipe is one of the most common methods used for diagnosing a pipeline, its main disadvantage is attributed to the high cost and downtime associated with the method, other than the fact that this method does not provide a comprehensive picture about the status of the whole pipe because of the variations in corrosion rate or deposit buildup found in different sections of a pipe. Therefore, many concepts of Pipe Inspection Robots (PIR) have been utilized as alternative methods [1-10]. In general, the main challenge that may pose technical difficulties to PIRs is the ability to navigate inside a pipe that has discrete internal cross-section e.g. segments that changes from a pipe of a large diameter connected to a pipe of a smaller diameter especially if the change in pipe diameter is considerably large e.g., more than double the smaller section. In these scenarios, when the robot needs to pass through such pipe segments, it has to be able to: a) find the hole location at the cross-section where the pipes mate (e.g. up, down, right or left of the center of the hosting pipe segment); b) steer into the hole direction; c) change its radial dimension to fit into the smaller pipe diameter; and d) propel/move autonomously forward and backward.

Several concepts of PIRs have been developed to propel horizontally through a pipeline [1, 2]. However, beside their inability to travel through vertical or curved paths, their propulsion mechanisms depend on friction force which is not always sufficient to cause the motion. Other mechanisms were designed to overcome those difficulties by utilizing sets of front and rear rollers/wheels which are pushed against the interior wall of the pipe and used to drive the robot [3-5]. Although some of those PIRs could travel only in pipes of constant diameters [3, 4].

Other mechanisms could handle only small changes in pipe diameters since their wheels are pushed radially against the pipe walls using compression springs of small strokes. Also their springs usually exert smaller forces towards the end of the spring stroke causing slipping of the wheels [5].

Moreover, inch-worm robots [6, 7] and snake-like robots [8] have been reported in the literature. Those robots utilize a serpentine motion to travel through horizontal and curved pipes, and their designs allow for limited radial and axial extension-contraction to provide clamping force against pipe walls. Also, pipe crawlers demonstrated good ability in dealing with the change in pipe diameters [9]. In spite of the advanced designs of those robots, their intricate designs showed deficiencies in meeting one or more of the intended tasks when navigating through a pipeline especially when dealing with sudden changes in pipe diameter.

In addition, magnetic wheels were utilized in [10] to provide a wall climbing robot that may be used for inspecting the interior of surfaces of gas-tanks in oversea ships. Since it lacks the ability to adapt to pipe diameter, this robot is unsuitable for in-pipe navigation.

The objective of the this project is to provide a sophisticated configuration of a pipe inspection robot that can efficiently overcome the difficulties and challenges that face other existing robots and add additional capabilities and functionalities needed for such robots. The novel design of the proposed robot enables it to navigate autonomously through a pipeline.

2. Robot Description

The robot consists of three portions, two identical end clamps and a middle segment, in addition to a sensor assembly to enable the hole-finding feature. The clamps enable the robot to firmly attach to the inner walls of the pipe while the middle segment provides means to propel and to steer the robot. (Figs. 1 and 2).

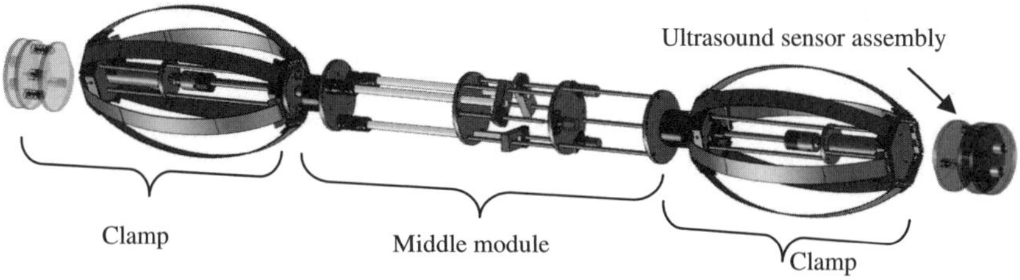

Fig. 1. A 3D CAD model of the robot showing the different segments.

2.1. *The Clamps*

The main function of the clamps (Fig. 3) is to provide temporary contact with the inner walls of the pipe. Each clamp assembly is built around a motor and consists of 6 flexible spring sheets connected at their ends to hexagonal flanges by hinges. The flanges are connected with a power screw and a DC motor which both are aligned along the clamp's axis. While one hexagonal flange is fixed at the end of the clamp, the other flange slides along the power screw when the motor rotates. This movement causes the flexible spring sheets to straighten or bend outwards thus causing the sheets to come in contact with the inner wall of the pipeline. The closer the flanges are, the stronger the clamping force to the pipe will be. However, since the robot is designed to propel through pipes of different diameters, this clamping force can be chosen to be fixed regardless of the pipe diameter. This can be achieved by installing a torque limiter between the power screw and the motor. The torque limiter is engaged, i.e., the motor directly drives the power screw as long as the torque is less than a preset value, which depends on the specifications of the torque limiter. Once the motor torque exceeds the limiter's preset torque value, the torque limiter disengages and disconnects the motor from the power screw causing the motor to continue spinning but not the power screw. Therefore, the

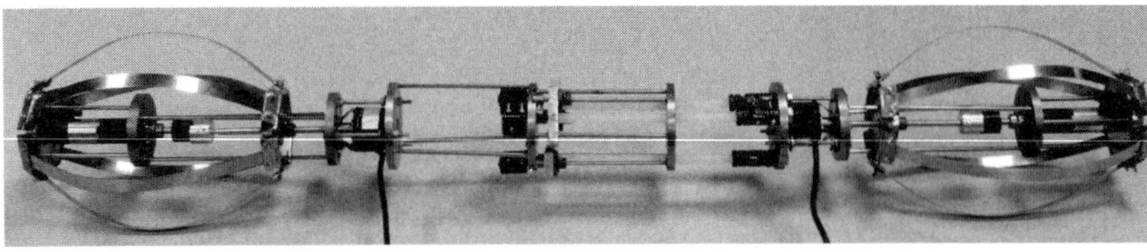

Fig. 2. Prototype of the PIR with inflated clamps for attach-to-pipe pose.

sliding hexagonal flange will stop at a certain distance along the power screw causing the spring sheets to apply a constant clamping force to the pipe regardless of the pipe diameter. The controller, in turn, receives two encoder readings, one from the motor encoder and the second from the encoder that is installed on the shaft of the power screw. When the motor is engaged with the power screw, both encoders will read the same angular velocity. But once disconnected, the power screw encoder stops rotating reading "zero" angular velocity thus, the controller realizes that the clamp is tightened with its maximum force to the pipe and ends the "clamping task".

Another feature of the clamp is the hole-finding sensor assembly (Fig. 4) which is mounted on the front of each clamp. The sensor assembly is designed to identify the existence of obstacles and/or the location of a hole during the movement of the robot when the robot passes through local variations like the sudden change in pipe diameter, eccentric sections, or connection into a network of pipes through T-junctions. The hole finding sensor assembly consists of three parallel circular discs with an ultrasonic sensors placed on the top disc, with the base plate attached to the nose of the clamp. Using a micro servo-motor, the top disc spins 360 degrees clockwise and counter-clockwise searching for a hole location while the robot is in motion along the path. The other two discs are separated by 3 helical springs separated by angles of 120 degrees to keep the plates parallel, and each is concentric with a contact sensor that can detect any compression in the spring if it occurs.

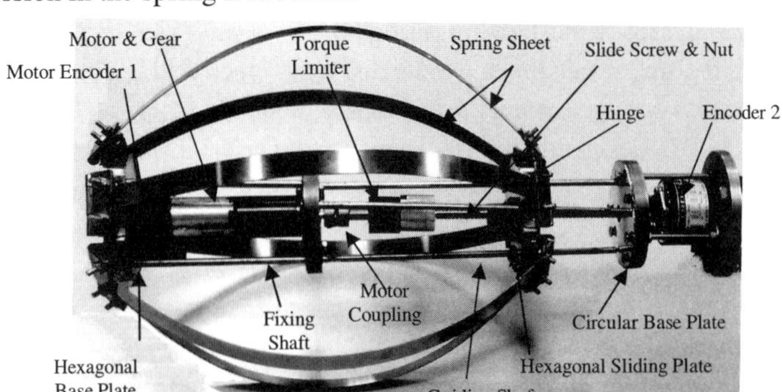

Fig. 3. Components of a "Clamp".

In case of approaching an eccentric pipe, e.g. moving through two pipe segments joined with a shift in the hole location, the ultrasound assembly sensor will try to find the location of the hole and feed it to the controller which commands the steering segment to change the direction of the propulsion (as explained in the next section). However, if the ultrasonic sensor fails to realize the hole location or the appropriate clear path, the robot nose will hit an obstacle causing one or two contact sensors to be depressed hence the robot controller can realize the obstacle-free direction or the hole location and steers the robot to propel it into that direction. The mechanism of propulsion and steering is explained in the next section.

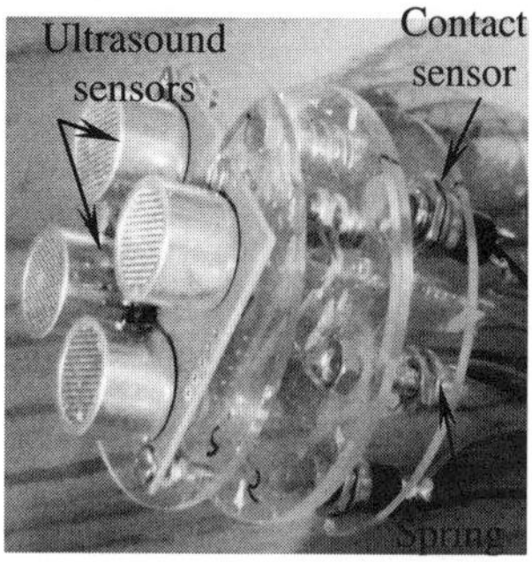

Fig. 4. Ultrasound assembly set for hole- finding feature

2.2. *Middle segment*

This segment (Figs. 1 and 2) is responsible for two important tasks, propulsion and steering. This portion consists of three nylon Polymeric rods connected from one end to universal joints which are fixed to an end flange, while at the other end, the nylon rods attach to aluminum blocks. Each block can travel along a power screw which is driven by a high-power micro-motor. During the movement of the blocks along the power screws, the nylon rods can extend and retract, i.e. they extend and emerge out of the middle segment assembly or retract and hide into it. When the micro-motors spin with the same speed, the nylon rods extend/retract with the same rate which pushes/pulls the end flange along a linear path. Since the end flanges are attached to the clamps, this enables the robot to propel along a linear path. On the other hand, if one of the rods extends slower than the other two, the nylon rods will bend with different radii causing the end flange to steer (Fig. 5). Therefore, the middle module enables the robot to propel and to steer to the correct direction as instructed by the controller based on the direction feedback of the ultrasound sensor assembly.

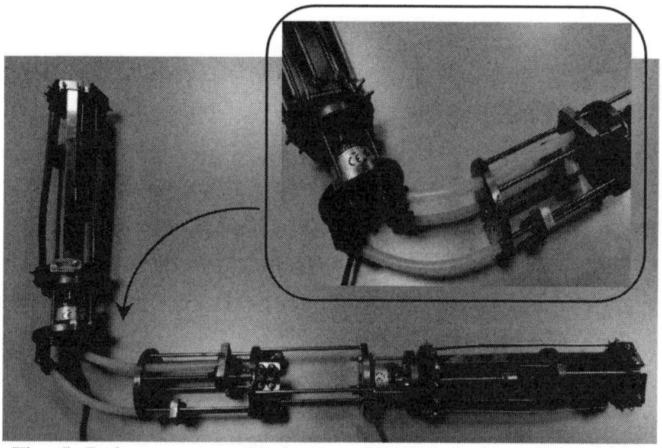
Fig. 5. Robot can turn 90 degrees. Exploded view of the middle segment showing the steering mechanism details. Clamps are in release-pipe pose.

2.3. *Motion sequence and pattern*

The three modules of the robot (two end clamps and middle segment) go through the following sequential movement pattern that causes the robot to propel.

Initially, the front and back clamps' spring sheets stick/attach to the inner walls of the pipe while the middle module nylon rods are retracted (Fig. 6).

The motion starts by relaxing the front clamp from the pipe. This can be done by relaxing the spring sheets of the clamp using the clamp's power screw. Then the nylon rods of the middle segment extend causing the front clamp to advance. Once the rods are fully extended, the front clamp springs attach to the pipe walls again.

The back clamp spring sheets relax using the back clamp power screw, and then the nylon rods of the middle module retract causing the distance between the clamps to shrink and thus the back clamp to advance to the front. After full retraction of the nylon rods, the spring sheets of the back clamp attach to the pipe walls again using the clamp's power screw.

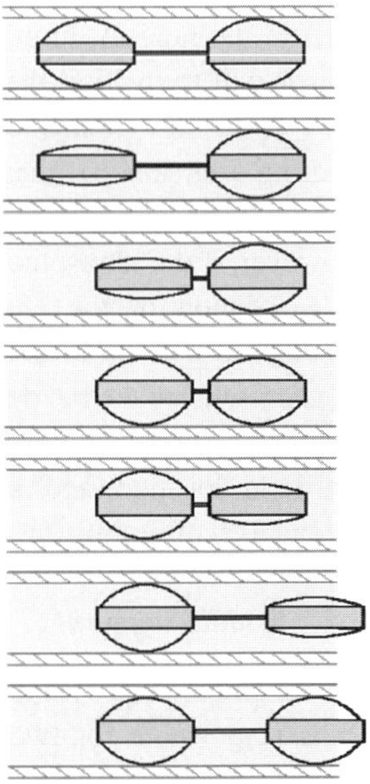

Fig. 6. Motion sequence of the robot adopts inch-worm motion pattern

Repeating that motion pattern can cause the robot to propel in one direction while reversing the sequence can reverse the motion direction.

3. Tests and Results

The robot was tested inside a 15 mm ID Plexiglas pipe in order to determine the capability of propelling along a linear path. The clamping force and the propulsion speed depend on the robot specifications and parameters. For instance, the clamping force to the walls can be adjusted by changing one or more of the following parameters/specifications: 1) the spring sheets' thickness, width and material; 2) the torque limiter's set-torque. The robot's current configuration uses 0.63 mm thick x 10 mm wide stainless steel spring sheets, and a torque setting of the selected torque limiter is 392 mN· m. moreover, the length of the spring sheets controls the robot's capability to handle bigger pipe diameters. In order to measure the clamping force, the clamps were attached to the pipe, and then the robot nose was hooked to a load cell and was pulled to move. The robot clamps together can provide a 76 N, i.e., 38 N/clamp. It is worth mentioning that the roughness of the inner pipe walls will affect these values.

The propulsion speed was calculated based on the total time required to complete the full cycle of an inch-worm motion pattern and found to be 2 mm/s. However, the speed can be improved by minimizing the waste in time. For instance, attaching/releasing the clamps to the pipe can be performed by spinning the clamp motor few spins sufficient to apply release the necessary force or to release the clamp. Also, this speed can be improved further by using micro-motors with higher RPM to drive the elastic rods of the middle segment and/or by selecting power screws with larger pitch.

Although the paper focuses on the locomotion aspect of the robot and its propulsion along a linear path, the robot's steering capability was tested and showed that it can steer to 90 degrees.

Table I summarizes the specifications of the robot based on its current configuration. Future work includes analyzing the robot performance when passing through turns and testing the capability of the robot's hole-finding feature in junctions.

Table I. Specifications of the robot.

Requirement considered	Specification
Weight	< 3.6 kg
Material	Brass and aluminum
Architecture	Modular: Clamp- steering/propulsion segment-Clamp
Motion type	Inch-worm
Power supply	Self powered , 8-AA batteries+ 2X 9V batteries
Dimensions:	
	Length : 97 cm (fully retracted) 105.5 cm (fully extended)
	Outer diameter : (min) 8 cm- fully collapsed clamps (max) 24 cm- fully expanded clamps
Max linear stroke	8.5 cm
Linear propulsion speed (tested in 15 cm ID pipe)	2 mm/s
Clamping force (tested in a 15 cm ID Plexiglas pipe)	76 N
Max turning angle	90 degrees
Operating mode	Autonomous

Conclusions

A novel in-pipe inspection robot has been designed and constructed. The robot's features enable it to navigate autonomously through a pipeline. Its clamping modules enable the robot to dynamically adjust its radial dimension and clamping force to attach to pipes of different inner diameters and overcome the obstacles associated with changes in pipe diameters, thus it can travel through pipelines that comprise segments of varying inner diameters. Its sophisticated middle module enables it to not only propel, but also to steer into the desired direction when approaching a junction or a bend. In addition, the robot is supported by a hole-finding capability that allows it to travel trajectory along turns and T-junctions.

Acknowledgments

This research was supported by Research Affairs at the United Arab Emirates University under Research Grant # *1605-07-01-10* (PI: Basem F. Yousef). Features of the design are reported and claimed in US provisional patent file No. 61577020

References

1. F. Kirchner and J. Hertzberg “A Prototype Study of an Autonomous Robot Platform for Sewerage System”, *Maintenance Autonomous Robots,* 1997; 4(4), p. 319-331.
2. HB. Kuntze and H. Haffner, “Experiences with the development of a robot for smart multisensoric pipe inspection”, *Proc. IEEE International Conf. on Robotic and Automation*, 1998. p. 1773-8.
3. HR. Choi and SM. Ryew “Robotic system with active steering capability for internal inspection of urban gas pipelines”, *Mechatronics*, 2002; 16(12), p. 713-36.
4. Z. Zhu and Z. Pan, “Miniature pipe robots”, *Industrial Robot: An International Journal*, 2003; 30(6), p. 575–83.
5. I. Hayashi, N. Iwatsuki, K. Morikawa and M. Ogata, “An in-pipe operation microrobot based on the principle of screw”, *International Symposium on Micromechatronics and Human Science*, 1997, p. 125-9.
6. T. Fukuda, H. Hosokai, M. Uemura, “Rubber gas actuator driven by hydrogen storage alloy for in-pipe inspection mobile robot with flexible structure”, Proc. IEEE International Conf. on Robotics and Automation, 1989, p. 1847–52.
7. S. Aoshima, T. Tsujimura and Yabuta “A miniature mobile robot using piezo vibration for mobility in a thin tube”, Transactions of ASME, Journal of Dynamic Systems, Measurements and Control, 1993, 115, p. 270-8.
8. KU. Scholl, V. Kepplin, K. Berns and R. Dillmann, “Controlling a multi-joint robot for autonomous sewer inspection”, Proc. IEEE International Conf. on Robotics and Automation, 2000, 2, p. 1701-6.
9. A. Zagler and F. Pfeiffer, “MORITZ” a pipe crawler for tube junctions”. Proc. IEEE International Conf. on Robotics and Automation, 2003, p. 2954-60.
10. W. Fischer, F. Tâche, and R. Siegwart, “Inspection system for very thin and fragile surfaces, based on a pair of wall climbing robots with magnetic wheels,” in Proc. IEEE/RSJ Int. Conf. Intelligent Robots and Systems, San Diego, CA, USA, 2007, pp. 1216-1221.

BASIC PERFORMANCE OF FLAT KICK-SWIMMING ROBOT POWERED BY IMPULSE FORCE GENERATOR WITH SPATIALLY-DEFORMABLE V-SHAPE ELASTICA

SYOUTA IMAI, HIROMI MOCHIYAMA

Department of Intelligent Interaction Technologies, University of Tsukuba, Tennoudai 1-1-1, Tsukuba, Ibaraki 305-8573, Japan
E-mail: motiyama@iit.tsukuba.ac.jp

In this paper, we show basic performances of a novel very flat under-water robot which swims by kicking surrounding water impulsively. This robot utilizes an impulse force generator based on snap-through buckling of a V-shape elastica whose ends are driven rotationally by two motors respectively. While synchronization of the two motors allows the robot move forward impulsively, phase difference of them brings quick turning motion to the robot. Under-water experiments show that the proposed robot can not only move forward impulsively at the maximum velocity of 0.74[m/s] corresponding to 2.18[BL/s] (body length per second), but also turn quickly at the maximum angular velocity of 289[deg/s].

Keywords: Flexible Robot, Impulse Force Generator, Swimming Robot, Continuum Robot.

1. Introduction

Real fish has much higher motion quickness than existing under-water vehicles. One of the most significant differences in motion quickness between them, especially in small sizes, is the ability to change traveling directions which we can see from fish in aquarium or from movies of swimming fish in natural environments. Robotic realization of such under-water motion quickness must be attractive especially for children, and can be utilized in the entertainment industry which is expected to grow to the service robot industry. The purpose of this research is to develop a compact swimming robot which has high swimming motion quickness comparable with real fish in small sizes.

Many fish-like swimming robots have been developed from the viewpoint of bio-mimetics.[1–3] These fish-like robots realized swimming motions very

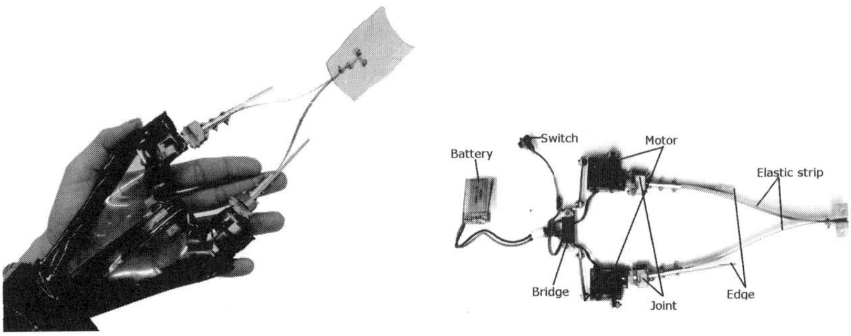

Fig. 1. (Left) Flaggyo: The flat kick-swimming robot. (Right) The impulse force generator based on V-shape closed elastica spatially deformable.

similar to those of real fish. However, only slow fish-like motions were mimicked. On the other hand, our research group has been trying to realize agile fish-like motions using impulse force in water which might be the key to achieve quick motions of real fish. Watari *et al.* developed a kick-swimming robot with length of 300[mm] which uses the impulse force generator based on snap-through buckling of the most basic type of closed elastica.[5] They succeeded to achieve an average traveling velocity of 1.45 body length per second by this robot which kicks water impulsively. However, this robot was not an under-water vehicle, but actually a boat on water although it kicks water with a paddle attached to the elastica. Yamada et al. developed a swimming robot which can turn 90 degree in 0.4[s]. This robot utilizes a special type of impulse force generator based on spatial snap-through buckling of an arch-shaped elastica spatially deformable.[4] However, in this robot, the impulse force generator is used only for turning motions, which means an additional thrust mechanism is needed. One of the main difficulties to achieve quick fish-like motions by a compact robot is the trade-off between large impulse force generation and reduction of drag force in water. Moreover, the device must be simple to be compact.

In this paper, we show basic performances of a novel very flat underwater robot (Fig. 1) which swims by kicking surrounding water impulsively. This robot, named Flaggyo,[a] utilizes an impulse force generator based on snap-through buckling of a V-shape elastica whose ends are driven rotationally by two motors respectively. While synchronization of the two motors allows the robot move forward impulsively, phase difference of them brings

[a]The word 'gyo' means 'fish' in Japanese. Therefore, 'Flaggyo' expresses 'flat fish'.

quick turning motion to the robot. Under-water experiments show that the proposed robot can not only move forward impulsively at the maximum velocity of 0.74[m/s] corresponding to 2.18[BL/s] (body length per second), but also turn quickly at the maximum angular velocity of 289[deg/s].

This paper is based on our previous paper written in Japanese and presented at a Japanese domestic conference in Japanese,[6] but not just a translated one. The robot mechanism has been improved, and the experimental data has been updated.

The organization of this paper is as follows. In Section 2, we explain the impulse force generator we developed for swimming robots. In Section 3, we explain a prototype of the flat kick-swimming robot which has our impulse force generator as a driving mechanism. In Section 4, we explain the underwater experiments in order to verify basic swimming performances of the robot. At the end, in Section 5, we summarize the results of this paper.

2. Impulse Force Generator with V-shape Elastica

2.1. *Mechanism*

The impulse force generator used for our swimming robot is a device for generating impulse forces by utilizing spatial snap-through buckling of a V-shape elastica spatially deformable. This device mechanically consists of a V-shape elastica, two rotational motors, two one-way wheels, a bridge link and two edge links. The V-shape elastica is just stacked two elastic strips whose one end is connected. Each motor is attached to each end of the V-shape elastica through the one-way wheel. The use of a one-way wheel, which transits rotations in one direction, but not does in the other direction, reduces the drag due to the motor at rapid deformation, thus increases quickness of the elastica deformation. The bridge link is used for jointing the two motors. It has a spatially elastic structure by connecting elastic strips, and therefore bridges the two motors flexibly. This elastic structure reduces loads acted on the axes of the motors. Each edge link is fixed to each connecting point between the motor and the one-way wheel. Appropriate length of the edge links allows the sharp point of the elastica not to collide with its own arch during the deformation.

By driving the two motors in the opposite direction each other, the elastica deforms spatially so that the sharp point of the elastica passes through its own arch, and goes back to the original shape. During the deformation, the elastica changes its shape drastically twice, i.e., snap-through buckling occurs twice in one cycle. If we continue to rotate the motors, we can obtain the snap-through buckling repeatedly.

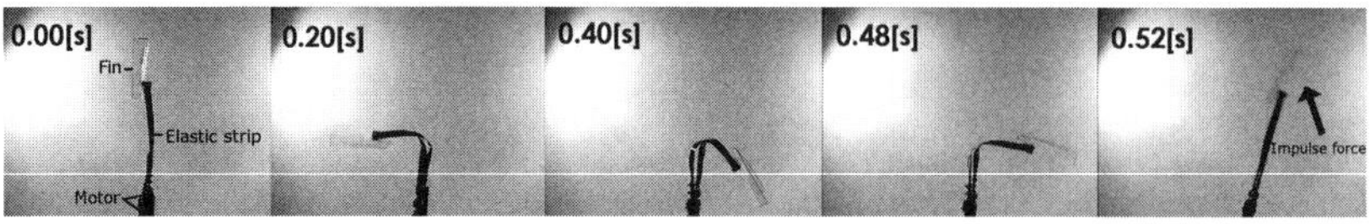

Fig. 2. The shape transition of the impulse force generator based on spatially-deformable sharp closed elastica.

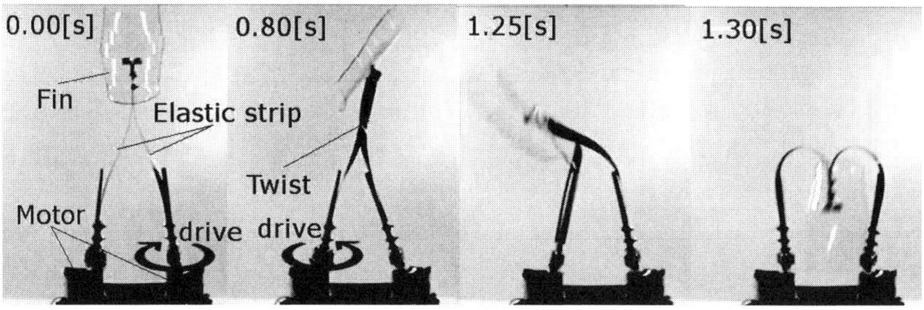

Fig. 3. The twisting motion of top of the elastic strip.

In our prototype of the impulse force generator, we use two elastic strips (JIS SK5-CSPH) for a V-shape elastica whose size is of $140\times 8.0\times 0.25$[mm], two RC servo motors (RB90, MiniStudio Inc., China), a 3.7[V] lithium ion battery (551717HS10C, FULLRIVER, China), and two one-way wheels (GP.387 Large-diameter one-way wheel WT, TAMIYA Inc., JAPAN). The body frame is made of aluminum. The distance and the angle between the motor axes are set to 60[mm] and 35[deg] respectively.

2.2. *Motion*

Fig. 2 is a series of photos of the deformation of the elastica from a lateral view. Note that time intervals between adjacent photos are not uniform. While the elastica bends slowly from 0.00[s] to 0.48[s], it changes the shape quickly in the short time from 0.48[s] to 0.52[s], which means the snap-through buckling occurs. We will utilize this phenomenon for a swimming robot to kick surrounding water impulsively.

We also found the phenomenon that the sharp part of the elastica moves sideways during the deformation. Fig. 3 is a series of photos of the deformation of the elastica from an anterior view when we make phase difference between the motors. In this case, we start to drive the right motor at 0.00[s] and the left motor at 0.80[s]. First, the elastica twists from 0.00[s] to 0.80[s],

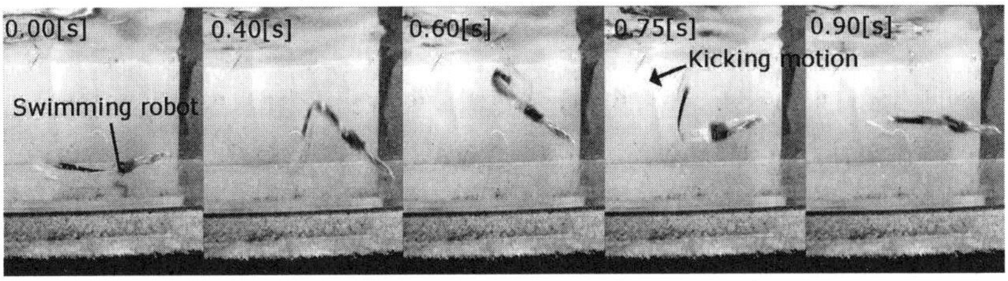

Fig. 4. The swimming motion of the swimming robot from the side view

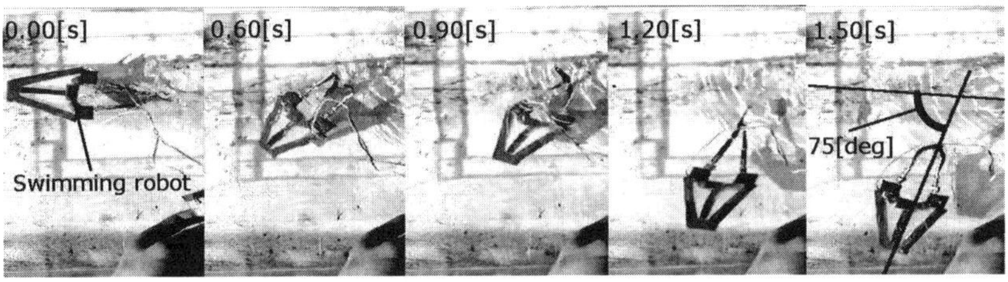

Fig. 5. The turning motion of the swimming robot from the top view

then it starts to bend. The elastica bends quickly in the short time from 1.25[s] to 1.30[s], which can be admitted as snap-through buckling. Here we should notice the following two points different from the no phase difference case. First, the sharp part of the elastica moves sideways when the snap-through buckling occurs. We will utilize this motion for a swimming robot to turn rapidly. Second, the snap-through buckling occurs just before the sharp part of the elastica passes across its own arch, which means that we have two times of snap-through buckling for one motion cycle. This first snap-through buckling is hidden in the no phase difference case. This second point motivates us to split the motion cycle into two motion phases appeared later.

3. Flat Kick-swimming Robot

The flat kick-swimming robot proposed here is a one which is very flat and swims by kicking surrounding water with the impulse force generator.

3.1. *Mechanism*

The flat kick-swimming robot consists of the upper body part and the lower body part. The upper body is made of two thin plastic films whose edges are taped and involves a battery inside. The lower part is the impulse force generator shown in the previous section at the sharp tip of which a fin is attached. The fin is curved for pushing water strongly and is slit up laterally for having appropriate elasticity. The robot is very flat, i.e., its height is of only 16[mm] compared with 340[mm] in length and 120[mm] in width. The mass is of 72[g] which corresponds to that of a size C battery (69[g]).

3.2. *Swimming Motion*

This robot has a motion cycle comprised of a pair of two motion phases, i.e., the turning phase and the traveling phase in turn. In the turning phase, the robot turns by kicking the surrounding water sideways with the first snap-through buckling in the motion cycle as shown in Fig. 3. In the traveling phase, it moves forward by kicking the surrounding water backward with the second snap-through buckling as shown in Fig. 2. The robot swimming is said to be the translational swimming if no rotation is generated in the turning phase, and the rotational swimming otherwise. If we rotate the motors in phase, we will obtain the translational swimming for the robot. Otherwise, i.e., if we make phase difference between the motors, we will have the rotational swimming. The phase difference time is hereinafter denoted by ϕ[s].

Fig. 4 is a series of photos showing the side view of the robot in the translational swimming from the time 0.00[s] to 0.90[s]. From 0.00[s] to 0.60[s], the robot bows underwater with paddling the fin downward. Then, its head is lifted while the fin passes through its own arch from 0.60[s] to 0.75[s]. After the robot kicks the water backward strongly at 0.75[s], the robot moves forward (rightward in the figure) with keeping the body along the traveling direction horizontally where only small drag forces are acted on the robot due to its flat shape. From the figure, we can see that the robot moves forward by kicking the surrounding water with deforming its body largely.

On the other hand, Fig. 5 is a series of photos showing the top view of the robot in the rotational swimming from the time 0.00[s] to 1.50[s]. From 0.00[s] to 1.20[s], the robot turns underwater with moving backward slightly. After the robot kicks the water backward at 1.20[s], the robot moves forward in the direction of about 75[deg.] right turn from the initial one. The robot can accelerates immediately after the turning motion due to the motion cycle with the turning phase and the traveling phase.

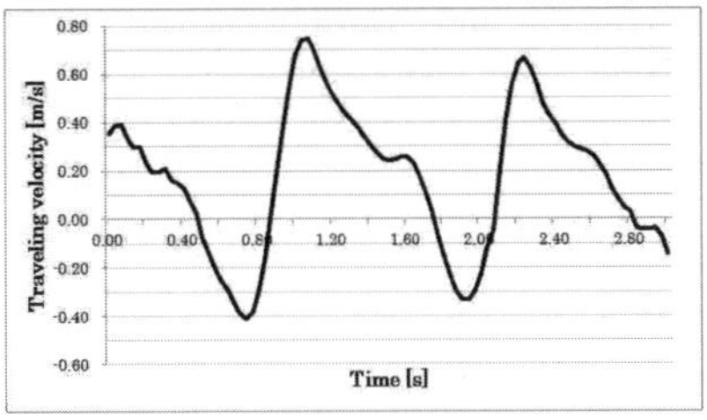

Fig. 6. Time-Traveling velocity graph of moving forward impulsively

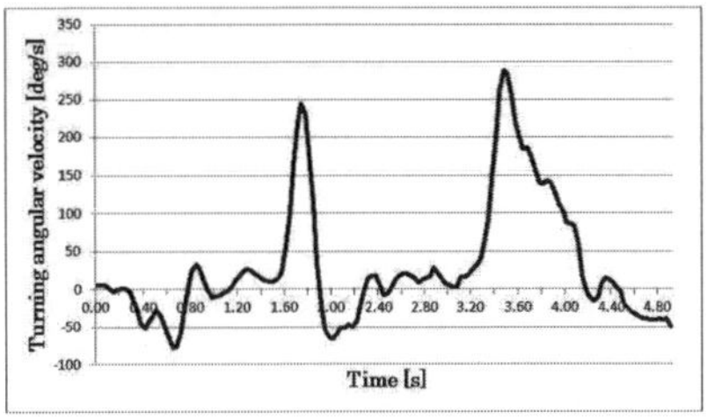

Fig. 7. Time-Turning velocity graph of turning motion

4. Basic Performance

In this section, we explain under-water experiments to verify basic swimming performances of the flat kick-swimming robot. In these experiments, the robot swims under water in a tank with size of 1800[mm] × 600[mm] × 600[mm], and its motion is recorded by a high-speed camera. Traveling distance and its velocity, turning angle and its angular velocity are obtained from the recorded movie by using a motion analysis software (TEMA lite).

4.1. *Translational Swimming*

Fig. 6 is a time graph showing the horizontal traveling velocity of the robot in the translational swimming, i.e., $\phi = 0$[s]. The horizontal axis means the time [s], while the vertical one denotes the velocity of the robot [m/s] underwater. The graph has two steep peaks after 0.40[s] which reflects rapid velocity increases of the robot by kicking the surrounding water impulsively. From this figure, we can see that the robot reaches the maximum velocity of 0.74[m/s], corresponding to 2.18[BL/s] (Body Length per second), from -0.40[m/s] in a moment of 0.33[s], which shows that the robot has the ability to move forward impulsively.

4.2. *Rotational Swimming*

Fig. 7 is a time graph showing the angular velocity of the proposed swimming robot around the vertical direction in the rotational swimming, i.e., $\phi \neq 0$[s]. The horizontal axis means the time [s]. Note that the vertical one dose not denote the velocity of the robot as the previous figure, but dose the angular velocity [deg/s] underwater in this case. Again the graph has two steep peaks after 1.20[s] which reflects rapid angular velocity increases of the robot by kicking the surrounding water sideways impulsively. From this figure, we can see that the robot achieves the maximum angular velocity of 289[deg/s] in a moment of 0.20[s], which shows the robot also has the ability to turn quickly.

5. Conclusions

In this paper, we showed basic performances of the flat kick-swimming robot powered by the impulse force generator based on snap-through buckling of a V-shape elastica. Under-water experiments showed that the proposed robot can not only move forward impulsively at the maximum velocity of 0.74[m/s] corresponding to 2.18[BL/s], but also turn quickly at the maximum angular velocity of 289[deg/s].

The robot is still far inferior to real fish in swimming ability. The future works includes further enhancement of the kicking force and decreasing the drag force of the robot.

This study was supported by a Grant-in-Aid for Scientific Research (B) (No. 22360100) from Japan Society for the Promotion of Science (JSPS).

References

1. Barrett D.S., M.S. Triantafyllou, D.K.P. Yue, M.A. Grosenbaugh and M. Wolfgang: Drag reduction in fish-like locomotion, J. Fluid Mech, 392, 183/212, 1999.
2. Hirata, K.: Development of experimental fish robot, Proc. of the Sixth International Symposium on Marine Engineering, 711/714, 2000.
3. Hu, H.: Biologically inspired design of autonomous robotic fish at Essex, Proc. fo the IEEE SMC UK-RI Chapter Conference 2006 on Advances in Cybernetic Systems, 2006.
4. Yamada, A., Y. Sugimoto, H. Mameda, H. Mochiyama and H. Fujimoto: An Impulsive Turning Mechanism for Swimming Robots Based on Repeated Snap-through Bucking, Proceedings of the 13th International Conference on Climbing and Walking Robots and the Support Technologies for Mobile Machines (CLAWAR2010), 253/260, 2010.
5. Watari, M., A. Yamada, H. Mochiyama and H. Fujimoto: A Robotic Catapult Based on the Closed Elastica with a High Stiffness Endpoint and Its Application to Impulsive Swimming Robot, Proceedings of the 11th International Conference on Climbing and Walking Robots and the Support Technologies for Mobile Machines (CLAWAR2008), 389/396, 2008.
6. Imai, S. and H. Mochiyama: Flat Kick-swimming Robot with Quick Turning Ability Powered by Impulse Force Generator, Proc. of the 12th SICE System Integration Division Annual Conference (SI2011), 0237/0240. *(in Japanese)*

"UNIBALL" A NOVEL METHOD OF MOVEMENT FOR PIPE CLIMBING ROBOTS

GURVINDER S. VIRK
University of Gävle
S-801 76 Gävle, Sweden

REZA ZANDIAN
University of Applied Science South Westphalia- Campus Soest,
Lübeckerring 2, 59494 Soest, Germany

Most of the proposed robots for pipe inspection are slow and include many mechanical parts leading to complex and unreliable operation. In this paper a novel moving robot for pipe applications, so called "UNIBALL" is proposed which overcomes these problems. Here, instead of wheels, just one ball coupled with electromagnets is used and all the movements are performed by driving the ball beneath the robot. The paper presents the robot design procedure and its detailed structure together with initial experimental test results.

1. Introduction

1.1. *Condition analysis*

Millions of miles of pipelines carry everything from water to crude oil but these pipes are vulnerable to attack by internal and external corrosion, cracking, third party damages and manufacturing flaws [3]. If a petroleum or chemical pipeline leaks, it can be a major environmental disaster and hence periodic inspections are performed to find potential flaws and damages before they become cause for concern. Robotic science can help to facilitate such processes of inspection and, as a result, several methods have been proposed for climbing robots suitable for pipe inspection tasks [4]. These methods are briefly introduced in the next section but their main disadvantages are that there is poor maneuverability, need for large numbers of complex mechanical parts, extra weight and limitation to move on just one surface of the pipe (inside or outside), etc. Hence new and more effective methods are needed to improve the overall efficiencies of climbing and inspecting robots used in such applications [5]. Some examples of climbing robots are shown in Figure 1.

(a) Snake type robot, Carnegie-Mellon University

(b) Special design, D-70/2 HELI-PIPE from Active Structures Laboratory

(c) Tank wheel inspector, Versatrax 150™, Inuktun

(d) Permanent magnet, Automated Ultrasonic Inspection

Figure 1. Various locomotion designs for pipe inspection robots

1.2. *New uniball-based method*

In the new method proposed here, the climbing robot moves on a ball (made of rubber, metal, etc, to suit different applications) which is driven via three rotating bars installed in different positions like the faces of a pyramid, so that the ball can move in any x, y and z direction (during the climbing mode). To keep the robot stable, four electromagnets are installed in the corners of the robot to control it tilt as well as supporting its weight during climbing tasks. The distance of the travel surface to each side of the robot is measured by four range sensors installed beneath the robot. The outputs of these sensors are fed to the electromagnets in a closed-loop control sense to adjust the robot's tilt angle. The electro-magnets are mounted on two links which can be adjusted according to the surface of the pipe being climbed. For inspection, a camera and an ultrasonic sensor are planned to be used for inspecting and detecting any problems in the pipe structure. In addition to these sensors, a GPS and a digital compass are to be included for effective localization and navigation.

1.3. *Why the uniball method?*

The benefits of the uniball method can be summarized as follows:

- Faster maneuverability due to the ball structure
- No limit in turning or changing direction

- Simplicity in the mechanical structure and fewer and simpler components needed leading to a lower level of maintenance and lower cost
- Ability of moving on both internal and external surfaces of pipes with no limitation in rotating or climbing.

2. Mechanical design

2.1. *Frame structure*

The frame of the robots has two plated segments; the lower plate holds the motors and the ball holder and the upper plate holds the circuits, sensors, the electrical magnets and batteries. Beside the plates, there are two movable links which are adjustable (to allow tuning of the sensors and the magnets) depending on the pipe diameter. Electro-magnets are installed on the upper plates perpendicularly to the surface of the pipe to control the robot tilt angle. Range sensors are also installed in the same configuration as the electro-magnets. Figure 2 shows the robot body frame and the position of some components.

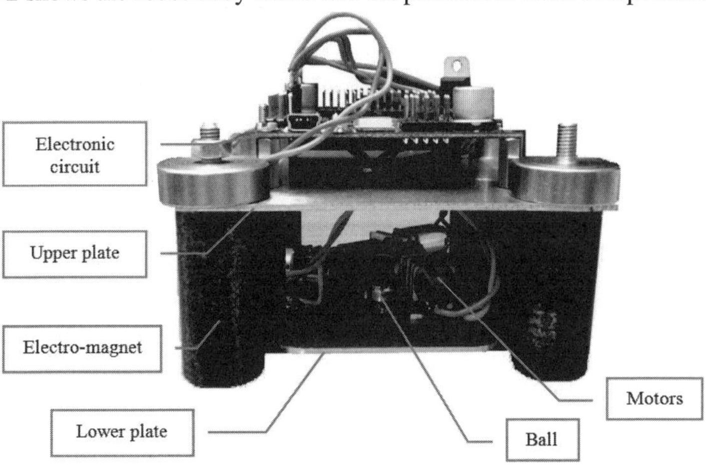

Figure 2. Uniball robot frame and location of some key components

2.2. *Motors and the uniball*

Three motors are used in the robot to drive the Uniball robot in a 2D plane to cove in the x and y directions on the pipe surface. Each motor is installed above the ball at 60° to the other motors so that they can drive the ball in any required direction by controlling the three motors to move the robot on any XY surface. It is also possible to achieve such 2D locomotion by using two motors to save cost but using three motors give us more drive power and finer level of control over the driven ball. The ball and the rolling bars are made of high friction rubber to have good locomotion capabilities and the overall configuration is shown in Figure 3.

Figure 3. Configuration of the motors and the driven ball

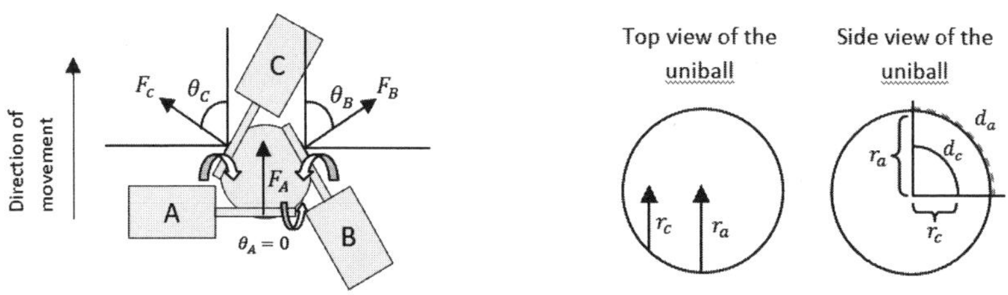

(a) (b)

Figure 4. (a) Analysis of forces on the driven ball; (b) Analysis of the distance for each motor

The amount of the drive power from the three motors can be calculated according to the force vectors shown in Figure 4(a). Based on the direction of the movement, an angle between the direction and the perpendicular bisector of each motor appears which has to be considered in calculating the total force and is given by the following:

$$F_{total} = F_A \cos\theta_A + F_B \cos\theta_B + F_C \cos\theta_C \quad (1)$$

Considering equal forces and $\theta_A = 0°$, $\theta_B = 60°$, $\theta_C = -60°$

$$F_{total} = F_A + \sqrt{3}F_A = (1+\sqrt{3})F_A \quad (2)$$

For each motion, it is necessary to consider the "wheel diameter", i.e., the diameter of the base disc that is driven by each motor. Considering Figure 4(a), 4(b) shows the displacement of the ball as it is driven along the full diameter of the ball by motor A and along a base radius that gives a quarter the circumference by motors B and C. These differences in drive radii demand the control system to calculate the speed and displacement of each motor independently based on the motion direction of the robot. To calculate the displacement for each motor, the following equations can be used:

$$d_x = 2\pi r_x \; and \; r_x = r\cos\theta \xrightarrow{yields} d_x = 2\pi r\cos\theta \qquad (3)$$

where r_x is the base radius of the ball along the drive path, θ is the angle between the direction of motion and the perpendicular bisector of each motor, r is the radius of the sphere and d_x is the displacement needed for that motor. The speed and displacement of each motor must be calculated to avoid contradictory movement of the motors against each other.

2.3. *Tilt management system*

To control the tilt of the robot, it is necessary to measure the gap on the sides of the robot and the travel surface so that it is controlled as needed; for the distance measurement, infra-red distance sensors are used which can measure distances in the range 5-60 mm. According to the characteristics of the sensors, their outputs are voltage which changes non-linearly with distance (see Figure 5(a)).

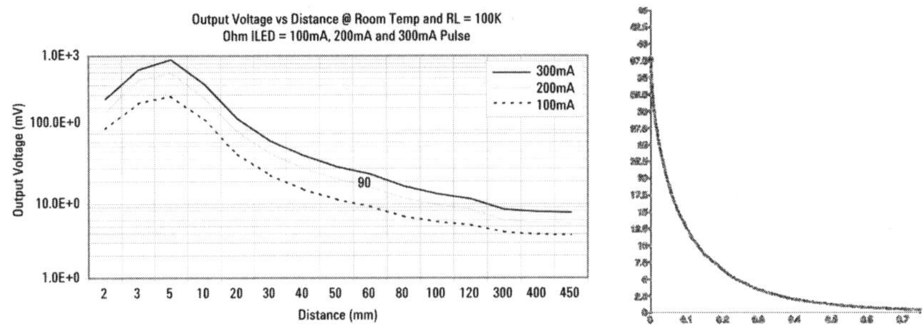

(a) (b)

Figure 5. (a) Output of the IR sensor vs. distance, (b) Force of the magnet vs. distance to surface

For the tilt control, electro-magnets are to be used; the design of these need to investigate the force and distance relationship in various configurations in which they may be used. The amount of the force available from the magnet is inversely proportional to its distance d_s to a ferrous surface and is given by equation 4 ([2]):

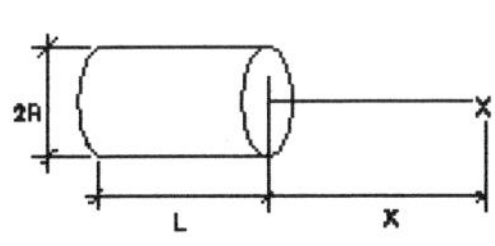

$$F \approx 0.58 B_r^2 L_m \sqrt{A}/x \qquad (4)$$

where F is the magnetic force, B_r is the residual induction, L_m is the magnetic path length, A is the cross-section area of the pole and x is the air gap to the metal surface [2]. The relation of the force versus the distance of the magnet to the surface based on this formula is shown in Figure 5(b). The magnetic force F is designed to be perpendicular to the travel surface by ensuring the robot tilt angle is controlled by the tilt management system. The control aspects are discussed in section 3.

3. Design of electronics and control sub-systems

The brain of the robot is a microcontroller able to sense the environment and control the drive motors and the electro-magnets in appropriate ways. The overall architecture of the system is shown in Figure 6 where a controller is implemented in the software which enables the microcontroller to make the necessary control actions.

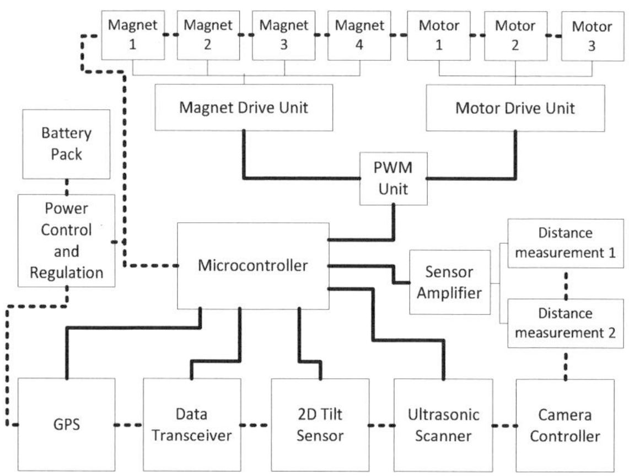

Figure 6. Block diagram of the overall Uniball system architecture

For the controller that is implemented via software, there are two main aims, namely, maintaining the required balance and providing the desired locomotion commands to drive the three drive motors. The control system in this case is a MIMO system with the following details:

- Tilt balance control system: 8 IR sensors measuring the air gap distances on all four sides of the robot and 4 electro-magnets to be controlled in maintaining the correct tilt angle. This system can be seen as a 2-input/2-output system to control the x-y tilt. However, because of the symmetry in the mechanical layout the tilt control system design can be carried out independently as 2 single input-single output control systems.
- Navigation system: 2 IR range sensors for obstacle detection to drive the 3 motors to move the robot as required for navigation purposes. This also is a 2-input/2-output control system to navigate the robot in the x-y plane.

The control algorithm for both control loops are designed to receive data, calculate the error and transfer it to a PID controller to correct the error in a feedback sense to achieve the desired effect. Figure 7 shows the block diagram of the tilt management system and Figure 8 presents the control block diagram of the locomotion system. The former shows just one of the two blocks available in the robot while the latter shows one of the three locomotion control systems.

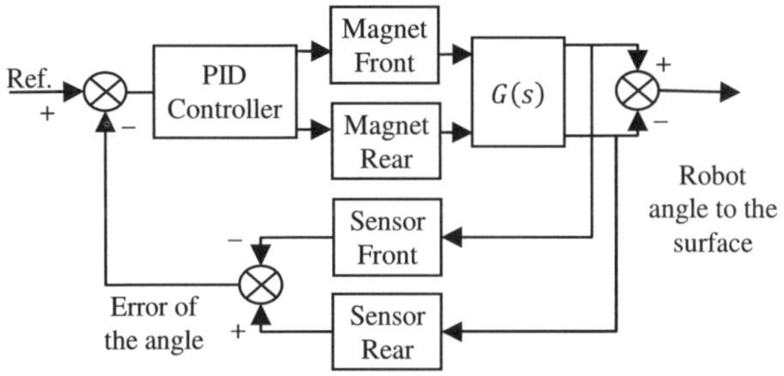

Figure 7. Block diagram of the tilt control system for one axis

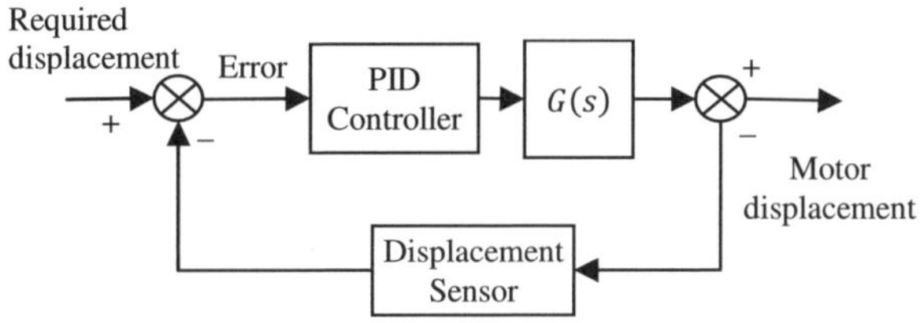

Figure 8. Block diagram of the locomotion control system for one motor

4. Test locomotion results

In order to test the performance of the locomotion system, various tests have been performed. These include the robot being programmed to move 1 metre at an angle of 70°; under this motion, it is seen that the robot rotates as it moves forward in the desired direction which causes it to lose its sense of direction. To solve this problem, a digital compass has been used to measure the unwanted rotations and calculate the changing direction to continually move the robot in the correct direction. To clarify this, Figure 9 show an example where the robot starts to move along the defined path facing forward (say); after a little motion the body has rotated but as the compass has detected this, it is possible to define the new angle for the controller of the motors to match the desired path. This rotation occurs continually and at the end of the path the body has turned 130° but by correcting the control direction, the robot is able to move on the defined path correctly.

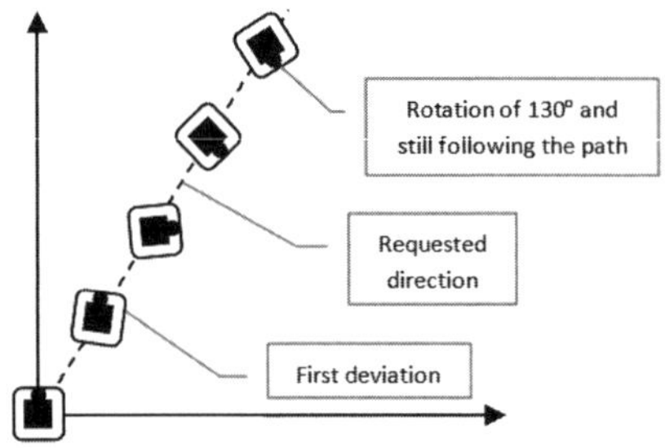

Figure 9. Test results of locomotion (moving an angle of 70°)

5. Conclusions and future works

The stability controller of the robot brings enough maneuverability to the robot which makes it an ideal device for moving in or outside of pipes. As well as that it is expected that we can use magnets to hold the weight of the robot on to the surface of the pipe even if the robot is moving on the ceiling of the pipe upside down. The results of locomotion tests have proved that the robot has a fast response to sudden movement requirements. The robot developed thus far uses a manual remote controller and the next area of focus for this work is to improve the robot to works autonomously using other controllers such as neural networks which have the ability to predict and control significant nonlinearities such as those in the Uniball robot. There are two main limitations of the concept, namely passing flanges over the pipe and contamination of the ball in dirty and greasy areas. These two problems will be addressed in the next stages of the planned research.

References

1. M Yorihisa Yamamoto, NXT Ballbot Model-Based Design, rev 1.0, 2009.
2. www.magnetsales.com/Design/DesignG.htm, 2000 Magnet Sales & Manufacturing Company, Inc (visited @ 01.2012).
3. Fabien Tache and the others, Adapted Magnetic Wheel Unit for Compact Robots Inspecting Complex Shaped Pipe Structures, 2007 Switzerland.
4. Hyouk Ryeol Choi and Se-gon Roh, In-pipe Robot with Active Steering Capability for Moving Inside of Pipelines, School of Mechanical Engineering, Sungkyunkwan University Korea, 2007.
5. M. M. Moghaddam, M. R. Arbab Tafti, Design, Modelling & Prototyping of a Pipe Inspection Robot, Proc 22nd ISARC, Ferrara, Italy, 2005.
6. Oliver Nelles, Non-linear system identification, Springer publication ISBN 3-540-67369-5, Germany, 2001.

MOBILE GROUND-MOVING AND WALL-CLAIMING (MGMWC) ROBOT*

IEM HENG

Computer Engineering Technology, New York City College of Technology, 186 Jay St Brooklyn, New York 11201, USA

ANDY S. ZHANG

Mechanical Engineering Technology, New York City College of Technology, 186 Jay St Brooklyn, New York, USA

The mobile ground-moving and wall-claiming robot is a unique robot that is capable of moving on a leveled ground or claiming on vertical walls. The MGMWC robot can climb vertical flat surface by using suctions cups through an innovative pneumatic system. The current prototype pneumatic system consists of two air compressors, one piston, two solenoid valves and four 7-inch diameter suction cups. This MGMWC robot is much larger than those typically found on the commercial market because a larger scale device can carry larger loads, and therefore provide more useful applications. Applications include the climbing of architectural exterior surfaces, civil engineering structures, the outer hulls of ships, and aircraft tails (while on the ground). Obvious uses include cleaning of windows or vertical surfaces, delivery of safety ropes or harness to humans, inspection of window joint seals, and inspection of the surface of dams or ships or aircraft surfaces for cracks and leaks. Further, on discovery of faults, this device has sufficient load capacity to perform repairs on a defected surface, whether temporary or permanents. In a real sense, the MGMWC robot could provide the necessary "finger in the dike" that averts disaster. And, as a robotic device, it will keep on its post indefinitely. It is controlled wirelessly, using ZigBee (XBee shield), through tele-operated mode with a programmable Arduino microcontroller.

1. Introduction

Living in New York City and using skyscrapers as an inspiration, the MGMWC robot is an alternative to clean the exterior walls and windows of skyscrapers. Window cleaning is considered the most dangerous job that causes many deaths and injuries to workers as shown in Fig. [1]:

Figure 1

* The work is partially funded by a grant from the National Science Foundation Advanced Technology Education Division. The award number is NSF ATE No 1003712.

Among the serious commercially available wall climbing robots found that are similar to MGMWC, several wall climbing robots are located in different regions around the globe. For instance, one is located in Switzerland, one in Japan, one is in Germany, and one is in Spain. The Swiss company, Serbot [2], appears to focus on self window cleaning as the primary function. The Japanese company, Urakami [3], has three models for window cleaning. The University Ailen in Germany, Max V [4], is a chain-driven climbing robot with vacuum chambers and focuses on window glass cleaning. The CSIC Madrid in Spain, Rest [5], is a six-legged welding robot with magnetic feet and appears to focus on wall cleaning. But those robots are self-regulated window cleaning without tele-operated mode. In addition, those window cleaning robots are small scale robots which will support a relatively smaller payload. Unlike those commercial robots, the MGMWC robot is a large size-robotic structure to withstand gravity and, used not just for window cleaning with tele-operated mode (human operator with remote control), it can be used to perform other tasks related to building exterior walls. And it is also capable of moving on the ground.

2. MGMWC Robot Overview

In this section, the developmental stages of MGMWC robot is discussed which includes mechanical and electrical components, microcontrollers, and suction cup lifting system.

2.1. *Construction Stages of MGMWC Robot*

The details of a conceptual design for the MGMWC robot as illustrated in the following Figure 2:

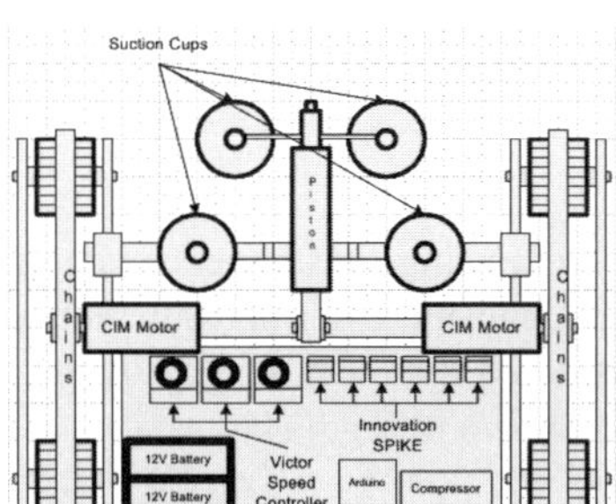

Figure 2. Conceptual Design of MGMWC Robot

Based on the conceptual design of MGMWC robot in Figure 2, this robotic device consists of four wheels, four suction cups, one piston, two CIM motors, two victor speed controllers, one utrasonic sensor, two 12V batteries, three solenoids, two spikes, two compressors and one Arduino microcontroller [6] with XBee shield [7]. Figure 3 shows different stages in building the MGMWC robot.

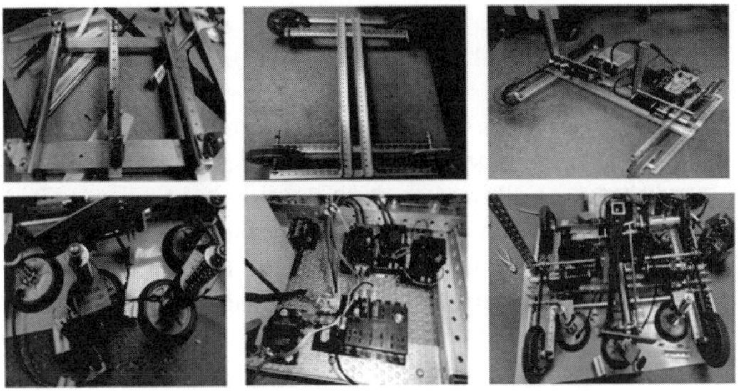

Figure 3: Design Stage of MGMWC Robot

2.2. *Prototyping of MGMWC Robot*

From the conceptual design in Figure 2 and development stages in Figure 3, a 3D model and a physical prototype of MGMWC robot were subsequently created as shown in Figure 4.

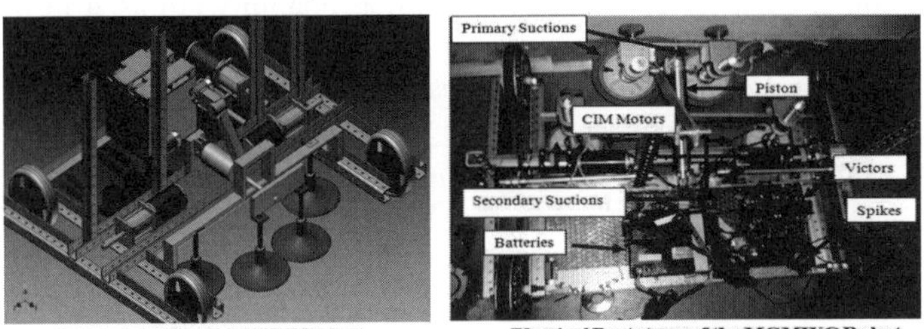

3D Model of the MGMWC Robot **Physical Prototype of the MGMWC Robot**

Figure 4: Prototyping of MGMWC Robot

Four 7-inch ANVER suction cups are being used. Two primary suction cups are connected to the lifting arm driven by a pneumatic piston located in the center of the MGMWC robot. Two secondary suction cups are placed on two sides of the lifting arm. Each suction cup can lift up to 200 pounds. Since suction cups work in pair, at any given time, there will be two suctions cups that are being used thus allowing the MGMWC to carry large load while still be able to climb a vertical plane. The lifting arm is made of a four-bar linkage, such that it moves in an articulated motion. One compressor is used to provide compressed air to the piston which is used to lift the arm attached to the primary suction cups. The second compressor is used to activate the two secondary suction cups. The two primary suction cups share one solenoid valve, while the secondary suction cups share another solenoid valve. The valves regulate the airflow direction.

The reason why the compressor is used instead of a vacuum pump is: Once the air pressure reaches a level of 45 psi, a "reversed" airflow of the compressed air away from the suction cup creates vacuum condition faster than using vacuum pump. This is critical for the continuous operation of the MGMWC robot.

Once the primary suctions are safely attached to the surface, the secondary suctions along with the main body are lifted to a higher position. Once they are fully positioned the air is allowed to enter the primary suction cups' cavity to release them. The secondary suction cups will remain attached until the primary suction cups move up and are fully attached again. Hence, the system alternates airflow to and from the primary to secondary suction cups for safely attaching and detaching, while climbing upward.

The entire pneumatic system requires eight spike relays [8]. They are used to control the vacuum pumps that provide the suction effect and the release of air, and to regulate the air flow of suction cups. The decision to utilize the spikes in substitution of a regular IC H-bridge system is its capacity to withstand up to 20 Amperes of current as oppose to 1 Amperes of an H-bridge IC chip.

The MGMWC robot uses four motors in total. Of the four motors, two are CIM motors [9] for the chassis system and two smaller motors for the pneumatic system. Each motor requires about 2A of current; therefore, two 12V batteries with a nominal capacity of 18Ah for extra dependability are used in the entire system. The two motors for the chassis are designated for the wheels in the structure and placed on opposite side of each other. One of the motors is used to lift the four- bar linkage connected to the piston. The remaining motor provides the movement of the four- bar linkage connected to the secondary suction cups for proper placement.

Due to the precise synchronization requirement of the pneumatics system and high current demand of the MGMWC robot, Victor speed controllers [10] are used in the MGMWC robot. Victors support a high current capacity with a minimal voltage drop. They are efficient in controlling speed and direction of motors. They also provide precise control of lifting systems. Victors use PWM signal to control the motion of the four motors in the robot. The PWM signal is provided by the Arduino Mega microcontroller, which is the central operating system of the MGMWC robot.

The Arduino Mega [11] is one of the many variations of ATMEL microcontrollers. It uses open-source software that can easily interface with electronics, with the proper electrical equipment. The Arduino board originally provides 5V, but allows external power sources. The best thing about this microcontroller is its ability to control large projects, with a relatively low cost.

2.3. *MGMWC Robot: Suction Cup Lifting System*

The MGMWC robot chassis is made of aluminum C-channels to reduce the weight. Two CIM motors and two 36:1 planetary gearboxes from BaneBots were used to drive the robot on the horizontal surface. Two RS 550 DC motors with two 256:1 planetary gearboxes were used to lift the robot. They provide enough driving force to lift the suction cup arms. Two suction cups attached to the mainframe will start to grab the surface using vacuum principle. Once the pressure inside the two suction cups reaches nearly vacuum condition, a pneumatic piston will raise the other pair of suction cups up by 10 inches. The raised suction cups will in turn to start to grab the surface of the wall. Once the raised suction cups are firmly grasped to the wall, the pneumatic piston will lift the robot up by 10 inches. One concern was that the planetary gearbox does not provide self-locking. Once the power was cut off, the arm that holds the suction cups might give in. To avoid this problem, we use two worm-gear motors to drive the suction cup lifting system.

To prevent falling due to turning moment of gravity, all suction cups are placed in the front of the MGMWC robot. Figure 5 is a free body diagram of the MGMWC robot in one of the typical vertical position.

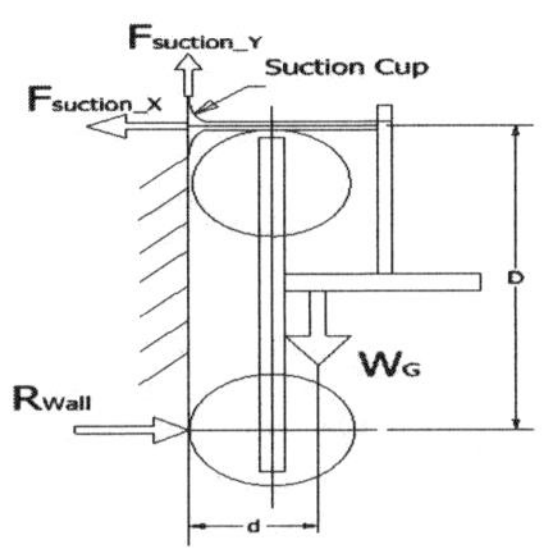

Figure 5: Free Body Diagram of MGMWC in vertical position

Where $F_{suction\text{-}x}$ = Suction force produced by suction cups in the horizontal direction,
$F_{suction\text{-}y}$ = Vertical frictional force created by vacuum condition,
R_{wall} = Reaction from the wall,
W_G = Weight of gravity of the MGMWC robot.

To prevent MGMWC robot from falling, the lifting moment $M_{lift} = F_{suction}$ x D has to be larger than the tilting moment $M_{tilt} = W_G$ x d. The net lifting moment becomes

$$M_{net\_lifting} = F_{suction\_x} D - R_{wall} d \quad (1)$$

3. Analysis and Testing Results

A wall climbing device must overcome forces that surface traveling devices need worry about. The normal force of gravity that holds a vehicle to the road is the force that will cause a wall climber to fall. Typical wall climbers have usually been quite small, relying on small suction cups or materials that adhere as a gecko foot pad does. While innovative and interesting, such devices cannot carry as large a payload, as the MGMWC robot can. What is unconventional in this design is the synthesis of parts intended for other purposes into a new innovative device that can be deployed easily.

Technically speaking, it is very difficult for a large-scale robot, with a weight of over 100 pounds to climb on a 90° vertical plane. Preliminary testing of a suction cup, as shown in Figure 6 below, proves that the MGMWC robot should be able to climb surfaces of materials such as glass, Plexiglas, metal, and other continuous surface areas. Nonetheless, gravity being an imminent force, the MGMWC robot would not be able to climb infinitely upward. The suction cups can hold up to 800 lbs while stagnate or 400 lbs while travelling, however as the structure keeps moving to extreme heights, physics still comes into effect and the force of gravity acted upon the robot becomes greater. This may be a factor to set a maximum height requirement on the MGMWC robot before it may become too much of a force. Nonetheless, the MGMWC robot does perform to its primary task which is to successfully travel on a vertical plane with distance depending on battery power and motor speed.

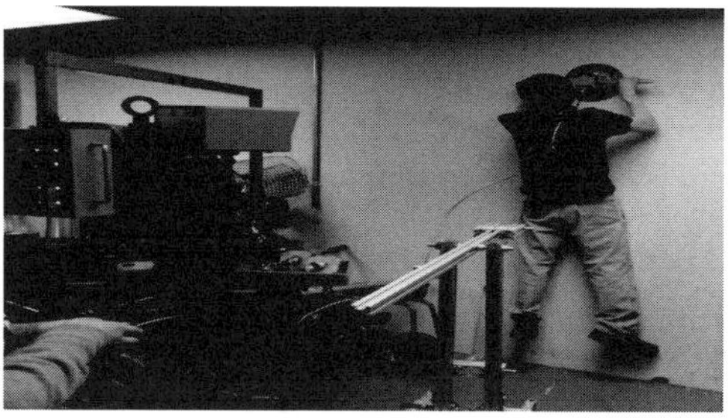

Figure 6: Suction Cup Testing on Dry Wall Material

By using industrial suction cups of large diameter and positioning the heaviest components close to the wall surface, the MGMWC robot is able to carry a significant load. As an illustration, a suction cup of 7-inch diameter has an area of 38.5 square inches. Once the atmosphere is evacuated, that will provide a holding force F_{max} of approximately 566 lb., when vacuum condition is reached ($p_{vacuum} = 0$) as shown in equation (2) below. Figure 7 below demonstrates how suction cup works.

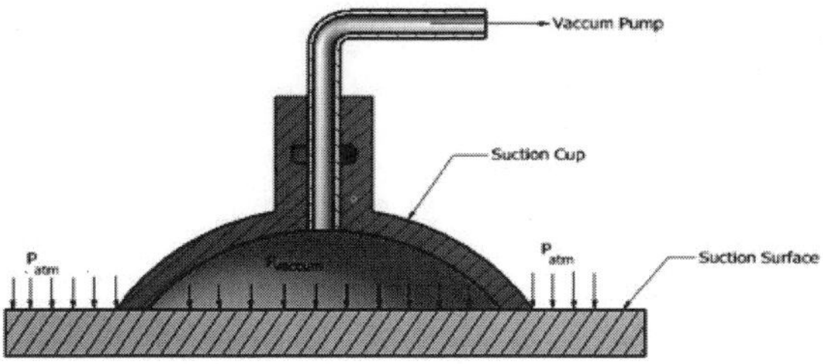

Figure 7: Suction Cup Working Principle

Given that there are four such pads, with minimum of two in contact at a given time, the device should carry a significant load with an adequate safety factor. One must admit, however that this depends in part on the surface condition. One instance is that a surface wetting agent might be employed to increase adhesion of the pad. This latest concept for the device is still in a progress stage of testing.

$$F = p_{atm} A_o - p_{vacuum} A_i \tag{2}$$

Where F = suction force (lb.),

p_{atm} = atmospheric pressure (psia),

A_o = circular area based on the outside diameter of the suction cup (in^2),

p_{vacuum} = suction pressure inside the vacuum cup cavity (psia),

A_i = circular area based on the inner diameter of the suction cup (in^2).

In our case, since the atmospheric pressure is p_{atm} = 14.7 psia, the outside circular area of the suction cup is $A_o = \frac{\pi D_o^2}{4} = \frac{\pi (7)^2}{4}$ = 38.5 in^2, and the circular area of the suction cup cavity is $A_i = \frac{\pi D_i^2}{4} = \frac{\pi (6)^2}{4}$ =28 in^2, if the pressure inside the cavity can reach p_{vacuum} = 10.0 psia (to be conservative), the suction force for each suction cup can be: $F = 14.7x38.5\text{-}10x28 = 286\ lb$. Since two suction cups are used at any given time, the total suction force would be twice of F ($F_{total} = 2F$). For above case, the lifting force would be:

$$F_{total} = 2x\ 14.7x38.5\text{-}10x28 = 572\ lb.$$

If ideal vacuum condition is reached (p_{vacuum} = 0 psia) (ideal vacuum condition), then $F = F_{max} = 14.7x38.5 = 566\ lb$. The total lifting force would be $F_{total} = 2x\ F_{max} = 2x\ 14.7x38.5 = 1132\ lb$. This guarantees that the lifting moment is quite larger than the tilting moments as indicated in equation (1).

To minimize the weight of the robot, Arduino microcontroller was selected. With appropriate extensions it can accommodate wireless control, a camera, and can function under user control as desired. Further, it provides ample capability to support the logic required to accomplish a "fail-safe" operation besides any other needed utilitarian functions.

4. Communication System Overview

The communication system for both MGMWC robot and user control device is shown in Figure 8 below. The user control device utilizes two joysticks to remote control the MGMWC robot, one Arduino microcontroller with Xbee shield to provide wireless communication between the robot and user control device, and one laptop to program the Arduino sketch language. The sketch language is similar structure as C/C++ language. Whereas, the MGMWC robot has accumulated of two CIM motors powered by 12V batteries, two victor speed controllers, one ultrasonic sensor, three solenoids, two spikes, two compressors, and one Arduino microcontroller with Xbee shield. The microcontroller is a vital source to communicate all the components between the MGMWC robot and user control device.

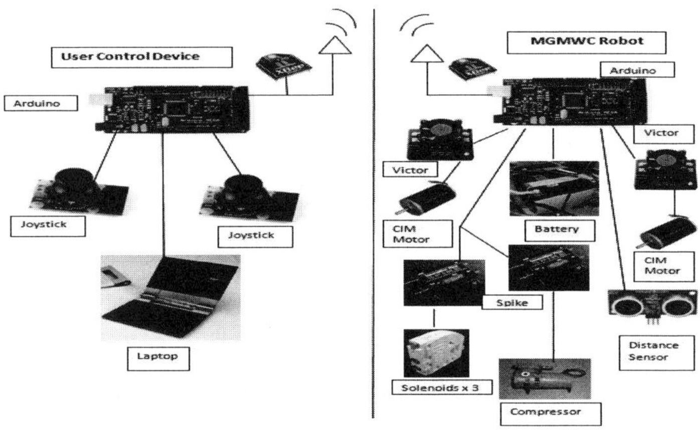

Figure 8: Communication System

5. Summary and Conclusion

In summary, the MGMWC robot uses four suction cups to provide suction force needed to keep the device on a smooth continuous vertical surface. To prevent falling due to turning moment of gravity, all suction cups are placed in the front of the MGMWC robot. Suction cups are used in pairs. Once the MGMWC robot reaches vertical position, two suction cups attached to the mainframe will start to grab the surface using vacuum principle. Once the pressure inside the two suction cups reaches nearly vacuum condition, a pneumatic piston will raise the other pair of suction cups up by 10 inches. The raised suction cups will in turn to start grabbing the surface of the wall. This action repeats itself to achieve climbing.

The current MGMWC robot prototype has the following innovative design features:

1. Two suction cups are used at any given time to keep the robot in place.
2. A "fail-safe" feature is incorporated to prevent movement if only one "cup" is holding the device to the surface.
3. Hold on a vertical surface and able to move up about 50feet and back down again without falling.
4. Perform the tasks above while carrying a load of at least 100 lbs.
5. MGMWC robot is capable of moving on the ground and wall claiming.

However, the current design of MGMWC robot can only move vertically in wall claiming. It does not allow the robot to turn while wall claiming. We hope to solve this problem by redesigning mechanisms that will allow the MGMWC robot to make turns while claiming up the wall.

Acknowledgments

The work is partially funded by a grant from the National Science Foundation Advanced Technology Education Division. The award number is NSF ATE No 1003712. The authors appreciate greatly the support from the NSF. In addition, the authors would like to thank Aidan Murphy (NYC College of Technology student) who is involved in making MGMWC robot.

References

1. http://www.csgcares.com/
2. http://www.serbot.ndswing05.ch/template_gray/robot_cleaning.html
3. http://www.urakami.co.jp/eng/index.html
4. K. Berns, C. Hillenbrand, T. Luksch. "Climbing Robots for Commercial Applications – a Survey", University of Kaiserslautern, 67653 Kaiserslautern, Germany (2006).
5. J.C. Grieco and M. Prieto and M. Armada and P. Gonzales de Santos. "A Six-Legged Climbing Robot for High Payloads", Proceeding of the 1998 IEEE International Conference on Control Applications (1998).
6. Igoe, Tom (2008). "Making Things Talk", O'Reilly Media, Inc., Sebastopol, California.
7. http://www.arduinofun.com/
8. http://www.robotcombat.com
9. http://www.robotmarketplace.com
10. http://www.ifirobotics.com
11. http://www.flickr.com

SECTION–6
LOCOMOTION

IMPEDANCE CONTROL FOR A BIOINSPIRED UNDERACTUATED LEG

JUAN CARLOS AREVALO, MANUEL CESTARI, DANIEL SANZ-MERODIO and ELENA GARCIA

Centre for Automation and Robotics, UPM CSIC
Carretera Campo Real km.0200, Arganda del Rey, Spain 28500,
E-mail: juan.arevalo@car.upm-csic.es

Bioinspired legged robots are a good option for field missions over complex terrain such as extraterrestrial landscape, disaster scenarios, etc. For these applications robots should exhibit good performance in terms of mobility, payload and endurance. In this work a control law to exploit the HADE leg dynamic features in order to save electrical energy by reducing the required mechanical work is presented. This control law is a force based impedance control scheme that considers the fact that the prototype is an underactuated mechanism. An analysis on the energy saving capabilities of the system, based on spring stiffness is theoretically performed and experimentally validated.

Keywords: Bioinspired robots, Impedance Control, Energy Saving

1. Introduction

Legged robots are a good choice for missions over complex terrain such as disaster scenarios, extraterrestrial exploration, etc, because of their superior adaptability to these terrains. Over the years, several legged robots have been built. However, very few are capable of agile locomotion required for the aforementioned missions.[1] On the other hand, new biologically inspired robots are emerging,[1–3] which attempt to take advantage of animals' key features that distinguish their dynamic behavior. One important characteristic of this kind of robots is the use of passive elements to simplify the control and to save energy as it seems to be the case of animals with their springy tendons.[2,4] Moreover, control methods are being developed in order to take advantage of the dynamic characteristics of these systems.[2,4] Most of these studies have been carried away with the general goal of increasing energy efficiency and therefore increasing autonomy and endurance. In this work, a force based impedance control scheme[5] is implemented on a bioinspired underactuated leg prototype (see Fig. 1). This prototype is a planar

3 DOF leg directly driven by two Series Elastic Actuators (SEA); one at the hip and other at the knee, while the fetlock joint is passively driven by a spring resembling the superficial digital flexor in horses. The impedance control was chosen motivated by the fact that animals seem to change their leg apparent stiffness in order to accommodate to the type of terrain.[6,7] On the other hand, underactuated robots have been subject of study in the

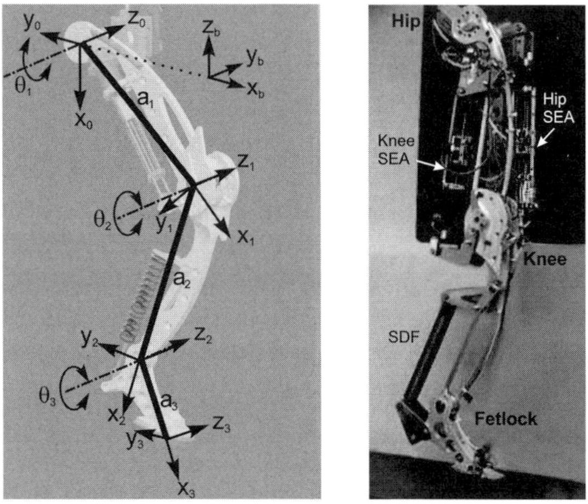

Fig. 1. The HADE Leg Prototype. Conceptual design (left). Real prototype (right)

robotics community for a number of reasons, and it has been demonstrated that the number of passive joints has to be lesser or equal to the number of active joints for the system to be controllable.[8,9] Control strategies have been proposed in order to control such systems by exploiting their intrinsic dynamic coupling[8,10,11] in both joint space and task space. Such strategies allow the passive joints to move freely until the desired joint position is achieved, and then the passive joint is blocked, allowing the other joints to move in order to achieve the overall desired position. In the case of the leg here presented, such strategies cannot be directly applied since the passive joint is not allowed to move freely because of the spring attached. In this work, a force based impedance control law is proposed and experimentally validated and the possibility of reducing the required mechanical power for its operation in function of the spring stiffness is analyzed and tested. This paper is structured as follows: Section 2 presents a control strategy in order to take advantage of the leg design and its energy saving capabilities are analyzed. In Section 3 an experimental validation of the control law and of its energy saving properties is presented. Finally, in Section 4 conclusions are stated and future work is outlined.

2. Force Based Impedance Control Strategy

The proposed control law attempts to achieve the desired position by controlling the exerted force in the corresponding coordinate (i.e. the vertical force drives the vertical coordinate and the horizontal exerted force drives the horizontal coordinate). This means:

$$F_r = \begin{bmatrix} F_{xr} \\ F_{yr} \end{bmatrix} = \begin{bmatrix} -K_x(x_r - x) - B_x(\dot{x_r} - \dot{x}) + F_{xd} \\ -K_y(y_r - y) - B_y(\dot{y_r} - \dot{y}) + F_{yd} \end{bmatrix} \tag{1}$$

where F_{xr} and F_{yr} are the reference forces for the x and y coordinates; x_r and y_r are the position references, x and y are the robot's position; K_x, K_y, B_x and B_y are the desired stiffness and damping of the respective coordinates; and F_{xd} and F_{yd} are the desired vertical and horizontal forces respectively. Given the force references in the task space, the necessary torque in each joint is given by the well known equation:

$$\tau_r = J^T F = \begin{bmatrix} F_{xr}J_{11} + F_{yr}J_{21} + F_{\phi r}J_{31} \\ F_{xr}J_{12} + F_{yr}J_{22} + F_{\phi r}J_{32} \\ F_{xr}J_{13} + F_{yr}J_{23} + F_{\phi r}J_{33} \end{bmatrix} \tag{2}$$

where J_{ij} are the corresponding terms of the Jacobian matrix with $i = 1, 2, 3$ and $j = 1, 2, 3$; and $F_{\phi r}$ is the desired moment for the leg orientation. However, since the fetlock is driven by a spring, its joint angle and torque cannot be controlled. For that reason, this leg can be treated as an underactuated manipulator with $\tau_3 = Kq_3$.

As mentioned before, the existent strategies for controlling underactuated robots cannot be directly applied to the leg here presented. Because, such strategies allow the passive joints to move freely until the desired joint position is achieved, and then the passive joint is blocked, allowing the other joints to move in order to achieve the overall desired position and that is impossible in this case because of the spring. Nevertheless, the main idea behind these control strategies, which is correcting the behavior of the passive joints with the active ones, can still be applied in our case. Following this idea, in order to compensate for the torque produced by the spring, the following task space force references are selected:

$$F_r = \begin{bmatrix} F_{xr} \\ F_{yr} \\ F_{\phi r} \end{bmatrix} = \begin{bmatrix} F_{xr} \\ F_{yr} \\ (\tau_3 - F_{xr}J_{13} - F_{yr}J_{23})\frac{1}{J_{33}} \end{bmatrix} \tag{3}$$

where τ_3 is the fetlock torque. Replacing (3) into (2) and J_{ij} with its corresponding values in this equation, and after some manipulation, the torque references (τ_{ri}) can be written as:

$$\tau_r = \begin{bmatrix} \tau_{r1} \\ \tau_{r2} \\ \tau_{r3} \end{bmatrix} = \begin{bmatrix} F_{xr}\hat{J_{11}} + F_{yr}\hat{J_{21}} - \tau_3 \\ F_{xr}\hat{J_{12}} + F_{yr}\hat{J_{22}} - \tau_3 \\ \tau_3 \end{bmatrix} \tag{4}$$

where, $\hat{J_{11}} = -l_1S_1 - l_2S_{12}$; $\hat{J_{12}} = -l_2S_{12}$; $\hat{J_{21}} = l_1C_1 + l_2C_{12}$ and $\hat{J_{22}} = l_2C_{12}$ This choice in the reference values allows the fetlock joint angle to vary freely without disturbing the global hoof force tracking, while its contribution is reflected in the other two joint torques. Moreover F_{xr} and F_{yr} can be chosen independently from τ_3 since its contribution is compensated by the other torque references. This leaves us with the control scheme depicted in Fig. 2.

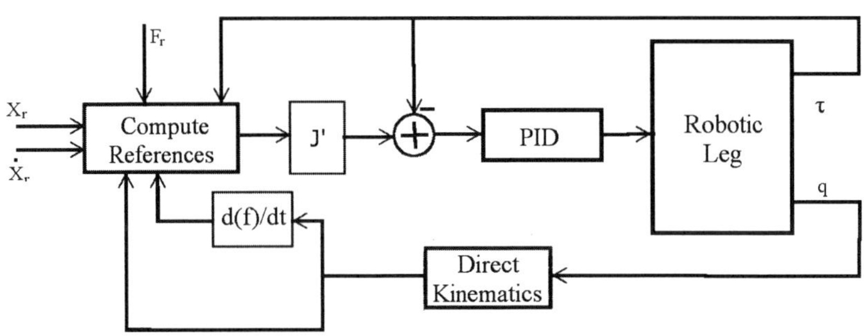

Fig. 2. Impedance Control Scheme

Now, if the articular positions q_1 and q_2 are fixed and the forces in the task space F_x and F_y are also fixed, τ_r is given by:

$$\tau_r = \begin{bmatrix} A - \tau_3 \\ B - \tau_3 \\ \tau_3 \end{bmatrix} \tag{5}$$

where A and B are constants. Moreover, assuming that the controller reaches the reference torques quickly enough, the forces F_x and F_y are given by:

$$F_x = 2\tau_3 l_1 C_1 + l_2 C_{12}(A - B) - B l_1 C_1 \quad (6)$$

$$F_y = -2\tau_3 l_1 S_1 + l_2 S12(B - A) + B l_1 S_1 \quad (7)$$

and the mechanical work is given by:

$$W = \begin{bmatrix} (A - \tau_3)q_1 \\ (B - \tau_3)q_2 \\ \tau_3 q_3 \end{bmatrix} \quad (8)$$

The fact that the force referenced are followed ensures the impedance control objective since these references reflect a relationship between the position and the force according to (3).

2.1. *Effect of spring stiffness on the legs mechanical work*

In direct driven robots, the mechanical work done by the joints in order to maintain a certain position is translated in electrical energy consumption, because in spite of the mechanical power being zero there is no mechanical element that can ensure the required torque level in order to maintain the needed position, as it is the case of highly geared machines with no backdriveability, since the transmission mechanically ensures that the leg configuration is mantained. For that reason, the energy needed to maintain a certain torque value implies current consumption on the actuators. In this section, the contribution of the spring stiffness in the total mechanical work done by the leg is analyzed.

Consider that the fetlock joint is driven by one spring with a stiffnes K_1 and then this spring is replaced by another spring with a stiffenss K_2 such as $K_2 = \alpha K_1$ with $\alpha > 1$. This means that:

$$\tau_3^2 = \tau_3^1 \Longleftrightarrow q_3^2 = \frac{q_3^1}{\alpha} \quad (9)$$

$$W_3^2 = W_3^1 \Longleftrightarrow q_3^2 = \frac{q_3^1}{\sqrt{\alpha}} \quad (10)$$

where q_3^1 and q_3^2 are the fetlock coordinates using springs with K_1 and K_2 respectively; τ_3^1 and τ_3^2 are the fetlock coordinates using springs with K_1 and K_2 respectively; and W_3^1 and W_3^2 are mechanical work due to the fetlock coordinate using springs with K_1 and K_2 respectively. It can be inferred from (9) and (10) that using a spring with higher stiffness, the work can be reduced. As long as:

$$\frac{q_3^1}{\alpha} \leq q_3^2 \leq \frac{q_3^1}{\sqrt{\alpha}} \quad (11)$$

for a stiffer spring, the mechanical work done by the actuators driving the hip and knee joints will be lesser or equal than the work done using a less stiff spring. Nevertheless, in our work since the proposed control law cannot fix the coordinates q_1 and q_2, in order to reduce the total work one has to rely on (6) and (7). Inspecting these equations it is clear that since $\tau_3 \leq 0$, the larger τ_3, the faster the desired value of F_y is reached and the same could be said about F_x. However, if F_y is set to zero the contribution of τ_3 needs to be counteracted which means increasing the work done by the other joints.

2.2. *Discussion*

The above condition to reduce the mechanical work with this control law seem to be restrictive. Nonetheless, this condition is naturally imposed to a leg in stance phase for any walking robot, i.e. to ensure the capability of the leg to propel the body a tangential force should be exerted. This implies that F_y should not be set to zero during stance phase. During the swing phase, this analysis also holds if there is no unexpected contact and if the spring is stiff enough to maintain a $q_3 = 0$ position without deviating too much from its equilibrium position. However, this does not represent any energy saving since the spring is not doing work. This analysis is valid only if the weights of the different springs are fairly similar, which imposes a practical limitation to the energy saving using this method. It is important to note that this control law is applicable only to directly driven legs of this type. Moreover, intuition tells us that the energy saving capabilities of this control law has certain limits. These limits however, are difficult to analyze because the equations are dependent on the three joint coordinates in a nonlinear way. However a practical corroboration is presented in the following section.

3. Experimental Setup and Results

In order to test our proposed control method and the conclusion that a stiffer spring will save power, the control law presented in this work (see Fig. 2) was implemented in the HADE Leg prototype (see Fig. 1). Each actuator of the leg includes encoders to measure its position and force, while the spring deflection is measured by another linear encoder. The control and data acquisition were implemented with an NI PXI-1042Q with an execution time of the control loop of 500 μ s per cycle. The experiments consist in placing the leg in an arbitrary initial position (with the leg in contact with the ground) and then commanding a desired force pattern in order to illustrate the considerations made in Section 2 using a spring with $K = 3113$ N/m. Then, replacing the spring for another with $K = 6670$ N/m leaving the control structure unaltered and placing the leg in the same initial position as before and again commanding a desired force pattern.

3.1. *Control law validation*

Figure 3 shows the hip, knee and fetlock tracking performance against time. In this figure it can be seen that even changing the spring the controller exhibits a good reference tracking performance. More importantly, the Fetlock tracking error is zero as the control law determines. Figure 4 shows

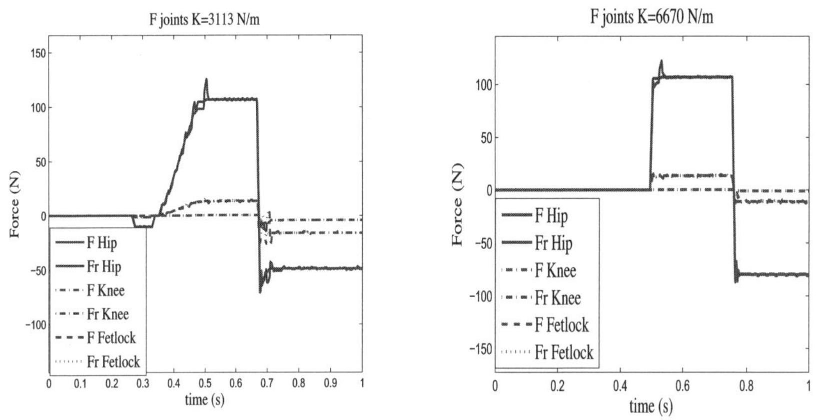

Fig. 3. Hip force (continuous red line), Hip reference force (continuous blue line), Knee force (dash dotted blue line), Knee reference force (dash dotted red line), Fetlock force (dotted blue line) and the Fetlock reference force (dotted red line) against time for K=3113 (left) and K=6670 (right)

the leg's force components F_x and F_y and its reference values F_{xr} and F_{yr} against time. In the figure, it can be observed that the tracking is also good. However, for the stiffer spring the task space force references exhibit less chattering.

3.2. *Mechanical power reduction*

Figure 5 shows the total work done by the leg with $K = 6670N/m$ (dotted red line) and with $K = 3113N/m$ (dashed blue line). As it can be seen in Fig. 5 for $F_x = -10$ N and $F_y = 0$ N the work done by the leg with the stiffer

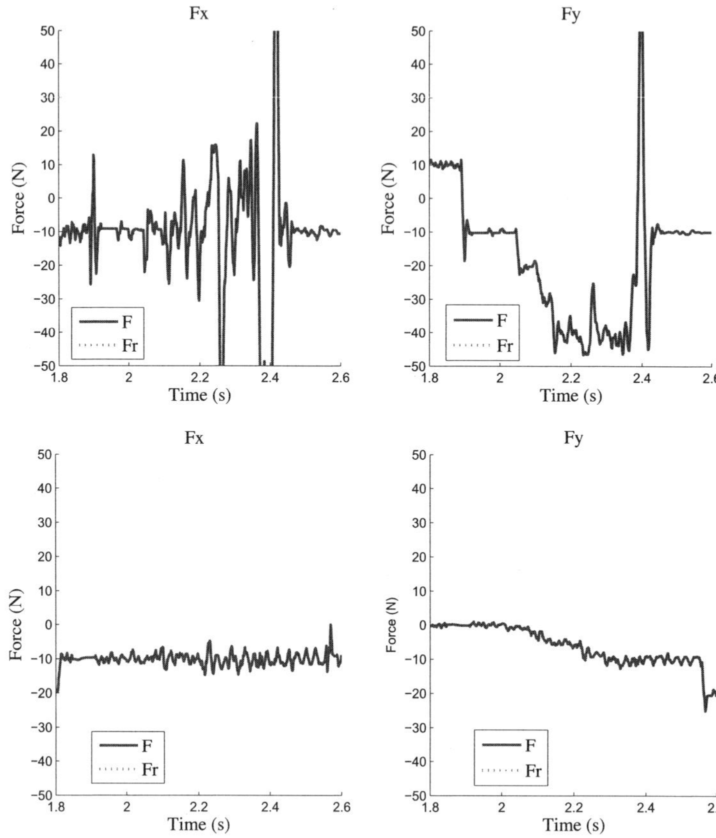

Fig. 4. Vertical and horizontal forces. Vertical (left) force (continuous red line) and vertical force reference (dotted blue line); Horizontal (right) force (continuous red line) and horizontal force reference (dotted blue line) vs. time for a spring of $K = 3113$ N/m (up) and for a spring of $K = 6670$ N/m (down)

spring is higher as was predicted in the previous section. When $F_y \neq 0$ N and $F_y = -10$ N ($t = 1.9$ s), the work of the leg with the stiffer spring is lower than the work done by the leg with the softer spring. Afterwards, in order to allow variations in q_1 and q_2 coordinates, the commanded force for the horizontal component with the softer spring was changed and then returned to $F_y = -10$ N, the work done by the leg with a softer spring become less than the work done by the leg with the stiffer spring as it showed in Fig. 5 at $t \approx 2.35s$. This behavior illustrates that the energy saving has a limit as it was intuitively predicted.

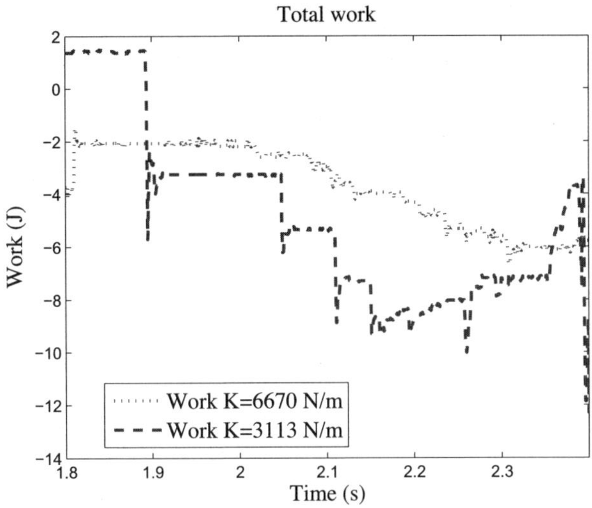

Fig. 5. Total Work performed by the leg. Net work done by the leg with a spring of $K = 6670N/m$ (dotted red line) and net work done by the leg with a spring of $K = 3113N/m$ (dashed blue line) vs. time

4. Conclusions

A force based impedance control scheme implemented on a bioinspired underactuated leg is presented and experimentally validated. This control law was designed to apply the knowledge on the control of underactuated manipulators. Moreover, analyzing the contribution of the fetlock torque in the resultant equations for the active joints' torque, a way of reducing electrical power consumption has been proposed. The aforementioned way of reducing the electrical power by reducing mechanical work has been theoretically analyzed with a concrete range of variation for the fetlock coordinate given, in the case that the articular positions were explicitly controlled. Moreover, a qualitative analysis was performed in order to provide conditions for also saving energy using a control law that does not explicitly control the joint coordinates. Experimental results corroborating the qualitative analysis are also presented.

An analysis on the optimal spring constant for power reduction with some optimization algorithm, as well as the extension of this analysis for a full gait are left for future work.

Acknowledgements

This work has been partially funded by the Spanish National Plan for Research, Development and Innovation through grant DPI2010-18702 which has funded personnel costs and by AECID through grant PCI-Iberoamerica A1/039883/11 which has funded equipment and consumable costs.

Mr. Juan Carlos Arevalo would like to thank the Spanish National Reserach Council, and Mr. Manuel Cestari would like to thank the Spanish Ministry of Economy and Competitiveness for funding their PhD research.

References

1. E. Garcia, J. C. Arevalo, G. Muoz and P. Gonzalez-de Santos, On the biomimetic design of agile-robot legs, *Sensors* **11** (2011).
2. J. Hurst and A. A. Rizzi, Series compliance for an efficient running gait, *IEEE Robotics & Automation Magazine* (2008).
3. R. Altendorfer, Rhex: A biologically inspired hexapod runner, *Autonomous Robots* **11** (2001).
4. J. E. Pratt and G. A. Pratt, Exploting natural dynamics in the control of a planar bipedal walking robot *Procedings of the Thirty-Sixth Annual Allerton Conference on Comunication, Control and Computing* (Monticello, Illinois, 1998).
5. J. C. Arevalo and E. Garcia, Impedance control for legged robots: An insight into the concepts involved, *IEEE Transactions on Systems, Man and Cybernetics: Part C* (to appear).
6. T. A. McMachon, The role of compliance in mammalian running gates, *Journal of Experimental Biology* **115** (1985).
7. P. G. Weyand, D. B. Sternlight, M. J. Bellizzi and S. Wright, Faster top running speeds are achieved with greater ground forces not more rapid leg movements, *Journal of Applied Physiology* **89** (2000).
8. H. Arai and T. Susumu, Dynamic control of a manipulator with passive joints in an operational coordinate space, in *Proceedings of the 1991 IEEE International Conference on Robotics and Automation*, (Sacramento, California, 1991).
9. M. Bergerman, M. Lee and Y. Xu, *Dynamic Coupling Measure of Underactauted Manipulators*, Tech. Rep. CMU-TR-RI-94-25, Carnegie Mellon University (1994).
10. H. Arai and T. Susumu, Position control of a manipulator with passive joints using dynamic coupling, *IEEE Transactions on Robotics and Automation* **7** (1991).
11. M. Bergerman and Y. Xu, Robust joint and cartesian control of underactuated manipulators, *Transactions of the ASME, Journal of Dynamic Systems, Measurements and Control* **118** (1996).

ON PASSIVE MOTION OF THE ARMS FOR A WALKING PLANAR BIPED

BASSEL KADDAR , YANNICK AOUSTIN and CHRISTINE CHEVALLEREAU

LUNAM IRCCyN, UMR 6597 C.N.R.S

Ecole Centrale de Nantes, Université de Nantes, BP 92101

1 rue de la Noë, 44321 Nantes cedex 03, France

E-mail: Bassel.Kaddar@irccyn.ec-nantes.fr, Yannick.Aoustin@irccyn.ec-nantes.fr

Christine.Chevallereau@irccyn.ec-nantes.fr

The energetic effects of passive motion of the arms and addition of torsion springs to the arms joints of a nine-link planar bipedal robot are studied for walking gaits. Only the locomotor system of the biped is actuated. Starting from a cyclic optimal motion of the biped with the arms stuck to the trunk, we explored the existence of a passive motion of the arms produced by the dynamics of the locomotor system of the biped. Multiples solutions including large amplitude of arms motion exist if the duration of walking step corresponds to the natural period of arms oscillation. The effect of torsion springs on the natural period and the cost functional is explored. Finally, costs functional of reference trajectories and trajectories with passive motions of the arms are compared.

1.1. *Introduction*

Several studies are done on the definition of walking gaits [1, 2]. For many researchers, trajectories of walking are achieved by using optimization [3]. Bipedal walking gaits with arms are also studied. Many of these studies interest with the generation of upper body motion aiming at improving the motion stability of the robot biped [4]; likely by the compensation of yaw moment [5]. Few studies and results are available to describe the effects of arms on bipedal walking gaits especially on the energy consumption. Aoustin and Formal'skii studied the optimal arm swinging of a planar biped [6]. Walking gaits are ballistic and torques are applied only during the impact. They showed that for a given period of the walking gait step and a length of the step, there is optimal amplitude of arm swing for which a cost functional is minimum. The effect of arms motions on the energy consumption, where the robot biped is fully actuated or only joints of the locomotor system are actuated, has not yet been shown.

The scope of this work is to investigate the dynamical effects of a passive motion of the arms on the energy consumption of a planar bipedal robot during its walking gait.

We first look for if a passive motion of the arms, due to the dynamic of actuated locomotor system, may exist for the walking gait. It is shown that large amplitude passive motion of the arms can be obtained for only an appropriate duration of the walking step. The introduction of torsion springs into shoulder joints is also studied to enable large passive motion of the arms with various walking periods. Finally, we show the dynamical effects of this passive motion of the arms on the energy consumption during walking of a planar bipedal robot.

1.2. *Presentation of the biped and walking gait*

The biped under study, presented in the figure 1, is a nine-link planar biped composed of two identical legs, two identical one-link arms and a torso. Each leg consists of a femur, a tibia and a rigid foot. All joints are revolute and have only one degree of freedom such that they can only move in the sagittal plane. Our biped is 2D equivalent of humanoid robot HYDROiD [7]. Desired gaits consist of single support phases separated by instantaneous and perfectly inelastic contact on the ground. This contact is supposed to be without impact.

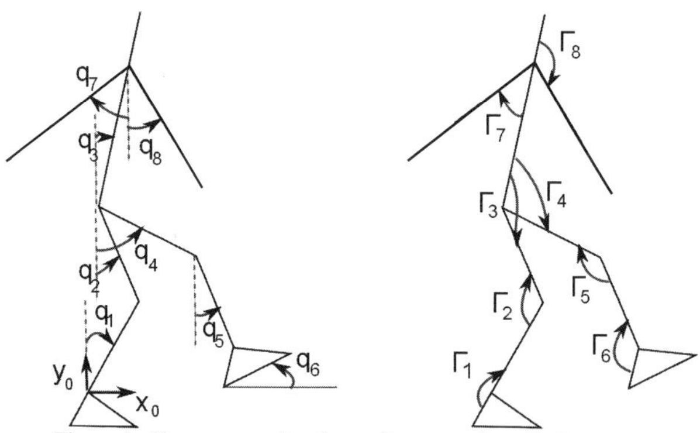

Figure 1. Planar generalized coordinates and applied torques.

The walking step starts with a single support phase and ends with the contact of the swing foot on the ground where the feet change their role, i.e. the stance foot becomes the swing foot and vice versa. The instantaneous contact on the ground is supposed to be with a flat foot and there is no rotation on the heel or toe of stance foot during swinging phase. The swinging foot also touches the ground with a flat contact. We consider that the stance leg lifts off the ground after the contact of the swing foot with the ground [6].

1.3. *Modeling of the robot biped*

The dynamic modeling is based on the assumption that the stance foot remains fix on the ground, i.e. there is no take off, no sliding and no rotation during the single support phase. As a result, stance foot is considered as base link of the robot biped.

Two models of our biped are used: the first takes into account that the arms are stuck to the trunk such that we consider a seven-link biped; the second considers revolute joints in the shoulders. The generalized coordinates of the biped are described by the vector of absolute angles $\mathbf{q}= [\mathbf{q}_1\ \mathbf{q}_2 \ldots \mathbf{q}_n]^t$. $\mathbf{\Gamma}= [\mathbf{\Gamma}_1\ \mathbf{\Gamma}_2 \ldots \mathbf{\Gamma}_6]^t$ is the joint torque vector of the locomotor system of the biped. Where $n=6$ or 8, depending on the biped model.

1.4. *Methodology*

The study is done by the following steps:

1.4.1. *Finding the natural period of arms oscillation:*

We consider the biped arm as a simple pendulum suspended from a frictionless pivot. For small oscillations, the pendulum swings with a natural period about $T_o \approx 1.19\ sec$. The natural period depends on the mass, length, and inertia and slightly on initial values of angular variable and angular velocity. The natural period of upper parts oscillation of the biped is chosen in order to coordinate arms motion with the motion of the locomotor system.

1.4.2. *Generation of optimal trajectories for the biped with the arms stuck to the trunk:*

The desired gait is cyclic and consists of a single support phase and an instantaneous impactless flat foot contact with the ground. We use a polynomial of fourth order for the evolution of the actuated joints' motions as a function of time to generate a reference trajectory which minimizes the following criterion:

$$C_\Gamma = \frac{1}{d}\int_0^T \mathbf{\Gamma}^t\,\mathbf{\Gamma}\,dt \tag{1}$$

where d is the step length and T is the time period of one step.
We used the *SQP* method (Sequential Quadratic Programming) [8] with the *fmincon* function of Matlab to solve the parametric optimization problem. Inequality constraints on the ZMP position, the torques and the contact between

the stance foot and the ground are considered to find the optimal reference trajectory.

1.4.3. *Generation of optimal trajectories with passive movement of the arms:*

Optimal reference trajectories, for the biped with the arms stuck to the trunk, are used for the biped with passive motions of the arms. The upper parts trajectories $\mathbf{q}_u(t)$ are due to the dynamic of the locomotor system and are calculated as following:

A system of equations (2) is obtained from the partitioning the matrices of the dynamic model:

$$\begin{bmatrix} \mathbf{A}_{ll} & \mathbf{A}_{lu} \\ \mathbf{A}_{ul} & \mathbf{A}_{uu} \end{bmatrix} \begin{bmatrix} \ddot{\mathbf{q}}_l \\ \ddot{\mathbf{q}}_u \end{bmatrix} + \begin{bmatrix} \mathbf{H}_l \\ \mathbf{H}_u \end{bmatrix} = \begin{bmatrix} \mathbf{B}_l \mathbf{\Gamma} \\ 0_{2\times 1} \end{bmatrix} \quad (2)$$

where the index u refers to the upper parts and the index l refers to the lower parts. $\ddot{\mathbf{q}}_u \in R^{2\times 1}$ is the accelerations vector of the upper parts of the biped, $\ddot{\mathbf{q}}_l \in R^{6\times 1}$ is the accelerations vector of the locomotor system of the biped, $\mathbf{A}_{ll} \in R^{6\times 6}, \mathbf{A}_{lu} \in R^{6\times 2}, \mathbf{A}_{ul} \in R^{2\times 6}$ and $\mathbf{A}_{uu} \in R^{2\times 2}$are submatrices of inertia matrix $\mathbf{A} \in R^{8\times 8}$, $\mathbf{H}_u \in R^{2\times 1}$and $\mathbf{H}_l \in R^{6\times 1}$ are submatrices of $\mathbf{H}$ which is a vector of Coriolis, centrifugal and gravity forces. $\mathbf{B}_l \in R^{6\times 6}$ is the actuation matrix.

Then, the upper parts trajectories can be deduced through the numerical integration of the equation (3) with the initial conditions of the angular configurations and velocities of the arms $\mathbf{q}_u(0), \dot{\mathbf{q}}_u(0)$:

$$\ddot{\mathbf{q}}_u = {\mathbf{A}_{uu}}^{-1}(-\mathbf{A}_{ul}\ddot{\mathbf{q}}_l \quad -\mathbf{H}_u) \quad (3)$$

where $\mathbf{q}_u = [\mathbf{q}_7 \, \mathbf{q}_8]^t$ (See figure 1).

By taking into account the exchange of the role of the arms between the final time *(t= T)* of a step and the initial time *(t= 0)* of the next step, the following equalities must be respected to satisfy the cyclicity conditions:

$$\mathbf{q}_u(0) = T_E \, \mathbf{q}_u(T), \;\; \dot{\mathbf{q}}_u(0) = T_E \, \dot{\mathbf{q}}_u(T) \quad (4)$$

where $\mathbf{q}_u(0)$, $\dot{\mathbf{q}}_u(0)$ vectors of angular and velocities variables of the arms at the initial time t=0 respectively. $\mathbf{q}_u(T)$, $\dot{\mathbf{q}}_u(T)$ vectors of angular and velocities variables of the arms at the final time t=T respectively. T_E is the permutation matrix which does the exchange of the role of arms.

To design the passive motion of the arms, it is necessary to find the solution $\mathbf{q}_u(t)$ of system of equation (3) with the boundary values satisfying the equalities (4). A Newton-Raphson method is used to solve the boundary problem. Then, trajectories with the passive motion of the arms can be optimized.

1.5. *Results*

1.5.1. *Time period is constant:*

The time period required such that the arms swing and then return to their initial positions is two times that of the period T of walking step. Therefore, optimal reference trajectories are obtained with a walking period T which is equal to half of the natural period T_o of arms oscillation which means $T = T_o/2 = 0.595s$. By using these reference trajectories for the biped with passive motion of the arms, two possibilities appear depending on initial conditions $\dot{q}_u(0)$, $q_u(0)$: a passive motion of the arms with large amplitude and a passive one with very small amplitude which is almost null. A passive motion with large amplitude is possible if and only if the period of walking stride (two steps) is close enough to the natural period of the arms oscillation.

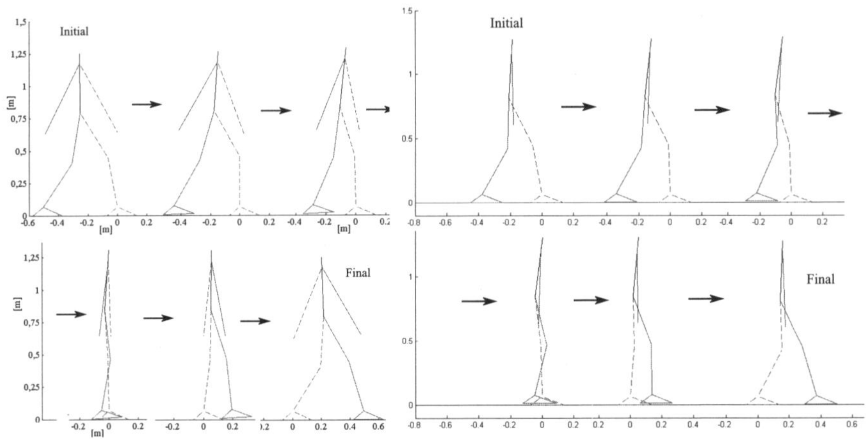

Large amplitude motion | Small amplitude motion

Figure 2. Walking gait as a sequence of snapshot figures is shown at walking speed V=0.8 m/s, T=0.57 s

For walking speed $V=0.8\ m/s$, the two types of passive motions are obtained only if the period T is between $0.56s$ and $0.61s$ which is close to $T_o/2$, this interval of T is related to walking speed. If the period T does not belong to the previous interval, only arms motion with small amplitude can be found. Figure 2 shows the two types of passive motion at walking speed of $V=0.8$ *m/s* and period $T=0.57$ *sec*. In the following, time period is considered as a variable of optimization.

1.5.2. *Time period is a parameter of optimization:*

When reference trajectories are optimized without fixing the time period, optimal time period will be significantly shorter, between *0.35s* and *0.4s*. Then, passive motions with large amplitude of the arms are not possible. The use of torsion spring in arms joints decreases the natural period and enables to obtain large amplitude of the arms. The time period is used as a parameter of optimization. Then, we found optimum reference trajectories for several walking speeds. Springs' coefficient k is used as a variable of optimization while finding optimal trajectories of the biped with passive motions of the arms. Criterion is calculated for optimal trajectories. Arms passive trajectories with large amplitudes have costs functional lower than these of reference trajectories for all walking speeds, figure 3.

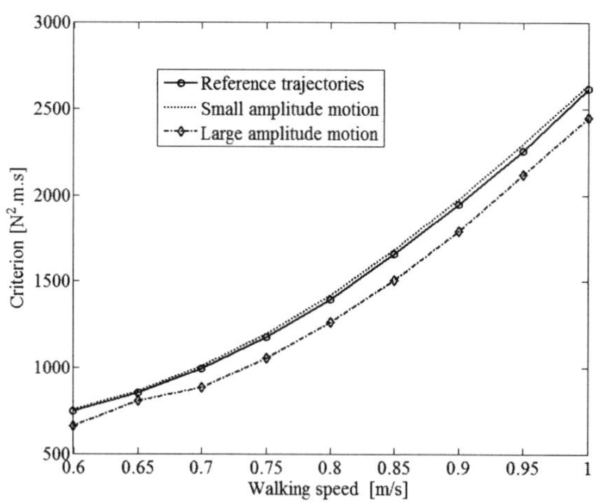

Figure 3. Evolution of energy criterion versus walking speed.

The optimal k is about [*8-12*] *N/rad* and differs slightly from one walking speed to another. To understand the effect of the springs, we take an optimal trajectory obtained at a walking speed *0.75 m/s* and we changed only the value of the springs' coefficient. The maximum amplitude of arms motion and the cost functional are obtained as functions of *k*. Figure 4, shows that large amplitude of the arms can be obtained when *k*=[*9.6-13.5*] N/rad. Cost functional is minimal when *k* is equal to *11.3 N/rad* for a walking speed of V=*0.75 m/s*. The desired amplitude of arms passive motion can be obtained by changing the springs' coefficient.

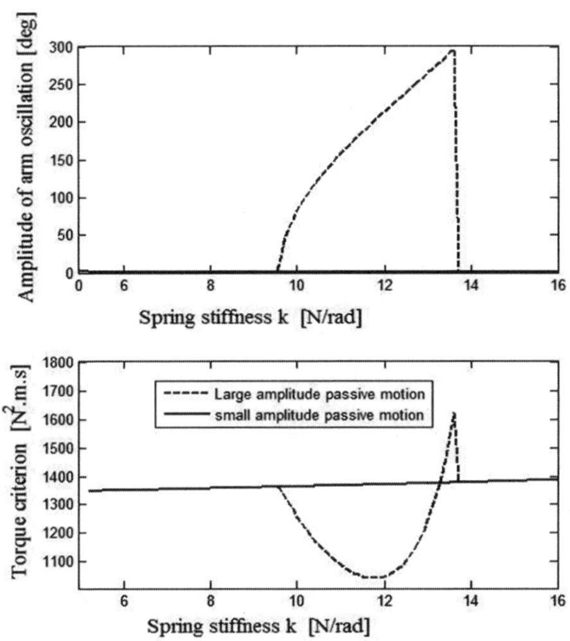

Figure 4. Evolution of energy criterion and maximum amplitude of arms motion versus *k* at V=0.75 m/s.

1.6. *Conclusion and Perspectives*

Two types of optimum passive motions of the arms were found during walking of the biped: one with very small amplitude (about one or two degrees) and one with large amplitude. Large amplitude motion can only be obtained if the time period of a step is close enough to half of the natural period of arms oscillation which can be changed by means a torsion spring applied in the arms joints. Cost functional is lower when arms swing with large amplitude passive motion in comparison with both optimal reference trajectories and small amplitude arms motion. It is worth to compare this passive motion with a case where the arms are actuated. In this work, we assumed that the swinging leg hits the ground with a flat foot and the support leg immediately leaves the ground. In future works, we study other types of walking gaits having double support phases and impacts besides single support phases such that the swing foot arrives the ground with non flat foot and do a rotation motion until the complete contact with the ground.

References

1. X. Mu, and Q. Wu, "A complete dynamic model of five-link bipedal walking". *Proceedings of the American Control Conference, 2003, 6,* pp. 4926–4931, 2006.
2. D. Tlalolini, Y. Aoustin, and C. Chevallereau, "Design of a walking cyclic gait with single support phases and impacts for the locomotor system of a thirteen-link 3d biped using the parametric optimization". *Multibody System Dynamics,23*, pp. 33–56, 2010.
3. C. Chevallereau, and Y. Aoustin, "Optimal reference trajectories for walking and running of a biped robot". *Robotica, 19*, pp. 557–569, 2001.
4. S. Kajita, F. Kanehiro, K. Kaneko, K. Fujiwara, K. Harada, K. Yokoi, and H. Hirukawa, "Resolved momentum control: humanoid motion planning based on the linear and angular momentum," *International Conference on Intelligent Robots and Systems*, vol. 2, pp. 1644 – 1650, Oct. 2003.
5. D. Xing and J. Su, "Arm/trunk motion generation for humanoid robot", *SCIENCE CHINA Information Sciences*, vol. 53, pp. 1603 – 1612, 2010.
6. Y. Aoustin and A. Formal'skii, "On optimal swinging of the biped arms", *International Conference on Intelligent Robots, IROS 2008.* pp. 2922– 2927, 2008.
7. S. Alfayad, "Robot humanoïde HYDROïD : Actionnement, Structure Cinématique et Stratégie de contrôle," *Thèse de Doctorat de l'Université Versailles Saint Quentin,* Paris, Novembre 2009.
8. M. Powell, "Variable metric methods for constrained optimization," *Lecture Notes in Mathematics.* Springer Berlin / Heidelberg, pp.62–72, 1977.

SOME ASPECTS OF THE MOVEMENT *MERO* MODULAR WALKING ROBOT ON UNARRANGED TERRAIN

ION ION[1], ADRIAN CURAJ [2], THEODOR BORANGIU[2], AURELIAN VASILE[3]
GRIGORE STAMATESCU[2]

[1]Technology of Manufacturing Department, [2]Automatic Control and System Engineering Department, [3]Mechanism and Robot Theory Department, POLITEHNICA University of Bucharest

The modular walking robots are used to unconventional displacement of the technological loads over the unarranged terrains. The walking robots moving on a terrain, strewn with many obstacles, which can be convex or concave, there is a danger that the position of these robots are not stable The modular constructions of the walking robots led to a more suppleness and very good adaptation to any terrain surface. These support surface obstacles during robot movement can lead to change direction, height of body movement and orientation in space and the corresponding reorganization and order tracking system of moving the legs. The performance of walking robots is closely related to the adopted gait Crab walking is very important to an omni directional walking robot In this article, the study will focus on some aspects of the gaits of a modular walking robot MERO 2.on the rough terrain

1. Introduction

In the area of level-walking gaits, gaits can be categorized into two types according to the terrain condition: gaits on perfectly smooth terrain and gaits on terrain which contains forbidden areas (areas which are not suitable for foot placement).

An important feature of the walking robots is that the ground's configuration never influences or affects their movement too much. Such a feature makes this locomotion type turn into an attractive solution for many applications that require movement on an unplanned land, namely having an uneven configuration. As uneven land covers an endless variety it is difficult to match all the types and cases of walking on such grounds.

To achieve and manage a walking robot it is necessary to know all the walking possibilities, because the selection of the legs number and its structure depends on the selected type of the gait. The selection of the type of gait is a very complicated matter, especially in the real conditions of walking on the rough terrain. Therefore, it is necessary that the terrain surface to be selected before the type of gait is chose.

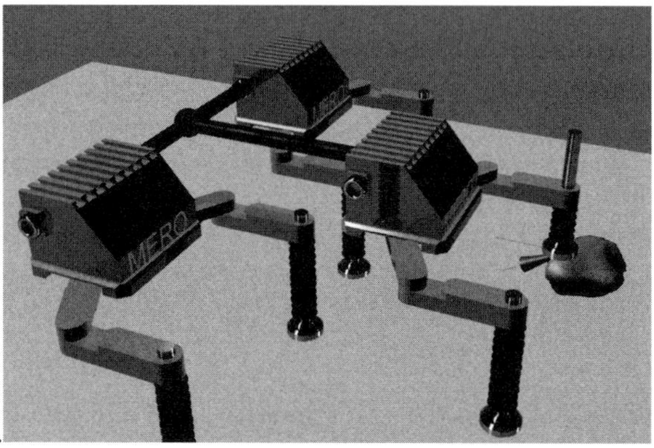

Figure 1. The modular walking robot MERO2-computer graphics model (in construction)

For example, some obstacle-crossing gaits of a hexapod and quadruped were studied in [1], [6], [3]..A computer-generated free gait was developed in [4],[1].A "discontinuous follow-the-leader gait" was developed and successfully implemented to control a hexapod in walking over uneven terrain [14]. A "continuous follow-the-leader gait" was formulated and studied in [2],[3] Two terrain-adaptable free gaits were developed to enable a quadruped to enable a quadruped to walk on a rough planar terrain [3]. [5].

The walking robots are able to move on a terrain with irregular surface. As a result, the leg mechanisms must have at least three degrees of freedom (Fig.1)

Among the above-mentioned gaits, the gaits which have the "follow-the-leader" feature seem to be most suitable for rough terrain in walking.

The body movements were inserted into two designed leg movements and the overall body motion was discontinuous [4]. [6].This type of walking mode provides good mobility on rough terrain. [1].

However, it suffers a slow speed due to the discontinuous body motion.

This may not be a problem for walking in severely rough terrain, but it certainly limits the performance of the walking machine in mildly rough terrain.

Compared to the "discontinuous follow-the-leader gait", a "continuous follow-the-leader gait" can reach a higher speed and maintain smoother body motion on mildly rough terrain due to its periodic nature [6], [2].

The MERO modular robot movement will be gone after a crab type that will materialize if the configuration of the gait hexapod 3 * 3 in the support phase, immediately after the transfer phase he moves the robot body.[6], [2], [1]. Fig.2

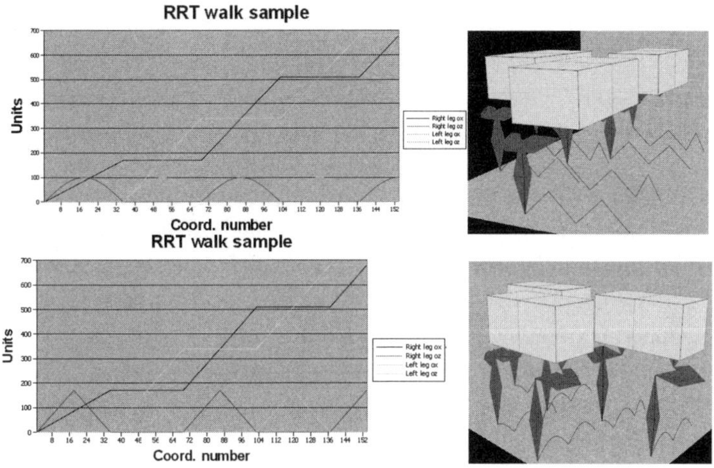

Figure 2. Gait type 3*3 for a walking robot MERO 2

2. Moving the walking robot over an obstacle isolated

To study the issue we need a simplification of the geometrical features of the real terrain. For this, we introduce the notion obstacle.[4], [6].

The obstacle's geometry is defined by one or two parameters and the main types of obstacles such as a slope, a ditch and an insulated wall.

The presence of the obstacle on the support area can lead to the change in the movement direction, in the height of the movement of the body of the walking robot or its orientation in space as well as to the appropriate redefining of the steps' order and sequence and in the regime of the legs' movement.

When the obstacle's dimensions allow the robot's movement, to simplify the walk's study it is required to swap it for an obstacle having a simpler geometry, to be defined by one parameter or two only. As the walking robot (Fig.1) has three or more modules each module has two legs, these moves have to be coordinated so that the move is ensured in conditions of the stability of the system.

In order to allow a theoretical approach of the gait of the walking robot it is necessary to define a lot of terms. Given that this work is studied a hexapod robot, will be presented only general concepts on static stability.

Static equilibrium of a robot walked under their own weight effect can be verified with the support polygon (Fig. 3). Which is the polygon formed by the horizontal projections of the points of support leg in support phase. [6],[4]

Walking is statically stable if, at any time, the vertical projection of center of mass G is inside the support polygon, being able to define the limits of: Stability: S stability limit is the minimum distance from the center of mass projection G'a the support polygon $S \geq 0$: Longitudinal stability limit is Sl where Sl = min (Sf, Ss). [2],[4] Sf -front stability limit is the distance from the center of mass projection G'a the front side of the polygon of support, measured in the direction of movement. Ss back stability limit is the distance from the center of mass projection G'a the back side of the support polygon, measured in the direction of movement. Let us consider a scheme of a walking hexapod robot (Fig 4) numbering each leg. Stability of walking robot's movement is influenced by some of the characteristics parameters of the robot walking can change a lot when used as a means of transport

For instance, an additional load on board, changes the barycenter's position and the inertia moments of the module's platform. Environmental factors such as the wind or other elements may bias the robot, and their influence is barely predictable. Such disturbances may cause considerable deviations in the robot's real movement, than expected.

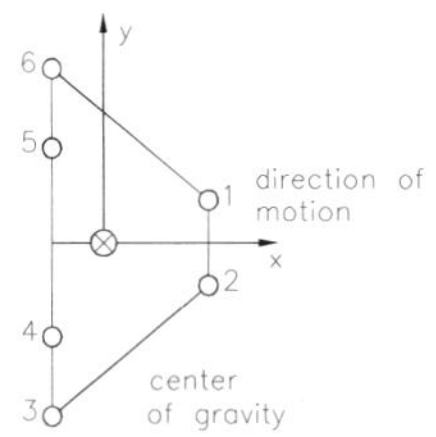

Figure 3. The longitudinal stability margin

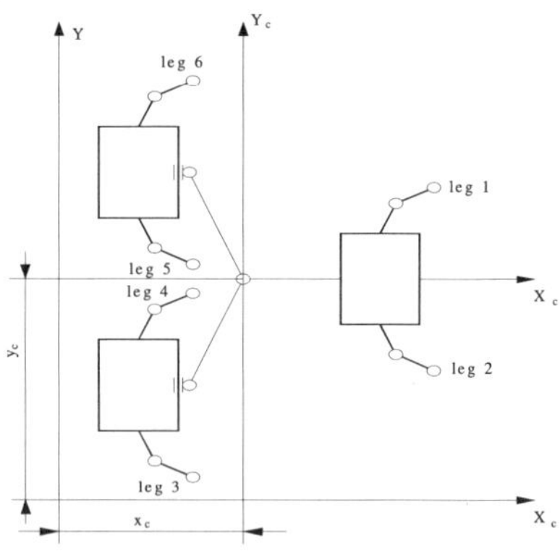

Figure 4. The model a modular robot hexapod

Working out and complete enough mathematical pattern for studying the movement of a walking robot, is interesting, both as regards the structure of its control system and verifying the simplifying principles and hypotheses, that the control program's algorithms rely on.

The upper control level analyzes the direction cosines' values and the speed's, for each leg. It defines each movement's initial and end moments.

While feet are lifting, it is checked the height of the feet's positions as against the support surface. Provided one of the heights is smaller than accepted, the foot is ordered to lift till the required height is reached, afterwards, the foot pursuits its movement forward. The movement stops with one of the following reasons:

1) the stride attains its utmost length limit
2) one of the driving pairs' variables has reached the limit of working area;
3) a "contact" signal was received from one of the feet's touch sensors.

In order to keep the platform in a horizontal position, we use the piece of information supplied by the horizontal and vertical stance transducers (fig.4) registering the deviation by the walking robot's platform from its horizontal position. Restoring the platform's horizontal line is accomplished by moving vertically different legs in the support phase, accordingly to the decision by the balance monitoring block.

The safety line is established in accordance with the prescribed value of the stability limit for a certain walking regime, or contingent of the reaction forces values, where the feet touch the support ground; these forces are not allowed to drop beneath the accepted values.

2.1 *The modular walking robot moves on a slope.*

The main difference between walking on a slope (Fig.5) and walking on a horizontal plane surface consists in the fact that the projection of the robot's

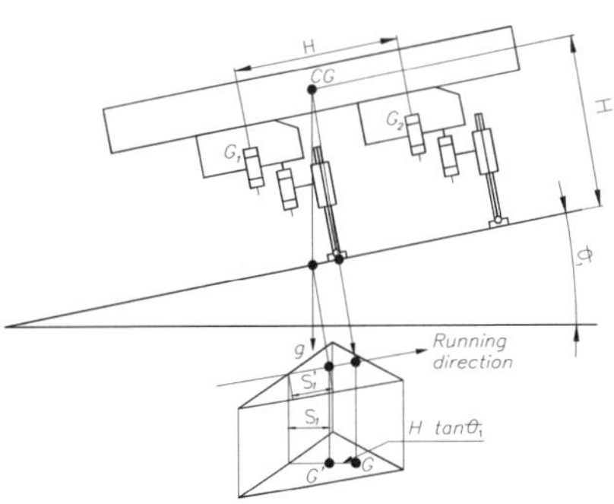

Figure 5. The slope's inclination angle

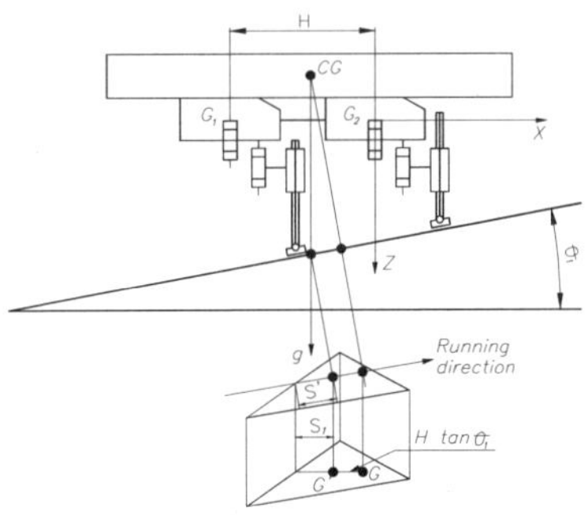

Figure 6. The modular robot walks on the slope and the body's inclination is diminished

center of gravity changes position against the sides of the support polygon, if the slope exceed a certain limit, the projection no longer lies inside the support polygon. When the robot walks on a slope when it goes both up and down, the limits of the longitudinal fore and back stability diminish accordingly, due to the change in the position of the projection of the center of gravity. A sensor is used to emphasize the walking robot's stance a sensor that measures the platform's inclination in two planes, a sagital and a frontal one.

To improve the robot's stability when it moves along a slope (Fig.6), there are two strategies such as:

-its body's height is diminished and its position adjusted;

-the step's length is also curbed.

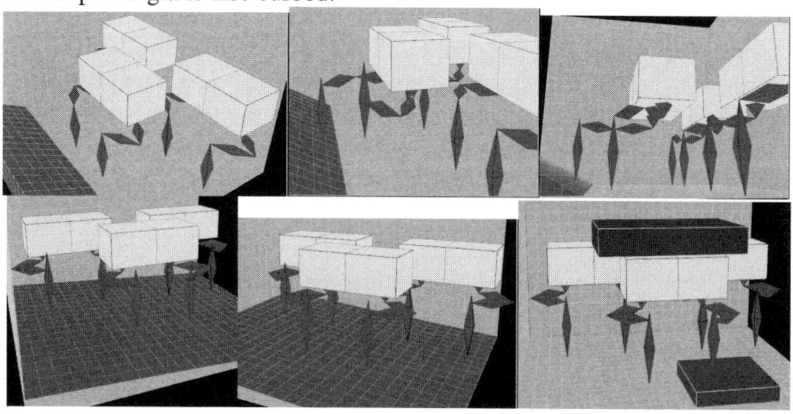

Figure 7. Stepping over isolated obstacles for a walking robot MERO 2
(Computer graphics simulation)

3. Results and conclusion

The application described here was developed with the purpose of easing the simulation and providing a better understanding of different walk strategies of two types of MERO modular robots (humanoid leg - RRR and crab-like leg RRP) in different configurations. Two main configurations were taken into consideration: RRP - 3 module assembly and RRR - 3 module assembly, one module in the front and two modules in the back, in a triangle shape.

There are two distinct windows that start when the application is launched: The control window (Fig.8) and the visual window (Fig. 9).

The control window supplies the user with every tool needed to simulate different aspects of the robot, visible in the visual window. The application allows zooming, axes rotation and free rotation inside de visual window in order to highlight different. There are three modes of working with this application.

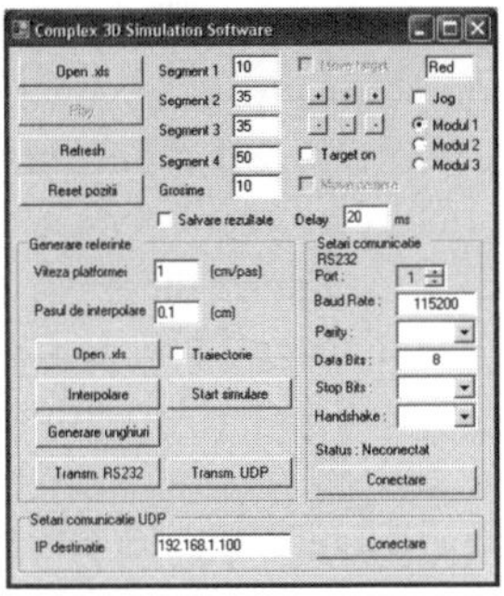

Figure 8. Control Window

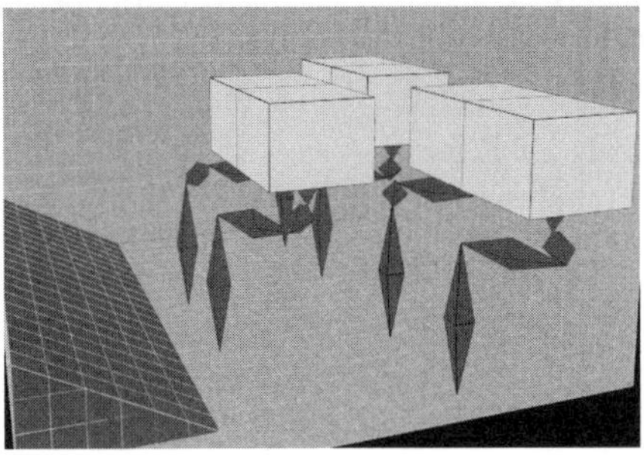

Figure 9. Visual Window

The first one and the simplest is the “Jog” mode. By enabling this mode the user can then select witch module he wants to jog and can do so by moving sliders that correspond to each degree of freedom of every joint of the selected robot.

The second mode allows loading an .xls (Microsoft Excel) file containing on each row a set of joint angles or distances in case of a translational joint. The file is analyzed by the application and simulated in the visual window line by line, the robot moving in the way described by the file. There are also functions of interpolating the points given in the file and saving the resulted trajectories of the robot’s leg in a different file.

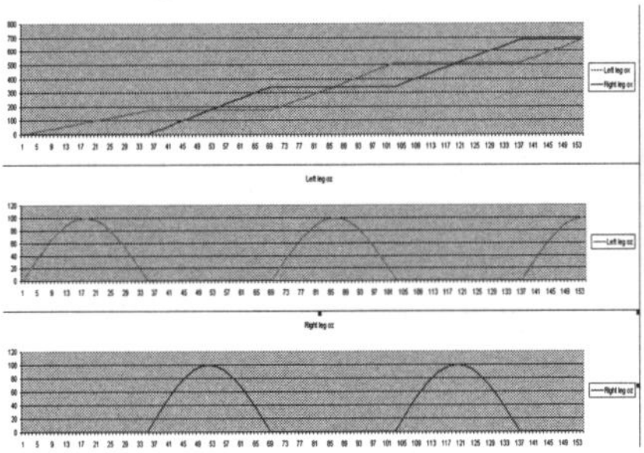

Figure 10. Trajectories for each leg projected on ox and oz axes

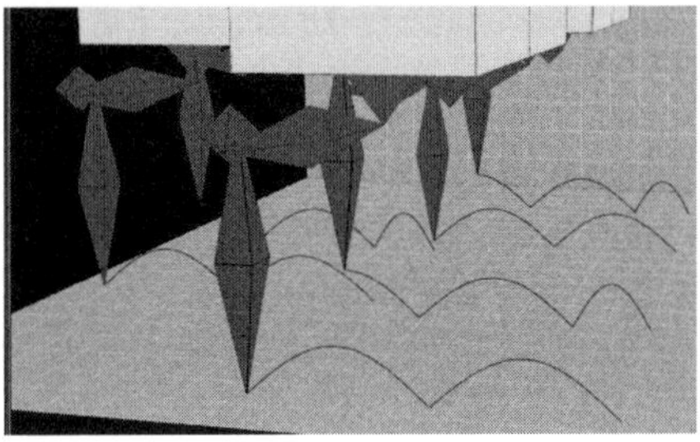

Figure 11. The actual trajectory simulated on the virtual robot

This simulation is done by applying direct cinematic and drawing each of the robot’s segment according to the file input. The user can also modify the lengths of each of the robot’s segments by modifying the values in the text-boxes on the control window. The last work mode of the application is somewhat the reversed of the previous one. The loaded file contains a trajectory for each leg of the robot. Such a trajectory is pictured in Fig.10 and Fig.11. The application, using the inverse kinematics (Denavit-Hartenberg), computes each of the joint's angles or displacements for each line in the file, in the end outputting these in new file.

These calculated values can be then transmitted to the real robot through Ethernet UDP frames or by RS232/RS485 in a proprietary protocol designed for MERO robots. As future feature, the application will also capture the movement of the real robot and compare it with the simulated one in terms of response and trajectory accuracy. An other feature that is being considered is a forth work mode that implies the initial robot's configuration and an unarranged virtual terrain, the application being used to find the best way to overcome the obstacles by choosing one or a combination of two or more previously loaded and learned walk strategies.

References

1. G., Figliolini, P., Rea, (2006) "Gaits analysis of six-legged Walking Robot when a leg failure occurs ",*Proceedings of 9th International Conference on Climbing and Walking Robots and support Technologies for Mobile Machines,* September12-14, Brussels, Belgium pp.276-283,
2. I., Ion, I Simionescu,., A., Curaj, A. Vasile (2011), *MERO* Modular Walking Robots, Solution for Displacing Technological Equipments on Irregular Terrains *Proceedings of the 13th World Congress in Mechanism and Machine Science,* Guanajuato, México, 19-25 June, 2011
3. A., Roennau., G Heppner., T Kerscher., R Dillmann., (2010)," A behavior-based gait inspired by a walking stick insect" *Proceedings of the 13th International Conference on Climbing and Walking robots and the Support Technologies for Mobile Machines CLAWAR 2010,* 31 August – 3 September 2010, Nagoya Japan,
4. S.M Song. and J.K. Waldron, "(1987). An analytical approach for gait study and its application on wave gaits", the *International Journal of Robotics Research,* **6**, 60-71,
5. W., Lewinger, R., Quinn, "Neurobiologically-based control system for an adaptively walking hexapod" *Proceedings of the 13th International Conference on Climbing and Walking robots and the Support Technologies for Mobile Machines CLAWAR 2010, 31 August – 3 September 2010,* Nagoya Japan,
6. K.J., Waldron (1991) Force and Motion Management in Legged Locomotion, *Proceedings of the 24th IEEE Conference on Decision and Control,* Fort Lauderdale, pp.12-17

A CLOSED-LOOP OPTIMAL CONTROL APPROACH FOR ONLINE CONTROL OF A PLANAR MONOPED HOPPER

SOHA POUYA, MOSTAFA AJALLOOEIAN, AUKE JAN IJSPEERT

Biorobotics Laboratory, Institute of Bioengineering, Ecole Polytechnique Federale de Lausanne (EPFL), CH 1015, Lausanne, Switzerland.

In this paper we present a closed-loop optimal control approach for the online control of a legged robot locomotion, particularly the hopping of a simulated monoped robot. Modeling is done based on the spring loaded inverted pendulum (SLIP) model suggested as the animal and human running gait template. The key idea is to efficiently inject energy to the system so that the monoped can track the desired apex height and forward velocity. The state of the system is observed in the Poincaré section at the apex point and the corresponding discrete dynamics is formulated by using available analytical solutions. The goal is then to synthesize an optimal control law which can bring the apex state at any step to the desired state at the next step. We show the controller performance in providing fast and accurate response in the presence of noise and through different scenarios while minimizing the control effort.

1. Introduction

Modeling and control of legged locomotion has been the field of interest and research from both biology and robotics viewpoints to understand the underlying locomotor mechanisms, as well as take inspiration to build efficient and powerful robotics platforms. Studies have been done at different levels of abstraction from very simplified models to sophisticated neuro-mechanical models (Ref. 1). For animal and human running gaits, the Spring-Loaded Inverted Pendulum (SLIP) has been proposed as an archetypal model, predicting correctly ground reaction forces and center of mass trajectories (Ref. 2). The SLIP model is a two-state hybrid dynamical system consisting of a point mass attached to a massless springy leg. Control laws can be added to this model to set the leg orientation at touchdown and can extensively improve the running stability (Ref. 3). In comparison to static robotic locomotion, this model presents dynamic and more efficient legged locomotion by exploiting the leg natural dynamics. The role of exploiting compliance in the legged locomotion has long been pointed out (Ref. 4). Raibert and his colleagues have successfully used this idea by using SLIP-based simple control rules for the control of their hopping and running robots (Ref. 5). Saranli et. al. (Ref. 6) proposed a deadbeat control approach using the inverse return map of SLIP model (Ref. 7) and compared the controller performance with the modified Raibert decoupled control laws. Their coupled control approach shows better performance in tracking the desired jumping height and forward velocity while not needing tedious tuning of controller parameters. In this paper, we show how one can use closed-loop optimal control methods for this tracking problem while minimizing the control effort (energy consumption). Open-loop optimal control methods have been used to produce stable and efficient periodic gaits (Refs. 8 and 9). Tedrake (Ref. 10) has suggested an optimal control approach to optimize the region of initial conditions from which the robot can recover. Our focus however is to enable the system to track the input commands in an efficient way. We use the same analytical return map as (Ref. 7) for the later comparison. The controller tracking performance is investigated in the presence of noise and through different scenarios including step, sinusoidal, complex and random commands. We first briefly explain the methods used for return-map modeling and optimal control formulation and finally present the results obtained for the different scenarios.

2. Mathematical Modeling of System Dynamics

The modeling of the leg dynamics is performed in the discrete form by degrading the continuous dynamics to one less dimension of Poincare map. Control of the system state variables is also performed in the discrete fashion. The key idea is to exploit the natural dynamics of the system and only manipulate the behavior by injecting energy at specific moments of the natural motion. This approach stands between the passive dynamic walkers where no actuation is involved and the static walkers such as Asimo with continuous and high accuracy control. Accordingly we base our modeling tools on the Poincare map also called *return map* formulation. Mathematically speaking we collapse the dimensions of the system by transecting the system at one characteristic point resulting a discrete map called Poincare map. The general form of such a system can be stated as follows:

$$x[k+1] = f(x[k], u[k]), \quad y[k] = h(x[k], u[k]) \tag{1}$$

where $x[k] \in \mathbb{R}^n$ is the state of the system at time k (an integer), $u[k] \in \mathbb{R}^p$ is the control input and $y[k] \in \mathbb{R}^q$ is the system output. Functions f and h are the mapping functions which should be extracted for the system. For our system, the characteristic point selected is the *apex* point where the

Table 1: physical properties and the initial values.

Physics		Initial States	
Hip Mass (m)	50.48 kg	$\theta_{lo}[0]$	25°
Gravity (g)	9.81 m/s^2	$E_{lo}[0]$	500 J
Rest Leg Length (l_0)	1 m	$\psi_{lo}[0]$	1
Control Bounds ($[u_{min}, u_{max}]$)		Initial Controls Guess	
$\theta_{td}[i]$	$[\pi/4,\ \pi/2]$ rad	$\theta_{td}[0]$	20°
$k_c[i]$	$[100, 500]$ $N.m^3$	$k_c[0]$	250 $N.m^3$
$k_d[i]$	$[100, 500]$ $N.m^3$	$k_d[0]$	250 $N.m^3$

leg reaches to the highest point in its trajectory and its vertical velocity is zero. The state of the system consists of the position and the forward velocity at the apex point. Closed form solutions for this system is not trivial in contrast to its simple dynamical model. We use the analytical solutions presented in (Ref. 7) together with a definition of energy consumption measurement to perform the task. We present the overall formulation and more detailed equations can be found in Ref. 6.

The state of the SLIP model is selected as $\mathcal{X}_l = \{X_l | X_l = [\bar{x},\ \bar{y},\ \bar{v_x}]\}$ with $\bar{x}$ and $\bar{y}$ being the positions and $\bar{v_x}$ the forward velocity of the apex point. State is selected such that it can give an intuitive description of the system behavior including the apex *jumping height* and *forward velocity.* The control variables are selected similar to the Raibert controllers as $\bar{\mathcal{U}} = \{\bar{u} | \bar{u} = [\theta_{td},\ k_c,\ k_d]^T\}$ with θ_{td} being the angle of attack at touchdown and k_c, k_d respectively the compression and decompression spring constants in the stance phase. The spring used for this formulation is an *air spring* with the potential energy formulated as $U(x) = \frac{K}{2}(\frac{1}{x^2} - \frac{1}{x_0^2})$. Closed-form formulation can be extracted in the liftoff coordinate system by transferring the state to the new set of state variables $\mathcal{Z}_{lo} = \{Z_{lo} | Z = [\theta, E, \psi]_{lo}^T\}$ respectively, the angle of attack, energy and ratio of forward to vertical velocity at liftoff. The apex states can then be calculated from the liftoff state variables as:

$$v_{x_{ap}} = \text{sign}(\psi_{lo})\sqrt{\frac{2E_{lo}\psi_{lo}^2}{m(1+\psi_{lo}^2)}}; \quad y_{ap} = \frac{E_{lo}}{mg(1+\psi_{lo}^2)} + l_0\cos\theta_{lo}$$

We use these equations to describe the system dynamics together with the control laws detailed in the next section to perform the desired tasks.

3. Closed-Loop Optimal Control

The role of the control algorithm is to find the suitable control input $u^* = [\theta_{td}^*,\ k_c^*,\ k_d^*]^T$ which can take the current apex state X_{ap} to the desired state X_{ap}^* while satisfying the problem constraints. Previous works have suggested simple decoupled controllers (Ref. 5) or using the approximated inverse map as the deadbeat control law (Ref. 7). We propose an optimal control approach which provides a coupled control of the state and in addition minimizes the required control effort. Moreover this approach can replace the tedious procedure of computing the inverse return map and particularly makes it more suitable for higher dimensional systems. We show the effect of taking the coupled dynamics of the system as a whole throughout the experiments in the section 4.

The optimal control problem is formulated as presented in the table 2. The goal is to minimize the control effort while satisfying the state and control constraints. This introduces a constrained optimization problem which can be solved by using the nonlinear programming (NLP).We formulate the *control effort* as the quadratic cost of change of the spring stiffness. In other words, we assume there is an operating point for the spring stiffness, k_0, and any change form that would need consuming energy. The cost function therefore is formulated as $J = (k_c[i+1] - k_0)^2 + (k_d[i+1] - k_0)^2$. *Constraints* of the problem are formulated regarding the physical constraints on the state variables (figure 1) and the control constraints due to the actuators saturation. The boundary for the angle of attack is [45, 90] and for the range of the air spring stiffness values we roughly rescale the spring potential energy from Ref. 7 resulting in the boundary of $[100,\ 500] N.m^3$. We use the quadratic errors of tracking the desired state trajectories as *nonlinear equality constraints*. The sequential quadratic programming (SQP) solver provided with MATLAB *fmincon* function is used to solve the problem. To be able to embed this optimization problem in the control loop, we use only one step horizon for measuring the cost function. While it works for this simple problem, for more complex problems using a bigger horizon (such as model predictive control methods) is suggested to improve the stability of the solutions. Finally this proposed algorithm uses the measured values of the state variables at each step to find the required control variables for the next step. In other words the control is performed in closed-loop helping the system to overcome the deviations in the presence of noise. We show the performance of the controller with a 10% noise on the sensor data.

4. Results and Discussions

The performance of the proposed control approach is evaluated through different scenarios assuming 10% noise on liftoff angle measurement. Figure 1 shows the results for selected scenarios each row showing one scenario:

Table 2: Pseudo-code for the optimal control formulation.

do for each step i

$$u[i]^* = \underset{u}{\text{argmin}} \left((k_c[i+1] - k_O)^2 + (k_d[i+1] - k_O)^2 \right)$$

$$u_{min} < u < u_{max}$$

$$y_{ap}[i+1] - y_{ap}^{desired}[i+1] = 0$$

$$v_{x_{ap}}[i+1] - v_{x_{ap}}^{desired}[i+1] = 0$$

$$Z_l[i+1] = f(Z_l[i], \tilde{N}.u[i]^*)$$

set $u_O[i+1] = u[i]^* + \epsilon$

control the system with $u[i]^*$ and measure state $\mathcal{Z}_{lo}$

until failure or maximum successful steps

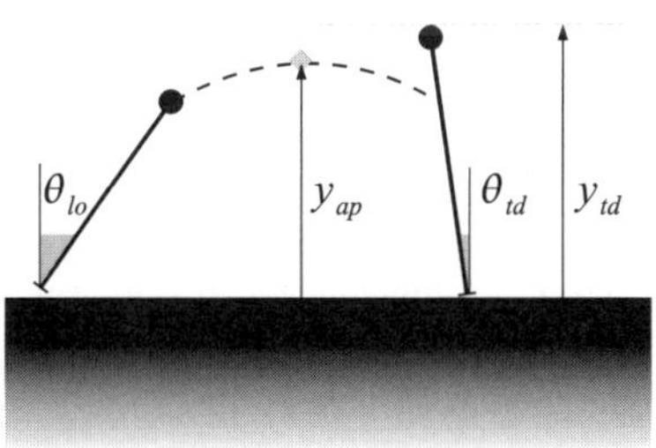

Fig. 1: Constraints imposed on the system state variables including feasibility of the apex height and liftoff angle (physiological constraints) and the system energy.

a-step, b-sine c-complex and d-random command inputs. The desired and controlled state is shown in the left and the control variables in the right graphs. It should be noted that the horizontal axis shows the steps (due to the discrete form of the system). Results show that for the first three cases the monoped is able to track the command accurately and after only a few first steps. The computation time for each step is about 50-100 milliseconds (using MATLAB on an Intel(R) quad-core PC) which is less than the flight duration and enables the online use of this algorithm.

(i) Response to step: As the first experiment we use the simplest scenario by imposing a step command for the forward velocity of the hopper while keeping the altitude fixed. This scenario also can be used for the qualitative comparison of our controller performance with the one from Ref. 7. Accordingly we use the same values including fixed altitude of $1.2m$ and step increase on forward velocity from 0.5 to 2.5 m/s. Figure 2-a shows the resulting behavior for following these commands. Similar to Ref. 7 we use the first 10 steps to bring the system to steady-state forward velocity of 0.5 m/s. At this stage the control problem is how to use the available actu-

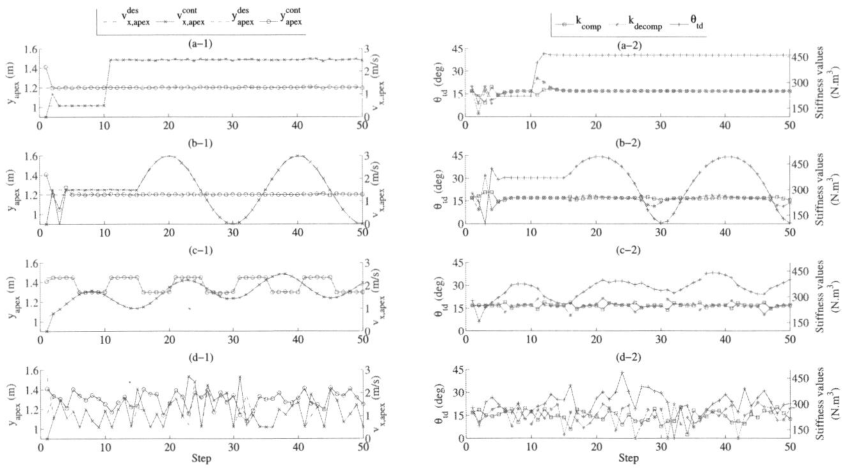

Fig. 2: Results of the designed scenarios (from top to down): a-step in forward velocity while fixing apex height, b-sinusoidal command for forward velocity while fixing apex height, c-complex commands for both and d-random commands for both. The desired and controlled state is shown in the left and the control variables in the right graphs.

ators to increase the velocity with 2 m/s in an efficient and responsive way. The nonlinear programming methods embedded in the control loop solve this control problem as an optimization problem and provide the control variables which optimize the control effort at the same time taking both the state and control constraints into account. The resulting control variables are represented in the graph (a-2). As expected the angle of attack plays the main role to increase the forward velocity and the stiffness values tend to keep the minimum-energy values for this gait. However they have an important role to stabilize the gait when the angle of attack is in the transient phase. This coupled dynamics which is neglected in the simplified decoupled control schemes can be exploited to improve the controllability of the hopper. When the angle of attack reaches to the steady state, the stiffness values converges to their rest value to minimize the control effort required for this task. It is worth-noticing that the controller is using only one step horizon to minimize the control effort. Increasing the horizon can result in more optimized control effort by taking to account the consequences of the current control on the N next step.

(ii) Response to sinusoidal trajectory: In the second experiment we use a more difficult scenario by imposing a sinusoidal desired trajectory for the forward velocity and keeping the altitude fixed. Using the same values as Ref. 7, we set the apex height fixed to 1.2 m and change the velocity

in a sinusoidal pattern with amplitude and offset both of 1.5 m/s. This scenario introduces a difficult task for the hopper since it should be able decelerate to zero speed and accelerate to $3m/s$ while still keeping stable hopping to a fixed height. Again the optimal control formulation is used to provide the right feasible control variables at each step. The results for the state and control variables are represented in figure 2, graphs (b-1) and (b-2) respectively. They clearly show the main role of the angle of attack in the speed control and at the same time the complementary role of the air springs in supplying the rest of the required energy to fulfill the task. Angle of attack follows a similar pattern to forward velocity. A closer look at the k_c and k_d patterns shows that while the hopper is speeding up, the stiffness has slightly higher values in the decompression than the compression phase which injects energy to the system ($\Delta E_U = U_{k_d}(r_0) - U_{k_c}(r_0) > 0$), and the opposite occurs while decelerating in order to remove the energy from the system. It should be noted that this task can not be performed successfully by using only the angle of attack taking its constraint into account.

(ii) Response to a complex pattern: In contrast to the two previous experiments, this experiment is designed to not only excite one of the state variables of the system but also to impose a coupled and complex pattern to both velocity and height of apex. As depicted in the figure 2 graph (c-1), forward velocity increases from zero to 2 m/s in a periodic fashion while the apex height value switches every five steps (between 1.3 and 1.45 m/s). This scenario is a good example of the situations where a more sophisticated control law is required to be able to take the coupled dynamics into account. As represented in the graph (c-2) of the figure 2 the angle of attack is not selected in accordance to the forward velocity pattern anymore since its value can derange the desired apex height. However this is still visible that the angle of attack is mostly contributing to forward velocity control and the change of spring stiffness is being used to inject or remove energy respectively in each step increase or decrease in apex height. While this has been already known and used in Raibert like control rules, we propose here a method to take their combinational effect into account.

(iv) Response to a random task: This scenario presents an ultimate situation where both the desired apex height and forward velocity are changing in random fashion. One can imagine it as hopping in an unstructured environment where the hopper should be quick and responsive to adjust its speed and jumping height. Graph (d-1) of figure 2 shows a sample of such desired apex height and speed. The results for this example shows that the hopper is more successful to follow the desired apex height than the

velocity. This is due to the actuator saturation that is assumed for this problem. The motor speed to adjust the angle of attack during one step is less than what is required for some of the speed jumps. It should be noted that the simulated values should be adapted for a new real setup and can drastically change the tracking performance.

5. Conclusion and Future Work

We proposed a closed-loop optimal control formulation for the online control of a single leg hopping gait. This work presents the basic approach and idea on a very simplified model. The approach however is scalable and can be used for higher dimensional models with not only the discrete return map but with their continuous dynamics. We used SQP solvers provided by MATLAB for academical applications while there have been great progress in faster and more robust solvers. We are currently extending this work by using the same approach for more realistic SLIP model (including energy dissipation) and more detailed dynamics modeling for 2D monoped and quadruped robots. In the optimal control formulation we use the horizon with only one step. For future works using more steps can improve the robustness of the solutions. We also plan to test the simulated algorithms on our newly designed prismatic leg.

Acknowledgement

This project has received funding from EPFL and from the European Community's Seventh Framework Programme FP7/2007-2013 Future Emerging Technologies, Embodied Intelligence, under the grant agreement No 231688 (Locomorph) and Challenge 2 Cognitive Systems, Interaction, Robotics under grant agreement No 248311 (AMARSi).

References

1. P. Holmes, R.J. Full, D. Koditschek, J. Guckenheimer, *SIAM Rev.* **48**, (2006).
2. R. Blickhan *J. of Biomechanics*, **22**, 1217–1227 (1989)
3. A. Seyfarth, Hartmut Geyer and Hugh Herr, *J. Exp. Biol.*, **206**, (2003).
4. T. A. McMahon *J. Exp. Biol.*, **115**, 263–282 (1985)
5. M. H. Raibert, *J. Exp. Biol.*, **29**, 499–514, (1986).
6. U. Saranli, J. Schwind, D. Koditschek, *ICRA*, (1998).
7. J. Schwind, PhD thesis, U. of Michigan, (1998).
8. K. D. Mombaur, et. al, *Robotics, Automation and Mechatronics*, **2**, 2004.
9. D. Remy, K. Buffinton, and R. Siegwart, *IORS*, (2011).
10. R. Tedrake, PhD thesis, MIT, (2004).

SYNTHESIS BIPED DYNAMIC WALKING BASED ON OFFLINE TRAJECTORY GENERATION

A. SAFONOV, F.PALIS, V. BILOUS

University of Magdeburg, Department Electrical Energy Systems, P.B. 4120, 39106 Magdeburg, Germany

A. TELESH, U. SCHMUCKER

Fraunhofer IFF, Department Virtual Engineering, Sandtorstr. 22, 39106 Magdeburg, Germany

The aim of this paper consists in development of a method for motion trajectory synthesis of a biped anthropomorphic robot on the basis of ZMP-concept. In this paper the suggestion is made to implement the process of dynamic walking of the biped robot in two stages: trajectory generation of a dynamic stable motion of the robot on the basis of his mathematical model in offline-mode and tracking of the received walking trajectories with use of a supplementary stabilizing (control) system.

1. Introduction

Anthropomorphic biped robots are complex mechanical systems with a great number of interconnected controlled and passive degrees of freedom (DOF). Organization and optimization of motions of such a system is quite a complicated problem which is connected with construction of a complex hierarchical control system. Motion trajectory of a biped robot is generated on the basis of ZMP-principle[1]. According to this principle, stable state of the robot is provided under the condition of situation of the zero moment point within the foot supporting plane. At this point the summation action of the moment of all forces acting on the foot from supporting surface is equal to nought [2]. In the present paper we propose to generate offline such trajectories by which robots motions relative to feet are free. Thus, during walking the moments in ankle joints are not applied, and the zero moment point is located strictly under ankle joint.

2. Synthesis of walking trajectories

The biped robot mechanism has a great amount of DOF that complicates the task of finding of joints motion trajectories. For the purpose of simplification, we introduce so-called reflexes. These reflexes describe motions dependences of some joints on other [8]. For influence on mechanisms state at the end of a step some impacts are made upon the system in a form of additional motions while leg and body displacement. These motions are executed during one step time and have zero values of derivatives at the beginning and at the end of a step (fig. 1).

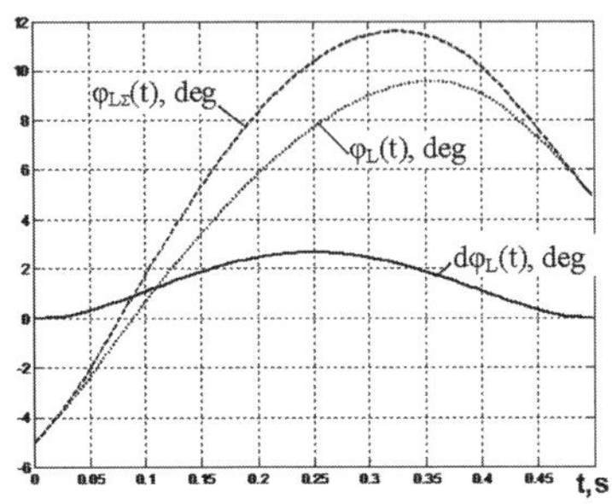

Fig. 1 – Additional leg movements

Robots walk has the following features:

- Moments in robots feet are not applied.
- Robots supporting leg is always in straightened state
- Robots body keeps a vertical position

Trajectories generation is implemented in offline mode on the basis of a mathematical model of the robot by means of Euler optimization algorithm. At the same time optimization of values of control actions K1, K2 is fulfilled for the purpose of receiving of necessary system final states (fig. 2).

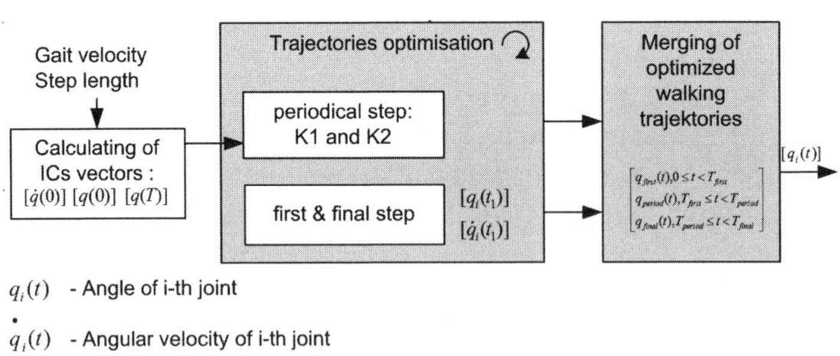

Fig. 2 – motion trajectory generation algorithm

There is possible to allocate three difference step phases in the walking process:

The first step - acceleration. During this step starts the robot moving from initial states (e.g. initial velocities of robot joints are zero). At the end of step has the robot such states, which are initial for the following (periodical) step.

The second phase - periodical steps. This is a step, which initial and final conditions of all joints are equivalent. Final step - deceleration. In this phase, robot stops.

Movement's trajectories of the robot are generated for each phase separately. At the first are generated moving trajectories of periodical steps of the robot at which movements of the mechanism's joints during a current step completely repeat movements during the previous step. The initial (final) conditions of final (first) step can be directly find from the necessary trajectories.

3. Design of control system

For ensuring of stable walking on imperfect surfaces an additional control system is used. Signals for its activation are trajectories of body bending in frontal and sagittal planes relative to vertical line. As feedback signals act real values of these angles, which are received from bending inertial sensors [3] (Fig. 3).The stabilization system influences the mechanism through the torques in ankle joints and body's movements. Simultaneous using of two control inputs increase robustness of control system and decrease of the time of transient processes.

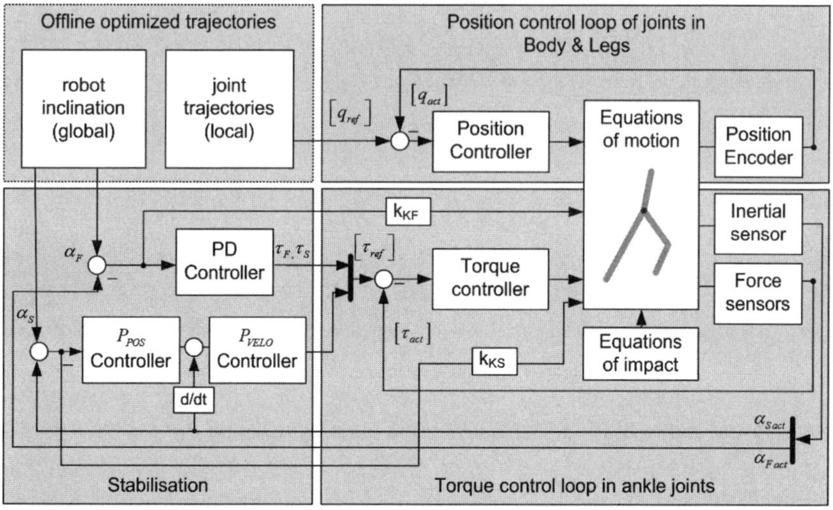

Fig. 3 – Control system of robot

The additional control actions are realized with body movements in sagittal and frontal plane. This additional control inputs are reduced necessary torques in the ankle joints and increased dynamical properties of control system (s. Fig 4.)

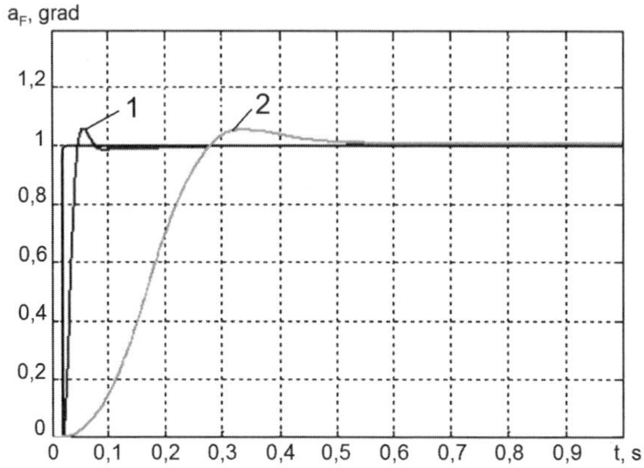

Fig. 4 – Step response of the control system with (1) /without (2) body's movement

4. Simulations results

Simulation of the robot control system while walking was carried out on the model with calculation of surface contact [4].

Figure 5 shows torques in ankle joint in sagittal and frontal plane.

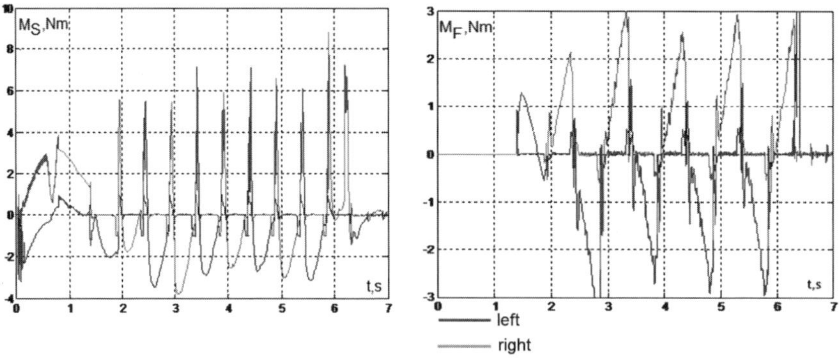

Fig. 5 – Torques in ankle joint in sagittal and frontal plane

Figure 5 shows the motion sequence of one step of the robot.

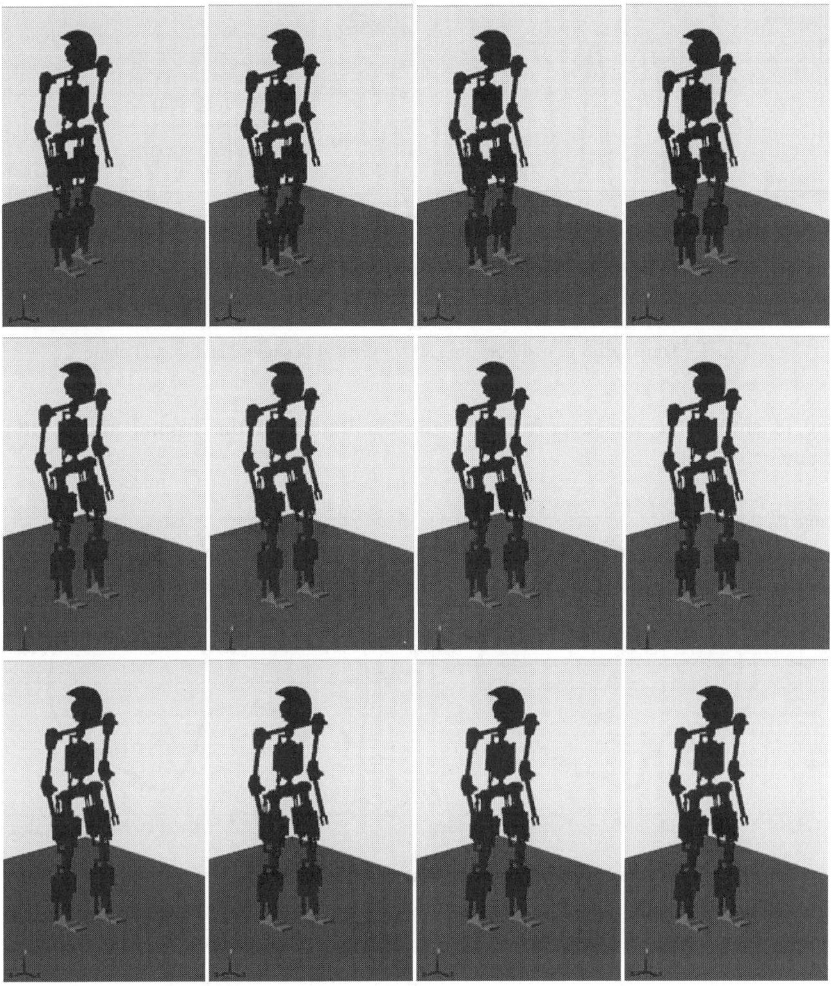

Fig.6 – Motion sequence of one step of the robot

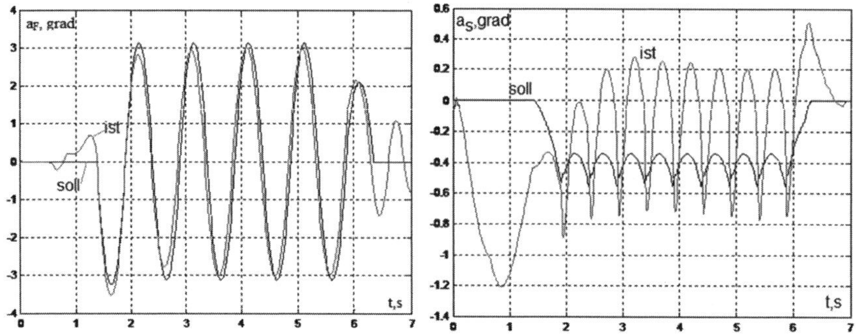

Fig. 7 – Trajectories of body bending above vertical in sagittal and frontal plane

Fig. 8 shows the phase portraits of the oscillation in the frontal and sagittal plane.

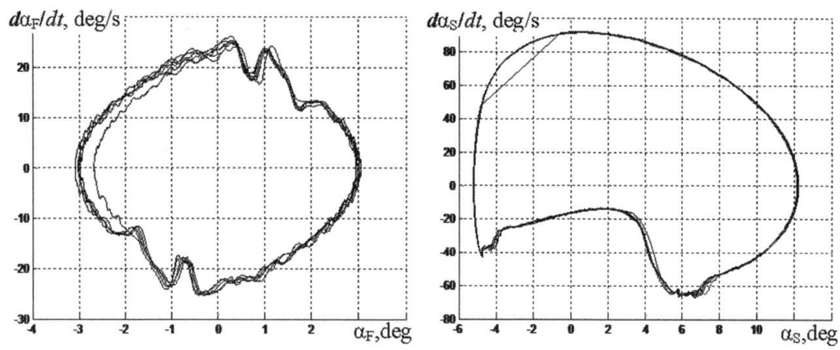

Fig. 8 – Phase portraits of the mechanism

Fig. 6, 7, 8 shows stable diagrams of periodical gate of biped mechanisms behavior while walking. Small differences among trajectories is caused by stochastic process of contact with virtual surface.

5. Experiments on robot ROTTO

Experimental investigations of the operation of the control algorithm for dynamic walking were carried out on the robot ROTTO. Information about the robot parameters, set of sensors, power sensors, communication and control system can be found in [3], [4], [6], [7].

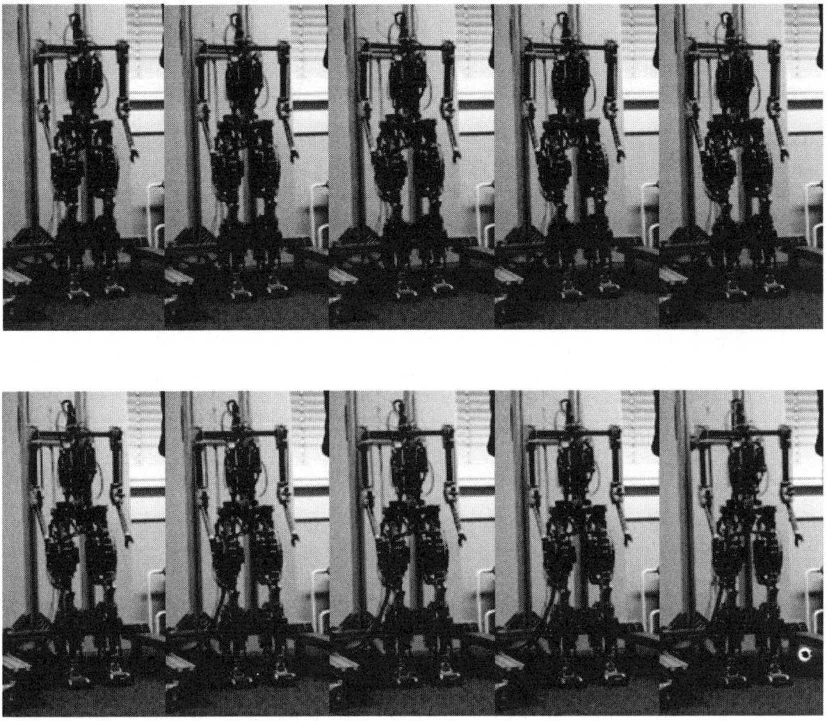

Fig. 9 – Motion sequence of one step of the robot

From the Figure 9 it is seen that the robot is moving its leg during the deviation in the frontal plane. At that the moving leg rises by 1 cm over the surface. The supporting led deflection from the vertical by 3 grad. The maximum forward walking velocity of the robot which was reached during the experiments was 0,3 km/h.

6. Conclusion

The advantage of the present algorithm of synthesis of biped robot motion trajectory consists in absence of need for dynamic calculation and kinematic transformations. Its drawback is trajectories rigidity, i.e. absence of an opportunity of their correction from the side of control system. In future we intend to upgrade the robot motions control system by means of introduction of feedback for motion trajectories generated for an ideal surface with application of ZMP-principle. This would enable to receive a more flexible robot motion control system with a view to adaptability to ambient conditions.

References

1. M. Vukobratovic. Shagayushie Roboti i antropomorfnie mehanismi. Mir, Moscow, 1976. In Russian
2. M. Vukobratovic and B. Borovac. Zero-moment point - thirty five years of its life. International Journal of Humanoid Robotics, 1(1):157-73,2004
3. Y. Zavgorodniy, „Konstruktion und Steuerung von Schreitrobotern mit ballistischem Laufverhalten“, Ph.D. Thesis. Magdeburg, 2009
4. Contact processing in the simulation of CLAWAR. Juhasz Konyev, Rusin, Schmucker. Proceedings of 10th International Conference (CLAWAR 2007), Singapore, 16-18 July 2007.
5. Peremeshhenie antropomorfnykh mekhanizmov. Formal'skij A.M. - M.: Nauka. Glavnaja redakcija fiziko-matematicheskojj literatury, 1982. - 368 s. (in Russian).
6. www.uni-magdeburg.de/ieat/robotslab/
7. Linear elastic actuator of a biped robot "ROTTO"/Melnykov, Andriy; Konyev, Mykhailo; Palis, Frank; Schmucker, Ulrich/ In: Emerging trends in mobile robotics. - Singapore [u.a.]: World Scientific, ISBN 981-432797-2, S. 588-595, 2010 Kongress: CLAWAR; 13 (Nagoya) : 2010.08.31-09.03
8. Feedback Control of Dynamic Bipedal Robot Lovomotion, Eric R. Westervelt, Jessy W. Grizzle, Christine Chevallereau, Jun Ho Choi, and Benjamin Morris. CRC Press. London - New York, 2007

GAIT MODIFICATION STRATEGY FOR A SIX-LEGGED ROBOT WALKING ON ROUGH TERRAIN*

Dominik Belter[‡]
Institute of Control and Information Engineering, Poznan University of Technology, 60-965 Poznan
ul. Piotrowo 3A,
[‡]*E-mail: {Dominik.Belter}@put.poznan.pl*

This paper presents the concept and selected results of a gait modification system for a six-legged robot walking on rough terrain. The robot modifies the sequence of leg's transfer phases according to the roughness of the terrain. On a flat terrain the robot uses tripod gait and according to the roughness factor the gait is modified to more secure cyclic wave and free gaits. Results of simulations with the motion planning algorithm on rough terrain are provided. The influence of the proposed strategy on the efficiency of the motion planning algorithm is presented.

1. Introduction

Efficient locomotion of walking robots on rough terrain can be achieved by various mechanical designs and control systems. The RHex robot is able to traverse rugged and obstacle ridden ground without complicated environment sensing and control algorithms[10]. It was created according to the model known as "rimless wheel". It is a one-degree-of-freedom mechanical system. The rimless wheel consists of a central mass with several spokes. At any time a single spoke touches the ground and the system behaves like an inverted pendulum. The RHex robot can traverse various types of terrains and climb stairs.

A reactive controller is used to obtain high locomotion capabilities of the BigDog robot on rough terrain[13]. Moreover, some existing 2D planning techniques are adopted to plan the motion in 3D environment[13].

Despite high locomotion capabilities these robot would fail in presence of some obstacles. Without a foothold selection algorithm the BigDog robot

*This work was supported by NCN grant no. 2011/01/N/ST7/02080

would fail during crossing a stream where some position of the feet (e.g. located on the stones) are acceptable and the rest would cause fall. Similar problems would have the RHex robot if its legs get stuck in between stones.

Some other robots represent deliberative approach. The Lauron III,[4] LittleDog[7] and Messor[12] robots should plan their motion. Without foothold selection and body motion planning they have limited capabilities to climb obstacles. However, with the proper motion planning algorithm they have potential to cover high and demanding obstacles. This article presents the influence of the gait type modification on the results of motion planning. The role of the foothold selection, body path planning and posture optimization are also pointed out.

2. Related work

Recent works with the stick insects[8] show that the animals constantly switch the gaits between cyclic tripod, tetrapod and non-cyclic free gait. This behavior was implemented on the biologically inspired Lauron robot. The robot can switch between fixed walking patterns. The sequence of the leg transfer strongly depends on the stability of the robot.

A different approach based on the primary/secondary gait concept is used on a quadruped robot[1]. The primary gait is the fixed sequence of the leg transfers defined for the wave gait. While walking on rough terrain the robot chooses the next leg to transfer. First, the robot considers the kinematic margin for all legs. The next transfered leg has the smallest value of the kinematic margin. The modified sequence of the leg transfers is called the secondary gait. If more than one leg has the same minimal value (e.g. while walking on flat terrain), the next transfered leg is set according to the defined fixed sequence for the wave gait – the robot uses the primary gait.

The free and wave gaits are mainly used while walking on rough terrain[6]. These gaits are secure (the support polygon is large) but the maximal speed of walking in these modes is low. However, the use of the tripod gait while walking on rough terrain is more risky. The support polygon (triangle) for the tripod gait is smaller than the support polygon for the wave and free gaits. During a real mission on rough terrain the robot should switch between cyclic tripod gait and free gait according to the roughness of the terrain. The tripod gait should be used on nearly flat terrain to walk with higher speed. The robot should switch to free gait to preserve higher stability on demanding and rough terrain.

3. Motion planning

3.1. *Control system*

Our six-legged robot Messor[12] is equipped with a Hokuyo laser rangefinder to measure the shape of the terrain[2]. The data obtained from the laser rangefinder are used to create the elevation map of the environment. The robot uses a Rapidly-exploring Random Trees-based algorithm to plan its motion on the elevation map[3]. A foothold selection algorithm is used to minimize the risk of slippage[2]. The robot also optimizes its posture during motion planning. The optimization algorithm tries to maximize kinematic and stability margin by searching for the optimal inclination and distance to the ground of the robot's platform.

In this research the simulator of the Messor robot is used. The dynamical properties of the robot are simulated by using Open Dynamics Engine[11]. The control software in simulator is the same as for the real robot. An advantage of the simulator is the knowledge about the robot state during the experiment. The same ground-truth is difficult to obtain on the real robot.

3.2. *Gait modification*

The planning algorithm allows to use any defined type of gait. To switch between various types of gait we proposed the following heuristic. First, the spherical variance ω is computed[9]. The grid map is converted to triangle mesh by using the considered point and its neighbors (the number of triangles n is 8). The vector normal to each triangle on the surface $\overrightarrow{\mathbf{N}_i} = (x_i, y_i, z_i)$ is computed. Then the module R of the vector set $\overrightarrow{\mathbf{N}_i}$ is computed as follows:

$$R = \sqrt{\left(\sum_{i=0}^{n} x_i\right)^2 + \left(\sum_{i=0}^{n} y_i\right)^2 + \left(\sum_{i=0}^{n} z_i\right)^2}. \tag{1}$$

The module R is divided by the number of vectors n to normalize the values into a range between 0 and 1. Finally, the spherical variance is computed as follows:

$$\omega = 1 - \frac{R}{n}. \tag{2}$$

The value of ω defines a "roughness" of the terrain. The robot computes the ω value for the current and desired position of the robot. To determine the "difficulties" of the transition between the current and the next desired position of the robot, the mean values of spherical variance ω and distance between current and desired position of the robot is computed. When this value is smaller than threshold (the terrain is easy to traverse) the robot uses tripod gait, elsewhere the robot uses a more secure pentapod gait.

The pentapod gait is generated according to the primary/secondary gait concept[1]. The sequence of leg transfers is fixed for the wave gait (primary gait). First, the robot transfers the fourth leg, then the third, the fifth, the second, the sixth, and the first leg (4,3,5,2,6,1).

4. Simulations

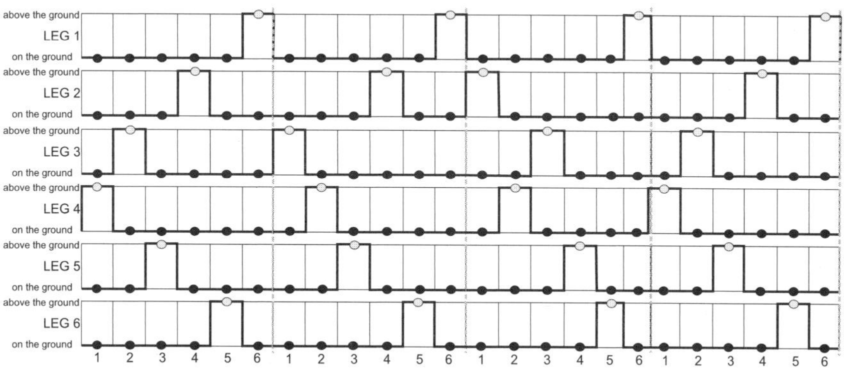

Fig. 1. Example modification of the gait cycle while walking on rough terrain

The results of the experiment with motion planning for the pentapod gait are shown in Fig. 1. The diagram represents four cycles of the gait. Only two states of the legs are presented in the graph – the position on the ground and the position above the ground. A gray circle represents the position of the leg above the ground, while a black circle represents the position of the leg on the ground. The first and fourth cycle are the same as for the wave gait. The default sequence changes when the robot walks on rough terrain. The sequence is modified because the kinematic margin varies for each leg. In the second cycle the sequence is 3,4,5,2,6,1, and in the third cycle the sequence is 2,4,3,5,6,1.

The modification of the gait sequence simplifies motion planning. It is easier to find a transition between two positions of the robot. Such an approach allows to avoid deadlocks which are caused by the lack of the kinematic and stability margin. For the tripod gait a transition is not always possible, because the robot can't fulfill the stability and kinematic requirements.

We conducted a verification simulation on rough terrain to determine the usefulness of the proposed strategy. The robot planned its motion on rough terrain (Fig. 2). The algorithm checked if the transition between two nodes was possible by using pentapod and tripod gait. In 272 trials the transition was possible only by using pentapod gait while the number of transitions using only tripod gait was 72. It shows that the efficiency of the more stable pentapod gait on rough terrain is higher.

We also verified the heuristic strategy of switching between gaits. The results are shown in Fig. 2A. On flat terrain the robot uses tripod gait. When the terrain is more rough (higher than the threshold) the robot uses pentapod gait. Moreover, length of the step is set to 75% of the nominal value to decrease the locomotion speed. By using this gait the risk of fall decreases. Moreover, the motion planning algorithm finds more easily the transition between two nodes of the RRT tree (steps).

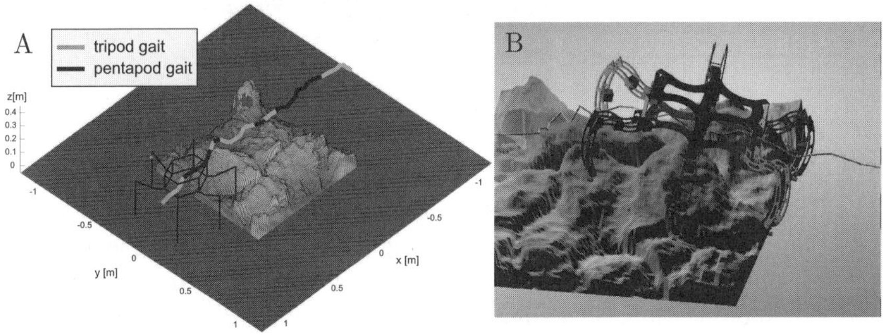

Fig. 2. Simulation on the rough terrain mockup: modification of the gait (A) and the simulated robot during motion execution (B)

Another problem solved by the proposed algorithm is presented in Fig. 3. This problem is similar to the problem of stepping stones while crossing a stream. There are some areas allowed for the feet (stones) and the robot has to balance its body to reach these areas. In the experiment the algorithm found a path from one side to the other side of the "stream". The results are shown in Fig. 3.

To show how each module of the motion planner increases the capabilities of the robot on rough terrain, a series of simulations were conducted[a].

[a]A video is available on http://lrm.cie.put.poznan.pl/clawar2012.avi

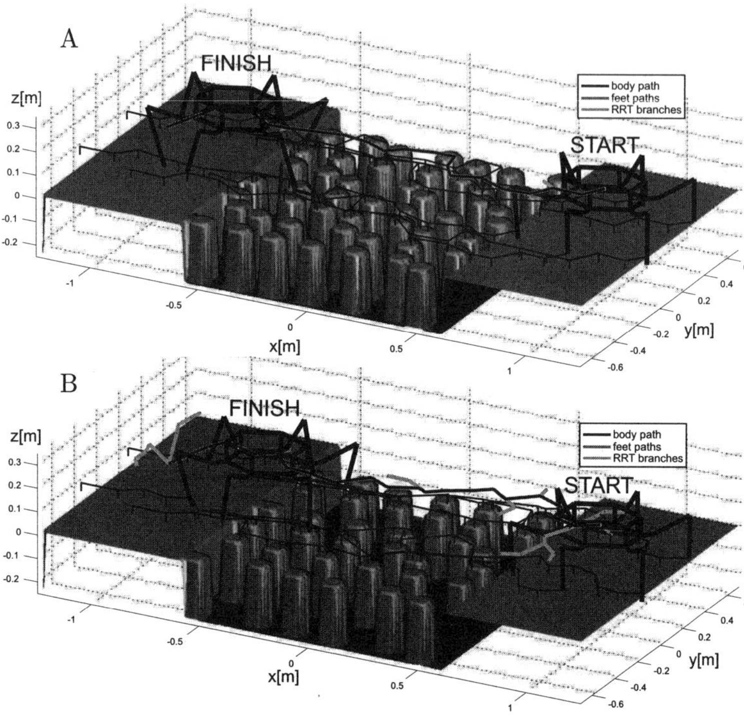

Fig. 3. Results obtained on artificial terrain – "crossing a stream" problem

The RRT-based planer is compared to the method which uses only foothold selection and the planned path is a straight line, RRT-based planer with posture optimization and RRT-based planner with posture optimization and gait modification. The simulations were conducted on the maps presented in Fig. 3. The number of stones in the gap is modified (25, 36, 49, 64, 81, 100, 121 stones in the following simulations). In Fig. 3A one can see a terrain with 49 stones, and a terrain with 36 stones in Fig. 3B. For each configuration of the environment a series of 10 simulations is conducted for all planning methods. The initial and goal position of the robot were randomly modified in the range of ± 5 cm. The relation between the number of stones and the success ratio is shown in Fig. 4.

The efficiency of the method which uses only the foothold selection algorithm decreases gradually (Fig. 4A). On the other hand the planning time is very short (about 0.2 s in Fig 4B). The path is straight so the robot is not always able to find proper footholds. When the robot plans the path, it can modify the position of the platform to reach proper footholds. In this

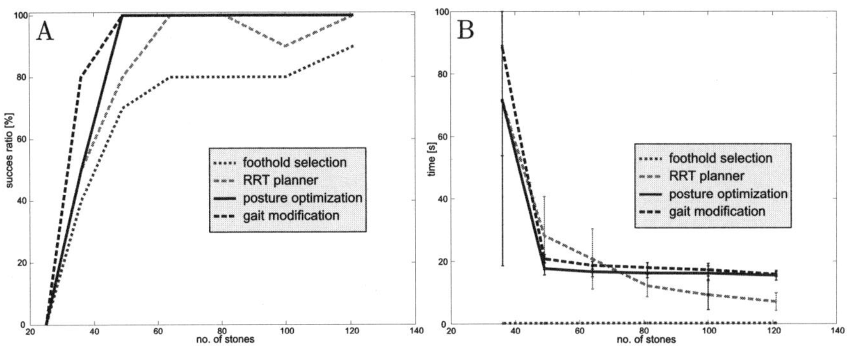

Fig. 4. Results of simulation – efficiency of motion planning for "crossing the stream" problem: success ratio (A) and time consumed for motion planning (B)

case also the planning time increases. When the module which optimizes the posture of the robot is added to the motion planner the efficiency rises to 100% for the ground configuration with the number of stones bigger than 49. However, the efficiency of the motion planning algorithm can be increased even more if the gait modification strategy is used. The success ratio for the number of stones bigger than 49 is 100% and for 36 stones is 80%. It is the best obtained value compared to other methods. The computation time for this approach increases only slightly. It should be also mentioned that the robot walks slower when it uses wave or free gait. The execution time of the planned motion increases.

5. Conclusions and future work

The article presents the strategy for gait modification for six-legged robots. The algorithm is based on the primary/secondary gait concept. The robot modifies the sequence of leg's transfer phases. It increases the stability margin during climbing more difficult obstacles. It was shown in simulations that the efficiency of motion planning increases when the proposed strategy is used (Fig. 4). Finally the robot is able to traverse more demanding obstacles. The computation cost increases slightly when the proposed strategy is used. This strategy imitates a behavior of real animals. The robot walks slower and makes shorter steps during climbing rough obstacles and increases speed on nearly flat terrain (by using tripod gait).

In the future we are going to implement the proposed strategy on the real robot. We are also going to use information about terrain and obstacle types obtained from a camera to modify the parameters of the motion planner and increase its efficiency.

References

1. Bai S, Low K.H, Zielińska T (1999) Quadruped Free Gait Generation Based on the Primary/Secondary Gait, Robotica, Vol. 17(4), 405–412
2. Belter D, Skrzypczyński P (2011) Rough terrain mapping and classification for foothold selection in a walking robot, Journal of Field Robotics, Vol. 28(4), 497–528
3. Belter D, Skrzypczyński P (2011) Integrated motion planning for a hexapod robot walking on rough terrain, 18th IFAC World Congress, Milan, Italy, 6918–6923
4. Celaya L, Albarral G (2006) Implementation of a hierarchical walk controller for the Lauron III hexapod robot, Proc. 9th Int Conf on Climbing and Walking Robots (CLAWAR), Brussels, Belgium, 409–416
5. Kalakrishnan M, Buchli J, Pastor P, Mistry M, Schaal S (2010) Fast, robust quadruped locomotion over challenging terrain, Proc. IEEE Int. Conf. on Robotics and Automation, Anchorage, USA, 2665–2670
6. Celaya E, Porta J.M (1998), A control structure for the locomotion of a legged robot on difficult terrain, IEEE Robotics & Automation Magazine, Vol. 5(2), 43–51
7. Rebula J, Neuhaus P, Bonnlander B, Johnson M, Pratt J (2007) A controller for the LittleDog quadruped walking on rough terrain, Proc. IEEE Int. Conf. on Robotics and Automation, Roma, Italy, 1467–1473
8. Roennau A, Heppner G, Kerscher T, Dillman R, A behaviour-based free gait inspired by a walking stick insect, Proc. 13th Int Conf on Climbing and Walking Robots (CLAWAR), Paris, France, 181–188
9. Sanctis L, Garrido S, Moreno L, Blanco D, Outdoor Motion Planning Using Fast Marching, Mobile Robotics: Solutions and Challenges, Tosun O. et al. (Eds.), World Scientific, Singapore, 2009, 1071–1080
10. Saranli U, Buehler M, Koditschek D (2001), RHex: A simple and highly mobile hexapod robot, International Journal of Robotics Research, vol. 20, 616–631
11. Smith R, Open Dynamics Engine, http://www.ode.org, 2012
12. Walas K, Belter D (20011) Messor – Versatile walking robot for search and rescue missions, Journal of Automation, Mobile Robotics & Intelligent Systems, 5(2), 28–34
13. Wooden D, Malchano M, Blankespoor K, Howardy A, Rizzi A, Raibert M (2010) Autonomous navigation for BigDog, Proc. IEEE Int. Conf. on Robotics and Automation, Anchorage, USA, 4736–4741

A QUADRUPED ROBOT WITH ON-BOARDING SENSING AND PARAMETERIZED GAIT FOR STAIR CLIMBING

T. BREWER, K.N. KAIPA, and S.K. GUPTA[†]

Department of Mechanical Engineering
University of Maryland, College Park, USA
[†]*E-mail: skgupta@umd.edu*

This paper presents the design of a quadrupedal robot that can automatically adapt its gait to, and climb, staircases of different configurations. This is accomplished by endowing the robot with a parameterized gait for stair climbing: First, a gait plan is synthesized that allows the robot to climb a stair of known dimensions. Second, the robot approaches a previously unseen stair and perceives its height and width by using an onboard vision system. Third, the synthesized gait plan is parameterized by the perceived estimates of height and width of the stair. Fourth, the robot executes the parameterized gait to climb the staircase; this thereby eliminates the need for a complex control system to achieve the same purpose. Whereas quadruped robots have previously demonstrated stair climbing, to the best of our knowledge, none have so far been capable of climbing stairs of variable height while simultaneously carrying all the needed perception, processing, and power modules on-board. Our work is one of the first successful attempts toward the above goal. Results with the robot climbing a variety of stair configurations demonstrate the effectiveness of our approach.

Keywords: Walking robots; Quadruped robots; Stair climbing; Parameterized gait.

1. Introduction

Design of robots that can automatically climb and negotiate man-made structures such as stairs and curbs is a difficult problem. This is evidenced by several research attempts to develop legged,[1–13] tracked,[14,15] and hybrid[16,17] robotic platforms for stair-climbing applications. Whereas stairs are not representative of naturally occurring terrain features, they are widely found in urban environments. Therefore, stair climbing machines can be used as first responders in urban search and rescue missions, mobile platforms for handicapped patients, and other service robots that must operate in urban settings. This, however, raises the question of how to endow

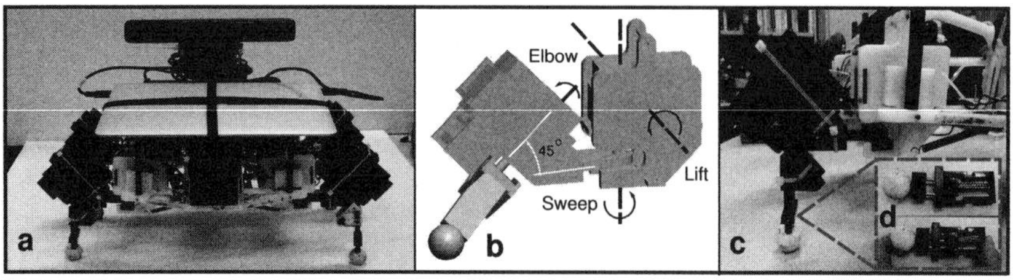

Fig. 1. (a) The physical quadruped robot. (b) A 3D model showing the leg joints. (c) A close-up view of the robot's leg. (d) Passive suspension system.

these robots with an ability to efficiently perceive pertinent features of the stairs in their surroundings and use such information for adjusting their gait to a variety of stair configurations.

In this paper, we present the design of a quadruped robot that can automatically adapt its gait to, and climb, staircases of different configurations. On-line gait planning for a given set of stairs is a computationally intensive task. This requires a high performance computer to be available on-board. Furthermore, significant time needs to be spent in planning the gait before it can be executed. In many search and rescue application scenarios, time is a critical parameter and delay of even few seconds could lead to failure. Hence, we have chosen a different approach. We have synthesized a parameterized stair-climbing gait off-line, which is defined using the parameters of the stair. Once the parameters of the stairs can be estimated during execution, a stair climbing gait can be instantiated instantaneously by using these parameters. This approach eliminates the need for on-line gait planning and the associated computation time.

The parameterized gait is implemented in four stages: First, a gait plan is synthesized that allows the robot to climb a stair of known dimensions. Second, the robot approaches a previously unseen stair and perceives its height and width by using an onboard vision system. Third, the synthesized gait plan is parameterized by the perceived estimates of height and width. Fourth, the robot executes the generated parameterized gait to climb the staircase. Quadruped robots with large sensing or extensive gait planning capabilities usually have the sensors mounted externally rather than having them onboard. Our main design goal is to create a robot which is capable of climbing stairs, while carrying all the needed sensing, processing, and power modules onboard. Results with the robot climbing a variety of stair configurations demonstrate the effectiveness of our approach.

Adaptive climbing of stairs up to a gradient of 20 degrees was first demonstrated by the quadruped robot PVII.[1] The quadruped TITAN VI[2] climbed ordinary staircases of 30 to 40 degrees. A dynamically stable step climbing was demonstrated by the quadruped Scout I.[3] Pack[5] developed a perception based control of a quadruped robot, which computed its posture relative to the staircase by matching an externally mounted camera's image of the staircase with an internal staircase model and used this information to move to, and position itself in front of, the stairs before climbing it. As this approach assumes that the staircase model is known a priori, it doesn't scale well across different types of stairs. However, in our approach, the stair parameters are explicitly estimated and used in real time in order to adapt the robot's movements to a variety of stairs.

Lee et al.[6] applied reinforcement learning to sequence foot placements for a quadruped robot and demonstrated its ability to climb a single step, double-angled steps, double-cornered steps, and four interleaved steps. Loc et al.[7] developed sensing and gait planning for a quadruped robot to operate in environments like slopes, convex corners, and concave corners. The environment was reconstructed by combining data from range and gyroscope sensors with the robot's internal state, and used by a gait planner to compute a motion plan for the robot. Even though our approach bears some similarity with their idea of combining sensing with gait planning, it differs by explicitly perceiving the stair parameters rather than attempting to reconstruct the robot's surrounding environment.

Ko et al.[8] performed trajectory planning for quasi-static stair climbing by a quadruped robot. Experimental results demonstrate that the robot is able to climb a natural stair by using their algorithm. However, the robot is not equipped with a feedback mechanism that helps it to sense the height and width of the stair. LittleDog is one of the recent quadruped robots that has demonstrated effective stair climbing.[9,10] However, the robot is constrained as its control computer and the motion capture system are externally situated.

2. Physical Robot

The robot consists of 16 degrees of freedom (Figure 1(a)). The leg configuration is shown in Figure 1(b). Each leg consists of three active joints and one passive joint: (a) *Lift:* The lift joints allow the robot to raise and lower its limbs. (b) *Sweep:* These joints allow the robot to sweep its limbs from forward to backward. (c) *Elbow:* The elbow joints allow the robot to appropriately place its feet on the stair. (d) *Passive suspension:* Each lower leg

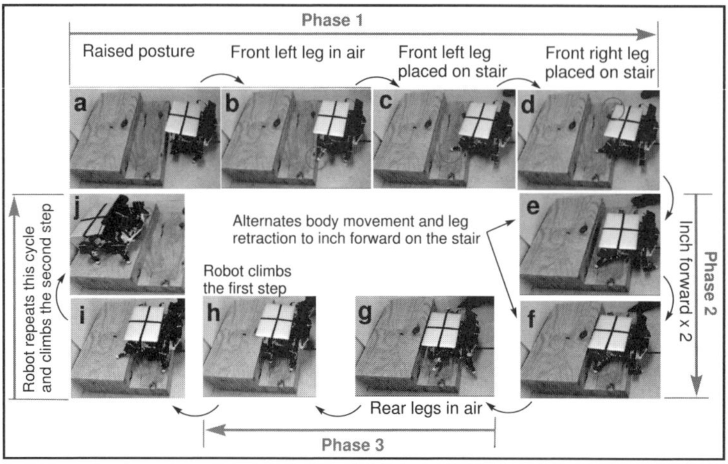

Fig. 2. The robot using a synthesized gait to climb a staircase of known dimensions

segment is coupled to its foot (rubber ball) by a spring with travel length = 0.5 inches and spring constant = 6 pounds/inch (Figure 1(d)). An onboard Microsoft Kinect sensor interfaced to an onboard laptop provides height and width perception of stairs.

3. Stair-climbing Gait Synthesis

The robot's stair-climbing gait is divided into three phases corresponding to a total of fifty discrete gait steps. A gait sequence for climbing a stair with a height of 5 inches is synthesized by defining the angles for all the twelve limb joints at each gait step.

Phase 1: Initially, the robot stands in front of a stair with each leg positioned at its zero position, in which the upper segment of each leg is horizontal and the lower segment is vertical. To begin climbing, the robot slides its feet inward, thereby lifting its body up (Figure 2(a)). Next, the robot lifts its front left leg and places its foot on the step (Figure 2(c)). Next, these motions are repeated by the robot's front right leg. In the resulting posture, the robot's front feet are rested on top of the stair and its body is slightly pitched toward the rear (Figure 2(d)).

Phase 2: In this phase, the robot inches forward until its CG is lying above the step and its rear legs are touching the side of the stair. The robot moves forward by rotating all its sweep and elbow joints backward (Figure

2(e)). Next, the robot retracts each leg and places it back on the ground. This movement cycle is repeated twice.

Phase 3: The previous phase ends with the front legs on top of the stair and the rear legs in contact with the side of the stair (Figure 2(f)). Next, the robot simultaneously raises the rear legs, resulting in lowering of the robot's rear until it rests on the edge of the stair (Figure 2(g)). The rear feet are then raised above the step and placed on the step. Once all four feet are on the same stair (Figure 2(h)), the robot repeats this cycle to climb the next step (Figure 2(j)).

4. Stair-climbing Gait Parametrization

The stair-climbing gait was synthesized for a 5 inch staircase. Therefore, the gait is parameterized to enable the robot to adapt to climbing previously unseen stairs. The height and width are decoupled in that they influence two mutually exclusive portions of the stair-climbing gait: (1) climbing a stair and (2) walking forward on a climbed stair.

Stair height parametrization: Lift is used to raise the foot off the ground and place the foot on the stair, thereby making it a function of stair height. However, the sweep and elbow angles remain constant. Therefore, parametrization is applied to the lift angles of all the legs. Let h represent the stair height estimated by the vision sensor, $i \in \{1, 2, 3, 4\}$ represent the ith leg, t_j represent the jth time step, and $\theta_5^{(i)}[t_j]$ represent the lift angle of the ith leg, synthesized for the 5-inch stair, at time step t_j. Let $\theta_h^{(i)}$ represent the lift angle sequence of ith leg for a stair of perceived height h. Now, the parametrization for $\theta_h^{(i)}$ at each discrete time step t_j is given below:

$$\theta_h^{(i)}[t_j] = \begin{cases} \theta_5^{(i)}[t_j] \times (\frac{h}{5-\epsilon}), & \text{if } t_j \in \{Phase1, Phase3\} \\ \\ \theta_5^{(i)}[t_j], & \text{otherwise} \end{cases} \tag{1}$$

The original gait was developed for a stair height of 5 inches. Therefore, a factor of $\epsilon = 0.5$ is used to provide for a margin of error. Lift angles for the left rear leg after parametrization for three different stair heights are shown in Figure 3(a).

Stair width parametrization: The robot covers roughly 4 inches using three iterations of its forward walking gait. If w is the sensed width of the stair, then the robot implements a walking gait for $\lceil 1.33 \times w \rceil$ iterations after climbing a stair.

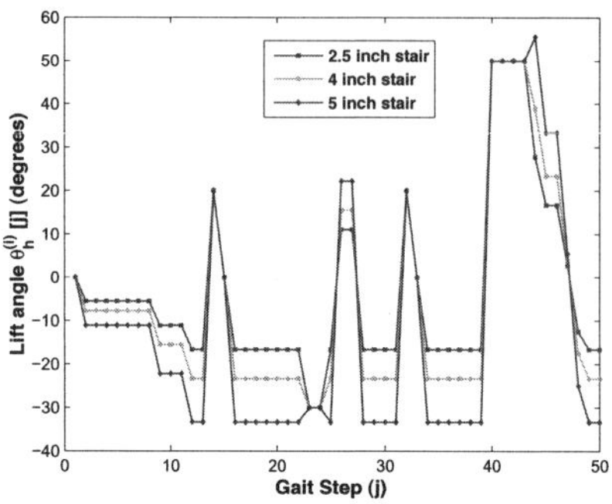

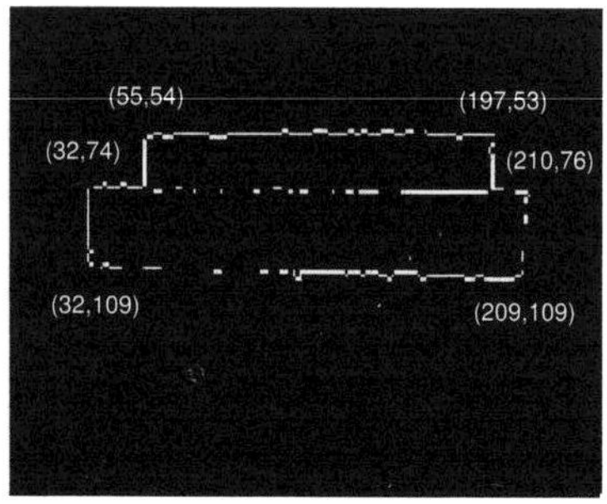

Fig. 3. (a) Parameterized left rear leg's lift joint angles. (b) Edge-detection applied to the depth image of the stair obtained from Kinect sensor.

5. Vision System for Height and Depth Perception

We develop a vision based algorithm for the height and width perception of stairs. First, a Microsoft Kinect sensor is used to obtain a depth image of the scene comprising the staircase. During camera calibration, the depth data was found to be nearly linear between 3 ft. and 9 ft. with a slope of 19.5 per foot. Second, a simple edge-detection of the depth image is performed (Figure 3(b)). Third, the resulting image is input to a corner-detection algorithm, which extracts L-type corners that are typical of stairs and filters out other irrelevant and spurious corners (The detected corners are shown in Figure 3(b)). The width of the stair-step is computed as the difference between the depth to the first step and the depth to the second step. The height calibration is achieved by placing objects of known height at measured distances from the sensor and counting the number of vertical pixels in the object's image. By using these measurements, a linear relationship between the number of pixels per inch and distance from the Kinect is obtained, which is then used for height measurement. All the height and width parameters were estimated with an accuracy of ± 0.15 in.

6. Results

We report results from experiments in which the robot used parameterized gait to climb three stairs: (a) h = 2.5 in., w = 12 in., (b) h = 4 in., w = 10 in., and (c) h = 5 in., w = 10 in. Even though the 5 inch stair was used for the purpose of gait synthesis, it was still used in this experimental regime as it involved testing the stair-climbing capability of the robot by

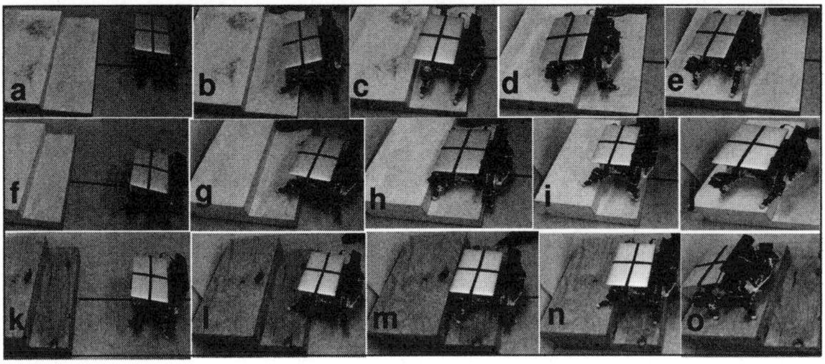

Fig. 4. Snapshots of video footage of the robot climbing three different staircases: (a-e) 2.5 inch stair. (f-l) 4 inch stair. (k-0) 5 inch stair.

using parameterized gait. The robot was placed at a distance of 2.5 feet in front of each stair and allowed to execute the gait without any human intervention. Figure 4 shows the robot successfully climbing all the three stairs. The rate of climb was found to be roughly 30 sec/step.

7. Conclusions

We presented the design of a stair climbing quadruped robot that can adapt to previously unseen staircases by using a parameterized gait. The robot used an on-board vision system to perceive critical stair parameters, tuned its gait based on these parameters, and successfully climbed stairs of varying configurations. We designed the simplest possible gait to ensure robustness. Experimental results showed that the gait is robust with respect to the sensor uncertainty. Currently, the robot approached the stairs head on. This can be extended by using depth data to compute the robot's relative orientation with respect to the stair and introducing it into the robot's gait plan.

References

1. S. Hirose and Y. Umetani, The basic motion regulation system for quadruped walking vehicle, in *Proc. ASME Int. Design Engineering Technical Conference (IDETC'80)*, (Beverely Hills, CA, 1980).
2. S. Hirose, B. Poneda, K. Arai, and T. Ibe, Design of prismatic quadruped walking vehicle TITAN VI, in *Proc. Int. Conference on Advanced Robotics*, (Pisa, Italy, 1991), pp. 723-728.
3. M. Buehler, R. Battaglia, A. Cocosco, G. Hawker, J. Sarkis, and K. Yamazaki, SCOUT: A simple quadruped that walks, climbs, and runs, in *Proc. IEEE*

Int. Conference on Robotics and Automation, (Leuven, Belgium, 1998), pp. 1707-1712.

4. S. Talebi, M. Buehler, and E. Papadopoulos, Towards dynamic step climbing for a quadruped robot with compliant legs, in *Proc. 3rd Int. Conference on Climbing and Walking Robots*, (Madrid, Spain, 2000).
5. D.J. Pack, Perception-based control for a quadruped walking robot, in *Proc. IEEE Int. Conference on Robotics and Automation*, (Minneapolis, MN, 1996), pp. 2994-3001.
6. H. Lee, Y. Shen, C-H. Yu, G. Singh, and A.Y. Ng, Quadruped robot obstacle negotiation via reinforcement learning, in *Proc. IEEE Int. Conference on Robotics and Automation*, (Orlando, Florida, 2006), pp. 3003-3010.
7. V-G. Loc, S-g. Roh, I.M. Koo, D.T. Tran, H.M. Kim, H. Moon, and H.R. Choi, *Rob. and Aut. Sys.* **58**, (2010).
8. C-C. Ko, S-C. Chen, C-H. Li, and P-C. Lin, Trajectory planning and four-leg coordination for stair climbing in a quadruped robot, in *Proc. IEEE/IRSJ Int. Conference on Intelligent Robots and Systems*, (Taipei, Taiwan, 2010), pp. 5335-5340.
9. K. Byl and R. Tedrake, Dynamically diverse legged locomotion for rough terrain, in *Proc. IEEE Int. Conference on Robotics and Automation*, (Kobe, Japan, 2009), pp. 1607-1608.
10. A. Shkolnik, M. Levashov, I. R. Manchester, and R. Tedrake, *Int. J. Rob. Res.*, **30**, 2 (2011).
11. E.Z. Moore, D. Campbell, F. Grimminger, and M. Buehler, Reliable stair climbing in the simple hexapod 'RHex', (Washington, DC, 2002), pp. 2222-2227.
12. J-S. Gutmann, M. Fukuchi, and M. Fujita, Stair climbing for humanoid robots using stereo vision, in *Proc. IEEE/IRSJ Int. Conference on Intelligent Robots and Systems*, (Sendai, Japan, 2004), pp. 1407-1413.
13. P. Michel, J. Chestnutt, S. Kagami, K. Nishiwaki, J. Kuffner, and T. Kanade, GPU-accelerated real-time 3D tracking for humanoid locomotion and stair climbing, in *Proc. IEEE/IRSJ Int. Conference on Intelligent Robots and Systems*, (San Diego, CA, 2007), pp. 463-469.
14. A.I. Mourikis, N. Trawny, S.I. Roumeliotis, D.M. Helmick, and L. Matthies, *Int. J. Rob. Res.*, **26**, (2007).
15. A. Kalantari, E. Mihankhah, S. Ali, and A. Moosavian, Safe autonomous stair climbing for a tracked mobile robot using a kinematics based controller, in *Proc. IEEE/ASME Int. Conference on Advanced Intelligent Mechatronics*, (Singapore, 2009), pp. 1891-1896.
16. S.A. Stoeter, P.E. Rybski, M. Gini, and N. Papanikolopoulos, Autonomous stair-hopping with scout robots, in *Proc. IEEE/IRSJ Int. Conference on Intelligent Robots and Systems*, (Lausanne, Switzerland, 2002), pp. 721-726.
17. S.D. Herbert, A. Drenner, and N. Papanikolopoulos, Loper: A quadruped-hybrid stair climbing robot, in *Proc. IEEE Int. Conference on Robotics and Automation*, (Pasadena, CA, 2008), pp. 799-804.

BENEFITS AND DRAWBACKS OF MRF KNEE DAMPING IN LEGGED ROBOTS

G. Muñoz*, M. Cestari, J.C. Arevalo, D. Sanz-Merodio and E.Garcia

Centre for Automation and Robotics, UPM-CSIC
La Poveda, 28500 Madrid, Spain
http://www.car.upm-csic.es/control
**E-mail: gustavo.munoz@csic.es*
**Site: http://www.gustavo.x10.mx*

In order to create efficient legged robots it is needed to care about energy consumption. During motion, energy is used in two different ways: positive energy to generate movement, and negative energy to brake movement. The use of passive components to dissipate the energy by friction could be a possible solution to avoid braking with active elements. This paper deals with the problem of determining experimentally the benefits of damping the knee by absorbing negative energy in MRF dampers, against actively braking using the actuators.

Keywords: agile quadruped, magnetorheological, damping, energetic efficiency

1. Introduction

Broadly speaking, legged locomotion can be considered as the cyclic combination of two phases: swing and stance, where negative and positive mechanical work is performed. Therefore, two different energy consumptions arise in the walking cycle: a positive consumption to generate movement, and a negative consumption to brake inertias and absorb perturbations. To perform an improvement in energy efficiency, some authors propose damping the knee using passive elements to dissipate energy, instead of active elements.[1]

However, when analyzing the damping requirements at the knee, both by simulation,[2] and by studying biological data, the conclusion is that the knee damping coefficient required is variable during the gait cycle. One of the more recent solutions is the use of magnetorheological fluids (MRF) devices which can modify its damping coefficient dynamically.

The main motivation of this work is to determine if the energy consumption is larger by dissipating it actively in the motor or by using an

MRF device to absorb part of it (considering its own energy consumption), working together with the motor (comparative energy balance of both devices). This article is focused on evaluating the benefits and drawbacks of using this kind of magnetic device in legged robots, testing them using the HADE2 robotic leg.[3]

Section 2 describes the magnetorheological theory, section 3 describes the test bench used to develop the experiments (HADE2 robotic leg) and the MRF knee device used, section 4 describes the experiments developed; and finally section 5 and 6 are the discussion on improvements and conclusions respectively.

2. MRF Damping

Magnetorheological Fluids (MRF) are smart fluids that can modify their viscosity thanks to the ferromagnetic particles in suspension, which react to magnetic fields adopting chain structures (Fig. 1-a).

An MRF rotary damper is composed of a rotor, inside a housing which contains embedded coils used to create the magnetic field (Fig. 1-b), with MR fluid between both parts. Depending on the magnetic field, the fluid has four possible states (Fig. 2):

- With no magnetic field, the fluid behaves as a non-Newtonian and temperature dependent fluid that exhibits an off-state viscosity.
- With magnetic field and below the yield stress, the fluid behaves as a viscoelastic material.
- With maximum magnetic field, the magnetic material behaves like a solid.
- With maximum magnetic field and over its maximum yield stress, the fluid shows a viscous behaviour which depends on speed.

MRF devices are designed to work in one of these three modes: flow mode (Fig. 3-a) or valve mode, shear mode (Fig. 3-b) and squeeze mode (Fig. 3-c).[4] This last working mode is not going to be described because no application is related to it in this article.

These modes depend on the mobility of the device parts relative to the fluid and, the fluid itself. The grey shaded parts represent the fixed parts, the dotted part represents the MR fluid and the white parts, the mobile parts.[5] The resistive strength depends on the off-state viscosity of the fluid itself, and on the magnetic field applied. The off-state viscosity depends on the MR fluid particle density, and the binding substance.

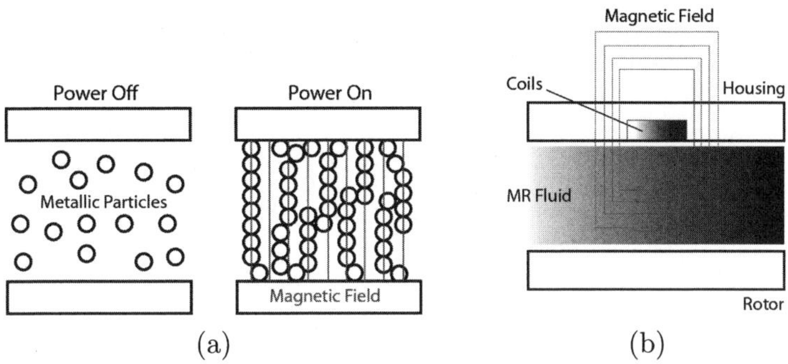

Fig. 1. Magnetorheological Fluid Behaviour. (a) Ferromagnetic Particles Behaviour under a Magnetic Field. (b) Mechanical Design of a Basic MRF Device

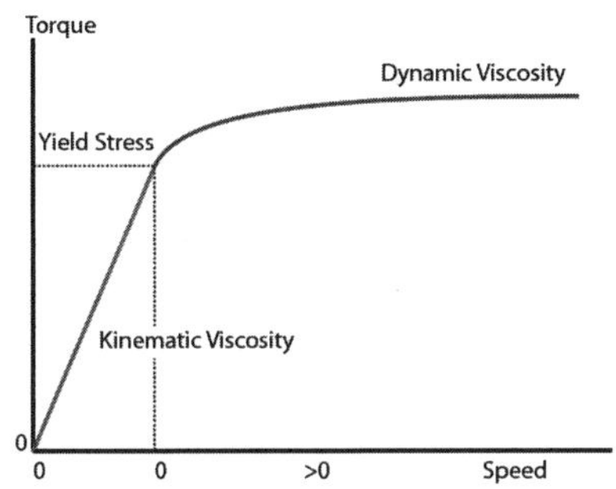

Fig. 2. MRF Damper Speed-Torque Curve

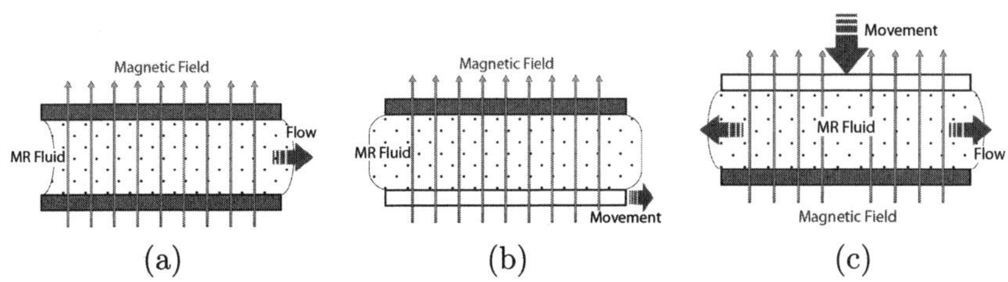

Fig. 3. Working Modes of the three Possible Configurations for MRF Devices. (a) Flow Mode or Valve Mode. (b) Shear Mode. (c) Squeeze Mode.

Flow mode or valve mode pressure drop is described in Eq. (1),

$$\Delta P = \Delta P_R + \Delta P_{MRF} = \frac{12 \cdot \eta \cdot Q \cdot l_g}{h_g^3 \cdot w_g} + \frac{f \cdot \tau_{MRF} \cdot l_g}{h_g} \qquad (1)$$

where ΔP_R is the pressure component depending on the rheological properties of the fluid, ΔP_{MRF} is the pressure component depending on the magnetorheological properties of the fluid, l_g, w_g and h_g are the dimensions of the gap throughout the fluid flows, τ_{MRF} is the yield strength, Q is the flow rate, η the dynamic viscosity and f that is a factor determined experimentally. This mode is present in linear dampers as the model shown in Fig. 4.

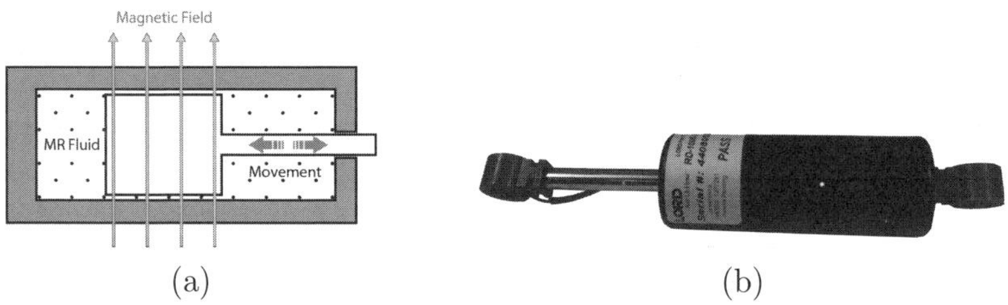

Fig. 4. Linear damper working in flow mode. (a) Schematic representation. (b) LORD MRF Linear Damper (RD-1005-3)

Shear mode force is described in Eq. (2). This mode is present in rotational brakes as the model shown in Fig. 5,

$$F = F_R + F_{MRF} = \frac{\eta \cdot S \cdot A}{h_g} + (\tau_{MRF} \cdot A) \tag{2}$$

where F_R is the force component depending on the rheological properties of the fluid, F_{MRF} is the force component depending on the magnetorheological properties of the fluid, S is the relative speed, h_g is the height of the gap throughout which the fluid flows, η the dynamic viscosity, τ_{MRF} is the yield strength developed in response to the applied magnetic field and $A = l_g \cdot w_g$ is the working interface area.

3. Experimental Set-up

3.1. *Robotic Leg Test Bench*

The HADE2 robotic leg is the second prototype of a research project for agile quadruped locomotion, and it is a three degree of freedom planar leg with two Series Elastic Actuators by Yobotics, one at the hip and one at the knee, and a spring for passively moving the ankle. The two SEA motors are composed of a linear ball screw moved by an electric motor and an elastic element between the motor and the load that allow us to measure force

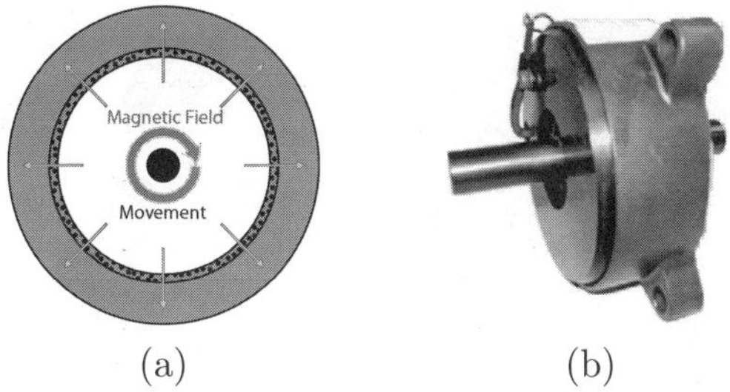

Fig. 5. Rotary damper working in shear mode. (a) Schematic representation. (b) LORD MRF Rotational Damper (RD-2087-01)

according to the Eq. (3), where k_{spring} is the elastic constant and x_{sensor} is the deflection of the elastic component.

$$F = -k_{spring} \cdot x_{sensor} \tag{3}$$

The knee joint has been designed to enclose a rotary MRF damper as seen in Fig. 6, because it has been demonstrated through animal observation and simulation studies (Fig. 7), that during most of the gait cycle, the energy consumed at the knee is negative, being used for braking instead of generating active movement. This is the reason to try to dissipate as much of this negative energy as possible into a passive device.

The locomotion controller runs on a PXI platform by National Instruments. The controller is based on a state machine which shifts from one gait phase to the following. The gate cycle of a leg is divided into three states as seen in Fig. 8.

Toe-off The leg applies force to leave the ground.
Swing The leg brakes the inertias and prepares itself to land again.
Stance The foot is on the ground again propelling the robot body forward and prepares itself to reach the next phase: toe off.

During each state, the articular positions of the leg are controlled using a compliance scheme shown in Fig. 9. In this control scheme, the compliance block is a viscous damping composed of a spring-damper system described in the Eq. (4), where k is the spring elastic constant, x is the spring deflection, b is the damping coefficient and F is the resultant force.

$$F = k \cdot x + b \cdot \frac{d(x)}{d(t)} \tag{4}$$

The PID block is a conventional proportional-integral-derivative controller that allow the leg reaches walking frequencies up to $1.3Hz$.

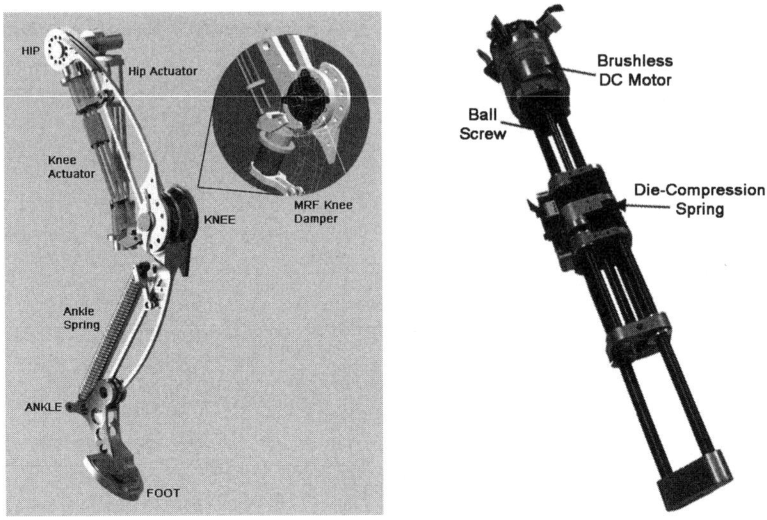

(a) (b)

Fig. 6. HADE2 Leg for an Agile Quadruped. (a) MRF Knee Damper Allocation. (b) Series Elastic Actuator Motor

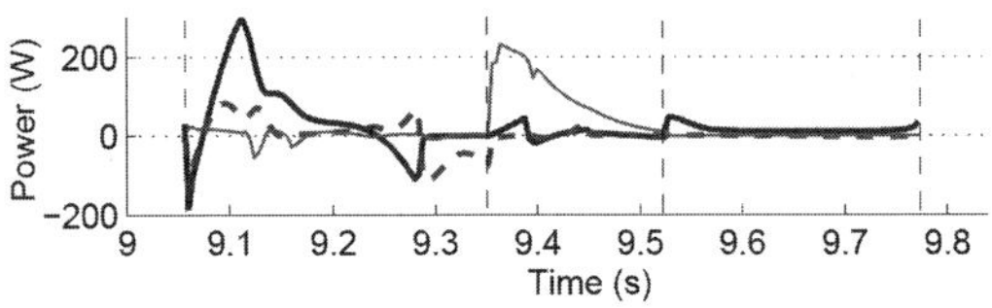

Fig. 7. Power during one Gait Cycle of an Agile Quadruped Running at 1.5 m/s Obtained from Simulation

4. MFR Knee Damper

MRF knee damping has been used mostly in prosthetic devices to adapt ground contact and make more realistic prosthetic limbs and increase stability when walking, damping the knee rotation in amputees,[6] or improving the energetic consumption in the gate cycle of legged robots.[7]

The MRF damper used (Fig. 5-b) is placed at the knee as a bearing to allow knee rotation (Fig. 6-a). This device is commanded with a current source (LORD WonderBox®) controlled by a continuous voltage signal between $0.5V$ and $5V$ that comes from a National Instruments PXI analogue output, to generate the magnetic fields in the MRF device coils.

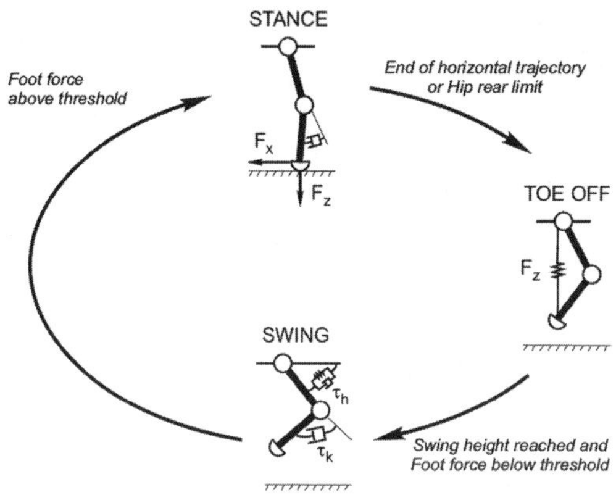

Fig. 8. Gait Cycle

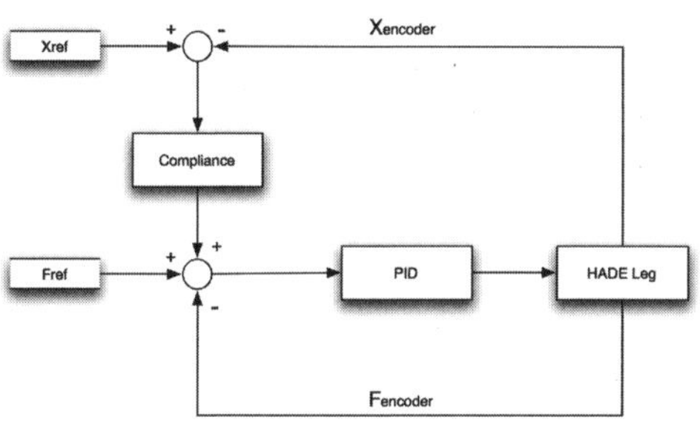

Fig. 9. Control Schema

5. Experiments

The experiments are conducted to obtain MRF damper and SEA power consumption in a gate cycle, in the knee joint. For these experiments a gait cycle is repeated at speeds between $1Hz$ and $1.2Hz$, and power is measured at the SEA, and at the MRF device. The compliance-controlled SEA acts as a spring-damper in two possible situations:

- Gait cycle without MRF damper $\Longrightarrow$ SEA energy.
- Gait cycle with MRF damper $\Longrightarrow$ SEA energy + MRF damper energy.

The idea is to use the MRF damper to substitute the SEA work when negative, trying to dissipate negative energy in a passive element by friction (replacing the emulated damper in the compliance scheme for a real device). This resultant torque is shown in Eq. (5).

$$T_{absolute} = T_{SEA} + T_{MRFdamper} \tag{5}$$

Where $T_{absolute}$ is the total torque against the movement applied to the joint, T_{SEA} is the braking torque generated by the SEA and $T_{MRFdamper}$ is the braking torque generated by the MRF damper.

The energetic improvement only takes place if the energy consumed by the MRF damper is lower than the energy consumed by the SEA doing the same braking work. These experiments are conducted to determine if this premise is true or not.

During the experiments the electric energy consumed by each device is measured. To measure the power consumed by the SEA, current through motor coils and voltage in motor terminals are measured and then processed using MatLab. The energy consumed by the MRF damper is calculated using the current signal and the resistance value of its coils with this equation: $P = V^2/R$.

The experimental results are represented in Fig. 10 along five locomotion cycles, where the peak consumption takes place during leg extension and the lowest consumption takes place during leg flexion as it is shown in the figure. The continuous and the dashed lines represent the consumption of the SEA with and without the action of the MRF damper, respectively. Due to the poor contribution of the damper, the energy consumed by the SEA is similar along the locomotion cycle in both cases. As expected, the starred-dotted line shows an increase of the total energy consumed when the damper consumption is added. This demonstrates that adding the action of the damper to the locomotion cycle increases the energy consumption of the leg.

6. Discussion on Improvements

The main problem of this MRF damper device is that it does not generate enough torque for a robot of this size. The torque at the knee due to inertias and weight are too much for the damper, and for torques in the range of the resistive damper torque, the energy consumed is similar or even greater than the one consumed in the motor, so the use of this kind of rotational devices is not efficient for these kind of applications. Some recent works[8] have improved the MRF device concept, increasing the share surface and of course the torque of the device, but at a cost of increasing the power consumption. Linear dampers have demonstrated to be stronger working

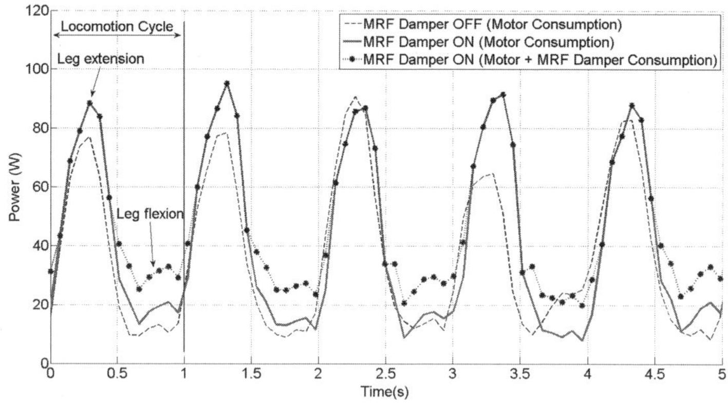

Fig. 10. SEA plus Damper Power Consumption

in flow mode,[9,10] so maybe the solution is to apply this working mode to new devices, instead of increasing the share surface, trying not to increase the power consumption.[11]

Another alternative solution may be to increase the resistive torque of the rotary damper through gearboxes. However the weight of the robot would increase too, yielding larger inertias to be damped. The farthest from the joint the weight is, the greater the inertias are, and even locating the weight in a good position, extra energy is needed to accomplish the movement, or even, bigger motors will be needed, adding even more weight.

7. Conclusions

New agile-robot limb designs are promising thanks to the new active and passive devices for movement and damping of the robots; but the current development of both is still insufficient for the specifications of the future robots. Weights and speeds are a problem trying to optimize energy consumption when working with these new devices, and in the future, these devices should be over dimensioned respect to the robot necessities, to avoid extra energy consumptions with respect to conventional devices.

The variable damping through magnetorheological devices is a possible line to develop more efficient robots because it is a natural way to act in biological organisms, through structures with variable stiffness; but for the moment, the development of these devices is limited to low scale devices with small torques, or for low amplitude vibrations. The low shear stress limit of the MR fluid in large-scale applications makes the damper being even more energetic inefficient than the motor alone, because it is working at maximum power to brake torques over its limits easily controlled by the motor itself, using less power than the MRF device, due to its structure.

New damper designs are needed for large-scale applications such as agile robots and exoskeletons.

Acknowledgements

This work has been partially funded by the Spanish National Plan for Research, Development and Innovation through grant DPI2010-18702 which has funded personnel costs and by AECID through grant PCI-Iberoamerica A1/039883/11 which has funded equipment and consumable costs.

References

1. A. Valiente, *Design of a quasi-passive parallel leg exoskeleton to augment load carrying for walking*, Tech. Rep. 2003, DTIC Document (2005).
2. E. Garcia, D. Sanz-Merodio, F. Sanchez, J. C. Arevalo and G. Muñoz, A lower-limb exoskeleton for gait assistance in quadriplegia, *System* , 1 (2011).
3. E. Garcia, J. Arevalo, G. Muñoz and P. Gonzalez-de Santos, Combining series elastic actuation and magneto-rheological damping for the control of agile locomotion, *Robotics and Autonomous Systems* **59**, 827(Octtober 2011).
4. M. R. Jolly, J. W. Bender and J. D. Carlson, Properties and Applications of Commercial Magneto-Rheological Fluids, *Journal of Intelligent Material Systems and Structures* **10**, 5(January 1999).
5. A. G. Olabi and A. Grunwald, Design and Application of Magneto-Rheological Fluid, *Materials & Design* **28**, 2658 (2007).
6. H. Herr and A. Wilkenfeld, User-adaptive control of a magnetorheological prosthetic knee, *Industrial Robot: An International Journal* **30**, 42 (2003).
7. S. a. Migliore, L. H. Ting and S. P. DeWeerth, Passive joint stiffness in the hip and knee increases the energy efficiency of leg swinging, *Autonomous Robots* **29**, 119(April 2010).
8. D. Senkal and H. Gurocak, Serpentine flux path for high torque MRF brakes in haptics applications, *Mechatronics* **20**, 377(April 2010).
9. M. Ahmadian and J. Norris, Experimental analysis of magnetorheological dampers when subjected to impact and shock loading, *Communications in Nonlinear Science and Numerical Simulation* **13**, 1978(November 2008).
10. F. Li, The modeling research of Magnetorheological Damper in advanced intelligent prosthesis, *2009 Chinese Control and Decision Conference* , 781(June 2009).
11. W. Li, G. Yao, G. Chen, S. Yeo and F. Yap, Testing and steady state modeling of a linear MR damper under sinusoidal loading, *Smart materials and structures* **9**, p. 95 (2000).

SURVEY OF SEMI-PASSIVE LOCOMOTION METHODOLOGIES FOR HUMANOID ROBOTS

Carlos Neves* and Rodrigo Ventura**

Institute for Systems and Robotics
Instituto Superior Técnico
1049-001 Lisboa, Portugal
**E-mail: cneves@isr.ist.utl.pt*
***E-mail: rodrigo.ventura@isr.ist.utl.pt*

Traditional methods for robotic biped locomotion employing stiff actuation display low energy efficiency and high sensitivity to disturbances. However, a recent trend based on the use of passive elements together with actuation has emerged, inspired by biological locomotion. This paper presents a survey of methodologies for semi-passive legged locomotion, focusing on foot design and ankle control solutions, granting adaptability to uneven terrains. The semi-passive, or hybrid, approach for biped locomotion allows for the (1) maximization of energy efficiency and (2) significant improvement of adaptability to external disturbances. By exploring the dynamics of Energy Storage and Return (ESAR) elements, like springs or rubber bands, it is possible to power joints by releasing stored energy instead of only using torque provided by electrical motors and to create gaits that minimize energy consumption. These elements also grant complacency to the robot which increases the gait's stability, specially on irregular terrains, and robustness to external disturbances.

Keywords: Humanoid robot, Legged Locomotion, Semi-Passive Locomotion, Semi-Passive Actuation, ESAR elements

1. Introduction

While most robotic tasks require the precision and speed that typical actuators provide, these mechanisms are not suitable for legged locomotion. The high torque values involved lead to high energy consumption and the high bandwidth behaviour results in sensitivity to small disturbances, increasing the natural instability of legged locomotion. Energy Storage and Return (ESAR) elements are usually elastic mechanisms, like springs, rubber bands or compliant sheets, that are able to deform and store energy that can be returned to power the joints, drastically reducing the power consumption of the actuators. Compliant actuation also transforms the joint into a low bandwidth system, which reduces the instability and absorbs vibrations from the foot-ground impacts.

2. ESAR Elements

Incorporating ESAR elements into the frame of a robot is a simple yet effective way of increasing stability and robustness of the platform, while reusing energy. Some groups have developed robots with passive locomotion, meaning that they require no actuation.[1,2] These robots are only able to walk down a slope and can't avoid obstacles, but some of the solutions can be adapted to actuated robots in order to improve energy efficiency.[3] Two candidate locations for these elements are the feet and knees of the robot. Some of the possible designs are:

- **Foot Pads** - Most mammals have some sort of padding underneath their feet, some in the form of fur others in the form of fat deposits. Adding a sole of rubber or springs to a robot's foot increases its stability by reducing the foot-ground forces generated on impact,[4,5] which also improves sensor reliability because of reduced vibrations. If coupled with sensors, the pads provide extra time for a controller to sense changes in ground slope and act on it, preventing a fall. The stiffness k of the padding should be such that it supports the weight of the robot when standing but also that is able to absorb the initial impact vibrations, suggesting a non-linear relation between force and displacement.[6]
- **Passive Toes** - The introduction of a compliant front section on a robot's foot, mimicking human toes, improves the stability and speed of walking gaits when compared to using a flat foot. Simulation results on HRP-2TJ[7] show an increased speed and a smoother gait due to a stage where the robot is balanced on the back foot toes previous to heel strike and HRP-2LT[8] is able to run at 3km/h and perform small jumps with no sensor feedback, since the compliance of the toes also provides an impulsive mechanism for the foot push off. Other robots are now being developed in order to accommodate this feature[9,10] in order to take advantage of the increased double support time.
- **Knee and Ankle Springs** - These joints are responsible for a large portion of energy consumption on a robot, both when standing and walking. Adding springs to these joints aims at mimicking the ligaments of the human body, providing both constrictions to the joint, such as not allowing lateral movement of the knee or limiting the leg swing, and supporting the muscles. Springs in the knee ensure that the motors will save energy

while the robot is standing and will store and reuse energy during the gait.[6] The ankle joint requires both the compliance and energy recycling that springs provide, in order to adapt to uneven terrain and increase the energy efficiency of the system. These approaches are useful not only for humanoid robotics but also for foot or knee prosthetics[a], where it is important to maintain a correct walking gait and reduce energy consumption.[11,12]

- **Body Compliance** - While body compliance does not affect locomotion directly, using a compliant exoskeleton, like Myon,[10] or adding a compliant vertebral column, like Acroban,[13] makes the robot more robust to external disturbances.

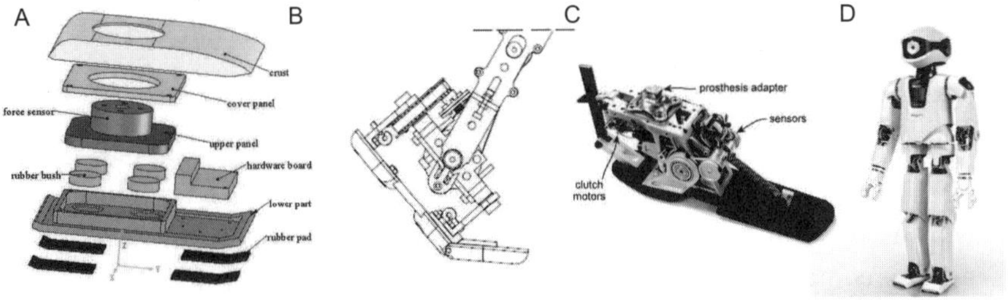

Figure 1. A: Flexible foot design[4] B: Passive Toes[9] C: Energy-recycling artificial foot[12] D: Myon Humanoid Robot[10]

3. Compliant Actuators

Although ESAR elements can reduce energy consumption and increase the stability and robustness of a humanoid platform, the actuators still play an important part on stability. Compliant actuators combine typical actuators with passive elements, allowing the absorption of small disturbances, improving the response time and increasing the energy efficiency of a walking gait. Some of the current actuators are:

- **Series Elastic Actuator (SEA)** - This is the simplest of compliant actuators, consisting in a spring in series with a stiff actuator, introduced in 1995.[14] By measuring the displacement L of the spring and by direct

[a]Knee Prosthetic: http://pofsea.org/programs/vietnam/local-prosthetic-and-orthotic-component-development/knee-joint/, retrieved in 09-04-2012

application of Hooke's Law it becomes possible to perform force control on a joint using position control of a typical actuator.

- **Antagonistic Controlled Stifness** - This family of actuators are based on a antagonistic setup, inspired by the human body, where each of the actuators pulls the load on a different direction. This allows to control not only the goal position of the actuator but also its stiffness, as long as the spring elements are non-linear. The Variable Stiffness Actuator (VSA-II)[15,16] and the Actuator with Mechanically Adjustable Series Compliance(AMASC)[17,18] are two of the approaches that can be found in literature.
- **Pneumatic Artificial Muscles (PAM)** - These actuators are usually employed in an antagonistic setup, where the inflation of a membrane will pull the load or joint, since the deflation of a membrane is more energy consuming. If both muscles are deflated, no torque is applied and the system is completely compliant; if both muscles are inflated, both apply force and the system is stiffer. The position of the joint is controlled by the difference in the inflation of the muscles. Although they can apply large torques and are naturally compliant, they have slow dynamics and the gas pressure control is rather complex, which explains why it is not a more common actuating method.[19–21]
- **Structure Controlled Stiffness** - The goal of this family of actuators is to change the compliance of a spring by altering its structure (width, length, number of active coils). Two members of this family are the Jack Spring Actuator[22] and the Mechanical Impedance Adjuster.[23,24] The Jack Spring takes advantage of the fact that the compliance of an helical spring's compliance depends on the number of active coils. By adjusting the position of a datum and the number of coils between the datum and the joint it is possible to control both position and compliance of the system (see Fig.3 B).
- **Mechanically Controlled Stiffness** - The Mechanically Adjustable Compliance and Controllable Equilibrium Position Actuator (MACCEPA)[25–27] and the VS-Joint[28] are two examples of a set of actuators where both equilibrium position and compliance are controllable. In this case, the compliance is controlled by changing the pretension of a spring. The MACCEPA actuator consists on three bars connected around a pivotal point: a fixed bar, a profile disc and an arm. The arm and the disc are also connected by a linear spring. The position of the disc and the pre-tension of the spring are controlled by two stiff actuators, allowing control of the position and compliance of the actuator independently.

The torque-angle curve is determined by the shape of the profile disc. It is possible to have a curve where the torque increases exponentially with displacement or even different curves, whether the displacement is positive or negative.

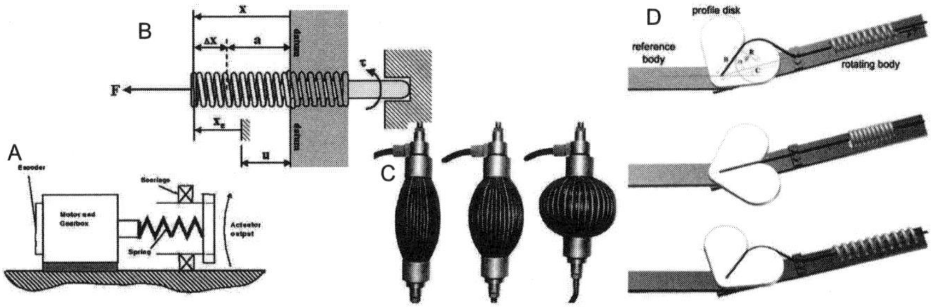

Figure 2. A: SEA working principle[14] B: The Jack Spring working principle[22] C: A PAM in different inflate stages[19] D:MACCEPA 2.0 with a profile disk in different position[27]

4. Control Methodologies

Tackling the legged locomotion problem using a semi-passive approach demands control methods that take into account the specificities of the humanoid platforms and the dynamics of the passive elements. Some control approaches that improve the benefits of the hardware changes are:

- **Adjusting Compliance** - By taking advantage of the dynamics of passive elements and of compliant actuators, the energy efficiency of a walking robot can be improved by jointly adjusting the compliance of the actuator and the walking speed of the robot.[29]
- **Attitude Estimation** - Walking locomotion displays a periodic movement, accompanied by several impacts, which add drift as well as vibrations and make attitude estimation difficult. Adapting an Extended Kalman Filter to work with ground-directed range sensors and the robot characteristics, the attitude estimations become both more robust and with faster convergence, which are then used to control locomotion in closed loop.[30]
- **Balance Recovery** - When facing external disturbances, a humanoid platform must be robust and able to react appropriately. Using Cognitive

Sensorimotor Loops, a well-proven sensorimotor control structure, it is possible to achieve and maintain a stable state for a set of joints by independently controlling each joint.[31]

- **Ground Slope Estimation** - Detecting ground slope is important in order to adapt the walking gait and maintain balance. If the robot lacks image processing, one common method to perform ground slope detection is to place force or pressure sensors in different points of the feet and control the ankle in order to obtain similar readings from all sensors.[32,33]

5. Conclusions

This paper presents a survey of current methods used for the semi-passive locomotion of humanoid platforms. The incorporation of ESAR elements and compliant actuators with adequate control methodologies allow these robots to improve energy efficiency, gain stability and become more robust to external disturbances.

Acknowledgments

Partially funded with grant SFRH/BD/80044/2011, from Fundação para a Ciência e a Tecnologia, and supported by the FCT [PEst-OE/EEI/LA0009/2011].

References

1. S. Collins and A. Ruina, A bipedal walking robot with efficient and human-like gait, in *Proceedings of the 2005 IEEE International Conference on Robotics and Automation, 2005*, April 2005.
2. D. Hobbelen and M. Wisse, Limit cycle walking, in *Humanoid Robots: Human-like Machines*, ed. M. Hanckel (AT: I-Tech Education and Publishing, 2007).
3. S. Anderson, M. Wisse, C. Atkeson, J. Hodgins, G. Zeglin and B. Moyer, Powered bipeds based on passive dynamic principles, in *2005 5th IEEE-RAS International Conference on Humanoid Robots*, December 2005.
4. J. Li, Q. Huang, W. Zhang, Z. Yu and K. Li, Flexible foot design for a humanoid robot, in *IEEE International Conference on Automation and Logistics (ICAL), 2008*, September 2008.
5. P. Zhang, C. Zhou, L. Zhang, Y. Tian and Z. Liu, Adaptive compliant control of humanoid biped foot with elastic energy storage, in *2009 IEEE/ASME International Conference on Advanced Intelligent Mechatronics*, 2009.
6. R. Alexander, Three uses for springs in legged locomotion, *International Journal of Robotics Research* **9**, 53(April 1990).
7. R. Sellaouti, O. Stasse, S. Kajita, K. Yokoi and A. Kheddar, Faster and smoother walking of humanoid HRP-2 with passive toe joints, in *Proceed-*

ing of the 2006 International Conference on Intelligent Robots and Systems IEEE/RSJ, 2006.

8. S. Kajita, K. Kaneko, M. Morisawa, S. Nakaoka and H. Hirukawa, ZMP-based biped runnning enhanced by toe springs, in *2007 IEEE International Conference on Robotics and Automation*, April 2007.
9. V. Santos, R. Moreira, M. Ribeiro and F. Silva, Development of a hybrid humanoid platform and incorporation of the passive actuators, in *Proceedings of the 2010 IEEE International Conference on Robotics and Biomimetics*, December 2010.
10. M. Hild, T. Siedel, C. Benckendorff, M. Kubisch and C. Thiele, Myon: Concepts and design of a modular humanoid robot which can be reassembled during runtime, in *Proceedings of the 14th Conference on Climbing and Walking Robots and the Support Technologies for Mobile Machines*, 2011.
11. J. Hitt, T. Sugar, M. Holgate, R. Bellman and K. Hollander, Robotic transtibial prosthesis with biomechanical energy regeneration, *Industrial Robot: an International Journal* **36**, 441 (2009).
12. S. H. Collins and A. Kuo, Recycling energy to restore impaired ankle function during human walking, *Public Library of Science ONE* **5** (2010).
13. O. Ly, M. Lapeyre and P. Oudeyer, Bio-inspired vertebral column, compliance and semi-passive dynamics in a lightweight humanoid robot, in *IEEE/RSJ International Conference on Intelligent Robots and Systems*, 2011.
14. M. M. Williamson, *Series Elastic Actuators*, tech. rep., Massachusetts Institute Of Technology - Artificial Intelligence Laboratory (1995).
15. G. Tonietti, R. Schiavi and A. Bicchi, Design and control of a variable stiffness actuator for safe and fast physical human/robot interaction, in *Proceedings of the 2005 IEEE International Conference on Robotics and Automation*, 2005.
16. R. Schiavi, G. Grioli, S. Sen and A. Bicchi, VSA-II: a novel prototype of variable stiffness actuator for safe and performing robots interacting with humans, in *Proceedings of 2008 IEEE International Conference on Robotics and Automation*, 2008.
17. J. Hurst, J. Chestnutt and A. Rizzi, *An Actuator with Mechanically Adjustable Series Compliance*, tech. rep., Robotics Institute, Carneggie Mellon University (2004).
18. J. Hurst, J. Chestnutt and A. Rizzi, The actuator with mechanically adjustable series compliance, *IEEE Transactions on Robotics* **26**, 597 (2010).
19. F. Daerden and D. Lefeber, Pneumatic artificial muscles: actuators for robotics and automation, *European Journal of Mechanical and Environmental Engineering* **47**, 10 (2000).
20. K. Inoue, Rubbertuators and applications for robots, in *Proceedings of the 4th International Symposium on Robotics Research*, (MIT Press, 1998).
21. F. Daerden, Conception and realization of Pleated Pneumatic Artificial Muscles and their use as compliant actuation elements, PhD thesis, Vrije Universiteit Brussel1999.
22. K. Hollander, T. Sugar and D. Herring, Adjustable robotic tendon using a 'jack spring', in *Proceedings of 9th International Conference on Rehabilitation Robotics, 2005*, 2005.

23. T. Morita and S. Sugano, Development of a new robot joint using a mechanical impedance adjuster, in *Proceedings of IEEE International Conference in Robotics and Automation*, 1995.
24. T. Morita and S. Sugano, Development of an anthropomorphic force-controlled manipulator WAM-10, in *Proceedings of 8th International Conference on Advanced Robotics*, 1997.
25. R. V. Ham, B. Vanderborght, M. V. Damme, B. Verrelst and D. Lefeber, MACCEPA: The mechanically adjustable compliance and controllable equilibrium position actuator for controlled passive walking, in *Proceeding of 2006 IEEE International Conference on Robotics and Automation*, 2006.
26. R. V. Ham, B. Vanderborght, N. G. Tsagarakis and C. S. andD. G. Caldwell, MACCEPA, the mechanically adjustable compliance and controllable equilibirum position actuator: Design and implementation in a biped robot, in *Proceeding of 2009 IEEE International Conference on Robotics and Automation*, 2009.
27. B. Vanderborght, N. G. Tsagarakis, C. Semini, R. V. Ham and D. G. Caldwell, MACCEPA 2.0: Adjustable compliant actuator with stiffening characteristic for energy efficient hopping, in *Proceeding of 2009 IEEE International Conference on Robotics and Automation*, 2009.
28. S. Wolf and G. Hirzinger, A new variable stiffness design: Matching requirements of the next robot generation, in *IEEE International Conference on Robotics and Automation ICRA2008*, 2008.
29. B. Vanderborght, B. Verrelst, R. V. Ham, M. V. Damme, P. Beyl and D. Lefeber, Development of a compliance controller to reduce energy consumption for bipedal robots, *Autonomous Robots* **24**, 419 (2008).
30. S. Singh and K. Waldron, Towards high-fidelity on-board attitude estimation for legged locomotion via a hybrid range and inertial approach, in *Experimental Robotics IX*, eds. M. Ang and O. Khatib, Springer Tracts in Advanced Robotics, Vol. 21 (Springer Berlin / Heidelberg, 2006) pp. 589–598.
31. M. Kubisch, C. Benckendorff and M. Hild, Balance recovery of a humanoid robot using cognitive sensorimotor loops (CSLs), in *Proceedings of the 14th Conference on Climbing and Walking Robots and the Support Technologies for Mobile Machines*, 2011.
32. K. Suwanratchatamanee, M. Matsumoto and S. Hashimoto, A simple tactile sensing foot for humanoid robot and active ground slope recognition, in *IEEE International Conference on Mechatronics (ICM) 2009*, April 2009.
33. K. Suwanratchatamanee, M. Matsumoto and S. Hashimoto, Haptic sensing foot system for humanoid robot and ground recognition with one-leg balance, *Industrial Electronics, IEEE Transactions on* **58**, 3174(August 2011).

SYSTEMATIC, BOTTOM-UP ROBOT DESIGN USING A BIOMECHANICAL EXPERIMENTAL METHODOLOGY

Jorgen Larsen, Kasper Stoy and David Brandt

The Maersk Mc-Kinney Moller Institute, University of Southern Denmark,
Campusvej 55, 5230, Odense, Funen, Denmark
E-mail: { jcla, kaspers, david.brandt }@mmmi.sdu.dk
http://www.sdu.dk/en/

Sten Grimmer and Martin Groß

Institut für Sportwissenschaft, Technische Universität Darmstadt,
Magdalenenstraße 27, 64289 Darmstadt, Deutschland
E-mail: { sgrimmer, gross }@sport.tu-darmstadt.de
http://www.lauflabor.de/

Using a bottom-up, model-free approach when building robots are often seen as a less scientific way, compared to a top-down model-based approach, because the results are not easily generalisable to other systems. We do, however, hypothesize that this problem may be addressed by using solid experimental methods. In this paper we will show how the use of well-known experimental methods from biomechanics are used to measure and locate weaknesses in our bottom-up, model-free implementation of a quadruped walker.

Keywords: LocoKit, Locomotion, Locomorph,

1. Introduction

Mechanical design of robots is often top-down, model-drive, that is, there is a specific kinematic model that the robot implements. This approach has the advantage that the resulting robot is controllable using powerful control methods. However, this also means that the robot is often unnecessarily complex and often poorly suited for physical interaction with its environment. Model-free, bottom-up mechanical design of robots is also possible and often leads to simpler designs and better interaction with the environment.[1] However, while research results from model-based systems can be generalized and used on robots with other kinematic structures, this is rarely the case for model-free robots because the designs are ad hoc. We believe, however, that this is not inherent to the model-free approach, but because that the experimental methods in robotics, with a few exceptions Ref. 2–4, tend to be poor, in particular, when it comes to understanding how specific elements of a mechanical system contributes to the whole.

To improve the understanding of robotic locomotion and to get deeper insight into details of the construction, we use an experimental approach from biomechanics suitable for analysing the kinematics and dynamics of the system. Such analysis measurements are well established in the field of animal and human locomotion for gait analysis. While these communities take advantage of these methods to understand the fundamental principles of gaits, it is also a valuable way for engineers to compare the performance of robots to that of animals.

The motivation for this paper is thus to understand how to combine a bottom-up, model-free approach to robot design with strong experimental methods taken from biomechanics in order to generate scientific results that are sound and generalizable to other platforms.

In section 2, we will introduce a robot called SpringyBot, which is inspired by ideas from the Scout, Ref. 5 by Poulakakis I. et al. and iSprawl, Ref. 1 by Kim, S. et al. In addition, we describe a typical gait analysis setup, which we use to evaluate the robot and to understand the mechanical constraints of our construction, which will be discussed in section 4. Section 3 will show results of our robotic gaits analysis, which we will conclude on in section 5.

2. Material and methods

This section describes firstly the robot SpringyBot that have been used during these experiments and hereafter the test set-up used to make measurements on the robot during the experiments.

2.1. *The SpringyBot robot*

We have used our robotic construction kit, LocoKit - Refs. 6,7, to build the quadruped robot that was used during all experiments - see Fig. 1(a).

The aim of the LocoKit construction kit is to enable us to build small, robust and light-weight legged robots. LocoKit should also provide the ability to make local adjustments on the skeleton of the robot, and, in general, allow us to build bottom-up, model-free robots with limited need for a mathematical model to describe the system beforehand.

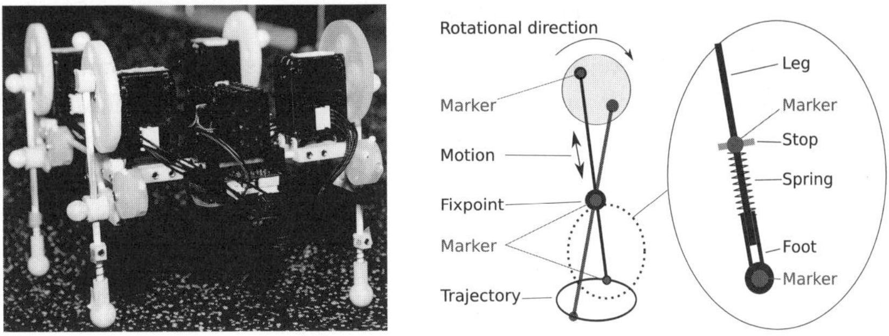

(a) The SpringyBot built from LocoKit.

(b) The leg set-up where the leg is driven by the top disk, and thereby generates the trajectory shown.

Fig. 1. In figure (a) the leg setup is shown. To be able to track the motion of the foot, IR markers (red) have been placed at the positions shown. The leg have been inspired by concepts found i the two robots; Scout Ref. 5 by Poulakakis I. et al. and iSprawl, Ref. 1 by Kim, S. et al. Our design is made as lightweight as possible in order to keep the mass of the leg to a minimum. In (b) the configuration of the tested robot is shown. This robot is inspired by the biomechanical SLIP model.

Our robot implements legs that utilise a setup, where each leg is run by one motor, that by a crank mechanism makes is work as shown in Fig. 1(b). By implementing a spring in the foot of the leg, we hope to get elastic leg behaviour similar to what is seen in a Spring Loaded Inverted Pendulum (SLIP) model, Ref. 8.

Due to the simplicity of the mechanics with only one motor per leg, the control scheme used to run the motors is only to ensure that the motors are running with a constant speed, and that there is a fixed synchronisation between the two pairs of motors - "RF-LH" and "LF-RH".[a] The controller makes a zero degree phase shift internally in a motor pair, and a 180 degree phase shit between each pair. This synchronisation between the motor-pairs gives the robot a trot gait pattern. The controller is running on a small ARM7 embedded micro controller on the robot and the robot is powered by batteries to make it autonomous. To be able to control it from a PC, an Xbee module has been used, both on the robot and on the computer side. We are able to send high-level commands to the robot, like stop/start and/or make changes to the synchronisation between the motors on the fly as well as changing the speed.

[a]R=Right, L=Left, F=Front, H=Hind

2.2. *Robotic gait analysis and data analysis*

Gait analysis was used to standardize and to permit us to make comparisons between the robot's gait and animal gaits. For the gait evaluation a track consisting of three force plates (two 60 cm x 50 cm, Typ 9260AA6; one 30 cm x 50 cm, Typ 9260AA3, Kistler, Winterthur, Switzerland) was introduced, see Fig. 2.

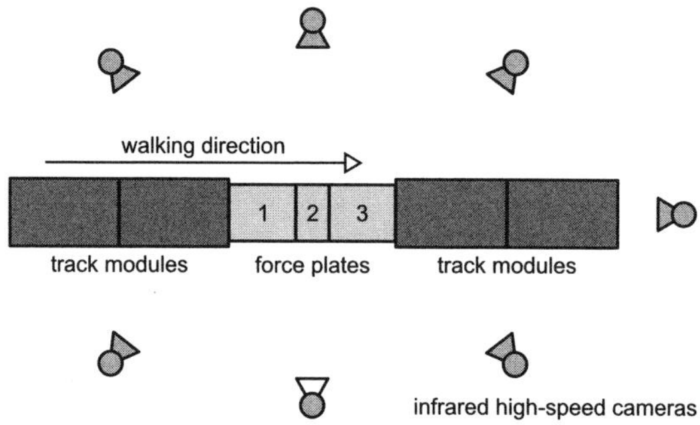

Fig. 2. The setup for the robotic gait analysis. The track consists of three force plates in row with two wooden track modules at both ends. Between two 60 cm x 50 cm force plates one smaller 30 cm x 50 cm force plate was installed. The force plates as well as the track modules were covered with carpet. For motion capturing 7 infrared high-speed cameras were installed around the track. One camera was recording high-speed videos of the robots in parallel to automatic motion capturing.

The kinematics were recorded using seven high-speed cameras (Oqus series, Qualisys, Gothenburg, Sweden) at 250 Hz, see Fig. 2. Force plates and cameras were triggerd for simultaneous measurements. In addition the data was synchronized due to time lag and time drift following the suggestions of Lipfert et al. Ref. 9.

From the force data and the kinematics, the touch-down and take-off events were separated in order to define the gait cycle, i.e. from one touch-down of one leg to the next touch-down of the same leg, as done by O'Connor et al. Ref. 10.

From the markers we calculated leg lengths with respect to the leg features spring and rest length as well as the leg angle, i.e. the angle of the leg with respect to the ground.

2.3. *Experiment protocol*

In our experiments we had the robot walking with a slow trot gait on the test track while being monitored by the cameras and force plates as shown in Fig. 2. For each walk over the track, a new set of measurements was taken. What we hope to see is a natural behaviour of the leg that is comparable to e.g. a dog in terms of phase plots of leg-length vs. leg-angle.

3. Results

In Fig. 3 the results on the leg behaviour are shown as a phase plot between the leg length and the leg angle with respect to ground. As seen, the trajectory of the legs does not show elastic behaviour, as observed on dogs, where the leg is shortened towards mid-stance, followed by a lengthening towards take-off (TO). In all cases, the leg length of SpringyBot is lengthened towards mid-stance and shortened towards take off, which is opposite of what we had expected.

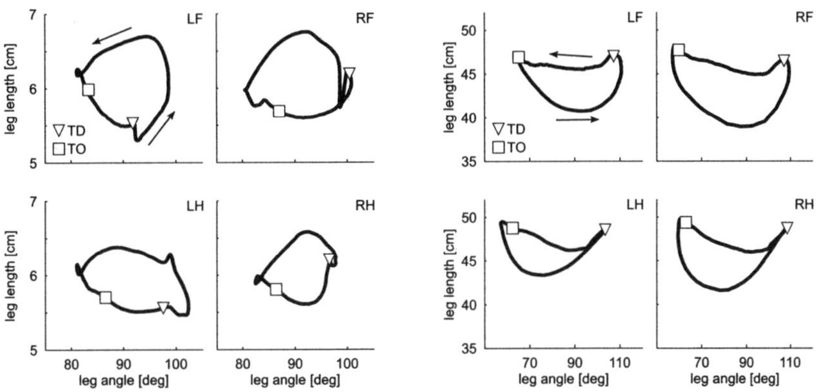

(a) Phase plot for SpringyBot

(b) Phase plot for trotting dog.

Fig. 3. Phase plot showing the length of the leg to a specific angle for SpringyBot and a trotting dog (TD: Toutch down, TO: Take off)

When measuring the lengths of the spring, leg-segment, and total leg directly as shown in Fig. 4, it is seen that the spring is compressed from the beginning of the stance phase to the end.

4. Discussion

Our goal from the beginning was to test our simple bottom-up approach in designing a walking robot with inspiration from the SLIP model and nature, and then see how this physical implementation performs compared

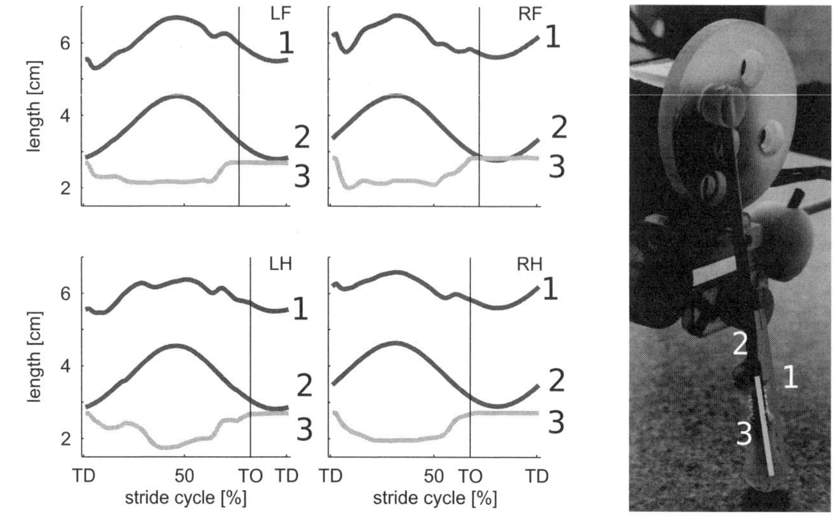

Fig. 4. Shows lengths of the specific parts of the leg. The green curve shows how the spring is fully compressed during most of the stance phase.

to similar experiments performed with real animals. The overall reason for doing this is: (1) To find out how well our implementation can be used as a simplification of real walking in animals where the mechanics is much more complex compared to our robot, and (2) To see if our bottom-up, model-free approach can, by the use of solid experimental methods, be a faster way of getting experimental data that can be used to improve the system, in contrast to ordinary model based design.

The robot used in this experiment was not built based on specific information regarding width, length or height, but based on intuition.

From the results in our first experiments in Fig. 3 it is seen that the cyclical pattern of the leg length vs. angle, is very different from what is observed on e.g. a dog. The difference is that when the leg of a trotting dog is getting shortened toward mid-stance, and lengthened towards take-off, the opposite is true in the robotics experiments. This could imply that the spring in the feet of our robot is either too short or too soft. This observation is also confirmed when looking at the data in Fig. 4. What we see here is three plots, each showing the length of different parts of the leg. The most interesting plot is the length of the spring during a complete stride (green (3)). What is seen is, that the spring is fully compressed during contact phase, and therefore it is not possible to get the desire behaviour where the leg is shortened towards mid-stance and lengthened towards take-off.

This can to some extend explain why the leg length in Fig. 3(a) did not follow the expected curve, because the motor will instead of compressing the spring during stance, push against the ground and therefore, due to the mechanical solution, lift the robot during stance.

By using the solid experimental methods described, we gained knowledge that helped us in coming to this results, which, without would have been more difficult. The methods described from biology and biomechanics are well tested and shows exactly what is important when dealing with locomotion e.g. leg and body dynamics. Even though these methods have been developed to be used on either animals or humans, the vital and interesting information about the kinematics and forces are still the same, and therefore it allows us to make direct comparisons to e.g. a dog in this case.

By the use of solid experimental methods we have gained more than easing the process of pinpointing the problems in the structure, we are also able to generalise our results to other platforms. This is because, that even though we are using and making experiments on ad-hoc designs, we are looking at specific elements as e.g. spring compression in the leg, which is interesting among different platforms. We hope that form our experimental data we will be able to suggest optimal springs for future leg designs and with the use of LocoKit as robotic platform, the suggestion can easily be tested in a robot.

5. Conclusion

In this paper we have shown how to use used a structured scientific method to test our bottom-up, model-free walking robot. Our results show, that by the use of this method it was easy for us to locate the week point of the mechanical design of the robot, and come up with possible solutions.

We feel that the experimental process with the use of solid experimental methods is highly beneficial for our future studies of walking robots, because we get systematic results regarding specific mechanical parts. This means, that even though our robot is build using a model-free, bottom-up approach we still can generalize our results to other robots and thereby make our results useful to other researchers as well.

6. Acknowledgments

The research leading to these results has received funding from the European Community's Seventh Framework Programme FP7/2007-2013 - Future Emerging Technologies, Embodied Intelligence, under grant agrement no. 231688.

References

1. S. Kim, J. E. Clark and M. R. Cutkosky, *Int. J. Rob. Res.* **25**, 903(September 2006).
2. B. Andrews, B. Miller, J. Schmitt and J. E. Clark, *Bioinspiration and Biomimetics* **6**, p. 026009 (2011).
3. C. Li, A. Hoover, P. Birkmeyer, P. B. Umbanhowar, R. S. Fearing and D. I. Goldman (2010).
4. K. Nakatani, Y. Sugimoto and K. Osuka, **23**, 483(March 2009).
5. *Experimentally validated bounding models for the scout ii quadrupedal robot* (April), (IEEE, 2004). Proceedings. ICRA04. 2004 IEEE International Conference on Robotics and Automation.
6. *Energy Efficiency of Robot Locomotion Increases Proportional to Weight* (0), 2011). Proceedings of the 2nd European Future Technologies Conference and Exhibition 2011 (FET 11).
7. *LocoKit - A Construction Kit for Exploration of Morphology of Legged Robots* 2011. Proceedings of The International Symposium on Adaptive Motion in Animals and Machines 2011.
8. R. Blickhan, *Journal of Biomechanics* **22**, 1217 (1989).
9. S. W. Lipfert, M. Günther and A. Seyfarth, *Journal of Biomechanics* **42**, 786 (2009).
10. C. M. OConnor, S. K. Thorpe, M. J. OMalley and C. L. Vaughan, *Gait and Posture* **25**, 469 (2007).

FEEDBACK CONTROL EXPERIMENTS WITH THE PARKOURBOT

A. Long[§] [*], A. Degani[⊥], and K. Lynch[§] [^]

[§] *Northwestern University, USA*
E-mail: {[*] awlong, [^] kmlynch}@northwestern.edu
[⊥] *Technion - Israel Institute of Technology, Israel*
E-mail: adegani@tx.technion.ac.il

The ParkourBot is a planar dynamic climbing biped robot and is capable of dynamically locomoting between two "vertical" walls in a reduced-gravity planar environment of an inclined air table. During flight, springy legs are compressed to store elastic energy. At impact, this potential energy is quickly converted to kinetic energy. In this paper, we demonstrate that simple feedback control strategies with a high-speed vision system can be used to control the ParkourBot to bounce in place, climb up and climb down.

1. Introduction

The ParkourBot is the planar dynamic climbing biped robot shown in Fig. 1(a). During flight, the springy legs are compressed to store elastic energy. At impact, this potential energy is quickly converted to kinetic energy. By controlling the leg angles and the injected energy, the ParkourBot is capable of dynamically locomoting between two "vertical" walls in a reduced-gravity planar environment of an inclined air table. In this paper, we demonstrate that simple feedback control strategies can be used to control the ParkourBot to locomote, similar to the cartoon in Fig. 1(b).

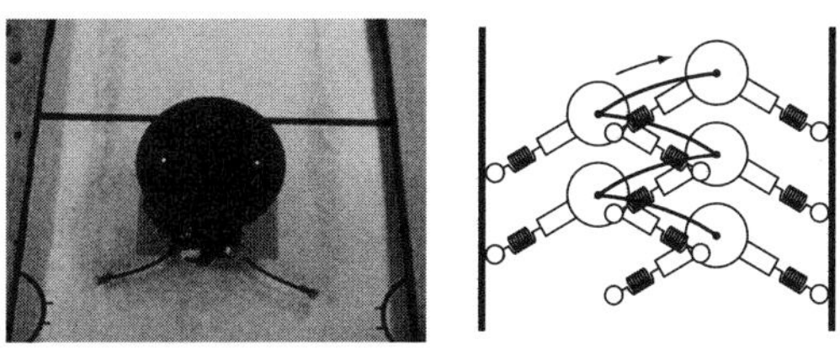

(a) (b)

Fig. 1. (a) Image of the ParkourBot, a planar biped climbing robot, on an inclined air table. (b) A cartoon of the ParkourBot ascending a chute.

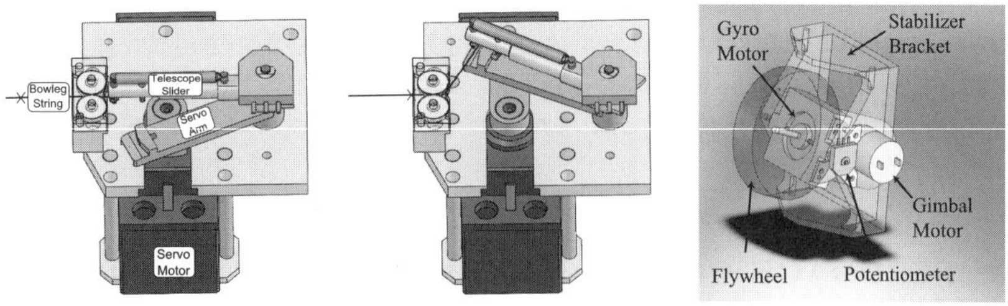

(a) (b) (c)

Fig. 2. The retract mechanism alternates between pushing the string up (a) and down (b) to store energy in the leg. (a) and (b) are reprinted from the work by Degani et al.[1] ©[2011 IEEE]. The gyrostabilizer (c) is used to minimize body rotation.

There are three different types of control for the ParkourBot: the leg angle control, the leg retract control and the body angle control. To minimize the body torque at impact, we fixed the leg angle control to place the foot, hip and center of mass along a line. With a high-speed vision system, we track the robot's position and orientation by recording the position of two LEDs mounted on the ParkourBot. As will be shown in this paper, by controlling the body angle with a gyrostabilizer and the injected energy with the leg retract control we can use simple feedback control strategies with the vision data to control the ParkourBot to bounce in place, climb up and climb down.

The outline of this paper is as follows. Sec. 2 discusses related work. The ParkourBot's design is detailed in the work by Degani et al.[1] and is reviewed briefly in Sec. 3. Sec. 4 motivates use of a gyrostabilizer with vision feedback to maintain the robot's body orientation. Sec. 5 describes our general feedback control strategy of the leg retract control. Finally, Sec. 6 demonstrates experimentally that feedback control with a vision system can be used to control the ParkourBot to follow a reference trajectory.

2. Related Work

In previous work, hopping and running has been achieved for many robots using both open-loop and closed-loop control. Several examples of experiments using open-loop control include Ringrose's monoped,[2] Cham and Cutkosky's dynamic Sprawl robots,[3] and Degani's et al. DSAC[4] and DTAR.[5] By relying on asymptotic limit cycles, one can apply a sequence of calculated open-loop controls repetitively to achieve a desired form of locomotion. Open-loop control has been applied to the ParkourBot;[1]

however, due to errors in modeling and implementation, with open-loop control it is difficult to obtain desired behavior such as foot placement and hopping height. Therefore, feedback control is required for improved locomotion.

Raibert and Hodgins' pioneering work on single leg hopping robots demonstrated that simple feedback control laws could be used to stabilize hopping on rough horizontal terrain.[6] The BowLeg hopper (predecessor to the ParkourBot) was also controlled with simple control laws to traverse a set of stepping stones.[7] Recently, an approximate solution to the well known spring-loaded inverted pendulum (SLIP) model has been proposed, resulting in a deadbeat controller for single leg hopping robots.[8] For the ParkourBot, we use a high-speed vision system to record the position of LEDs mounted on the body to obtain the orientation and position of the robot during locomotion. In this paper, we demonstrate that simple feedback control with a high-speed vision camera can be used to control the ParkourBot to bounce in place, climb up and climb down in a rigid chute.

3. Mechanical Design

This section reviews the mechanical design of the ParkourBot (presented in more detail in the work by Degani et al.[1]). The ParkourBot consists of two springy BowLegs and a body with a gyrostabilizer. The body is mounted on a 6 inch disc to float on an inclined air table. This forces the ParkourBot to operate in the plane of the table and reduces the effect of gravity. In this paper, we set the inclined table to approximately 10% of true gravity.

The BowLeg is fabricated with unidirectional fiberglass and is capable of storing large amounts of elastic energy. The BowLeg operates similar to an archer's bow. During flight, energy is stored in the leg by retracting a string attached to the foot. At impact, the string goes slack and a clutch disengages the string, allowing the leg to extend quickly. This converts the stored potential energy to kinetic energy, causing the robot to 'kick' off the surface. For each leg, note that the retract mechanism alternates between pushing the string upwards and downwards as shown in Fig. 2. For calibration purposes, this means that we need to determine two different controls for each leg that produce the same amount of retraction. An angle mechanism changes the leg angle with respect to the body.

The gyrostabilizer shown in Fig. 2(c) is utilized on the ParkourBot to minimize body rotation in the plane. It consists of a fast-spinning flywheel mounted on a single-axis gimbal. When torque is applied to the robot such as when a foot hits a surface, instead of the robot changing orientation the gyro pivots to a new angle to accommodate this change in angular momentum. An additional small motor and a potentiometer is attached to this pivot axle to allow for feedback control of the ParkourBot's body angle. In this paper, the motor powering the flywheel will be referred to as the gyro motor and the small pivot motor will be referred to as the gimbal motor. As will be shown in this paper, this feedback is necessary to maintain orientation since the gimbal drifts during operation of the gyro.

4. Body Angle Control

In this section, we motivate the use of active control with the gyrostabilizer. The trajectory in the plane of the robot's position without the gyrostabilizer and with constant retract controls can be seen in Fig. 3(a). The plot of the body angle as a function of time shown in Fig. 3(b) indicates that the robot's orientation changes significantly over only a few bounces, ultimately leading to falling. With a passive gyrostabilizer (gyro motor turned on and gimbal motor turned off), the robot does not tip over as quickly as shown in Fig. 3(c) and Fig. 3(d); however, the robot's orientation drifts over time. Therefore, gimbal motor correction is necessary to maintain the orientation.

We implemented a feedback controller to keep the body orientation within a range of $+/-3$ degrees. When the body angle drifts to the limits of this range, we apply a series of short pulses of fixed voltage and duration, determined experimentally, which has the effect of bringing the orientation to the other range limit. This results in the body orientation drifting from one limit to the other limit and quickly returning to the first limit. The trajectory in the plane and the body orientation as a function of time using this strategy can be seen in Fig. 3(e) and Fig. 3(f), respectively. This simple control maintains the body angle for many bounces but cannot maintain it indefinitely because when the gyro motor reaches one of the limits of the stabilizer bracket the gimbal motor can no longer change the gimbal angle in that direction. When this occurs, the robot usually falls over.

5. Injected Energy Control

This section describes our feedback control strategy for the injected energy control. The retract mechanism stores energy in the springy legs by pulling on a string attached to the foot. Ideally we would control the energy stored in the legs, but our actual control is a servo motor value. Therefore, we implemented a strategy with an inner loop and an outer loop as shown in Fig. 4(a). The outer loop calculates an injected energy to be applied on the

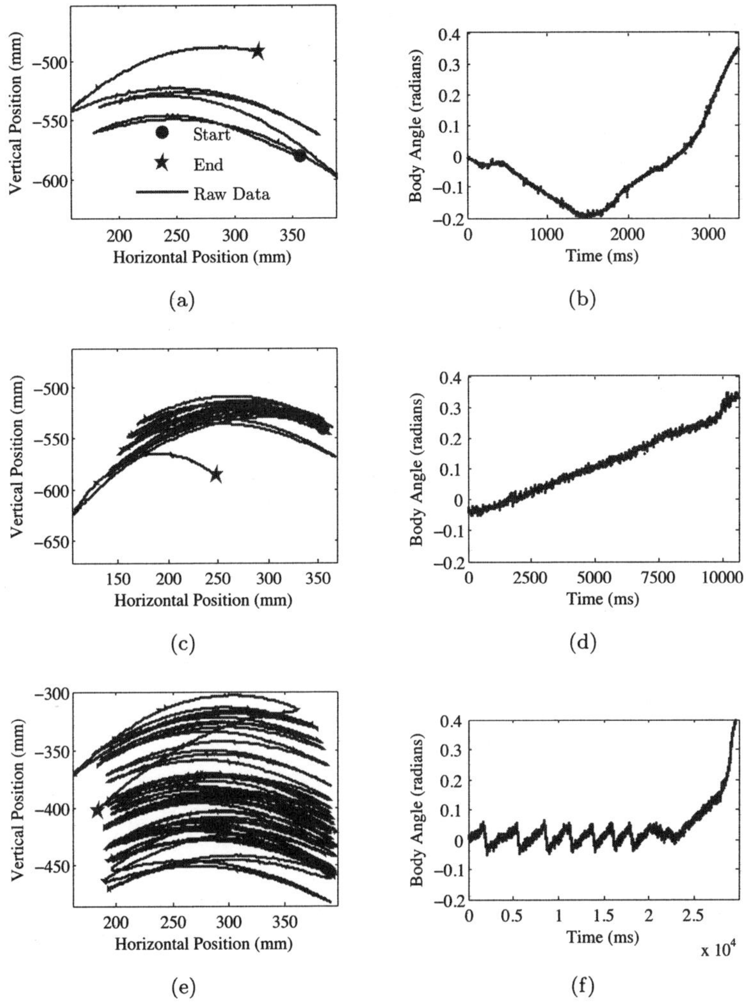

Fig. 3. Plots of the trajectory in the plane and the body angle as a function of time, of the ParkourBot hopping five times with no gyrostabilizer (a) - (b), 20 times with a passive gyrostabilizer (c) - (d) and 60 times with an active gyrostabilizer under vision feedback control (e) - (f).

next impact based on the difference between the next desired peak height and the recently estimated peak height. The inner loop computes the servo retract control based on the desired injected energy and an estimate of the previous injected energy. For each leg, the retract mechanism alternates

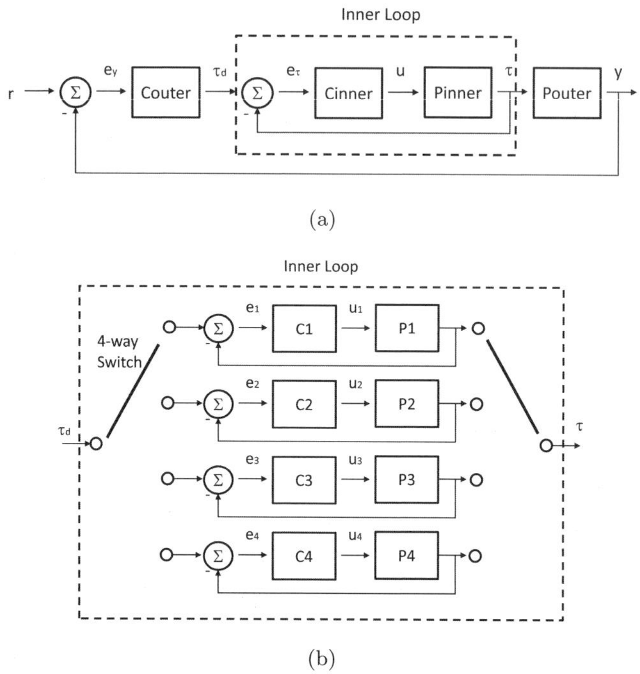

Fig. 4. (a) Feedback control block diagram with inner and outer control loops. (b) Feedback control inner loop for four switching controllers.

between pushing the string upwards and downwards as shown in Fig. 2(a) and Fig. 2(b), respectively. Therefore, our feedback strategy is to switch between four different inner control loops as shown in Fig. 4(b). Our strategy for estimating the peak state and the injected energy is based on the work by Long et al.[9] and a recursive least squares filter.[10]

To demonstrate the effect of the inner loop, we ran an experiment until the robot fell over with a constant desired peak height (bouncing in place) with no inner loop and a proportional-integral outer loop controller. Without the inner loop, the calculated injected energy control converges to a period-4 pattern as seen in Fig. 5(a). The peak height oscillates around the desired height as shown in Fig. 5(b).

With a proportional-integral controller for both the inner loop and outer loop, the peak heights can be seen in Fig. 6(b). The calculated injected energy control appears to be converging to a single injected energy value as shown in Fig. 6(a) which is expected for a bouncing in place behavior. The initial height was much lower resulting in the the relatively large overshoot.

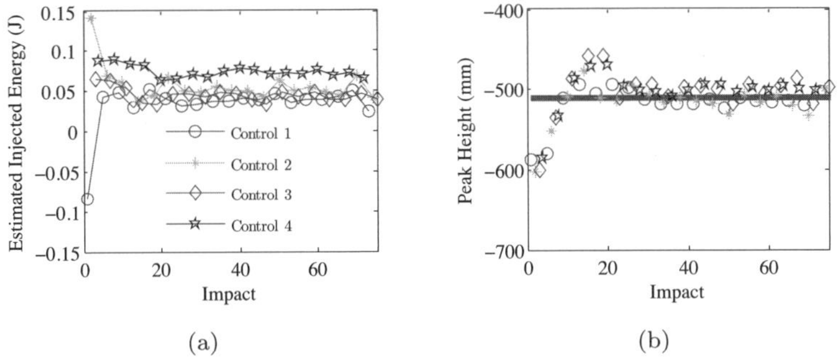

Fig. 5. Feedback Experiment with No Inner Loop: (a) has the calculated injected energy for each impact. (b) contains the estimated peak heights for each flight.

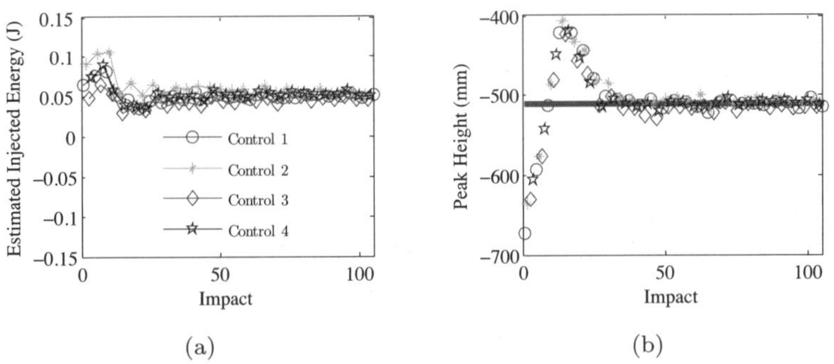

Fig. 6. Feedback Experiment with Four Inner Loops:(a) has the calculated injected energy for each impact. (b) contains the estimated peak heights for each flight.

6. Feedback Control Experiment

A goal of our experiments for the ParkourBot is to demonstrate that we can control it to bounce in place, climb up and climb down in a rigid chute. Fig. 7 contains the estimated peaks heights with a reference trajectory using the feedback control strategies described in Sec. 4 and Sec. 5. To reduce the overshoot after the climbing phases, we implemented minimum jerk reference trajectories to have the robot gradually reduce its momentum when approaching the bouncing in place phases.

7. Conclusion

This paper describes simple feedback controllers to control the ParkourBot to bounce in place, climb up and climb down on an inclined air table. Active control of the gyrostabilizer is used to maintain the body orientation

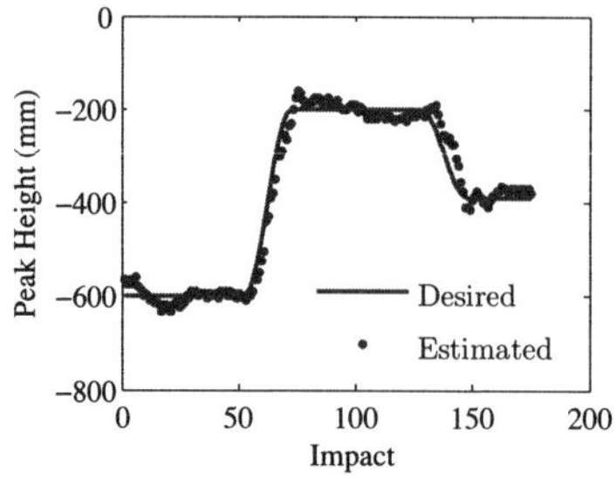

Fig. 7. The estimated peak heights as a function of flight number for a bounce in place, climb up, bounce in place, climb down and bounce in place reference trajectory.

during locomotion. Feedback control with an inner and outer loop allows us to control the peak heights of the robot's trajectory. In future work, we would like to control the ParkourBot to navigate through a complex set of footholds and these experiments are a first step towards this goal.

Acknowledgments

This work was supported by NSF IIS-0803734 and Andrew Long's NDSEG fellowship. The authors thank Adam Barber, Tsahai Phillip, Nelson Rosa, Ben Brown and Siyuan Feng for discussions about the ParkourBot.

References

1. A. Degani, S. Feng, H.B. Brown Jr., K. Lynch, H. Choset, and M. Mason, The ParkourBot - A Dynamic BowLeg Climbing Robot, in *IEEE Proc of Intl Conf on Robot and Auto*, 2011.
2. R. Ringrose, Self-stabilizing running, PhD thesis, MIT1997.
3. J. Cham and M. Cutkosky, *J of dynamic systems, measurement, and control* **129**, 275 (2007).
4. A. Degani, H. Choset and M. Mason, DSAC-dynamic, single actuated climber: Local stability and bifurcations, in *IEEE Proc of Intl Conf on Robot and Auto*, 2010.
5. A. Degani, H. Choset and M. Mason, *IEEE Trans on Robot* **27**, 360 (2011).
6. J. Hodgins and M. Raibert, *IEEE Trans on Robot and Auto* **7**, 289 (1991).
7. G. Zeglin and B. Brown, Control of a bow leg hopping robot, in *IEEE Proc of Intl Conf on Robot and Auto*, 1998.
8. U. Saranli, O. Arslan, M. Ankarali and O. Morgul, *Nonlinear Dynamics* **62**, 729 (2010).
9. A. Long, R. Gregg and K. Lynch, The Simplest Parkour Model: Experimental validation and stability analysis, in *Proc of the 14th Intl Conf on Climbing and Walking Robots*, 2011.
10. S. Haykin, *Adaptive filter theory* (Prentice Hall, NJ, 2003).

FOOT FORCE CRITERION FOR ROBOT STABILITY

MAHDI AGHELI, STEPHEN S. NESTINGER[†]

Mechanical Engineering Department, Worcester Polytechnic Institute
Worcester, MA 01609, USA

Stability analysis is a key component to controlling multi-legged robots especially when traversing irregular terrain. A variety of stability criteria exists in the field of multi-legged and multi-wheeled robotics. However, no criterion presents a quantitative stability extent that measures how close or far the robot is to the unstable or the maximum stable state. This paper presents the Foot Force Stability Margin which is applicable to all types of multi-legged and multi-wheeled robots. A modified extension of the Foot Force Stability Margin is provided and compared to a widely used stability margin through a numerical simulation. The comparison results show the accuracy, simplicity, sensitivity, and efficiency of the Modified Foot Force Stability Margin making it practical for use in on-line and real-time controllers.

1. Introduction

In terms of control, it is essential to monitor the stability of the robot at each instant through the use of a stability criterion while walking, moving or manipulating objects. There are several widely used stability criteria in the field of multi-legged and multi-wheeled systems. These criteria can be divided into static and dynamics-based criteria, and further classified into five categories based on the stability metric: distance, angle, energy, moment, and force.

Distance-based criteria [1-4] use either the distance between the support polygon and projection of the center of gravity [1-3] or the distance between the support polygon and the net force vector acting at the center of gravity [4] to measure the stability of the system. Angle-based criteria [5-8] use the angle between the support polygon and the net force acting at the center of gravity to represent the stability of the system. Energy-based criteria [9-13] look at the difference between the energy of the robot in two different situations: one in the current situation and one when tipping over. Most of the stability criteria fall under moment-based criteria [14-22] which calculate the moments about all possible tip over axes. Force-based criteria [4, 23-25] focus on the ground contact forces of all the feet of the robot. When tipping over, the foot forces,

[†] Corresponding Author.

except those defining the tip over axis or point (in planar robots [4]), become zero. Although there were some attempts to use this concept for robot control [23-25], there is no distinctly defined stability margin or extent in this category.

To the authors' knowledge, no criterion presents a quantitative stability extent that measures how close or far the robot is to the unstable or the maximum stable state. This paper presents the Foot Force Stability Margin (FFSM), a force-based criterion capable of presenting the stability of the whole system in an extent bounded by the unstable state and the maximum stable state. A Modified Foot Force Stability Margin (MFFSM) is provided to consider top-heaviness and geometrical sensitivities. A few studies comparing multiple stability criteria have been completed [21-22,26-28]. However, no optimal criterion exists for complex situations that include irregular terrain, inertial forces, and manipulation forces [26,28]. The accuracy and sensitivity of the FFSM and MFFSM are compared to Force Angle Stability Margin (FASM) [6-7] through a simulated scenario involving mobile manipulation. Although FASM is not the most suitable margin and is computationally more expensive than some other criteria [28], it is widely adopted (based on citations), providing a basis for comparison.

2. Foot Force Stability Margin (FFSM)

Although the FFSM can be adapted for use with systems that walk on walls or ceilings, this paper focuses on manipulation and walking on irregular terrain.

2.1. *Concept and Definitions*

A robot with n supporting legs has n foot force magnitudes, one per supporting leg. It is assumed that all foot force magnitudes are measurable. Under walking conditions, the foot force vectors $\boldsymbol{F}_i$, as shown in Figure 1, have nonnegative (zero or strictly positive) magnitudes, F_i. The critical foot force magnitude of each leg is zero which implies that the leg has lost contact with the surface. When a foot force magnitude is strictly positive, the normal component of the foot force vector, $\boldsymbol{f}_i$, applied by the contact surface to the foot point, is perpendicular to and points away from the object surface at the contact point. This implies that the leg is in contact with the surface.

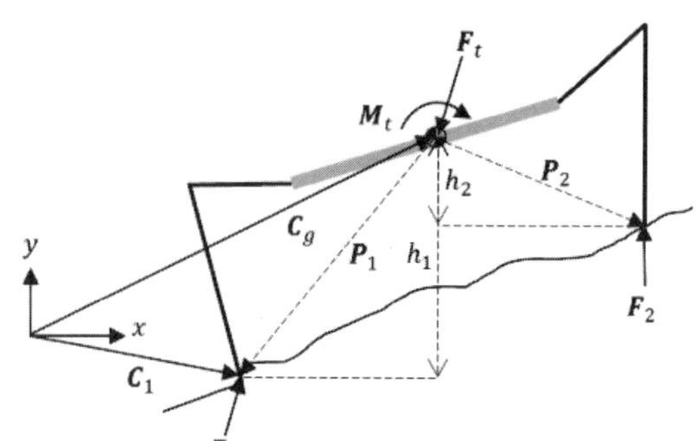

Figure 1. General planar robot.

The other two components of the foot force vector correlate with friction. All contacts between the legs and contact surfaces are assumed to be point contacts.

Based on the FFSM definitions and assumptions, a relation between stability and foot forces is given in Theorem 1.

Theorem 1: A spatial multi-legged walking robot with n *supporting legs* $(n \geq 3)$ *is stable at time* t *if and only if there are at least three legs with strictly positive normal foot forces at time* t.

Proof: Given a multi-legged $(n \geq 3)$ robot with two nonzero forces, indicating that only two legs have contact with the ground, there exists only one single axis connecting the two contact points such that the body can easily tip over about that axis. Therefore, the robot has to have at least three legs with strictly positive foot forces to be considered stable.

2.2. *Measure of FFSM*

Based on the theorem above, the FFSM at time t for a multi-legged robot with n supporting legs is defined as follows:

$$FFSM = \frac{f_1 f_2 \cdots f_n}{\bar{f}^n} \qquad 0 \leq FFSM \leq 1 \tag{1}$$

where n is the number of supporting legs with strictly positive foot forces, f_i is ith normal foot force magnitude and $\bar{f} = \frac{1}{n}\sum_{i=1}^{n} f_i$ is the average of all normal foot force magnitudes. Equation (1) provides a magnitude between zero and one $(0 \leq FFSM \leq 1)$ which indicates how close the system is to unstable or maximum stable state, respectively. As expected, Eq. (1) indicates that a more even distribution of foot forces enhances the stability of the whole system. The maximum stability $(FFSM = 1)$ occurs only when the foot forces are evenly distributed i.e. the standard deviation of foot force magnitudes is zero.

Let m be the number of legs which are going to lose contact with the surface where $m \leq n - 3$. If $n \geq 4$ and m legs lose ground contact, which generally occurs on irregular terrain, Eq. (1) would indicate a zero stability margin given the initial n. However, the system may still be stable with $n - m$ supporting legs. For example, this may occur when a robot changes from a quadruped $(n = 4)$ to a tripod $(n = 3)$. Therefore, at each iteration, n should be updated accordingly $(n \leftarrow n - m)$.

FFSM does not provide instability recovery, nor does it handle dynamic balancing of a system with at most two legs in contact with the ground. However, since all of the effects of gravity, external forces, inertial forces and disturbances can be observed in foot forces [7], the dimensionless FFSM represents the dynamic aspects of stability. In a practical stability controller, once the foot forces are known, the FFSM can be calculated using Eq. (1). The simplicity and expressiveness are attractive characteristics of FFSM. Furthermore, the FFSM can be modified for enhanced sensitivity to any desired system characteristic.

3. Modified Foot Force Stability Margin (MFFSM)

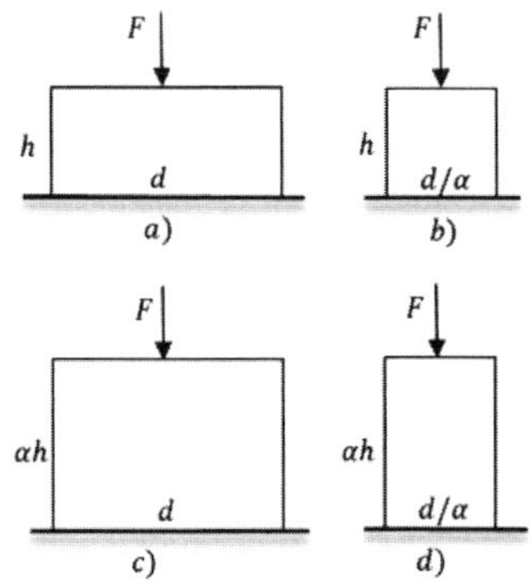

Figure 2. Example of parameter sensitivity.

Consider the cross section of a legged robot as shown in Figure 2. Assuming that $\alpha > 1$ is a constant, the FFSM is the same for all four cases. However, this is not conceptually true from the point of view of system energy since some have a higher tip over potential in the presence of lateral noise. For a specific instance, suppose a lateral force is acting as noise. Thus, conceptually, Figure 2a and Figure 2d have the lowest and highest geometrical tip over potential, respectively. On the other hand, in Figure 2a-d, the tip over potential is reduced if force F is replaced with αF. This represents sensitivity to top-heaviness. However, the FFSM, for all situations, gives the same stability margin without considering the Center of Gravity (CG) height, heaviness, or surface contact positions of the robot. Since these parameters directly affect the tip over potential of the system, a Modified FFSM (MFFSM) is developed to enhance the sensitivity of the stability margin to these parameters. Conceptually, to enhance the stability from a physical perspective, the CG height of the robot should be reduced as much as possible while the distance between the foot point and the center of gravity should be spread out. For this purpose, consider a spatial n-legged robot traversing irregular terrain as shown in Figure 3, where the support polygon is not restricted to be planar. In Figure 3, $\boldsymbol{M}_t$ is the net moment vector acting at the CG, and $\boldsymbol{F}_t$, which is the net force vector acting at the CG, can be derived from $\boldsymbol{F}_t = -\sum_{i=1}^{n} \boldsymbol{F}_i$ where $\boldsymbol{F}_i$ is ith foot force vector whose normal component is $\boldsymbol{f}_i$ with magnitude of $f_i = \|\boldsymbol{f}_i\|$. Hence, $\bar{f} = \frac{1}{n}\sum_{i=1}^{n} f_i$ directly correlates with the amount of

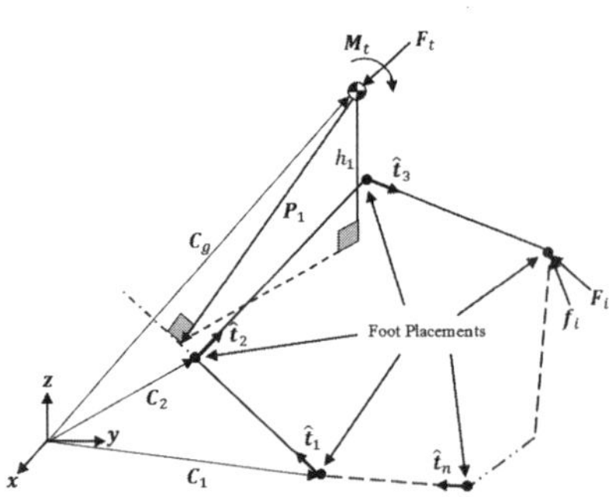

Figure 3. A general n-legged robot.

the net force and could be multiplied by FFSM for top-heaviness sensitivity [7]. On the other hand, the stability enhancement due to height reduction and foot placement spread could be directly taken into consideration by multiplying FFSM by $\frac{P_i}{h_i}$. However, this is true as long as the CG lies within the support polygon. Otherwise, $\frac{P_i}{h_i}$ should be replaced with $\frac{1}{P_i \cdot h_i}$. Finally, the MFFSM is defined as follows:

$$MFFSM = FFSM \cdot \frac{(P_i)^j}{h_i} \cdot \bar{f} \tag{2}$$

where $j = 1$ if the projection of the CG is inside or on the support polygon, $j = -1$ if it is outside of the support polygon, h_i is defined as the height of the CG with respect to the tip over axis aligned with the gravity vector which is the vertical portion of the tip over axis normal vector ($\boldsymbol{P}_i$) given by

$$h_i = \boldsymbol{P}_i \cdot (-\widehat{\boldsymbol{k}}), \tag{3}$$

where $\widehat{\boldsymbol{k}}$ is the unit vector of $\boldsymbol{z}$-axis of the global coordinate system. In Eq. (2), $P_i = \|\boldsymbol{P}_i\|$ where $\boldsymbol{P}_i$, the tip over axis normal vector, is the vector created by connecting the CG perpendicularly to the tip over axis. In other words, $\boldsymbol{P}_i$ is that portion of the vector $(\boldsymbol{C}_{i+1} - \boldsymbol{C}_g)$ which is perpendicular to the tip over axis where $\boldsymbol{C}_i$ is the position vector of ith ground contact point and $\boldsymbol{C}_g$ is the position vector of CG. This indicates that $\boldsymbol{P}_i$ can be obtained by subtracting that portion of the vector $\boldsymbol{C}_{i+1} - \boldsymbol{C}_g$ which is along the tip over axis as follows:

$$\boldsymbol{P}_i = (\boldsymbol{C}_{i+1} - \boldsymbol{C}_g) - [(\boldsymbol{C}_{i+1} - \boldsymbol{C}_g) \cdot \hat{\boldsymbol{t}}_i]\hat{\boldsymbol{t}}_i. \tag{4}$$

where $\hat{\boldsymbol{t}}_i = \boldsymbol{t}_i / \|\boldsymbol{t}_i\|$ is the unit vector of $\boldsymbol{t}_i$, ith tip over axis vector, given by

$$\boldsymbol{t}_i = \boldsymbol{C}_{i+1} - \boldsymbol{C}_i, \qquad i = 1, \ldots, n-1. \tag{5}$$

In Eq. (5), when $i = n$, $\boldsymbol{C}_{i+1}$ will turn into $\boldsymbol{C}_1$.

In a spatial robot, tipping over occurs about an axis called tip over axis. The tip over axis is chosen as the axis created by connecting the two surface contacts which have the largest foot force magnitudes amongst all foot forces. The subscript i in Eqs. (2-5) relates to that tip over axis. This reduces the calculations to only one axis. However, if there are two such axes with the same set of foot forces at the same time, the measure of MFFSM could be different for them since $\frac{(P_i)^j}{h_i}$ varies. In this case, a lower MFFSM indicates a higher tip over potential about the tip over axis and vice versa. In a very specific case, when the MFFSM is the same for two axes, tip over will occur about the single foot contact point with maximum foot force instead of an axis. One example is a tripod robot with a lateral force applied along the bisector of the supporting isosceles triangle. When using MFFSM, one should simultaneously consider

FFSM to take advantage of a measure between 0 and 1 that provides knowledge of how far or close the robot is to the most stable and unstable states.

4. Simulated Comparison Experiment

Figure 4. A tripod robot with a tool and a cylindrical rod.

One of general applications of legged robots is manipulation. Assume there is a hexapod robot with a gripper tool and a cylindrical rod as shown in Figure 4. While having three legs on the ground, the robot is commanded to reduce its height $(0 < t < 10\ \text{sec})$, pick up the cylindrical rod $(10 < t < 11\ \text{sec})$, increase its height $(11 < t < 26\ \text{sec})$, and move forward with the cylindrical rod $(t > 27\ sec)$. Figure 5 depicts the changes in height of the CG, external force acting at the gripper of the robot, and the lateral movement of the CG with respect to the time. During the manipulation process, the changes in the stability of the system are shown in Figure 6 based on FFSM, MFFSM, and FASM.

4.1. *Discussion of the Results*

All margins react $(t = 10\ and\ t = 26)$ and go through instability at the same time (point O in Figure 6). This shows the accuracy and correctness of FFSM and MFFSM. The results indicate that the sensitivity of FASM to the height of the robot is higher than the others $(0 < t < 10\ sec)$. However, MFFSM is more sensitive to top-heaviness as compared to the FASM between $t = 10$ and 11 seconds, when the robot picks up the rod. Lifting the rod increases the weight and moment about the CG. Therefore, the MFFSM increases sharply. However, the robot then increases the CG height, which decreases the stability. At $t = 26$, the robot approaches a height larger than at $t = 10$ and destabilizes, but the MFFSM is larger than it was at instant $t = 10$. The sensitivity of the MFFSM to the top-heaviness is the main reason of this phenomenon. Also the graph shows that under some specific situations, the MFFSM is more conservative than the FASM and less under some other situations. Figure 6 shows that MFFSM is more sensitive to lateral movements of the robot $(t > 27\ sec)$ which is one of the most common reasons for the tipping over of multi-legged robots due to instability.

The provided example demonstrates the simplicity, accuracy, and sensitivity of MFFSM and shows that it is reliable under static/quasi-static situations.

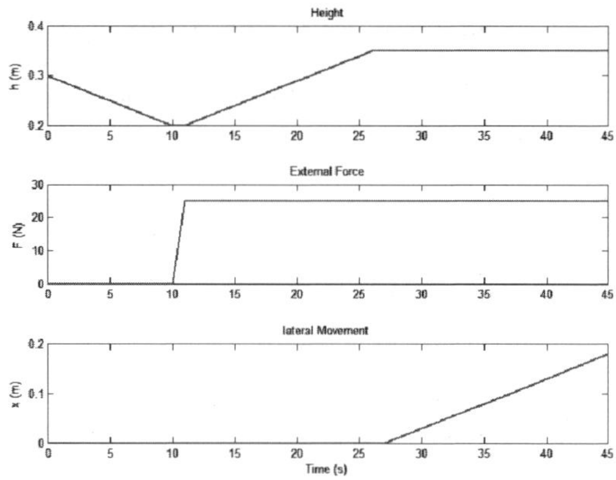

Figure 5. The changes in height, external force, and lateral movement during manipulation.

5. Conclusion

This paper presented the Foot Force Stability Margin (FFSM). It was shown that the FFSM is bounded by the unstable and maximum stable states. The Modified Foot Force Stability Margin (MFFSM) was derived to consider stability sensitivity to top-heaviness, robot height, and foot positions. To verify FFSM and MFFSM, a numerical comparison with FASM was completed. The results showed that FFSM and MFFSM are reliable under static situations.

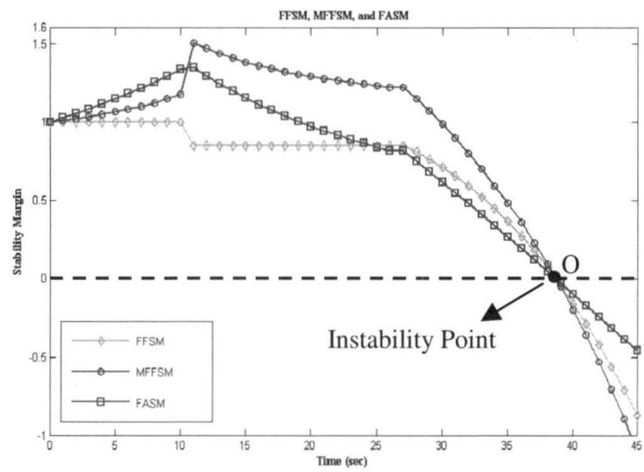

Figure 6. Comparison results of FFSM, MFFSM and FASM.

References

1. R. B. McGhee and A. A. Frank, On the Stability Properties of Quadruped Creeping Gaits, *Math. Bioscience*, **3**, 331 (1968).
2. R. B. McGhee and G. I. Iswandhi, Adaptive Locomotion of a Multilegged Robot over Rough Terrain, *IEEE Trans. Syst. Man. Cybern.*, SMC-**9**(4), 176 (1979).
3. Z. Wang, X. Ding, A. Rovetta and A. Giusti, Mobility analysis of the typical gait of a radial symmetrical six-legged robot, *Mechatronics*, **21**, 1133 (2011).
4. S. V. Sreenivasan and B. H. Wilcox, Stability and Traction Control of an Actively Actuated Micro-Rover, *J. Robot. Syst.*, **11**(6), 487 (1994).
5. S. Hirose, H. Tsukagoshi and K. Yoneda, Normalized Energy Stability Margin and its Contour of Walking Vehicles on Rough Terrain, in *IEEE Int. Conf. Robot. Autom.*, 181 (2001).
6. E. G. Papadopoulos and D. A. Rey, A New Measure of Tipover Stability Margin for Mobile Manipulators, in *IEEE Int. Conf. Robot. Autom.*, 3111 (1996).
7. E. G. Papadopoulos, and D. A. Rey, The force-angle measure of tipover stability margin for mobile manipulators, *Vehicle Syst. Dyn.*, **33**, 29 (2000).

8. A. Diaz-Calderon and A. Kelly, On-Line Stability Margin and Attitude Estimation for Dynamic Articulating Mobile Robots, *Int. J. Robot. Res.*, **24**, 845 (2005).
9. D. A. Messuri and C. A. Klein, Automatic Body Regulation for Maintaining Stability of a Legged Vehicles During Rough-Terrain Locomotion, *IEEE J. Robot. Autom.*, RA-**1**(3), 132 (1985).
10. A. Ghasempoor and N. Sepehri, A measure of Machine Stability for Moving Base Manipulators, in *IEEE Int. Conf. Robot. Autom.*, 2249 (1995).
11. S. Hirose, H. Tsukagoshi and K. Yoneda, Normalized energy stability margin: Generalized stability criterion for walking vehicles, in *Int. Conf. Climbing and Walking Robots*, 71 (1998).
12. E. Garcia and P. A. G. de Santos, A New dynamic energy stability margin for walking machines, in *Int. Conf. Adv. Robot.*, 1014 (2003).
13. J. K. Davidson and G. Schweitzer, A mechanics-Based Computer Algorithm for Displaying the Margin of Static Stability in Four-Legged Vehicles, *J. Mech. Des.*, **112**(4), 480 (1990).
14. B.S. Lin and S.M. Song, Dynamic modeling, stability and energy efficiency of a quadrupedal walking machine, in *IEEE Int. Conf. Robot. Autom.*, 367 (1993).
15. S. Sugano, Q. Huang and I. Kato, Stability Criteria in Controlling Mobile Robotic Systems, in *IEEE/RSJ Int. Conf. Intell. Robots. Syst.*, 832 (1993).
16. D. Kang, Y. Lee, S. Lee, Y. Hong and Z. Bien, A Study on an Adaptive Gait for a Quadruped Walking Robot under External Forces, in *IEEE Int. Conf. Robot. Autom.*, 2777 (1997).
17. Q. Huang, K. Tanie and S. Sugano, Coordinated Motion Planning for a Mobile Manipulator considering Stability and Manipulation, *Int. J. Robot. Res.*, **19**, 732 (2000).
18. J. Kim, W. K. Chung, Y. Youm and B. H. Lee, Real-time ZMP Compensation Method using Null Motion for Mobile Manipulators, in *IEEE Int. Conf. Robot. Autom.*, 1967 (2002).
19. J. Lee, J. Park and B. Lee, Turnover prevention of a mobile robot on uneven terrain using concept of stability space, *Robotica* **27**, 641 (2009).
20. K. Yoneda and S. Hirose, Tumble Stability Criterion of Integrated Locomotion and Manipulation, in *IEEE/RSJ Int. Conf. Intell. Robots. Syst.*, 870 (1996).
21. S. A. A. Moosavian and K. Alipour, Stability Evaluation of Mobile Robotic Systems using Moment-Height Measure, in *IEEE Conf. Robot. Autom. Mechatronics*, 1 (2006).
22. S. A. A. Moosavian and K. Alipour, On the dynamic tip-over stability of wheeled mobile manipulators, *Int. J. Robot. Autom.*, **22**(4), 322 (2007).
23. A. Meghdari, D. Naderi and M. R. Alam, Neural-network-based observer for real-time tip overestimation, *Mechatronics,* **15**, 989 (2005).
24. J. Wang and Y. Li, Static Force Analysis for a Mobile Humanoid Robot Moving on a Slope, in *IEEE Int. Conf. Robot. Biomimetics*, 371 (2008).
25. Y. Liu and G. Liu, Interaction Analysis and Online Tip-Over Avoidance for a Reconfigurable Tracked Mobile Modular Manipulator Negotiating Slopes, *IEEE/ASME Trans. Mechatronics,* **15**(4), 623 (2010).
26. E. Garcia, J. Estremera and P. G. de Santos, A comparative study of stability margins for walking machines, *Robotica*, **20**, 595 (2002).
27. P. R. Roan, A. Burmeister, A. Rahimi, K. Holz and D. Hooper, Real-World Validation of Three Tipover Algorithms for Mobile Robots, in *IEEE Int. Conf. Robot. Autom.*, 4431 (2010).
28. P. G. de Santos, E. Garcia and J. Estremera, Quadrupedal Locomotion: An Introduction to the Control of Four-Legged Robots, *Springer-Verlag, Berlin* (2006).

A DYNAMIC MODEL OF RUNNING WITH A HALF-CIRCULAR COMPLIANT LEG

M. M. ANKARALI

Dept. of Mechanical Engineering,
Johns Hopkins University, Baltimore 21218, MD, USA
E-mail: mertankarali@jhu.edu

E. SAYGINER*, Y. YAZICIOGLU§, A. SARANLI‡ and U. SARANLI†

**Dept. of Information Systems, §Dept. of Mechanical Engineering,*
‡Dept. of Electrical and Electronics Eng., †Dept. of Computer Engineering,
Middle East Technical University, Ankara 06800, Turkey
*E-mail: *sayginer@metu.edu.tr, §yigit@metu.edu.tr, ‡afsars@metu.edu.tr,*
†saranli@ceng.metu.edu.tr

Substantial research on legged robotics focuses on the design and morphology of leg structures. In this context, the advantages of passive compliance for reliability and simplicity have long been recognized. For example, composite leg designs with a half-circular profile used on later versions of the RHex hexapod were found to support a rich set of dynamic behaviors. However, the complex geometry and compliance properties of these legs have been difficult to model, preventing the use of dynamic simulations. In this paper, we present a simple dynamic model for this leg design for a planar monopod, taking into account both the kinematics of rolling contact and the nonlinear compliance of the geometry. We show through simulations that realistic predictions for system trajectories can be generated and leave experimental validation of this model for future work.

Keywords: RHex, Half-Circular Compliant Leg, Castigliano's Theorem

1. Introduction

Legged morphologies admit a wide range of behavioral alternatives through behaviors such as running, self-righting and climbing. The RHex hexapod[1] has been a good example of how biological inspiration can help improve mobility, motivating more recent instantiations of similar ideas.[2] A distinguishing feature of recent versions of RHex are its half-circular, compliant leg designs that were found to have robust mechanical properties and good dynamic performance but have been difficult to model accurately. Consequently, existing behavioral simulations to study this robot were mostly based on a radially compliant, telescoping leg structure,[3,4] closer to what has been used on monopedal or bipedal robot designs. In this paper, we present a new model that incorporates both the kinematics of rolling contact and the nonlinear compliance associated with RHex's composite half-circular legs, extending on previous modeling attempts towards a uniform model to support complete dynamic simulations of the robot.[5] An alternative approach to solve this modeling problem is through the use of numerical simulations based on finite-element methods.[6] However such methods are not only computationally demanding, but they also do not provide any insight into the underlying dynamics of the compliant mechanism to support the analysis and control of the overall system.

Some of the components for the model we present in this paper were introduced by earlier work.[7] However, this paper provides a more detailed kinematic model of the leg as well as additional novel components such as the formulation of the no-slip constraint for ground contact and hybrid transitions for running dynamics. For clarity, we present our leg model in the context of a lossless monopedal runner, the C-Pod , which has simple but sufficiently rich dynamics to illustrate basic properties of our model. The morphology of this system is similar to the TD-SLIP model from our earlier work,[8] incorporating both damping and hip actuation but using the a compliant half-circular leg morphology instead of a telescoping design.

2. Phases, Constraints and Kinematics of C-Pod

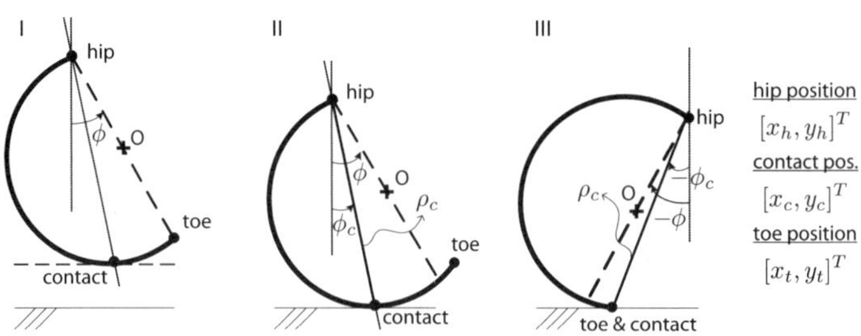

Fig. 1. Phases for C-Pod running: I - flight, II - rolling stance and III - toe stance. Point O marks the center of the leg circle. Dashed lines show the diameter of the arc passing through the hip, whose angle with the vertical defines the leg angle ϕ.

Similar to all legged systems, the C-Pod is a hybrid dynamical system. As illustrated in Fig. 1, the model goes through three phases during running:

I. Flight , II. rolling stance and III. toe stance. During flight, the leg is assumed to be massless and the body behaves as a point-mass acted upon by gravity. Unlike other commonly used monopedal models,[8] two different stance phases are necessary for C-Pod to capture the effects of the limited arc length of the circular leg. The dynamics associated with these two phases are the primary focus of this paper. The vector $[x_c, y_c]$ denotes the point on the leg closest to the ground, with $y_c = 0$ whenever the system is in stance. During stance, the rolling phase during stance is in effect whenever $x_c \neq x_t$, with ground contact falling on the circular part of the leg. Otherwise, the contact point is at the toe and the system is in its toe stance phase. The absolute horizontal location of the leg and the kinematics of rolling during stance are determined when the leg touches the ground based on the associated arc-length l_{td} and the contact position $x_{c,td}$, where l denotes the length of the arc between the contact and hip position and sub-script "td" stands for touchdown.

Our half-circular leg model is based on two basic assumptions:

(1) The leg always maintains a circular shape parametrized through its radius but preserves its total arc length,
(2) The leg undergoes pure rolling motion during stance with no slippage, corresponding to the constraint

$$l - l_{td} = x_c - x_{c,td}$$

to be maintained throughout the rolling stance phase.

The realization of these assumptions is one of the main contributions of this paper, allowing us to formulate system kinematics during rolling stance based only on the coordinates of the hip joint and the touchdown parameters l_{td} and $x_{c,td}$. Kinematically, the leg angle ϕ can be found during stance through the solution to

$$y_h(\sin\phi - \pi + \phi) + (1 + \cos\phi)(x_h - x_{c,td} + l_{td}) = 0 \, , \qquad (1)$$

based on which all other kinematic stance variables can also be found as detailed in Table 1.

Since C-Pod is a hybrid dynamical system, kinematic boundaries must be defined for phase transitions. Transition from flight to stance happens when the fully extended circular leg comes into contact with the ground. In contrast, two transitions are possible during the rolling phase. If the ground reaction force becomes zero, the leg lifts-off and the system transitions into

Table 1. Kinematic solutions during the rolling stance phase

Quantity	Derivation	Description
ϕ_c	$= \phi/2$	Angle of the "virtual spring"
r	$= y_h/(1 + \cos\phi)$	Leg angle and radius
ρ_c	$= 2r\cos\phi$	Angle of the "virtual spring"
l	$= (\pi - \phi)r$	Arc length of the compressed section
x_c	$= l - l_{td} + x_{c,td}$	Horizontal contact position

flight. If, however, the arc-length of the compressed section becomes equal to the arc-length of the leg, the system transitions into toe stance.

During toe stance, the leg acts as a nonlinear radial spring with a fixed toe. Since we know the contact position, the length and angle of this "virtual spring" can be computed as

$$\rho_c = \| \mathbf{v}_c \|_2, \quad \phi_c = atan2(\mathbf{v}_c(1), -\mathbf{v}_c(2)) \ ,$$

where $\mathbf{v}_c = [x_c, y_c]^T - [x_h, y_h]^T$ is defined as the virtual leg vector. The hip angle ϕ during toe-stance is then found by numerically solving

$$\pi + 2(\phi_c - \phi) - 2l_0 \cos(\phi_c - \phi)/\rho_c = 0,$$

where l_0 is the arc length of the whole leg. Remaining kinematic parameters during toe-stance are found using direct algebraic relations.

3. Stance Dynamics of the Half-Circular Compliant Leg

The dynamics of the C-Pod system during stance are governed by the force generated by the compressed half-circular leg, which we approximate using an adaptation of Castiglianos theorem.[9] In the absence of a hip torque and with a massless leg, the leg force must act along $\mathbf{v}_c$ (i.e. along the line connecting the hip and contact point). Our method relies on finding the total strain energy in the compliant leg as a function of the ground reaction force, whose derivative with respect to the force yields the instantaneous deflection of the leg. This can then be used to compute the instantaneous stiffness as

$$k(r,l) = \frac{2EI/r^3}{-3\sin(l/r) + (\cos(l/r) + 2)(l/r)} \ , \tag{2}$$

which only depends on the radius and arc length of the compressed leg segment, making it applicable to both stance phases. Consistent with our assumption above, we consider the rest leg length $\rho_{c,0}$ to be the distance between the current contact point and the hip for an uncompressed leg, which can be computed using

$$\rho_{c,0} = 2r_0 \sin\left(\frac{l}{2r_0}\right) , \tag{3}$$

where r_0 is the radius of the uncompressed leg. The definition of (3) is structurally different than previously proposed definitions,[7] which were prone to unrealistic discontinuities in leg forces during hybrid transitions between rolling and toe stance phases. Our model ensures continuity of leg forces during such transitions.

Subsequently, we use the deflection $\delta = \rho_{c,0} - \rho_c$ to compute the leg force as $F = k(r,l)\delta$. Finally, this leg force and the fact that the leg force acts along $\mathbf{v}_c$ can be used to finalize the dynamics of the C-Pod body as

$$\begin{bmatrix} \ddot{x}_h \\ \ddot{y}_h \end{bmatrix} = -\frac{\mathbf{v}_c}{\rho_c} F/m + \begin{bmatrix} 0 \\ -g \end{bmatrix} ,$$

where m and g corresponds to the body mass and gravity, respectively.

4. Simulation Results

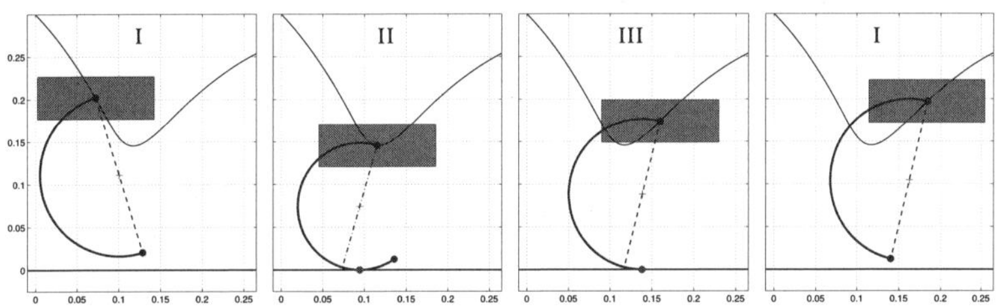

Fig. 2. Snapshots from an example simulation within a single stride. Black curve is the the body trajectory for the whole stride.

Fig. 2 illustrates four snapshots of a simulated C-Pod from a single stride, with the system phase indicated in roman numerals for each snapshot. Our results show that the trajectory and the evolution of all system variables are qualitatively realistic, noting that an experimental validation of these results is still necessary to reach any quantitative conclusions about their accuracy. In the absence of such a validation, which we intend to perform in the near future on a recently completed monopedal platform, the only qualitative evaluation we can perform is an evaluation of whether the compliance model achieves conservation of energy for a lossless system structure.

In this context, we simulated a single stride of the C-Pod model using a range of different initial conditions and we observed relative energy differences after a full stride. Simulations were done with $m = 1.5kg$ (mass per each leg for RHex), $r_0 = 0.19m$, $EI = 2.7859N$ (E and I are the modulus of elasticity and moment of area of the leg cross-section, respectively), selected to closely match physical properties of our RHex robot. The range of initial conditions we considered were $y_h \in [r_0, 2r_0]m$, $\dot{x}_h \in [-10r_0, 10r_0]m/s$ and $\phi_{td} \in [-45^o, 45^o]$. For each simulation, we evaluated the percentage energy change in a single stride, defined as

$$PE := 100\frac{|E_k - E_{k+1}|}{E_k}$$

Table 2 summarizes our results. We observed that the mean energy difference after a full stride across all simulations was approximately 2%, which is reasonable given that energy losses will be on the order of %30 – %50 for physical robotic systems with non-negligible damping.[10,11] Interestingly, energy changes for strides without the rolling phase (i.e. only toe-stance) were negligible, showing that our rest length and non-linear stiffness definitions provide a fully conservative spring model. Consequently, small changes in energy occur mainly due to the rolling phase, which is not unexpected since the kinematics during the rolling phase inherently depend on our first assumption of the leg preserving its circular shape, potentiall resulting in approximation errors.

Table 2. Apex to apex percentage energy changes for a single stride with the C-Pod model.

	$\mu \pm \sigma$ for PE	$PE_{\max}$
All strides considered	1.88 ± 1.66	6.03
Strides with only rolling stance	2.16 ± 1.48	5.92
Strides with only toe-stance	0.00 ± 0.00	0.00
Strides with both stance phases	2.35 ± 1.59	6.03

In addition to the results presented in Table 2, we illustrate single stride energy changes as a function of initial height in Fig. 3. In a somewhat expected manner, an increase in initial height results in an associated increase in the energy errors. This is mainly because the amount of leg compression also increases with larger initial apex heights, making the impact of our approximations and assumptions more significant. Most mathematical models for compliant systems assume small deflections[12] and Castigliano's theorem underlying our model is no exception. For these reasons, the positive correlation between initial apex height and energy errors we observed in our simulations are reasonable and expected. Fortunately, the maximum height that RHex can reach even for its fastest speeds is well below $2r_0$, suggesting that energy errors exhibited by our model should not be unreasonable under normal operating conditions.

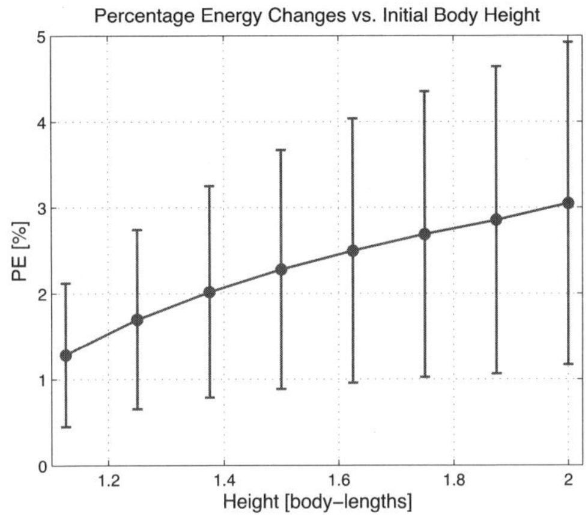

Fig. 3. Percentage energy changes as a fucntion of initial apex height. The vertical bars and solid markers represent the corresponding standard deviation and mean.

5. Conclusion

In this papaer, we proposed a dynamic model of running with a half-circular compliant leg, which is one of the core properties of the recent versions of the RHex robots. We had taken into account both the complex kinematics of rolling contact under the assumption of no-slippage over the ground and the nonlinear compliance of the geometry. We show through simulations that realistic predictions for system trajectories can be generated. We also observed that energy differences of the lossless and non-actuated C-Pod are negligible compared the energy losses observed in real robots. However in order to test the accuracy and reliability of our model, we have to perform an experimental validation that we leave as our future work.

6. Acknowledgments

We would like to thank Yasemin O. Aydin for her support in our comparisons with alternate models. This project was partially supported by TUBITAK project 109E032.

References

1. U. Saranli, M. Buehler and D. E. Koditschek, RHex: A simple and highly mobile robot, *International Journal of Robotics Research* **20**, 616(July 2001).
2. D. Goldman, H. Komsuoglu and D. E. Koditschek, March of the sandbots, *IEEE Spectrum* **46**, 30(April 2009).
3. M. Ankarali and U. Saranli, Control of underactuated planar pronking through an embedded spring-mass hopper template, *Autonomous Robots* **30**, 217 (2011).
4. U. Saranli, Dynamic locomotion with a hexapod robot, PhD thesis, University of Michigan, (Ann Arbor, MI, USA, 2002).
5. Y. O. Aydin, K. C. Galloway, Y. Yazicioglu and D. E. Koditschek, Modeling the compliance of a variable stiffness C-shaped leg using Castigliano's theorem, in *ASME Conference Proceedings*, (44106)2010.
6. G. Oral, Flexible multibody dynamic modelling and simulation of RHex hexapod robot with half circular compliant legs, PhD thesis, Middle East Technical University, (Ankara, Turkey, 2008).
7. E. Sayginer, Modelling the effects of half circular compliant legs on the kinematics and dynamics of a legged robot, PhD thesis, Middle East Technical University, (Ankara, Turkey, 2010).
8. M. M. Ankarali and U. Saranli, Stride-to-stride energy regulation for robust self-stability of a torque-actuated dissipative spring-mass hopper, *Chaos* **20**(Sep. 2010).
9. J. E. Shigley and C. R. Mischke, *Mechanical Engineering Design* (McGraw-Hill, 1989).
10. M. Ahmadi and M. Buehler, The ARL monopod II running robot: Control and energetics, in *Proceedings of the International Conference on Robotics and Automation*, 1999.
11. J. Hurst and A. Rizzi, Series compliance for an efficient running gait, *IEEE Robotics Automation Magazine* **15**, 42 (september 2008).
12. U. Saranli, O. Arslan, M. M. Ankarali and O. Morgul, Approximate analytic solutions to non-symmetric stance trajectories of the passive spring-loaded inverted pendulum with damping, *Nonlinear Dynamics* **62**, 729 (2010).

A RESISTIVE FORCE MODEL FOR LEGGED LOCOMOTION ON GRANULAR MEDIA*

CHEN LI, TINGNAN ZHANG, and DANIEL I. GOLDMAN

School of Physics, Georgia Institute of Technology, 837 State St NW
Atlanta, GA 30332, USA

Compared to agile legged animals, wheeled and tracked vehicles often suffer large performance loss on granular surfaces like sand and gravel. Understanding the mechanics of legged locomotion on granular media can aid the development of legged robots with improved mobility on granular surfaces; however, no general force model yet exists for granular media to predict ground reaction forces during complex limb intrusions. Inspired by a recent study of sand-swimming, we develop a resistive force model in the vertical plane for legged locomotion on granular media. We divide an intruder of complex morphology and kinematics, e.g., a bio-inspired robot L-leg rotated through uniform granular media (loosely packed ~ 1 mm diameter poppy seeds), into small segments, and measure stresses as a function of depth, orientation, and direction of motion using a model leg segment. Summation of segmental forces over the intruder predicts the net forces on both an L-leg and a reversed L-leg rotated through granular media with better accuracy than using simple one-dimensional penetration and drag force models. A multi-body dynamic simulation using the resistive force model predicts the speeds of a small legged robot (15 cm, 150 g) moving on granular media using both L-legs and reversed L-legs.

1. Introduction

A variety of desert animals including insects, lizards, and mammals encounter granular surfaces like sand and gravel in their natural environments and appear to move across them with ease [1]. In contrast, man-made vehicles, most of which move on wheels and tracks, can become stuck on granular media [2] because these substrates yield and flow under sufficient load. Understanding how legged animals generate sufficient ground reaction forces without sinking deeply on granular media will help improve robotic mobility on these challenging substrates [3]. However, unlike fluids for which forces can be computed by the Navier-Stokes equations [4], no comprehensive force model yet exists for granular media. While classical terramechanics has guided the design and

* This work was supported by The Burroughs Wellcome Fund, Army Research Lab MAST CTA, and the Army Research Office.

performance tests of large, off-road, wheeled and tracked vehicles [5, 6], the empirical and computational methods were developed specifically for wheel/track-terrain interfaces and thus it is not clear if they apply to small legged locomotors [5, 6].

Despite recent progress in modeling limb-ground interactions during legged locomotion on granular media [1, 7, 8], there is need to develop a force model to accurately and quickly predict forces on intruders of arbitrary shapes and trajectories moving through granular media. These recent studies [1, 7, 8] modeled the ground reaction forces in granular media with one-dimensional force models, such as the force on a vertically penetrating disc oriented horizontally [9], or the force on a horizontally dragged plate oriented vertically [10]. While these studies reveal some basic mechanisms of locomotion on granular media, these models do not fully capture forces on intruding limbs that are complex in both morphology and kinematics [1, 7, 8, 11, 12]. The discrete element method (DEM) which calculates forces on all interacting particles and the intruder [13] is accurate in predicting forces, but it is computationally expensive and limited in the number of particles that can be simulated, thus restricting the size of the granular bed or the size of particles [13].

2. Development of resistive force model

Our resistive force model for legged locomotion was inspired by a resistive force model developed in the study of a small lizard swimming horizontally within granular media [14]. The resistive force model divides an intruder of complex morphology and kinematics into small segments, each generating forces that are assumed independent. Summation of segmental forces over the intruder gives the net forces on the intruder. This approach is valid when the intruder moves slowly enough (< 50 cm/s, below which particle inertial force only contributes to < 10% of the total force) such that granular forces are dominated by friction (speed-independent) and the inertia of the accelerated particles is small [14].

To develop a resistive force model to predict ground reaction forces for legged locomotion on granular media (Fig. 1), we considered an intruder of complex morphology and kinematics moving within the vertical plane, e.g., a bio-inspired L-shaped leg (L-leg) of a hexapedal robot (Fig. 1a) rotated through granular media during locomotion. At each instant of time during rotation, each leg segment (gray bars in Fig. 1b) has a particular depth, orientation, and direction of motion (arrows in Fig. 1b), which may change at the next instant.

Therefore, we measured segmental lift f_z and drag f_x as a function of depth $|z|$, angle of attack β, and angle of intrusion γ (Fig. 1c). We used a robotic arm to move a limb "segment", a small aluminum plate of area $A = 3.8 \times 2.5$ cm^2, 0.6

cm thick, at a slow speed of 1 cm/s, as a force sensor measured forces during both intrusion and extraction. The granular medium (~ 1 mm diameter poppy seeds) was prepared by a fluidized bed into a uniform, repeatable, loosely packed state (volume fraction = 0.58) before each intrusion and extraction [7].

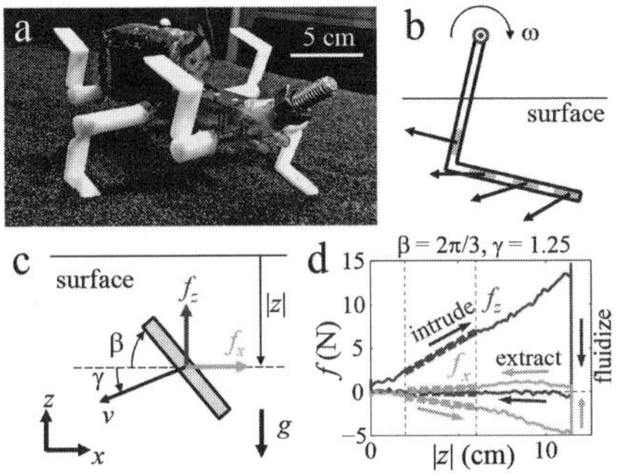

Figure 1. Development of the resistive force model for legged locomotion on granular media. (a) A hexapedal robot with L-legs standing on a granular substrate (~1 mm poppy seeds). (b) Representative leg segments (gray) have various depth, orientation, and direction of motion (arrows) as the L-leg rotates about the hip (red) through granular media within the vertical plane. (c) Segmental lift f_z and drag f_x on a plate depend on depth $|z|$, angle of attack β, and angle of intrusion γ. (d) f_z and f_x vs. $|z|$ for representative $\beta = 2\pi/3$ and $\gamma = 1.25$.

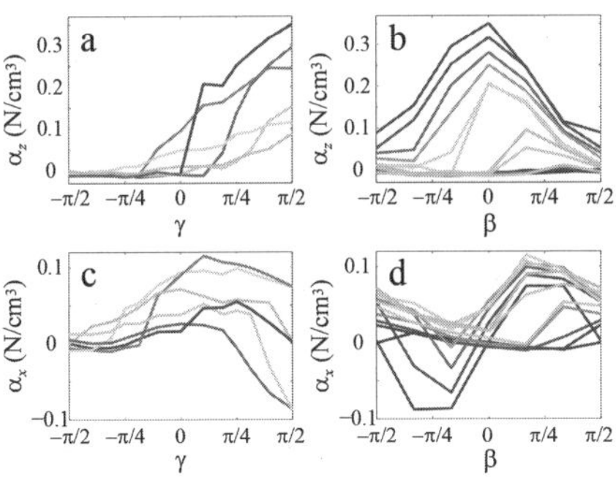

Figure 2. Vertical stress α_z (a,b) and horizontal stresses α_x (c,d) as a function of angle of intrusion γ and angle of attack β. In (a,c), colors from red to blue correspond to β = 0, 0.52, 1.05, π/2, 2.09, 2.62, and π. In (b,d), colors from red to blue correspond to γ = π/2, 1.25, 0.98, π/4, 0.59, 0.32, 0, –0.32, –0.59, –π/4, –0.98, –1.25, and –π/2.

We found that for all β and γ, both f_z and f_x were approximately proportional to $|z|$ during both intrusion and extraction (Fig. 1d). We performed a linear fit with zero intercept, $f_{z,x} = \alpha_{z,x}A|z|$ (where A is the segment area), to segmental forces vs. depth data over the steady state regions (dashed lines in Fig. 1d) to determine vertical and horizontal stresses $\alpha_{z,x}$ as a function of β and γ. Both α_z and α_x depended sensitively on β and γ (Fig. 2). For all β, both α_z and α_x had larger magnitudes (absolute value) for positive γ than for negative γ (Fig. 2a,c); i.e. it was generally harder to push the plate into granular media than to extract the plate. For all γ but $\gamma = 0$ and $\pi/2$, the magnitudes of both α_z and α_x were asymmetric about $\beta = 0$ (Fig. 2b,d); i.e., to move a plate with vertically mirroring orientations along the same direction generally required different forces. These asymmetries in stresses are due to the symmetry breaking by gravity in the vertical plane and the finite yield stress of granular media [15].

3. Predictive power of resistive force model

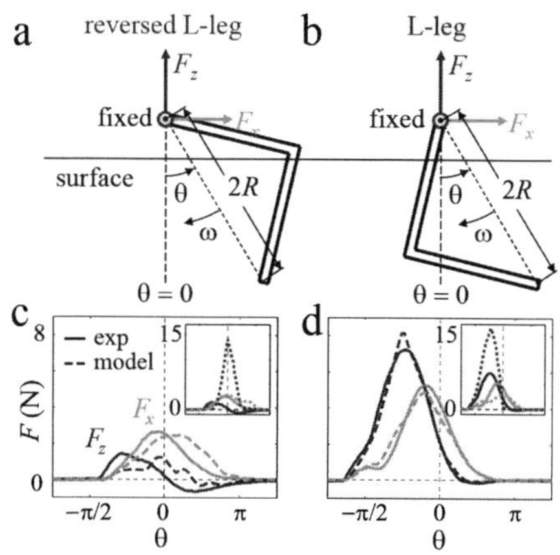

Figure 3. Predictive power of the resistive force model. A reversed L-leg (a) and an L-leg (b) were rotated about a fixed axle (red) through granular media as net lift F_z and drag F_x were measured as a function of leg angle θ (c,d). Solid and dashed curves in (c,d) are experimental measurements and resistive force model predictions. Insets: comparison of predictions from one-dimensional vertical penetration and horizontal drag force models (dotted) with experimental measurements (solid).

To test the power of the resistive force model for predicting net forces on an intruder of complex morphology and kinematics, we examined the net lift F_z and drag F_x on the an L-leg ($2R$ = 7.6 cm, width = 2.5 cm, made of aluminum) rotated about a fixed axle through granular medium, in both normal and reversed configurations (Fig. 3a,b). We used the robotic arm to perform rotation at angular velocity = 0.2 rad/s (tip speeds ~ 1 cm/s) as the force sensor measured forces as a function of leg angle θ [11]. The fluidized bed prepared the granular medium to a uniform loosely packed state before each rotation [7]. We calculated F_z and F_x from the resistive force model by summing segmental forces over the intruder using $F_{z,x} = \int \alpha_{z,x}(\beta_s, \gamma_s)|z|_s dA_s$ (where s denotes a segment), and compare them with experimental measurements.

Without any fitting parameters, the model prediction of $F_{z,x}$ vs. θ on both the L-leg and reversed L-leg (Fig. 3c,d, dashed curves) matched experimental measurements (Fig. 3c,d, solid curves) to within 15%, capturing both the magnitudes and asymmetric profiles. As a comparison, the $F_{z,x}$ vs. θ predicted by similar integration using the one-dimensional vertical penetration model, i.e., $\alpha_z(\beta, \gamma > 0) = \alpha_z(0, \pi/2)$ and $\alpha_z(\beta, \gamma < 0) = \alpha_z(0, -\pi/2)$ [9], and horizontal drag force model, i.e., $\alpha_z(\beta, \gamma) = \alpha_z(\pi/2, 0)$ [10] (Fig. 3c,d, inset, dotted curves) did not agree: it over-predicted F_z by 800% and under-predicted F_x by 50% for the reversed L-leg, and over-predicted F_z by 200% for the L-leg. The discrepancy in the exact shape of $F_{z,x}$ vs. θ between experiment and resistive force model for the reversed L-leg may be because particles became contained more easily within a concave leading surface [15].

4. Predicting legged locomotor performance

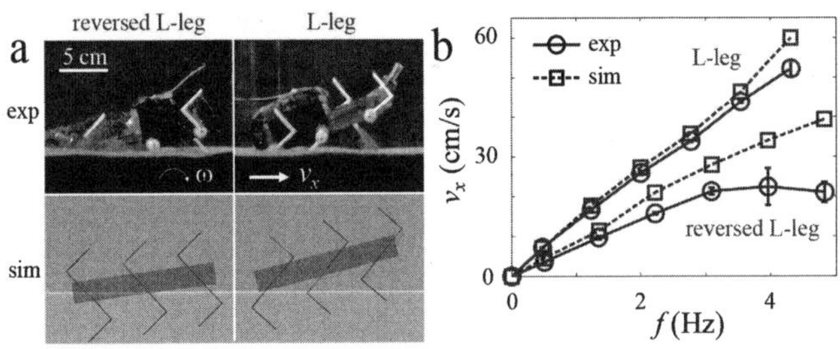

Figure 4. Legged locomotor performance in experiment and simulation. (a) Side views of a hexapedal robot (top) and a simulated robot (bottom) moving on granular media at stride frequency $f \approx 2$ Hz using both reversed L-legs (left) and L-legs (right). (b) Average forward speed v_x as a function of stride frequency f. Solid and dashed curves in (b) are experimental measurements and simulation predictions using the resistive force model.

To test whether the resistive force model could be used to predict legged locomotor performance, we studied in both experiments and simulation the average forward speed v_x of a small, bio-inspired, hexapedal robot (15 cm, 150 g) moving on granular media using both L-legs and reversed L-legs (Fig. 4). We used high speed videos to record both dorsal and lateral views as the physical robot ran across granular media (Fig. 4, top). The robot legs were made from plastic with similar friction coefficient (0.36) to that of aluminum (0.40) relative to poppy seeds. The granular medium was prepared into a uniform loosely packed state by a fluidized bed trackway before each run [7]. A multi-body dynamic simulation of the robot was created using MBDyn [16] (Fig. 4a, bottom), which used the resistive force model to calculate ground reaction forces and torques from the robot morphology and kinematics. This simulation required ~ 10 s to simulate 1 s of locomotion. The stride frequency f of the robot was varied between 1 Hz and 5 Hz in both experiment and simulation, and v_x was measured.

The robot in experiment and simulated robot displayed similar kinematics (Fig. 4a) during locomotion. We found that the robot always moved faster at any f with L-legs than with reversed L-legs (Fig. 4b). The simulation prediction of v_x vs. f (Fig. 4b, dashed curves) matched experimental measurements (Fig. 4b, solid curves) to within 15% for all f tested using both L-legs, and to within 30% for reversed L-legs up to f = 3 Hz (Fig. 4b, solid curve). The plateau and larger variation in v_x for f > 3 Hz using the reversed L-leg was a result of the pitch-instability of the robot observed in experiment, which was not found in simulation. As a comparison, similar simulation using the one-dimensional vertical penetration [9] and horizontal drag [10] force models over-predicted v_x vs. f by 50% for the L-leg and by 100% for the reversed L-leg.

5. Leg shape variation test

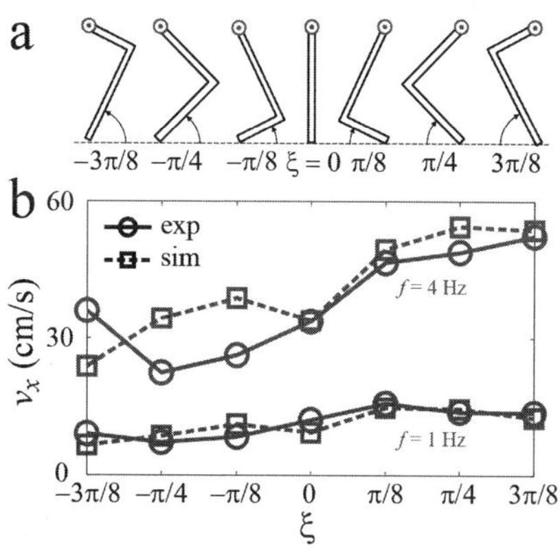

Figure 5. Leg shape variation test. (a) Definition of the angle ξ which parameterizes the shape of the L-leg. (b) Average forward speed v_x vs. ξ for two representative stride frequencies, f = 1 and 4 Hz. Solid and dashed curves in (b) are experimental measurements (interpolated from data in Fig. 4b) and simulation predictions using the resistive force model.

With the resistive force model and multi-body dynamic simulation, we can quickly perform design and performance tests for legged locomotors on granular media. To demonstrate this capability, in both experiment and simulation we changed the shape of the L-legs by varying the angle ξ which defines the location of the point adjoining the two straight pieces of an L-leg (Fig. 5a), and examined whether the simulation could predict how changes in leg shape affected the average forward speed v_x of the robot.

We found that v_x was higher using legs of positive ξ than using legs of negative ξ (Fig. 5b). The simulation (Fig. 5b, dashed curves) captured the measured v_x vs. ξ to within 10% for positive ξ and to within 50% for negative ξ. In addition, agreement was better (within 30%) at $f = 1$ Hz and worse (within 50%) at $f = 4$ Hz. As a comparison, similar simulation using the one-dimensional vertical penetration [9] and horizontal drag [10] force models over-predicted v_x by up to 200% at $f = 1$ Hz and by up to 100% at $f = 4$ Hz. The larger discrepancy between experiment and simulation using the resistive force model for negative ξ may be a consequence of particles being contained within concave leading surfaces. The larger discrepancy at higher f may be because as the robot legs moved faster, inertial effects became significant. Both these effects are not yet captured by the resistive force model.

6. Conclusions

We developed a resistive force model in the vertical plane for legged locomotion on granular media. Without any fitting parameters, the resistive force model demonstrated superior predictive power over previously used one-dimensional vertical penetration and horizontal drag force models in predicting the net forces on an intruder with complex morphology and kinematics. Furthermore, a dynamic simulation using the resistive force model predicted the locomotor performance of a small legged robot on granular media using L-legs of a variety of shapes, with significantly better accuracy than that using vertical penetration and horizontal drag force models. By contrast, in a previous study where the vertical penetration force model was used to explain the quasi-static locomotion of a similar robot, two fitting parameters were required to obtain agreement between experiment and model.

Our resistive force model provides a promising candidate for a general force model to calculate forces accurately and quickly for legged locomotion on uniform granular media. By contrast, a discrete element method simulation using same granular bed would contain 5 million poppy seed particles and take 30 days to simulate 1 s of locomotion. Aided by experimental and computational tools, the resistive force model promises to accelerate the design and performance tests for future legged robots on granular media.

Acknowledgments

We thank Yang Ding, Nick Gravish, Paul Umbanhowar, Gareth Meirion-Griffith, and Hal Komsuoglu for discussion, Jeff Shen for help with robot modification, and Pierangelo Masarati for MBDyn support.

References

1. C. Li, S. T. Hsieh, and D. I. Goldman, *J. Exp. Biol.* (in review).
2. J. Kumagai, *IEEE Spectrum* **6**, 44 (2004).
3. R. Pfeifer, M. Lungarella, and F. Iida, *Science* **318**, 1088 (2007).
4. S. Vogel, *Life in Moving Fluids: The Physical Biology of Flow* (1996).
5. M. G. Bekker, *Off-the-road locomotion, research and development in terramechanics* (1960).
6. J. Y. Wong, *Terramechanics and Off-Road Vehicle Engineering* (2010).
7. C. Li, P. B. Umbanhowar, H. Komsuoglu, D. E. Koditschek, and D. I. Goldman, *Proc. Natl. Acad. Sci.* **106**, 3029 (2009).
8. N. Mazouchova, N. Gravish, A. Savu, and D. I. Goldman, *Biol. Lett.* **6**, 398 (2010).
9. G. Hill, S. Yeung, and S. A. Koehler, *Europhys. Lett.* **72**, 137 (2005).
10. N. Gravish, P. B. Umbanhowar, and D. I. Goldman, *Phys. Rev. Lett.* **105**, 208301 (2010).
11. C. Li, P. B. Umbanhowar, H. Komsuoglu, and D. I. Goldman, *Exp. Mech.* **50**, 1383 (2010).
12. C. Li, A. M. Hoover, P. Birkmeyer, P. B. Umbanhowar, R. S. Fearing, and D. I. Goldman, *Proc. SPIE* **7679**, 1 (2010).
13. R. Maladen, Y. Ding, P. B. Umbanhowar, A. Kamor, and D. I. Goldman, *J. Roy. Soc. Int.* **8**, 1332, (2011).
14. R. D. Maladen, Y. Ding, C. Li, and D. I. Goldman, *Science* **325**, 314 (2009).
15. Y. Ding, N. Gravish, and D. I. Goldman, *Phys. Rev. Lett.* **106**, 028001 (2010).
16. G. Ghiringhelli, P. Masarati, and P. Mantegazza, *Nonlinear Dynamics* **19**, 333 (1999).

DESIGN AND CONTROL OF RANGER: AN ENERGY-EFFICIENT, DYNAMIC WALKING ROBOT

Pranav A. Bhounsule, Jason Cortell and Andy Ruina

Biorobotics and Locomotion Laboratory,
306 Kimball Hall, Cornell University,
Ithaca, NY 14853, USA
E-mail: pab47@disneyresearch.com, jbc2@cornell.edu, ruina@cornell.edu

We present the design and control of an energy-efficient, knee-less, essentially planar, four-legged bipedal robot called Ranger. In separate trials, Ranger: 1) walked a 40.5 mile ultra-marathon on a single charge and without human touch, setting a robot distance record; and 2) walked stably at Total Cost Of Transport (TCOT= total energy used per unit weight per unit distance travelled) of 0.19, apparently less than that of any other legged robot to date. Key design features are: a light weight and high strength box body, low-inertia leg design for fast and efficient swing, foot actuation that combines toe-off and ground clearance, a steering mechanism that enables turning of this essentially planar robot, and a low-power modular networked electronics hardware system. The model-based control approach uses a simplified offline trajectory optimization with a reflex-based feedback controller for stabilization. Ranger's reasonable success suggests that these design and control ideas could be extended to the development of an energy-efficient higher degree of freedom, 3-D bipedal robot.

1. Introduction and Motivation

Present legged robots are either highly energy-efficient like passive dynamic walkers[5] but fall down frequently, or like PETMAN,[1] BigDog,[6] and ASIMO[8] are more robust but guzzle energy. For legged robots to be useful they will have to be energy-efficient as well as robust. Here we present some progress made in this direction: the design and control of a simple planar robot with just 4 degrees of freedom that is highly energy-efficient and reliable at the limited task of steady level walking. We believe that aspects of the design and control approach we present here can scale up to more complex machines and to more versatile tasks.

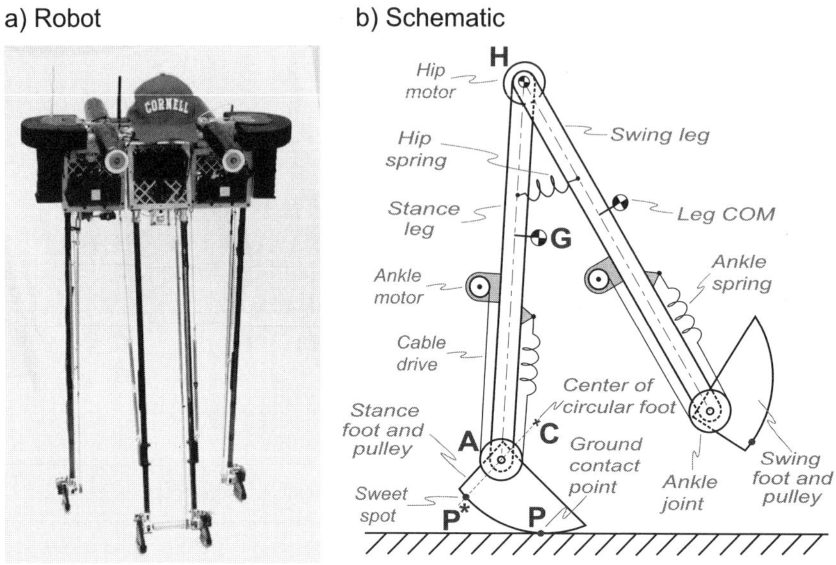

Fig. 1. **(a)** Ranger. **(b)** 2D schematic. The fore-aft cylinders with 'eyes' and the foam 'ears' (both visible in photo) are only for shock absorption in case of falls. The hat is decorative (hollow). The closed and rigid aluminum lace boxes, conceptually shown as point H, house all of the motors and gearing, various pulleys for the ankle cable drives, and most of the electronics (on the drawing the hip motor location is only schematic). There are two boxes connected by a hinge: an outer box, shaped like an upside-down U, rigidly connected to the outer legs, and an inner box, filling the space in the U, holding the inner legs (each of which can twist for steering). The hip spring, which aids leg swing, is shown schematically as symmetric between the two legs but shows as a diagonal cable and spring in the photograph. The feet are shaped so that toe-off is possible, so that no torques are needed during single stance, and so that ground clearance during swing can be achieved (by rotating the toe towards the hip).

2. Ranger

The Cornell Ranger is a four-legged knee-less biped (figure 1 a). It is about 1 m tall and has a total mass, including batteries of 9.9 kg.

It is autonomous in that all sensing and computation is on board, batteries are on board, and it has no booms, tethers or cable connections. It is not autonomous in that, at least so far, it needs to be started manually, and steering is done with a model plane type radio control.

Hardware. Leg swing is powered by a DC brushed hip motor. Each of the ankle pairs (one inner pair and one outer pair) has a motor along the hip axis that actuates the ankle via a one-way (toe-off) cable drive. Foot lifting, for ground clearance, is powered by a return spring on each ankle. A fourth motor twists the inner legs about a vertical axis with each step in order to steer the robot; the amount and direction of steering is governed by the remote control. The motors and electronics are powered by seven 25.9 V lithium-ion batteries with a total capacity of about 493 watt-hours.

Electronics. The main control loop runs, with no supervisory operating system, on an ARM9 microcontroller. Four ARM7 processors on custom boards monitor and control the three main joints (the outer ankles are counted as a single joint) and the steering. Two more ARM7 processors supervise the on-board communications network (CAN), the Bluetooth data reporting, and the onboard data display and lights. The Inertial Measurement Unit (IMU) also contains a proprietary microprocessor board. The multi-processor bus-based architecture (CAN) was chosen to facilitate design evolution, to simplify overall wiring (e.g., so a new sensor could be added without new wiring into the main processor) and to compartmentalize the control software (high-level on ARM9, low level on ARM7s). Sensors for each motor include an optical encoder, a voltage sensor, and a current sensor. In addition, each joint has an absolute angle sensor. Each foot has an optical strain gauge for measuring foot distortion (and hence foot contact). From the 3D IMU, Ranger's control only uses the sagittal plane angular rate sensor. The top-level control loop runs at 500 Hz on the ARM9 processor; data is sent to and from the satellite ARM7 processors once per loop execution; the motor current controllers, and their associated sensors operate at 2 kHz on the ARM7 processors.

Software. The total custom control code is about 10,000 lines of C and C++ code, the bulk of which is associated with low-level measurement, low-level control and communications protocols. The main control loop is based on hierarchical concurrent finite state machines. Control and estimation tasks are coordinated by a simple cooperative-multitasking scheduler, while low-level input-output, such as from motor encoders, uses processor interrupts.

For debugging and development, walking data is viewed and logged via wireless system. Although control parameters can be adjusted wirelessly mid-walk, during attempts at walking distance records autonomy is maintained by sending to the robot only steering signals and requests for data (e.g., cumulative number of steps, battery voltage) and not sending any walking control nor any changes of walking control parameters.

3. Control

Our model-based robot control algorithm is presented next. We summarize the key aspects of the control. For more details please see.[3,4]

Model for simulation. The 2D model (figure 1 b) consists of two rigid legs characterized by a mass, center of mass and polar moment of inertia about the center of mass. The feet are massless and round and are assumed to roll without slip when on the flat, level ground. The robot is powered by three motors; one for the hip and one for each ankle. The ankle motors are connected to the feet through a linear, torsional spring ('series elastic actuation'). Also, there is a zero-free-length hip spring pulling the legs to their parallel position. We assume a walking step to consist of the following concatenation of phases: single stance → heel-strike → double stance. The foot-to-ground collisions are assumed to be instantaneous, hard, with no slip and bounce, and with continuity in the position by discontinuity in the velocities. The power to the DC motor model is the sum of two parts: mechanical work, and electrical dissipation in the motor resistance and brush contact. The output torque from the DC motor is proportional to the current to the motor minus the frictional torque, which we found to be load-dependent.

Energy-optimal trajectory control (Fine grid). We try to find that periodic motion which minimizes the Total Cost Of Transport (TCOT), defined as the power used per unit weight per unit speed. The optimization variables are the time to take one step, all initial angles, and velocities in single stance. The control functions to be optimized are motor currents, which are parameterized as piecewise linear functions of time. We also enforced various constraints: current should be within actuator bounds, tensional contact with the ground is forbidden, and the swinging leg's foot should have sufficient ground clearance.

Simplifying the optimal trajectory. In order to implement the energy-optimal trajectory control solution presented above we proceed in two stages; first, we simplify the description of the near-optimal trajectory, and then we add a reflexive (event-based) feedback. These are discussed in the next two sections.

(i) Optimizing the coarse-grid description. First, we use a simpler coarse-grid representation that has few parameters. We did this by looking at the general shape of the fine-grid optimization and choosing intervals where we could use a simpler controller (see figure 2). The coarse-grid form is chosen manually so as to best catch the features of the fine-grid optimization but with fewer terms. Our goal is to achieve a gait with close to the same TCOT, informally doing a simultaneous optimization of both energy use and simplicity.

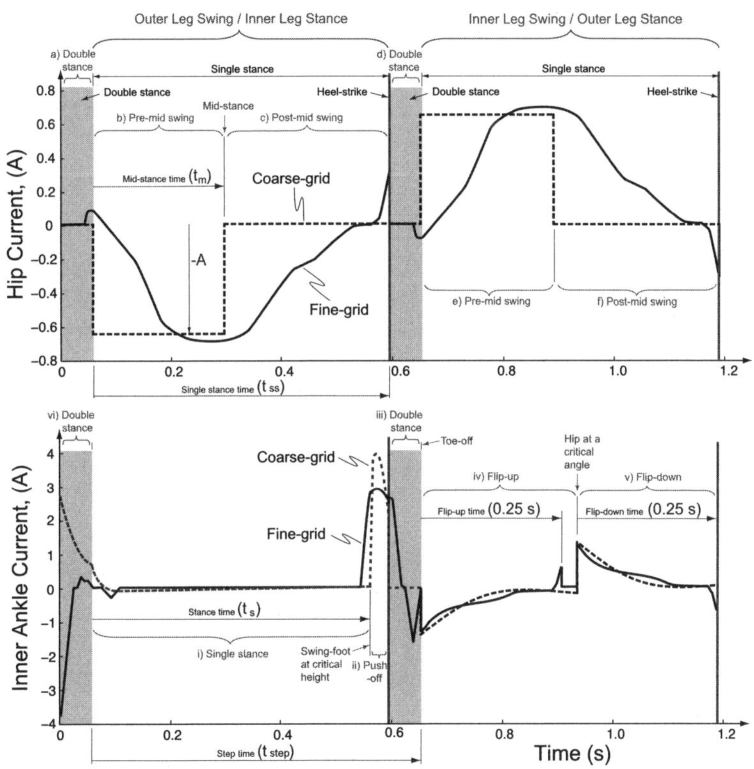

Fig. 2. Numerically, the fine-grid optimum (solid red) is obtained by solving the energy-optimal control problem (see energy-optimal trajectory control section). The current obtained was a piecewise linear function of time involving about 126 parameters, with TCOT $= 0.167$. The coarse-grid optimal (dashed blue) motor currents are the simplified version of the fine-grid solution (see optimizing the coarse-grid description above). Here, we first divide an entire step into phases (e.g., pre-mid swing, foot flip-up). Next, in each phase we use a compliant controller that involves selecting a constant, a stiffness, and a damping term (i.e. $I = I_0 + K\theta + C\dot{\theta}$, where I is current, θ is angle and I_0, K, C are constants). Our coarse-grid description has only 15 parameters and a TCOT $= 0.18$. Thus, in going from a fine grid to a coarse grid we have reduced the parameters by almost a factor of 8, but have increased the cost by only 14%.

(ii) Reflex feedback: discrete intermittent feedback control. Next, to stabilize the gait we wrap a linear feedback controller around this nominal trajectory. If the control output is U (e.g., current), then we decompose the control into two parts: a trajectory generator part (coarse-grid optimization described above) and a stabilizing controller part (described here).

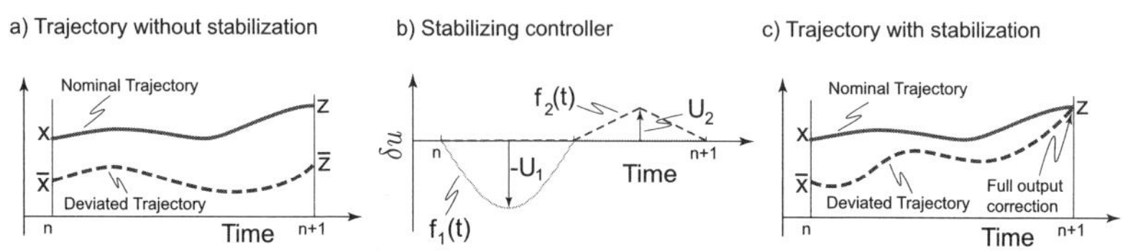

Fig. 3. A schematic example of our reflex-based discrete intermittent feedback controller

We illustrate the idea with a schematic example. Consider the nominal trajectory of a second-order system shown as a continuous red color line in figure 3a. Let n and $n+1$ be instances of time at which we are taking measurements from sensors. Let us assume that we take two measurements, $x = [x_1 \;\; x_2]'$ at time n (e.g., a position and velocity). We are interested in regulating two outputs, z_1 and z_2 at time $n+1$.

Suppose that due to an external disturbance the system deviates from its nominal trajectory, and thus follows the deviated trajectory shown as a dashed blue line in figure 3a. Now, the sensors read $\bar{x}$ at time n. In the absence of any feedback correction, the output values would become $\bar{z} = [\bar{z}_1 \;\; \bar{z}_2]'$.

The stabilizing controller measures deviations at time n ($\delta x_n = x - \bar{x}$) and uses actuation to minimize the deviations in output variables ($\delta z_{n+1} = z - \bar{z}$). For illustration we choose two control actions, $\delta u_n = [U_1 f_1(t) \;\; U_2 f_2(t)]'$, a half sinusoid and a hat function, each active for half the time between time $n+1$ and n. This is shown in figure 3b. We adjust the amplitudes of the two control functions U_1 and U_2, based on measured deviations δx_n, to regulate the deviated outputs δz_n. For example, with a proper selection of the amplitudes of the two functions it is possible to fully correct the deviations in the output variables, as shown in figure 3c.

In order to compute the amplitudes U_1 and U_2 we use a linear controller as follows: We linearize the system about the nominal trajectory (actually, we only linearize the section to section map). The sensitivities of the dynamic state to the previous state and the controls $\delta U_n = [U_1 \;\; U_2]'$ are: $\mathbf{A} = \partial x_{n+1}/\partial x_n$, $\mathbf{B} = \partial x_{n+1}/\partial U_n$, $\mathbf{C} = \partial z_{n+1}/\partial x_n$ and $\mathbf{D} = \partial z_{n+1}/\partial U_n$.

Thus we have $\delta x_{n+1} = \mathbf{A}\delta x_n + \mathbf{B}\delta U_n$ and $\delta z_{n+1} = \mathbf{C}\delta x_n + \mathbf{D}\delta U_n$. Using the above two equations, we set up a discrete linear quadratic regulator (DLQR) that is to find δU_n from δx_n. Our resulting linear controller has the form $\delta U_n = -\mathbf{K}\delta x_n$.

On Ranger. In order to implement the stabilizing controller described above, we used the following heuristics: we measure the robot state x_n during mid-stance, i.e., when the stance leg is vertical; then we try to regulate the robot velocity at the next mid-stance z_{n+1}. Our control actions U_n are as follows: If the robot velocity at mid-stance is slower than the nominal, then the robot has lower energy than the nominal at mid-stance and so we inject energy by pushing off harder (similar to walking uphill). If the robot velocity at mid-stance is faster than nominal, then the robot has higher energy than the nominal at mid-stance and so we extract out energy by increasing the step length (similar to walking downhill). We use the linearized version of the equation presented above to do the stabilizing control.

4. Results

(i) Energy-efficiency record. Table 1 lists the various gait parameters and energetics for the fine-grid trajectory control problem, the coarse-grid control representation and the experimental robot data.

For the energy-optimal trajectory control problem we had 126 parameters in all and a TCOT of 0.167. With our coarse-grid representation we reduced the parameters to 15 while increasing the TCOT to 0.18, a 7% increase in cost. However, the coarse-grid representation could not realize stable walking and had to turn on the stabilizing controller. We see that in going from the optimal trajectory control of the robot to our coarse-grid control with stabilizing control, we have reduced the parameters from 126 to 30 and added gait reliability, but at the cost of increasing the TCOT from 0.167 to 0.19, a 14% increase. The TCOT of 0.19 makes our robot perhaps slightly more energy-effective than the Collins walker, which had a TCOT of about 0.2.[2]

(ii) Long distance walking record. The broad goal of the Ranger project[7] is to develop a reliable robot capable of walking long distances on minimal amounts of energy. We set ourselves the goal of making Ranger walk a marathon distance of 26.2 miles or 42.2 kilometers, without falling down, without stopping, and without recharging.

On 1-2 May 2011, before we had optimized the energy-effective controller presented earlier in this paper, Ranger walked 40.5 miles or 65 kilometers, non-stop, and on a single battery charge. Ranger took 186,076 steps

Gait Parameter	Fine-grid	Coarse-grid	Experiment
Total COT	0.167	0.180	0.190
Motor COT	0.087	0.100	0.110
Overhead COT	0.083	0.080	0.080
Hip COT	0.019	0.018	0.030
Ankle COT (push-off)	0.029	0.052	0.046
Ankle COT (foot-flip)	0.039	0.029	0.034
Step Length	0.38	0.39	0.38
Step Velocity	0.64	0.66	0.62
Step Time	0.60	0.60	0.61
Double Stance (%)	9.5	5.0	3.0
Control Parameters	126	15	30

at a leisurely pace of 2.12 kilometers per hour or 1.32 miles per hour to set this distance record. The total energy consumption for Ranger for this walk was 493 watt-hours and it had a TCOT of 0.28 (as noted, this was later reduced to TCOT = 0.19).

References

1. Boston Dynamics. PETMAN - BigDog gets a big brother. [available online]. `http://www.bostondynamics.com/robot_petman.html`, April 2012.
2. S.H. Collins and A. Ruina. A bipedal walking robot with efficient and human-like gait. In *Proceeding of 2005 International Conference on Robotics and Automation, Barcelona, Spain*, 2005.
3. P. A. Bhounsule, J. Cortell, B. Hendriksen, J. G. D. Karssen, C. Paul, A. Ruina, A robot that can walk far using little energy, *The International Journal of Robotics Research (in preparation).*
4. P. A. Bhounsule, *A controller design framework for bipedal robots*, PhD Thesis, Cornell University, 2012.
5. T. McGeer, Passive dynamic walking, *The International Journal of Robotics Research*, 9(2):62–82, 1990.
6. M. Raibert, K. Blankespoor, G. Nelson, R Playter and others, *Proceedings of the 17th International Federation of Automatic Control*, 2008.
7. A. Ruina et al. Cornell ranger 2011, 4-legged bipedal robot. [available online]. `http://ruina.tam.cornell.edu/research/topics/locomotion_and_robotics/ranger/Ranger2011/`, April 2012.
8. Y. Sakagami, R. Watanabe, C. Aoyama, S. Matsunaga, N. Higaki, and K. Fujimura, The intelligent ASIMO: System overview and integration, *In Proc. of International Conference on Intelligent Robots and Systems, Lausanne, Switzerland*, 3:2478–2483, 2002.

IMPLEMENTATION AND ANALYSIS OF FUZZY-ZMP-WALKING CONTROL IN THE GIMBIPED

FERDI PERDANA-KUSUMAH†, JOSÉ-LUIS PERALTA-CABEZAS‡,
TOMI YLIKORPI and AARNE HALME

Department of Automation and Systems Technology, School of Electrical Engineering, Aalto University, Otaniementie 17, 00076 Aalto, Espoo, Finland
E-mail: †ferdi.kusumah@aalto.fi, ‡joseluis.peralta@aalto.fi

This paper presents the results of a Fuzzy-ZMP-Walking control implementation on Aalto University's GIMBiped robot. The robot was restricted to move only on sagittal plane, avoiding in this way any control for sideways balancing. The GIMBiped robot was originally designed to perform controlled Limit Cycle Walking (LCW) and the results of these experiments are oriented for future comparison with the realization of 2D LCW-based controllers in the same robot. The membership functions for the Fuzzy-ZMP-controller were implemented such to control the ZMP to be in a region of stability based on adjustments done in the walking speed. Results show relatively poor performance for several 2D generated trajectories with different fuzzy calculation rate, ZMP stability region size and placement. The previous issue indicates that ZMP strategy is not suitable for the current design of GIMBiped and if a dual (ZMP and controlled LCW) approach wants to be applied, redesign is needed. Some main points for the new design are also presented in this paper.

1. Motivation & Problem Statement

In Aalto's *Automation Technology Laboratory* (ATL) and in the *Center of Excellence* (CoE) in *Generic Intelligent Machines* (GIM), where this study was conducted, one of the main research topics are *Field and Service Robotics* (FSR) [1]. *Field Robotics* target the study of intelligent mobile machines in unstructured and dynamic environments (*e.g.* construction, forestry, agriculture, mining, military, and space applications) [2], and *Service Robots* are aimed to provide services to humans instead of manufacturing purposes [3]. These intelligent mobile machines shall autonomously carry out monotonous, dangerous or tedious tasks; therefore crucial prerequisites for FSR are *mobility* and *autonomy*. On flat surfaces, wheels are most efficient however for rough terrain or in human-like environments leg locomotion is more advantageous [4]. Within legged locomotion, humanoids-bipeds are the most attractive platform for the FSR potential applications, given that the ultimate goal in bipedal robotic research for FSR is to serve as locomotion platform for a complete humanoid robot, which could replace, if desired, the human presence in the future work-site [5].

Researchers have been studying biped robots for more than forty years [6] and, while many advances in the field have been accomplished and successfully ported to real applications, specially in the area of human prosthetics, there still remain the challenge to transfer the existing solutions into real applications for FSR [7]. The foremost issue is related to what was mention above as the main requirement for FSR, *mobility* and *autonomy*. In *mobility*, biped robots have evolved greatly since its beginning, in the late 60s, nevertheless they are still far from what a human can do in the work-site. Similarly, autonomy of biped platforms has been tackled on several different grounds, but its core problem remains, and it is associated to energy issues [8]. Because of this energy issues, lately the main attention has been redirected to the long time autonomy of these robots. For that, much effort has been made to develop new more efficient biped robots.

The motivation of being part of this not yet conquered investigation field and, the need to evolve into the next generation of work-site partners, gave birth to the GIMBiped project [7,9,10]. The objective of this project is to study and develop dynamic and energy efficient bipedal robotic waking.

1.1. *Energy Issues in Biped Service Robots*

State-of-the-art humanoid robots, such as Honda ASIMO [11], have not taken the benefit of the fluent gait inherent for humans and animals. Instead, the traditional approach has been to control each joint angle to achieve desired posture in terms of Center of Mass or Zero Moment Point (ZMP). These paradigms require plenty of energy to perform and a better result in terms of energy efficiency can be achieved by the *Limit Cycle Walking* (LCW) [8]. The idea of LCW is to correctly adjust the size and mass of the limbs such that with the proper initial joint angles and angular speeds the legs will eventually cycle back to the same initial conditions. The cycle can then start again from the beginning resulting in a smooth human-like gait. The next generation of LCW robots includes actuators in one or more joints enabling locomotion on a level terrain and up-hill [8]. However, so far the few LCW bipeds that have actually been implemented in hardware can only perform simple control maneuvers with limitations in mobility and robustness [8], making those poor in versatility.

Because of that in [7] we proposed to separate the main operational states of bipedal robots as *Walking* and *Standing*. Both of these states have

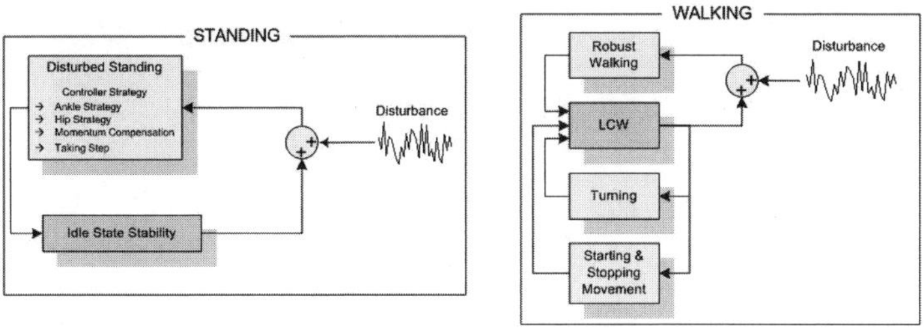

Fig. 1. Main states in bipedal robots operation and their respective efficient sub-states.

an energy efficient sub-state, called *Limit Cycle Walking* and *Idle State Stability* (ISS) (Fig. 1). But there are also other sub-states, which in the case of *Walking*, does not involve just walking, but also to overcome disturbances and perform other tasks (turning, starting and stopping movement, squatting, etc.). One possibility for the implementation of the sub-state *Robust Walking* is to apply some kind of ZMP approach, which has proven to be highly robust and versatile [11]. Here we propose then the use of Fuzzy-ZMP-walking for the robust walking state of our GIMBiped robot, which at the same time can serve for comparison and evaluation of energy consumption with the future development of LCW-based controllers.

2. Methodology

2.1. *The GIMBiped*

Until now mainly rotational motors have been used to drive the robot's joints; however, their gearbox reduction, needed in this case to reach the desired torque in the active phase, does not allow the free natural movement of the swinging leg in the passive stage of walking. Because of that much energy is wasted in the continuous position control of the leg in both phases. On [9] we purposed the use of permanent magnet synchronous linear motors to efficiently address LCW. They do not have gear boxes and for that reason nothing breaks the natural movement in the swinging phase of the leg's trajectory. Also, they can be easily controlled by torque because of its direct force/current relation thus avoiding the overuse of position control. However, the previous compliance is a disadvantage when ZMP approach wants to be applied, and that is the motivation to study our robot's performance under ZMP control in this paper.

The main processing units are two PC104 embedded computers. One, named Motor Hub, is in charge of sending positioning commands to the

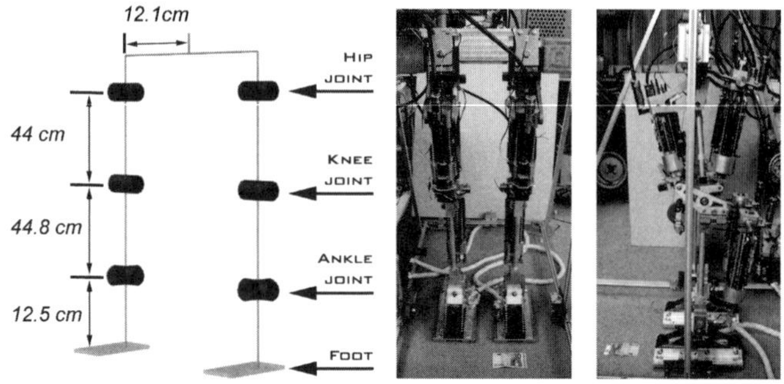

Fig. 2. GIMBiped platform. Left: kinematic layout; Center: front view; Right: left view

Table 1. Fixed Parameters for all the Setups in Experiments with the GIMBiped.

Max.[1] command skipped	Robot's Mechanical Param.						Walking Traj. Param.			
	Hip width	Thigh length	Shin length	Foot height	Foot width	Foot length	Number of steps	Step length	Step height	Double[2] Support %
18	12.1	44	44.8	12.5	16	31	10	10	8	10

[1] The maximum number of commands skipped to influence the transfer rate of predefined trajectory positions. By changing the transfer rate, the walking speed of the robot can be made faster or slower

Note: All values apart from [1] and [2] in cm.

linear motor's controller using CAN protocol, and the other PC104, named Sensor Hub, is responsible to collect force sensor data from both feet. Both computers are connected to a network for remote access. A specially design network architecture called GIMNet [12] is used to control the robot. The robot is connected to a frame which allows only sagittal move, the mechanical parameters are presented in Table 1 and ilustrated in Fig. 2.

2.2. *Trajectory Planning, Sensor System & Controller*

2.2.1. *Trajectory planning*

In the trajectory planning, smooth cubic polynomial interpolations are used to generate the desired position of the feet and hip, based on some walking parameters like step length, step height, and others shown in Table 1. Inverse kinematic using damped least square method is then applied to obtain the needed joint angles. These angles were converted to strokes in order to be used by the linear motors. The embedded linear motor controller's PID was then used to reach the desired position [9]. The trajectories were generated off-line and also used to compute the reference area for fuzzy control calculations, these technique is explained in Section 2.2.3 below.

Table 2. Parameters for the positioning controller in the Linear Motors.

Modulus	Modulus	PID parameters								
Max. Vel.	Max. Acc.	P $[A/mm]$			I $[A/(mm/s)]$			D $[A/(mm*s)]$		
$[m/s]$	$[m/s^2]$	Ankle	Knee	Thigh	Ankle	Knee	Thigh	Ankle	Knee	Thigh
0.2	10	4	2	6	2	0	3	4	2	6

2.2.2. *Sensor System*

Four one-dimensional force sensors, based on strain gages, were used in each foot to calculate the ZMP position [13]. The force signals from the sensors were amplified 1000 times by a differential amplifiers. The signals were then converted to a digital form by a 16 bit ADC and transferred to the Sensor Hub through serial communication. The client software then obtain the data through the network and calculate the corresponding ZMP position and perform Fuzzy control. The linear motor has built in sensors which are capable of measuring motor's slider position and velocity, temperature, motor voltage and current. The linear motors' controller have their own PID positioning control, its parameters are presented in Table 2 and they remain constant during the experiments. These controllers receive control commands and send motors variables data through CAN interface that can be accessed from the Motor Hub.

2.2.3. *Controller*

Fuzzy control was the approach taken here to maintain robot's walking stability. The control is performed in a client program using data obtained from the Sensor Hub through the GIMNet. The controller uses two inputs, which are: the difference between the current ZMP and reference area position (ZMP error), and its derivative (ZMP error rate). The *reference area* is defined as a specific area inside the foot boundary in which the robot is consider to be stable if the ZMP is within it, that means the ZMP position error is zero and no action is required by the controller. The *reference area rate*, is the velocity of the movement of the reference area. The controller was applied based on the initial assumption that the feet always remain flat to the ground.

The controller changes the walking speed online based on a stability index calculated from the ZMP error and ZMP error rate. The fuzzy rules map every possible combination of input/output which drive changes in the number of command skipped to maintain the ZMP inside the boundary of the foot. The controller changes the position commands resolution by

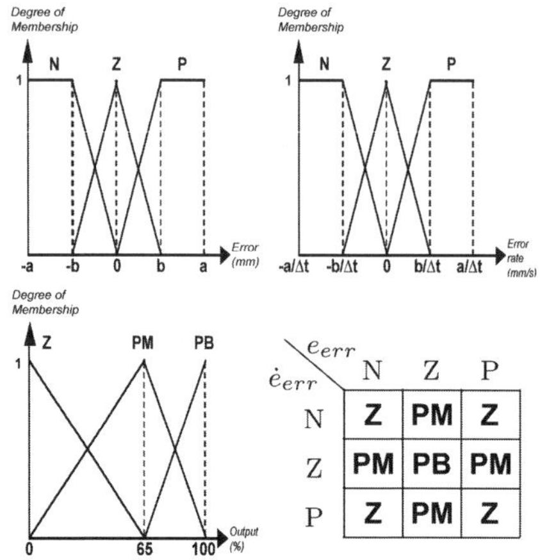

$\dot{e}_{err}$ \ e_{err}	N	Z	P
N	Z	PM	Z
Z	PM	PB	PM
P	Z	PM	Z

Fig. 3. Membership functions of the Fuzzy-ZMP. Left Up: input error MF; Right Up: input erro rate MF; Left Down: output MF; Right Down: Fuzzy Rule Table

including or skipping some trajectory commands to change the walking speed. The membership functions of the controller are shown in Fig. 3.

The universe of the fuzzy error input is from $-a$ to a, with a is the half dimension of the foot length. For the error rate input, the universe contains the value from $-a/\Delta t$ to $a/\Delta t$, with Δt the time span between each data sample. The b value is half of a. The output membership function produce crisp outputs in percentage format. The linguistic variable consists of N for $Negative$, Z for $Zero$, P for $Positive$, PM for $PositiveMedium$ and PB for $PositiveBig$. The inference method was root sum square and the weighted average approach was used in defuzzyfication process.

One issue to notice is that the configuration of the robot's feet are different than in humans, since the ankle joint's position is located almost at the center of the foot. This creates a critical case where the ZMP is located almost at the center of the foot when the robot is in the standing position. This affects the responsiveness of the controller when the robot is performing its steps since the ZMP just starts to move when the robot is about to fall forward. To sense the instability faster, the reference area in the stance phase is shifted to the heel to produce the error and error rate required to slow down the step speed. Also, a delay is applied in the fuzzy calculation to reduce the controller reaction when needed. The movement rate of the reference area is calculated offline by the trajectory planning.

Table 3. Variable Parameters for different Setups in experiments with the GIMBiped.

Setup	Reference Area			Fuzzy
	Size[1] $[mm]$	Initial mov. rate $[mm/s]$	Final mov. rate $[mm/s]$	calculation delay $[\mu s]$
Set 1[2]	–	–	–	–
Set 2	20	-200	200	5000
Set 3	10	-200	200	5000
Set 4	50	-200	200	5000
Set 5	20	0	0	5000
Set 6	20	-100	100	5000
Set 7	20	-200	200	0
Set 8	20	-200	200	10000

[1] Actual size of reference area is twice the value (+ and − from the center of the foot)
[2] Without control, following trajectory on open loop

3. Experiments, Conclusions and Future Work

The experiments were conducted with eight different sets of control parameters. Each experiment was performed 10 times and the final values of the results were averaged. The different set of parameters are shown in Table 3.

An overview of the numerical results for all the different sets, showing total walking time, power consumption and the motors' temperature difference between the beginning and the end of the walking, are shown in Table 4

Without applying any control (Set 1), the robot consumed plenty of power. The robot foot did not slow down when it touches the ground and created more loads at the moment of impact, specially for the motors with large PID values (hip and ankle motors). The PID controller tries to keep all motors in their desired positions all the times during sequence of the walking. By not applying control, both feet usually lift or bend the toes during the walking, therefore they did not always land parallel to the ground. In this situation, the walking can only be analyzed visually since the information of the pressure sensor is not useful. The foot mostly landed on its toe specially the right foot because the motor's speed and acceleration could not keep up with the given position commands. The previous also affects the total walking time which was 21.23 s, and ideally it should be 16.67 s.

In the subsequent experiments the controller tried to stabilize the robot by changing the resolution of the trajectory points generated by the trajectory planner, thus the step speed became variable. The shortest step time that the controller can produce is 1.667 s and the longest is 30 s.

In terms of power consumption, it can be seen from the table that the fuzzy controlled robot used less power than the previous experiment

Table 4. Walking Analysis Overview.

Setup	Total time [s]	Power [Watt]							Δ Temperature[°C]					
		LA[1]	LK[2]	LT[3]	RA[4]	RK[5]	RT[6]	Tot[7]	LA	LK	LT	RA	RK	RT
Set 1	21.23	338.1	297.6	159.1	386.2	361.3	176.4	1710.7	1.27	7.56	0.18	2.24	9.53	0.49
Set 2	29.59	263.3	301.6	127.8	302.1	362.2	104.2	1461.2	1.28	9.71	0.4	2.73	13.2	0.1
Set 3	30.12	259.4	283.9	128.5	295.4	367.6	106.1	1440.9	1.37	10.1	0.4	2.84	13.3	0.3
Set 4	25.11	267.8	229.7	133.9	358.6	354.9	143.9	1488.8	1.89	7.17	0.48	2.66	11.9	0.3
Set 5	28.43	283.3	293.2	119.1	304.5	371.3	106.7	1478.1	1.17	9.49	0.39	2.76	13.4	0
Set 6	28.93	290.1	295.2	123.3	305.3	371.2	103.6	1488.7	1.55	10.1	0.6	2.37	13.6	0
Set 7	27.57	291.9	278.1	131.8	370.9	384.4	110.8	1567.9	1.47	8.82	0.4	2.72	14.6	0.5
Set 8	29.75	265.6	298.1	120.6	270.3	338.7	97.51	139081	1.07	10.1	0.3	2.44	12.5	0.3

[1]=Left Ankle; [2]=Left Knee; [3]=Left Thigh
[4]=Right Ankle; [5]=Right Knee; [6]=Right Thigh; [7]=Total Power Consumption

(without control). The effect of slowing down the movement of the robot is the reduction of its moment, therefore with a suitable control, it is less likely that the foot will hit the ground with high speed and create a huge load for the corresponding leg motors. In the current control scheme, reducing the step speed means increasing the resolution of the trajectory points. The position commands sent will be closer to each other so the motor speed and acceleration can keep up with the commands. In the fuzzy control experiments, the foot still landed on the toe with small inclination angle. The problem was most probably resulted from PID accumulation error on each motor and inaccuracy of robot's dimension parameters in the angle to stroke conversion equations.

It can be seen from Table 4 that the total energy consumption of Set 3 is smaller than the two previous experiments. In Set 3 the size of the reference area was changed and, by reducing its size, the calculated error will grow and the fuzzy control will increase the command resolution to reduce the walking speed. On the contrary, increasing the reference area too much can lead to smaller error calculation and this can make the controller less responsive to the ZMP movement. The result will be worse if the ZMP mostly stay in the reference area, and no error is produced leading the controller to move the robot at full speed abandoning robot's stability.

From Sets 2, 5 and 6, the lowest power consumption was achieved by the Set 2. Set 2 has the largest gradient values for the reference area rate (± 200 mm/s). This affects the area movement speed by quickly moving it to the heel. The error rate produced is increased and the controller acts faster in slowing down the walking speed.

The experiment with the lowest power consumption was achieved by Set 8. The fuzzy controller used 10000 *ms* delay in its computation. By reducing the data taking and calculation rates, the controller becomes less sensitive to noise and rapid changes, allowing the controller to better follow the general movement of the ZMP. In the case with 0 *ms* delay (Set 7), the fuzzy control will take and calculate the data as fast as possible while the time scale of the ZMP movement is much larger, so the fuzzy will not only account the noise in its calculations but also see the slow ZMP movement as a series of temporary stops. This can be seen from the Table 4 where the power intake in Set 7 is the largest of the experiments with control because the control system failed to reduce the impact on the robot feet.

The outcome has shown us important guidelines for future improvement in our mechanical design and control techniques for dual approach (ZMP+LCW). Is seems slower walking speeds helps maintain stability, however slow speed makes the robot spend more time and energy to reach its destination. Unfortunately, the current robot mechanical design does not allow the knee to work intensely. As Table 4 shows, the temperature changes on both knees are higher than in the hip or ankle motors. This indicates that they worked outside the area of their nominal operation.

Since our main problem is the knee motor overload, changes in the prototype should consider a lighter motors in the shank to ease the load of the knee, more powerful knee actuators and better cooling mechanism. Also more proficient ankles with springs and dampers are needed, together with different type of feet. Furthermore the controller should be perfected based on additional sensor feedback (inclinometers, accelerometers and IMU).

References

1. J. Saarinen *et al.*, *Proc. World Congress,* The Concept of Future Worksite-Towards Teamwork-centered Field Robotic Systems, (2011).
2. A. Kelly *et al.*, Field and Service Robotics: Results of the 7th Int. Conf., **62**, (2010).
3. S. Lee *et al.*, Recent Progress in Robotics: Viable Robotic Service to Human (2009).
4. D.J. Todd, Walking Machines: An Introduction to Legged Robots, (1985).
5. K. Yokoi *et al.*, *Robotics Research* A Tele-operated Humanoid Operator, **25** (2006).
6. M. Vukobratović *et al.*, *Biomedical Eng.* On the Stability of Biped Locomotion, **1** (1970).
7. J.-L. Peralta *et al.*, *Proc. CLAWAR2010,* Idle State Stability, Limit Cycle Walking & Regenerative Walking: Towards Long Time Autonomy in Bipeds, (2010).
8. S. Collins *et al.*, *Science,* Efficient Bipedal Robots Based on Passive-dynamic Walkers, (2005)

9. J.-L. Peralta *et al.*, *Proc. Int. Conf. on Humanoid Robots,* Novel Design of Biped Robot Based on Linear Induction Motors, (2009).
10. J.-L. Peralta *et al.*, *Proc. Int. Conf. on Robotics and Aut.*, Assessment of Limit-cycle-based Control on 2D Kneed Biped, (2011).
11. Y. Sakagami *et al.*, *Proc. Int. Conf. Intelligent Robots and Systems,* The Intelligent ASIMO: System Overview and Integration, (2002).
12. J. Saarinen *et al.*, *Proc. 2nd Wrorshop in Generic Intelligent Machines,* Philosophy of GIMnet, (2009).
13. F. Kusumah, *M.Sc. thesis,* Foot Force Sensor Implementation and Analysis of ZMP Walking on 2D Bipedal Robot with Linear Actuators, (2011).

THE EFFECT OF SPRAWL ANGLE AND WALL INCLINATION ON A BIPEDAL, DYNAMIC CLIMBING PLATFORM

J. D. Dickson and J. E. Clark*

Department of Mechanical Engineering,
Florida A & M University - Florida State University College of Engineering,
Tallahassee, FL 32310, USA
**E-mail: clarkj@eng.fsu.edu*
www.eng.fsu.edu/stride

Animals have shown the ability to climb vertical surfaces with high speed and stability. Utilizing the underlying dynamics of these animals, robotic platforms have been developed that climb vertical surfaces with similar speed. It is hypothesized that the pre-incidence angle of the legs, commonly referred to as sprawl angle, for these robotic platforms can significantly affect vertical velocities, efficiency, and stability as well as passively controlling body oscillations. To date, little empirical work has been conducted on the effect of sprawl angle and wall inclination on the performance of dynamic climbing platforms. This paper presents initial research utilizing a biologically inspired dynamical climbing platform to understand the effect of sprawl angle and wall inclination on dynamic climbing. Simulations have shown that a sprawl angle of 30° maximizes vertical velocity overall, while experimental results show that a sprawl angle of approximately 10° maximizes vertical velocity, while in both increasing sprawl angle increases lateral velocities over all wall inclinations.

Keywords: Dynamic climbing; sprawl angle; bio-inspired robotics

1. Introduction

Animals have shown the ability to climb vertical surfaces with high speed and stability. Utilizing the underlying dynamics of these animals, robotic platforms have been developed that climb vertical surfaces with similar speed. The bio-inspired, dynamic scansorial platforms have shown capabilities of faster vertical velocities than previous, quasi-static climbers through active management of their kinetic energy.

Dynamic climbing platforms have drawn their inspiration from biological analogs including geckos and cockroaches. While these animals are vastly different morphologically, their ground reaction forces and velocity profiles generated during climbing were found to be remarkably similar.[1,2] This discovery led to the development of the Full-Goldman Template[2,3] which utilizes two legs to generate similar force and velocity profiles as its biological progenitors. It is hypothesized that the pre-incidence angle of the legs, commonly referred to as sprawl angle, can significantly affect vertical velocities, efficiency, and stability as well as passively controlling body oscillations.[2,3]

Current biological data indicates that cockroaches transition their legs from pushing during horizontal running to pulling for vertical climbing.[4] To date, little work has been conducted on the effect of wall inclination on the performance of dynamic climbing platforms.

While biological data suggests that animals utilize an effective sprawl angle of approximately 10° on vertical surfaces,[2] only a cursory examination of the effect of sprawl angle on dynamic climbing platforms has been conducted.[5,6] This paper presents the development of a small-scale, bipedal, dynamic climbing platform based on the Full-Goldman template with an adjustable shoulder sprawl angle. The effect of varying sprawl angle on vertical and lateral velocity is experimentally investigated on a variety of climbing inclinations.

2. Simulation

In order to study the effects of sprawl angle and wall inclination on dynamic climbing, a new simulation was developed in Working Model 2D capable of variable sprawl angle and wall inclination based on the Full-Goldman template.[2] Considerations for the development of the physical platform led to a simulation mass of 200g. The dynamic properties of the Full-Goldman template were preserved through utilization of dynamic similarity. A thorough discussion of dynamic scaling is presented by Alexander[7] and Clark *et al.*[3] Scaling of the Full-Goldman template from 2g to 200g led to a simulation that utilizes a leg stroke of 5.1cm and runs at 4.2Hz.

The simulation utilizes a geared transmission system in which a single motor supplies torques to both gears. Each gear serves as a crank for the crank-slider mechanism which drives the linear arms angled at a prescribed sprawl angle. A wrist spring of $131Nm^{-1}$ is incorporated into the system connecting the foot to the end of the arm. The foothold is approximated as a pin joint during stance and turned off during flight. The stance conditions are calculated based on the angular position of the gears driving the crank-slider mechanism. A torque-limited speed controller was implemented in the simulation in addition to the simple frequency controller.

The primary motivation for the addition of the torque-limited speed controller to the 2D simulation is the analysis of the effects of friction drag on the climbing surface when the wall inclination is reduced from vertical. The frictional force due to drag on the carpet was implemented as a force opposing vertical motion equal to the normal force of the robot, expressed as gravitational constant (9.81) multiplied by the mass of the robot and the cosine of the angle of the wall (ψ), on the wall multiplied by the coefficient of friction, as shown in (1). The coefficient of friction, μ, was empirically calculated for the carpeted climbing wall as 0.2650.

$$F_f = \mu(9.81 * mass) * cos(\psi) \tag{1}$$

2.1. *Simulation Results*

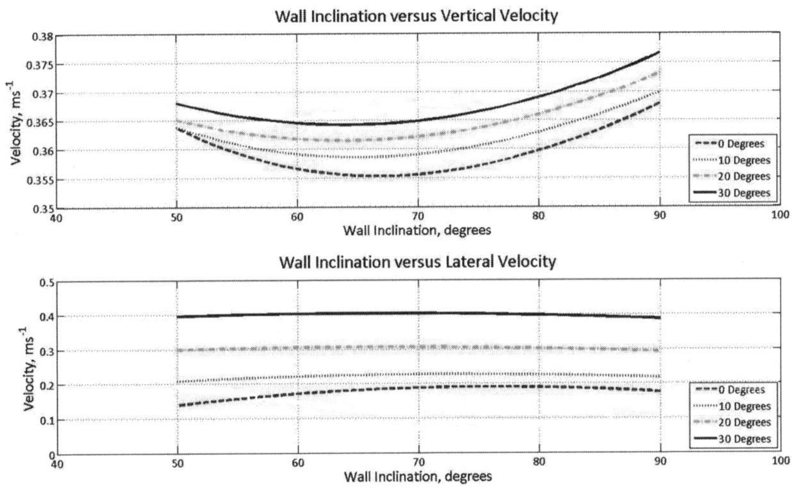

Fig. 1. Vertical and lateral velocities of the bipedal dynamic climbing simulation utilizing a mass of 200 g, a torque limited speed controller operating at 4.2 Hz, and wrist spring stiffness of $131Nm^{-1}$ run on varying wall inclinations between 90° (vertical) and 50° incrementing by 10° with sprawl angle between −10° and 30° incrementing by the same amount.

The velocities of the simulation, shown in Figure 1, show perceptible trends with the alteration of hip sprawl angle. The simulation shows a maximum vertical velocity utilizing a sprawl angle of 30° for most wall inclinations, as previous simulation studies conducted by Clark *et al.*[5] and

Lynch *et al.*[6] predict. For sprawl angles between 0° and 30° a minimum vertical velocity is shown at a wall inclination of approximately 70°. This wall inclination corresponds to a maximum resistive force on the platform through the combination of the gravitational force and drag force of the wall. Overall, the variation in vertical velocities is minute when compared to their overall magnitude. The lateral velocity shows an expected upward trend in maximum lateral velocity with increasing sprawl angle.

The trends shown in the simulation of sprawl angle show similarities to previous studies on the effect of sprawl angle on dynamic climbing. It is expected that a sprawl angle of 30° will maximize vertical velocity and max lateral velocity on the experimental platform. It is also expected that a sprawl angle of 0° will minimize the lateral velocities.

3. Platform Fabrication

Fig. 2. (Left) Bipedal dynamic climbing platform. The single actuator rotates both gears which are physically locked 180 out of phase to move the linear guide rails. (Top Right) Mechanism for compliant wrist. (Bottom Right) Adjustable shoulder to allow variable sprawl angle.

The miniature bipedal dynamic climber utilized the same scale and drive system as the Working Model 2D simulation and borrowed design considerations from the initial dynamic climbing platform, Dynoclimber.[8] A single DC motor was utilized to provide actuation for both arms leading to an under actuated system for reduction of mass. The platform physically locks the arms 180° out of phase. Based on torque requirements determined utilizing the 2D simulation, a 6 V Faulhaber DC motor (# 1331 006 SR) with a Faulhaber Series 15/3 6.3:1 ratio spur gear head was chosen to drive the climber at the design target of 200g. The final physical platform weighed 187g.

A crank-slider mechanism utilizing 5 cm gears, leading to an additional gear reduction of 3.75:1, coupled with miniature linear guide rails and bearing blocks (Misumi #SSEB6-40) provide the necessary linear motion required based on the scaled Full-Goldman template. A slotted shoulder mount and guide block provide the ability to alter the sprawl angle at the shoulder and lock it during experimental runs. Marks on either side of the shoulder indicate pre-defined sprawl angles incrementing by 10°. The platform utilizes commercially available wrist springs of $131Nm^{-1}$ with a maximum extension of 2 cm. Simple hooks are rigidly attached to the slider block of each wrist mechanism, shown in Figure 2.

4. Experimental Setup and Procedure

To evaluate the performance of the platform's scansorial motion, a 1.22m x 2.5m carpeted wall was built and oriented vertically, the wall is designed to be capable of adjusting its inclination between 90° and 50°. A motion capture system was implemented for tracking the location and orientation of the platform using two LEDs and a Casio Exilim Pro EX-F1 high-speed camera at 300 fps.

The platform was run 10 times for each sprawl configuration varying between 0° and 30° by 10° on the given wall inclination. The wall inclinations varied between 90° (vertical) and 50° incrementing by 10°. Each run extracted three strides at the top of the run to ensure steady state operation for numerical analysis utilizing MATLAB.

5. Experimental Results

The platform was first compared to the 2D Working Model dynamic simulation discussed in Section 2. The simulation utilized the same physical constraints as the platform including mass, wrist spring, frequency, and torque limitations. It was found that at a sprawl angle of 10° on a vertical wall the simulation predicted the platform's trajectory and velocity profiles closely. Figure 3 shows the comparison of the platform run on a vertical wall with a sprawl angle of 10° and the Working Model simulation

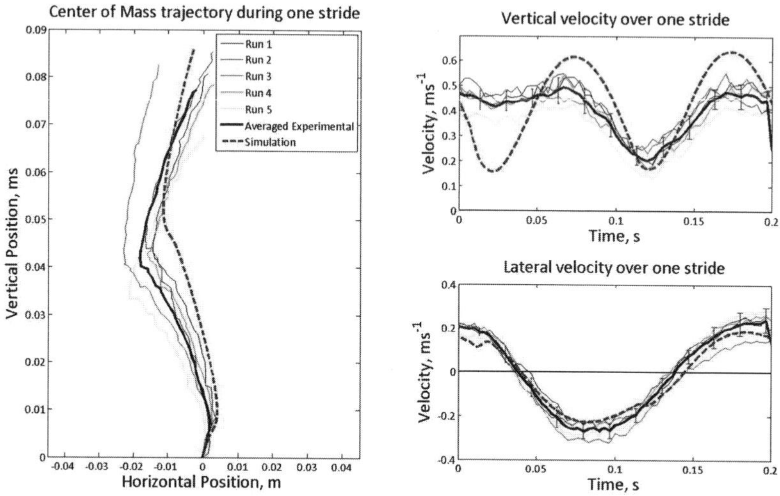

Fig. 3. Comparison of the center of mass trajectory and velocity profiles of the Working Model 2D simulation and the physical platform. The solid line represents the averaged values of the individual runs. The dotted line represents the simulation prediction of the platform's behavior utilizing the same parameters as the physical robot.

run with the same parameters. A limping gait is apparent in the vertical velocity of the platform, most likely attributable to a slight asymmetry in the drive mechanism of the platform. The experimental platform was shown to climb at a vertical velocity of $37.0 \pm 3.4cms^{-1}$ approaching the scaled velocity of a cockroach climbing at the same mass and exceeding the predicted simulation velocity of $36.6cms^{-1}$. This vertical velocity sets this miniature dynamic climbing platform as the second fastest climbing robot following the original dynamic climber, Dynoclimber, capable of velocities up to $66cms^{-1}$.[8]

After comparing the simulation to the physical platform, 200 individual climbs were run over the number of sprawl angles and wall inclinations. The results of these runs are shown in Figure 4a and Figure 4b.

The overall trends found on the physical platform show a maximum vertical velocity at a sprawl angle of 10° over all wall inclinations, as Figure 4a illustrates. Between 90° and 50° the maximum vertical velocity increases with decreasing wall inclination, as Figure 4b shows. As predicted by the simulation, and shown in Figure 4a, with increasing sprawl angle the maximum lateral velocity increases.

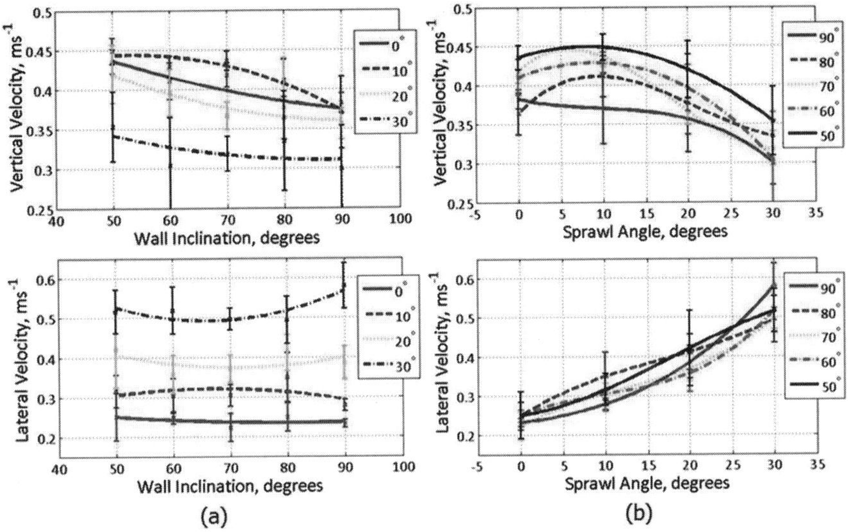

Fig. 4. Comparison of sprawl angle effects on vertical and lateral velocities over a range of wall inclinations. (a) Comparison of all wall inclinations plotted over sprawl angle. **(Top)**Sprawl angle versus vertical velocity plotted over all wall inclinations **(Bottom)**Sprawl angle versus maximum lateral velocity plotted over all wall inclinations. (b) Comparison of wall inclination versus vertical velocity and max lateral velocity plotted over all sprawl angles. **(Top)** Wall inclination versus vertical velocity plotted over all sprawl angles. **(Bottom)** Wall inclination versus max lateral velocity plotted over all sprawl angles.

6. Conclusions and Future Work

A comparison of the Working Model 2D simulation results with previous simulation studies[5,6] on the effect of leg sprawl on velocities shows similar trends including increasing vertical and lateral velocities with increasing sprawl angle and a 30° angle maximizing both.

While the simulation captures the effect of leg sprawl and slope for lateral velocities on the physical platform, it over predicts the vertical speed for large sprawl angles. The experimental data shows a maximum vertical velocity attained utilizing a 10° sprawl angle over all wall inclinations. The discrepancy in vertical velocity appears to be caused by increased lateral forces imparted to the linear guide block bearings as sprawl angle increases. The friction of the guide block increases as lateral force on it increases, thus reducing the overall vertical velocity. The sprawl angle of 10° represents the point where increased vertical velocity due to sprawl angle can still overcome the increased friction due to the guide block, where as the benefit of increased sprawl angle on vertical velocity is lost after 10°. For wall inclinations of 90° to 50° the ideal sprawl angles do not seem to shift.

In order to address these discrepancies, the simulations utilized for dynamic climbing will need to incorporate a friction model for the legs and methods of reducing this friction should be investigated for the robot.

Future work will focus on analyzing the effects of sprawl angle and wall inclination on climbing efficiency and ground reaction forces utilizing simulation and experimental studies over wall inclinations between 0° and 90°. These studies will be compared to animals such as the cockroach *Blaberus discoidalus* and the gecko *Hemidactylus garnnotti* to gain insight on how biological systems utilize leg sprawl angle during high speed vertical running.

References

1. K. Autumn, S. Hsieh, D. Dudek, J. Chen and C. Chitaphan, *The Journal of Experimental Biology* (2006).
2. D. Goldman, T. Chen, D. Dudek and R. Full, *The Journal of Experimental Biology* (2006).
3. J. Clark, D. Goldman, T. Chen, R. Full and D. Koditschek, Toward vertical dynamic climbing, in *Proceedings of the 9th International Conference on Climbing and Walking Robots*,
4. J. Schmitt and S. Bonnono, *Journal of Theoretical Biology* (2009).
5. J. Clark, D. Goldman, P. Lin, G. Lynch, T. Chen, H. Komsuoglu, R. Full and D. Koditschek, Design of a bio-inspired dynamical vertical climbing robot, in *Proceedings of Robotics: Science and Systems*,
6. G. Lynch, L. Rome and D. Koditschek, *Applied Bionics and Biomechanics* (2011).
7. *Principles of Animal Locomotion* (Princeton University Press, 2003).
8. G. Lynch, J. Clark and D. Koditschek, A self-exciting controller for high-speed vertical running, in *Proceedings of the International Conference on Intelligent Robots and Systems*,

ATRIAS 2.0, A NEW 3D BIPEDAL ROBOTIC WALKER AND RUNNER*

Alireza Ramezani and J.W. Grizzle

Department of Mechanical Engineering and
Department of Electrical Engineering and Computer Science
Control Systems Laboratory, University of Michigan
‡*E-mail: {aramez,grizzle}@umich.edu*

This paper introduces ATRIAS 2.0, a new platform for the study of bipedal locomotion in robots. One of the purposes of the robot is to further explore the role of compliance in achieving energetically efficient and agile locomotion. A second purpose is to inspire the development and experimental validation of analytical feedback control algorithms for dynamic 3D locomotion. A third purpose is to address the challenge of tightly integrating hardware and software to achieve extreme robustness to unknown ground-height variations while walking or running.

Keywords: robot; bipedal; control; hybrid; gait; zero dynamics; optimization.

1. Introduction

ATRIAS 2.0, shown in Fig. 1, is a collaborative effort of Oregon State University, Carnegie Mellon University, and the University of Michigan. The robot has been conceived for energy efficiency, speed, and robustness with respect to natural terrain variations, without over-reliance on external sensing, such as vision. In addition, the robot will be untethered, with all computation and actuation onboard, using electric motors, batteries, and mechanical springs for cyclic gait-energy storage. ATRIAS 2.0 is being designed and constructed at Oregon State University, under the leadership of J. Hurst. Single-leg (monopedal) prototypes of the robot are currently being tested.[1] This paper describes the robot and reports initial work on developing a dynamic model and a feedback controller for walking on level ground.

*Supported by DARPA Contract W91CRB-11-1-0002.

2. Overview of the Mechanism

ATRIAS 2.0 is a 3D bipedal robot comprised of rigid links joined by revolute joints to form a torso, left and right hips, and left and right legs, as shown in Fig. 1. The left and right hip are connected to the torso through revolute joints with a common axis aligned in the sagittal plane. Each hip is independently actuated in the frontal plane by a brushless DC-motor connected through a gear ratio of 18:1. These motors are mounted on the torso.

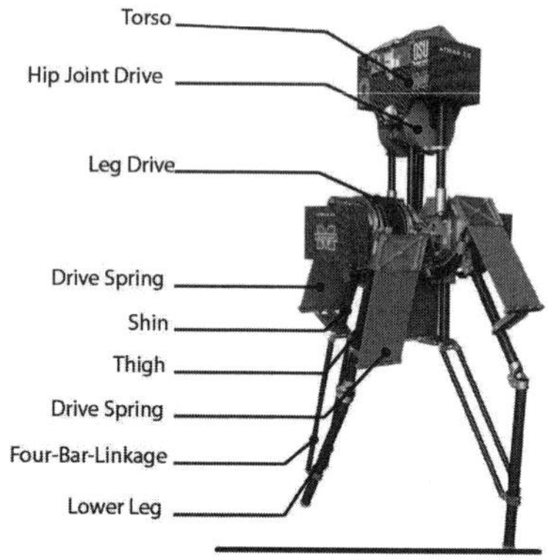

Fig. 1: SolidWorks rendering of ATRIAS 2.0, a 3D bipedal robot that has a mass of 60 Kg and 1 m long legs.

The right (resp. left) hip houses the motor and gear drive assembly for the right (resp. left) leg. Each hip is approximately 10 cm wide, so that the total width of the hips when the legs are parallel is 20 cm. Each leg is formed by a planar 4-bar linkage (i.e., formed by revolute joints with parallel axes), and with the links sized to form a parallelogram, as shown in Figures 1 and 2. The common axis of the top joints of the right 4-bar linkage is aligned with the axis of the right hip; the left leg is similar. At the current time, the legs are terminated in point feet. Nontrivial feet will be added later.

The top two links of each 4-bar linkage are independently driven with a series-compliant actuator. Specifically, each link is connected in series with a large spring, a 50:1 harmonic drive, and a DC brushless motor. A conceptual representation of the actuator is shown in Fig. 2-(a).

It follows that in general position, meaning no contact with the environment, bipedal ATRIAS 2.0 has 16 degrees of freedom (DOF) and 6 independent actuators. Six DOF are associated with the torso. There are 2 DOF where the hips attach to the torso, and each leg has 4 DOF, with 2 DOF arising from the 4-bar mechanism and an additional two DOF due to the springs used in the series-compliant actuators.

3. Generalized Coordinates

Coordinates are defined for the robot in general position, and when the robot is in single support, appropriate holonomic constraints can be applied

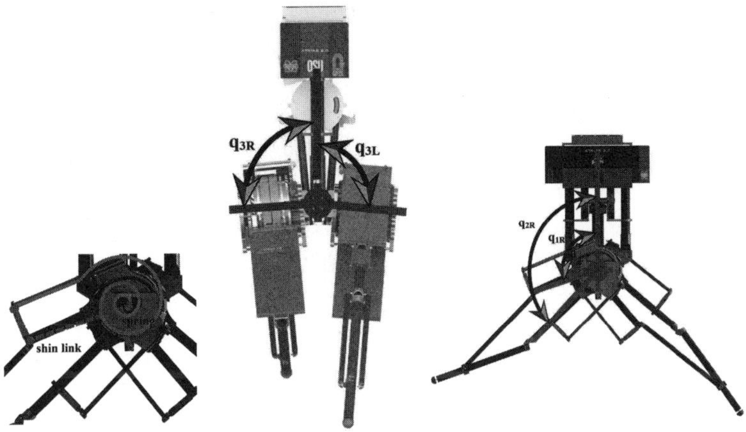

(a) Series-compliant actuation (b) Front view (c) Side view

Fig. 2: (a) Series-compliant actuators, consisting of a DC brushless motor, a 50:1 harmonic drive, and a large spring, are used to power the legs in the sagittal plane. Each hip is formed by the assembly of two such actuators. (b) Coordinates q_{3R} (resp. q_{3L}) is the relative angle between the torso and the right (resp. left) hip in the frontal plane. (c)q_{1R} (resp. q_{2R}) is the relative angle between the hip and the top-front (resp. top-rear) bar of the 4-bar linkage of the right leg in the sagittal plane; similar for the left leg.

to arrive at a reduced set of generalized coordinates. The coordinates are defined independent of which leg is eventually the stance leg; this is different from the approach followed in reference.[3] All relative angles are positive in the direction that respects the right-hand rule.

Let a world frame be defined and attach a Cartesian coordinate frame (x_T, y_T, z_T) to the torso, oriented so that when the torso is upright, the z-axis points upward and the y-axis points forward. TaitBryan Z-X'-Y" (Euler) angles are used to parameterize the orientation of the torso with respect to the world frame. These angles are denoted q_z, q_x, and q_y and are called yaw, pitch, and roll, respectively.

The angles of the right and left hips relative to the torso are denoted q_{3R} and q_{3L}, respectively, as shown in Fig. 2-(b)-(c). Because the four links in each leg form a parallelogram, only two coordinates are need to parameterize the four links. The angles of the top two links relative to the hip are denoted q_{1R} and q_{2R} for the right leg, and q_{1L} and q_{2L} for the left leg. This completes the coordinates for the body of the robot.

Turning to the actuators, because the motors at the hips are connected to the body through a fixed gear ratio, the angles of the rotors relative to the torso are uniquely determined from q_{3R} and q_{3L}. On the other hand, the motors driving the legs are connected through springs, and hence additional coordinates are needed. For each motor, the angle of the output shaft of the harmonic drive is denoted q_{gr}. These angles are used instead of the angles of the rotors of the motors because q_{gr1R}, q_{gr2R}, q_{gr1L}, and q_{gr2L} are in the "same coordinate frame" as the link angles of the legs.

4. Dynamic Model Of Walking

A hybrid dynamic model of walking has been derived under the assumptions used in reference.[3] The walking gait consists therefore of alternating phases of single support (one leg on the ground) and double support (2 legs on the ground). During single support, the end of the stance leg acts as a pivot; in particular, it does not slip and the leg end is free to rotate in pitch, roll, and yaw, though viscous friction is included to reduce the amount of yaw about the stance leg end. The walking surface is flat and rigid. The double-support phase is therefore instantaneous.[4] Impacts with the ground are assumed transversal, meaning that the vertical component of the leg-end velocity is strictly negative. In steady state, the gait is symmetric with respect to the left and right legs.

The mechanical data of the various rigid bodies[†] comprising the robot (i.e., links, gears, and motor rotors) are posted on the web.[2] The values are either estimates obtained from the SolidWorks model of the robot or from manufacturer data sheets.

When in single support, the Cartesian coordinates of the torso are redundant and can be eliminated. Hence, for developing the single-support model, with either the left or the right leg in stance, the configuration variables are $q_s = (q_z, q_x, q_y, q_{1R}, q_{2R}, q_{1L}, q_{2L}, q_{gr1R}, q_{gr2R}, q_{3R}, q_{gr1L}, q_{gr2L}, q_{3L})$. Note that the last 6 coordinates are independently actuated whereas the first 7 coordinates are unactuated. Let $Q_s \subset SO(3) \times S^{10}$ be a connected open subset giving the feasible set for the configuration variables.

A separate dynamic model was developed for each leg in single support using the method of Lagrange, and then expressed in state variable form

$$\dot{x} = f_s(x) + g_s(x)u, \tag{1}$$

[†]The mass of batteries and power amplifiers are lumped in the torso. The mass of each motor and rotor are lumped with the body on which it is mounted. The rotor inertias of the motors used in the series-elastic actuators are treated separately.

where $x = (q_s, \dot{q}_s)$ is the 26-dimensional state vector and u is the vector of 6 actuator torques. Impacts occur when the end of the swing leg contacts the ground. Let $p_z : Q_s \to \mathbb{R}$ denote the vertical height of the swing leg above the ground so that the impact surface is

$$\mathcal{S} = \{(q_s, \dot{q}_s) \in TQ_s \mid p_z(q_s) = 0\}. \tag{2}$$

The impact is modeled as a contact of two rigid bodies and is expressed

$$x^+ = \Delta(x^-), \tag{3}$$

where x^- is the value of the state just before the impact and x^+ is its value just after the impact.

Combining the single support and the impact models for the left (L) and right (R) legs results in the hybrid model

$$\begin{aligned} &\Sigma_L : \begin{cases} \dot{x} = f_{s,L}(x) + g_{s,L}(x)u, & x^- \notin \mathcal{S}_L \\ x^+ = \Delta_{L\to R}(x^-), & x^- \in \mathcal{S}_L \end{cases} \\ &\Sigma_R : \begin{cases} \dot{x} = f_{s,R}(x) + g_{s,R}(x)u, & x^- \notin \mathcal{S}_R \\ x^+ = \Delta_{R\to L}(x^-), & x^- \in \mathcal{S}_R. \end{cases} \end{aligned} \tag{4}$$

5. Gait Optimization and Feedback Control Design

Walking gaits were designed using the method of virtual constraints and hybrid zero dynamics as presented in references,[3,7] adapted to the morphology of ATRIAS 2.0. In this method, one holonomic constraint per actuator is proposed in the form of an output that, when zeroed by a feedback controller, enforces the constraint. The constraints are written as

$$y = h(q_s, \alpha) = h_0(q_s) - h_d(\theta, \alpha), \tag{5}$$

where $h_0(q_s)$ specifies the vector of variables to be controlled, $h_d(\theta, \alpha)$ is the desired evolution of the actuated variables as a function of $\theta(q_s)$, and α is a vector of real parameters to be chosen. Because θ is used to replace time in parameterizing a periodic motion of the biped, it is selected to be strictly monotonic (i.e., strictly increasing or decreasing) along a walking gait.

The virtual constraints are used to design a parameterized family of low-level feedback laws that create a parameterized smooth surface Z_α that is invariant under the closed-loop hybrid model and attractive.[3,5] Hybrid invariance creates a low-dimensional hybrid subsystem of the high-dimensional closed-loop hybrid system. The low-dimensional hybrid subsystem is called the Hybrid Zero Dynamics (HZD), and solutions of it are (exact) solutions of the high-dimensional hybrid system.

In each single-support phase, the zero dynamics takes the form

$$D_{zero}(q_{zero}, \alpha)\ddot{q}_{zero} + H_{zero}(q_{zero}, \dot{q}_{zero}, \alpha) = 0, \tag{6}$$

where, if the virtual constraints (5) have vector relative degree two,[7] the dimension of q_{zero} equals the dimension of q_s minus the number of independent actuators. Hence, for the model used here, dim $q_{zero} = 7$. The process of deriving (6) also provides a closed-form expression[3,5,8] for the the control input $u^*_\alpha(q_{zero}, \dot{q}_{zero})$ achieving

$$0 = h(q_s, \alpha) \Leftrightarrow h_0(q_s) = h_d(\theta, \alpha),$$

that is, the control input that zeros the virtual constraints and creates Z_α.

Selection of the virtual constraints and optimization of the hybrid zero dynamics was performed as detailed in reference,[3] with modification suggested by the study of MABEL, a planar (2D) biped with series-compliant actuation.[6] Cost functions were defined as

$$J_1(\alpha) = \frac{1}{d}\int_0^{T_I} \|u^*_\alpha(t)\|_2^2 \, dt \tag{7}$$

and

$$J_2(\alpha) = \frac{1}{gM_{tot}d}\int_0^{T_I} \sum_{i=1}^{6} [p_i(t)]_+ dt \tag{8}$$

where T_I is the step duration, d is the distance traveled by the center of mass in one step, g is the gravitational constant, M_{tot} is the total mass of the robot, $p_i(t)$ is actuator power, and $[p]_+ = p$ when $p \geq 0$ and equals zero otherwise. The parameter vector α was chosen to minimize one of the two cost functions subject to the gait being periodic and symmetric, feasible ground reaction forces, etc. The cost J_1 represents electric motor energy per distance traveled, while J_2 is the dimensionless cost of mechanical transport.

As shown in references,[3,5,8] the computation of the virtual constraints leads to a variety of feedback controllers for rendering the periodic orbit (i.e., walking gait) exponentially stable. Details for ATRIAS 2.0 will be given elsewhere due to space limitations.

6. Results

Walking gaits for speeds of 0.9 m/s, 1.0 m/s, and 1.4 m/s have been computed via optimization of the hybrid zero dynamics. The optimization at

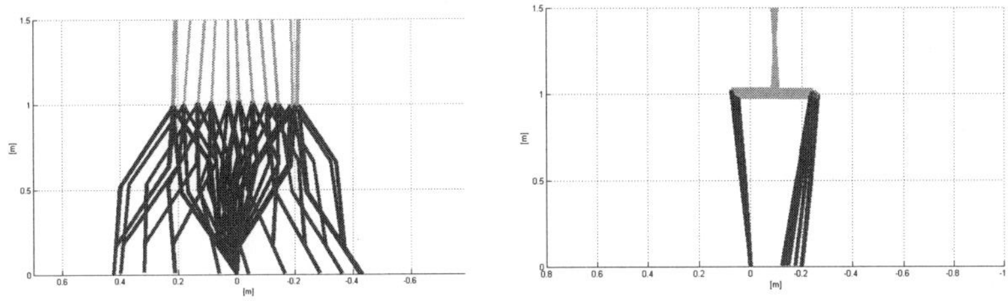

(a) (b)

Fig. 3: Stick diagram of the walking gait of ATRIAS 2.0. Green lines indicate the torso and the hip, the blue and red lines indicate the right and left legs, respectively.

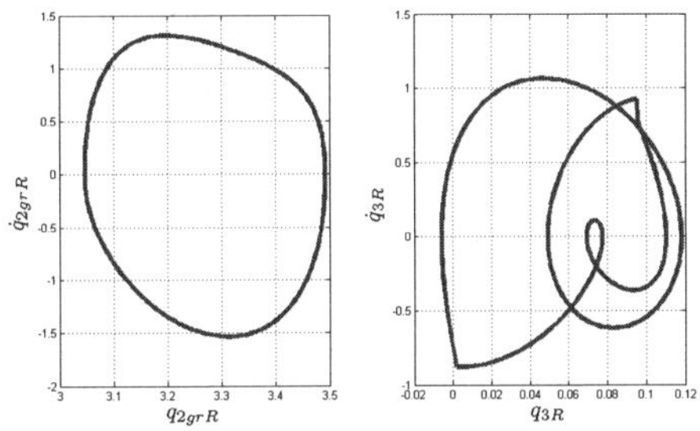

Fig. 4: The plots indicate that a periodic gait is achieved.

Actuator	**Positive Work (J)**	**Negative Work (J)**	**Peak Power (W)**	**Peak Torque (Nm)**
q_1 **stance**	0.5	12.6	-105.6	2.8
q_2 **stance**	16.3	0	98.5	-2.0
q_3 **stance**	2.3	4.7	-48.6	3.1
q_1 **swing**	1.0	0.6	-10.5	0.2
q_2 **swing**	1.8	2.3	20.1	0.9
q_3 **swing**	3.2	0.4	12.2	-1.7

Table 1: After optimizing $J_2(\alpha)$, the energy breakdown over one step is shown for walking at 1 m/s with a step length of 0.8 m (i.e., center of mass travels $d = 0.4$ m) and step duration of 0.4 s.

this point does not take into account the power losses at the harmonic drives, which are expected to have an efficiency of approximately $\eta = 70\%$.

A stick figure representation of the 1.0 m/s gait is shown in Fig. 3. The gait has been optimized with respect to J_2. Fig. 4 shows that a periodic orbit is obtained.

Table 1 breaks down the energy budget for the 1.0 m/s gait, showing how power cycles through the actuators over each step on the periodic orbit. The dimensionless cost of mechanical transport, given by (8), is $CMT = 0.107$, when the harmonic drives are assumed lossless. With the same gait and assuming 70% for the harmonic-drive efficiency, $CMT = 0.142$.

7. Acknowledgements

Ching-Long Shih and Christine Chevallereau contributed to the Lagrangian model of ATRIAS 2.0.

References

1. J. Hurst, Dynamic Robotics Laboratory, Oregon State University. `http://mime.oregonstate.edu/research/drl/`, March 2012.
2. Dynamic Robotics Laboratory - open source project site. `http://beaversource.oregonstate.edu/projects/drl`, March 2012.
3. C. Chevallereau, J.W. Grizzle, and C-L. Shih. Asymptotically stable walking of a five-link underactuated 3D bipedal robot. IEEE Transactions on Robotics, 25(1):37–50, February 2009.
4. Y. Hürmüzlü and D. B. Marghitu. Rigid body collisions of planar kinematic chains with multiple contact points. International Journal of Robotics Research, 13(1):82–92, 1994.
5. B. Morris and J. W. Grizzle. Hybrid invariant manifolds in systems with impulse effects with application to periodic locomotion in bipedal robots. IEEE Transactions on Automatic Control, 54(8):1751–1764, August 2009.
6. Koushil Sreenath, Hae-Won Park, Ioannis Poulakakis, and Jessy W. Grizzle. Compliant hybrid zero dynamics controller for achieving stable, efficient and fast bipedal walking on MABEL. International Journal of Robotics Research, 30(9):1170–1193, August 2011.
7. E. Westervelt, J.W. Grizzle, and D.E. Koditschek. Hybrid zero dynamics of planar biped walkers. IEEE Transactions on Automatic Control, 48(1):42–56, January 2003.
8. E. R. Westervelt, J. W. Grizzle, C. Chevallereau, J.-H. Choi, and B. Morris. Feedback Control of Dynamic Bipedal Robot Locomotion. Control and Automation. CRC Press, Boca Raton, FL, June 2007.

COMPARING ENERGY EFFICIENCY OF HOPPING SYSTEMS WITH LINEAR AND PIECEWISE LINEAR SPRINGS

F. FARIVAR, H. YAGHINI BONABI, M. H. H. KANI, M. J. YAZDANPANAH, M. NILI AHMADABADI

Control and Intelligence Processing Center of excellence, School of ECE, Faculty of Engineering, University of Tehran

In this paper, double-mass hopping systems with piecewise linear and linear springs are compared in terms of energy consumption. The hoppers' compliance specifications are set to achieve zero-energy-loss hopping on a flat surface with equal initial energy. These behaviors are set as the desired behavior of the hopping systems in presence of surface irregularity, carrying load, and error in the sensory system. A sliding-mode controller is designed for each hopper to follow its desired behavior. Simulation results show that the two systems are equivalent in the designed situations. However, linear system consumes less energy when the surface is irregular or when the hopper carries a load. Nevertheless, piecewise linear spring system is more energy efficient in presence of combination of realistic conditions; i.e. the error in the sensory system, bumpy surface, and carrying a load.

1. Introduction

In theoretical and practical studies, it is illustrated that using compliant elements in robots –especially in legged ones- results in interesting emergent behaviors, higher energy efficiency, and better stability in addition to lower control effort [1-9]. While most of the existing compliant legged-robots have linear springs in their joints, natural legged systems benefit from nonlinear compliant characteristics of their tendon, ligament and muscles [10, 11]. Even though it is shown that complex nonlinear springs are efficient in legged robots [12], benefits of using such multifaceted elements over the hardship of making and using them are not clear yet. Moreover, the complaint behavior of natural legged systems are close to piecewise linear and utilizing piecewise linear springs is straightforward. Therefore, here we compare the role of linear and piecewise linear springs in energy efficiency of a double-mass hopping system.

This paper is organized as follows. In Section 2, hopping systems with linear and nonlinear spring specifications are described. Also, hybrid dynamic equations are presented in this section. Sliding mode control is designed in

Section 3 to control hopping system behaviors. In Section 4, simulation results are presented. Discussions and conclusions are mode in Section 5.

2. Hopping Systems

A double-mass hopping system consists of a body, a foot, a spring system, and an actuator. Body and foot are modeled by two point masses. The spring system and the actuator are embedded between the two point masses. Two hopping systems with the piecewise linear and linear spring systems are shown in Fig.1 (a) and Fig.1 (b) respectively. The ground bumpiness is modeled by an ideal prismatic joint that moves the ground up and down.

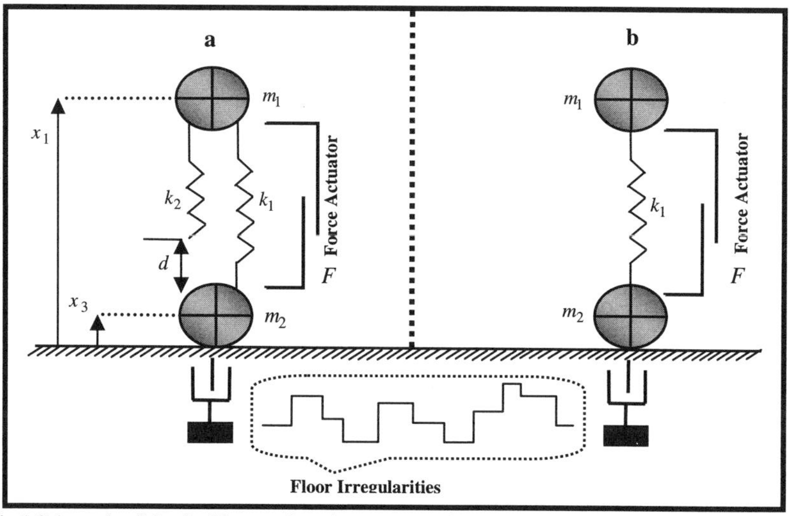

Fig.1 Schematic of the double-mass hopping systems with, (a) piecewise spring, (b) linear spring.

In Fig. (1), $m_{i(=1,2)}$ and $k_{i(=1,2)}$ denote the masses and the stiffness of the spring elements, respectively. Variable $x_{i\,(=1,3)}$ is the displacements of masses. $D_{i(=1,2)}$ is the rest length of springs. Notice that d is a gap between the second mass-spring in Fig. 1(a). Actuator force is denoted by F.

These two systems are designed to hop with an initial energy and without any energy lost. The trajectories of zero-energy-lost systems are set as the desired trajectories for the corresponding system. Phase portrait trajectories corresponding to the designed hopping systems with piecewise linear and linear spring specifications are shown in Fig. 2 (a) and Fig. 2 (b), respectively.

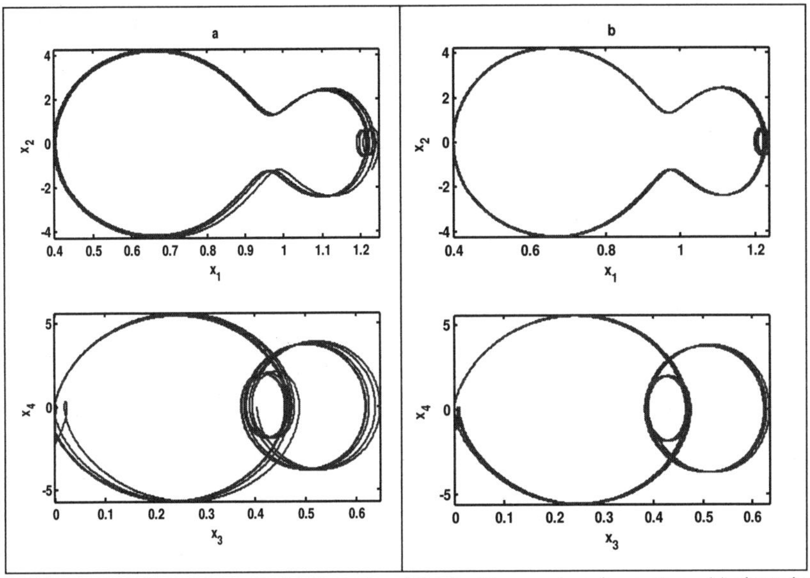

Fig.2 Zero-energy-lost phase portrait trajectories of the double-mass hopping system, (a) piecewise case, (b) linear case.

Sticking impact is assumed for the foot touch-down. The hybrid model of the hopping system with the piecewise linear spring specifications is obtained as Eq. (1).

$$\left(\textit{Fligt Phase}\ \&\ k_2\ \textit{is active}\right)$$

$$\begin{cases} \dot{x}_1 = x_2 \\ \dot{x}_2 = -g - \dfrac{k_1}{m_1}(x_1 - x_3 - D) - \dfrac{k_2}{m_1}(x_1 - x_3 - D + d) \\ \dot{x}_3 = x_4 \\ \dot{x}_4 = -g + \dfrac{k_1}{m_2}(x_1 - x_3 - D) + \dfrac{k_2}{m_2}(x_1 - x_3 - D + d) \end{cases} \tag{1}$$

$$\left(\textit{Stance Phase}\ \&\ k_2\ \textit{is active}\right)$$

$$\begin{cases} \dot{x}_1 = x_2 \\ \dot{x}_2 = -g - \dfrac{k_1}{m_1}(x_1 - x_3 - D) - \dfrac{k_2}{m_1}(x_1 - x_3 - D + d) + \dfrac{F}{m_1} \end{cases}$$

where g is the gravitational force. The equations for the linear spring case are achieved by omitting d and k_2 on Eq. (1). Notice that the actuator force is applied on the stance phase.

State trajectories of the designed hopping systems with an initialized energy are shown in Fig. 3.

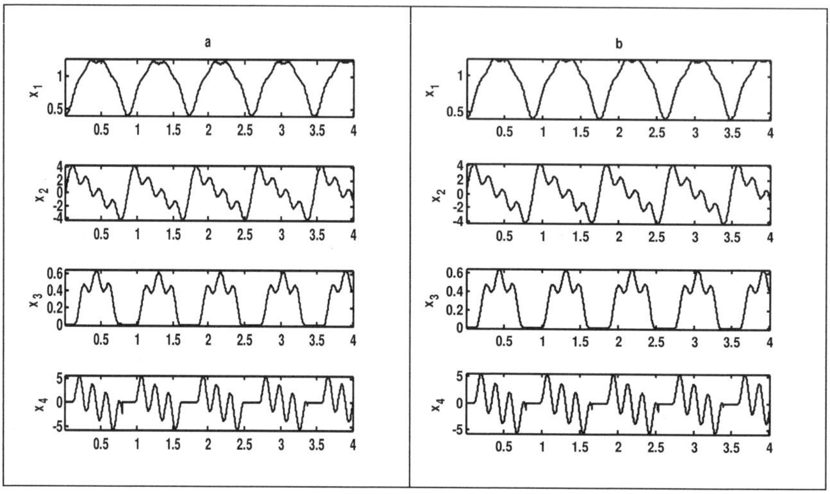

Fig.3 State space trajectories of the double-mass hopping systems, (a) piecewise case, (b) linear case.

In next section, a classic nonlinear control is designed to control the hopping system behaviors.

3. Sliding Mode Control

In this section, a nonlinear control system based on sliding mode control (SMC) is designed. Control effort is an actuator force F, introduced in previous section. Behaviors of designed hopping systems without any actuator force are the target models for controlled systems. State trajectories of the designed systems without any actuator force are considered as desired trajectories. Hence, states of controlled hopping systems should track the target model states. Let us define the tracking error as $e_1 = x_1 - x_{d-1}$ and $e_2 = x_2 - \dot{x}_{d-1}$, where x_{d-1} is the first state of the target model. The sliding surfaces is defined as $S(t) = e_2(t) + \lambda e_1(t)$, where $S(t) \in R$ and λ is a real positive constant parameter. So, actuator force is obtained as Eq. (2).

$$F = m_1\left(-\eta S - f_1(x_1, x_3) - \lambda e_2 + \ddot{x}_{1d}\right) \tag{2}$$

where $f_1(x_1, x_3) = -g - \frac{k_1}{m_1}(x_1 - x_3 - D) - \frac{k_2}{m_1}(x_1 - x_3 - D + d)$, and η is a real constant parameter.

Remark. The actuator force is applied on the stance phase.

4. Simulation Results

Hopping systems with piecewise linear and linear spring specifications are tested under conditions as; (1) Variation of m_1 is considered as a symbol of carrying load. (2) Floor irregularities are generated as the up and down steps by normal distributions with $\mu = 0$ and different variances. (3) Error in the position and velocity sensory data. Errors are modeled by normal distributions with $\mu = 0$ and different variances. (4) Combination of conditions (2) and (3).

Parameters of hopping systems are specified as follows:

$m_1 = 5kg; m_2 = 1.5kg; k_1 = 1300\left(N/m\right); k_2 = 100\left(N/m\right); D_1 = 0.7m; d = 0.2m$.

Tables (1-4) show the energy consumptions on all tests. Examples of system behaviors are shown in Fig. (4).

Table 1. Energy consumption of the hopping systems in presence of floor irregularities.

	Nonlinear System $m_1 = 5kg$	Linear System $m_1 = 5kg$	Ground Profile characteristics	The Number of Iteration
Actuator Energy (Average)	1.4574×10^3	1.4367×10^3	$\mu = 0; \sigma = 0.015$	$n = 10$
	2.7769×10^3	2.7135×10^3	$\mu = 0; \sigma = 0.03$	$n = 10$
	4.7628×10^4	3.4093×10^4	$\mu = 0; \sigma = 0.06$	$n = 10$
	5.1672×10^4	3.8584×10^4	$\mu = 0; \sigma = 0.09$	$n = 10$
	1.2003×10^5	7.2480×10^4	$\mu = 0; \sigma = 0.12$	$n = 10$

Table 2. Energy consumption of the hopping systems with variations in the body mass in presence of floor irregularities.

	Nonlinear System $m_1 = 6kg$	Linear System $m_1 = 6kg$	Nonlinear System $m_1 = 7kg$	Linear System $m_1 = 7kg$	Ground Profile characteristics	The Number of Iteration
Actuator Energy (Average)	8.4216×10^4	8.3438×10^4	1.7311×10^5	1.8114×10^5	$\mu = 0$ $\sigma = 0.015$	$n = 10$
	7.7032×10^4	7.7262×10^4	1.6625×10^5	1.7728×10^5	$\mu = 0$ $\sigma = 0.03$	$n = 10$
	3.3145×10^4	4.3277×10^4	1.3998×10^5	1.8311×10^4	$\mu = 0$ $\sigma = 0.06$	$n = 10$
	3.5420×10^4	1.0892×10^5	7.2574×10^4	4.7165×10^4	$\mu = 0$ $\sigma = 0.09$	$n = 10$
	1.1452×10^5	7.0755×10^4	5.8139×10^5	1.2376×10^5	$\mu = 0$ $\sigma = 0.12$	$n = 10$

Table 3. Energy consumption of the hopping systems in presence of sensory errors.

	Nonlinear System $m_1 = 5kg$	Linear System $m_1 = 5kg$	Position Sensor Error Characteristics	Velocity Sensor Error Characteristics	The Number of Iteration
Actuator Energy (Average)	1.5559×10^3	5.9375×10^3	$\mu = 0; \sigma = 0.002$	$\mu = 0; \sigma = 0.003$	$n = 20$
	2.0299×10^3	3.9072×10^3	$\mu = 0; \sigma = 0.001$	$\mu = 0; \sigma = 0.005$	$n = 20$

Table 4. (a) Energy consumption of the hopping systems in presence of sensory errors and floor irregularities.

	Nonlinear System $m_1 = 5kg$	Linear System $m_1 = 5kg$	Position Sensor Error Characteristics	Velocity Sensor Error Characteristics	Ground Profile characteristics	The Number of Iteration
Actuator Energy (Average)	1.7439×10^3	2.5902×10^3	$\mu = 0$ $\sigma = 0.002$	$\mu = 0$ $\sigma = 0.003$	$\mu = 0$ $\sigma = 0.015$	$n = 10$
	1.9938×10^3	1.9351×10^3	$\mu = 0$ $\sigma = 0.002$	$\mu = 0$ $\sigma = 0.003$	$\mu = 0$ $\sigma = 0.03$	$n = 10$
	2.9163×10^4	8.9423×10^4	$\mu = 0$ $\sigma = 0.002$	$\mu = 0$ $\sigma = 0.003$	$\mu = 0$ $\sigma = 0.06$	$n = 10$
	5.8163×10^4	7.7820×10^4	$\mu = 0$ $\sigma = 0.002$	$\mu = 0$ $\sigma = 0.003$	$\mu = 0$ $\sigma = 0.09$	$n = 10$
	7.4415×10^4	7.6552×10^4	$\mu = 0$ $\sigma = 0.002$	$\mu = 0$ $\sigma = 0.003$	$\mu = 0$ $\sigma = 0.12$	$n = 10$

Table 4. (b) Energy consumption of the hopping systems in presence of sensory errors and floor irregularities.

	Nonlinear System $m_1 = 5kg$	Linear System $m_1 = 5kg$	Position Sensor Error Characteristics	Velocity Sensor Error Characteristics	Ground Profile characteristics	The Number of Iteration
Actuator Energy (Average)	1.1761×10^3	1.5232×10^3	$\mu = 0$ $\sigma = 0.001$	$\mu = 0$ $\sigma = 0.005$	$\mu = 0$ $\sigma = 0.015$	$n = 10$
	2.5470×10^3	2.9604×10^3	$\mu = 0$ $\sigma = 0.001$	$\mu = 0$ $\sigma = 0.005$	$\mu = 0$ $\sigma = 0.03$	$n = 10$
	1.4323×10^4	3.3213×10^4	$\mu = 0$ $\sigma = 0.001$	$\mu = 0$ $\sigma = 0.005$	$\mu = 0$ $\sigma = 0.06$	$n = 10$
	2.3565×10^4	4.4585×10^4	$\mu = 0$ $\sigma = 0.001$	$\mu = 0$ $\sigma = 0.005$	$\mu = 0$ $\sigma = 0.09$	$n = 10$
	6.2667×10^4	6.5690×10^4	$\mu = 0$ $\sigma = 0.001$	$\mu = 0$ $\sigma = 0.005$	$\mu = 0$ $\sigma = 0.12$	$n = 10$

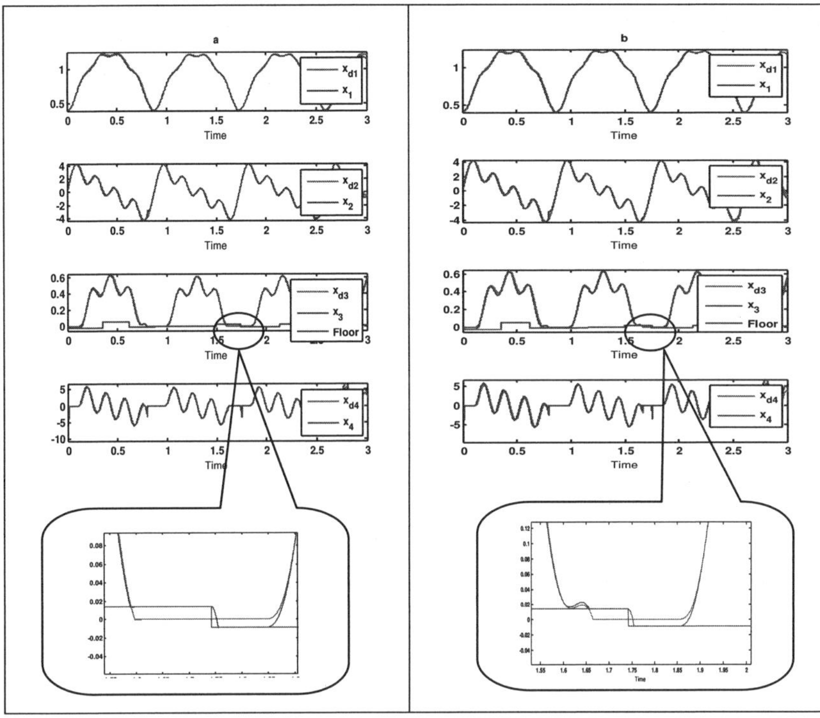

Fig.4 State space trajectories of the double-mass hopping systems, (a) piecewise case, (b) linear case.

5. Discussions and Conclusions

A summary of the obtained results are as follow:

(1) In presence of floor irregularities, the energy consumption of the hopping system with the linear spring is less than the energy spending of the hopping system with the piecewise linear spring (See Table.1). The difference between the energy efficiency of the two systems increases with the surface bumpiness.

(2) In presence of the body mass variations (m_1) on the rather flat floors, the hopper with piecewise linear spring has not any advantage over the system with the linear spring (See Table.2, The first two rows).

(3) In presence of moderate floor irregularities and variations of m_1 , the piecewise linear system has a higher energy efficiency (See Table.2 rows 3 and 4). The linear system shows a better energy efficiency when the ground is too bumpy (See Table.2 the last row).

(4) In the presence of sensory error with or without floor irregularities, the piecewise linear system has a higher energy economy (See Table. 3 and Table. 4). The difference between the two systems is higher for moderate sensory error and ground bumpiness.

As the results show, hopping system with the piecewise linear spring has not any energy advantage over the linear one if the systems have precise sensors and suitable control systems. Nevertheless, in realistic situations when the floor is moderately bumpy, sensory system is erroneous - especially in low cost robots-, and there are modeling errors –like in the body mass-, the piecewise linear system shows a considerably higher energy efficiency.

The results suggest that, benefits of using piecewise linear springs in legged robots should be studied more seriously. Beside, characteristics of piecewise and linear spring systems should be compared when a simpler control system is applied.

References

1. M. Raibert, "Legged Robots That Balance", *MIT Press*, 2000.
2. G. Pratt, M. Williamson, P. Dillworth, J. Pratt, and A. Wright, "Stiffness isn't everything", *Experimental Robotics IV*, pp. 253–262, 1997.
3. J. Bongard and R. Pfeifer, "How the Body Shapes the Way We Think: A New View of Intelligence", *A Bradford Book*, 2006.
4. B. Vanderborght, "Dynamic Stabilisation of the Biped Lucy Powered by Actuators with Controllable Stiffness", *Springer*, 2010.
5. S. A. Migliore, E. A. Brown, and S. P. DeWeerth, "Biologically inspired joint stiffness control", *IEEE International Conference on Robotics and Automation (ICRA)*, pp. 4508–4513, 2005.
6. H. J. Bidgoli, A. Vafaei, A. Sadeghi, and M. N. Ahmadabadi, "Learning Approach to Study Effect of Flexible Spine on Running Behavior of a Quadruped Robot", *13th International Conference on Climbing and Walking Robots (CLAWAR)*, 2010.
7. M. Hutter, C. D. Remy, M. A. Hoepflinger, and R. Siegwart, "ScarlETH: Design and control of a planar running robot", *IEEE/RSJ International Conference on Intelligent Robots and Systems (IROS)*, pp. 562–567, 2011.
8. K. W. Buffinton, C. D. Remy, K. W. Buffinton, and R. Y. Siegwart, "Energetic of Passivity Based Running with High-Compliance Series Elastic Actuation", *12th Mechatronics Forum Biennial International Conference*, 2010.
9. M. Hutter, C. D. Remy, M. Hopflinger, and R. Siegwart, "SLIP running with an articulated robotic leg", *IEEE/RSJ International Conference on Intelligent Robots and Systems (IROS)*, pp. 4934–4939, 2010.
10. R. Shadmehr and M. A. Arbib, "A mathematical analysis of the force-stiffness characteristics of muscles in control of a single joint system", *Biological cybernetics*, Vol. 66, No. 6, pp. 463–477, 1992.
11. P. Provenzano, R. Lakes, T. Keenan, and R. Vanderby, "Nonlinear ligament viscoelasticity", *Annals of Biomedical Engineering*, Vol. 29, No. 10, pp. 908–914, 2001.
12. J. D. Karssen and M. Wisse, "Running with improved disturbance rejection by using non-linear leg springs", *International Journal of Robotics Research*, Vol. 30, No. 13, pp. 1585-1595, 2011.

STARLETH: A COMPLIANT QUADRUPEDAL ROBOT FOR FAST, EFFICIENT, AND VERSATILE LOCOMOTION*

MARCO HUTTER, CHRISTIAN GEHRING, MICHAEL BLOESCH, MARK A. HOEPFLINGER, C. DAVID REMY, and ROLAND SIEGWART

Autonomous Systems Lab, Institute of Robotics and Intelligent Systems, Swiss Federal Institute of Technology (ETHZ), Zurich, Switzerland, mahutter@ethz.ch

This paper introduces Starl*ETH*, a compliant quadrupedal robot that is designed to study fast, efficient, and versatile locomotion. The platform is fully actuated with high compliant series elastic actuation, making the system torque controllable and at the same time well suited for highly dynamic maneuvers. We additionally emphasize key elements of a powerful real time control and simulation environment. The work is concluded with a number of experiments that demonstrate the performance of the presented hardware and controllers.

Keywords: compliant quadruped, legged locomotion, torque control, SEA

1. Introduction

Inspired by locomotion in nature, research in legged robotics is currently undergoing substantial changes. In contrast to the early walking machines [1] which were built stiff and controlled kinematically, more and more compliant systems are nowadays finding their way into robotics; thereby offering an entirely new field of research. Due to their mechanical softness in terms of back-drivability and low system inertia, these systems are perfectly suited to continuously interact with their environment. They can cope with (unexpected) ground contact collisions and are at the same time inherently save for hand-in-hand collaboration with human operators. Similar to nature, where compliant elements in muscles and tendons largely contribute to the energetic efficiency [2,3], springs can be exploited in robotics to support non-linear oscillations.

*This research was supported by the Swiss National Science Foundation through the National Centre of Competence in Research Robotics, as well as by the Hans Eggenberger Foundation.

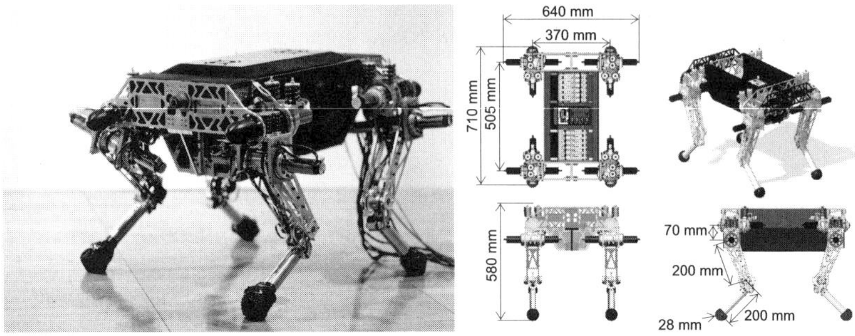

Fig. 1. Starl*ETH*, a compliant quadruped developed at the Autonomous Systems Lab to combine *efficient* and *versatile* locomotion. (Foto: François Pomerleau)

The transition from classic industrial robots to such novel platforms is accompanied by a paradigm shift from position control to torque control strategies. To give an example, research done within the DARPA learning locomotion challenge [4] has impressively shown, how far position control strategies can be pushed through precise kinematic trajectory planning and execution. However, at the same time this challenge clearly demonstrated the limitations of stiff position control, since detailed knowledge, not only about the internal states of the robot, but also about the environment was required. As a solution to cope with these limitations, new torque control approaches (e.g. whole body control [5] or inverse dynamics [6]) allow to modulate the system's behavior in a sophisticated and adaptive manner.

To combine the advantages of *efficiency* and *versatility* in a single device, we developed Starl*ETH* (Springy Tetrapod with Articulated Robotic Legs) that pushes the state of the art in the field of compliant quadrupeds. This paper introduces the hardware prototype and reports on tests in several static and dynamic experiments. We address the setup of a simulation and control environment in detail and show how a proper integration of soft- and hardware can lead to an extremely powerful tool for development. While we will briefly outline possible control approaches and give precise references to existing literature in the corresponding fields, a full description of the employed controllers is not within the scope of this article.

2. Hardware Description

Starl*ETH* is based on four identical legs that are arranged in an X configuration. With linear dimensions of about 0.6 m and a total weight of 23 kg, it has the size of a medium-sized dog. Every leg has a total of 3 degrees of

Table 1. Key data of Starl*ETH*

total leg length	488.5	mm	max./min. hip AA[1]	-30/30	°
leg segment length	200	mm	max./min. hip FE[2]	-40/80	°
total mass	23	kg	max./min. knee FE	-145/0	°
payload	25	kg	max. hip AA,FE speed	600	°/s
max. hip AA,FE torque	24	Nm	max. knee FE speed	350	°/s
max. knee torque	40	Nm	motor power	200	W

Note: [1] abduction/adduction, [2] flexion/extension

freedom, one each for hip abduction/adduction, hip flexion/extension, as well as knee flexion/extension. The robot's design is based on earlier studies that we conducted to optimize a planar running leg [7,8]. This leg was extended with an abduction degree of freedom. The key data of the mechanics as well as the actuation are summarized in Table 1. We put emphasis on keeping the inertia of the moving segments minimal by concentrating all actuators at the main body through the use of chain and cable pulley systems. This is beneficial to ensure fast swing leg motion and to reduce impact losses at the intermittent ground contact. The lightweight main body is fabricated as a carbon fiber sandwich monocoque with aluminum front- and back connectors. It contains well protected all electronic parts that are cooled through active air circulation. The key elements of Starl*ETH* are the high compliant series elastic actuators that are implemented in all joints using linear compression springs in a pre-compressed setup (antagonistically at the hip joints). Springs decouple the motor (*Maxon EC-4pole* 200 W) and gearbox (*Harmonic Drive CSG-14*, 1:100 reduction) from the joint. This setup ensures robustness against impacts, allows for energy storage to improve the efficiency [7], and provides full torque controllability. All joint angles, motor angles, and spring deflections are precisely measured in every joint [8]. The lightweight ball feet (70 g, Figure 3(b)) are based on air filled racquet balls that provide amble cushioning with minimal weight while having a high contact friction coefficient. To detect changes in the contact force, we tightly integrated force sensing resistors (FSR). The proprioceptive sensor equipment is completed with a *X-sense MTi IMU* that includes an accelerometer and gyroscopes.

3. Simulation and Control Framework

Controlling highly complex and inherently unstable systems like Starl*ETH* with 12 actuated joints and 6 unactuated (floating base) degrees of freedom in an often unknown environment requires a very powerful and reliable

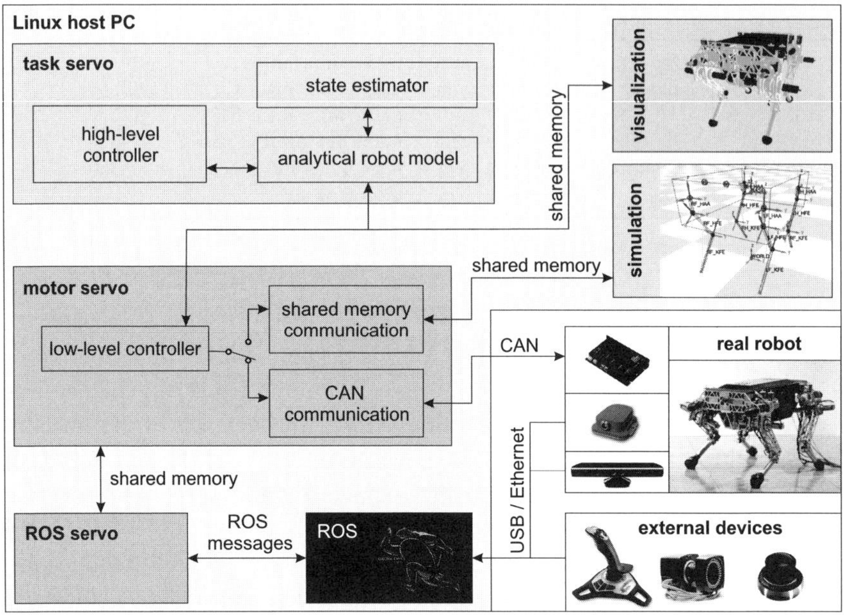

Fig. 2. The real-time simulation and control environment is based on SL [9] and analytical global kinematics and dynamics. The framework allows an immediate exchange of simulation and actual robot, as well as a direct integration into ROS.

control setup. After several iterations through different configurations, we realized a framework as depicted in Figure 2. The setup is based on a single centralized host-PC that is responsible for the entire high-level control part. The software contains the real-time simulation and control environment SL [9], which is used as a multi-body dynamics engine, for visualization of the robot and its environment, as well as for handling the internal messaging between the different servos through shared memory.. It additionally provides tools for data logging and post-processing.

As a core element, the *motor servo* is responsible for timing and coordination of the individual actuators. It contains all low level regulators (position, velocity, torque, ...) and signal filters. Using either shared memory to access the simulation, or a CAN gateway to communicate with the actual robot, the user has the possibility to run simulations and experiments (even in parallel) with exactly the same controller implementation. This is highly advantageous for controller development, since it avoids code reimplementation, simplifies debugging and testing, and ensures certain safety if the controllers show good robustness in the simulation.

The *ROS servo* provides an interface to the *robotic operation system* that can not meet the hard real-time constraints of SL itself. This connection opens the door to a huge set of existing interfaces to different sensors and existing software packages. In our current setup, this includes the IMU, the Microsoft Kinect sensor, as well as several external devices such as a 3D joystick for control or a vicon motion capture system to track the main body position/orientation.

Finally, the *task servo* decouples the controller entirely from the multibody simulation and provides the global kinematics and dynamics (robot model) of the system in analytical form. Therefore, we developed proNEu [10], a MATLAB framework that uses the Symbolic Math toolbox[a] to translate a relative kinematic tree into global kinematics using Euler rotations, and to derive the system dynamics using projected Newton-Euler equations. In addition to speed and simplicity of the implementation, this separation between the model used in the controller and the simulation has several benefits. First of all, the control engineer has user friendly access to any kind of updated kinematic or dynamic parameters. Second, sensor noise, imperfect state estimation, or (intentionally) introduced differences between simulation and controller allows to test the robustness of the controller against modeling errors and limits the problematic of instable plant inversion. The state estimator which is implemented on top of the robot model fuses kinematic information with the IMU sensor signals. It is based on an Extended Kalman filter [11] and estimates not only the base coordinates but is also able to generate a sparse ground elevation map using the contact feet.

In practice, the rate-limiting factor in our framework is the data connection of the CAN bus system between the host PC and the motor controllers. Although we managed to parallelize 4 channels (1 per leg), the bit rate of 1 MBits/s limits the overall loop cycle time to about 400 Hz.

4. Experimental Results

In the first part of the experimental validation, we are focusing on hardware performance. First of all, walking and running machines need to provide sufficient torque that enables them to conduct various maneuvers. A representative scenario is standing up from a completely crouched position with maximal payload (Figure 3(a)). Due to the high power per weight ratio, we are able to stand up with a payload of 25 kg. This is more than the total weight of the actual robot and considerably better than for systems

[a]www.mathworks.com/products/symbolic

(a) standing up with 25kg payload

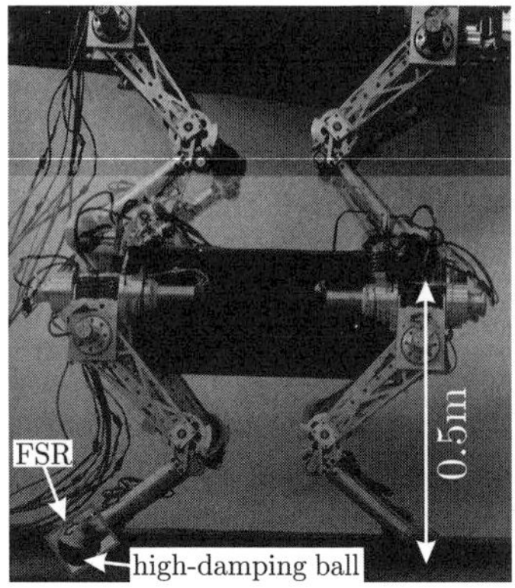

(b) falling down from 0.5m

Fig. 3. The torque capability (a) was experimentally validated with 25kg payload. Its passive compliance (b) protects the actuators from large peak forces at landing and allows for joint torque control.

such as Little Dog [12] or ALoF [13]. In theory, the peak torque at the knee of 40 Nm with a lever arm of 0.2 m results in a maximal total lift force of 800 N or a maximal stand-up payload of ≈50 kg. However, these peak loads should only occur in highly dynamic maneuvers and were hence not experimentally validated for safety reasons.

To be capable to perform dynamic maneuvers, it is crucial that the system can cope with impacts. In contrast to a stiff system (again c.f. Little Dog [12] or AloF [13]), pre-impact kinetic energy should not be entirely dissipated in an collision that harms the mechanical structure. Instead, the system is able to either damp out a certain amount of energy or to store it as potential energy in the actuator springs. To demonstrate both, we dropped Starl*ETH* from a height of ≈0.5 m. The passive mechanical compliance in the actuators, as well as the high damping feet elements protect the system from the impact at landing. This ensures that, even if all motors are blocked, none of the gearboxes gets damaged through high peak forces. The high compliant springs lead to large oscillations (red-dotted line in Figure 4) including a number of short rebounces (grey background) after landing. This can be avoided through active damping (black-solid line in Figure 4) using a virtual model controller that exerts the force $\mathbf{F}_{mb}$ and torque $\mathbf{M}_{mb}$ at the main body

$$\mathbf{F}_{mb} = k_p\left(\mathbf{r}_{des} - \mathbf{r}_{act}\right) + k_d\left(\dot{\mathbf{r}}_{des} - \dot{\mathbf{r}}_{act}\right) \tag{1}$$

$$\mathbf{M}_{mb} = k_p\left(\varphi_{des} - \varphi_{act}\right) + k_d\left(\dot{\varphi}_{des} - \dot{\varphi}_{act}\right) \tag{2}$$

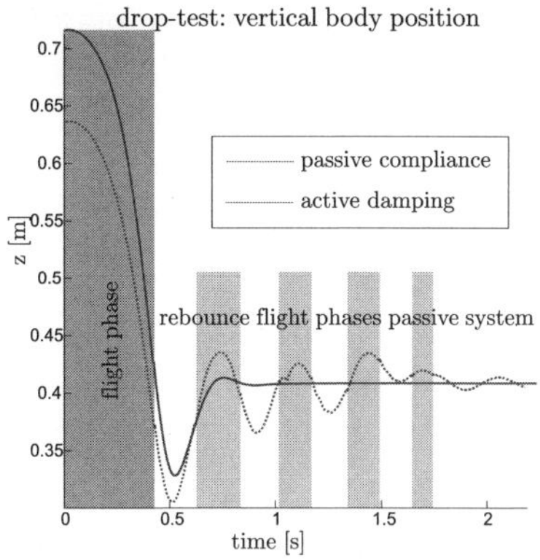

Fig. 4. The vertical main body position of the system dropped from $\approx 0.5m$ was estimated using IMU values during flight and kinematic information during contact phase: the red-dotted curve shows the high oscillations due to the passive compliance with intermittent flight phases (grey area), while active damping (blue-solid) using virtual model control ensures nearly critical damping of the system.

with $\mathbf{r}, \boldsymbol{\varphi}$ representing the main body position and orientation and the spring and damping constant k_p, k_d. In combination with additional gravity compensation $\mathbf{F}_j^g = -m_j\mathbf{g}$ for the $N_p = 4 \cdot 3 + 1$ bodies, the force is distributed onto the different contact forces $\mathbf{F}_i^c$ ($N \leq 4$):

$$\begin{pmatrix} \mathbf{F}_1^c \\ \vdots \\ \mathbf{F}_N^c \end{pmatrix} = \begin{bmatrix} \mathbf{I} & \dots & \mathbf{I} \\ \tilde{\mathbf{r}}_1^c & \dots & \tilde{\mathbf{r}}_N^c \end{bmatrix}^+ \begin{pmatrix} \mathbf{F}_{mb} - \sum_{j=1}^{N_b} \mathbf{F}_j^g \\ \mathbf{M}_{mb} - \sum_{j=1}^{N_b} \mathbf{r}_j^g \times \mathbf{F}_j^g \end{pmatrix} \tag{3}$$

These contact forces are subsequently translated into joint torques $\boldsymbol{\tau}$ using Jacobian-transposed mapping with the relative Jacobians at contact points $\mathbf{J}_i = \frac{\partial \mathbf{r}_i^c}{\partial \mathbf{q}} = \frac{\partial\left(\mathbf{r}_i^{foot} - \mathbf{r}_i^{base}\right)}{\partial \mathbf{q}}$ and center of gravity $\mathbf{J}_j = \frac{\partial \mathbf{r}_j^g}{\partial \mathbf{q}} = \frac{\partial\left(\mathbf{r}_j^{CoG} - \mathbf{r}_i^{base}\right)}{\partial \mathbf{q}}$:

$$\boldsymbol{\tau} = \sum_{i=1}^{N} \mathbf{J}_i^T \mathbf{F}_i^c + \sum_{j=1}^{N_b} \mathbf{J}_j^T \mathbf{F}_j^g \tag{4}$$

5. Conclusion

The paper introduced Starl*ETH*, a compliant quadruped robot that was developed to study *fast*, *efficient*, and *versatile* locomotion. This lightweight system is driven by high compliant series elastic actuation that makes it fully torque controllable, robust against impacts, and hence perfectly suited for sophisticated dynamic maneuvers. A framework based on SL [9] in combination with an analytical global kinematics and dynamics implementation serves as a powerful real-time simulation and control environment. It gives the control engineer a compact and user friendly tool to develop and test control algorithms. Starl*ETH* was successfully tested in various experiments that demonstrated a high torque capability, its compliant protection against impact collisions, and full torque controllability. Hence, the presented platform in combination with the control and simulation environment is perfectly suited for dynamic maneuvers and sophisticated model based torque control strategies. As a first step, successful walking experiments were conducted using the presented virtual model controller (1)-(4) in combination with kinematic swing leg control [14]. This holds as a basis for future work towards more dynamic maneuvers.

References

1. P. Gonzalez de Santos, E. Garca and J. Estremera, *Quadrupedal locomotion: an introduction to the control of four-legged robots* (Springer, 2006).
2. G. A. Cavagna, F. P. Saibene and R. Margaria, *J Appl Physiol* **19**, 249 (1964).
3. R. M. Alexander, *International Journal of Robotics Research* **9**, 53 (1990).
4. P. D. Neuhaus, J. E. Pratt and M. J. Johnson, *The International Journal of Robotics Research* **30**, 216 (2011).
5. R. Philippsen, L. Sentis and O. Khatib, An open source extensible software package to create whole-body compliant skills in personal mobile manipulators, in *IROS*, 2011.
6. M. Kalakrishnan, J. Buchli, P. Pastor, M. Mistry and S. Schaal, Fast, robust quadruped locomotion over challenging terrain, in *ICRA*, 2010.
7. M. Hutter, C. D. Remy, M. A. Hoepflinger and R. Siegwart, High compliant series elastic actuation for the robotic leg scarleth, in *CLAWAR*, 2011.
8. M. Hutter, C. D. Remy, M. A. Hoepflinger and R. Siegwart, Scarleth: Design and control of a planar running robot, in *IROS*, 2011.
9. S. Schaal, *The SL Simulation and Real-Time Control Software Package*, tech. rep., USC (2009).
10. M. Hutter, C. Gehring and R. Siegwart, *proNEu: Analytical Kinematics and Dynamics*, tech. rep., Autonomous Systems Lab, ETHZ (2011).
11. M. Bloesch, M. Hutter, M. A. Hoepflinger, C. D. Remy, C. Gehring and R. Siegwart, *submitted to RSS Robotics Science and Systems* (2012).
12. M. P. Murphy, A. Saunders, C. Moreira, A. A. Rizzi and M. Raibert, *The International Journal of Robotics Research* (2010).
13. C. D. Remy, O. Baur, M. Latta, A. Lauber, M. Hutter, M. A. Hoepflinger, C. Pradalier and R. Siegwart, *Industrial Robot: An International Journal* **38**, 264 (2011).
14. M. Bloesch, State estimation and robust control for a quadruped robot (2011).

IMPROVED EFFICIENCY IN LEGGED RUNNING USING LIGHTWEIGHT PASSIVE COMPLIANT FEET*

MARCO HUTTER, CLAUDE HOLENSTEIN, DARIO FENNER, C. DAVID REMY, MARK A. HOEPFLINGER, and ROLAND SIEGWART

Autonomous Systems Lab, Institute of Robotics and Intelligent Systems, Swiss Federal Institute of Technology (ETHZ), Zurich, Switzerland, mahutter@ethz.ch

This paper investigates the mechanical benefits of employing a passive foot segment to improve energetic efficiency in legged running. The proposed lightweight design significantly reduces impact and damping losses, while simultaneously allowing for a natural-looking stance configuration. Actuator input and ankle spring properties were optimized in simulation and successfully tested in 2D running experiments.

Keywords: compliant leg design, passive ankle joint, efficient hopping, SEA

1. Introduction

Locomotion in nature is highly effective in terms of *speed* and energetic *efficiency*. To achieve both, elastic elements in muscles and tendons play a highly important role by introducing a passive support to the vertical oscillation of the main body, as it is found in all dynamic gaits of animals and humans [1]. Inspired by these principles, research has tried to transform the idea of elastic energy storage into robotic devices using series elastic actuation. Springs in each joint can not only be used as force elements to accurately control the joint torque, but also act as passive elements that enhance the natural dynamics of the system. The springs can store energy during the deceleration phase after landing and subsequently release it before lift-off. In the optimal case, the actuator must never produce negative work, but uses the entire stance phase to introduce energy into the system [2].

While optimal control strategies allow exploiting the natural dynamics in a very efficient manner [3], major limitations are given by the mechanical

*This research was supported by the Swiss National Science Foundation through the National Centre of Competence in Research Robotics.

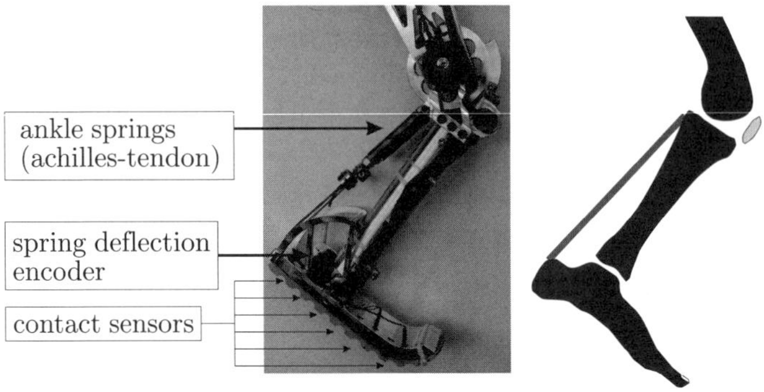

Fig. 1. The robotic foot is built similar to a human leg. Springs emulate the behavior of the Achilles tendon and Soleus muscle. An orthopedic foot sole ensures a soft impact, and sensors allow contact detection and precise joint torque measurement.

structure of the hardware system. For example, the capability to store energy depends heavily on the damping and friction effects in the joint springs. In addition to this, a second substantial drain of energy in dynamic maneuvers can be attributed to impact collision at landing. In articulated legs such as Starl*ETH* [4], these energy losses largely depend on the configuration at impact and on the individual segment mass. Although swing leg retraction [5] or pre-impact speed adaptation potentially allow to lower these losses, their implementation in real robotic devices is often difficult to realize.

In this paper we describe the development and mechanical optimization of a passive foot element (Figure 1). This is done in theory, simulation, and on a real robotic system, with the overall goal of increasing the efficiency of one-legged running. Optimization strategies are not only used to improve motor control policies, but also to optimize the mechanical properties of the leg. Inspired by biology, the foot is equipped with springs emulating the behavior of the Achilles tendon and Soleus muscle (Figure 1), while, at the same time, the sole of an orthopedic foot absorbs the impact shocks.

2. Impact Analysis of Segmented Legs

The dynamics equation of a system colliding with its environment can be described by $\mathbf{M}\left(\dot{\mathbf{q}}^+ - \dot{\mathbf{q}}^-\right) + \mathbf{J}_s^T \mathcal{F}_s = \mathbf{0}$ with the generalized velocities before ($\dot{\mathbf{q}}^-$) and after ($\dot{\mathbf{q}}^+$) impact, the mass matrix $\mathbf{M}$, the support Jacobian $\mathbf{J}_s$, and the contact impulsive force $\mathcal{F}_s$. Since the contact point is immediately brought to zero velocity $\dot{\mathbf{r}}_s^+ = \mathbf{J}_s\dot{\mathbf{q}}^+ = \mathbf{0}$, the instantaneous change in generalized velocities is given as

$$\Delta\dot{\mathbf{q}} = \dot{\mathbf{q}}^+ - \dot{\mathbf{q}}^- = -\mathbf{M}^{-1}\mathbf{J}_s^T\left(\mathbf{J}_s\mathbf{M}^{-1}\mathbf{J}_s^T\right)^{-1}\mathbf{J}_s\dot{\mathbf{q}}^- = -\bar{\mathbf{J}}_s\mathbf{J}_s\dot{\mathbf{q}}^-, \tag{1}$$

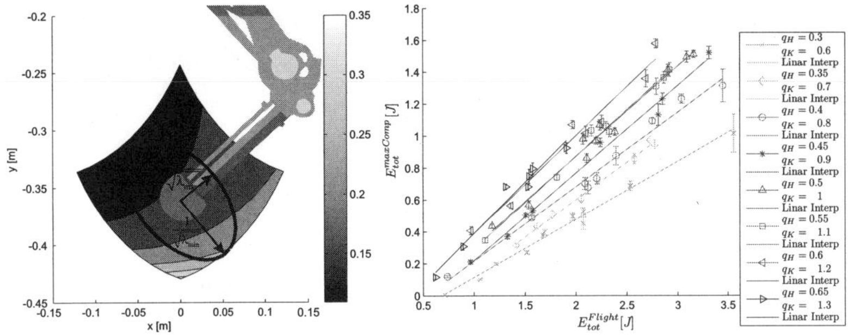

(a) GIE for Scarl*ETH* (b) Drop-Test energy loss

Fig. 2. A GIE analysis (b) indicates that the pre-impact velocity should be perpendicular to the shank in order to minimize the impact losses. Theoretic analysis (a) and experimental validations (b) show that a crouched position results in lower energy losses.

with the support point inertia $\mathbf{\Lambda}_s := \left(\mathbf{J}_s\mathbf{M}^{-1}\mathbf{J}_s^T\right)^{-1}$ and the dynamically consistent generalized inverse of the support Jacobian $\bar{\mathbf{J}}_s = \mathbf{M}^{-1}\mathbf{J}_s^T\mathbf{\Lambda}_s$ [6]. The energy loss in such a collision can be written as

$$E_{loss} = \Delta E_{kin} = -\tfrac{1}{2}\Delta\dot{\mathbf{q}}^T\mathbf{M}\Delta\dot{\mathbf{q}} \tag{2}$$

This can be expressed in task space using $\Delta\dot{\mathbf{r}}_s = -\dot{\mathbf{r}}_s^- = \mathbf{J}_s\Delta\dot{\mathbf{q}}$ as

$$E_{loss} = \Delta E_{kin} = -\tfrac{1}{2}\Delta\dot{\mathbf{r}}_s^T\mathbf{\Lambda}_s\Delta\dot{\mathbf{r}}_s = -\tfrac{1}{2}\dot{\mathbf{r}}_s^{-T}\mathbf{\Lambda}_s\dot{\mathbf{r}}_s^- . \tag{3}$$

Since only part of the derivation was given in [7], the entire proof for (2) and (3) is compactly outlined in (A.1) and (A.2).

The complex impact mechanics can be analyzed through an eigenvalue decomposition of the foot point inertia $\mathbf{\Lambda}_s$. The highest impact energy loss occurs if the pre-impact speed $\dot{\mathbf{r}}_s^-$ is aligned with the eigenvector corresponding to the greatest eigenvalue. In [8] and [7] a visual interpretation is proposed by plotting the generalized inertia ellipsoid (GIE) with the perpendicular axis $1/\sqrt{\lambda_{1,2}}\mathbf{e}_{1,2}$.

3. Implications on a 2-Segmented Leg

Investigation of different landing configurations for the 2-segmented leg Scarl*ETH* [2,4] based on the GIE (Figure 2(a)) indicates that impact energy losses are minimal if the pre-impact velocity $\dot{\mathbf{r}}_s^-$ is nearly perpendicular to the inertia main axis of the contact segment. For in-place hopping, this implies that energy losses are minimized for a crouched positions, ideally with the shank segment being close to horizontal (dark area, $< 15\%$ energy loss). On the other hand, more elongated leg configurations (bright area,

$> 25\%$ energy loss) have a higher end-effector inertia in vertical direction and hence significantly more energy drain with the same vertical pre-impact speed.

To confirm these theoretical results on an actual robotic device, we conducted several experiments with Scarl*ETH*. In reality, a contact collision is not an instantaneous process. It induces extremely high accelerations at the different segments which is very difficult to capture in measurements. Additional losses due to friction or damping effects that are impossible to model require to evaluate efficiency in a different way. As a very intuitive solution, we propose to measure how much of the center of gravity (CoG) energy during flight phase

$$E_{tot}^{-} = mg\left(y_{CoG} - y_0\right) + \tfrac{1}{2}m\dot{\mathbf{r}}_{CoG}^{2} \tag{4}$$

can be preserved until the mid-stance point. At that point, the velocity of all segments is zero and the knee spring (characteristic $c(\delta)$ identified in [4]) is maximally compressed

$$E_{maxComp} = mg\left(y_{CoG} - y_0\right) + \int_0^{\delta_{max}} c\left(\delta\right) d\delta. \tag{5}$$

Since energy is constant during flight phase, all available measurements between apex (highest point in flight curve) and touchdown are used to estimate (4). The energy at maximal compression is accurately determinable as we do not have to account for kinetic energy that is prone to noise.

The results of multiple dropping tests are depicted in Figure 2(b). Taking the height relative to the leg length at impact ensures that only the actual jumping height contributes to the potential energy. For a constant landing configuration the dependency between the energy level (4) and the energy loss (3) is linear, with an increasing slope for less crouched configurations. This is in agreement with the theory: Since a different starting height changes only the norm of $\dot{\mathbf{r}}_s^- = \dot{\mathbf{r}}_{CoG}^-$ but not its direction, (3) reduces to $-\frac{1}{2}\lambda(\dot{\mathbf{r}}_s^-)^2$ with the geometric scalar λ, and the relation $\frac{\Delta E_{kin}}{E_{tot}^-} = \frac{-1/2\lambda(\dot{\mathbf{r}}_s^-)^2}{1/2m(\dot{\mathbf{r}}_s^-)^2} = -\frac{\lambda}{m}$ becomes constant. To keep λ small, it is indispensable to minimize the contact segment mass and to align the inertia main axis perpendicular to the impact direction - requirements that seek for an additional foot element.

4. Passive Compliant Foot Segment

Inspired by biomechanical studies (e.g.[9]) that pointed out multiple advantages of 3-segmented legs, we developed the foot element depicted in

Figure 1. This segment is built as lightweight as possible, since the contact inertia $\mathbf{\Lambda}_s$ is proportional to the mass of the most distal segment. Hence, the structurally optimized foot was manufactured as a single high-tensile aluminum part that has a total weight of only 125 g including a high resolution joint encoder (80'000 qc/rev), 10 digital distributed ground contact switches, and an orthopedic foot sole. In the current setup, the two modular ankle springs weight 56 g and produce a rotational stiffness of $c_a = 10\,\mathrm{Nm/rad}$.

Previous studies conducted in our group aimed at finding optimal periodic gaits for a given robot model [3]. Periodicity in this example means that the system defined through parameters $\mathbf{p}$ passes through the same states $\mathbf{x}$ at every apex transit (highest position in flight phase), respectively that the stride-to-stride transfer function P maps the periodic states onto themselves $\mathbf{x}^{k+1} = P^*\left(\mathbf{x}^k, \mathbf{p}, \mathbf{u}\right) = \mathbf{x}^*$. In the present work, we extended this and simultaneously optimized the actuator input $\mathbf{u}(t)$ and ankle joint stiffness c_a by minimizing the positive mechanical actuator work (joint torque $\boldsymbol{\tau}$)

$$\underset{c_a, \mathbf{u}(t)}{\text{minimize}} \qquad f\left(\mathbf{x}^*, \mathbf{p}, \mathbf{u}(t)\right) = \int_0^{T_{hop}} \max\left(0, \boldsymbol{\tau}\dot{\mathbf{u}}\right) dt \tag{6}$$

subject to the motion periodicity and hopping height constraints

$$P^*\left(\mathbf{x}^*, \mathbf{p}, \mathbf{u}(t)\right) - \mathbf{x}^* = 0 \tag{7}$$

$$\max\left(y(t)\right) - h = 0. \tag{8}$$

In this optimization problem, we fixed the period time of motion T_{hop}, the hopping height h, as well as all mechanical parameters $\mathbf{p}$ except the ankle joint stiffness c_a. The motor actuation $\mathbf{u}(t)$ is described as a time based Fourier series with 3 coefficients:

$$\mathbf{u}(t) = \sum_{i=0}^{3} a_i \sin\left(i\frac{2\pi}{T}t\right) + b_i \cos\left(i\frac{2\pi}{T}t\right) \tag{9}$$

This has the benefit that we can account for the motor bandwidth limitations by keeping the number of Fourier coefficients small. The solution of the optimization problem is depicted in Figure 3(a): The bars indicate the value of the cost function (positive actuator work), the solid line the impact loss of the 3-segmented leg in comparison to the 2-segmented leg (dashed line). The actuator work as well as the total impact loss is reduced by increasing the stiffness. Up to about $10\,\mathrm{Nm/rad}$ the ankle spring is compliant enough to allow for a double contact phase (heel makes contact), above that, only the toe touches the ground. The obtained results are close to the theoretical predictions [10] for a massless leg which showed that $c_{ankle} = \frac{l_{foot}}{l_{thigh}} c_{knee}$ is required for a symmetric bending (Figure 3(b)).

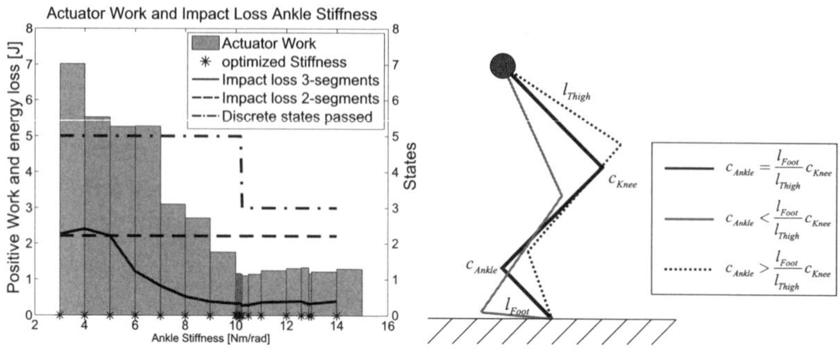

(a) ankle stiffness optimization (b) massless 3-segmented leg

Fig. 3. The positive actuator work minimization (6) shows minimal energy consumption for a stiffness close to the point where heel strikes ground. (b) The ankle stiffness of a massless 3-segmented leg structure has a theoretical optimum [10].

These results indicate that in terms of efficiency, there is no benefit to make the ankle so stiff that heelstrike is fully avoided. However, considering controllability of the system, double contact is quite advantagous: since the ankle is not actuated, proper torque control (e.g., using virtual model control strategies [4]) is only applicable during this phase. In the time between toe and heel contact, a considerable amount of energy is stored in the ankle springs and the vertical heel speed is reduced such that the second impact at the heel is comparably small. In summary, the stiffness should be chosen to be just below the critical stiffness.

5. Experimental Validation

A simplified control structure was implemented in the 3-segmented leg to experimentally test the leg behavior in single legged running experiments for a varying ankle joint stiffness. The application of a passive compliant foot shows a big improvement of motor energy consumption (Figure 4(a)): a compliant ankle spring (4.0 Nm/rad (red-dotted), 2.85 J) requires more energy compared to the stiffer springs (7.2 Nm/rad (blue-solid), 10.5 Nm/rad (black-dashed), both 2.45 J). Evaluating the hopping height (Figure 4(b)) shows the same tendency with a much higher hopping height (0.175, 0.185m) for the stiffer springs and a huge gap to the more compliant setup with a hopping height of only 0.04m. Hence, the optimization results could be entirely approved: as long as the heel still touches the ground, increasing the stiffness results in a remarkable improvement. As soon as the critical stiffness (only toe touches the ground) is reached, making the ankle stiffer does not improve the overall performance.

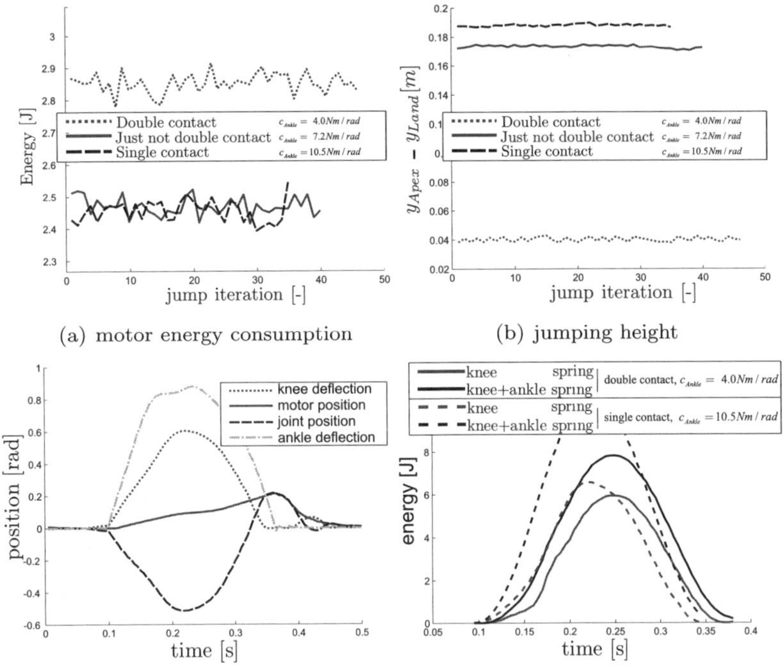

(a) motor energy consumption

(b) jumping height

(c) position distribution

(d) energy distribution

Fig. 4. The foot segment was tested in single legged running experiments using a varying ankle joint stiffness.

The hopping experiment with an ankle stiffness of 10 Nm/rad (Figure 4(c)) shows that the proposed control strategy provides only positive actuator power (joint velocity $>$ 0 during the entire stance phase) and the actual motor travel distance is only about 20% of the actual joint deflection.

The energetic improvement can be entirely assigned to the ankle. Comparing the stiffest and the most compliant ankle joint shows, that the energy stored in the knee remains the same while the total energy stored in the ankle and knee spring can be drastically increased (Figure 4(d)).

6. Conclusion

In the presented paper we showed theoretically as well as experimentally that the efficiency in single legged running with a 2-segmented leg is mechanically limited to a large extent due to energy losses by impacts at landing. As a biologically inspired solution, a highly optimized foot segment with joint torque and contact sensing capabilities was developed. A parameter and actuator optimization framework allowed to choose the best ankle joint stiffness and actuation input for highly efficient planar running. Due to the lower impacts at landing, the new flight configuration is less crouched than with the 2-segmented leg. A thorough experimental evaluation in planar running confirmed the theoretical expectations.

Appendix A. Proof of Kinetic Energy Loss

$$\begin{aligned}
\Delta E_{kin} &= \tfrac{1}{2}\dot{\mathbf{q}}^{+T}\mathbf{M}\dot{\mathbf{q}}^{+} - \tfrac{1}{2}\dot{\mathbf{q}}^{-T}\mathbf{M}\dot{\mathbf{q}}^{-} = \tfrac{1}{2}\left(\dot{\mathbf{q}}^{+} + \dot{\mathbf{q}}^{-}\right)\mathbf{M}\left(\dot{\mathbf{q}}^{+} - \dot{\mathbf{q}}^{-}\right) \\
&\overset{\dot{\mathbf{q}}^{+}=\dot{\mathbf{q}}^{-}+\Delta\dot{\mathbf{q}}}{=} \tfrac{1}{2}\left(\Delta\dot{\mathbf{q}} + 2\dot{\mathbf{q}}^{-}\right)\mathbf{M}\Delta\dot{\mathbf{q}} \\
&\overset{\Delta\dot{\mathbf{q}}=-\bar{\mathbf{J}}_s\mathbf{J}_s\dot{\mathbf{q}}^{-}}{=} -\dot{\mathbf{q}}^{-T}\mathbf{M}\bar{\mathbf{J}}_s\mathbf{J}_s\dot{\mathbf{q}}^{-} + \tfrac{1}{2}\Delta\dot{\mathbf{q}}^{T}\mathbf{M}\Delta\dot{\mathbf{q}} \\
&\overset{\mathbf{J}_s\bar{\mathbf{J}}_s=\mathbf{I}}{=} -\dot{\mathbf{q}}^{-T}\mathbf{M}\bar{\mathbf{J}}_s\mathbf{J}_s\bar{\mathbf{J}}_s\mathbf{J}_s\dot{\mathbf{q}}^{-} + \tfrac{1}{2}\Delta\dot{\mathbf{q}}^{T}\mathbf{M}\Delta\dot{\mathbf{q}} \\
&\overset{\mathbf{M}\bar{\mathbf{J}}_s\mathbf{J}_s=\left(\bar{\mathbf{J}}_s\mathbf{J}_s\right)^T\mathbf{M}}{=} -\dot{\mathbf{q}}^{-T}\left(\bar{\mathbf{J}}_s\mathbf{J}_s\right)^{T}\mathbf{M}\bar{\mathbf{J}}_s\mathbf{J}_s\dot{\mathbf{q}}^{-} + \tfrac{1}{2}\Delta\dot{\mathbf{q}}^{T}\mathbf{M}\Delta\dot{\mathbf{q}} \\
&\overset{\bar{\mathbf{J}}_s\mathbf{J}_s\dot{\mathbf{q}}^{-}=\Delta\dot{\mathbf{q}}}{=} -\tfrac{1}{2}\Delta\dot{\mathbf{q}}^{T}\mathbf{M}\Delta\dot{\mathbf{q}} \qquad \text{(A.1)} \\
&\overset{\Delta\dot{\mathbf{q}}=\mathbf{M}^{-1}\mathbf{J}_s\boldsymbol{\Lambda}_s\Delta\dot{\mathbf{r}}_s}{-} \tfrac{1}{2}\Delta\dot{\mathbf{r}}_s^{T}\boldsymbol{\Lambda}_s\underbrace{\mathbf{J}_s\mathbf{M}^{-1}\mathbf{M}\mathbf{M}^{-1}\mathbf{J}_s^{T}}_{\boldsymbol{\Lambda}_s^{-1}}\boldsymbol{\Lambda}_s\Delta\dot{\mathbf{r}}_s \\
&= -\tfrac{1}{2}\Delta\dot{\mathbf{r}}_s^{T}\boldsymbol{\Lambda}_s\Delta\dot{\mathbf{r}}_s \qquad \text{(A.2)}
\end{aligned}$$

References

1. C. T. Farley, J. Glasheen and T. A. Mcmahon, *Journal of Experimental Biology* **185**, 71 (1993).
2. M. Hutter, C. D. Remy, M. H. Hoepflinger and R. Siegwart, High compliant series elastic actuation for the robotic leg scarleth, in *CLAWAR*, 2011.
3. C. D. Remy, Optimal expoitation of natrual dynamics in legged locomotion, PhD thesis2011.
4. M. Hutter, C. D. Remy, M. H. Hoepflinger and R. Siegwart, Scarleth: Design and control of a planar running robot, in *IROS*, 2011.
5. J. G. D. Karssen, M. Haberland, M. Wisse and S. Kim, The optimal swing-leg retraction rate for running, in *ICRA*, 2011.
6. O. Khatib, *IEEE Journal of Robotics and Automation* **3**, 43 (1987).
7. J. P. Schmiedeler and K. J. Waldron, *Advances in Robot Kinematics* , 129 (2000).
8. H. Asada, *Journal of Dynamic Systems Measurement and Control-Transactions of the Asme* **105** (1983).
9. R. Blickhan, A. Seyfarth, H. Geyer, S. Grimmer, H. Wagner and M. Guenther, *Philosophical Transactions of the Royal Society* **365**, 199 (2007).
10. A. Seyfarth, M. Guenther and R. Blickhan, *Biological Cybernetics* **84**, 365 (2001).

ENVIRONMENT-DEPENDENT SELECTION OF LOCOMOTION PATTERNS WITH KAIRO-II

L. Pfotzer*, S. Bohn, G. Heppner, A. Rönnau, R. Dillmann

Forschungszentrum Informatik an der Universität Karlsruhe (FZI)
Haid-und-Neu-Str. 10-14, 76131 Karlsruhe, Germany
**E-mail: pfotzer@fzi.de*
www.fzi.de

This paper describes a method of detecting different ground conditions with a modular wheel-driven robot by measuring the motor currents. Depending on the roughness and strength of the terrain a suitable locomotion pattern will be selected to optimise the locomotion of the robot. We have evaluated the implemented methods on different indoor and outdoor surfaces.

Keywords: Ground Condition Sensing, Locomotion Switching, Biologically-Inspired Locomotion, Modular Robots

1. Introduction

Using different biologically-inspired locomotion methods is an common way to overcome deficits of wheel-driven robots in unstructured environments.[1] Many approaches concentrate on improving and optimising the locomotion methods itself. Starting the locomotion, respectively switching between different locomotion methods, usually requires an operator to decide which method is suitable for the current surface. With an autonomous robot we want to use different locomotion methods for different environments without the need for a supervising operator manually adapting the control parameters. The robot should know about its locomotion possibilities and decide by itself which is the best way of moving through the current environment.

There already exist some methods of sensing and detecting the ground. Yang et al.[2] developed a tactile sensor measuring frictional forces. Their results look very promising, but using tactile sensors seems inappropriate for wheeled robots. Another approach investigates different visual terrain classification methods for mobile robots.[3] Using visual sensors for ground detection is computationally expensive and it is often not possible to get a significant value describing the suitability of the ground for wheeled robots.

In this work we focus on detecting different environments by measuring the current of the driving wheel motors. Moreover, we define two different ground estimation values: The *ground roughness* and the *ground strength*. Combining these values, it is possible to automatically choose a suitable locomotion method. With our approach we want to optimise the autonomous locomotion of biologically-inspired robots like our modular inspection robot KAIRO-II.

2. The Multi-Segmented Inspection Robot KAIRO-II

Fig. 1: The multi-segmented inspection robot KAIRO-II.

KAIRO-II is a multi-segmented sensor carrier, consisting of drive segments with wheels which are connected through joint segments (see Fig. 1). Each drive segment has two DOF and each joint segment has three DOF. All joint angles are measured with optical encoders, giving both absolute and incremental data.

The hierarchical control software is implemented in the robotic framework MCA2 (Modular Controller Architecture)[4] and includes base control, motion execution and maneuver control. The complete base control for regulating motors and reading sensors is implemented by a UCoM[5] (Universal Controller Module) in every segment. This custom made circuit board has an integrated motor current measuring component to determine the currents of the wheel motor. The software layer above the base control handles sensor data and calculates the joint angles of the robot kinematics. At the top level the maneuver control plans the motion of the robot using a virtual rail. There are several locomotion maneuvers defined, such as wheel-driven, inchworm, caterpillar and sidewinder locomotion.[1] Every maneuver is implemented as a collection of states inside a finite state machine.[6]

3. Ground Condition Sensing

To measure and detect the ground conditions, we use the internally measured wheel motor currents. We want to have an estimation value to examine which locomotion method is suitable for the ground, rather than classifying the ground for example into grassy, sandy or stony ground. For wheel-driven robots the drivability is more important than the type of the ground.

3.1. *Ground Roughness*

In a first step, we defined a roughness value R for continuous estimation of the terrain roughness. This value can be calculated using the *Moving Standard Deviation (MSTD)* which is well known in the finance sector, measuring the market volatility. The *MSTD* is calculated in two steps. First, the *Moving Average* (M_{avg}) of the motor currents X_i of one segment is computed for a given time period Δt as shown in Eq. 1.

$$M_{avg} = \frac{X_1 + \cdots + X_n}{n}, \text{where}\, n = \text{amount of current values in}\, \Delta t \qquad (1)$$

Next, the *MSTD* δ of one segment is calculated with Eq. 2 using the previously calculated *Moving Average* (M_{avg}).

$$\delta = \sqrt{\frac{(X_1 - M_{avg})^2 + (X_2 - M_{avg})^2 + \cdots + (X_n - M_{avg})^2}{n}} \qquad (2)$$

The resulting roughness value R combines the *MSTD* values of all robot segments by a simple mean filter. To receive an up-to-date value while driving with the robot, we calculate R continuously over a short time window Δt (see Fig. 2). This *MSTD* window is variable and influences the considered count of motor current measurements. While a shorter window size Δt leads to a faster changing roughness, a longer window size results in a smoother roughness function. We achieved good results with a *MSTD* window size of $\Delta t = 300ms$. This corresponds to an amount of $n = 10$ measurements for every calculation cycle.

3.2. *Ground Strength*

As a second step, we investigated how to measure the strength S of the ground. This is necessary because the robot could sink into soft soil, which

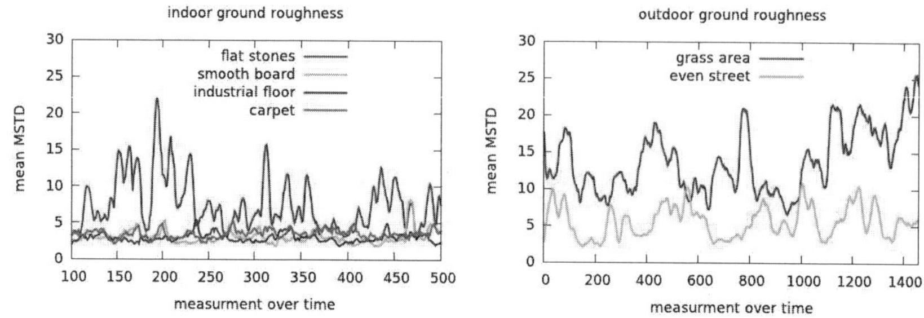

Fig. 2: Continuously calculated ground roughness R on different types of indoor and outdoor surfaces.

is difficult to detect with the motor currents. Thus, we can run a special ground test, if the robot stops. This test consists of the following moving sequence: The robot raises its head segment and activates the wheel motor of this segment. As the head segment is lowered, the motor current will increase when the wheels reach the ground. Depending on the gradient of increase, we can infer the ground strength S. This gradient of increase σ can be calculated by fitting a least square line into the measured motor current data (see Fig. 3). The gradient of the least square line is easy to compute. A high gradient implies a solid ground, a flat gradient a soft ground. On soft grounds the motor current rises much slower than on solid grounds, because the rotating wheels dig into the soft soil instead of blocking up immediately.

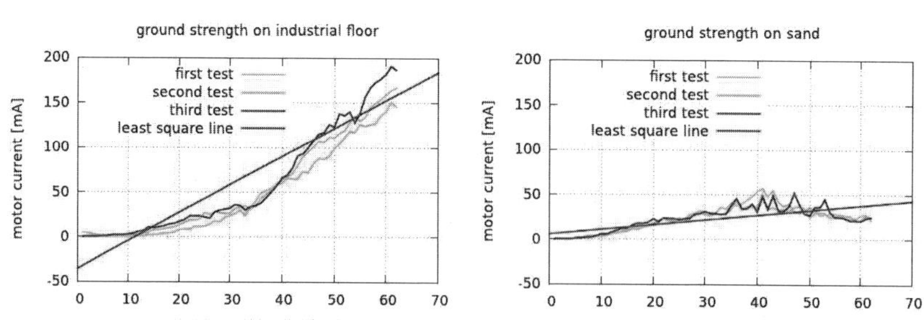

Fig. 3: Measured motor currents and least square line on two different surfaces: industrial floor and sand. The ground strength value corresponds to the gradient of the calculated least square line.

4. Locomotion Selection

The gathered values for ground roughness and ground strength can be used for selecting a suitable locomotion method. First of all, we have to consider which locomotion methods are available and for which ground conditions they are suitable. Currently, there are four different locomotion methods implemented for KAIRO-II: Wheel-driven, inchworm, caterpillar and sidewinder locomotion. The evaluation of these locomotion methods[1] indicated that wheel-driven is the preferred locomotion because of the highest speed. In addition, inchworm is suitable for little rough grounds, caterpillar is best for very rough grounds and sidewinder is well for soft soils. If the roughness value is between a lower and an upper threshold value, the locomotion selection switches to the inchworm locomotion. This motion is faster than the caterpillar locomotion, which is used if the ground roughness exceeds the upper threshold value. The locomotion selection switches to the sidewinder locomotion if the ground strength indicates soft soil by a very small gradient. Every special locomotion is executed only for a specific count of cycles. Then, the robot is back at its home position and the locomotion selection reruns the ground strength test.

5. Implementation

Realising the computation of the ground estimation values, we implemented two new modules in the control software of KAIRO-II: *TerrainRoughness* and *TerrainStrength*. Both modules are placed in the sensor group and receive the wheel motor currents as inputs. The terrain roughness module continuously computes the roughness value R as described above. In contrast to the roughness module, the strength module is only active if the ground detection maneuver has been enabled. This maneuver is implemented as a new state *StateGroundTest* in the maneuver control of KAIRO-II and executes the ground detection sequence described above. While running this maneuver, the *TerrainStrength* module records the motor currents of the segment selected for the test. After finishing the moving sequence by a minimum lowering height or a maximum motor current, the ground strength S is calculated from the recorded motor currents. The determined ground strength stays valid until a new measurement sequence is started.

To implement the locomotion selection, we added a new module *MotionChooser* at the top level of the control software. Based on the calculated roughness and strength values, the motion chooser decides which locomotion is suitable. Currently we have implemented this decision by expert knowledge but we are working on examining further selection methods. Figure 4 gives an overview to the overall locomotion selection system. This figure only shows the newly added or modified modules of the control software.

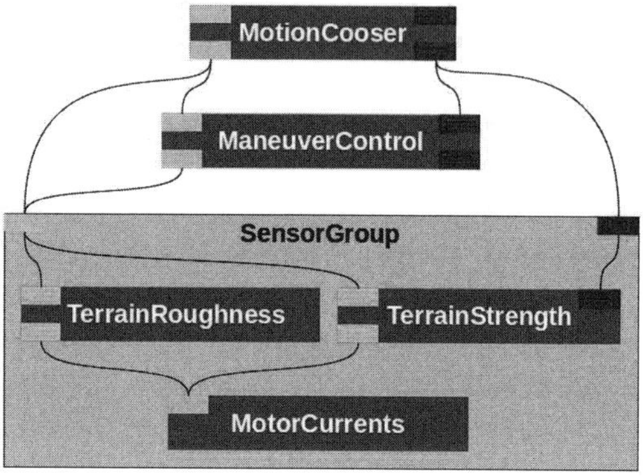

Fig. 4: Implemented software structure of the locomotion selection system of KAIRO-II. Only the newly added and modified MCA2 modules are shown in this figure.

6. Evaluation and Results

Evaluating the above described methods, we performed different tests. First, we evaluated the calculation of the roughness value. For this purpose, we recorded the ground roughness R, while driving on different indoor surfaces like a smooth board, industrial floor, carpet and flat stones (see Fig. 5). We also tested our implementation in outdoor terrains like grass and even street. If we compare the measured ground roughness values R in Fig. 2 we can see, as expected, that the roughness on stones is much higher than on smooth surfaces.

It is possible to evaluate the new ground test maneuver and the calculation of the ground strength at the same time because they are tightly related. Thus, we used grounds with different strengths like industrial floor and sand (Fig. 6). Figure 3 shows the recorded motor currents of three different tests and the fitted least square line of one test. It is noticeable that there is a great difference of the gradient between the industrial floor and sand. Evaluating the repeating accuracy, we rerun the tests three times with each ground, obtaining good results.

(a) smooth board

(b) industrial floor

(c) carpet

(d) flat stones

Fig. 5: Measurement of ground roughness with KAIRO-II at four different terrains: smooth board, industrial floor, carpet and flat stones.

(a) industrial floor

(b) sand

Fig. 6: Measurement of ground strength with KAIRO-II at two different terrains: (a) industrial floor and (b) sand.

Our final evaluation with an indoor test environment showed that the overall system consisting of ground detection and locomotion selection, work as expected. Depending on the detected ground roughness, the *mMotionChooser* switches correctly to the inchworm or the caterpillar locomotion. Likewise the switching to the snake-like locomotion could be observed, if the ground strength increases.

7. Conclusion and Outlook

Wheel-driven robots with the ability of using alternative locomotion patterns like caterpillar or snake locomotion often lack an automated method for choosing a ground condition dependent locomotion. This is why we developed a system to select a suitable locomotion for the detected type of ground. To achieve this, we defined the two ground estimation values roughness and strength, which can be calculated by monitoring the motor currents. Furthermore, we implemented a system which determines these ground values and automatically selects suitable locomotion patterns. The evaluation of the ground estimation and the overall system showed that it is possible to select suitable locomotion patterns based on different ground conditions.

We are going to continue our work testing the stability of our approach in further outdoor environments. In addition, we are going to investigate different locomotion selection methods.

References

1. L. Pfotzer, S. Bohn, C. Birkenhofer and R. Dillmann, Biologically-inspired locomotion with the multi-segmented inspection robot KAIRO-II, in *Proceedings of CLAWAR2011, 14th International Conference on Climbing and Walking Robots and the Support Technologies for Mobile Machines*, 2011.
2. Y. Lü, S. Ma, B. Li and L. Chen, Ground Condition Sensing of a Snake-like Robot, in *Proceedings of the 2003 IEEE, Intemational Conference on Robotics,Intelligent Systems and Signal Processing, Changsha, China*, 2003.
3. K. B. Y. N. Khan, P. Komma and A. Zell, Grid-based visual terrain classification for outdoor robots using local features, in *IEEE Symposium on Computational Intelligence in Vehicles and Transportation Systems (CIVTS)*, 2011.
4. K. Uhl and M. Ziegenmeyer, MCA2 – An Extensible Modular Framework for Robot Control Applications, in *Proceedings of CLAWAR2007, 10th International Conference on Climbing and Walking Robots*, 2007.
5. K. Regenstein, Modulare, verteilte Hardware-Software-Architektur für humanoide Roboter, PhD thesis, Karlsruher Institut für Technologie. 2010.
6. C. Birkenhofer, Adaptive Steuerung eines mehrsegmentigen Inspektionsroboters, PhD thesis, FZI Karlsruhe, 2010.

SOFTER LEGS ALLOW A MODULAR HEXAPOD TO RUN FASTER

J. Sastra*, S. Revzen and M. Yim

University of Pennsylvania, School of Engineering and Applied Science, Philadelphia, PA 19104

We describe a bioinspired hexpedal robot that uses only 7 motors and is built from CKBot modular robot modules. The robot has elastic 2-DOF legs whose stiffness falls in the biologically relevant range, allowing it to run dynamically. We used gradient search to optimize the running gait parameters on two sets of legs with different stiffness. Surprisingly, the softer legs run 16% faster at a gait frequency 11% lower – contradicting the simple design intuition that more rigid legs and higher gait frequencies are key to faster running.

Keywords: Gait optimization; leg stiffness; modular robots

1. Introduction

In the last decade we have witnessed an enormous increase in the interest in legged robots, primarily centering around the ability of legged platforms to traverse terrain that is difficult or impossible for wheeled vehicles. For quasi-static legged systems, the governing mechanics are well understood[1]. The situation is starkly different for dynamically moving legged robots, where the combination of hybrid transitions, surface interaction and robot deformation often proves challenging to analyze.

Modular robots offer the chance to quickly prototype and explore the parameter space of a given robot design. Such an approach has been the subject of ongoing work, primarily with a bio-inspired focus[2]. Our work on rapidly prototyping robots with the CKBot platform[3], lead us to explore the possibility of dynamical locomotion with these modules[4].

Motivation for our design — Among the fastest autonomously powered legged robots are the Sprawl[5] and the RHex[6] robot families. Both are bioinspired, applying design principles and parameter ranges obtained from studies of cockroach (*Blaberus sp.* and *Periplaneta sp.*) locomotion. We present a robot of the same overall type – a hexapod with elastic legs, whose mass lies in between that of the largest Sprawl and smallest RHex[4].

This robot was designed with the goal of achieving dynamic running using a modular robot platform.

2. Robot design

Unlike the cockroach inspired robots whose leg motions are primarily in the saggital plane, our robot also borrowed from the body-plan of centipedes (*Scolopendridae sp*) by having a laterally bendable body and laterally projecting legs, each of which was a fiberglass cantilever plate mounted on a servo allowing its roll angle relative to the body to be controlled. We attached a "foot" (a normally offset rubber pad) on each side of the leg cantilever. The design achieves an effective 2-DOF motion for each of its 6 feet with only 7 motors. We ran the robot in a motion tracking arena (Vicon, 16 cameras, 100fps) while tracking thc rigidly attached markers.

Gait optimization — The gait parameters we optimized were frequency, leg roll amplitude, leg yaw amplitude (achieved via body bending), sweep angle and duty cycle (relative duration) of the stance period. These were tuned by a Nelder-Mead simplex based gradient search using a python program that ran the robot back and forth in the 5m x 5m tracking arena. The robot was kept on track by a human operator using a joystick. Pressing a button would start and end a trial. The optimization attempted to minimize time divided by distance between robot position at start and end of each trial by changing the gait parameters.

Data analysis — After optimization we collected tracking data of all markers while running the robot back and forth. A total of 52 trials were conducted with an average of 372 frames and an average distance of 3.0m (minimum of 303 frames and 2.1m). From one end to the other end of the arena counts as one trial. Trials were segmented by thresholding velocity of the robot. We removed marker association errors using a combination of manual and automated tools. An average of 1.4% of markers were occluded in a trial and were linearly interpolated. We used the fore-aft motions of the legs to reconstruct the dynamical phase of the gait limit cycle using the Phaser algorithm[7], allowing us to reinterpret our data as a function of phase and to construct Fourier series models of the marker trajectories superior to those that can be achieved with footfall based phase estimates.

3. Results

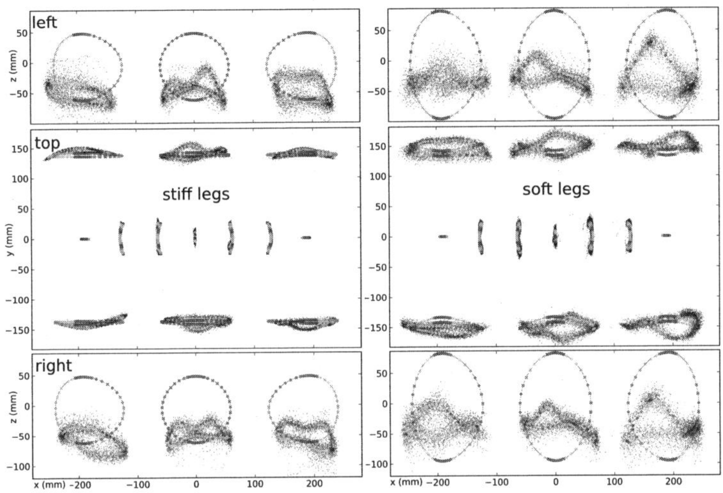

Plots show trajectories of body and leg corner markers, with colors indicating speed in 3D space (red faster, blue slower). We plot the positions predicted by a rigid body model (square color markers on solid green line), the observed positions (point cloud of black points) and a Fourier series model of the observed positions (round color markers).

Our optimizations terminated with the following gait parameters results

Leg	Freq. Hz	Roll deg	Yaw deg	Sweep deg	Duty %	Speed cm/s	K_{tri} N/m	L mm	H mm
Soft	3.1	33	11	22	19	78	331	152	68
Stiff	3.5	22	12	20	18	67	1950	137	43

We observed that soft legs give 16% higher speed with a 11% lower gait frequency. This difference cannot be accounted for by either frequency-length scaling, which predicts soft legs would be slower, or by estimating ground speed of the feet, which predicts soft legs should be much faster than observed in practice.

3.1. *Biomechanical approach*

Biomechanists[8,9] have developed methods to compare animal gaits despite large variability in scale and body structure. A key observation is the presence of dynamic similarity between running in various organisms, which expresses itself in the existence of a common parameter, the "Froude Number" $\mathfrak{f}$, governing the transition between running and walking ([9] fig 1).

We argue that for our robots, the appropriate leg length to use is not the unloaded hip-height [8], because very substantial compression occurs even when standing still. Since our robot runs in an alternating tripod gait, we use as our vertical length l the equilibrium height of the body when standing on a single tripod of legs. At a weight of $15.7\,N$, this implies $l = 21\,mm$ and $\mathfrak{f} = 3.0$ with soft legs; $l = 35\,mm$, $\mathfrak{f} = 1.3$ with stiff legs. We conclude that with the soft legs, the robot achieves a running gait dynamicaly similar to animal running ([9] fig 1), while with the stiff legs, the robot is (dynamically speaking) merely executing a rapid walk.

4. Discussion

Our results suggest that in the biologically relevant parameter range of leg stiffness, the performance of dynamically running robots is difficult to anticipate from first principle arguments, or at least those arguments that rely on rigid body models and no-slip contact conditions. That the results are counter-intuitive in laboratory conditions on a flat and uniform floor, speaks to the baffling complexity that must appear as we attempt to design robots to run in uncontrolled environments. The combination of simplicity and flexibility provided by using a modular plaform affords us the opportunity to explore performance optimization through careful empirical study – at least until such a time when we have better theory to guide our designs.

References

1. Y. Or and E. Rimon, *IJRR* **29**, 3 (2010).
2. J. C. Larsen, Locomotion through morphosis: development of the modular robotic toolkit - locokit, Master's thesis, South Denmark University (2011).
3. S. Revzen, J. Sastra, N. Eckenstein and Yim, Ckbot platform for the icra 2010 planetary challenge, in *Workshop "Modular Robots: The State of the Art", Proceedings of IEEE ICRA conference*, 2010.
4. J. Sastra, W. G. Bernal-Heredia, J. Clark and M. Yim, A biologically-inspired dynamic legged locomotion with a modular reconfigurable robot, in *ASME Dynamic Systems and Control Conference*, Octtober 2008.
5. J. G. Cham, J. K. Karpick and M. R. Cutkosky, *Int J Rob Res* **23**, 141 (2004).
6. R. Altendorfer, N. Moore, H. Komsuolu, M. Buehler, H. B. Brown, D. McMordie, U. Saranli, R. J. Full and D. E. Koditschek, *Autonomous Robots* **11**, 207 (2001).
7. S. Revzen and J. M. Guckenheimer, *Phys Rev E* **78**, p. 051907 (2008).
8. T. A. McMahon and G. C. Cheng, *J Biomech* **23**, 65 (1990).
9. R. Alexander and A. Jayes, *Journal of zoology* **201**, 135 (1983).

FOOT DESIGN EVALUATION FOR A SIX-LEGGED WALKING ROBOT

A. Roennau, G. Heppner, L. Pfotzer, R. Dillmann

FZI Research Center for Information Technology,
Interactive Diagnosis- and Servicesystems (IDS),
Haid-und-Neu-Str. 10–14, D-76131 Karlsruhe, Germany
E-mail: {roennau, heppner, pfotzer, dillmann}@fzi.de, www.fzi.de

The stability, velocity and energy efficiency of walking robots strongly depend on the quality of the established ground contact. Foothold planning is a good method to optimise the ground contact by selecting stable places to set the foot while walking in rough terrain. Besides the foothold, the design of the foot has great influence on the ground contact quality and therefore on the overall walking performance. In this paper we propose a set of basic design principles for feet of multi-legged robots. These design principles are evaluated with the six-legged walking robot LAURON by testing several different foot designs in various environments. The relevance of these proposed design principles is demonstrated by comparing the walking velocity and energy consumption in a series of experiments.

Keywords: foot design, walking velocity, energy consumption, slippage, ground contact, six-legged robot, hexapod

1. INTRODUCTION

Robotic walking is an interesting and challenging field of research due to the required mechatronic complexity for smooth motion and the challenge to control and ensure the stability of the walking locomotion at all times. In bipedal walking many robots make use of sophisticated foot designs to optimise the ground contact during the stance phase and to increase the robots' overall walking capabilities. Some of these footed robots have made impressive improvements in the field of dynamic walking in the last years (e.g. PETMAN of BostonDynamics, HRP-4c[1] or WABIAN-2R[2]). Although the foot seems to be inherent important for fast and stable bipedal walking, the prevalent design of feet for multi-legged robots are rather simple ball-like constructions. Taking a closer look at the biological counterparts of multi-legged robots, one will find very complex kinematic feet that improve the ground contact with great abilities: reduce ground collision impact, minimise slippage, adapt to surface irregularities and many more. For example the foot of a small stick insect, like Carausius morosus, has multiple degrees of freedom (claws, Arolium, 4 pairs of Euplantulae) to ensure stable walking in rough terrain.[3] Like many other walking robots, the kinematic structure and control system of the six-legged walking robot LAURON was inspired by an insect (stick insect). But during the design and construction of LAURON IVc the impressive kinematics and great complexity of the tarsus of the stick insect has been reduced to a simple industrial rubber stop. If we have a closer look at current multi-legged walking robots, we will see that the majority of these robots is equipped with a simplified ball-like foot tip (e.g.[4],[5]). Nevertheless, there are a few multi-legged robots using sophisticated foot designs (e.g. TITAN VIII,[6] SpaceClimber[7]). The advantages of these specialised feet confirm the importance of the foot design.

Fig. 1. Bio-inspired walking robot LAURON IVc with a simple rubber stop as foot tip.

This paper wants to draw attention to the relevance of the foot design for multi-legged robots. Fast and stable walking can only be achieved with the right leg coordination, a foot hold planning mechanism and reliable ground contacts. Among these important elements, the contact of the foot with the ground during the stance phase is the most relevant environmental interaction for legged locomotion. Therefore, a sophisticated and optimised design of a foot can increase the walking capabilities significantly by reducing slippage, increasing velocity and improving the energy efficiency.

A fundamental set of design principles will be presented and benchmarked with the help of an evaluation of some basic foot designs for the six-legged walking robot LAURON. Although, all these feet have an astonishing simple design, their effect on the walking process is remarkable. The designs will be rated according to their influence on the walking speed and the energy consumption on different terrain types. A final discussion at the end of this paper will assess the presented foot design principles and conclude the foot design evaluation.

2. APPROACH

2.1. *Six-legged Walking Robot LAURON*

The walking robot LAURON IVc (see Fig. 1) has a kinematic structure that was inspired by the stick insect. The orientation of the legs connected to the main carbon fibre body and the leg segments' length ratios are very similar to the stick insect's morphology. Each leg has 3 joints and a spring-damper foot with an additional passive degree of freedom. At the end of the foot tip LAURON has an industrial rubber stop with a screw mounting. More details on the robot in general, the sensor system and the control architecture can be found in.[8]

2.2. *Foot Design Principles*

During legged locomotion the foot will be stressed repeatedly: collisions with the ground at the end of each swing phase to establish ground contact, high force transmissions to maintain reliable ground contact in stance phase and possible collisions with obstacles in swing phase. Hence, a foot for a walking robot should have a robust mechanical design and fulfil some basic design requirements.

Compliance: To reduce peak stress induced by collisions with the ground and obstacles it is advisable to include some compliance in the design. A spring-damper system or deformable materials like rubber can absorb collision shocks and reduce the mechanical stress of the entire robot.

Adaptability: It is desirable to establish a nondetachable connection with the ground during the stance phase. By adapting to the terrain structure and increasing the contact area it is possible to improve the load capacity (see left side of Fig. 2). With the help of elastic elements the foot can achieve this terrain adaptability. Although the Coulomb friction model says that the size of the contact area does not influence the friction, a bigger contact area can also have a positive influence on the slippage (many micro mechanical couplings).

Friction: A high friction coefficient between the two involved materials (foot tip and ground surface) can reduce the slippage significantly. The

Fig. 2. (left) Adaptability as an important foot design principle, (right) Kinematic structure with 3 DOF per leg restricts orientation of the foot tip: A ball like foot tip can reduce the negative influence of this kinematic restriction, but is also minimises the overall contact area.

ground surface material cannot be changed in real application scenarios for walking robots, therefore the material of the foot is an important design criteria.

Ground Foot Alignment: The orientation of the foot tip towards the ground has a big influence on the contact area. A kinematic structure with 3 joints per leg restricts the orientation of the foot tip. This typical hexapod morphology creates a permanently changing foot tip orientation during the stance phase. The right part of Fig. 2 illustrated the orientation of the foot tip at the beginning and end of this phase. A good solution to minimise this problem can be found in Nature: Add an additional (passive or active) joint to the foot. A rather simple technical suboptimal solution is the usage of a ball like foot tip design. In this case the contact area is nearly the same during the stance phase, but the area is also smaller than a flat foot design.

2.3. *Foot Design Evaluation Methodology*

The foot designs used for the evaluation are designs that have been used on LAURON (e.g. Reference - current foot design, see Fig. 3) or were developed to evaluate the relevance of the proposed design principles. Of course, not all of the presented foot designs are suitable for all terrain types. The left part of Fig. 3 shows the foot designs that are being evaluated.

The evaluation of each foot design will be performed in a similar experimental setup. LAURON walks an identical distance in a fixed tripod gait over three different surfaces: solid industrial floor, thin carpet on industrial floor and a patch of soil covered with grass (shown in the right part of Fig. 3). All control parameters like body height or stride length are kept the same in every experiment. The evaluation criteria will be the walking speed and the energy consumption. To accurately measure the speed of LAURON we use a LMS100 laser scanner as external tracking system, which continuously

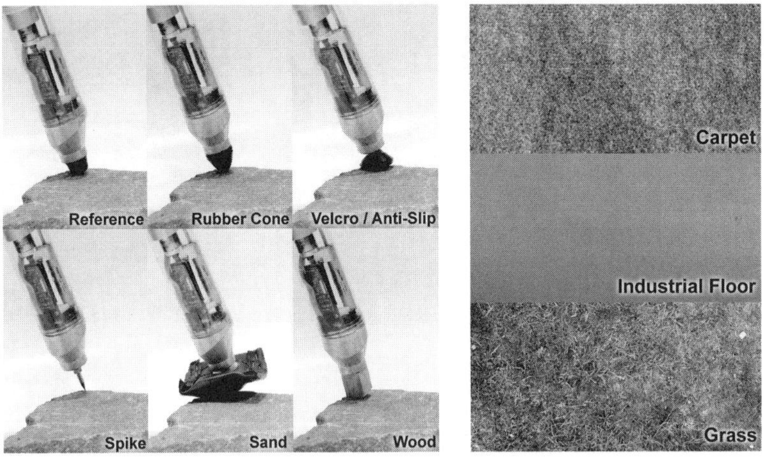

Fig. 3. (left) Overview of seven evaluated foot designs, (right) evaluated surfaces: carpet, industrial floor, grass

measures the distance to LAURON and can therefore calculate its speed. The scanner is positioned on top of the mobile power supply of LAURON (see Fig. 4). By attaching a perfect reflector to LAURON's back we are able to accurately pinpoint the distance and radial position. The reflector is big enough to provide a stable tracking target, even during walking in the outdoor tests. The scanner is connected to an external computer which pre-processes the distance values to reduce noise. The measured distance and calculated speed are then sent to LAURON via WLAN for on-board evaluation. The overall mean speed of a complete measurement is calculated by using the distances at the beginning and end of a test run, as well as the total run time. LAURON's energy consumption is measured with a precise on-board energy measurement system.[9] Although this measurement system is able to determine the energy consumption online, we will only present the total energy required for each test run.

3. EXPERIMENTS AND RESULTS

For the evaluation process each foot design was tested at least three times on every ground, only considering successful runs. All together over 100 individual experiments were conducted. During evaluation the varying influences of the design principles became evident. Compliance in the foot design not only damps shocks, but can also have a negative effect on the stability of the foothold, if the compliance becomes too high. LAURON uses

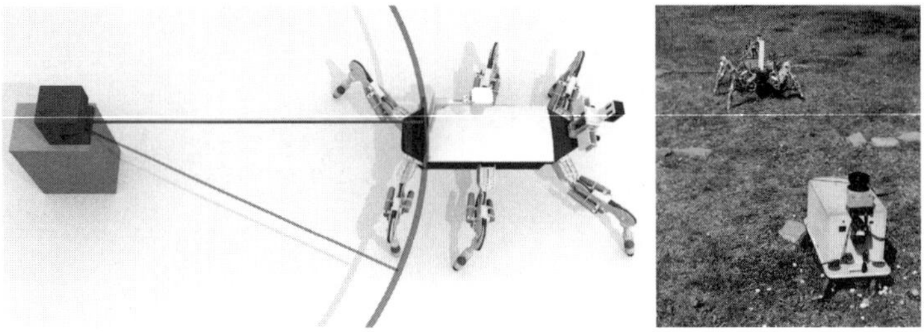

Fig. 4. External velocity measurement system with SICK LMS100

a spring-damper system to absorb strong shocks. Additionally, the carpet ground and the soft soil also act as dampers. The rubber cone foot (foot no.2) is made of a very soft rubber, hence the compliance is increased even further. As a result the small tip of this foot deforms under LAURON's weight and destabilises the support of the body. This leads to a poor performance and especially a high variance of the results, independent of the ground (illustrated in both Fig. 5 and 6). The wooden foot (foot no.3) on the other hand performs very well on the carpet. This directly arises from the solid structure of wood, which provides the needed traction while the carpet and the damper add the necessary compliance.

However, the wooden foot (foot no.3) violates the Ground Foot Alignment principle. This is visible in the industrial floor data. While standing the wooden feet are capable of supporting the entire weight of the robot. But, during locomotion only the edges of the foot touch the ground (see Fig. 2). As consequence the small contact area is not able to create a sufficient amount of traction and the legs slip. The Anti-Slip foot (foot no.6), equipped with a ball bearing, achieves a better Ground Foot Alignment and uses its complete contact area.

Friction is a key aspect on solid floors like the industrial floor. Without the micro mechanical couplings with a rough material like carpet the feet slip away easily. The feet made of rubber proved to be the most successful on this floor (see middle of Fig. 6). Too much friction on the other hand can lead to negative effects. Very high forces were needed to detach the Velcro foot (foot no.4) from the carpet. The spike foot (foot no.7) in the outdoor tests showed similar problems. In some steps the connection was so strong that LAURON was not able to lift its feet at all.

The most adaptable foot design is the sand filled foot (foot no.5). Even though other feet were faster or more energy efficient, the sand filled feet made a good impression in all experiments. This observation can be verified with the data through the small variance of the speed values. For example, the reference foot design (foot no.1) attains the highest overall speeds on the outdoor ground. Nevertheless, the achieved speeds vary significantly (shown in right part of Fig. 5). The higher energy consumptions as well as slower speeds, compared to the reference, mainly result from the high weight and the large shape of the foot design.

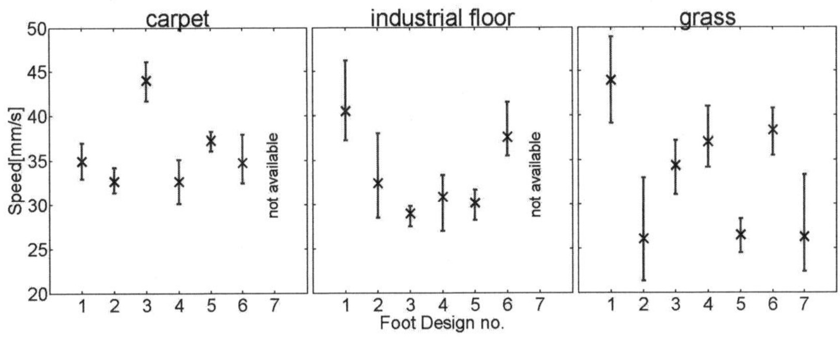

Fig. 5. Experiments: Walking speeds with different foot designs on carpet (left), industrial floor (middle) and soil with grass (right) - (1: reference rubber foot, 2: rubber cone foot, 3: wooden foot, 4: ball head foot with Velcro, 5: elastic sand foot, 6: ball head foot with anti-slip, 7: spike foot)

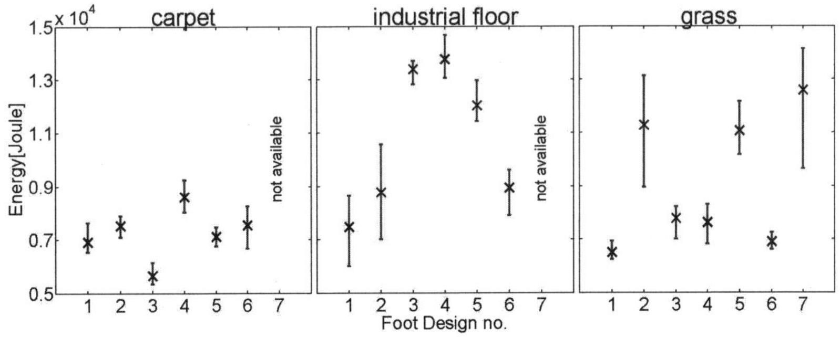

Fig. 6. Experiment: Energy consumption for different foot designs on carpet (left), industrial floor (middle) and soil with grass (right) - (1: reference rubber foot, 2: rubber cone foot, 3: wooden foot, 4: ball head foot with Velcro, 5: elastic sand foot, 6: ball head foot with anti-slip, 7: spike foot)

4. CONCLUSIONS AND FUTURE WORKS

In this paper we have presented basic design principles for foot designs and their impact on the performance of walking robots. Different aspects of these principles where addressed by simple foot designs that were evaluated by a great number of experiments with the walking robot LAURON. By comparing the speed and energy consumption of LAURON in tests on different terrains we were able to show the huge significance of the foot design for multi-legged locomotion. As humans wear special shoes on the beach or in the mountains it became clear that different foot designs are necessary for different environments.

In the future we want to investigate the influence of various foot designs in additional environments like unstructured terrain. These insights will help to better understand the walking process and lead to better foot designs for multi-legged walking robots.

References

1. K. Miura, M. Morisawa and et al., Human-like walking with toe supporting for humanoids, in *Intelligent Robots and Systems (IROS), 2011 IEEE/RSJ International Conference on*, 2011.
2. Hyun-jin. K., K. Hashimoto and et al., Realization of biped walking on uneven terrain by new foot mechanism capable of detecting ground surface, in *Robotics and Automation (ICRA), 2010 IEEE International Conference on*, 2010.
3. H. Cruse, V. Duerr, M. Schilling and J. Schmitz, Principles of insect locomotion, in *Spatial Temporal Patterns, COSMOS 1, pp. 4396. Springer*, 2009.
4. J. Rebula, P. Neuhaus, B. Bonnlander, M. Johnson and J. Pratt, A controller for the littledog quadruped walking on rough terrain, in *Robotics and Automation, 2007 IEEE International Conference on*, 2007.
5. C. Semini, N. G. Tsagarakis, E. Guglielmino, M. Focchi, F. Cannella and D. G. Caldwell, *Proceedings of the Institution of Mechanical Engineers, Part I: Journal of Systems and Control Engineering* , 831 (2011).
6. S. Ohtsuka, G. Endo, E. Fukushima and S. Hirose, Development of terrain adaptive sole for multi-legged walking robot, in *Intelligent Robots and Systems (IROS), 2010 IEEE/RSJ International Conference on*, 2010.
7. S. Bartsch, T. Birnschein and et al., Spaceclimber: Development of a six-legged climbing robot for space exploration *41st International Symposium on and 2010 6th German Conference on Robotics (ROBOTIK)* 2010.
8. A. Roennau, T. Kerscher and R. Dillmann, Design and kinematics of a biologically-inspired leg for a six-legged walking machine, in *3rd IEEE RAS and EMBS BioRob 2010*, (Tokyo, Japan, 2010).
9. A. Roennau, F. Sutter, T. Kerscher and R. Dillmann, On-board energy consumption estimation for a six-legged walking robot, in *14th International Conference on Climbing and Walking Robots*, (Paris, 2011).

EFFECTS OF LEG CONFIGURATION ON RUNNING AND WALKING ROBOTS*

MIKHAIL S. JONES† and JONATHAN W. HURST‡

School of Mechanical, Industrial, and Manufacturing Engineering
Oregon State University, 204 Rogers Hall
Corvallis, Oregon 97331, USA
† E-mail: jonesmik@engr.orst.edu
‡ E-mail: jonathan.hurst@oregonstate.edu
mime.oregonstate.edu/research/drl/

This paper investigates the relations between leg configuration and performance during running and walking operations. Specifically, we use standard kinematic techniques to visualize impact forces and kinetic energy loss during inelastic collisions with the environment in order to gain fundamental insights into robotic leg design. We show that for a two-link planar leg with revolute joints, a slightly crouched stance with a backward-facing knee (in contrast to humans) is optimal for both impact rejection and energy economy.

Keywords: legged, locomotion, collisions, dynamic, running

1. Introduction

Legged robots intended for energetic and dynamic locomotion encounter a repetitive series of collisions with the ground during every stride as the robot transitions from flight phase to stance phase. These collisions become especially prevalent in dynamic, unknown, or treacherous environments where unexpected and larger-than-normal impact forces often occur. At walking speeds, the severity and frequency of collisions can be minimized through knowledge of the environment and cautious control strategies, however at greater speeds, collisions become unavoidable due to the ballistic nature of running gaits. Assuming the leg actuators have sufficient bandwidth or series impedance to isolate actuator inertia, such collisions still have several undesirable traits as a result of leg inertia, including energy loss due to the

*This work was supported by the School of Mechanical, Industrial and Manufacturing Engineering at Oregon State University.

inelastic nature of the collisions and large end-effector impulses which could result in mechanical damage or failure.

We contend that the severity of these inelastic collisions with the environment can be minimized by strategically choosing the leg configuration during impact. The primary motivation for this work stems from our observations of the monopod robot Thumper, which showed a distinct preference for backward-hopping (in contrast to humans).[1] During stable backward-hopping experiments, Thumper was more energy economic, produced less ground impact noise, and was less susceptible to falling as compared to forward-hopping. Building on these observations, we investigate the implications of leg configuration on running and walking operations. Specifically, we find that for a two-link planar leg with revolute joints, a slightly crouched stance with a backward-facing knee (in contrast to humans) is optimal for both impact absorption and energy economy.

2. Methods

In order to quantitatively evaluate the kinetic energy loss and forces incurred during an impact with the environment, we borrow techniques commonly used for robotic manipulator analysis.[2] Note that the expressions given by Eq. (1) through Eq. (5) are standard in dynamics literature; they are provided here for completeness.

2.1. *Joint Space Kinematics*

We begin by reviewing the forward kinematic relationships between the joint velocities $\dot{q}$ and the corresponding end-effector velocity

$$\dot{x} = [J(q)]\dot{q}, \tag{1}$$

where $J(q)$ is the Jacobian matrix associated with the coordinate transformation from cartesian end-effector velocity to rotational joint velocities. Because we are interested in changes of kinetic energy, we proceed by considering the inverse kinematics problem relating the instantaneous change in the end-effector velocity $\Delta\dot{x}$ to the instantaneous change in the joint velocities

$$\Delta\dot{q} = \bar{J}(q)\Delta\dot{x}, \tag{2}$$

where $\bar{J}(q)$ is the dynamically consistent generalized inverse of the Jacobian matrix associated with the solution that minimizes the instantaneous kinetic energy.[3,4] For non-redundant manipulators, $\bar{J}(q)$ is simply the inverse of the Jacobian, however, for redundant manipulators

$$\bar{J}(q) = H^{-1}(q)J^T(q)[J(q)H^{-1}(q)J^T(q)]^{-1}, \tag{3}$$

where $H(q)$ is the symmetric positive-definite inertia tensor of a series of rigid bodies with respect to the joint coordinates. Substituting Eq. (3) into Eq. (2), allows us to determine the instantaneous change in joint velocities associated with a given instantaneous change in end-effector velocity.

2.2. *Kinetic Energy Loss*

For the purposes of this paper, we assume that collisions are perfectly inelastic, with no slip, in which neither the robot nor the environment deforms at the contact point. This is a reasonable assumption as we want the end-effector to remain planted on the ground to prevent chatter and slip. By using this criterion and representing the environment as a infinitely massive, immovable object with zero velocity, we can deduce that the end-effector velocity after impact becomes zero. We can investigate the kinetic energy loss ΔT associated with this instantaneous change in velocity through

$$\Delta T = \frac{1}{2}\Delta\dot{q}^T H(q)\Delta\dot{q}. \tag{4}$$

Because the instantaneous change in joint velocities is not typically known succeeding a collision, we choose to express the kinetic energy loss corresponding to an instantaneous change in end-effector velocity. Substituting Eq. (2) into Eq. (4) yields

$$\Delta T = \frac{1}{2}\Delta\dot{x}^T \bar{J}(q)^T H(q)\bar{J}(q)\Delta\dot{x}, \tag{5}$$

where $\bar{J}(q)^T H(q)\bar{J}(q)$ is denoted as the symmetric generalized inertia tensor of a series of rigid bodies $G(q)$.[5,6] Equation (5) allows us to conveniently calculate the kinetic energy loss of a manipulator given a instantaneous change in end-effector velocity.

3. Analysis

The primary goal of this investigation is to not only evaluate the performance of a particular leg configuration, but also to quantitatively and visually compare different configurations. Doing so will allow us to gain fundamental insights into robotic leg design and develop an intuition for why one configuration may be superior to others.

3.1. *Two-Link Model*

While the methods presented in this paper are applicable to a wide range of systems, we will focus on a particular model, the two-link planar manipulator with revolute joints. As illustrated in Fig. 1, the system consists of two rigid bodies hinged together by a pin joint allowing for two degrees of freedom relative to a fixed hip. In order to determine the kinetic energy

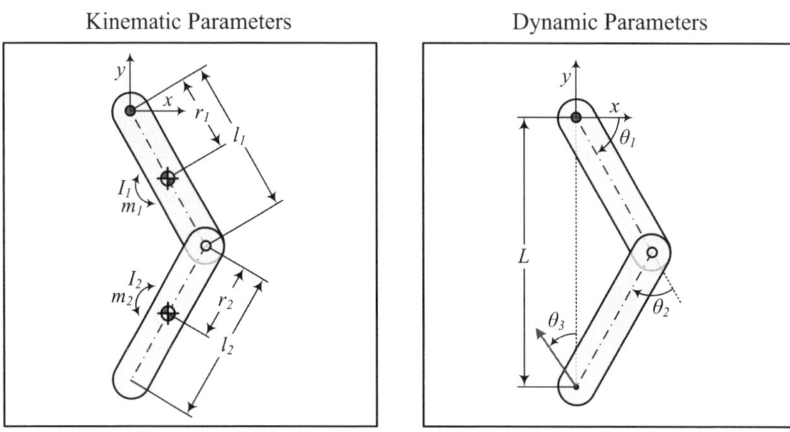

Fig. 1. Illustration of the kinematic and dynamic parameters for the two-link planar manipulator with revolute joints. Left: mass, inertia and dimensional constants are depicted. Right: the joint coordinates are shown relative to the fixed coordinate system at the hip, the impact vector (solid arrow) is shown extending from the toe.

loss associated with a configuration, we first need to derive the kinematic relationships for the two-link model. We proceed by deriving the inertia tensor with respect to the joint coordinates from Lagrange's equations.

$$H(q) = \begin{bmatrix} \alpha + 2\beta c_2 & \delta + \beta c_2 \\ \delta + \beta c_2 & \delta \end{bmatrix} \tag{6}$$

where

$$\alpha = I_{z1} + I_{z2} + m_1 r_1^2 + m_2(l_1^2 + r_2^2) \tag{7}$$

$$\beta = m_2 l_1 r_2 \tag{8}$$

$$\delta = I_{z2} + m_2 r_2^2 \tag{9}$$

Additionally, we determine the Jacobian matrix by differentiating the forward kinematic mapping.

$$J(q) = \begin{bmatrix} -l_1 s_1 - l_2 s_{12} & -l_2 s_{12} \\ l_1 c_1 + l_2 c_{12} & l_2 c_{12} \end{bmatrix} \tag{10}$$

where $s_i = \sin\theta_i$, $s_{ij} = \sin\theta_{i+j}$, and similarly for c_i and c_{ij}.

From the illustration on the right in Fig. 1, we can see that every configuration is represented by the unique combination of three parameters; θ_1 and θ_2, which determine the shape and orientation of the leg, and θ_3, which represents the impact angle relative to the generalized coordinate system. For convenience, the problem is simplified recognizing that θ_1 and θ_2 are dependent on the virtual leg length L due to the non-redundant nature of the two-link model. By then aligning the generalized coordinate system to the virtual leg length, we can essentially eliminate one degree of freedom and more concisely describe the system with only two parameters, L and θ_3. This allows us to independently visualize the effects of leg configuration and impact angle on performance.

3.2. *Kinetic Energy Loss*

Using the formulation in Eq. (5), the kinetic energy loss is calculated relative to the leg length and impact angle, generating the contour surface depicted in Fig. 2. The prominent feature of this plot is the valley starting at $\theta_3 = 0$, $L = 0$ stretching up to $\theta_3 = \pi/2$, $L = 1$. This region represents the optimal combinations of leg lengths and impact angles for the given kinematic parameters. In this case, the upper and lower leg segments are equally proportioned with evenly distributed mass throughout.

4. Discussion

The results shown in Fig. 2 reveal several fundamental insights into robotic leg design as it pertains to dynamic locomotion.

4.1. *Impact Angle*

During a typical running gait, the foot is extended in front of the body, and the resulting impact with the ground occurs predominately in the vertical direction with a small horizontal component. As illustrated in Fig. 3, by rotating the system as to align the impact vector with the vertical, two unique cases appear.

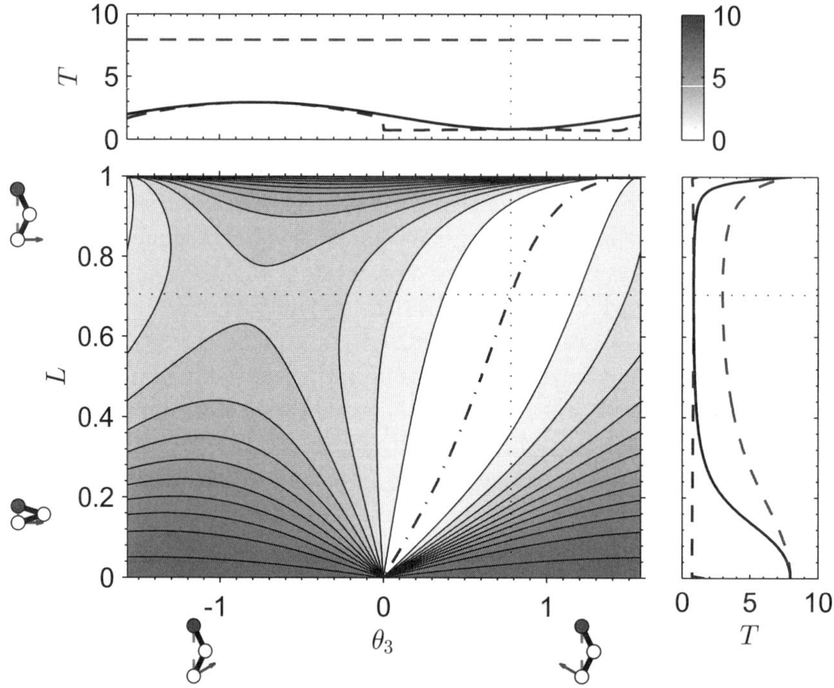

Fig. 2. The contour plot illustrates kinetic energy loss T as a function of virtual leg length L and impact angle θ_3 with the optimal (minimum energy loss) at varying leg lengths indicated by the dot-dashed line within the white shaded region. The top and side panels present the kinetic energy loss (solid black lines) through a two-dimensional slice (dotted black lines) of the three-dimensional contour surface. For a qualitative comparison, the minimum and maximum kinetic energy loss (dotted lines) along each dimension are also provided. (Because of end-effector singularities, the magnitudes of the energy loss function are scaled by an arctangent function.)

Assuming the leg is extended in front of the body we deduce the direction of travel for each case. We can see that values of $\theta_3 < 0$ correspond to a forward-facing knee (as compared to humans), while values of $\theta_3 > 0$ correspond to a backward-facing knee (in contrast to humans). Examining these two regions in Fig. 2, we can see that the vast majority of the low energy loss area (depicted as the filled white area) falls in the region of $\theta_3 > 0$. This is an interesting result that suggests a backward-facing knee is optimal for impact rejection and energy economy.

This can be intuitively explained by the difference in the generalized inertia matrices. With a backward-facing knee, the toe impact causes an

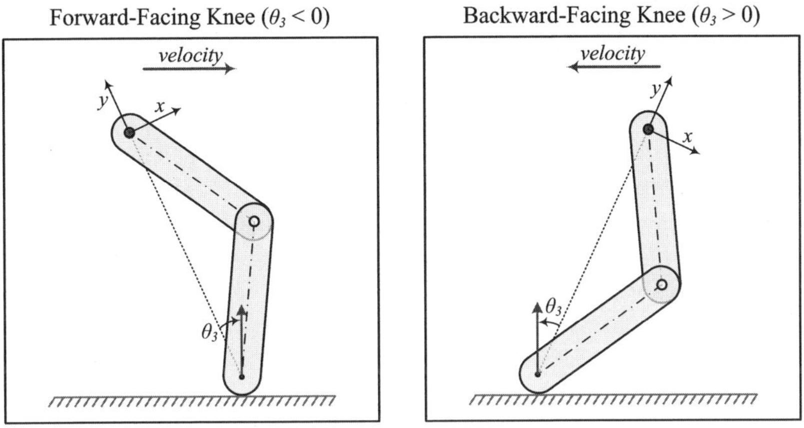

Fig. 3. Illustrations of the two unique leg configurations associated with a typical running gait impact. Left: the case of $\theta_3 < 0$, analogous to a forward-facing knee (as compared to humans). Right: the case of $\theta_3 > 0$, analogous to a backward-facing knee (in contrast to humans).

instantaneous rotation of the shin while a forward-facing knee results in an instantaneous translation of the shin, which, requires a greater impulse. This is in direct alignment with our observations made of the monopod robot Thumper which exhibited a preference for backward-hopping.[1]

4.2. *Leg Configuration*

The effect of leg configuration on kinetic energy loss is substantial as observed in Fig. 2. The optimal virtual leg length for this particular case that guarantees the lowest energy loss independent of impact angle occurs at $L = 0.71$ (as indicated by the dotted horizontal line in the side panel of Fig. 2). Leg lengths in close proximity to this optimal value ($L = 0.5$ to 0.9) show only slight increases in energy loss while lengths further away result in drastically increased energy loss. We can see that configurations in the vicinity of a singularity result in the largest energy loss and thus also the largest impact forces. This is especially true when the impact angle is aligned with virtual leg length. These findings agree with natural intuition and experience; when an impact is imminent we often slightly bend our knees or elbows to reduce risk of injury. This result suggests that a slightly crouched stance allows for better impact rejection and energy economy.

5. Conclusions

In this paper, we discussed the effects of leg configuration on running and walking robots from an impact perspective. We conclude that during walking and running operations, a slightly crouched posture allows large impact forces to be more readily rejected with minimal energy loss. We can also show that force generation at the end-effector can be maximized by extending the leg joints towards a singularity which may be beneficial in static situations such as standing. More interesting, is the difference between forward and backward running configurations. We show that a backward-facing knee (in contrast to humans) more efficiently rejects impacts then a forward-facing knee (as compared to humans). These results are intuitive yet have significant implications in robotic leg design.

References

1. J. W. Hurst, The Role of Compliance in Legged Locomotion, phd, Carnegie-Mellon University2008, p. 102.
2. R. M. Murray, Z. Li and S. S. Sastry, *A Mathematical Introduction to Robotic Manipulation* (CRC Press, 1994).
3. O. Khatib, *IEEE Journal of Robotics and Automation* **3**, 43 (1987).
4. O. Khatib, *International Journal Of Robotics Research* **13**, 19 (1995).
5. H. Asada, *Robotics and Automation Proceedings 1984 IEEE International Conference on* **1**, 94 (1984).
6. H. Asada and K. Ogawa, *Robotics and Automation Proceedings 1987 IEEE International Conference on* **4**, 751 (1987).

ADAPTIVE LOCOMOTION ON VARYING GROUND CONDITIONS VIA A RECONFIGURABLE LEG LENGTH HOPPER

FARRUKH IQBAL SHEIKH†, ROLF PFEIFER

Artificial Intelligence Laboratory, University of Zurich, Andreastrasse 15
Zurich, 8050, Switzerland

In this paper, we present the concept of adapting to changes in ground conditions like stiffness, damping and friction, using a novel two degree of freedom reconfigurable leg length hopping robot with a fixed passive compliance. In such a robot, the change in the dynamics of the single legged hopper can be induced by the change in coupled stiffness and damping of the system, i.e., stiffness and damping of the ground coupled with the stiffness and damping of the robotic leg. It is experimentally shown by in-place hopping of a robotic leg on various grounds (stiff, less stiff and soft) that the leg can effectively adapt to changes in coupled stiffness and damping by the rate and the amplitude at which the leg length changes. This is true, while the leg hops in-place as the role of ground friction is negligible. However, in forward motion where the ground friction dominates, a change in initial effective leg length, i.e., shortening or lengthening can provide an additional support to the hip motor in overcoming even large variations in ground friction. This is demonstrated through a planar locomotion experiment on different ground surfaces. The overall results provide strong support for this concept.

Keywords: Varying ground conditions, changeable leg length.

1. Introduction

Animals and humans are capable to adapt to varying ground conditions (stiffness, damping and friction) while maintaining their balance on irregular surfaces [1]. This kind of adaptive behaviour is a great source of inspiration for designing legged robots. In general, ground surfaces can be characterized by their stiffness, damping and friction properties. In legged robot locomotion, one approach to adapting to varying ground conditions is by physically changing parameters of the robotic leg to counter-act the overall change. For example, in [2] mechanically adjustable compliance in the robotic leg was introduced to adapt to the change in stiffness of the underlying surface. However, it is not entirely clear how these changes can be incorporated in practice during fast running of a legged robot. Except for the pioneering work of Marc Raibert [1],

† Work supported by the European Commission Seventh Framework Program, Theme ICT- 2007.8.5 as part of the project LOCOMORPH, under a grant no 231688. Video can be found at this link.

most of the electrically actuated legged robots developed in the past were designed to exploit the potential of passive compliance [3], [4], together with the oscillatory motion of the robotic leg. However, due to exploiting limited degrees of freedom in [3], [4], the performances of these systems were mainly restricted to stiff grounds. In this study, we demonstrate that, in the context of adaptive legged robot locomotion, all three ground properties, namely stiffness, friction and damping, are equally important. In other words, efficient adaptation should consider all three ground conditions in a unified framework.

We developed a bio-inspired 2-DOF robotic leg whose design of [5] follows the spring loaded inverted pendulum (SLIP) model, also referred to as the bouncing motion [5]. In [5], the altering leg length feature was first introduced. In this paper, we propose a systematic approach, which is based on the concept of embodiment [6], in order to test the framework of [7] for varying ground conditions. According to [6], the dynamic coupling of the robot's body with its controls and the physical environment is important to investigate the overall behavior of the robot. By employing this concept, the effects of altering leg length are practically investigated on number of different grounds. Experimental results demonstrate that the reconfigurable leg length approach is suitable to efficiently adapt to varying ground conditions both for in-place and planar hopping.

The remainder of this paper is structured as follows: Section 2 describes the mechanical structure and the control of the robotic leg. Section 3 explains the proposed mathematical model. Experiments and results are provided and discussed in Section 4. Finally, Section 5 draws some conclusions and details the future research direction of this work.

2. Mechanical Design and Control

The 2-DOF reconfigurable leg length hopper (RLLH) module, as shown in Fig. 1, was designed and constructed based on the SLIP model [7].

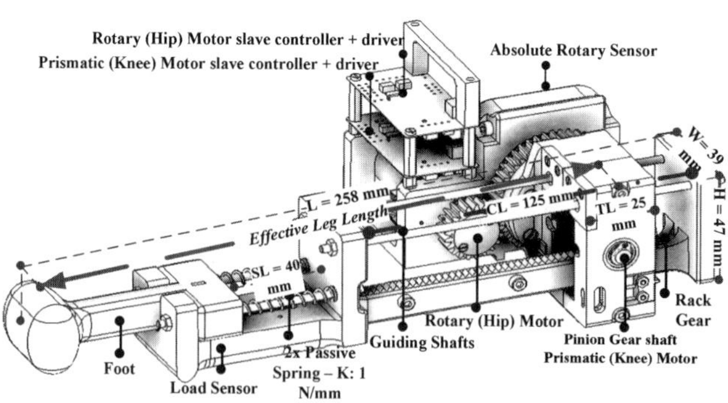

Figure 1. Reconfigurable Leg Length Hopper (RLLH) module. Leg segment dimension *(LxWxH: 258x39x47 mm³)*; total weight (including the weight of the boom rod, which can be seen in Fig. 4 (a): *0.778 ± 0.001 kg*. Purple line shows the range of reconfigurable leg length (CL). Blue line indicates the effective leg length (i.e., It is a distance measured from the center of rotation to the ground).

However, in contrast to the standard SLIP model, the leg has an additional active DOF, i.e., a linear joint which works in series with the passive compliance, similar to a muscle-tendon mechanism in a biological system [8]. This additional DOF provides supplementary advantages over the SLIP model, i.e., it regulates the energy into the passive-spring by altering the amplitude and the rate at which the leg length changes. In addition, adjusting the initial leg length to the various settings, i.e., shifting the initial set position of the leg in order to make it shorter or longer, can now be achieved online. In order to make the design lighter, two electric DC brushed motors are used. The first DC motor drives the linear motion for changing the leg length while the second one allows for a rotary motion of the hip joint.

The control of the RLLH [7] consists of a two layer framework. At the lower level, two types of positional PID motor control loops, one each for the rotary and the linear motion, were implemented. The gains of each PID were experimentally tuned for good performance. On the higher level, an open-loop sinusoidal control scheme was programmed in a master controller. It computes the desired trajectory of each joint by processing control functions (see section 2.2) at the rate of *125 ± 2.4E-05 Hz*. The results from this computation are then transmitted to the low level PID control for the execution of joint motion.

2.1. *Control law*

The oscillator force F_m, due to the motion of the reconfigurable linear joint, for a given operating time t is defined as follow:

$$F_m = -\frac{I_l}{d^2}\left(\Delta l \omega^2{}_p\right)\sin\left(\omega_p t + \varphi_p\right), \tag{1}$$

where, I_l is the moment of inertia of the leg, d is the radius of the pinion gear, Δl is the change in leg length (amplitude), $\omega_p = 2\pi f_p$ is the angular frequency of the oscillator, and φ_p is the phase shift in the oscillator. Furthermore, $v_p = \omega_p * d$ and $\Delta l = v_p * t$. F_m acts such that, when the leg length increases during the first phase of the control signal, the body moves up consequently performing positive work at the passive spring. This work is done by pushing the leg against the ground thus storing some energy in the spring. During the next phase of the control signal, when the applied torque is in the counter-clockwise direction, the leg length reduces by effectively taking the energy from the system (the negative work). Hence, by applying a simplified oscillatory control signal of various amplitudes and frequencies, the required energy in the passive spring can be regulated.

Similarly, the torque τ_r produced by the rotary joint during the oscillatory motion in the sagittal plane, is defined as

$$\tau_r = I_l * \ddot{\theta}_r = -\Delta\theta_r \omega_r^2 \sin(\omega_r t + \varphi_r), \tag{2}$$

where, $\ddot{\theta}_r$ is the joint angular acceleration, $\Delta\theta_r$ is the amplitude of the rotary motion, ω_r is the angular frequency of the oscillator, and φ_r is the phase offset of the oscillations. By adjusting the parameters $\Delta\theta_r$, ω_r and φ_r in the above equation, the forward and backward motion of the leg is controlled.

3. Mathematical Model

The dynamics of the in-place hopping of the RLLH can be described by the mass-spring-damper model, as shown in Fig. 2. By conducting a force analysis of the model, a differential equation describing the motion of the center of mass of the robotic leg, when the ground stiffness is high, is obtained as,

$$m\ddot{y} + D_l\dot{y} + k_l y = F_m + mg, \tag{3}$$

where, y is the motion of the body in the vertical axis, D_l is the damping while k_l is the stiffness constant of the leg, F_m is the oscillatory force produced by the motion of the prismatic joint, as defined in (1), and $m.g$ is the weight of the module. According to (3), when the ground surface is highly stiff, the natural frequency of the system can be defined as,

$$f_0 = \frac{1}{2\pi}\sqrt{\frac{k_l}{m}}, \tag{4}$$

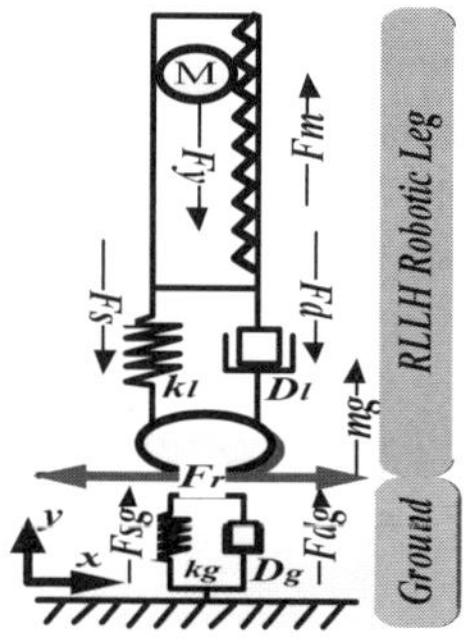

Figure 2. A mass-spring-damper representation of the robotic leg during in-place hopping.

However, for varying ground conditions, the resulting natural frequency f_0 of the system is not only a function of the stiffness and the damping of the robotic leg but it becomes a function of the coupled properties of the two, i.e., stiffness and damping of the ground coupled with the stiffness and damping of the robotic leg (see Fig.2, a model of RLLH (orange) in connection with the model of the ground (green)). These coupled properties affect the overall system dynamics when the leg is in contact with the varying ground conditions. In this case, the coupled stiffness k_c is defined as,

$$k_c = \frac{k_l * k_g}{k_l + k_g}, \tag{5}$$

where, k_l is the leg stiffness and k_g is the ground stiffness. The resulting natural frequency of this coupled interaction can be represented as,

$$f_0 = \frac{1}{2\pi}\sqrt{\frac{k_c}{m}}, \tag{6}$$

In case of very stiff grounds, i.e., $k_g \gg k_l$, the natural frequency of the system (6) mainly depends on k_l, because the coupled stiffness k_c reduces to k_l.

On the other hand, when the ground is soft, i.e., $k_g \ll k_l$, the ground stiffness dominates the system behavior and the coupled stiffness k_c reduces to the ground stiffness k_g. Thus, in this case k_g defines the natural frequency of the coupled system. Since, for soft grounds, k_g is lower than that of k_l, the required operating frequency of the actuator in order to adapt to surface changes will also be lower than in the case of stiff ground. This is advantageous, since, the electric actuators such as the DC brushed motors typically have a limited operational bandwidth. At the lower operating frequency, such motors can potentially be more efficient.

The presence of the dissipative forces, such as friction is very common in real systems. It causes a decrease in mechanical energy during the motion. This effect is modeled as a damper in our system that depends on the speed of the body. In a coupled system, its effect will be additive and can be compensated by injecting more energy (see Equation 1).

These two ground properties (stiffness and damping) lead to a change in the natural frequency and the amplitude of the power consumption of the system for varying grounds, which is shown in number of experiments (see Section 4.1).

The ground friction comes into play during the forward motion of the robotic leg. Due to varying nature of the ground, friction also varies and one way to compensate this change in friction is to reduce the length of the leg such that the rotary joint requires less torque to efficiently negotiate with the ground friction. This relation of the rotary joint torque τ_r and the leg length is defined as

$$\tau_r = F_r * \{L_0 + \Delta l \sin(\omega_p t + \varphi_p)\}, \tag{7}$$

where, L_0(initial leg length), Δl (change in leg length), and ω_p (rate of change). Thus, the shorter leg can theoretically negotiate better both the high and the low friction grounds compared to the longer leg. However, a longer leg can be useful in situations where an increased locomotion speed is required. The latter can be achieved by increasing the leg length during the steady state locomotion when the effect of the static friction becomes smaller. This is experimentally verified in section 4.2.

4. Experiments and Results

To test the role of the reconfigurable leg length for different ground conditions, two types of dynamic motions of the RLLH were studied: one, when the leg hops in-place on various grounds (see Section 4.1) and second, when the leg ran in hopping gait over different grounds, while being fixed to a boom (see Section 4.2).

4.1. *Hopping in-place*

This experiment was conducted while the robot was in an upright posture with respect to the ground. Hopping was achieved by actuating the hip motor at a constant position (i.e., the hip angle is fixed at $\pi/2$ *rad*). In-place hopping was performed by varying two parameters, the amplitude (Δl) and the frequency (f_p) in equation (1) that controls the force produced by the prismatic motion. For each change in amplitude and frequency, the total electrical power consumption of the RLLH was measured. The same experiment was repeated on three types of ground surfaces: stiff (force plate), less stiff (gym training mat) and soft (foam), as shown in Fig. 3 (a). In total, *152* combinations including *4* different amplitudes *(4-28 mm in steps of 8 mm)* and *38* different frequencies *(0.5-10 Hz in steps of 0.25 Hz)* were experimentally tested per ground surface.

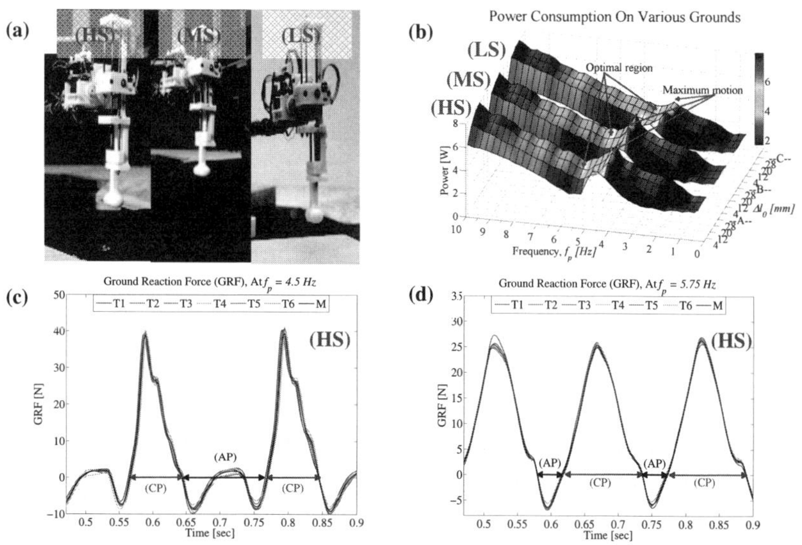

Figure 3. Hopping in place experiments. **(a)** Photographs taken during the experiment over (HS) very stiff ground (force plate), (MS) medium stiff ground (gym training mat), and (LS) soft, low-stiff ground (foam). **(b)** The power consumption plot for HS, MS, and LS. The x-axis represents the frequency ranging from *(0.5-10Hz in steps of 0.25Hz)*, the y-axis the amplitude of change in leg length *(4-12-20-28mm)* relative to the initial effective leg length of *135.5mm*, and the z-axis represents the electrical power consumption averaged over *6 trials* per control parameter. The duration of each trial was *7 sec*. **(c)** The ground reaction force (GRF) measured at a forced frequency (f_p) of *4.5 Hz* on ground (HS). At this frequency the robotic leg achieved the maximum aerial phase (AP) – see the duration of contact phase (CP) and aerial phase (AP) in Fig. 3 (c). This reflects the increase in the ground impact, thus the power consumption- see Fig. 3 (b), (HS). **(d)** The GRF measured at the optimal forced frequency *5.75 Hz* on ground (HS), where the lowest power consumption was observed. At this point the aerial phase decreases to the value shown in Fig. 3 (c). These results from c) and d) also indicate that the duration of the contact phase (CP) and the aerial phase (AP), can be actively modulated by the rate of change in leg length to adapt to varying ground conditions. As we formulated in (5) that on stiff ground such as (HS), the coupled stiffness k_c reduces to the leg stiffness, i.e. $k_c = k_l = 1$ *N/mm* – see Fig. 1. By substituting this value of k_l and *m: 0.778 ± 1E-3* k_g in equation (5), the natural frequency f_0 is computed, which yields *5.70 Hz*. This theoretical f_0 approximately matches the optimal frequency of (HS), which is *5.75 Hz,* where the robotic leg is energy efficient in maintaining steady-state hops.

It can be observed in the respective plot that the power consumption over different grounds (HS), (MS), and (LS) follows nearly the same pattern. On each of these surfaces, the power consumption is low at lower frequencies and it increases proportionally with the frequency. Interestingly, a sharp peak in the power curve is observed just after the frequency, where the in-place hopping started (see Fig. 3 (b), (HS) it is *4.25 Hz*). This peak represents the point where the contact phase (duration on ground) is less than the aerial phase (duration in air), i.e., high ground clearance, as can also be seen in Fig. 3 (c). Right after that, the power consumption sharply drops again as the frequency increases further and the contact phase (CP) increases compared to the aerial phase (AP) (see Fig.3 (d)). This frequency is the optimal frequency of the coupled system as it consumes least power. By further increasing the frequency above the optimal frequency, the power consumption increases too due to the decrease in aerial phase (AP). This shows that by altering the rate of change in leg length caused by the change in the frequency (f_P) of the actuation control signal in equation (1), the passive spring can be actively tuned to work in two different modes: First, in the energy efficient mode (optimal region), where the system can consume the least power and, second, an energy inefficient mode (maximum motion), where the system consumes more power but achieves high ground clearance. These modes are essential in designing the optimal control by exploiting the use of the passive compliance for locomotion.

As can be observed further, the optimal frequency for the soft ground (LS) shifted to a lower frequency as compared to the stiffer grounds (HS) and (MS). This shift in the frequency of overall system was caused by the change in the coupled stiffness and damping properties of the robotic leg with the ground surface, as formulated in equation (6). However, the proposed design is potentially able to adapt to it by varying the rate of change in leg length (see in Fig. 3 (a) In addition, it was observed that to efficiently adapt to the soft ground by mitigating the effect of deformation, higher ground clearance will also be needed. This can be obtained by increasing the amount of change in leg length at the optimal frequency.

4.2. *Planar locomotion in a hopping gait*

This experiment was performed by operating both the rotary (at the hip) and prismatic motors (to change the leg length) at the frequency of *5 Hz*, which has been previously obtained from the in-place hopping experiment (see Fig. 3). For the planar locomotion, the amplitude of the oscillation of rotary hip joint was kept constant at 0.1745 ± 0.035 *rad*. In addition, the phase shift between the rotary and the prismatic actuation signals was set to *0.92* rad with the change in leg length (i.e., Δl_0) to *10 mm and 18 mm*, during all trials. Only the initial effective leg length was altered (see Fig. 1), thereby the initial distance between the hip and the ground was changed (i.e., simulating different heights of the leg). In each trial, the angular velocity ω_s (see Fig. 4 (a)) about the boom fixed coordinate system was measured using the gyro and it was then converted into planar velocity.

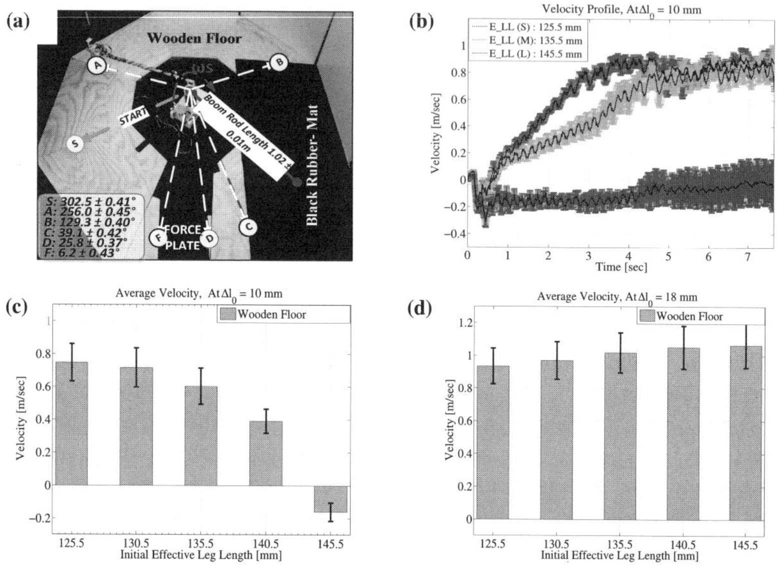

Figure 4. Forward hopping experiment. **(a)** A top view of the experimental-setup. Each trial ends after completing one revolution about the fixed boom axis. **(b)** The velocity profile with respect to the different initial leg lengths *(125.5 (short), 135.5(medium) and 145.5(long) in mm)*. For each leg length settings 5 trails were performed; the shaded region indicates the error about the mean value. **(c)** The measures of the change in forward velocity on wooden grounds with respect to the different initial leg settings. **(d)** The measures of change in forward velocity on wooden ground, when the amplitude of change in leg length Δl_0 increases to *18mm* than Fig. 4 (c).

As shown in Fig. 4 (b) and (c), the velocity at the smaller leg length (red curve) increases faster to the maximum than to the medium (green) and longer leg length (blue). Since the starting point for all these experiments were on wooden floor, we showed that the robotic leg in the short leg configuration was able to recover faster from the static-friction, than in the larger leg length condition, as described by equation (7). However, when the amplitude of change in leg length increases to 18 mm than 10 mm, then it enables the robotic leg to recover faster from the effect of friction, maintaining an increase in overall average locomotion speed with respect to increase in leg length, as shown Fig. 4 (d).

5. Conclusion and Future Work

We have practically demonstrated, that the reconfigurable second DOF of the RLLH that works in series with the passive compliance, is potentially useful to adapt to the changes in ground conditions like stiffness, damping and friction. It can be simply achieved by adapting the amplitude of the linear leg movement and its rate of change in the leg length. Both changes can be rapidly applied to compensate for the change in the overall system behavior, i.e., the change in coupled stiffness and damping between leg and ground. In addition, shortening and lengthening the leg makes this design equally suitable for compensating the varying ground friction. Moreover, the speed of the locomotion, which also depends on the leg length, is also controllable. Furthermore, we highlighted in the in-place hopping experiments that our robotic leg can function in two modes, the optimal energy efficient mode *(aerial phase < contact phase)* and the high bounce energy in-efficient *(aerial phase > contact phase)* mode.

In the near future, these results will be incorporated in a closed loop control for the RLLH. Similar concept will be extended towards controlling a state of the art robot called DTAR that consists of 4 of the presented robotic legs (RLLH).

Acknowledgment

This work is supported by the European Commission Seventh Framework Program, Theme ICT- 2007.8.5 as part of the project LOCOMORPH, under a grant no 231688.

References

[1] M. H.Raibert, "Legged Robots," *Communications of the ACM,* vol. 29, no. 6, June 1986.

[2] Jonathan W.Hurst, et al., "Series Compliance for an Efficient Running Gait," *IEEE Roboitcs & Automation Magazine,* pp. 42-51, 2008.

[3] Fumiya Iida, et al., "Exploiting Body Dynamics for Controlling a Running Quadruped Robot," pp. 229-235, 2005.

[4] Ioannis Poulakakis, et al., "On the Stability of the Passive Dynamics of Quadrupedal Running with a Bouncing Gait," *IJRR,* pp. 669-687, 2006.

[5] R.Blickhan, "The Spring-Mass Model For Running And Hopping," *Journal of Biomechanics,* vol. 22, pp. 1217-1227, 1989.

[6] Rolf Pfeifer, et al., "Self-Organization, Emodiment, and Biologically Inspired Robotics," *Science,* vol. 318, p. 1088, 2007.

[7] Farrukh Iqbal Sheikh, et al., "SLIP-Model-Compatible and Bio-Inspired Robotic Leg with Reconfigurable Length," in *Adaptive Motion in Animals and Machines (AMAM)*, 2011.

[8] R.M.Alexander, "Elastic Mechanisms in Animal Movement," *Cambridge University Press,* 1988.

SWING LEG CONTROL FOR ACTUATED SPRING-MASS ROBOTS*

Hamid Reza Vejdani* and Jonathan W. Hurst
Dynamic Robotics Laboratory, Oregon State University, Corvallis, Oregon, USA
**E-mail: hrvn@engr.orst.edu*

In this paper we present a control strategy for spring-mass running robots that maintains a consistent running gait on uneven terrains, while prioritizing a limit on the peak forces on the leg. The peak forces are a problem for real machines, potentially exceeding the peak forces of an actuator and leading to a fall, or even breaking robot components. Our control strategy relies on an actuated spring-mass model which is described in section 2. Our controller chooses a leg angle and a leg length during the flight phase, relying entirely on passive dynamics during the stance phase to have symmetric gait and not suffer from high leg forces during the stance phase.

Keywords: Actuated SLIP model, Running robot.

1. Introduction

The planar spring loaded inverted pendulum (SLIP) has been widely used in literature as a model for walking[1,2] and running.[3,4] The stability of this simple model in running[5] explains how animals can run robustly and efficiently in real world. Schmitt et. al.[6] added a very simple controller to the passive SLIP model and made the system more robust.

Very recently Ernst et. al.[4] proposed a flight phase control strategy for running on uneven terrain that leads to steady state running. In their method, the leg angle is chosen such that a steady state running is produced during the stance phase. This control strategy does not require any work during the stance phase, but the leg force is increased dramatically.

Our motivation for this study comes from the response of animals to hidden disturbances[7] as is shown in Figure 1. For them, too, peak leg force appears to be a concern. Ground-running birds carefully limit their leg peak forces when encountering unexpected drop perturbations by extending their

*Supported by grant RGY0062/2010 of the HFSP (Human Frontier Science Program).

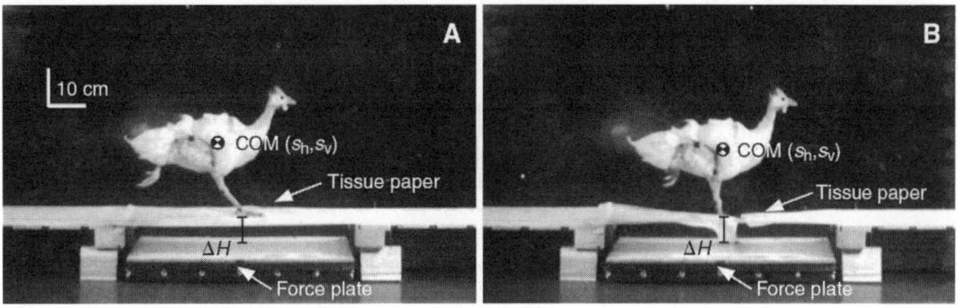

Fig. 1. By extending their leg length and adjusting their leg angles, guinea fowls reject hidden disturbances without suffering from high leg forces.[7]

legs and adjusting their leg angles[7] (Figure 1).

Inspired by the behavior of animals in running, we intend to investigate the effect of the leg length on the dynamics of running. Therefore, a leg actuator is added to the SLIP model[8] to control the length of the leg during the flight phase (Figure 2). The observation of animals' behavior shows that in the level running, the CoM trajectory is close to a symmetric path. When they encounter hidden drops in the ground, their leg length increases to fill the unknown hole height and their leg angle is adjusted to reject the disturbance robustly and efficiently. The steady state running (equilibrium gait) for the SLIP model can be obtained when the leg touch-down angle is the same as the leg lift-off angle and also the CoM velocity components are the same.

2. Methods

The model that we use here is an actuated version of the SLIP model which is shown in Figure 2. The leg actuator is in series with the spring to control the leg length during the flight phase. We keep the motor locked during the stance phase, therefore the dynamics of the system will be entirely passive (SLIP model) in stance phase. The motor inertia and maximum motor torque are considered for the leg actuator to model a realistic electric motor. In addition to the leg length control we assume, like previous studies,[4–6] that the leg angle can also be controlled during the flight phase.

Previously, Ernst et. al.[4] proposed a method to have symmetric gait (steady state running) by only adjusting the leg angle during the flight phase. Figure 3 shows the CoM trajectory of the SLIP model with this control policy. To implement this control policy on a real robot, we need to generate a look-up table which gives us the appropriate touch-down angle

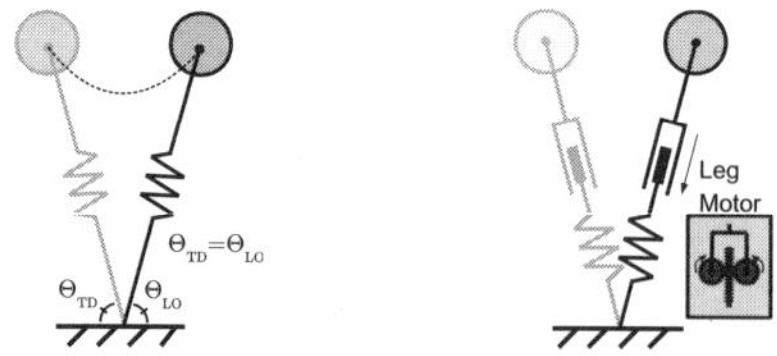

Fig. 2. **Left:** The passive SLIP model **Right:** The actuated SLIP model.

with respect to the vertical component of the CoM velocity. Unfortunately, the peak leg force during the drop increases too much using this control policy. Simulations show that the increase in the peak force in drop gait is about 33% of the steady state value. This increase may break the leg or the transmission of the robot.

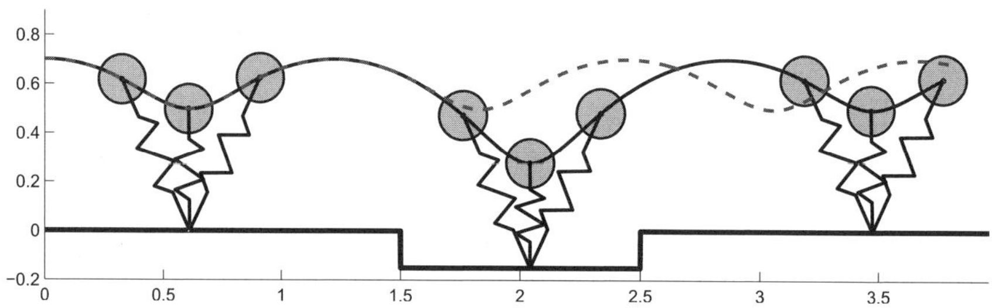

Fig. 3. CoM trajectory of the SLIP model adjusting the leg angle with constant leg length to have symmetric gait.[4]

The control policy used by animals uses the leg extension in addition to leg retraction. Therefore, at each instant (means constant vertical velocity during the flight phase) we look for the relation between the leg length and the leg angle that leads to a symmetric gait for SLIP model. Moreover, we monitor the change in the peak leg force in the stance phase. We need to expand the previous look-up table that Ernst et. al.[4] proposed. In the new look-up table different leg lengths should also be included. Therefore, the required leg angle in each instant is obtained based on the falling time (or vertical velocity) and the current leg length by interpolating among the look-up table data. Since the system is purely passive in stance phase, the well known SLIP model equations of motion are used in stance phase.[2,6]

In Figure 4 the relation between the leg length and the leg angle is shown for different vertical velocities to have equilibrium gait. Each point (pair of leg angle and leg length) on the lines leads to an equilibrium gait for the corresponding vertical velocity. The numbers on the curves show the peak leg forces at those points. It can be seen that to have symmetric gait, the peak force is nearly constant for different leg lengths (numbers along each curve in Figure 4). It means if the leg length extends while the leg angle is being adjusted concurrently (like animals do), the peak force in the stance phase does not vary too much. The whole point is that the leg extension compensates for the hole height and therefore, a symmetric path with nearly constant peak force is generated during the stance phase.

In summary we can design the controller as follows: if the vertical velocity passed the usual value at touch-down; the leg actuator should extend the leg towards the ground. Based on the current leg length at each moment the leg angle should be adjusted using the curves in Figure 4 or the look-up table that was mentioned earlier.

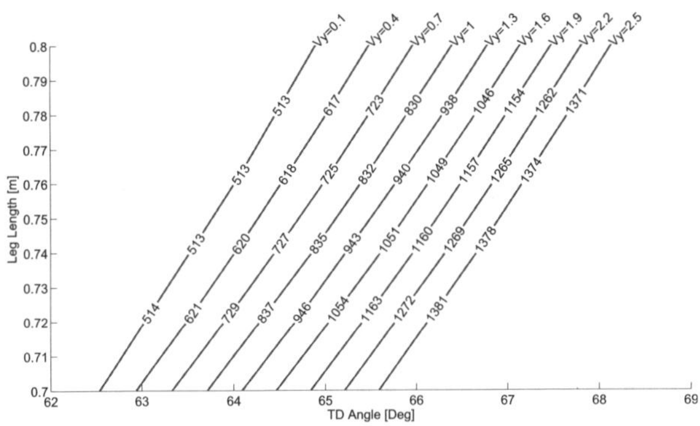

Fig. 4. Leg length vs leg angle for equilibrium gait policy with different vertical velocities. Numbers on the lines show the peak leg force [N] during the stance phase at that point. The peak forces are nearly constant along each curve.

Two different scenarios that may happen during the running are shown in Figure 5. When the vertical velocity of the CoM is less than or equal to the usual value of the vertical velocity at touch-down, the motor does not work and only the leg angle is adjusted to generate symmetric path in stance phase (if there is any step up in the ground). If the vertical velocity of the CoM passes the usual value of the vertical velocity at touch-down it means the robot encounters a step down in the ground. Therefore, the leg actuator extends the leg towards the ground. Because of the existence of the motor inertia and maximum torque, the toe reaches the ground with some delay. Meantime, the leg angle is adjusted based on the current leg length to have symmetric gait in stance phase whenever it hits the ground.

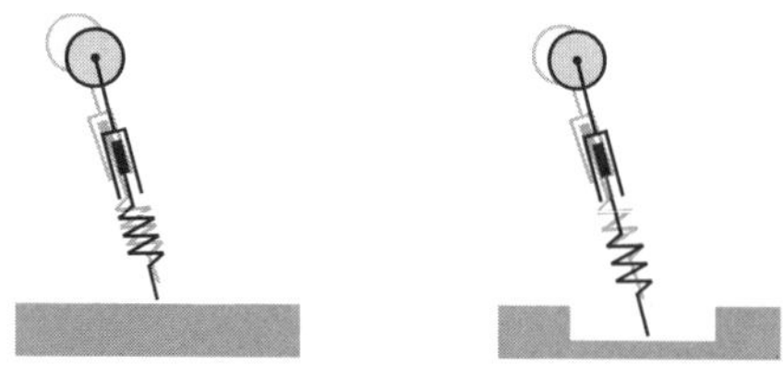

Fig. 5. **Left:** The leg actuator does not work on even ground (no leg extension, only leg angle adjustment during the flight phase) **Right:** The leg actuator extends the leg, the leg angle is adjusted concurrently based on the current leg length.

In our simulations, we assume that the motor has inertia and maximum torque limit. Any leg mass can also be included in motor inertia. In this case, the important physical limitations are considered in the results. To accomplish the simulations, following characteristics are chosen for the model.

Parameter	**Description**	**Value**
m	robot mass	$30.0kg$
k_{leg}	leg spring stiffness	$4500\frac{N}{m}$
T_{max}	maximum motor torque	$850N.m$
I	motor inertia	$2.78kg.m^2$
G	Gear ratio	$50:1$
v_{0x}	initial horizontal velocity	$2.5\frac{m}{s}$
h_0	initial CoM height	$70cm$
δ_{gnd}	ground disturbance	$-15cm$

3. Results

The CoM trajectory of the robot controlled with the policy described in section 2 is shown in Figure 6. When the vertical velocity of the CoM becomes greater than the vertical velocities at touch-down in previous strides, the leg actuator starts extending the leg towards the ground. At each moment, based on the new leg length and new vertical velocity the required touch-down angle is calculated from the look-up table. Figure 7 shows the leg force profiles for three cases. As can be seen in the figure, the increase in the peak force with the proposed control policy is only 11% more than the undisturbed case. This increase is due to the small increase in the vertical velocity of the CoM while the motor tries to hit the ground. The increase of the peak force due to the Ernst et. al.[4] method is about 33% more than the undisturbed case.

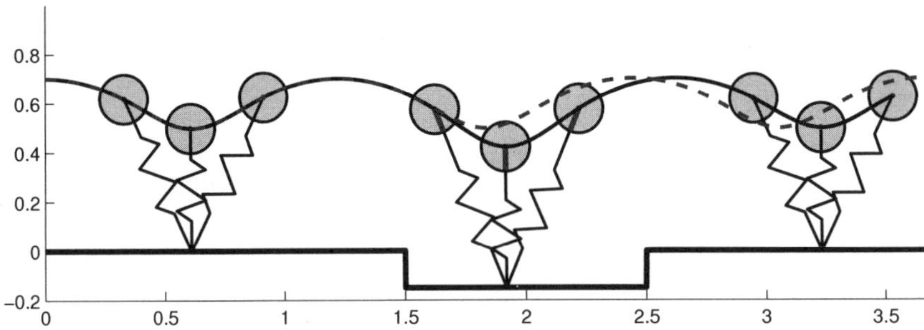

Fig. 6. The actuated SLIP model rapidly extends the leg and adjusts the leg angle concurrently. The red line on the leg at the drop gait is the increased leg length.

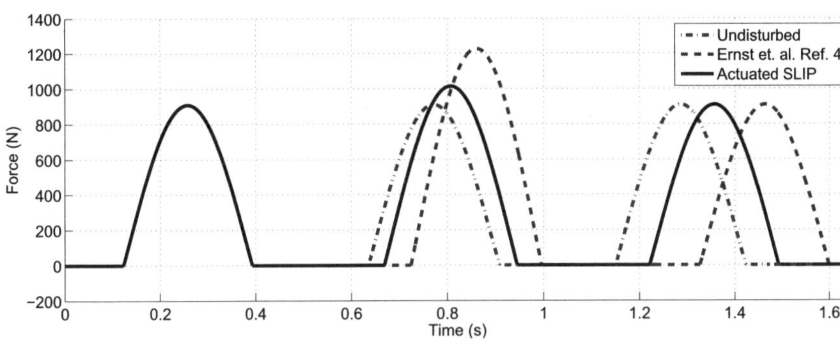

Fig. 7. Leg force profiles show the peak force increases only 11% if the leg is rapidly extended. This increase is due to the physical limitations of the motor like motor inertia. Without leg extension, increase in leg force would be 33%.

4. Conclusion

In this paper we proposed a bio-inspired control strategy for the flight phase that leads to steady state running but more importantly keeps the leg force nearly constant in the presence of disturbances. The proposed control policy comes from the fact that if the leg length extends while the leg angle is being adjusted appropriately (like animals do), the peak force in the leg does not increase too much. It means the leg extension partially compensates for the drop and the leg angle is updated to generate the symmetric path. It should be noted that the small increase in the leg force in our simulation is due to the physical limitations of the motors that prevent them to act instantaneously.

References

1. F. Parietti and H. Geyer, Reactive balance control in walking based on a bipedal linear inverted pendulum model, in *Robotics and Automation (ICRA), 2011 IEEE International Conference on*, may 2011.
2. H. Geyer, A. Seyfarth and R. Blickhan, *Proceedings of the Royal Society B* **273**, 2861 (2006).
3. R. Blickhan, *Journal of Biomechanics* **22**, 1217 (1989).
4. M. Ernst, H. Geyer and R. Blickhan, Spring-legged locomotion on uneven ground: a control approach to keep the running speed constant., in *Proc 12th Int Conf on Climbing and walking Robots (CLAWAR)*, 2009.
5. R. M. Ghigliazza, R. Altendorfer, P. Holmes and D. Koditschek, *SIAM Review* **47**, 519 (2005).
6. J. Schmitt and J. Clark, *Bioinspiration & Biomimetics* **4**, p. 17(pp) (2009).
7. M. A. Daley, J. R. Usherwood, G. Felix and A. A. Biewener, *The Journal of Experimental Biology* **209**, 171(November 2005).
8. D. Koepl and J. Hurst, Force control for planar spring-mass running, in *IEEE International Conference on Intelligent Robots and Systems*, 2011.

RUNNING ON SOFT GROUND: SIMPLE, ENERGY-OPTIMAL DISTURBANCE REJECTION*

CHRISTIAN M. HUBICKI† and JONATHAN W. HURST

Dynamic Robotics Laboratory
Oregon State University
Corvallis, Oregon, USA
†*E-mail: hubickic@engr.orst.edu*
mime.oregonstate.edu/research/drl

Soft ground is a ubiquitous hazard for legged locomotion and has yet to be conquered in a robust, dynamic, and economical manner. In search of a controller to meet these demands, we found that a simple force controller is energy optimal for spring-loaded running on unknown ground dissipation. The simplicity of this optimal controller suggests a fundamental insight into legged locomotion.

Keywords: legged, running, terrain, optimal control, force control, SLIP

1. Introduction

Handling uncertain terrain is an enduring challenge in legged locomotion, especially when running. We believe the solution lies in fundamental principles of spring-mass running. Some controllers for the Spring-Loaded Inverted Pendulum (SLIP) have been successful in rejecting potholes and slopes by adjusting leg posture mid-flight. However, negotiating soft terrain, such as soil or sand, requires thoughtful actuation while in brief contact with the ground. This challenge is further complicated as the SLIP model is nonlinear, hybrid-dynamical, high-dimensional, underactuated, and analytically unsolvable. This investigation discovered a simple feedback force controller that completely rejects soft ground in a single step for minimum energy cost.

In simulation, we demonstrate that the controller navigates even extremely soft ground without any estimates or foreknowledge of the ground

*Supported by the DARPA: Maximum Mobility and Manipulation Program (W91CRB-11-1-0002) and the Human Frontier Science Program (RGY0062/2010).

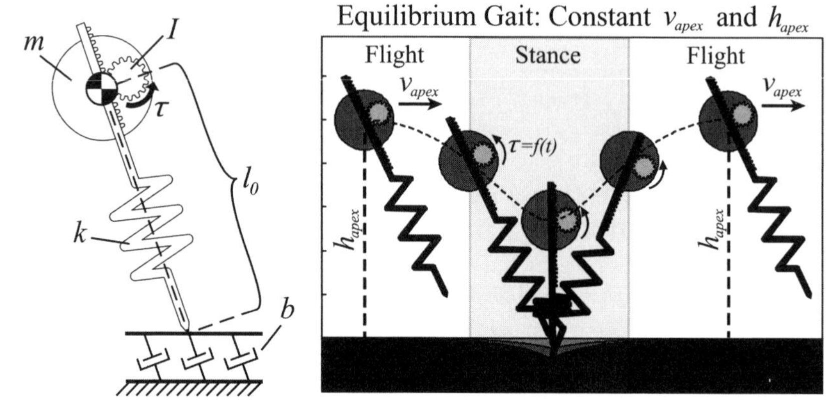

(a) (b)

Fig. 1: (a) An actuated variant of the Spring-Loaded Inverted Pendulum (SLIP) model, including a motor that regulates axial spring compression during stance, and a linear damping ground model acting inline with the leg. (b) A visualization of an equilibrium gait, where apex velocity and height remain constant across flight phases. On soft ground, this is only possible if the actuator intervenes during the stance phase, as shown above.

properties. Further, we present evidence from numerical optimizations that this controller is, in fact, the energy-optimal solution for handling any dissipative terrain. We conclude that these features make the force controller well-suited for real-world locomotion and realizable on hardware that approximates SLIP-model running.

Solving an optimal control problem with such mathematical barriers required an unconventional approach. We use trajectory optimization as a starting point, manually inspecting numerical optimal trajectories for similarities. Informed by these similarities, we hypothesized a general optimal controller and, as validation, compared it to the performance of optimal trajectories. We call this process *optimization-inspired* controller synthesis.

2. Background

The research trail of running robot control was blazed by Raibert,[1] whose simple and effective controllers inspired a wave of stable, energy-based running controllers.[2] Hybrid-Zero Dynamics, while having largely been applied to walking,[3] has recently yielded robust and efficient running gaits[4] on the spring-legged robot, MABEL. For the idealized, energetically-conservative

SLIP model, Ernst, Geyer, and Blickhan[5] (EGB) developed a feedforward steady-state flight-phase controller. While airborne, EGB control selects a leg angle which guarantees the same flight speed on the next step (an *equilibrium gait*) even without detecting the impending ground. However, guarantees made by EGB control and its predecessors assume a rigid landing surface. This investigation seeks a stance-phase controller complementary to EGB control which rejects the effect of soft surfaces while performing minimal actuator work. Such a controller, as Fig. 1b illustrates, would ensure economical, robust control while running on surfaces like soil, sand, and snow.

3. Methods

3.1. *Trajectory Optimization*

Using the actuated SLIP model shown in Fig. 1a, a trajectory optimization problem was formulated. Given **A)** an initial apex condition, **B)** a leg angle at touch down[a], and **C)** a ground-damping coefficient, find a torque trajectory[b] which **1)** results in an equilibrium gait[c] (as defined in Fig. 1b), and **2)** minimizes actuator work[d].

Using Sequential Quadratic Programming[e], energy-optimal torque trajectories were computed for the SLIP model[f] at various initial conditions and assorted surface consistencies. However, any individual trajectory is only useful for its respective narrow case and, at 10^3 seconds per optimization, are cripplingly complicated to compute in real time. While these trajectories are impractical on their own, we suspected that inspecting them all for similarities could yield insight into a simple, general solution.

---

[a]On rigid ground, these leg angles at touch down would produce equilibrium gaits, and are computed using EGB control.

[b]The input trajectory is a piecewise-linear, time-scheduled, torque applied to the leg-length control motor at the moment of touch down.

[c]An "equilibrium gait", similar to a "limit-cycle" gait, "periodic gait", or "steady-state gait", has equivalent state variables from step to step at the apex of flight.

[d]The objective function is defined as the time integral of the unsigned derivative of actuator work, resulting in minimal motor intervention in the passive dynamics.

[e]Using `MATLAB`'s constrained minimization algorithm, `fmincon`, with an equality constraint.

[f]SLIP model parameters are estimates from our experimental monopod, ATRIAS 2.0: $m = 32kg$, $l_0 = 0.90m$, $k = 3.0\frac{kN}{m}$, $\tau_{max} = 25Nm$, motor inertia $I = 0.003kgm^2$, transmission ratio: 50:1, and ground damping coefficient, b, is experiment dependent.

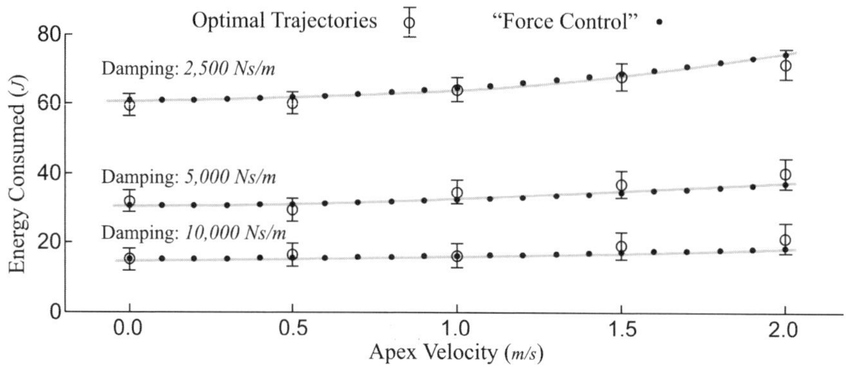

Fig. 2: Energy consumption of "force control" matches optimized trajectories, strongly suggesting that force control is an optimal solution to handling soft terrain. Energy error bars correspond to a 1% solver constraint tolerance.

3.2. *General Optimal Control*

Despite the optimal torque trajectories being widely varied, the axial ground reaction forces were remarkably similar across all tested surfaces. Even when a particularly soft surface caused the robot to sink 10% of its leg into the ground, conditions which could be likened to a sand dune or snow drift, the axial forces were nearly identical to that of rigid ground. This surprising similarity among energy-optimal trajectories led us to formulate a hypothesis: *"a controller that replicates the axial forces experienced on rigid ground is an energy-optimal control policy on soft ground."*

To test this hypothesis, we compared the energy economy of an axial-force controller to optimal energy consumption. To facilitate the necessary closed-loop control of the ground-reaction forces, a PI controller[g] was implemented which tracks the axial force trajectory experienced on rigid ground (dubbed *force control* for brevity).

4. Results and Conclusions

We compared the energy consumption and disturbance rejection of both force control and optimal trajectories. As shown in Fig. 2, when applied to the same situations as the optimization, force control preserved the equilibrium gait[h] on soft ground for the same energy cost as the optimized

[g]Integral control was necessary to eliminate steady-state error; Gains of $K_P = 10m$, $K_I = 20\frac{m}{s}$ were manually tuned.

[h]Equilibrium gaits were reliably produced within a 1% state variable tolerance.

trajectories[i]. We present this as numerical evidence which confirms our hypothesis that *force control is the energetically optimal solution to running on soft ground.*

We believe this solution is a fundamental insight into control of the SLIP model. By using numerical optimizations to inspire and test hypotheses about optimal control, we were able to bypass the mathematical complexity inherent to the model. We believe this approach, coined *optimization-inspired* controller synthesis, can apply to a broad class of problems.

Force control also has many practical advantages. The policy is simple, tracking easily computed force profiles instead of expensively computed optimal input trajectories. Further, it requires no estimation of the ground dynamics in order to be robust or optimal. This feature combines well with EGB flight control, since EGB requires no knowledge of the ground geometry, as demonstrated in a recent simulation study.[6] These features allow force control to be tested on practical surfaces with robotic hardware that can approximate SLIP-model dynamics.

References

1. M. H. Raibert, *Legged Robots That Balance* (MIT Press, Cambridge, 1986).
2. M. Ahmadi and M. Buehler, A control strategy for stable passive running, in *IEEE Conf. on Intelligent Systems and Robots*, 1995.
3. K. Sreenath, H.-W. Park, I. Poulakakis and J. W. Grizzle, *Int. J. Rob. Res.* **30**, 1170(August 2011).
4. K. Sreenath, H.-W. Park and J. W. Grizzle, *The International Journal of Robotics Research* (submitted, 2011).
5. B. R. Ernst M., Geyer H., *Proc 12th Int Conf on Climbing and Walking Robots (CLAWAR)* , 639 (2009).
6. D. Koepl and J. Hurst, Force control for planar spring-mass running, in *Intelligent Robots and Systems (IROS), 2011 IEEE/RSJ International Conference on*, sept. 2011.

[i]Small variance in the SQP performance is likely due to local minima caused by trajectory representation and the state variable tolerance allowing slightly "lazier" solutions.

THE DESIGN OF ATRIAS 1.0 A UNIQUE MONOPOD, HOPPING ROBOT*

Jesse A. Grimes† and Jonathan W. Hurst‡

Dynamic Robotics Laboratory, Oregon State University
204 Rogers Hall, Corvallis, Oregon, USA
†*E-mail: grimesje@engr.orst.edu*
‡*E-mail: jonathan.hurst@oregonstate.edu*
mime.oregonstate.edu/research/drl

ATRIAS 1.0 is a spring-legged, monopod robot designed and built as a prototype towards a human-scale 3D biped. The monopod has to meet certain requirements concerning locomotion dynamics and energy efficiency to meet the goal of a biped that can autonomously walk and run efficiently and robustly outdoors, untethered over realistic (non-ideal) terrain. The design of ATRIAS 1.0 includes adequate control authority for robust locomotion as well as incorporating the idea of passive dynamics for high energy economy. Towards this effort, the passive dynamics of ATRIAS 1.0 are designed to match the key features of the Spring Loaded Inverted Pendulum model: a massless leg, mass centered at the hip joint, and a series spring between the ground and the mass at the hip joint. In this paper the authors discuss the key features of this unique robot design.

Keywords: robot; legged; monopod; hopping; walking; running.

1. Introduction

ATRIAS is a prototype spring-legged monopod robot designed as a precursor to a human-scale biped that is intended to efficiently and robustly walk and run outdoors, untethered, over realistic (non-ideal) terrain. Legged robots that can currently locomote outdoors do so at significant energy expense.[1,2] Conversely, robots that walk or run efficiently do so at the expense of robustness.[3] Although no robot currently realizes both features, we believe these are not mutually exclusive goals, and that ATRIAS will bridge the gap.

*Supported by HFSP grant number RGY0062/2010 and DARPA grant number W91CRB-11-1-0002

Fig. 1. **Left:** ATRIAS version 1.0, a Monopod prototype for a bipedal walking, running robot. **Right:** CAD model of the biped configuration.

ATRIAS is designed to match a simple mathematical model called the Spring Loaded Inverted Pendulum (SLIP), a model that approximatly describes the steady-state legged locomotion for most legged animals.[4] By matching this established model, we can take advantage of existing robust and efficient control theory that is built on this foundation.[5–7] Successful application of these controllers on ATRIAS relies on close approximation of the SLIP model's passive dynamics. The key elements of the SLIP that we have approximated in ATRIAS are a massless leg, mass centered at the hip joint, and a series spring between the mass and the ground.

2. Background

Big Dog and LS3 are quadruped robots representing the state of the art in legged locomotion.[1,8] These robots can walk outside in a wide range of non-ideal, real world environments. Utilizing gasoline engines to power hydraulic actuators, these robots have ample power to allow for their controllers to be developed and implemented. Each have a cost of energetic transport rating of over ten times that of a human. Legged robots that currently locomote effectively outdoors do so at high energy cost.

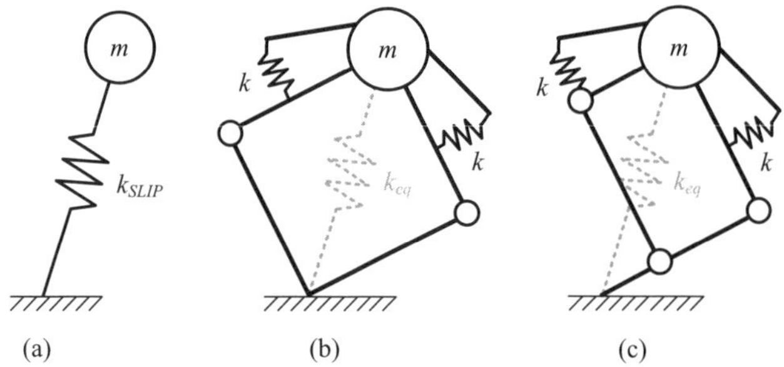

Fig. 2. **A:** Spring Mass model **B:** Simplified ATRIAS model **C:** ATRIAS model. This figure shows the similarity between the Spring Mass model and the ATRIAS model. The dynamics of the SLIP and ATRIAS are equivalent. The equivalent SLIP model is superimposed over the ATRIAS models.

Efficient legged robots locomote at the expense of robustness. Robots like Cornell Biped, Ranger and T.U. Delft's Denise use little or no actuation and rely on the mechanism design to achieve a stable walking gait.[3] They rely largely on their passive dynamics showing us that an understanding of this concept can lead to greater energy economy in legged robotics.

A combination of the two approaches, combining active control with passive dynamics, can result in an energetically efficient and robust legged robot. As an example MABEL is a planar biped built with passive dynamics to match the SLIP model and has proven to be a robust walker with a cost of transport of about twice that of a human.[7]

3. Robot Model and Mechanical Design

The goal of the mechanical design of our robot is to closely match the key features of the SLIP model. As with the SLIP model, our robot operates in a plane, constrained by a support boom allowing the robot to run in a circle in our laboratory while allowing un-restricted vertical motion.

A four-bar mechanism composed of lightweight carbon fiber tubes is used to meet the requirement of a massless leg as close as physically possible. This leg follows the same idea of high-speed pick-and-place robots featuring low-inertia limbs with simple pin joints, driven by motors placed above the limbs.[9] While a massless leg is impossible, this four-bar leg is only 10% of the total full monopod robot weight.

Placement of the motors at the hip joint also helps to match the SLIP-model feature of robot mass concentrated at the hip joint. These two brushless DC motors paired with low-friction, zero-backlash cable pulley

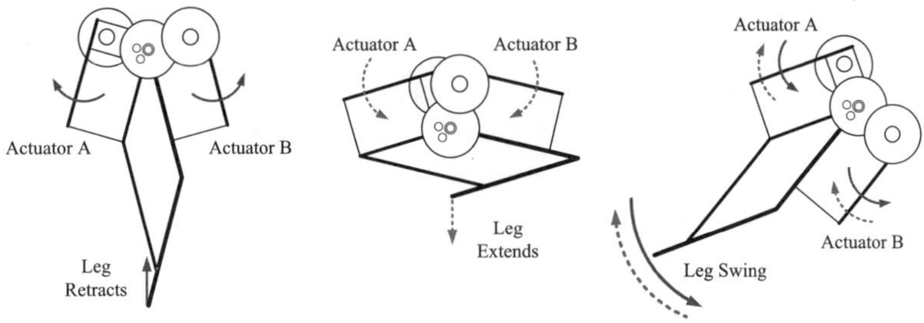

Fig. 3. Reaction torque is canceled during leg extension and retraction but not for leg swing. Both motors contribute to the torque for leg extension, in contrast to a design with one motor controlling leg length and second for leg angle. **Right:** Solid, red arrows indicate leg swing in one direction; dashed, blue arrows indicate leg swing in the opposite direction.

transmissions drive the two topmost members of the four-bar leg through a large series spring.[7,10] Figure 3 shows this configuration where both motors contribute to the leg length and the leg angle, combining torques to extend the leg. Both motors contribute to the large forces exerted by the leg during the stance phase of running or jumping. This allows the motors to be smaller and lighter compared to a configuration where one motor drives the leg length and the other the leg angle.[a]

A series spring is placed between each motor and its driven carbon fiber linkage in the form of a fiberglass leaf spring. Figure 5 shows deflection in the springs, which is measured with high resolution absolute encoders allowing for very precise force control.

The configuration of the four-bar leg, with the mass centered at the hip joint and series springs pictured in Fig. 2c, is dynamically equivalent to the SLIP model. The SLIP model generally has a linear spring stiffness, k_{SLIP} (Fig. 2a), but this is primarily for mathematical convenience. Due to ATRIAS's leg configuration, linear leaf springs result in an overall leg stiffness, k_{eq} (Fig. 2c), that is a non-linear, softening spring. To make ATRIAS a 3D-capable robot, a hip actuator is included to control the angle of the leg to the sagittal plane of the robot, allowing the robot to side step. This is a divergence from the 2D SLIP model but is a necessary extension for a 3D biped configuration, and is also useful to counter the angular changes imposed on the monopod by the robot support boom. This actuator is composed of a considerably smaller motor from the leg motors, includes a series spring and ball screw, driving the leg with respect to the body. The robot's

[a]MABEL is an example of this configuration.[7]

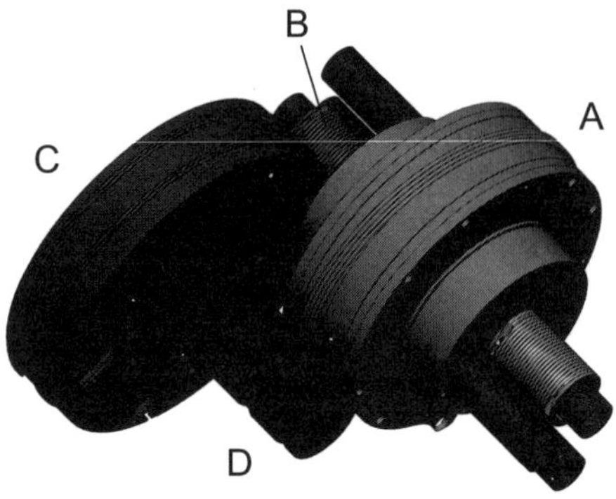

Fig. 4. ATRIAS 1.0 Cable Drive Transmission for one motor (transmission for second motor is removed for clarity). Four Pulleys make up the planetary gear system for one of the two force controlled actuators on the robot. D: Motor housing, fixed to body (non-rotating) A: Motor output spindle (input to transmission) A & C First stage of transmission, 5.38:1 C & D Second stage of transmission 3.73:1 C: Pulley, rotates around motor housing (D) with the transmission output. An overall transmission ratio of 20:1 is achieved.

body is composed of this actuator and a battery pack arranged such that the center of mass of the body is located at the hip joint. This maintains the desired mass concentration of the robot and the requirement set by the SLIP model.

4. Discussion and Future Work

While the cable pulley transmission is highly efficient, its system of pulleys require a large workspace when moving the leg. This is problematic when the leg has a large range of motion including that which is necessary for side stepping. The robot necessarily becomes very bulky in order to keep the pulley workspaces from intersecting.

The four-bar leg is a lightweight, low-inertia physical approximation of the massless leg on the SLIP model. Because one motor applies torques between the body and the thigh, and the other motor applies torques between the body and the shin, both motors must work together for coordinated leg movements. This has two implications: the reaction forces of the motors accelerating, and how they apply force and work.

Due to the counter-rotation of the motors during leg extension, motor reaction torques are canceled; however, co-rotation of the motors during leg swing these torques are added. For a monopod configuration this can cause undesired pitching of the robot body, but, for a biped configuration these

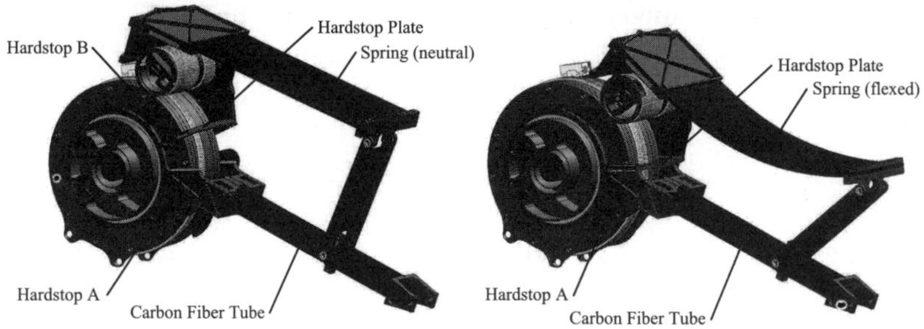

Fig. 5. Section view of one of ATRIAS's leg actuators. The carbon fiber tube is driven by the motor through the leaf spring and linkage forming a series elastic actuator.

reaction torques can be canceled by counter-rotating the two legs. This divergence from the SLIP model must be accounted for in any dynamic modeling, as a modification of the actuated SLIP model.

The four-bar leg has a distinct advantage in the application of force, and a disadvantage in conservation of work. For force application, both motors apply opposing torques to lift the robot off of the ground. However, when running forward at speed, both motors are moving in the same direction as the leg sweeps through the stride. Because the forces are applied in opposite directions while the speed is in the same direction, one motor is doing positive work while the other is doing negative work. In effect, energy is cycling between the motors and repeatedly encountering losses in the transmissions. This effect will reduce the robot's energy economy when running at high speeds, and whether or not it is a worthwhle tradeoff depends on the specfic gaits or behaviors.

The large leaf springs provide excellent energy storage and series compliance for the two leg actuators responsible for extending, retracting and swinging leg. However, series compliance in the hip actuator, responsible for side stepping, allows for good force application during stance but leads to complications in controlling the system during flight. The complication arises when the hip actuator transitions from applying forces during stance to supporting the leg during flight. This immediate change in controller goals results in undesirable oscillations during flight phase, complicating the task of orienting the leg for touchdown. With our monopod prototype, we essentially ignored the springs and the forces, and did a rough position control based on the kinematics of the boom, to keep the leg upright as it hopped. This design will not work well for a 3D, tether-free biped.

Toward the goal of making a biped robot walk and run in a realistic, non-ideal, 3D environment using SLIP model controllers, ATRIAS 1.0 is an important first step in the design process. With a very basic, intuitive controller, this robot has achieved a hopping gait. We demonstrated that the four-bar linkage leg configuration will work well, the leaf springs work well, and the kinematic structure of ATRIAS is good. In addition, of course, the software and electronics and many small engineering details were tested. Some things will change, including the hip actuator for lateral stepping, and the cable drive transmission, due to the large size. With what we have learned from ATRIAS 1.0, future revisions of ATRIAS paired with proper controllers will be capable of truly efficient, robust 3D walking and running.

References

1. Boston Dynamics Inc, http://www.bostondynamics.com/.
2. Honda Corp., Asimo humanoid robot, http://world.honda.com/asimo/.
3. S. Collins, A. Ruina, R. Tedrake and M. Wisse, *Science Magazine* **307**, 1082 (2005).
4. R. Blickhan, *Journal of Biomechanics* **22**, 1217 (1989).
5. M. Ernst, H. Geyer and R. Blickhan, Spring-legged locomotion on uneven ground: a control approach to keep the running speed constant., in *Proc 12th Int Conf on Climbing and walking Robots (CLAWAR)*, 2009.
6. D. Koepl, K. Kemper and J. Hurst, Force control for spring-mass walking and running, in *Advanced Intelligent Mechatronics (AIM), 2010 IEEE/ASME International Conference on*, july 2010.
7. K. Sreenath, H.-W. Park, I. Poulakakis and J. W. Grizzle, *Int. J. Rob. Res.* **30**, 1170(August 2011).
8. M. Raibert, K. Blankespoor, G. Nelson, R. Playter and the BigDog Team, Bigdog, the rough-terrain quadruped robot, in *Proceedings of the 17th World Congres (IFAC'08)*, July 2008. Seoul, Korea.
9. The ABB Group, IRB 360 FlexPicker robot Packaging Magazine(July, 2008).
10. Barrett Technology Inc, Wam arm, http://www.barrett.com.

MOBILITY OF LEGGED ROBOT LOCOMOTION WITH ELASTICALLY-SUSPENDED LOADS OVER ROUGH TERRAIN*

JEFFREY ACKERMAN, XINGYE DA and JUSTIN SEIPEL†

School of Mechanical Engineering, Purdue University
585 Purdue Mall, West Lafayette, IN 47907, USA

Elastically-suspended loads can reduce the energetic cost of legged robot locomotion over flat ground. In this paper, we studied the effect of elastically-suspended loads on the mobility of legged robot locomotion over rough terrain. The power input and speed of a simple hexapod robot running over rough terrain with an elastically-suspended load and a rigidly-attached load was compared. On average, the robot ran 16% faster and consumed 8.9% less power while carrying an elastically-suspended load versus a rigidly-attached load over rough terrain. Therefore, the experiments suggest that elastically-suspended loads may increase the mobility of legged robot locomotion over rough terrain.

1. Introduction

Legged locomotion offers superior mobility characteristics over wheeled or tracked vehicles for locomotion in unstructured environment [1]. The RHex robot is one of the premier legged robot platforms for locomotion over rough terrain [2-4]. The tuned spring-legs of the RHex coupled with the robust alternating tripod gait inspired by the cockroach enable the RHex robot to efficiently traverse rough terrain [5-8]. Besides the use of spring-legs, the principle of elasticity is not utilized elsewhere in the robot design. The inherent mass of the batteries and electronics that power the RHex robot represent a significant proportion of the total mass and are usually rigidly-attached to the robot chassis. In prior work, the authors have shown that elastically-suspending such inherent mass from the main robot chassis with a passive compliant suspension system can significantly increase the energy efficiency of legged robot locomotion by decoupling the vertical motion of the load from that of the main robot chassis [9, 10]. However, these experiments were conducted over flat ground; actual legged robots, like the RHex, need to be able to operate in

* This work was supported by NSF CMMI-1131423 and the Winkelman Fellowship at Purdue University.

† To whom all correspondences should be addressed: jseipel@purdue.edu

challenging environments. Therefore, the authors set out to test the effect of elastically-suspended loads on the mobility of legged robot locomotion over rough terrain. In this paper, we define mobility as the capability of the robot to traverse over rough terrain quickly and efficiently.

2. Robot Platform

A simple hexapod robot based on the RHex was modified to elastically-suspend a battery load from the main robot chassis (Figure 1). The robot used a single DC motor and an all-mechanical power transmission system to rotate two pairs of three spring-legs in an alternating tripod configuration. The suspension system used to elastically-suspended the load was inspired by the Christie suspension system [9]. It utilizes a bell crank mechanism to convert the vertical motion of the load to a largely horizontal motion of the suspension spring. The load was attached to one arm of the bell crank and a compliant linear extension spring was attached to the other arm. This compliant suspension system elastically decouples the vertical motion of the load from the vertical motion the main robot chassis. Fixing the bell crank to the main robot chassis with a bolt enables the robot to switch to a rigidly-attached load.

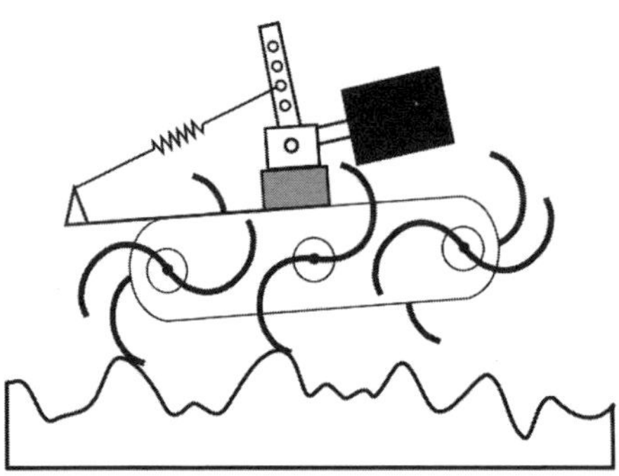

Robot Mass	275 g
Load Mass	125 g
Robot Dimensions	15.2 cm long 7.6 cm wide
Effective Leg Stiffness	6000 N/m
Effective Load Suspension Stiffness	37 N/m

Figure 1: The simple hexapod robot with an elastically-suspended load.

3. Experimental Methods

3.1. *Rough Terrain*

Rough terrain was simulated by randomly distributing and fixing about 150 foam pieces on a plastic track that measured 142 cm long by 47 cm wide (Figure 2). Each foam piece was about 3cm wide by 6 cm long by 1.67 cm high. Since the highest clearance of the robot was about 1.5 cm, these pieces were chosen to represent significant obstacles for the robot to surmount. The foam pieces were cut from a large foam block and each piece was quite stiff so no significant deformation occurred while the robot ran over the track.

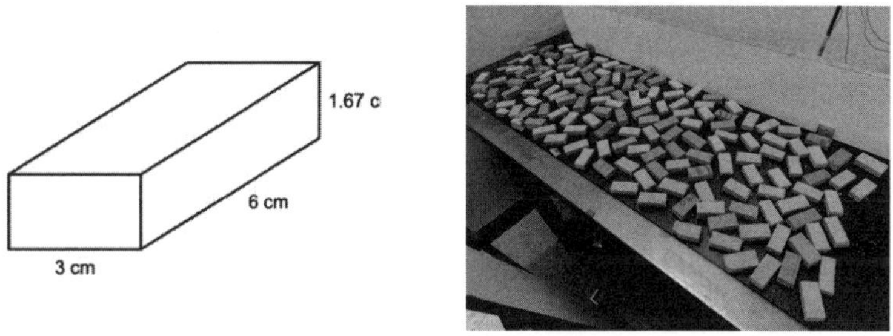

Figure 2: The rough terrain block dimensions (left) and track setup (right).

3.2. *Experimental Setup*

The power and speed of the robot with a rigidly-attached load versus an elastically-suspended load was measured to compare the relative mobility of the robot in each configuration. To determine power, the voltage across the motor terminals of the robot was measured with an ADC and the current flowing through the motor was measured with a hall-effect sensor. An analog low pass filter was used to reduce measurement noise and the robot was powered by a regulated power supply at a constant voltage during each trial. The data was sampled at 1 kHz using a USB4 DAQ. The first 0.75 seconds of data were ignored to remove the initial power spike from the comparison. The robot was started from the same point and the legs were checked for the proper alternating-tripod configuration before every trial. To measure the speed of the robot in each configuration, a laser tripwire at the end of the track was used to stop the data collection, yielding the time the robot took the traverse the known track distance. High-speed video was also taken at 240 fps to qualitatively observe the dynamics of the system in each configuration.

We also sought to determine the effect of the rough terrain height on the mobility of the robot over rough terrain. The average power and speed of the robot was measured while running it for 10 trials over a short track with flat ground and varying obstacle heights. Multiple thin books were stacked on top of each other to simulate increasing rough terrain height (Figure 3). Although the track length was shorter, the power and speed of the robot was measured using the same method discussed previously.

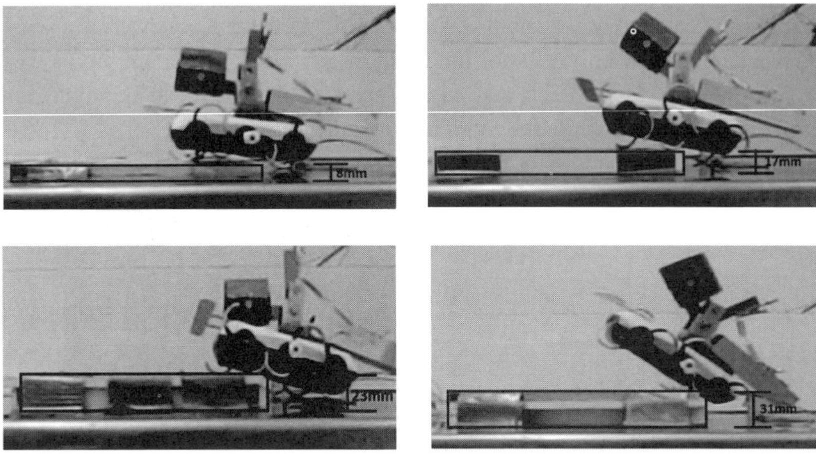

Figure 3: The speed and power of the robot was measured while running over flat ground (not shown), an 8 mm tall obstacle (top left), a 17 mm tall obstacle (top right), a 23 mm tall obstacle (bottom left), and a 31 mm tall obstacle (bottom right) to determine the effect of terrain height on the robot's mobility. The 31 mm tall obstacle proved to be too tall for the robot to reliably surmount.

4. Results and Discussion

Twenty trials were conducted for the robot traveling over the rough terrain with a rigidly-attached load and with an elastically-suspended load. On average, the robot with an elastically-suspended load traversed the rough terrain 16% faster while consuming 8.9% less power than with a rigidly-attached load (Figure 4). Furthermore, the robot with a rigidly-attached load failed to travel

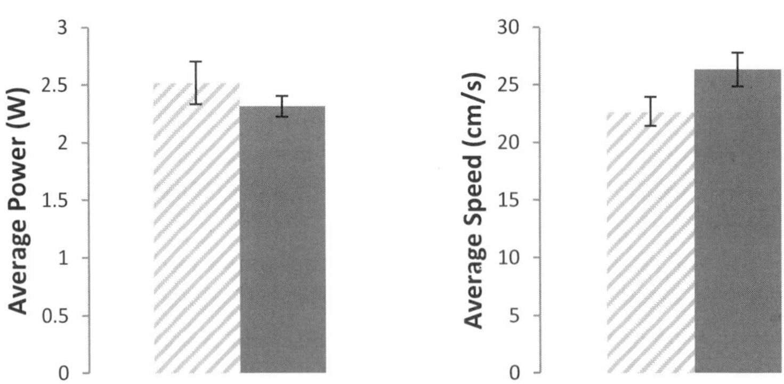

Figure 4: The average power and average speed for the robot with a rigidly-attached load (hatched) and an elastically-suspended load (shaded). The error bars show the standard deviation over 20 trails (not including the failed trails where the robot got stuck or hit the wall).

the entire track length four times because it got stuck twice and hit the right wall twice. The robot with an elastically-suspended load only failed once by getting stuck. The randomness of the rough terrain resulted in a relatively high standard deviation, but the overall trend is clear. The data shows that the robot was able to traverse the rough terrain noticeably faster with slightly less power consumption while carrying an elastically-suspended load versus a rigidly-attached load. Therefore, these experiments indicate that elastically-suspended loads could improve the mobility of legged robots traversing rough terrain.

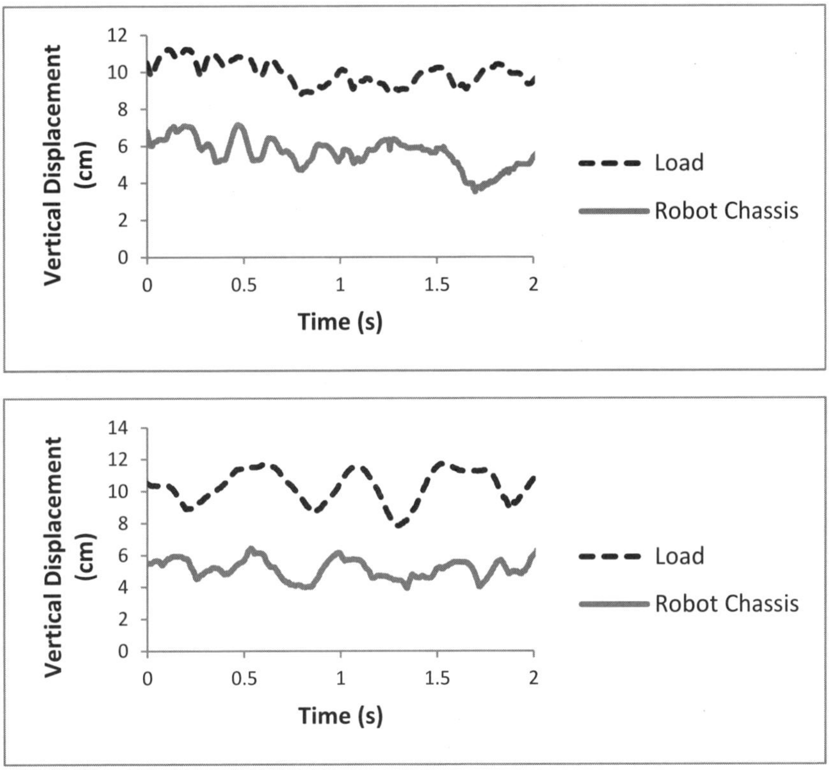

Figure 5: Data from high-speed video recorded at 240 fps showing the behavior of the robot running over rough terrain with a rigidly-attached load (top) and an elastically-suspended load (bottom). The point tracking was performed using the open source software Tracker Video.

High-speed video from the experiments showed that elastically suspending a load decouples the vertical motion of the load mass from the motion of the main robot chassis (Figure 5). The decoupling of these motions may enable the robot to resist the high-frequency perturbations that result from running over the rough terrain, enabling it to traverse the rough terrain more quickly and efficiently. The authors hypothesize that the complaint suspension essentially acts as a low-pass filter. High-frequency perturbations do not affect the legged robot system with a compliant load suspension as much as when the robot is carrying a rigidly-attached load. The robot is essentially lifting less effective mass while it traverses over the rough terrain due to the compliant load suspension. The motion of the elastically-suspended load mass also seems influence the motion of the main robot chassis, possibly enabling the robot to surmount rough terrain perturbations more effectively.

During the experimental trials, the robot traversed rough terrain with many perturbations due to the randomly dispersed blocks that were approximately 1.67 cm high on average. To better understand the effect of such rough terrain on the mobility of the robot, we tested the effect of a single perturbation of varying heights on the average power and speed of the robot.

The data shows that regardless of whether the robot was traveling over flat ground or a 2.3 cm high perturbation, the robot ran on average approximately 5.7% faster while consuming approximately 3.3% less power with an elastically-suspended load versus a rigidly-attached load (Figure). The robot was also tested while running over a 3.1 cm perturbation, but it was unable to get over this obstacle consistently because of the orientation of the suspended load mechanism (Figure 3).

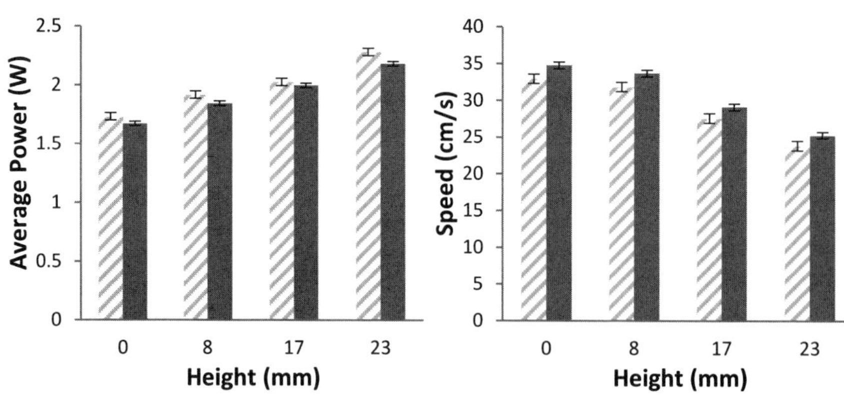

Figure 6: The regardless of the rough terrain height, the robot ran over the short track approximately 5.7% faster while consuming approximately 3.3% less power on average with an elastically-suspended load versus a rigidly-attached load. The error bars show the standard deviation over 10 trials.

The robot was able to travel over rough terrain with numerous randomly distributed perturbations 16% faster with 8.9% less power consumption while carrying an elastically-suspended load versus a rigidly-attached load, but was only able to travel 5.7% faster with 3.3% less power consumption while running over a single perturbation of the same height as the average height of the rough terrain. Since the robot was only run over a single perturbation in the latter experiments, the differences between the mobility of the robot with an elastically-suspended load versus a rigidly-attached load were not as pronounced. The data shows that the numerous perturbations of the rough terrain increase the difference between the mobility of the robot with a rigidly-attached load and an elastically-suspended load. Over a longer stretch of rough terrain, these differences may be further magnified. It is still unclear what effect height has on the mobility of legged robot locomotion over a long stretch of rough terrain. We plan to investigate the effect of terrain height and frequency on mobility in the future.

5. Conclusion

The mobility of legged robot locomotion over rough terrain could be increased by elastically-suspending part of the inherent mass of a legged robot with a compliant suspension system and decoupling the vertical motion of the load from the vertical motion of the robot chassis. This has numerous potential applications for legged robots because batteries, electronics, fuel, vision systems, or sensitive components could be elastically-suspended from the robot body to increase the mobility of legged robot locomotion over rough terrain.

Acknowledgments

The authors would like to thank Cameron Cecil and Yuhang Che for their help.

References

1. Kar, D., *Design of statically stable walking robot: A review.* Journal of Robotic Systems, 2003. **20**(11): p. 671-686.
2. Saranli, U., M. Buehler, and D.E. Koditschek, *Rhex: A simple and highly mobile hexapod robot.* The International Journal of Robotics Research, 2001. **20**(7): p. 616.
3. Altendorfer, R., et al., *Rhex: A biologically inspired hexapod runner.* Autonomous Robots, 2001. **11**(3): p. 207-213.
4. Moore, E., et al. *Reliable stair climbing in the simple hexapod'RHex'.* 2002: IEEE.

5. Buehler, M., et al., *Dynamic locomotion with four and six-legged robots.* 2000, DTIC Document.
6. Blickhan, R. and R. Full, *Similarity in multilegged locomotion: Bouncing like a monopode.* Journal of Comparative Physiology A: Neuroethology, Sensory, Neural, and Behavioral Physiology, 1993. **173**(5): p. 509-517.
7. Sponberg, S., *Neuromechanical response of musculo-skeletal structures in cockroaches during rapid running on rough terrain.* Journal of Experimental Biology, 2008. **211**(3): p. 433-446.
8. Spagna, J., et al., *Distributed mechanical feedback in arthropods and robots simplifies control of rapid running on challenging terrain.* Bioinspiration & biomimetics, 2007. **2**: p. 9.
9. Ackerman, J. and J. Seipel, *Energetics of Bio-Inspired Legged Robot Locomotion with Elastically-Suspended Loads*, in *IEEE/RSJ International Conference on Intelligent Robots and Systems.* 2011, IEEE: San Francisco, CA.
10. Ackerman, J., D. Xingye, and J. Seipel, *Energy Efficiency Of Legged Robot Locomotion With Elastically-Suspended Loads Over A Range Of Suspension Stiffnesses*, in *ASME International Design Engineering Technical Conferences.* 2012, ASME: Chicago, IL.

A FORCE THRESHOLD-BASED POSITION CONTROLLER FOR LEGGED LOCOMOTION OVER IRREGULAR TERRAIN

MAYUR PALANKAR AND LUTHER PALMER III*

Department of Computer Science & Engineering, University of South Florida, Tampa, Florida 33620, USA
**E-mail: palmer@cse.usf.edu*
www.cse.usf.edu/~palmer

Taking inspiration from biological studies on animal locomotion, a Force Threshold-based Position (FTP) controller has been developed for cyclic legged locomotion over irregular terrain. The controller uses feed-forward position control and force feedback in the form of threshold comparison to generate joint torques during the stance phase of each leg cycle. The FTP controller decouples the control of each leg using only localized feedback such that no multi-leg Jacobian needs to be computed. The controller is robust to terrain elevations without extracting any information about the terrain or inertial sensing of the body. The algorithm is applied to a simulated hexapod walking on irregular terrain, and an experimental system is being developed. The FTP controller is also applied to bipeds, quadrupeds and other biologically-inspired forms.

1. Introduction

Animal legs appear in many arrangements and morphologies, and enable multiple locomotion modes. Despite the versatility in arrangement seen in nature, every leg operates under the same intermittent paradigm of flight and stance phase. Foot forces are applied to the ground to propel and stabilize the body during the stance phase, and the leg is returned from its posterior position to a forward position during the flight phase.

Biologists comparing legged systems of differing morphology have observed general ground reaction force profiles during the stance phase while animals walk, run, jump and climb.[1] Full et al.[2] measured the three dimensional ground reaction forces produced by individual legs of the cockroach while running. Similar force patterns normal to the terrain were also observed by Goldman et al.[3] while recording forces of vertically-climbing cockroaches and geckos. Fig. 1 shows the template for vertical ground reaction force. At each step, a vertical ground reaction force greater than the body weight of the animal must be produced by the active legs. Based on the support provided by each leg, the force profile could be divided into three states (Fig. 1). This is further explained in the next section.

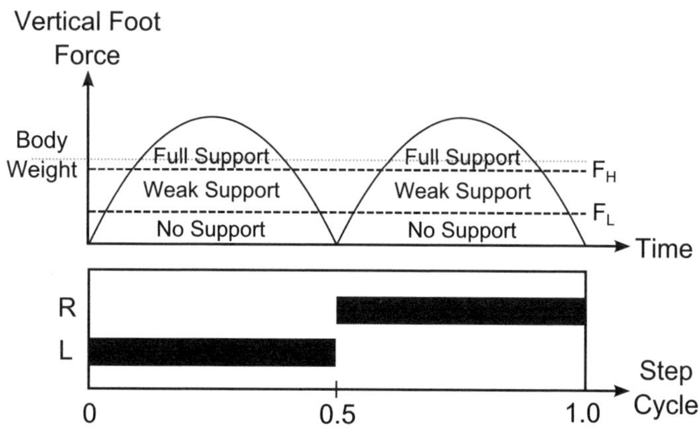

Figure 1.: A spring-mass model template[3] for vertical ground reaction force. Dashed line indicates body weight of the animal. The black bar represents the stance period and white space the flight period.

Joint positions during locomotion steps have also been archived for a significant number of biological systems. The position of the foot with respect to the body can be kinematically determined from the cyclic body dynamics observed during stable walking in animals. These foot force profiles and position trajectories will serve as the foundation for the FTP controller.

2. Force Threshold-based Position (FTP) Controller

The Force Threshold-based Position (FTP) controller attempts to replicate biological vertical force profiles for each leg of the system such that the leg can propel and stabilize the body at every step. This requires ground contact to be made at every step regardless of the terrain. However, trying to replicate the exact vertical force profile for each leg in a distributed system is not only computationally intensive but can also lead to instability. Therefore, the controller tries to achieve the vertical force profile using specific thresholds to control the position of each leg. This not only provides a stable walking pattern on flat terrain but also works well on irregular terrains.

2.1. *Position Control*

For many years, position control has been used to control walking systems.[4] Once joint angles are chosen for the leg touchdown position and the desired body motion during stance is known, the desired joint angles throughout the stance phase can be computed kinematically. A strength of position control is the inherent motion control and tilt stability that arise from well-designed trajectories. Because these trajectories are often kinematically-computed from stable and smooth body motions on well-modeled terrain, legs moving through their desired position can result in smooth body motions without the use of vestibular sensing of the body or information about the terrain. On complex terrain too difficult to model, such as natural terrain, and in systems that do not sense terrain irregularities, the interaction of the foot with the ground becomes unpredictable and can lead to a loss of stability when only feed-forward trajectories dictate the desired foot position. The FTP controller achieves the benefits of position control and overcomes its primary weakness by incorporating force feedback in the way of threshold comparison.

2.2. *Using Force Thresholds*

For position control to work properly on irregular terrain, the trajectory of the leg should be based on the ground height at the current step. If the ground is lower than expected, the controller must depress the foot to reach the ground and maintain the foot depression till the end of the stance phase. If the ground is higher than expected, the controller must reduce the foot depression during the stance phase so as not to lift the body higher than expected. Using this approach, the stance phase can be divided into a series of states dictating whether the foot should be depressed or elevated.

The different states and corresponding input conditions are given in Fig. 2. The low (F_L) and high (F_H) force thresholds and preferred foot depression (D_{PRE}) determine the state of the leg and its intended operation. The foot force generated by the leg at the previous step is compared with thresholds and based on the current foot depression, the foot is either depressed or elevated. The thresholds, shown in Fig. 1, correspond to the level of support the leg provides when generating a particular foot force.

Even though every animal generates foot forces equivalent to its body weight, the foot force generated by each leg is different and dependent on the gait and number of legs on the ground. Most of the walking gaits correspond to two sets of legs doing stance and flight phases alternatively

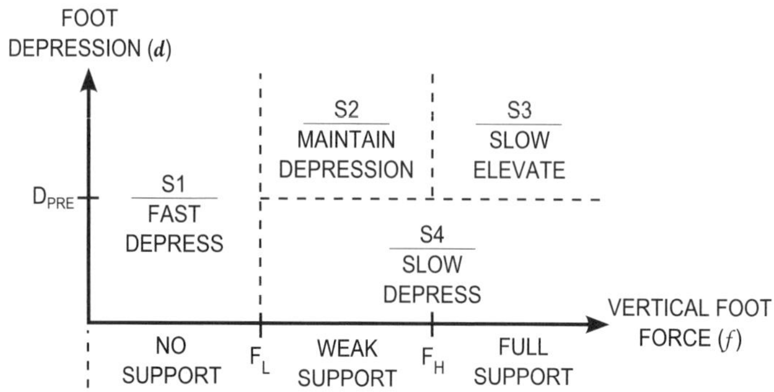

Figure 2.: FTP controller states: During stance, the foot is commanded to depress further or elevate upwards based on the foot force feedback and current foot depression.

with an overlap of the stance phase for stability. The F_L and F_H thresholds are used by the FTP controller to distinguish these phases; no support (NS), weak support (WS) and full support (FS). A particular leg or set of legs are in *no support* when the other set of legs are in *full support* and can maintain the body weight. The transition or overlap between the stance phases during which most of the legs are on the ground is the *weak support* phase. The preferred foot depression (D_{PRE}) is the maximum foot depression achieved by a position controller when walking on flat terrain while maintaining a particular body height from the ground. The FTP controller compares the computed foot depression to D_{PRE} which serves to stabilize stabilize body height such as observed with a position-only controller. No sensing of body tilt is needed either as the position aspect of the FTP controller stabilizes these motions innately.

The states shown in Fig. 2 for stance phase are explained below:

State	Condition	Explanation
$S1$	$f < F_L$	No ground contact Depress foot at a high constant rate (Δ_{FD})
$S2$	$(f > F_L) \wedge$ $(f < F_H) \wedge$ $(d > D_{PRE})$	Foot too far depressed but not supporting body Maintain foot depression
$S3$	$(f > F_H) \wedge$ $(d > D_{PRE})$	Foot too far depressed and fully supporting body Elevate foot at a slow constant rate (Δ_{SE})
$S4$	$(f > F_L) \wedge$ $(d < D_{PRE})$	Foot too far elevated and supporting body Depress foot at a slow constant rate (Δ_{SD})

3. System and Results

The body of the hexapod is modeled similar to a stick insect. Each leg of the hexapod has three degrees of freedom (Fig. 3a). Leg links l_2 and l_3 form a rigid angle of 90° between them. The first joint θ_1 controls the fore-aft motion of the leg while the other two (θ_2 and θ_3) controls the foot elevation and depression. The preferred foot depression D_{PRE} is 4.25 cm. The dimensions of the system are given in Fig. 3b. Simulation of the FTP algorithm was done using the RobotBuilder[5] simulation environment (Fig. 3c).

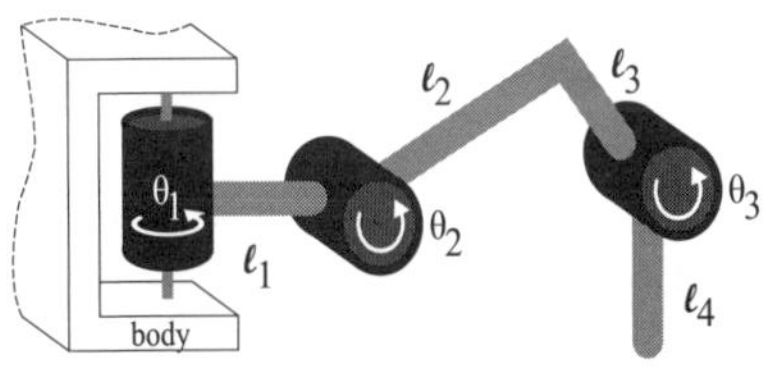

(a) Hexapod leg: Each leg has 3 actuators and described by 4 link lengths

Product	Centimeters
Body Width	7.8
Body Length	35.5
Body Height	2.5
Leg: Link Length 1 (l_1)	3.7
Leg: Link Length 2 (l_2)	4.5
Leg: Link Length 3 (l_3)	2.3
Leg: Link Length 4 (l_4)	8.8

(b) Hexapod dimensions

(c) Hexapod in simulation

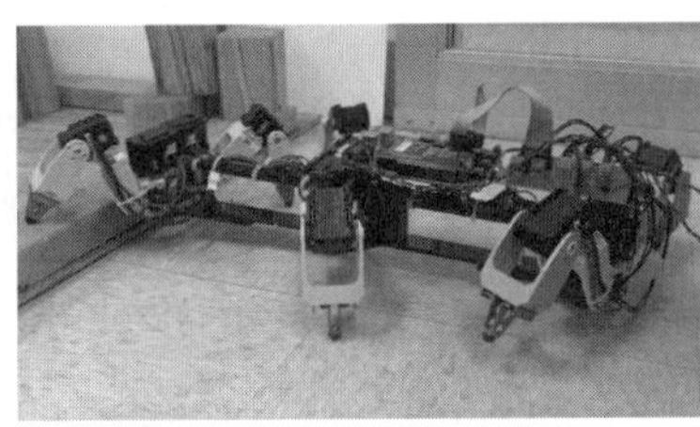

(d) Experimental hexapod system

Figure 3.: Hexapod System

An experimental hexapod has been built (Fig. 3d). The hardware hexapod system measures 50 cm in length and 30 cm in breadth. The onboard processing is done by a 200 MHz ARM920T CPU on a compact Single Board TS-7260 Computer (Technologic Systems). The joints θ_2 and θ_3 are controlled using a Dynamixel AX-12 (Robotis Inc.) servomotor, while the joint θ_1 is controlled using a Dynamixel RX-28 (Robotis Inc.) servomotor.

Figure 4 shows the comparison between the FTP controller and a position controller as the hexapod walks on a random terrain (deviation of ± 6 cm). Although other algorithms exist for legged locomotion on uneven terrain, a pure position control algorithm is the only control algorithm known to the authors which does not require vestibular sensing of body state or terrain information, and therefore, is the only appropriate com-

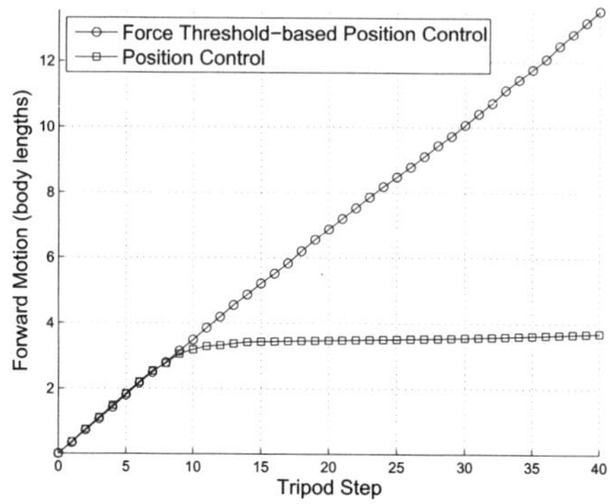

Figure 4.: Comparison between the FTP controller and the position controller walking over a randomly generated terrain.

parison for the FTP controller. The figure shows the displacement of the hexapod in terms of body length (Fig. 3b). After some steps, the hexapod under position control arrives at a location in which the feet do not reach the terrain and results in sluggish motion. The FTP controller, however, constantly seeks the ground resulting in forward motion at each step.

Figure 5 shows the body height of the hexapod walking on different terrains. The first two subplots show the body height as the hexapod climbs over an obstacle and a gap in the terrain while the latter two show the body height as it climbs up and down a single step. The height of the obstacle (or gap) and step is the same at 2.5 cm. The body height is measured from the geometric center of the hexapod. In the first two subplots, the body height remains fairly stable as each set of legs passes over the gap or the obstacle. In the latter two subplots, once the hexapod crosses over the single step, each leg tries to reach the preferred foot depression (D_{PRE}) which affects the body height. The choice of the slow depress (Δ_{SD}) and slow elevate (Δ_{SE}) rates dictate the number of steps required to return to the normal stepping height.

Figure 6 shows the foot depression and foot force of the front right leg as the hexapod walks over an obstacle (2.5 cm). The body height shown is measured from the front of the hexapod to the terrain. The figure shows three steps taken by the front right leg before, during and after the obstacle. In each of the three steps, the leg enters the $S1$ state and depresses the foot at a constant rate (Δ_{FD}). At 2.3 sec, when the leg contacts the ground early, the leg enters the $S4$ state and depresses the foot using Δ_{SD}. The

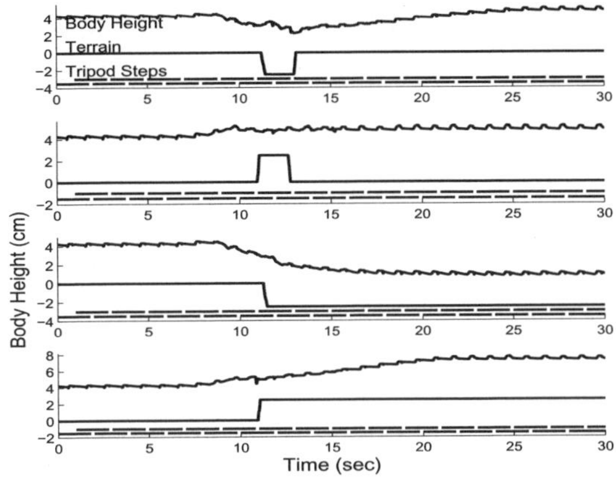

Figure 5.: Body height of the hexapod walking on different terrains.

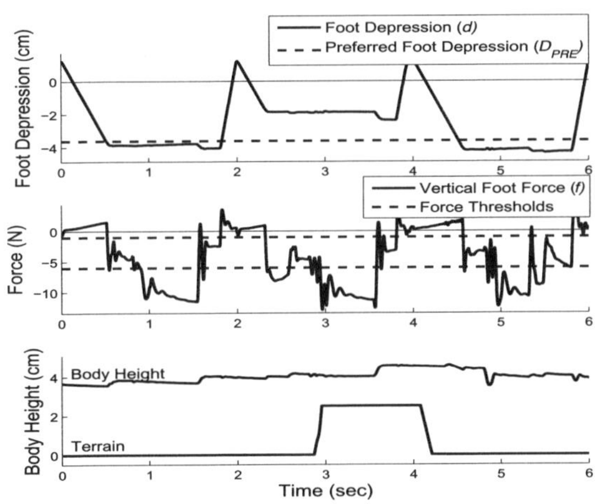

Figure 6.: Foot depression and foot force of the front right leg of the hexapod as it passes over an obstacle on the terrain.

foot depression doesn't reach D_{PRE} but increases the body height slightly. As the front right leg lifts the body (state $S4$), the middle left leg (not shown in figure) looses ground contact. Due to this, the middle left leg enters the $S1$ state again to depress the foot and regain ground contact. At the next step, the front right leg returns to normal cyclic pattern. This behavior can be verified using the FTP controller states in Fig. 2.

4. Conclusion and Future Work

A Force Threshold-based Position (FTP) controller for legged locomotion over irregular terrain has been presented along with simulation results. The controller achieves a cyclic locomotion without the use of visual sensors, a priori information about the terrain, inertial feedback or interleg feedback for controlling the joints. The controller decouples the control of each leg and stabilizes the body using only local feedback. The controller has been compared with position controller which also doesn't use a priori information about the terrain and inertial feedback. Additional testing must still be done in order to verify the potential of the algorithm and testing on the experimental system will reveal limitations under real-world constraints.

Currently, the algorithm is being implemented on an upright quadruped system[6] and will also be implemented on the experimental hexapod system. The basic operation of both algorithms is same with some minor changes to account for the morphology of the quadruped system. Future work will also include turning and steering of the body based on biological data. Also, the decoupling of the control system could be taken a step further by implementing a distributed controller at the joint level.

Acknowledgments

The work was funded in part by DARPA Maximum Mobility and Manipulation (M3) Program.

References

1. R. J. Full, *Energy Transformation in Cells and Animals* , 175 (1989).
2. R. J. Full, R. Blickhan and M. S. Tu, *Journal of Experimental Biology* **158**, 369 (1991).
3. D. I. Goldman, T. S. Chen, D. M. Dudek and R. J. Full, *Journal of Experimental Biology* **209**, 2990 (2006).
4. S. M. Song and K. J. Waldron, *Machines that Walk* (MIT Press, Cambridge, MA, 1989).
5. S. J. Rodenbaugh, RobotBuilder: a graphical software tool for the rapid development of robotic dynamic simulations, Master's thesis, The Ohio State University (2003).
6. L. R. Palmer III and C. Eaton, Toward innate leg stability on unmodeled and natural terrain: Quadruped walking, in *Proceedings of the IEEE/RSJ International Conference on Intelligent Robots and Systems*, submitted, March 2012, available at http://www.cse.usf.edu/~palmer/DOP/palmer_2012.pdf.

DESIGN OF A VARIABLE STIFFNESS LEG USING SHAPE MEMORY POLYMER COMPOSITES

DUNCAN W. HALDANE
Mechanical Engineering, University of California, Berkeley
JONATHAN E. CLARK
Director, STRIDe Lab, Florida State University

1. Introduction

Biomechanical studies have shown that animals modify their leg stiffness to adapt to changes in their running environment[1]. A recent empirical study investigating the performance of mechanically variable stiffness legs on a dynamic hexapedal robotic platform, Edubot, has shown that optimal performance was measurably increased by adaptively modulating the leg stiffness[2]. While effective, the resulting mechanism was too heavy, complex, and fragile for extended field operation.

The aim of this study is to develop an electrically activated variable stiffness leg for Edubot using a class of polymeric smart materials called shape memory polymers. Shape Memory Polymers (SMPs) experience a several order of magnitude drop in modulus as they approach their glass transition temperature. By fabricating a composite using a SMP as the resin matrix, a shape memory polymer composite is formed. A shape memory polymer composite (SMPC) developed in this manner should be capable of sustaining the high bending loads required of a locomotion appendage. By modulating the leg stiffness, the robot will then be capable of adaptation to variations in terrain and payload.

We present a design for a variable stiffness leg that utilizes a polymeric smart material to modify the leg compliance without any external mechanism. In section 2 the styrene-based shape memory polymer material used in the legs design is described and characterized. The design of and fabrication of the robot legs using this material are detailed in section 3, and the experimental testing of the legs is reported in section 4. Section 5 describes the results and directions for future work.

2. SMPC Material and Characterization

In order to develop a leg with smart material based stiffness modulation, a source of shape memory polymer resin had to be secured and a shape memory polymer composite material had to be developed. In the absence of an available commercial shape memory polymer resin, we synthesized a styrene-based chemically crosslinked co-polymer resin with shape memory properties. We designed the polymer to have a glass transition temperature of 75°C as well as a large glass transition zone to give a wide range of temperatures suitable for the control of leg stiffness. Unreinforced shape memory polymer, however, lacks the requisite mechanical properties (high strength, fracture toughness and recovery force) passively compliant robotic legs require. Recent research[3-5] aimed at incorporating continuous fiber reinforcement in an effort to create a shape memory polymer composite has shown that using continuous fiber reinforcement can significantly increase critical mechanical properties. The unanimous selection for the reinforcing fiber in these studies was carbon fiber.

To characterize the effect of replacing the current epoxy-based composite legs used in Edubot, three different composites using 3k T-300 carbon fiber were fabricated. The mechanical properties of the SMPC developed in this study were evaluated in comparison to composites made with two commercial resins. The styrene-based shape memory polymer resin was tested alongside a commercial epoxy resin (US Composites 635), as well as a commercial vinyl ester resin (IVEX 400).

These composites were evaluated at room temperature (the nominal case for a zero power hold SMPC mechanism), in accordance with ASTM standards. The SMPC had comparable properties to the epoxy composite and exceeded the composite made with IVEX-400 (another polymer resin) in every tested category. Material properties for composites all three resins are given in Table 1.

Table 1: Mechanical properties of tested composites

		SMP Resin	**USC 635 Epoxy**	**IVEX-400**	
Ultimate Tensile Strength	UTS (σ_{UTS})	335.0 (24.5)	350.0 (20.8)	202.7 (29.7)	MPa
Elastic Modulus	E (σ_E)	31.75 (1.53)	31.83 (0.91)	29.86 (2.29)	GPa
Strain at Failure	ε_f (σ_{ε_f})	11.4 (1.46)	11.2 (0.78)	6.1 (0.19)	$\varepsilon \times 10^{-3}$
Short Beam Shear Strength	τ_{SBS} ($\sigma_{\tau_{SBS}}$)	28.05 (1.77)	24.06 (0.81)	23.27 (2.16)	MPa

Although SMP appears to be an adequate substitute for epoxy in these legs, in order to achieve variable stiffness behavior, it was not required to fabricate the leg entirely from SMPC. We found that more reliable mechanical properties could be attained in the SMPC by fabricating it in smaller geometries. Therefore, a segment of SMPC was included in the leg at the effective center of deflection, 15% of the leg length away from the fixed attachment point. To increase the strength of the bond between the SMPC section and the epoxy leg, we used a methacrylate adhesive to bond the two composites at the fiber level. The completed leg, shown in Fig. 1, has the same nominal properties of the original leg with a diameter of 11.4 cm and a nominal stiffness of 1,100 N/m measured at 10% deflection of the leg.

3. Design

We considered several methods of activation for the composite leg for electronic control of the leg such as fabricating the composite using an electroactive polymer, or heating through the fiber matrix of the composite itself.

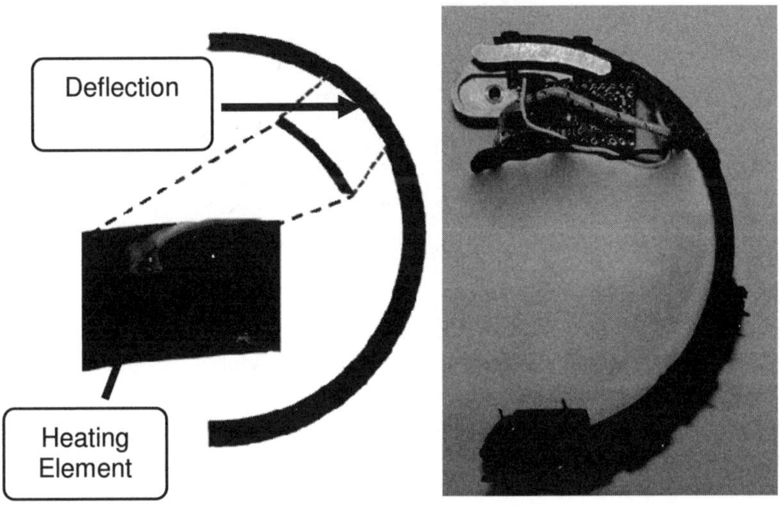

Figure 1: Variable stiffness leg, showing SMPC section (left), and complete assembly (right)

We elected to use Joule heating as our actuation method, wherein thermal power is dissipated into the leg from a resistive heating element. A Nichrome wire heating element and thermocouple were embedded in a thin fiberglass scrim and applied to the SMPC section of the leg. A miniature microcontroller, complete with an ATmega328P processor, MAX6674 cold junction compensated thermocouple ADC and motor driver was attached to the leg itself. This controller is designed as an independent unit with its own power source to eliminate the need for a rotor and stator as used in the alternative design[2]. The resulting leg weighs 25 grams and is 70% lighter than the previous, mechanical variable stiffness design. The most suitable heating element for this geometry was found to be an 'S' shaped curve.

4. Results

We first evaluated the performance of the electro-activation hardware. Figure 2 is an infrared photograph, showing that the 'S" shape heating coil configuration chosen results in an even heat distribution across the shape memory composite specimen.

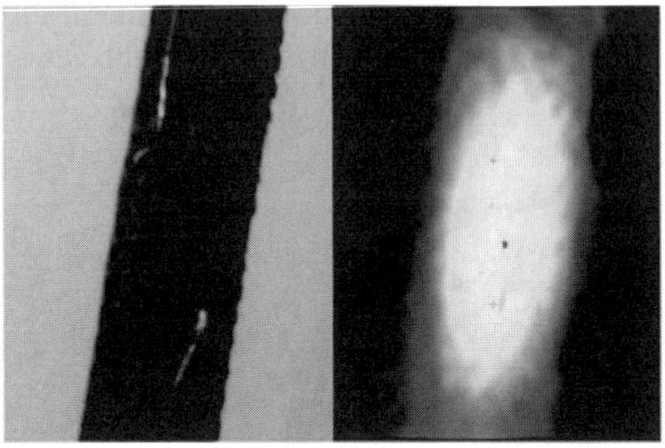

Figure 2: Heat distribution across the SMPC specimen

The heat activation system was designed to draw less than 2.5W when operating at its maximum temperature (85°C) to allow for comparable battery life between the temperature control circuits and Edubot. Steady state power consumption was measured for the temperature control unit and the peak power consumption falls within our design parameters, as shown in Figure 3. The time constant for heating actuation was found to be fairly low, on the order of 8s, with the selected heating element design. However, the cooling and stiffening of the leg requires heat transfer based on ambient conditions, and is much slower. Therefore, this design of leg is well suited for adapting to changes in steady state operation, as the mechanically variable stiffness leg has been used.[2]

With the temperature control circuit in place, a prototype leg was attached to the loading head of a MTS Insight testing system (Fig. 4) and compressed onto a linear stage to measure the effective radial stiffness of the variable stiffness leg. The leg was tested at room temperature and in 5°C increments from 40°C to 85°C. As shown in Fig. 5, the leg stiffness ranged from a maximum value of 2310 N/m at room temperature to 650 N/m at 85°C, a change of approximately 350%.

5. Discussion and Future Research

In this work, we synthesized a shape memory composite material and developed an integrated temperature-based stiffness control policy. Based on this material, we fabricated a new shape memory composite mechanism and characterized its radial stiffness characteristics. The variable stiffness leg developed in this study was capable of over a 70% decrease in stiffness, a range encompassing the expected operational states for Edubot. Further optimization is required before this leg can be implemented on the robot operating in the field. The mechanical strength and fatigue life need to be improved for an extended operational lifetime of the leg. A new layup, using a higher-grade carbon fiber, is expected to extend the lifetime of the component by increasing resistance to the micro-buckling failure mode. Once these improvements have been achieved, testing on a mobile robotic platform will demonstrate the utility of the new leg.

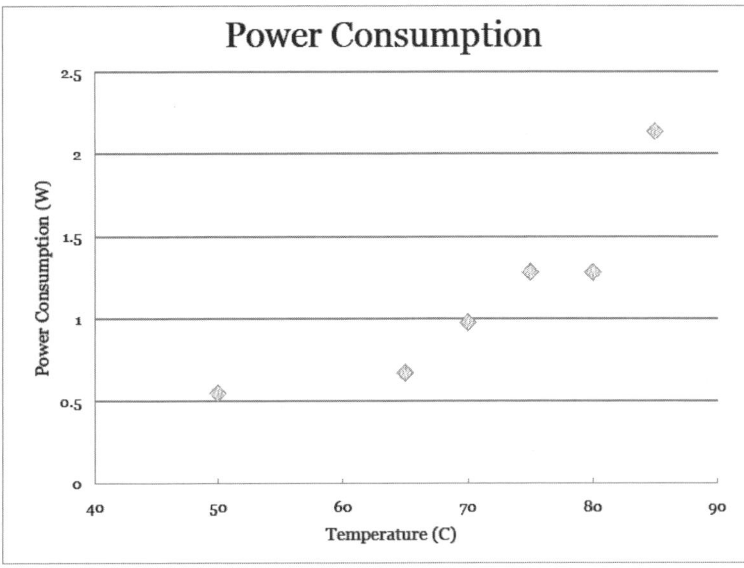

Figure 3: Power consumption of the variable stiffness leg as a function of specified flexure temperature.

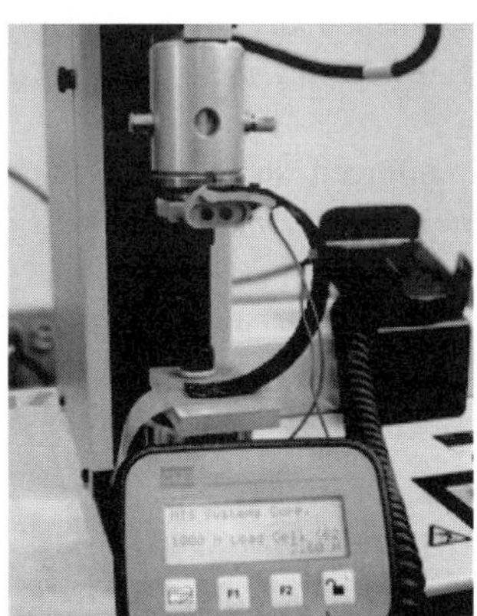

Figure 4: Stiffness testing equipment

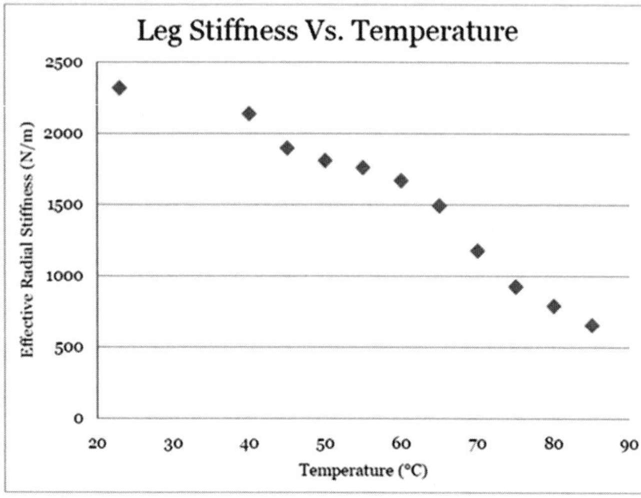

Figure 5: Experimental performance of the variable stiffness leg

References

1. D. P. Ferris, M. Louie, and C. T. Farley, "Running in the real world: adjusting leg stiffness for different surfaces," *Proceedings of the Royal Society B: Biological Sciences*, vol. 265(1400), 1998.
2. Galloway, K.C., Clark, J.E., Yim, M. & Koditschek, D.E. Experimental Investigations into the Role of Passive Variable Compliant Legs for Dynamic Robotic Locomotion. *Mechanical Engineering* 1243-1249 (2011).
3. Francis, W., Lake, M.S., Mallick, K., Freebury, G.E. & Engineering, G.E. AIAA 2003-1496 Development and Testing of a Hinge / Actuator Incorporating Elastic Memory Composites *44 Structures, Structural Dynamics , and Materials Conference 2003* .
4. Lan, X. et al. Fiber reinforced shape-memory polymer composite and its application in a deployable hinge. *Smart Materials and Structures* **18**, 024002 (2009).
5. Zhang, C. & Ni, Q. Bending behavior of shape memory polymer based laminates. *Composite Structures* **78**, 153-161 (2007).

MECHANICS OF CRAWLING ROBOT WITH SLIDING SEAL *

V. GRADETSKY, V.CHASHCHUKHIN

A. Yu. Ishlinsky Institute for problem in mechanics RAS, prosp. Vernadskogo 101, block 1, Moscow, 119526, Russia

Mechanical model of miniature robot with sliding seal is considered. Robot moves along the incline plane surface. Conditions of motion without sliding of the wheels are derived. These conditions are assigned design of control system of the robot. Suggested method of motion control is based on the combination control by the vacuum pump module providing pressing force and driving wheels module providing translation motion and rotation of the robot. The motion conditions without sliping and oscillations were established depends on the structure of robot's coupling mechanism contacted with surface.

1. Introduction

Robots with sliding supports are perspective type of wall-climbing robots and they attract attention of the researchers and designers [1-3]. Robot is designed at A. Yu. Ishlinsky Institute for problem in mechanics RAS and is presented on Fig. 1. It is arised special interest for miniature robots design. Such robots are characterized as multilinking as specific conditions of the controlled motion over inclined surfaces and wall. Specific conditions are linked with contradictiens friction forces influence. Friction forces between sliding seal and surface have to direct against streamline of motion. The module of forces has to be minimal value but the seal must save the air-tightness during the motion. On the other hand friction forces between the band of wheels and surface have to be greater and provide reliable cohesion with the surface and secure controlled motion. At the same time the pressing of all system has to be restricted by means of force acting on the body. Pressing force is arisen through the vacuum generation inside the body. This force is equals to the product of pressure difference on effective body's area. The control of this pressing force is implemented through changing the number of turns of fan block's engine in the robot's body.

* This work is supported by grant RFBR 12-08-00374 and 11-01-92002-HHC_a.

Fig 1. Robot with sliding seal.

Previously performed investigations [1-6] were refer to develop such robot's constructions and to select the working mods of operation but don't include the general mechanical processes influence on dynamics of control motion.

The conditions of secure control motion for the crawling wheel robot are delivered in this paper. Operative modes of the motion with sliding seal are analyzed. These modes provide pressing the robot's body to the surface of motion under various surface conditions.

An idea of using a special valve with pressure sensor has been proposed in [**Error! Bookmark not defined.**]. Including a vacuum pressure sensor satisfy a control generation of vacuum in robot's system by means of negative feedback. The robot motion conditions of nonslipping and nonvibration mode of operations were established.

2. Mechanics of robot motion and problem statement

Lets consider the mechanics of crawling robot direct motion over inclined wall. Robot (Fig. 2) consists of body 1, vacuum pump module 5 (engine with turbine wheel), two driving wheels 3, operating by dc motors 2, sliding seal 4 (a hoop attaching to the body's bottom), and vacuum chamber inside of the body. The vacuum pump module creates a necessary value of vacuum that equal to pressure difference ΔP between the atmospheric pressure P_0 and the pressure in the chamber P_1, $\Delta P = P_0 - P_1$. Lets consider a forces and moments acting on the mechanical system (Fig. 3). It is suppose that wheels moving without sliding. Rolling friction with the vertical surface is included in the consideration. The sliding friction forces direct inversely to the rolling friction forces [6]. Air drag is neglected. Motion is uniform. Wheels are identically.

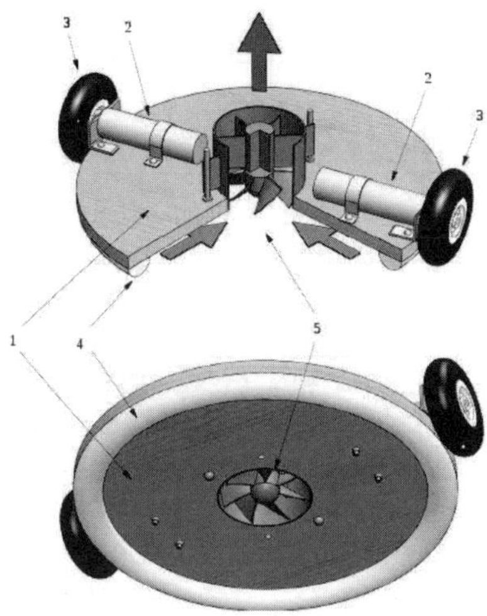

Fig. 2. Crawling robot with sliding seal.

The following external forces and moments are affected on system: G – weight of the robot applied to centre of mass, N_w – normal reaction of the surface forces applied to the wheels, F_w - friction force of the wheels (driving force). The reaction application point is displaced relatively wheel contact point on the value which is equal to rolling friction coefficient f_r. r – radius of the wheel, d_2 and d_1 – external and internal diameters of the sliding seal (Fig. 4), h – distance between centre of mass and surface, F_p – pressing force conditional by vacuum pump, forces of the surface reaction on sliding seal are is follows: F_1, F_2, N_1, N_2.

Let us suppose that the pressure on the sliding seal is $\boldsymbol{p}_s(\varphi)$, where φ - is a angle with the axle of the left wheel passing through the center of the seal ring (Fig. 4) The surface reaction force equal to $\boldsymbol{p}_s(\varphi)dS$ and it acts on element with area dS. Lets $n_s(\varphi)$ is the normal to surface projection of $\boldsymbol{p}_s(\varphi)$, and $f_s(\varphi)$ is a projection of $\boldsymbol{p}_s(\varphi)$ on the surface plane. Normal surface reaction forces are the following that acting on the seal

$$N_1 = \frac{d_2^2 - d_1^2}{8} \int_0^{\pi} n_s(\varphi) d\varphi, \; N_2 = \frac{d_2^2 - d_1^2}{8} \int_{\pi}^{2\pi} n_s(\varphi) d\varphi.$$

Tangential surface reaction forces (sliding friction forces) acting on the seal are the following:

$$F_1 = \frac{d_2^2 - d_1^2}{8} \int_0^{\pi} f_s(\varphi) d\varphi, \; F_2 = \frac{d_2^2 - d_1^2}{8} \int_{\pi}^{2\pi} f_s(\varphi) d\varphi.$$

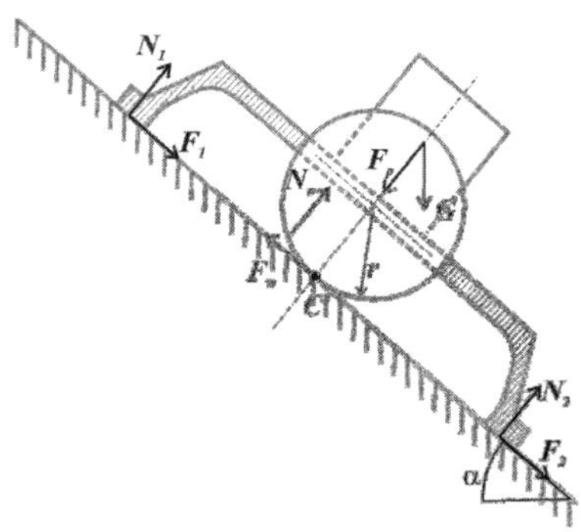

Fig. 3. Forces acting on robot.

The pressing force F_p depends on pressure drop in the chamber. Area under the chamber is $\pi d_1^2/4$. If the pressure in the chamber is p_1 than
$F_p = (p_1 - p_0)\pi d_1^2/4 = \Delta p \; \pi d_1^2/4$, where p_0 is an atmospheric pressure.

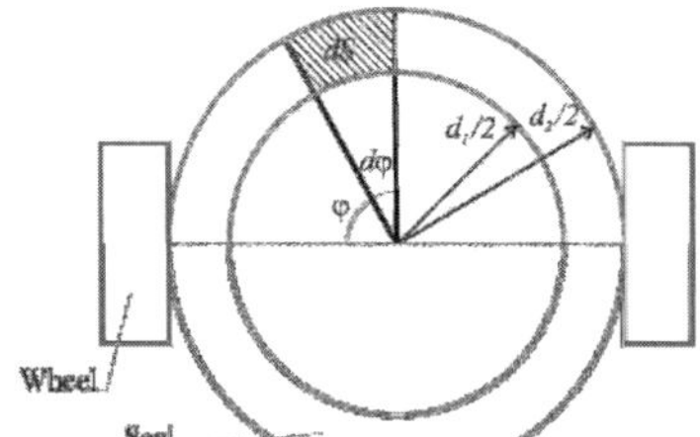

Fig. 4. Ring of the sliding seal.

The following problems are necessary to solve for motion robot conditions analysis on vertical wall and for developing of the control system:

– Calculate conditions to satisfy secure attachment to wall surface.

– Produce the influence of rolling and sliding friction forces on motion parameters.

– Determine the engine's driving forces on predetermined displacement *s* in case of uniform motion.

3. Secure conditions to satisfy robot's wheels attachment to the surface

Assume that mass of the robot is *m* and *x,y* are center of mass coordinates. Put an abscissa axis along the motion direction of the robot and ordinate axis perpendicularly to the surface plane. The motion equations of robot are represented as

$$\begin{cases} m\ddot{x} = F_w - F_1 - F_2 - G\sin\alpha; \\ m\ddot{y} = N_1 + N_2 + N_w - F_p - G\cos\alpha. \end{cases}$$

Suppose that robot moves uniform along the wall than $\ddot{x} = 0,\ \ddot{y} = 0$. The sliding friction force is coulomb, in other worlds $F_w \le \mu_w N_w$, where μ_w is a friction coefficient of the wheel, $F_1 = \mu N_1$, $F_2 = \mu N_2$, (μ – friction coefficient of the seal), i.e. the seal is sliding during the motion.
Equilibrium condition for moments relatively the axis passing through the contact point *C* of the wheel (Fig. 3) is

$$N_w f_r + N_1 R_1 + Gh\sin\alpha = N_2 R_2.$$

The moments of seal's normal reaction forces are characterized by following expressions:

$$N_1 R_1 = \frac{\left(d_2^2 - d_1^2\right)\left(d_2 - d_1\right)}{16} \int_0^{\pi} n_s(\varphi)\sin\varphi\, d\varphi,$$

$$N_2 R_2 = \frac{\left(d_2^2 - d_1^2\right)\left(d_2 - d_1\right)}{16} \int_0^{\pi} n_s(\varphi)\sin\varphi\, d\varphi.$$

In addition the requirements to be N_1, N_2, N_w nonnegative, have to be completed.
We obtain following system

$$\begin{cases} N_w = F_p - \dfrac{1}{\mu}F_w + \left(\cos\alpha + \dfrac{1}{\mu}\sin\alpha\right)G; \\ N_1 = -\dfrac{f_r}{R_2+R_1}F_p + \left(\dfrac{R_2+f_r}{\mu(R_2+R_1)}\right)F_w - \left(\dfrac{f_r}{R_2+R_1}\cos\alpha + \dfrac{f_r+h\mu+R_2}{\mu(R_2+R_1)}\sin\alpha\right)G; \\ N_2 = \dfrac{f_r}{R_2+R_1}F_p + \left(\dfrac{R_1-f_r}{\mu(R_2+R_1)}\right)F_w + \left(\dfrac{f_r}{R_2+R_1}\cos\alpha + \dfrac{f_r+h\mu-R_1}{\mu(R_2+R_1)}\sin\alpha\right)G; \\ N_1 \geq 0, \quad N_2 \geq 0, \quad N_w \geq 0, \quad F_w \leq \mu_w N_w. \end{cases}$$

4. Secure conditions to satisfy robot's wheels attachment to the surface

We have received four limitations for F_w, that presented by system (1). The equation determine the influence of rolling and sliding friction forces on motion parameters. Following to (1), it is possible to find the detachment preventions of different parts of robot body and reliable attachment of the robot wheels to the wall. Analysis of the system inequalities (1) permits to calculate limitations for driving forces (Fig. 5) and conditions of nonsliping (Fig. 6)

$$\begin{cases} F_w \geq \dfrac{f_r\mu}{R_2+f_r}F_p + \left(\dfrac{f_r\mu}{R_2+f_r}\cos\alpha + \dfrac{f_r+h\mu+R_2}{R_2+f_r}\sin\alpha\right)G; \\ F_w \geq -\dfrac{f_r\mu}{R_1-f_r}F_p - \left(\dfrac{f_r\mu}{R_1-f_r}\cos\alpha + \dfrac{f_r+h\mu-R_1}{R_1-f_r}\sin\alpha\right)G; \\ F_w \leq \mu F_p + (\mu\cos\alpha + \sin\alpha)G; \\ F_w \leq \dfrac{1}{2}\mu_w F_p + \dfrac{1}{2}(\mu_w\cos\alpha + \sin\alpha)G. \end{cases} \tag{1}$$

First inequality implementation corresponds to $N_1 \geq 0$, i.e. prevention the detachment of front part of body. We attempt to prevent the body's seal lifting from the surface because it will lead to pressing force decrease. Second inequality corresponds to $N_2 \geq 0$, i.e. the detachment prevention of rear part of body. Third inequality corresponds to $N_w \geq 0$, i.e. detachment prevent of the wheel from the surface. Fourth inequality corresponds to condition $F_w \leq \mu_w N_w$, i.e. prevention of the sliding of the wheel. On the Fig.5 the area of feasible solutions of the system (1) is showed for $F_p = 3G$, $R_2 = 1.1R_1$, $h = 0.5R_1$, $f_r = 0.1R_1$, $\mu_w = 0.6$, $\mu = 0.3$. If $F_w \leq 0$ than the drive engine must work in braking mode.

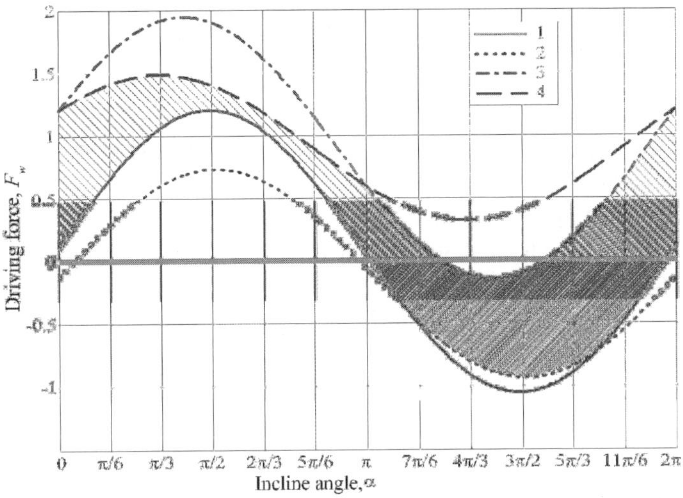

Fig. 5. Limitations for driving force.

5. Secure conditions to satisfy robot's wheels attachment to the surface

Following Fig. 5 it is possible to determine the engine's driving forces depends on angle α values and acting forces. The lines 1-4 on Fig. 5 denotes four values right side of (1).

When $F_w \geq 0$ than the driving motors work in advance mode. This condition is completed when

$$\begin{cases} \mu F_p + (\mu \cos\alpha + \sin\alpha) G \geq 0; \\ \frac{1}{2}\mu_w F_p + \frac{1}{2}(\mu_w \cos\alpha + \sin\alpha) G. \end{cases}$$

This implies

$$\begin{cases} F_p \geq -\left(\cos\alpha + \frac{\sin\alpha}{\mu}\right) G; \\ F_p \geq -\left(\cos\alpha + \frac{\sin\alpha}{\mu_w}\right) G. \end{cases}$$

This condition is completed when

$$F_p \geq \sqrt{\frac{\mu^2+1}{\mu^2}} G$$

The robot motion conditions without slipping characterizes the absence of oscillations because of the vibration acting is the result of very small attitude motion of sliding supports in both directions. Suppose, that moment M is applied to the wheel and the wheel rolls without sliding, then $M=N_w f_r + F_w r$. A condition $F_w \leq \mu_w N_w$ is the result in limitation on driving force, conditioned by driving moment, applied to wheel

$$F_w \leq \frac{\mu_w M}{f_r + \mu_w r}.$$

Conclusion

Suggested approach will be usefull for designer of climbing robots with sliding seal. The conditions were calculated to satisfy the reliable attachment robot's wheels to the surface. The analysis was performed to determine the rolling and sliding forces influence on robot motion parameters. Engine's driving forces were calculated in the case of uniform robot's motion. Taking into consideration given equations for forces and moments, it is possible to consider that motion conditions without slipping and vibrations are valid for miniature robot design, including limitations for driving forces.

References

1. Hirose S., Nagakubo A., Toyama R. Machine that Can Walk and Climb on Floors, Walls and Ceilings, Proc. Of Int. Conf. Of Advanced Robotics, Italy, 1991.
2. Sujiyama S., Nitoh S., Sato K. and Ozaki N Wall Climbing Vechicle with Magnetic Legs or Vacuum Legs., Proc of 16th ISIR, Brussels, Belgium, 1986.
3. Gradetsky V. Mechatronic System for Mobile Wall Climbing Robot, Proc. Of Mechatronic Design and Production conf., Tubitak-Modisa, Turkey, 1993, pp. 43-47
4. Gradetsky V., Knyazkov M., Fomin L., Chashchukhin V. Mechanics of Miniature robots. Nauka Publisher, Moscow., 2010, pp. 271 (in Russian).
5. Chernousko F., Bolotnik N., Gradetsky V., Manipulation Robots: Dynamics, Control and Optimization, CRC Press, Boca Raton, USA, 1994, 267 pp.
6. Bat' M.I., Dzhanelidze G.Yu., Kelzon A.S. Theoretical mechanics at examples and problems, v. 2. Nauka Publisher, 1975, 606 p (in Russian).

SMOOTH WALK OF A SKID-STEERING VEHICLE CONSISTING OF FEET JOINTED TO EACH WHEEL RIM

T. OKADA[1] S. TEZUKA[1] and H. WADA[2]

[1]*Faculty of Engineering, Niigata University, Niigata, 950-2181, Japan*
[2]*DOUBLE Research and Development Co.,LTD*
http://okada.eng.niigata-u.ac.jp

In general, when rolling on a flat surface, vehicles move smoothly as if they collect surface profiles. However, legged robots tend to swing vertically because of the front to back kicking of the legs. Therefore, walking on feet that are jointed to a wheel rim requires a decrease in vertical swing and speed unevenness. We propose a computational method to determine the angular variables of feet that are synchronously driven with the wheel rotations. In addition, we propose a structural model of a four-wheeled skid-steering vehicle (4WSSV) with multiple feet on each wheel. Next, we explain the difficulty of achieving a uniform vehicle speed when the wheels rotate with a constant velocity. To cope with this problem, we consider a control method that minimizes the speed unevenness. We derive an appropriate dynamic wheel rotation range for the case with feet landings. We simulate a walk based on a model for a flat surface with a slope. Finally, the approach is verified with walk demonstrations demonstrating that the vehicle behaves as designed, i.e., without a vertical swing and a close to even speed.

Keywords: Skid-Steering Vehicle, Wheel with Feet, Wheelchair, Vertical Swing, Speed Control

1. Introduction

Legged robots have several advantages when it comes to for moving and working on uneven terrain. For example, they can easily step onto or over obstacles on an irregular terrain which is difficult for wheeled and tracked robots. These advantages have prompted much research on the topic of the robot walk. For instance, a hybrid design with legs and wheels was developed.[1,2] A biped gait generation has been proposed for moving on rimless wheels.[3] A six-wheeled vehicle having only three flat spoke bars in each wheel was constructed for robust propelling.[4] The robot ASGUARD is designed for the use in a harsh outdoor environment with a focus on security and outdoor surveillance.[5] An adaptive, bio-inspired control approach is taken in to consideration to control the ASGUARD movement.

The absence of an explicit steering assembly makes the 4WSSV mechanically simple and robust. It moves with ease as an all-terrain response vehicle and has excellent maneuverability. Moreover, 4WSSVs move smoothly on a flat ground. However, inflexible wheeled legs/spokes are liable to generate a vertical swing while walking because of front to back leg kicking. Particularly, walking on feet that are jointed onto a wheel rim[6] requires a decrease in the swing and speed unevenness.

This study describes procedures to determine the control variables of 4WSSVs combining multiple feet to keep the vertical swing small and a suitable propelling speed to move the feet. First, we propose a structural model of a walk on feet jointed on a wheel rim and develops a computational method to determine the angular variables of the feet mechanism on wheels. Then, we explain why it is difficult to keep the vehicle speed uniform when the wheels rotate with a constant speed, that is, the wheel rotation speed needs to be changed to keep the vehicle speed constant. We also discuss an appropriate dynamic range of the wheel angle for a feet landing with minimal speed unevenness. The walk based on the model for a flat ground with a slope is made under normal and advanced controls for adapting the feet to the ground. The experimental motion verifies that the walk is carried out while satisfying the designated body height and speed.

The 4WSSV is designed strong enough to propel itself over certain obstacles without being disturbed. The wheels and feet in front and back are combined mechanically to execute the same motions. In addition, the right and left mechanisms have the same driving design. Its motion is slow but powerful and moves robustly on unstructured terrain. Therefore, motions of the four hip joints are possible, and we will focus on the front right hip joint. To reduce the number of drawings and the complexity of the analysis we will mostly deal with the hip joint.

2. Structural model of the 4WSSV combined with feet

The 4WSSV can independently control all feet on one side of the vehicle with two motors. The feet on the right or left can be unified to be parallel or crossed between odd-numbered and even-numbered groups.[6] Walks with these controls are shown on the left and right of Fig.1. We denote the first-type of the movement as a *normal walk* and the second-type as an *advanced walk*. The following parameters are defined for analyzing the static configurations in walking.

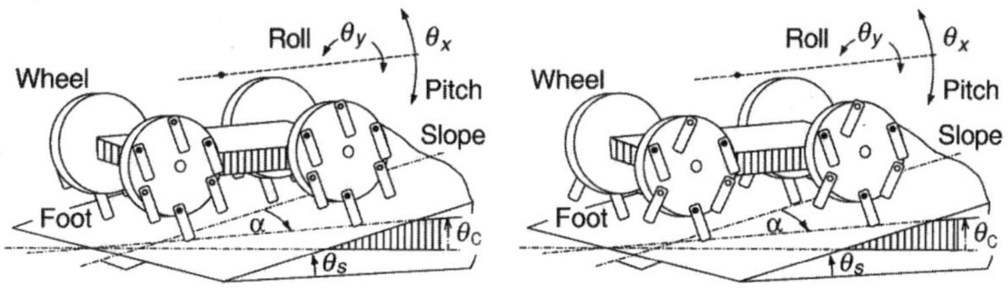

Fig. 1. Sketches of the normal walk (left) and advanced walk(right).

Symbol	Definition
J	J_h; Hip joint(=Wheel axis), J_a; Foot joint
R	Wheel radius
r	Distance between J_h and J_a
ℓ	Foot length
δ	Center angle of neighboring J_as
θ_r	Wheel rotation angle from the robot front
β	Foot angle from the gravity direction
θ_x	Pitch angle of the vehicle; cross angle between body center axis and horizontal plane
θ_y	Roll angle of the vehicle; cross angle between body plane and vertical plane
θ_s	Slope angle of a flat ground; cross angle between horizontal and body planes
θ_c	Slope angle of a propelling body axis; angle of a body axis from a horizontal plane
α	Traverse angle of a slope; cross angle between vehicular propelling and ground elevation directions

3. Normal walk with unified feet directions

During the normal walk, all the feet are unified to face a certain direction and only a wheel rotation is sufficient to walk. However, the hip joint position swings up and down in synchronization with the alternate foot landings. Therefore, an appropriate change of β is important to keep the height constant while θ_r changes.

At this moment, ℓ is generally shorter than r. In addition, ℓ must be smaller than the distance between two neighboring J_as to keep the free rotation from bumping. Therefore, it is evident that $\ell < 2r\sin(\delta/2)$. By using n, i.e., the number of feet, we obtain

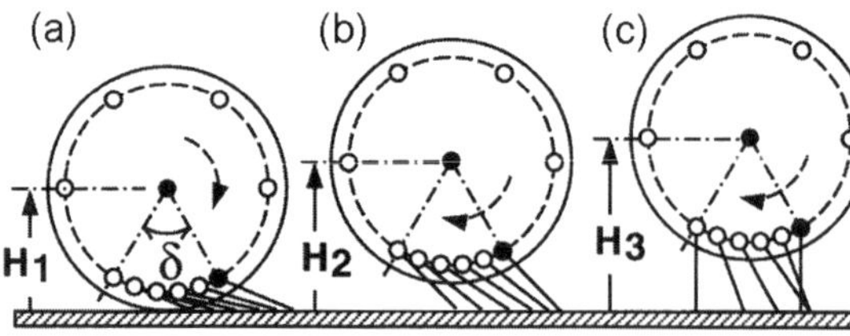

Fig. 2. Overwriting of feet configurations at heights of H_1, H_2, and H_3.

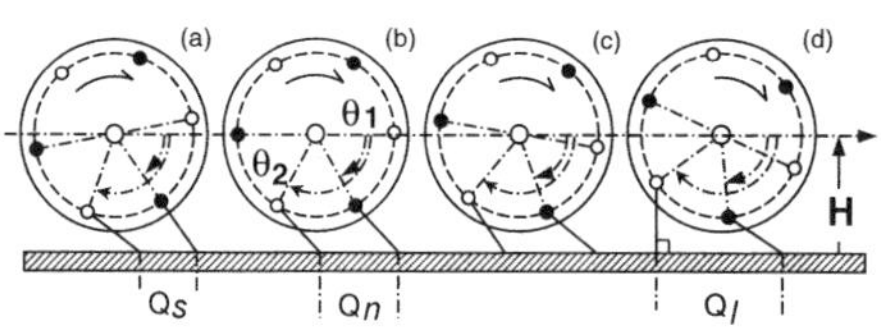

Fig. 3. Explanation of feet landing in advanced walk.

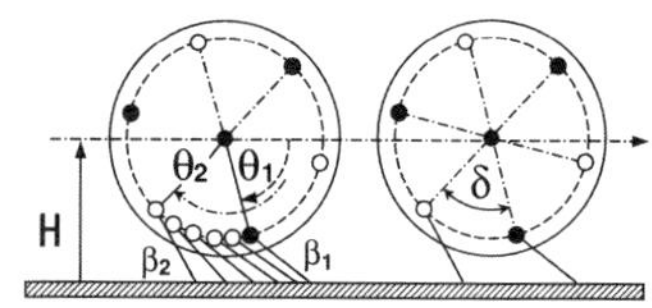

Fig. 4. Possible range of θ_r for landing in advanced walk.

$$\begin{aligned} \delta &= 2\pi/n \\ &= 2\cos^{-1}\{(R - \ell\sin\beta)/r\} \end{aligned} \tag{1}$$

The feet configurations need to be adjusted when the 4WSSV changes its body height (see Fig.2). In the critical standing configuration with two feet ready to alternate, a parallelogram is formed and used to support the body for stability. Normally, J_h always moves forward. However, Fig.2(c) reveals that J_h moves back because of the height H_3. Moreover, a constant propelling speed is not guaranteed because the wheel speed cannot be changed during the time one pivot foot alternates with the other. Therefore, controllability of the body height is considered at a lower level by changing the foot angle β.

4. Advanced walk with adjusted feet to the ground

Note that multiple feet are controlled independently between odd-numbered and even-numbered groups in all angular ranges. Therefore, we can consider the configurations in Figs.3 and 4. In the normal walk, a stride is determined as the length between two neighboring joints of J_a. However, in the advanced walk, the stride varies even when the body height is maintained. The symbols Q_s, Q_n, and Q_l in Fig.3 stand for short, normal and long strides, respectively. Symbols θ_1 and θ_2 show the start and end angles of

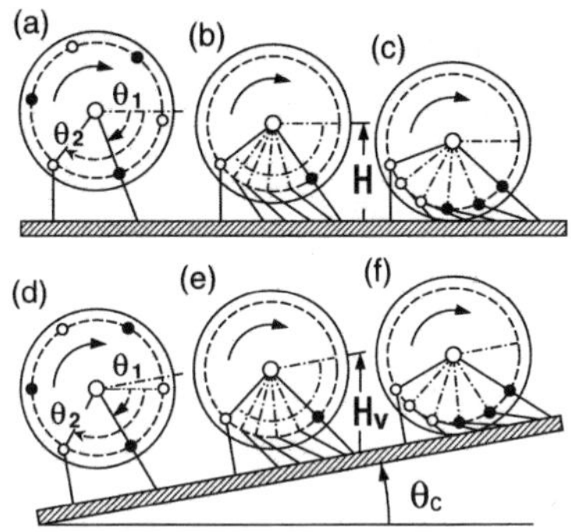

Fig. 5. Possible range of δ_r to keep a body height at different levels.

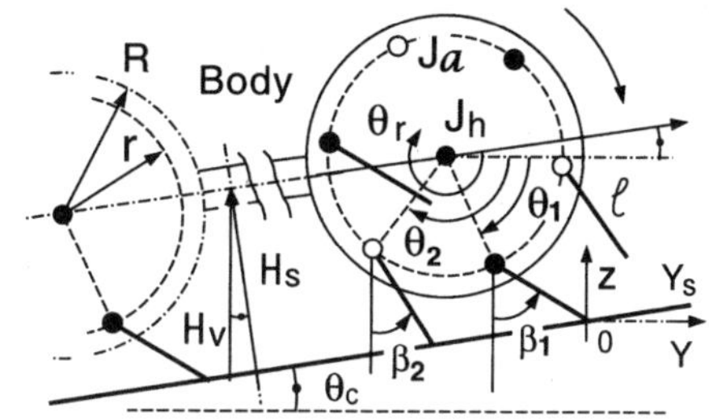

Fig. 6. Angular definitions of θ_r, θ_1, θ_2, β_1 and β_2.

the range δ (C.W. from the front horizontal direction), respectively. δ is $\pi/3$ radian in Fig.2 since n is 6.

The feet configurations to keep the body height maximal on horizontal flat and sloped grounds are shown in Fig.5(a) and (d), respectively. The form aligning a foot radially shows the minimal value of θ_1, and the form aligning a foot vertical to the ground shows the maximal value of θ_2. Figures (b),(c),(e), and (f) suggest that θ_1 and θ_2 are flexibly selected when θ_2-θ_1 equals δ.

Consider that h_{max} stands for the maximum H, then

$$h_{max} = \left\{(r+\ell\,)K_1 + \sqrt{K_1^2 - K_2(4\ell^2 - 3r^2)}\,\right\} \,/\, (2K_2) \tag{2}$$

where

$$K_1 = \ell(r + 2\ell) \tag{3}$$
$$K_2 = r^2 + r\ell + \ell^2 \tag{4}$$

It follows that β is calculated using θ_r and H_v (see Fig.6).

$$\beta = \sin^{-1}\left\{\,(\,-K_5 + \sqrt{K_5^2 - K_5K_7}\,)\,/\,K_6\right\} \tag{5}$$

where

$$K_5 = \ell K_8 \tan\theta_x \tag{6}$$
$$K_6 = \ell^2(1 + \tan^2\theta_x) \tag{7}$$
$$K_7 = K_8^2 \;-\; \ell^2 \tag{8}$$
$$K_8 = r(\sin\theta \;+\; \cos\theta\tan\theta_x) - H_v \tag{9}$$

At this moment, relations among θ_s, θ_c, and α are understandable from Fig.7. Coordinate transformation by θ_x and θ_y makes the area A_0 on the

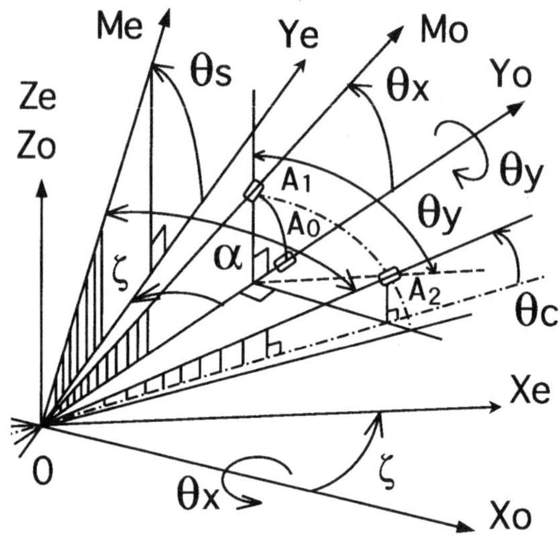

Fig. 7. The vehicle can recognize θ_s, α, and θ_c by sensing θ_x and θ_y.

X_0-Y_0 plane move to A_2 via A_1. The area A_2 exists on the plane including $\overrightarrow{OX_e}$ and $\overrightarrow{OM_e}$. Cross angles between X_0 and X_e axes, and between Y_0 and Y_e axes are same to become ζ. These satisfy $\theta_c < \theta_x < \theta_s$, in general. Angles of θ_s, θ_c and α are calculated uniquely because θ_x and θ_y are measured with external inclination sensors. In fact,

$$\theta_s = \begin{cases} 0, & \text{if } |\theta_x| \cong 0 \\ f_s, & \text{if } \theta_x > 0 \\ -f_s, & \text{if } \theta_x < 0 \end{cases} \tag{10}$$

$$\theta_c = \begin{cases} 0, & \text{if } |\theta_x| \cong 0 \\ f_c, & \text{if } \theta_x > 0 \\ -f_c, & \text{if } \theta_x < 0 \end{cases} \tag{11}$$

$$\alpha = \begin{cases} 0, & \text{if } |\theta_x| \cong 0 \\ f_a, & \text{if } \theta_y > 0 \\ -f_a, & \text{if } \theta_y < 0 \end{cases} \tag{12}$$

where functions f_s, f_c, f_a are

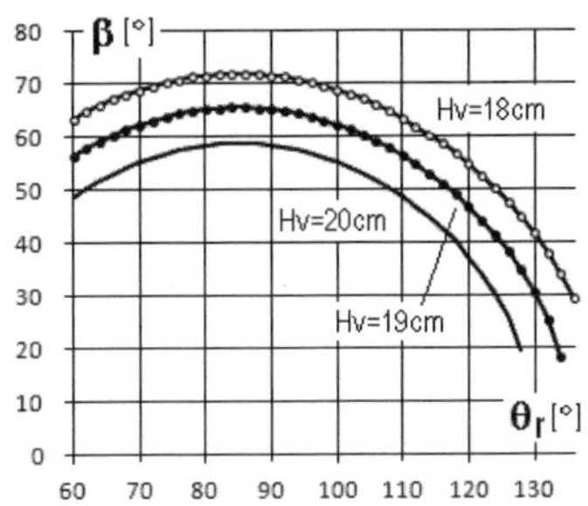

Fig. 8. β versus θ_r when H_v=18, 19, 20 under θ_x=5°.

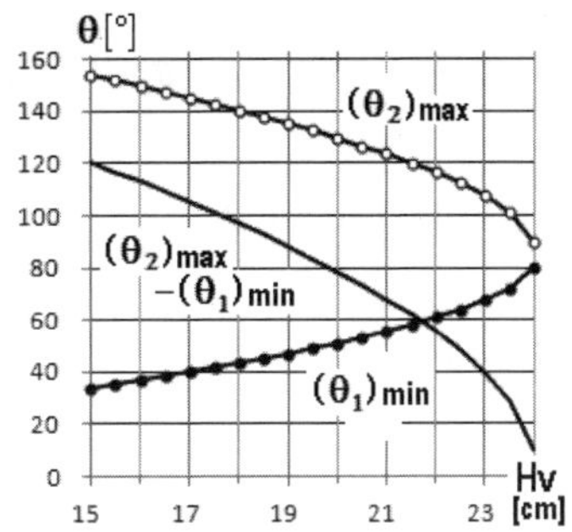

Fig. 9. Available range of θ between $(\theta_1)_{min}$ and $(\theta_2)_{max}$.

$$f_s = \cos^{-1}\left(\cos\theta_x \cos\theta_y\right) \tag{13}$$

$$f_c = \cos^{-1}\sqrt{\sin^2\theta_x \sin^2\theta_y + \cos^2\theta_x} \tag{14}$$

$$\begin{aligned} f_a = \cos^{-1}\{ & \sin\theta_x \cos\theta_y \sin\theta_s + (\cos\zeta\cos\theta_x \\ & - \sin\zeta\sin\theta_x\sin\theta_y)\cos\theta_s \} \end{aligned} \tag{15}$$

The angle ζ in (15) is written as

$$\zeta = \begin{cases} 0, & \text{if } |\theta_x| \cong 0 \\ f_z, & \text{if } \theta_x\theta_y > 0 \\ -f_z, & \text{if } \theta_x\theta_y < 0 \end{cases} \tag{16}$$

where

$$f_z = \cos^{-1}\left\{\left|\sin\theta_x\right| \Big/ \sqrt{\sin^2\theta_x + \cos^2\theta_x \sin^2\theta_y}\right\} \tag{17}$$

The angle β is uniquely determined under the condition that the feet bend forward as is the case for a human. The characteristics of β versus H_v are shown in Fig.8. In addition, we obtain the minimum value of $0 < \theta_1 < \pi/2$ and the maximum value of $\pi/2 < \theta_2 < \pi$ from the following equations.

$$\begin{aligned} (\theta_1)_{min} = \sin^{-1}\{ & (H_v\cos^2\theta_x \\ & + \sin\theta_x\sqrt{(\ell+r)^2 - H_v^2\cos^2\theta_x}\ )/(\ell + r\ )\} \end{aligned} \tag{18}$$

$$\begin{aligned} (\theta_2)_{max} = \sin^{-1}\big[& \{(H_v - \ell)\cos^2\theta_x \\ & + \sin\theta_x\sqrt{r^2 - (\ell - H_v)^2\cos^2\theta_x}\ \}/r\big] \end{aligned} \tag{19}$$

In the actual case where ℓ=10cm, r=14cm, θ_x=5°, the relations of $(\theta_1)_{min}$, $(\theta_2)_{max}$ and $(\theta_2)_{max} - (\theta_1)_{min}$ versus H_v are shown in Fig.9.

5. Calculation of θ_r and β with their speeds

Let us consider that (Y_h,Z_h) stands for the position of the front J_h in Fig.6. Then we have

$$Y_h = (-r\cos\theta - \ell\sin\beta)/\cos\theta_x \quad (20)$$

$$Z_h = r\sin\theta + \ell\cos\beta \quad (21)$$

In addition, we have

$$H_v = r\sin\theta + \ell\cos\beta + (r\cos\theta + \ell\sin\beta)\tan\theta_x \quad (22)$$

When the vehicle speed along the ground surface is denoted by $V_s(=\dot{Y}_h/\cos\theta_x)$ with a vehicular body height H_v, control variables of $\dot{\theta}_r$, β and $\dot{\beta}$ are given depending on θ_r.

Differentiating Eqs.(20) and (22) yields

$$r\sin\theta_1\ \dot{\theta}_1 - \ell\cos\beta_1\ \dot{\beta}_1 - \dot{Y}_h\cos\theta_x = 0 \quad (23)$$

$$r(\cos\theta_1 - \sin\theta_1\tan\theta_x)\ \dot{\theta}_1 - \ell(\sin\beta_1 - \cos\beta_1\tan\theta_x)\ \dot{\beta}_1 = 0 \quad (24)$$

Eliminating $\dot{\beta}_1$ from Eqs.(23) and (24) results in

$$r\cos(\theta_1+\beta_1)\ \dot{\theta}_1 + (\sin\beta_1 - \cos\beta_1\tan\theta_x)\dot{Y}_h\cos\theta_x = 0 \quad (25)$$

Similarly, we obtain

$$r\cos(\theta_2+\beta_2)\ \dot{\theta}_2 + (\sin\beta_2 - \cos\beta_2\tan\theta_x)\dot{Y}_h\cos\theta_x = 0 \quad (26)$$

Equations (25) and (26) clarify the relations of $\dot{\beta}$ and $\dot{\theta}_r$ versus θ_r shown in Fig.10 in the actual case where ℓ=10cm, r=14cm, θ_x=0° and $\dot{Y}_h$=15cm/s. Figures 11 and 12 show $\dot{\beta}$ and V_s versus θ_r, respectively. Figure 13 shows the error of V_s depending on the selection of θ_m i.e., $(\theta_1+\theta_2)/2$, under θ_x=5° and $\dot{\theta}_r$=1.0rad/s, for instance. Obviously, the optimal value of θ_m is found as a minimum to reduce the unevenness in the speed while keeping the vehicle body at the designated level. Actually, the values (95°, 91°, 87°) of θ_m correspond best to the values (16, 18, 20cm) of H_v, respectively.

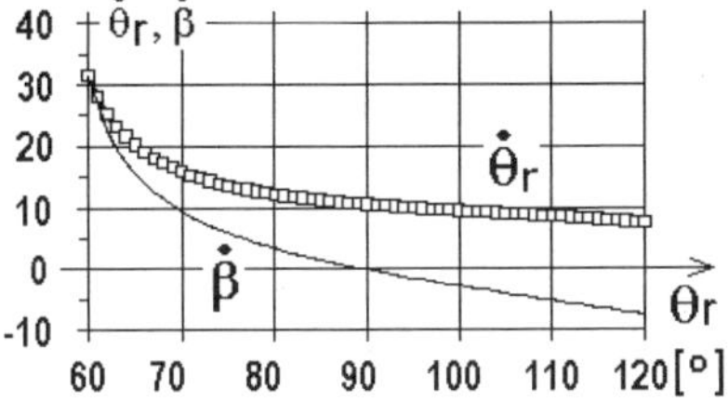

Fig. 10. $\dot{\theta_r}$ and $\dot{\beta}$ depending on θ_r.

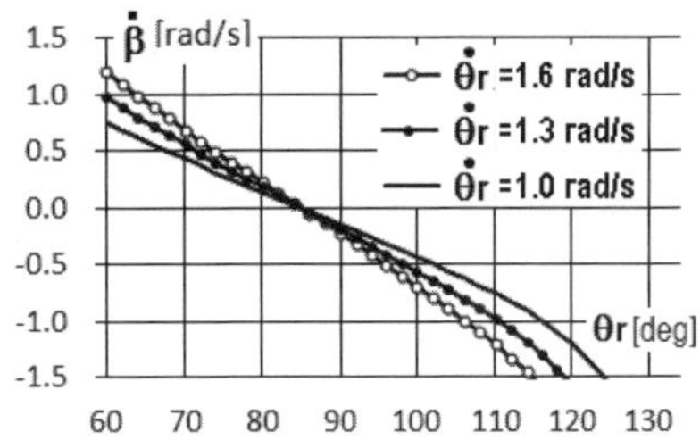

Fig. 11. $\dot{\beta}$ versus θ_r under H_v=19cm and θ_x=5°.

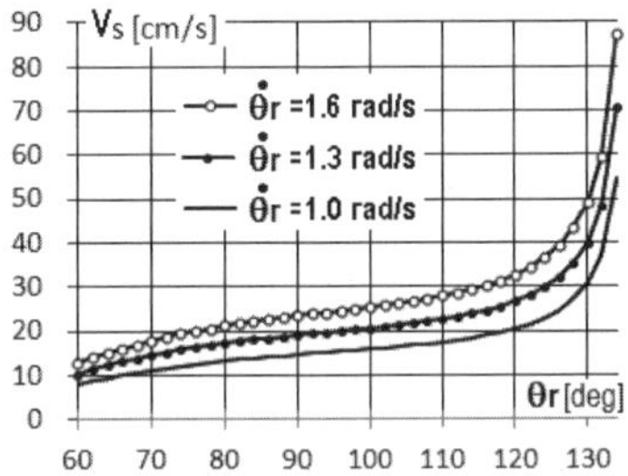

Fig. 12. V_s versus θ_r when $\dot{\theta_r}$=1.0, 1.3, and 1.6 rad/s under H_v=19cm and θ_x=5°.

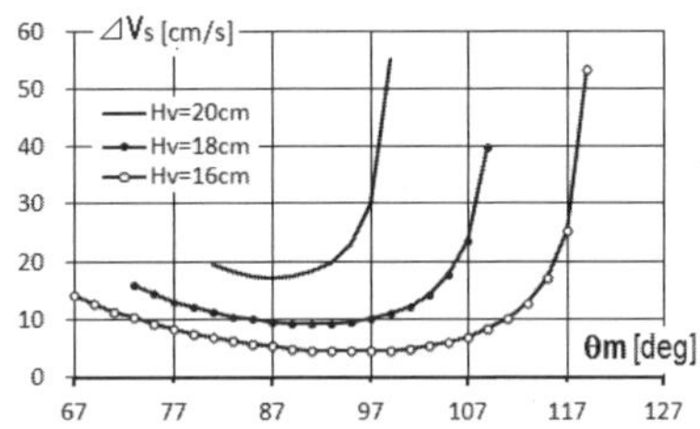

Fig. 13. Error of V_s influenced by the selection of θ_m.

Fig. 14. Walking of the robot, PEOPLER-IV on the campus corridor.

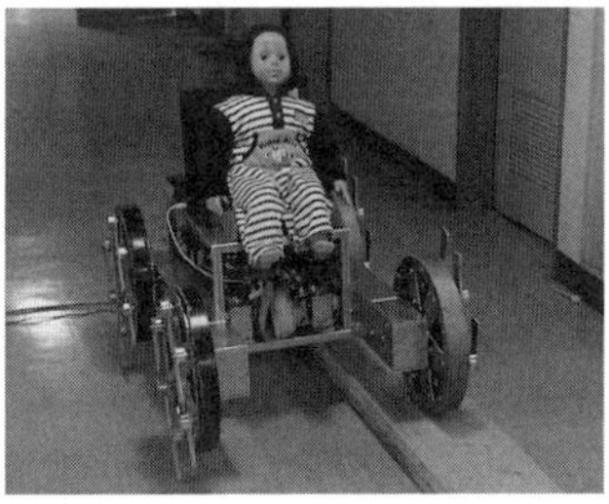

Fig. 15. Stable traversing on a sloped floor.

6. Experiments

The advanced walks of the PEOPLER-IV have been demonstrated on the campus floor. Figure 14 shows the photographs of an advanced robot walk controlled at a 15cm body height. Figure 15 also shows an advanced robot walk but with a left side without feet landing to keep the body plane horizontal. The left side is 8cm higher than the right.

7. Conclusion

To improve the mobility and performance of 4WSSVs by positioning the body height at a designated level and reducing the unevenness of the speed, we proposed to control the feet and wheels cooperatively. This study clarifies the relationships among the angular variables of the wheels and the feet in relation to the body height, wheel speed, and vehicle speed on a flat surface. The cooperative control of the feet and wheels is useful for controlling a hybrid wheelchair to carry a human smoothly on a muddy or sandy ground. The 4WSSVs might also be useful as robust carriers in a hazardous environment. Finally, this study was supported by KAKENHI(C21560259).

References

1. K. Suwannasit and S. Laksanacharoen: "A Bio-Inspired Hybrid Leg-Wheel Robot", Proc. of TENCON2004, pp.495-497, 2004.
2. N. Koyachi, K. Maruyama, Y. Kawai, F. Tomita and Y. Morikawa: "Step Over Motion of Four Wheeled and Four Legged Flexible Personal Robot", Proc. of ROBIO2009, pp.616-621, 2009.
3. F. Asano and Zhi-Wei Luo: "Asymptotically stable biped gait generation based on stability principle of rimless wheel", J. Robotica, Vol.27, pp.949-958, 2009.
4. A.S. Boxerbaum, J. Oro and R.D. Quinn: "Introducing DAGSI Whegs^{TM}: The latest generation of Whegs^{TM} robots, featuring a passive-compliant body joint", Proc. of ICRA2008, pp.1783-1784, 2008.
5. M. Eich, F. Grimminger, S. Bosse, D. Spenneberg, and F. Kirchner: "AS-GUARD: A Hybrid-Wheel Security and SAR-Robot Using Bio-Inspired Locomotion for Rough Terrain", Proc. of RISE08, 2008.
6. T. Okada, S. Tezuka and Y. Sasaki: "Sensor-Based Control of a Skid-Steering Mobile Robot Combined with Feet for Moving on All Terrain", Proc. of 14th CLAWAR, Paris, pp.666-675, 2011.

CONTROL SYSTEM OF OMNI-DIRECTIONAL MOBILE ROBOT WITH CASTER DRIVE MECHANISM TO DEVELOP REDUCTION OF MOTOR OUTPUT

Y. UENO*, and K. TERASHIMA
Department of Mechanical Engineering, Toyohashi University of Technology,
1-1, Hibarigaoka, Tempaku-cho, Toyohashi, Aichi 441-8580, Japan
**E-mail: ueno, terashima@syscon.me.tut.ac.jp*

H. KITAGAWA
Department of Electrical Control Engineering, Gifu National College of Technology
Kamimakuwa, Motosu, Gifu, 501-0495, Japan
E-mail: hkita@gifu-nct.ac.jp

K. KAKIHARA
KER CO., LTD.
Yutakagaoka, Toyokawa, Aichi, 442-0808, Japan
E-mail: kkakihara@ker-jp.com

Caster-drive mechanism is an omni-directional high mobile mechanism compared with other mechanisms by such the reason that the usual tire can be used. In this paper, an omni-directional mobile platform is built, which consists of three caster-drive wheels. In the experiment of measuring motor output, the increase of motor output is observed such the case that robot rotates different from translation movement. From the result of torque analysis, it is estimated that this problem is caused by offset structure of caster-drive mechanism. In this paper, the mechanism on the occurrence of this unfavorable phenomenon is discussed, and a novel method to improve this problem is proposed.

Keywords: Caster-drive mechanism; Omni-directional mobile robot.

1. Introduction

The omni-directional mobile robot is effective at narrow or crowded area. Several kinds of mechanisms are used to realize the omni-directional mobile robot. These mechanisms are classified into three kinds of types: free-roller mechanism, ball-wheel mechanism[1] and caster-drive mechanism[2]. Caster-drive mechanism is one that can realize an omni-directional movement by using usual wheels such as a pneumatic tire. Therefore, high mobility capability such as overcoming a difference in level and driving at the rough terrain can be achieved. However, in conventional caster-drive mechanism, since two motors are separately used for driving or steering a wheel, motor capacity is increased and efficiency of motor is decreased.

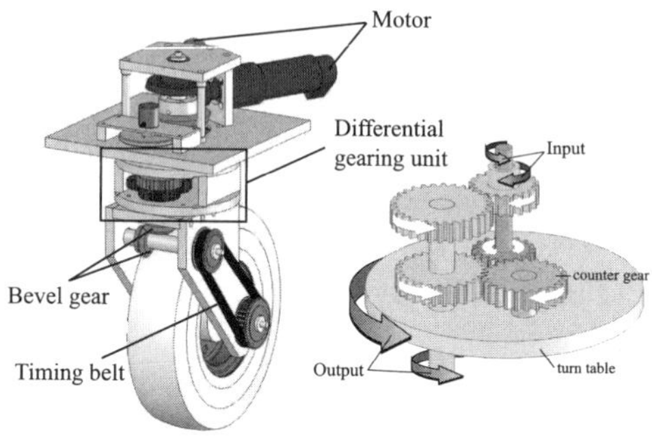

Fig. 1. Illustration of the proposed DDSS

Fig. 2. Overview of three-wheeled platform

Table 1. Specification of robot

Size (Diameter×Height)	ϕ680×900 (mm)
Weight	60 (kg)
Motor power	150 (W) × 6
Max. velocity	1.12 (m/s)
Max. acceleration	2.0 (m/s^2)

In order to improve this problem, novel mechanism named Differential Drive Steering Mechanism (DDSS) which is shown in Fig. 1 has been developed by authors[3]. In the past research, simulation results and experimental results show that the motor capacity can be reduced in the case of translation movement. However, while measuring exactly motor output when the robot rotates, the increase of motor output has been discovered.

Hence, the aim of this research is to clarify the key factor to induce this phenomenon, and to develop a control system in order to reduce the loss of motor output. Furthermore, past research has considered four-wheeled omni-directional mobile robot for stability and high-power. However, it has redundancy due to the excessive freedom. Then, we built a novel three-wheeled omni-directional mobile robot in this paper.

2. Novel Power-assist Mobile Robot with Three-wheels as Platform of Omni-directional Mobile Robot

In this research, we built an omni-directional mobile platform which consists of three DDSS wheels as shown in Fig. 2. Since the number of the minimum wheels which moves without a free caster on a plane floor is three wheels, the three wheels platform is built. Table 1 shows the specifications of the omni-directional mobile robot. Configuration of control unit is shown in

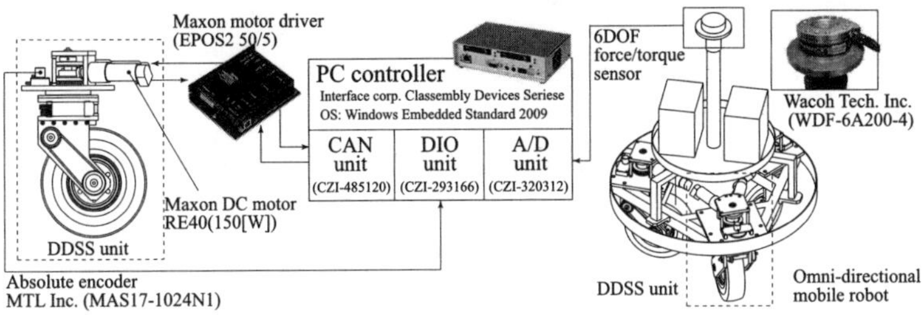

Fig. 3. Configuration of control system

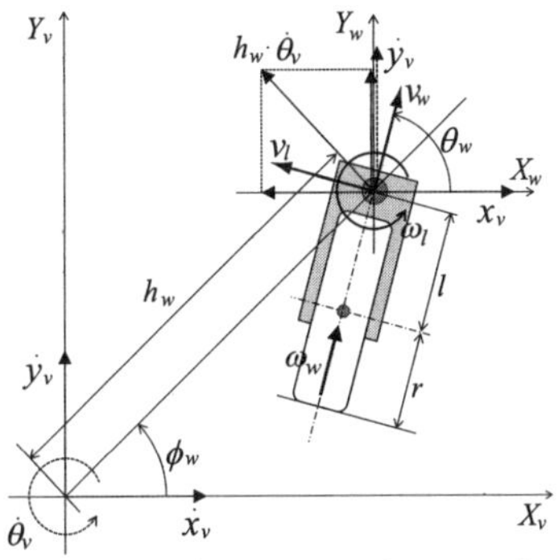

Fig. 4. Model of the caster-drive mechanism

Fig. 3. Angular velocity of motor is controlled by feedback of encoder data according to reference velocity which is given by PC controller through CAN-bus network. Reference velocity of motor is derived to PC controller from the input velocity of the robot and the wheel angle which is detected by absolute encoder. Although this robot can be controlled automatically, operation by human is also possible by the power-assist system. In order to detect input force from human, 6-DOF force and torque sensor is fixed at the top of the robot. By transforming to reference velocity from these input forces, a person can control this mobile robot by adding a force to the sensor. Detail of the control system of the power-assist system is omitted because it is written in the previous paper[4].

3. Principle of Mechanism

Figure 4 shows the model of the caster-drive mechanism in the coordinate system of vehicle. By intentionally making an offset between the grounding point of wheel and the steering axis, two independent velocities (v_w, v_l) are generated on the steering axis as shown in Fig. 4.

The inverse kinematic equation which produces driving and steering angular velocity $u_w = [\omega_w, \omega_l]^T$ from input vector in each axis of the robot

$x_v = [x_v, y_v, \theta_v]^T$ can be described as

$$u_w = B_w^{-1} \dot{x}_v \quad (1)$$

where

$$B_w^{-1} = \begin{bmatrix} C\theta_w/r, & S\theta_w/r, & h/r\,(S\theta_w \cdot C\phi_w - C\theta_w \cdot S\phi_w) \\ -S\theta_w/l, & C\theta_w/l, & h/l\,(S\theta_w \cdot S\phi_w + C\theta_w \cdot C\phi_w) \end{bmatrix}$$

Here, r is radius of wheel, l is offset between steering axis and grounding point of wheel, h is distance between steering axis and center of vehicle, θ_w is steering angle of wheel and ϕ_w is angle between X_v axis and h. For brevity, we use S- and C- instead of sin(-) and cos(-).

Although the caster-drive mechanism has high mobility performance, increase of motor capacity and decrease of motor efficiency are caused, because two motors in the conventional mechanism are separately used for driving and steering a wheel. When the mobile robot is in the steady motion, such as a straight running, only a motor for driving a wheel is used and a motor for steering becomes idle. In order to improve this problem, we developed a novel mechanism named as DDSS in the past research[3]. Differential gearing mechanism is attached to the caster-drive mechanism. The DDSS controls two outputs by contributing two motors output and then distributing their two outputs into rotation of wheel and steering through the differential gearing mechanism. Detail of the DDSS is omitted because it is written in the previous paper[3].

4. Experiment of Measuring Motor Output

Experiments which measures angular velocity and torque of motors when the robot moves forward and rotates at the center axis, have been conducted. Motor output is derived by product of angular velocity and torque. The experimental results are shown in Fig 5 and Fig 6. The result of velocity of the mobile robot shows that the robot is controlled along the reference velocity, in each case. A huge difference exists on the result of motor torque and output between two experiments. In the case of rotational movement, motor torque becomes higher than the translation movement.

The reason why motor output becomes higher value in the rotational movement is that the wheel is dragged in the case of rotation. Kinematic equation shows that the angular velocity of the robot is distributed to tangential direction of rotational trajectory to each wheel as $h_w \cdot \dot{\theta}_w$. Fig. 7 shows the state of the wheel in rotational movement. Since the wheel angle is fixed at tangential direction by the kinematic model, the trajectory of grounding point of wheel becomes outer than the trajectory of steering axis, and slippage of wheel is caused. In order to produce the slippage of

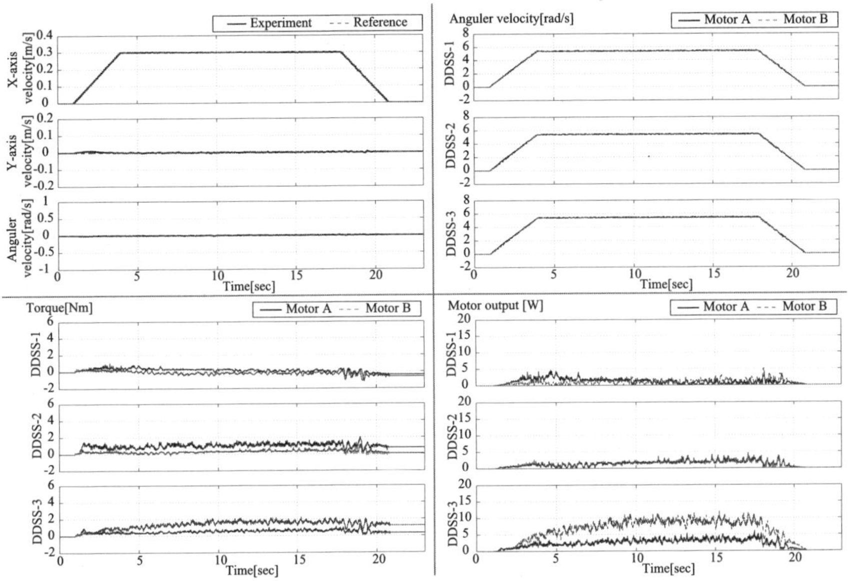

Fig. 5. Experimental result of robot's velocity at translatory movement

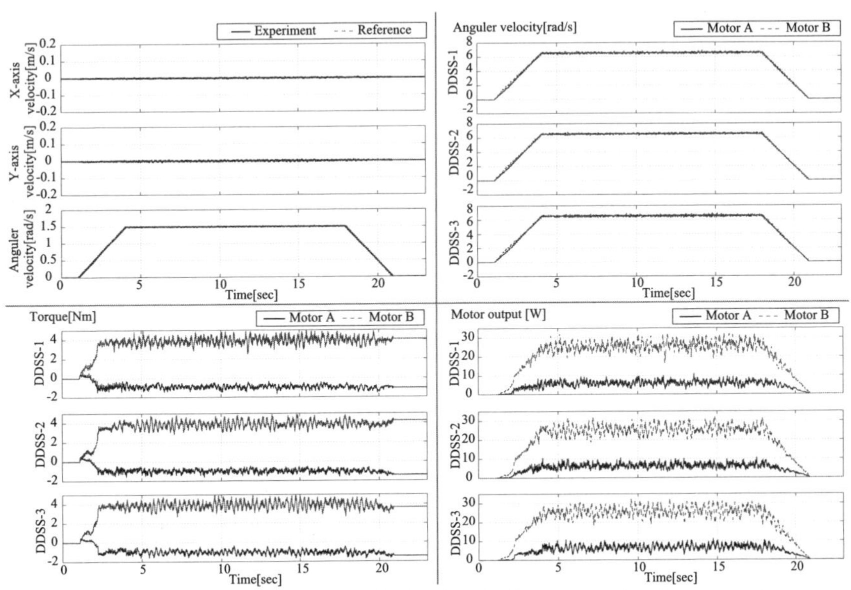

Fig. 6. Experimental result of robot's velocity at rotational movement

wheel, large torque is needed. It can be said that this problem is a problem of caster drive mechanism, and it is not mentioned in the past research. Therefore, it is necessary to solve this problem.

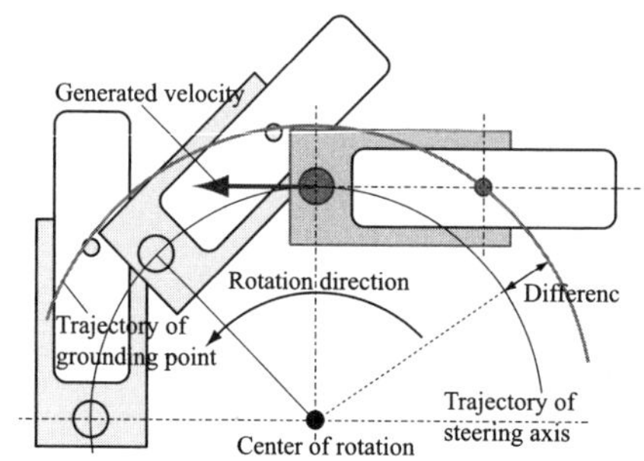

Fig. 7. The situation where the wheel is dragged

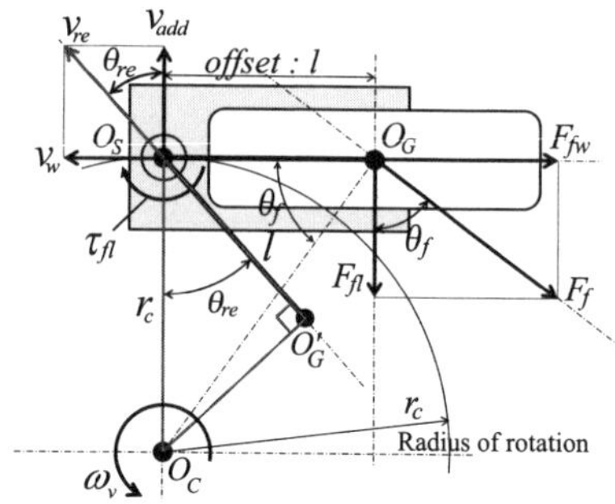

Fig. 8. Effect of the frictional force

5. Proposed Method for Improvement

In this section, we propose a novel method to improve the problem.

Firstly, a grounding point to reduce the influence of friction by the wheel is considered. The model of influence of friction is shown in Fig. 8. Here, O_C, O_G and O_S is respectively the rotation center, grounding point of wheel and steering axis. r_c is radius of rotation. F_f, F_{fw} and F_{fl} is respectively the frictional force, force component which affects wheel rotation and force component which affects steering rotation. τ_{fl} is a torque in the steering axis generated by F_{fl}. θ_f is the angle between line segments O_SO_G and O_GO_C. It seems that F_{fl} influences on the increase of motor output because motor has to generate a torque opposing to τ_{fl}. Therefore, it is thought that the motor output can be reduced by decreasing F_{fl}.

Secondly, a method to reduce the motor output is considered. Figure 8 shows that the angle θ_f which makes F_{fl} the minimum is $\pi/2$[rad]. In order to revise θ_f, we suggest a method to apply the velocity v_{add} into radial direction. Here, in Fig. fig:effectfriction.eps, O'_G is the grounding point after revising the wheel angle, and θ_{re} is the angle between line segments O_CO_S and $O_SO'_G$. Furthermore, v_w is the velocity of tangential direction to rotation given by $r_c \cdot \omega_v$, and v_{re} the velocity of after revision. The velocity v_{add} to generate v_{re} is given by

$$v_{add} = \left|v_w / \tan\theta_{re}\right| = \left|(v_w l) \Big/ \left(\sqrt{r_w^2 - l^2}\right)\right| \tag{2}$$

Note that v_{add} takes an absolute value because the direction of v_{add} is fixed regardless of the rotational direction.

Thirdly, derivation method of additional velocity of the DDSS to the arbitrary velocity input of the robot is considered. Figure 9 shows the model of rotation. Rotational center $O_c(x_c, y_c)$ is given by

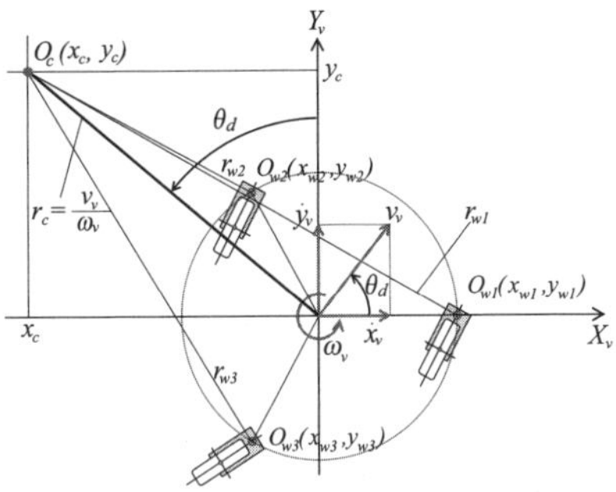

Fig. 9. Model of rotation center

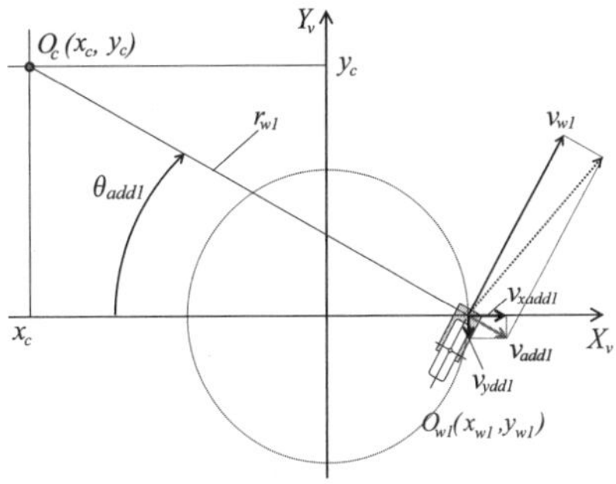

Fig. 10. Model of adding velocity

$$x_c = -r_c \sin\theta_d = -r_c\dot{y}_v/v_v = -\dot{y}_v/\omega_v \tag{3}$$

$$y_c = r_c \sin\theta_d = r_c\dot{x}_v/v_v = \dot{x}_v/\omega_v \tag{4}$$

The distance between rotational center and steering axis of each wheel $O_{wi}(x_{wi}, y_{wi}), i = 1, 2, 3$ is also given by

$$r_{wi} = \sqrt{(x_{wi} - x_c)^2 + (y_{wi} - y_c)^2}\ ,\quad i = 1, 2, 3 \tag{5}$$

Finally, v_{addi} is derived from Eq. (2). As seen in Fig. 10, the velocity v_{addi} is replaced to the elocity of DDSS to caster-drive wheel coordinate system (v_{xaddi}, v_{yaddi}) as follows:

$$v_{xaddi} = v_{addi}\cos\theta_{addi} = v_{addi}\left(x_{wi} - x_c\right)/r_{wi} \tag{6}$$

$$v_{yaddi} = v_{addi}\sin\theta_{addi} = v_{addi}\left(y_{wi} - y_c\right)/r_{wi} \tag{7}$$

By adding v_{xaddi} and v_{yaddi} to the reference velocity which is derived by conventional kinematic equation, wheel angle is revised.

Experimental Evaluation

In order to verify the effectiveness of proposed method, experiments with the same condition with Fig. 6 are conducted. Figure 11 shows an experimental result. This figure shows that the motor output of A, B in DDSS-1,2,3 is reduced in comparison with the result of Fig. 6, although the velocity of the mobile robot is the same with that. Furthermore, when the average value of a motor output for 4[sec] to 17[sec] is calculated, in the case of rotating by the previous method, the output of motor A and B is then 6.33[W] and 25.50[W], and by the proposal method, the output of motor A and B is 10.71[W] and 1.39[W]. It is confirmed that the output of motor B is decreased drastically. In contrast, the rise of motor output about 4[W] is found in motor A. It is because torque of motor A is close to zero in the experiment by previous method, but it is higher in the experiment by proposal method. On the other hand, torque of motor B is close

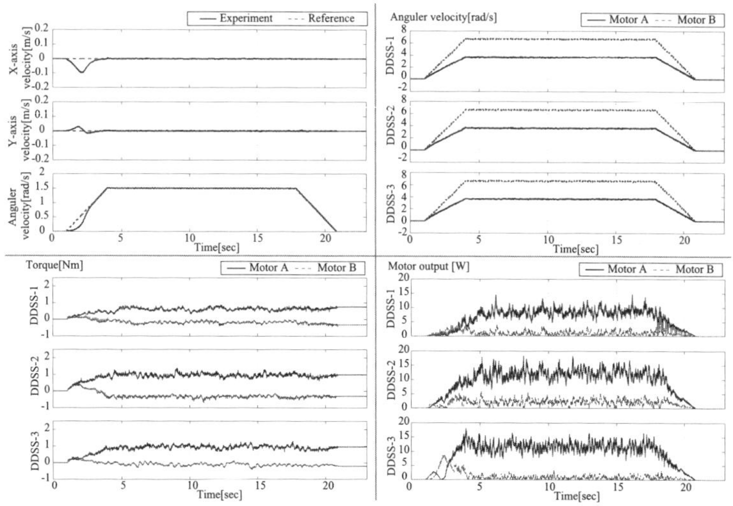

Fig. 11. Experimental result of robot's velocity at rotational movement

to zero in the experiment by proposal method. Thus, the output of motor B becomes smaller than the result by previous method. However, we could totally achieve the reduction of motor output up to 30%.

6. Conclusion

In this paper, novel power-assist mobile robot with three-wheels is built as platform of omni-directional mobile robot. Furthermore, it is confirmed that the motor output becomes higher in the case of rotational movement than the case of translation movement by experiments. Therefore, improvement method is proposed from the view point of the analysis of wheel's friction. Then, large reduction of motor output is verified from comparing with experiment by conventional method.

References

1. M. West, et al., "Design of Ball Wheel Mechanism for Omnidirectional Vehicles with Full Mobility and Invariant Kinematics", Journal of Mechanical Design, 1997, Vol. 119, pp.153–161.
2. M. Wada, et al., "Holonomic and omnidirectional vehicle with conventional tires," Int'l Conf. on Robotics and Automation, 1996, pp.3671-3676.
3. Y. Ueno, et al., "Development and Experimental Evaluation of a Novel Omnidirectional Wheel Mechanism, 13th CLAWAR, 2010, pp.817-824.
4. H. Kitagawa, et al., "Fuzzy power assist control system for omni-directional transport wheelchair", Int'l Conf. on Intelligent Robots and Systems, 2004, pp.1580-1585.

PERFORMANCE OF THE ADAPTIVE GAIT FOR THE LEG-WHEEL ROBOT, RT-MOVER

Shuro Nakajima and Kazuhisa Ietomi

The Dept. of Advanced Robotics, Chiba Institute of Technology, Narashino, Chiba 275-0016, Japan. Email: shuro.nakajima@it-chiba.ac.jp, www.nakajima-lab.it-chiba.ac.jp

We have developed a robot, called RT-Mover, which shows adequate mobility performance on targeted types of rough terrain. It has four drivable wheels and two leg-like axles but only five active shafts. A strength of this robot is that it realizes both a leg mode and a wheel mode in a simple mechanism. We are now researching and developing an adaptive gait algorithm targeting terrain on which random areas of raised terrain exists. In "adaptive gait" only the wheel in contact with the raised terrain performs a leg motion and movement continues. In this research, we numerically analyzed the range of raised terrain which can be traversed using adaptive gait, and verified movement ability through experimentation. From those results, we have found that adaptive gait allows motion from an arbitrary posture in an environment with arbitrarily positioned raised terrain if the height of the terrain is within about 0.1[m].

Keywords: Adaptive Gait, Leg-wheel Robot, Rough terrain, Mobile Robot

1. Introduction

Robots that can move over rough terrain with active body leveling are now in strong demand. Such a robot can have a variety of uses, such as carrying packages, assisting people who have difficulty in walking, and safety monitoring outdoors.

These robots can be generally classified into the following three categories: legged robots like ASV,[1] Wheeled robots like CRAB,[2] and leg-wheel robots like Epi.q-1.[3]

Leg-motion and wheel-motion differ in principle, being continuous surface motion and discontinuous surface motion, respectively. Both functions are required, but in the environment in which humans live, the majority of surfaces can be traversed with wheels, while some areas such as steps require leg motion. For this reason, this research targets leg-wheel robots with both leg and wheel functions. We developed the RT-Mover(Fig.1),[4] aiming to realize the functionality of allowing both leg and wheel motion while maintaining the constant horizontal position of the load bearing portion with the minimum number of drive axles. So far, we have already proposed a motion control method for wheel mode as well as a step gait that raises and lowers four wheels in sequence in order to move up and down steps. Also, we are investigating adaptive gait, which raises and lowers the required wheel like a leg in response to raised terrain (To be announced in a separate report).

This paper is positioned to inspect the range of surfaces including raised terrain in which adaptive gait can allow move. In other words, this research inspects the movement ability of adaptive gait. In Section 3, we perform numerical calculations regarding the height of raised terrain which can be traversed, and then discuss the applicable range of adaptive gait. In Section 4, we perform outdoor experiments and verify movement ability in an actual environment.

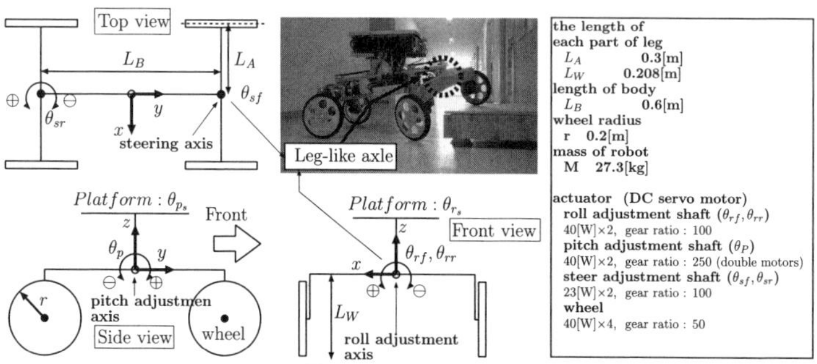

Fig. 1. RT-Mover

2. Adaptive gait for the leg-wheel robot, RT-Mover

Adaptive gait targets terrain upon which raised terrain exists at random, as shown in Fig.2. It continues movement by moving only the wheel in contact with rough terrain in a leg motion. Fig.3 is a flowchart which shows basic operations (I) and complementary operations (II). Basic operations are composed primarily of determination of contact with raised terrain, confirmation of stability and leg movement. Complementary operations are composed mainly of an operation to correct the situation if it is determined that stability or stride length are insufficient at the start of a leg motion and an operation to pull back if the above operation fails to perform the leg motion. This paper considers the range of raised surfaces which can be traversed using the adaptive gait described above. Information regarding adaptive gait itself will be described in a separate report.

3. Discussion on the performance of the adaptive gait

Through adaptive gait, wheels are moved in a leg motion as necessary in response to raised surfaces, making it possible to move over terrain on which raised surfaces exist at random. Here, we will investigate the maximum height of such raised surfaces which can be traversed by adaptive gait.

3.1. *Inspecting static stability with regard to raised surface height*

We will numerically analyze static stability when a leg is raised in adaptive gait. In adaptive gait, static stability at the moment the leg is raised is one condition for the leg motion capability. This is because in most cases, if static stability at the movement a leg is raised can be secured, static

(b)

Fig. 2. Sceans of the adaptive gait

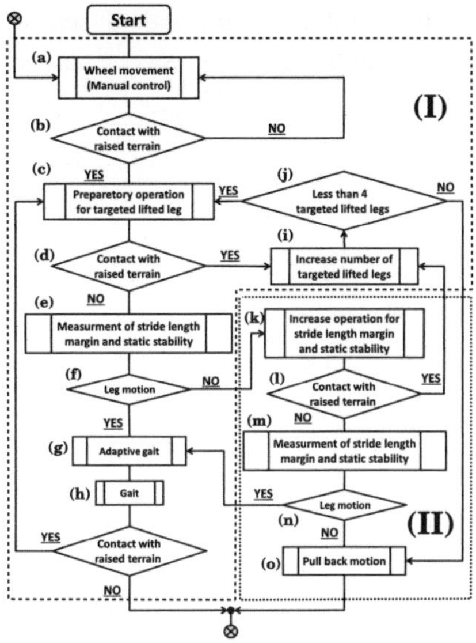

Fig. 3. A flow of the adaptive gait

stability can be maintained throughout the leg movement due to the mechanical characteristics of the RT-Mover. Thus, we will numerically analyze the static stability at the moment a leg is raised from a position. Dividing the states, we find the following three cases.

(1) Situations when raising 1 leg while 1 other leg is on a step (Fig.6)
(2) Situations when raising 1 leg while 2 other legs are on steps (Fig.8)
(3) Situations when raising 1 leg when 3 other legs are on steps (Fig.10)

The position being investigated is assumed to be in a condition in which the support leg steering angle is stopped before the static stability reaches maximum ($\pm\theta_{max}$, Fig.4). When the moving leg is in front, the front steering angle is the angle $\pm\theta_f = \pm\theta_{max} \mp \theta_{step}$ (Fig.4) which leaves stride length and lowers stability most. In other words, if the lifted leg is the left leg it is $-\theta_f$, if the lifted leg is the right leg it is θ_f. When the lifted leg is a rear leg, the rear steeling is $\pm\theta_{sm}$ as shown in the reference.[4] The reasons for this are described in papers related to adaptive gate, but if the lifted leg is the left leg it is θ_{sm}, if the lifted leg is the right leg it is $-\theta_{sm}$. Here, we will explain the calculation of the roll angle θ_r and the wheel pitch angle θ_p when the leg is on a step of height h[m]. Using these, we calculated the following stability. If L_A is the distance from the center of rotation of the roll adjustment axis to the wheel as shown in Fig.5, then $\theta_r = \sin^{-1}(\frac{h}{2L_A})$. Using body length L_B and the height h_p of the rise of the front of the body, the body pitch angle is calculated to be $\theta_p = \sin^{-1}(\frac{h_p}{L_B})$. The height of the rise of the body height is $h_p = L_A \sin\theta_r$.

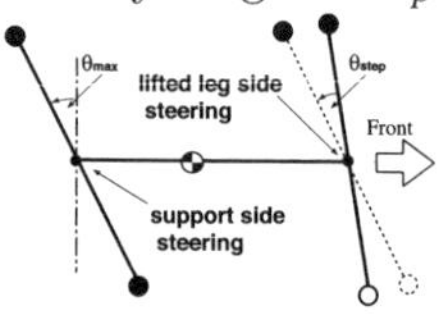

Fig. 4. Position of investigating stability

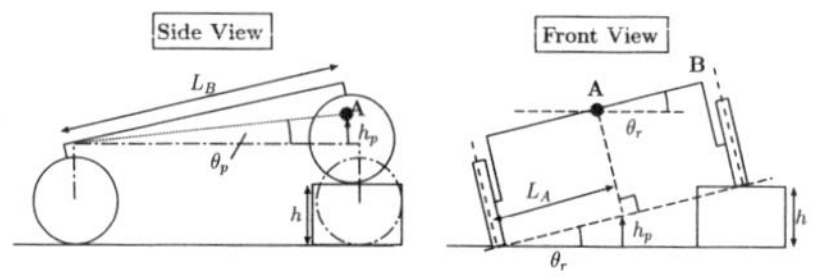

Fig. 5. Model in calculation with regard to step height

3.1.1. *Situations when raising 1 leg while 1 other leg is on a step*

We will numerically calculate the static stability when 1 leg is lifted when 1 other leg is on raised terrain of height h[m]. There are cases such as those shown in Fig.6(a)∼(f). Using a residential environment as the presumed usage environment, the maximum of height h[m] was set to 0.2[m], the step height of a sidewalk.

Fig.7(a)∼(c) show the static stability when raised terrain height h[m] is changed in the situations in Fig.6(a)∼(c). We can see that static stability falls below 0 at the lowest raised terrain height 0.106[m] in Fig. 7(c). The reasons stability decreases in Fig.7(c) are because through the body pitch angle tilting, the projected center of gravity moves to the rear, away from the center of the support polygon, and because the support polygon changes by the slope of the front roll adjustment axis.

Fig.7(d)∼(f) show the static stability when raised terrain height h[m] is changed in the situations in Fig.6(d)∼(f). Static stability falls below 0 at the lowest raised terrain height 0.097[m] in Fig.7(e). The reasons stability decreases in Fig.7(e) are also because through the body pitch angle tilting, the projected center of gravity moves to the rear, away from the center of the support polygon, and because the support polygon changes by the slope of the front roll adjustment axis(Fig.6(e)).

Here, as in Fig.6(b) for example, the projected center of gravity moves away from the center of the support polygon like in Fig.6(c), but as the front roll adjustment axis is in the opposite direction, static stability actually increases.

3.1.2. *Situations when raising 1 leg while 2 other legs are on steps*

Figure 9(a)∼(c) show the static stability when raised terrain height h[m] is changed in the situations in Fig.8 (a)∼(c). Static stability falls below 0 at the lowest raised terrain height 0.099[m] in Fig.9(b). In Figure 8(b) the body pitch angle is horizontal, but static stability lowers as the support polygon changes due to the slope of the front and rear roll adjustment axes.

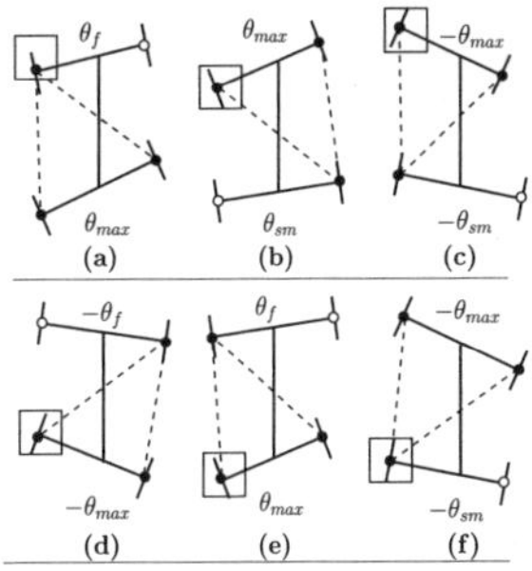

Fig. 6. Situations when 1 leg is on a step

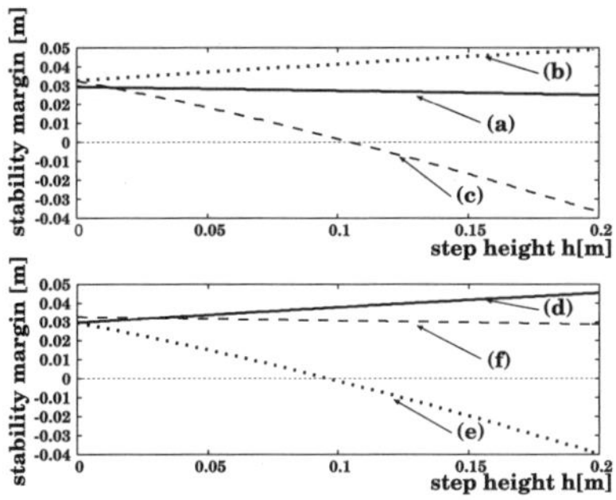

Fig. 7. Static stability when 1 leg is on a step

Figure 9(d)∼(f) show the static stability when raised terrain height h[m] is changed in the situations in Fig.8(d)∼(f). Static stability falls below 0 at the lowest raised terrain height 0.11[m] in Fig.8(e). In Figure 8(e) the body pitch angle does not tilt, but static stability falls as the support polygon changes due to the slope of the front and rear roll adjustment axes.

Here the front and rear roll axes incline in Fig.8(c) and (f), but because the rotation direction is opposite in front and behind, the support polygon does not change so as to lower static stability.

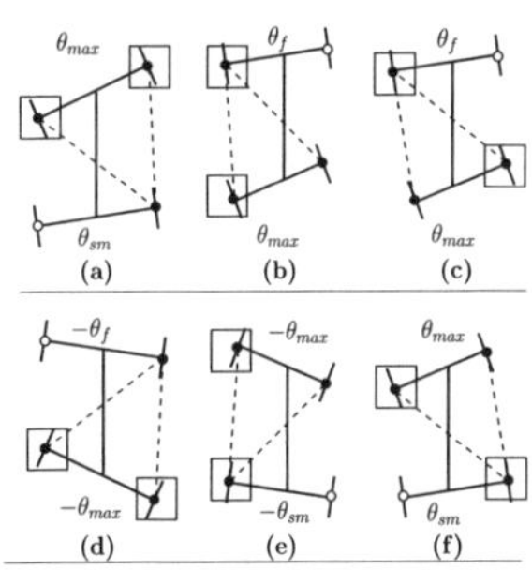

Fig. 8. Situations when 2 legs are raised on steps

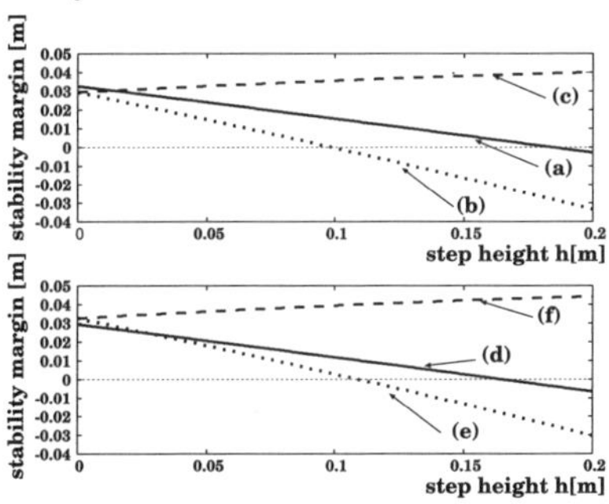

Fig. 9. Static stability when 2 legs are on steps

3.1.3. *Situations when raising 1 leg while 3 other legs are on steps*

We will consider static stability when 1 leg is lifted when 3 other legs are on raised terrain of height h[m] (Fig.10(a) and (b)). From the results of numeric calculation, we see that static stability does not fall below 0 in Fig.11 (a) and (b).

From the investigations results above, the lowest step height at which static stability falls below 0 is the condition shown in Fig.7(e), and the raised terrain height at that time is 0.097[m], roughly the same as the wheel radius. In other words, in adaptive gait, raised surfaces with heights of up to approximately 0.1[m] can be traversed regardless of a configuration

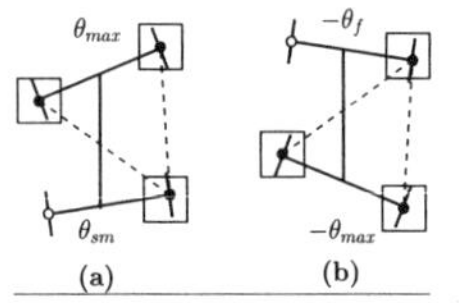

Fig. 10. Situations when 3 legs are raised on steps

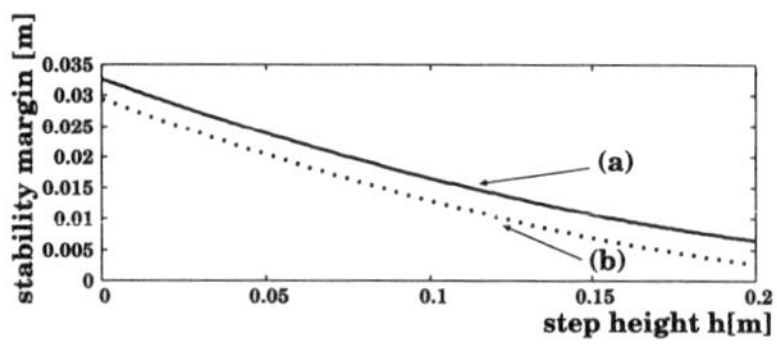

Fig. 11. Static stability when 3 legs are raised on steps.

of the robot body or a positional state of the raised terrain. On the other hand, as we can see from the calculation results, there are cases in which raised terrain surfaces of heights of even 0.2[m] can be traversed depending on a configuration of the robot body or a positional state of the raised terrain. Considering that the frequency of high raised surfaces decreases with height, and that it is rare to enter a position in which it is difficult to raise a leg, we can say that we can move in most places where raised surfaces of 0.15[m]~0.2[m] are scattered in an actual environment.

4. Evaluation of the adaptive gait through experiments

We carried out movement experiments in an actual environment to inspect the movement ability of adaptive gait. We chose a place that often exists in a city environment as an actual environment. The experiment conditions are the same as separate reports relating to adaptive gait. The route direction and wheel command speed in wheel mode were given by remote control.

4.1. *Adaptive gait on a narrow path*

Figure 12(a) shows the movement on a path which has surfaces of different heights on left and right and on which raised terrain exists in front. We see that movement can be continued even in this kind of environment. First, the robot moved forward in wheel mode, and the right-front leg made contact with a step. As a result, adaptive gait with 1 lifted leg was initiated (Fig.12(b)-4[s]). In Fig.12 (e)-4[s]~5[s], rear steering was turned to the maximum in a counter-clockwise direction to preserve static stability in order to lift the right-front leg. In Fig.12(d)-5[s]~7[s], as the leg capability conditions have been fulfilled, the right-front leg is lifted onto the step by leg motion. In Fig.12(e)-10[s]~14.5[s], front-rear steering is returned to 0[deg] by the gait completion motion. After that, when the robot moved forward in wheel mode, the right-rear leg made contact with a step, and adaptive gait with 1 lifted leg was initiated (Fig.12(b)-17(s)). In Fig.12(e)-17(s)~18.5(s), rear steering is set to θ_{sm}, and by turning front steering to the maximum, leg capability conditions were fulfilled and the right-rear leg was lifted onto the step through leg motion.

After that, front and rear steering angles are returned to 0[deg] by the gait completion motion and adaptive gait is completed.

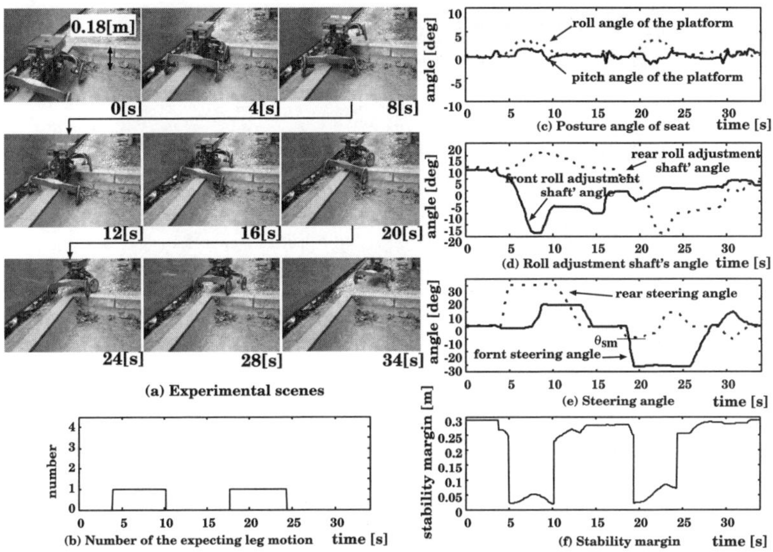

Fig. 12. Adaptive gait on a narrow path

4.2. *Adaptive gait on continuous steps*

Figure 13 shows movement on continuous steps of a wide depth. In particular, the height of the first step, Step A, is the same as the wheel diameter 0.2[m], but we can see that movement is possible. First, the robot collides with Step A in wheel mode and the right-front leg makes contact with Step A (Fig.13-2.5[s]). It enters adaptive gait with 1 lifted leg, and the right-front leg is lifted onto Step A through leg motion (Fig.13-7.5[s]). After the adaptive gait operation has completed, the robot moves forward in wheel mode (Fig.13-22.5[s]), the right-rear leg climbs Step A by the wheel function. The left-rear leg cannot climb with the wheel function, and climbs Step A through adaptive gait with 1 lifted leg (Fig.13-25[s]). After the adaptive gate operation has completed, the robot moves forward in wheel mode, and the front legs and right-rear leg climb Step B through the wheel function (Fig.13-25.5[s]). The right-rear leg, as before, cannot climb Step B with the wheel function and climbs Step B through adaptive gait (Fig.13-32.5[s]). After the adaptive gait operation has completed, the robot climbs Step C through the wheel function, at which point it has climbed all stairs (Fig.13-45[s]).

When frictional conditions are favorable as in this experiment, the probability that the robot can move in wheel mode alone increases. However, when the front of the body climbs the step and the body tilts backwards, it becomes difficult for the rear wheels to climb the step with the wheel function. At this time, it enters adaptive gait mode and can climb the step.

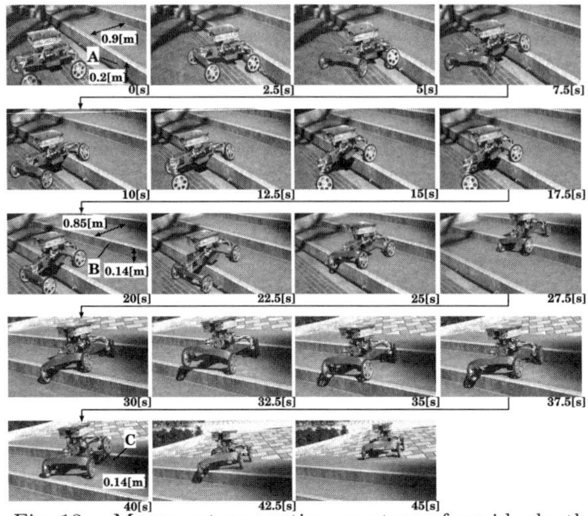

Fig. 13. Movement on continuous steps of a wide depth

5. Conclusion

In this report, we considered the range of surface on which the RT-Mover can move using adaptive gait, analyzing raised surface height in particular. In addition, we performed an outdoor experiment and inspected the movement ability of the RT-Mover. We will discuss stability during the leg motion in more detail and the effect of raised terrain after movement as inspections of adaptive gait movement ability as future tasks. In addition, we want to add further verification by adding outdoor experiments in other environments. Also, a future task for adaptive gait is support for moving down steps and grooved terrain by integrating external sensor information into the proposed gait which is based on internal sensor information.

Acknowledgement: This work was partially supported by the Program for Promoting Fundamental Transport Technology Research from the Japan Railway Construction, Transport and Technology Agency (JRTT).

References

1. S.M.Song and K.J.Waldron,"Machines That Walk:The Adaptive Suspension Vehicle", MIT Press,1989.
2. T.Thueer, et al.,"CRAB-Exploration rover with advanced obstacle negotiation capabilities",Proceedings of the 9th ESA Workshop on Advanced Space Technologies for Robotics and Automation, pp.1-8, 2006.
3. G.Quaglia,etal.,"The Epi.q-1HybridMobileRobot",The International Journal of Robotics Research, 29 (1): 81-91, 2010.
4. Shuro Nakajima,"RT-Mover: a rough terrain mobile robot with a simple leg-wheel hybrid mechaism", The International Journal of Robotics Research, Vol.30 Issue 13, pp.1609-1626, 2011.

TAIL ASSISTED DYNAMIC SELF RIGHTING

Aaron M. Johnson*, Thomas Libby†, Evan Chang-Siu†
Masayoshi Tomizuka†, Robert J. Full†, and D. E. Koditschek*

** University of Pennsylvania, 200 S. 33rd St, Philadelphia, PA 19104*
E-mail: {aaronjoh,kod}@seas.upenn.edu
† University of California Berkeley, 1099 Valley Life Science Bldg, Berkeley, CA 94720
E-mail: {tlibby,evancs,tomizuka,rjfull}@berkeley.edu

In this paper we explore the design space of tails intended for self-righting a robot's body during free fall. Conservation of total angular momentum imposes a dimensionless index of rotational efficacy upon the robot's kinematic and dynamical parameters whose selection insures that for a given tail rotation, the body rotation will be identical at any size scale. In contrast, the duration of such a body reorientation depends upon the acceleration of the tail relative to the body, and power density of the tail's actuator must increase with size in order to achieve the same maneuver in the same relative time. Assuming a simple controller and power-limited actuator, we consider maneuverability constraints upon two different types of parameters — morphological and energetic — that can be used for design. We show how these constraints inform contrasting tail design on two robots separated by a four-fold length scale, the 177g Tailbot and the 8.1kg X-RHex Lite (XRL). We compare previously published empirical self-righting behavior of the Tailbot with new, tailed XRL experiments wherein we drop it nose first from a 2.7 body length height and also deliberately run it off an elevated cliff to land safely on its springy legs in both cases.

Keywords: Tails, Angular Momentum, Motor Selection, Legged Robots

1. Introduction

Tails play a variety of roles in animals, from fat storage[1] to communication.[2] Perhaps more useful to mobile robotics is their ability to stabilize dynamic locomotion. By swinging their tails, geckos can self-right in less than a body length after a fall, or reorient through zero net angular momentum maneuvers.[3] Fast-running lizards use their tails to briefly store angular momentum after a perturbation, enabling feedback-driven attitude control during short leaps.[4] The effectiveness of this mechanism inspired Tailbot (Figure 1(a)), a robot with an active tail which enabled disturbance regulation[4] and other dynamic behaviors, including air-righting and traversing

(a) (b)

Fig. 1. (a) Tailbot[4, 5] (b) XRL[7, 8] with a new tail, and with approximately sized image of Tailbot inserted.

rough terrain.[5] The stabilizing function of tails appears to operate effectively over a wide range of size scales in natural systems, from 1g geckos to 10 kg lemurs,[6] and possibly beyond. In this paper we explore the efficacy of a stabilizing tail on robots across a range of size scales and provide insight into the design choices. As a second robotic example, we design a tail for X-RHex Lite (XRL)[7, 8](Figure 1(b)), a relative of the RHex hexapedal robot,[9] over 60 times more massive than Tailbot.

While tails can provide many benefits to mobile robots, here we focus exclusively on aerial self-righting. In general, if survivability or required performance is very sensitive to orientation, an inertial tail will be beneficial, and we treat on the problem of determining criteria for and then assessing empirically the capabilities of a "good" inertial tail. Designing a robot with a tail, or adding one to an existing design, has many costs, including the extra mass, volume and extended body envelope as well as the added complexity and new opportunities for failure. Weighing the penalties associated with these multi-faceted disadvantages against the benefits of increased maneuverability lies far beyond the scope of this paper whose contribution is to address the much narrower question of how to parametrize the design space and then how to select within it a design for a self-righting tail.

In general, we would consider a tail-like design to be one which adds an appendage to a robot, specialized for inertial manipulation (as opposed to say a flywheel[5]). Even then, potential designs can still be quite varied in attachment and mass distribution. Given that the robots in question are intended to navigate through unstructured environments, a tail that trails behind the body and presents few opportunities for entanglement seems preferable to more exotic alternatives. In this case, concentrating mass at

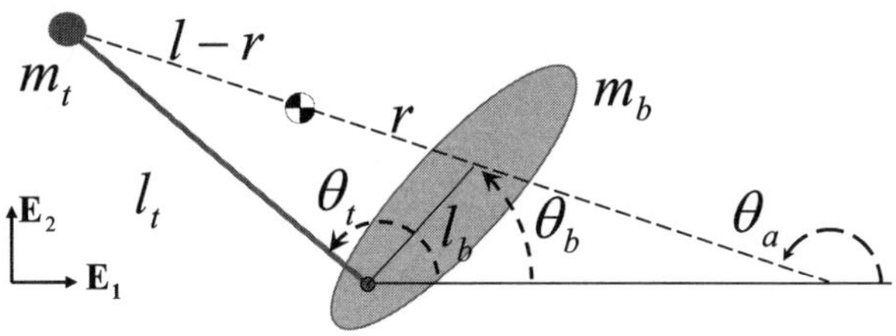

Fig. 2. Planar two body model.

the tail's tip generates the largest moment of inertia per unit mass. In this paper we will consider a point-mass tail at the end of a massless rod; the derivation provided is easily adaptable to other situations by adding the tail's moment of inertia about its center of mass.

The paper's organization roughly parallels the sequence of design and assessment steps associated with the natural stages of tail development. Section 2 proposes a morphological design space (the tail's kinematic and dynamic parameters) and imposes upon it a scale-independent measure of kinematic efficacy arising from conservation of angular momentum. Section 3 introduces the maneuverability task space (the elapsed time and body angle parameters) along with a power train design space (the tail actuator's motor and gearing parameters) and uses the trajectories resulting from a simple (linear) dynamical system model to bind these three parameter spaces together in a constraint that affords the expression of contrasting design criteria[a]. Section 5 presents the empirical results on XRL.

2. Kinematics

Consider a simple planar model of a tailed robot in an aerial maneuver, representing the tail by a point mass held by a massless rigid rod, and the rest of the vehicle represented by a single rigid body (Figure 2). The body and tail centers of mass are separated by a length l with angle θ_a, which is dependent on the joint angle $\theta_r = \theta_b - \theta_t$. Without loss of generality we assume that the origin is at the COM of the system and therefore the constant total angular momentum, H_0, about the system's center of mass,[4]

$$H_0 = I_a\dot{\theta}_a + I_b\dot{\theta}_b; \qquad I_a = \frac{m_b m_t}{m_b + m_t} l_t^2, \tag{1}$$

[a]We contrast with this simplified linear design model the kinematics (and some behavioral implications) of the general nonlinear case, but for the sake of space leave the detailed derivations in this and other sections to a technical report [12].

imposes a non-holonomic constraint on the dynamics (see the technical report[10] for full derivation)[b]. If the tail pivots at the body center of mass ($l_b = 0$), then $\dot{\theta}_a = \dot{\theta}_t$ and $l = l_t = const.$, and the constraint is linear in segment angular rates. We choose a time scale γ (units of $1/s$, see further discussion in the next section) such that $t^* = \gamma t$, where the * indicates dimensionless values. Substituting the dimensionless derivatives $\dot{\theta}_i = \gamma \dot{\theta}_i^*$, simplifying and dividing by I_b yields a dimensionless version of Eqn. (1),

$$\bar{H}_0 = \varepsilon\, \dot{\theta}_t^* + \dot{\theta}_b^*; \qquad \varepsilon = \frac{I_a}{I_b}; \tag{2}$$

where we define ε to be tail *effectiveness* (generalizing the previous definition[4]) and $\bar{H}_0 = H_0/(\gamma I_b)$ is a dimensionless momentum.

The effectiveness directly governs performance in two distinct tasks: zero angular momentum righting ($\bar{H}_0 = 0$), where ε is the ratio of segment speeds, and orientation regulation after an impulse, where a controller tries to maintain a stable body angle ($\dot{\theta}_b^* = 0$) with a tail velocity ($\dot{\theta}_t^* = \bar{H}_0/\varepsilon$). Because effectiveness is dimensionless, isometrically[11] scaled robots are kinematically similar — for a given tail rotation, the body rotation will be identical at any size scale. It is important to note that motor and gearbox selection have no impact on the effectiveness, ε, though they will impact righting duration and power requirements, hence this notion of effectiveness decouples the mechanics of tail design from the energetics of motor and gearbox selection.

When the tail pivot is not at the center of mass of the body ($l_b \neq 0$), the kinematics change with θ_r, leading to the nonlinear version of (1),[10]

$$\bar{H}_0 = \varepsilon(1 - \lambda \cos\theta_r)\dot{\theta}_t^* + (\varepsilon(\lambda^2 - \lambda\cos\theta_r) + 1)\dot{\theta}_b^*; \qquad \lambda = l_b/l_t. \tag{3}$$

This equation governs both stabilization and zero angular momentum maneuvering. In the latter case, $\bar{H}_0 = 0$, and by maintaining our definition of effectiveness in righting as the ratio of segment speeds, the configuration-dependent *non-linear effectiveness* is,

$$\varepsilon_n = -\frac{\dot{\theta}_b^*}{\dot{\theta}_t^*} = \frac{\varepsilon(1 - \lambda\cos(\theta_r))}{\varepsilon(\lambda^2 - \lambda\cos(\theta_r)) + 1}. \tag{4}$$

Because we have limited our dynamical analysis, below, to the (linear) case $\lambda = 0$, and only report XRL experiments for a tail mounted as close to this condition as could be readily implemented we merely suggest in the

---

[b] I_a is conventionally called the *reduced mass* moment of inertia. To consider a non-point mass tail, another term $I_t\dot{\theta}_t$ must be added to H_0.

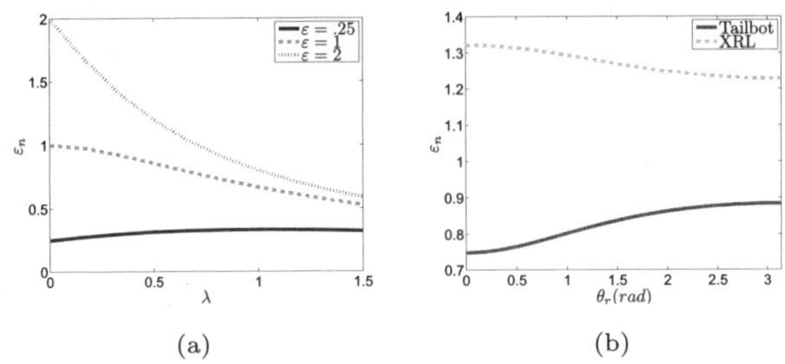

Fig. 3. Variation of non-linear effectiveness with changing λ when $\theta_r = \pi$ (a), and with respect to θ_r for the two robots (b).

numerical plots of Figure 3(a) the somewhat unintuitive manner in which ε_n varies in λ and θ_r. The effect of joint offset is relatively small — for $\varepsilon = 1$, $\lambda = 0.5$, ε_n decreases by only 15-33% depending on tail angle. For robots like Tailbot and XRL with with $\varepsilon \approx 1$ ($\varepsilon_{Tailbot} = 1, \varepsilon_{XRL} = 1.29$) and $\lambda < 0.5$ ($\lambda_{Tailbot} = 0.43, \lambda_{XRL} = 0.14$), the change in effectiveness with θ_r is small ($< 25\%$) further motivating the linear approximation.

3. Dynamics and Power Scaling for a Free Fall Task

While all isometrically scaled robots will maneuver with similar kinematics given enough time, a real terrestrial robot is constrained by the duration of its aerial phase (fall, leap, or other dynamic behavior) and this imposes a new set of requirements on a new set of parameters that specify the tail actuation power train.

Consider a maneuverability task specified by the requirement to reorient the body through a fixed angle $\theta_b = \theta_0$, in a desired time $t = t_0$. We develop in the technical report[10] the linear (assuming $\lambda = 0$) dynamics associated with a bang-bang style of control (i.e., accelerating and then decelerating the tail with maximal available torque) imparted by a conventionally power-limited actuator (i.e., whose maximal available torque must decrease linearly with speed.[12] Here we restrict our task to a simple reorientation maneuver through a fixed angle $\theta_b = \theta_0$, in a desired time $t = t_0$. The closed form trajectory of the body angle is,[10]

$$\theta_b(t^*) = \frac{\omega_m}{\gamma(1 + \frac{1}{\varepsilon})}(-1 + t^* + \exp(-t^*)) \tag{5}$$

where the dimensionless state variables are denoted with a *, for some time scale $t^* = \gamma t$, with $\gamma = \frac{4P(1+1/\varepsilon)}{I_b \omega_m^2}$, ω_m the motor no-load speed (after the gear box), and P the motor's peak rated mechanical power.

Thus for a fixed system and time, $t^* = \gamma t_0$, we see that the robot body has rotated $\theta_b(\gamma t_0)$. Conversely if we desire a body rotation of θ_0, we must solve the implicit function $t^*(\theta_0)/\gamma$ to find the time required.

We can also turn this problem around and ask for a given task specification, a θ_0 body rotation in t_0, what are the constraints on the system parameters? If the gearing, ω_m, is chosen to minimize the power, P, then,[10]

$$\theta_0 = \frac{2t_0^{3/2}P^{1/2}k_1}{(1+1/\varepsilon)^{1/2}I_b^{1/2}} \tag{6}$$

for constant $k_1 \approx 0.402$. Therefore the minimum power required is,

$$P = \frac{\theta_0^2}{4t_0^3k_1^2}(1+1/\varepsilon)I_b \tag{7}$$

but of wider interest may be power density, $P_d = P/m$.

This relationship reveals an important constraint on dynamic tail reorientation: the effect of robot size. Consider a robot isometrically scaled by a length scale L. Then mass m we will scale by the cube of length L and $I_b \propto L^5$. If the robot were required to reorient through the same angle in the same time regardless of size, then by substitution into Eq. (7) we would require power density $P_d \propto L^2$. However, a larger robot will fall slower relative to its length. Considering a free falling distance $h \propto L$ implies that the time available $t \propto L^{1/2}$. Therefore, from Eq. (7) the power density,

$$P_d \propto \frac{1}{L^3}\frac{1}{L^{3/2}}L^5 = L^{1/2} \tag{8}$$

scales as the square root of length. This indicates that inertial reorientation gets more expensive at large size scales; larger robots may suffer reduced performance, or must dedicate a growing portion of total body mass to tail actuation. However, the robots in this paper span a characteristic length range of almost four fold without dramatic differences in ability; in this case, variance in motor power density may trump scaling.

4. XRL Tail Design

In accord with the scale independent features of the mechanical design space introduced in Section 2, the XRL tail is an appropriately scaled approximation to that of Tailbot,[5] both comprising an approximate point-mass made of brass at about 1/10th body mass, attached to a carbon fiber tube of about one body length (see Table 1 for exact values). While Tailbot was a special-built machine, the tail for XRL must be added to an existing platform as a modular payload,[8] and as such the range of motion is

Table 1. Comparison of Physical Properties.

Attribute	Tailbot	XRL + Tail
Body Dimensions (cm)	12x9.3x7.0	51x40x10
Body Mass (kg)	0.16	8.1
Tail Mass (kg)	0.017	0.6
Body Inertia (kgm^2)	1.54x10^{-4}	0.15
Tail Range of Motion	225°	155°
Linear Effectiveness, ε	1	1.29
Tail Offset, λ	0.43	0.14
Peak Motor Power (W)	27	342

significantly lower than Tailbot's, especially given a 7.5° safety margin to avoid collision with the body. To compensate, the XRL tail design targets a slightly higher effectiveness so as to afford the same 90° body correction capability as Tailbot.

To mitigate the integration task, both Tailbot's and XRLs tail actuators are chosen to be the same as their wheel/leg motors. But to maximize the performance of the tail, a more careful study is needed. As a slight variation from the previous section, here the power is given for each motor and we now seek to determine the minimal completion time as a function of peak power (parametrized by morphology) rather than the inverse function as above. The time requirement t_0 from the previous section can be thought of a constraint, while here for those motors that meet that constraint we want to consider the fastest completion time as a metric. The optimal no load speed (after gear ratio) and resulting completion time functions are (see the technical report[10] for full derivation),

$$\omega_m = \left(\frac{\theta_0 \beta_0}{k_1 k_0^{3/2}}\right)^{1/3} ; \qquad t = \left(\frac{\theta_0^2}{4Pk_1^2}(1 + 1/\varepsilon) I_b\right)^{1/3} \tag{9}$$

Other metrics to consider are physical (size, mass), electrical (current and voltage available), and thermal. The thermal cost of a tail for inertial self-righting is in general small due to the very small time scales, however some motors may still overheat. Now, following,[13] whose numerical optimization step does not require the restriction to the linear dynamics ($\lambda = 0$) used to derive (9) and which can incorporate these additional metrics, the performance of all commercial motors[14] can be compared, or used as a constraint

Fig. 4. XRL self-righting in a fall.

in the gearbox optimization. Out of the 1,546 motors considered, 82 of them met the length (< 30mm), weight (< 200g), and minimum completion time (< 0.5s) requirements. Of those, the chosen motor was the third fastest, only 22% slower than what would be the optimal motor. The optimal gear ratio for our motor would be 27:1, the 28:1 gear ratio used is the closest commercially available[c].

5. Experiments

As an empirical validation of the foregoing scaling arguments, we conducted a series of initial tests (Figure 4) to see how large a body rotation can be achieved by a relative tail rotation of about 155°, which is limited by geometry. The robot was dropped nose first from a height of 1.36m (over 8 times the standing height and 2.7 times the body length). The body angle was measured from an IMU and regulated to horizontal by a simple PD controller. This test used the entire 155° range of relative tail motion, rotating the body a maximum of 89.7° before hitting the hard stop. From these two final positions we can calculate an average $\varepsilon_n = 1.38$, which matches our simulated result in Fig 3(b) to within 7% and is reasonable considering the errors involved in measuring inertia[15] and manufacturing. As a comparison, the robot was also dropped with no tail activation, causing the front two legs to snap as well as some minor internal damage. Thus if the robot tasks requires a fall from this height, it is definitely survivable assuming it successfully reorients to land within about 5° of level[d].

To demonstrate this new ability for XRL in a practical task, the second set of experiments was conducted outdoors, running along and then away from a 62 cm (3.8 times the hip height or 1.2 body-length) cliff. The robot's

[c]This analysis could be made more accurate by considering a current limit and more complicated controller.
[d]This is an empirical bound still subject to further tests.

Fig. 5. XRL surviving a run off a cliff outdoors.

inertial sensors detect the cliff upon initial body pitch, then actuate the tail according to the previously described closed loop control policy, and the robot lands on its feet (Figure 5). Another test with XRL running from a cliff with a *passive* tail confirmed that it would land nose first.

6. Conclusion

Robots at a range of sizes could benefit from a tail for inertial self-righting. Here we have shown that the kinematics will scale isometrically and are decoupled from the motor/gearbox selection. We have addressed the power considerations for different size robots, but also demonstrated the performance of tails on two robots with vastly different masses and morphologies. In the future we will explore other tail based behaviors.

A fascinating question that also lies largely beyond the scope of this paper surrounds the relative efficacy of appendage (here, a tail) vs. core (body trunk, as in Mather et al.[16]) actuation for the self-righting task. The primary advantages relative to the addition of a tail are that back bending preserves the overall morphology (diminishing the expansion of volume and body envelope) and essentially separates the body into two chunks with much lower moment of inertia, while a tail extends the body envelope and should see a larger effective inertia, depending upon where it is mounted with respect to the body's mass center. However, the penalty in bending is that the final orientation of both segments is important if the legs of the robot are to hit the ground simultaneously.[16] Further, the benefit of a more compact overall body envelope in a new design actually represents a significant deterrent for improving existing robotic platforms like RHex, whose core body structure cannot be substantially altered (without a major redesign[8]) but whose distal appendages can be relatively easily changed, added, or subtracted. We have shown that a tail can enable rapid rotations of up to 90° with relatively low added mass (10% – 20% of body mass), and that the effect of offsetting the tail joint can be relatively low[e].

[e]Although this preliminary design study simplifies the dynamics and diminishes the effective inertia by mounting the tail near XRL's mass center, our analysis suggests that relocating the mount at the body's extreme end would still result in a viable design.

Acknowledgments

This was supported primarily by the ARL/GDRS RCTA and the NSF CiBER-IGERT under Award DGE-0903711. The authors would like to thank Praveer Nidamaluri for building the X-RHex tail, as well as David Hallac, Justin Starr, Avik De, Ryan Knopf, Mike Choi, Joseph Coto, and Adam Farabaugh for help with the robot and experiments.

References

1. M. Aleksiuk, *Journal of Mammalogy* **51**, pp. 145 (1970).
2. E. Palagi, *Journal of Comparative Psychology* **123**, p. 1 (2009).
3. A. Jusufi, D. T. Kawano, T. Libby and R. J. Full, *Bioinspiration & Biomimetics* **5**, p. 045001 (2010).
4. T. Libby, T. Y. Moore, E. Chang-Siu, D. Li, D. J. Cohen, A. Jusufi and R. J. Full, *Nature* **481**, 181 (2012).
5. E. Chang-Siu, T. Libby, M. Tomizuka and R. J. Full, A lizard-inspired active tail enables rapid maneuvers and dynamic stabilization in a terrestrial robot, in *Intelligent Robots and Systems*, (San Fransisco, CA, USA, 2011).
6. D. C. Dunbar, *American Journal of Primatology* **16**, 291 (1988).
7. K. C. Galloway, G. C. Haynes, B. D. Ilhan, A. M. Johnson, R. Knopf, G. Lynch, B. Plotnick, M. White and D. E. Koditschek, *X-RHex: A Highly Mobile Hexapedal Robot for Sensorimotor Tasks*, tech. rep., University of Pennsylvania (2010).
8. G. C. Haynes, J. Pusey, R. Knopf, A. M. Johnson and D. E. Koditschek, *Laboratory on Legs: An Architecture for Adjustable Morphology with Legged Robots*, in *SPIE Unmanned Systems Technology XIV (8387)*, 2012, p. (in press).
9. U. Saranli, M. Buehler and D. E. Koditschek, *The International Journal of Robotics Research* **20**, 616 (2001).
10. A. M. Johnson, T. Libby, E. Chang-Siu, M. Tomizuka, R. J. Full and D. E. Koditschek, *Tail Assisted Dynamic Self Righting: Full Derivations*, tech. rep., University of Pennsylvania (2012).
11. T. Blackburn and K. Gaston, *Trends in Ecology & Evolution* **9**, 471 (1994).
12. P. Gregorio, M. Ahmadi and M. Buehler, *IEEE Transactions on Systems, Man, and Cybernetics, Part B* **27**, p. 626634 (1997).
13. A. De, G. Lynch, A. Johnson and D. Koditschek, Motor sizing for legged robots using dynamic task specification, in *IEEE International Conference on Technologies for Practical Robot Applications*, (Boston, MA, USA, 2011).
14. Maxon Motors, Maxon Catalog (2010–2011).
15. P. Ringegni, M. Actis and A. Patanella, *Measurement* **29**, 63 (2001).
16. T. W. Mather and M. Yim, Modular configuration design for a controlled fall, in *Intelligent Robots and Systems*, (St. Louis, USA, 2009).

SECTION–7

MISCELLANEOUS APPLICATIONS

D.R.O.P. THE DURABLE RECONNAISSANCE AND OBSERVATION PLATFORM

CLIFFORD MCKENZIE
Mechanical & Aerospace Engineering Dept., North Carolina State University
Raleigh, NC 27695, US

AARON PARNESS
Robotic Platforms Group, Jet Propulsion Laboratory, California Institute of Technology
Pasadena, CA 91109, US

The Durable Reconnaissance and Observation Platform (DROP) is a prototype robotic platform with the ability to climb concrete surfaces up to 85° at a rate of 25cm/s, make rapid horizontal to vertical transitions, carry an audio/visual reconnaissance payload, and survive impacts from 3 meters. The platform uses a two-wheel, two-motor design that delivers high mobility with low complexity. DROP extends microspine climbing technology from a linear to rotary implementation, providing improved transition ability, increased speeds, and simpler body mechanics while maintaining microspines' ability to opportunistically grip rough surfaces. Various aspects of the prototype robot's design and performance are discussed, including the climbing mechanism, body design, and impact survival.

1. Introduction

Figure 1. DROP weighs only 300g and is capable of climbing 85° rough surfaces at 25cm/s, move across flat ground at 45cm/s, and make rapid horiz-vert transitions enabled by a new rotary implementation of microspine technology.

Current remote presence needs of the U.S. Department of Defense and NASA are underserved by the state of the art in robotics. Generally, currently deployed platforms are large, restricted to flat terrain, have limited portability, and lack stealth (i.e. Talon [1], Packbot [2]). Conversely, many research platforms have become so small that the limitations on range, terrain, and payload reduce their utility in the field (i.e. DASH [3], Waalbot [4], MEDIC [5], etc.). The Durable Reconnaissance and Observation Platform (DROP), shown in Figure 1, seeks to fill a niche between these paradigms as a platform that can scale vertical surfaces, can carry a 100 gram reconnaissance payload (2.4 GHz video camera and microphone), has a functional range of >100m, but is light enough to be crashproof.

Inspired by climbing insects and spiders, microspine technology enables climbing by using arrays of 10s to 1000s of independent sharp steel hooks to grip microscopic asperities (bumps, holes, and slopes) on rough surfaces [6,7]. Microspines attach to porous, often dusty surfaces such as stone, wood, and concrete where most other attachment mechanisms fail. Microspines have been used successfully on these materials by the RiSE robot [8], Spinybot II [9], and perching airplanes [10], [11], as well as by human climbers [12]. Until now the use of microspines in these applications has required a linear engagement path, motivating legged architectures that are often complex, lack durability, and limit top speed. These legged architectures, such as those implemented on RiSE and Spinybot II, result in speed limitation due to inertial changes when the legs must cycle back and forth. In addition, because of a relatively high level of complexity, numerous points of failure are inherent to these robots should they slip during climbing, making climbing without a safety line a significant risk.

Using a rotary implementation of microspine technology, the Durable Reconnaissance and Observation Platform is able to climb near-vertical surfaces at much higher speeds than legged platforms. The simplicity of the microspine climbing mechanism also provides the platform with a significant weight reduction and increase in durability as compared to previous microspine climbing robots, eliminating the need for safety lines and slip recovery behaviors.

ReconRobotics' SCOUT platform is based on the same two-wheeled architecture as DROP and has demonstrated the benefits of an impact resistant robot for reconnaissance based on novel deployment strategies like throwing the robot into windows, onto rooftops, or into dangerous areas [13]. DROP seeks to improve upon this architecture by enabling vertical climbing and curb mounting without the loss of durability and horizontal maneuverability. The current DROP prototype is capable of traversing concrete surfaces up to 85° from the horizontal at a rate of 25 cm/s and surviving free-fall impacts from 3 meters. On horizontal surfaces it features high maneuverability, such as turn in place capability, and a top speed of 45 cm/s. The two-wheeled architecture also enables the robot to transition from horizontal to vertical surfaces more rapidly and reliably than its legged predecessor, RiSE [14], one of the few climbing robots able to perform such a maneuver at all.

2. Design

2.1 *Climbing Mechanism*

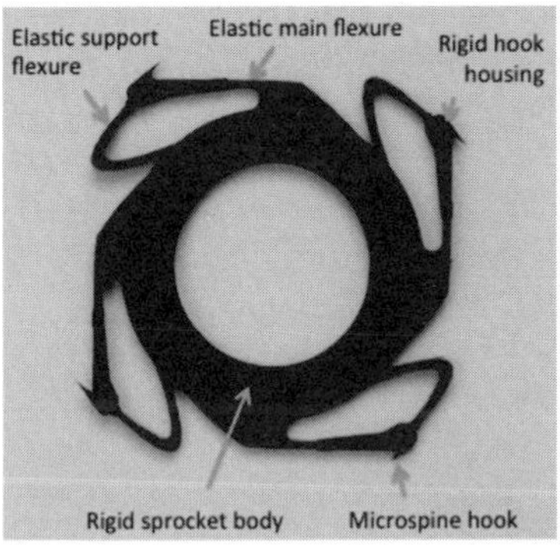

Figure 2. Four microspine hooks are shown arrayed on a Rotary Microspine, each attached by a thin, stretchable flexure. Multiple holes in the frame allow arrays of Rotary Microspines to be configured at different angles with respect to neighboring Rotary Microspines along the width of the robot.

DROP was developed around a rotary microspine climbing mechanism that provides continuous and secure engagement during climbing. Similar to previously demonstrated linear microspines [6], rotary microspines are produced using shape deposition manufacturing. Each microspine features elastomeric flexures which provide suspensions for each steel hook, allowing them to engage the surface independent of adjacent hooks. This flexure further allows each microspine hook to remain engaged through a large angular displacement (approximately 30°) as the mechanism turns. Figure 2 shows the current iteration of the rotary microspine design. Approximately 40 of these rotary microspine mechanisms are arrayed along the width of the body. This construction provides numerous opportunities for microspine engagement, resulting in secure engagement and allowing for climbing on concrete slopes up to 85°.

Previous research has shown that the probability of microspine engagement is largely dependent on hook size, engagement angle, and surface conditions [15, 7]. To increase the probability of desirable surface engagement on a variety of surfaces, each microspine is positioned to maintain an angle theta [Fig. 3] of between 30 and 45 through a prolonged period of rotation of the mechanism. Disengagement from the surface is the result of a decrease in this angle during climbing, eventually approaching 0, which results in detachment from the surface.

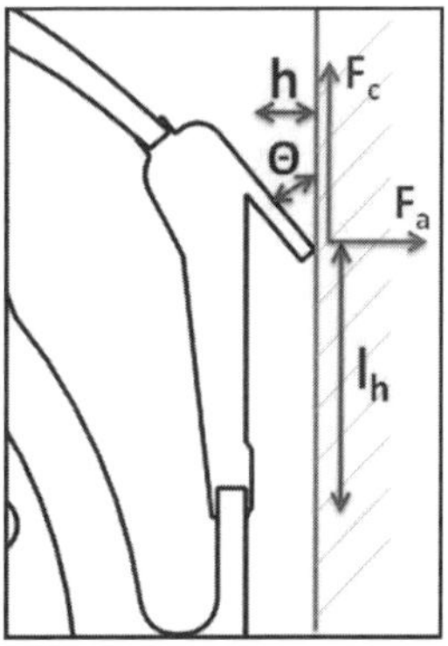

Figure 3. The Rotary Microspine was designed for a hook engagement angle 30°<θ<45°. As the rotary microspine spins, the hook will disengage naturally as θ approaches 0°.

To function properly, rotary microspines require the use of thin, rigid divider disks (0.15mm thick) to separate adjacent microspines. These dividers limit microspine motion to a single plane, preventing hook entanglement and providing a recess for microspines that do not engage, thereby eliminating detrimental forces normal to the climbing surface. In addition, the dividers provide a smooth contact surface for improved horizontal travel.

2.2 *Body*

Developed around the climbing mechanism, the body [Fig 4] is divided into two main components: 1) a central section, which houses two brushed motors, and 2) a tail section which contains the batteries, RF receiver, and microprocessor unit. The center and tail sections are connected by alternating sections of rigid and elastic materials, modeled loosely off vertebrae in a spine. This tail structure allows the main sections to bend and twist relative to one another, aiding in energy dissipation during impact. Selective laser sintered (SLS) high-elongation polyamide-based materials are used to construct the primary body components. The use of Shore 20A-60A hardness polymers in select areas, such as the motor mounts, also provide impact absorption capability.

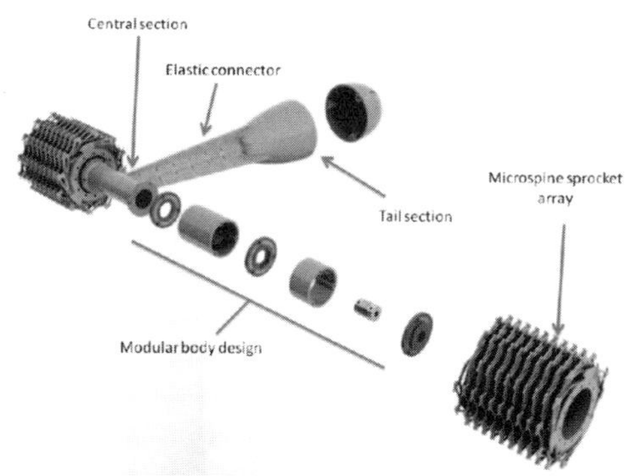

Figure 4. The body of DROP was designed to be modular and durable. High impact resiliency plastic (via SLS) and flexible elastomeric material (via SDM) were used in the construction.

The two-wheeled architecture chosen for DROP requires a balance between adhesive forces created by the microspines and the reactive force created at the tail section. For the quasi-static case shown in Figure 5 it is necessary that the adhesive force (F_a) be greater than the normal reaction at the tail (F_r) produced by the motor torque (T) and pitch-back moment. This can be expressed as:

$$F_{a,maximum} \geq \frac{(F_{mg} * r) + T}{l_t}$$

The radius of the microspine mechanism (r) is assumed constant. Therefore, the body length (l_t) becomes critically important. Seeking a balance of horizontal and vertical performance, it is necessary to constrain l_t. An initial body length of 180 mm was chosen based on expected adhesive forces. Ultimately, however, these adhesion forces are highly dependent on many factors including robot mass, surface conditions, and quality of spine engagement. Future testing of the robot with varied tail lengths is planned, along with several modifications of the rotary microspine design.

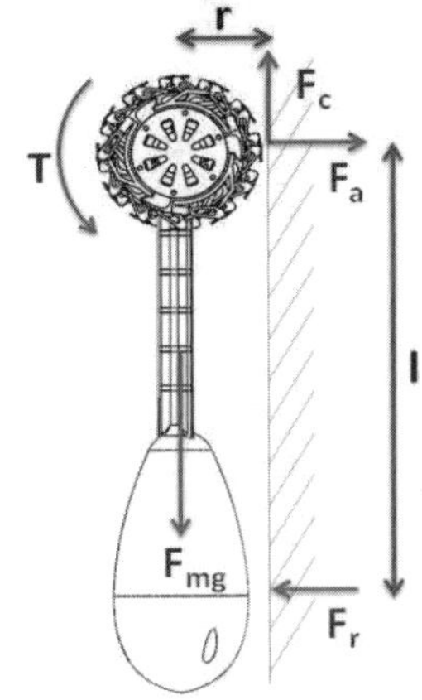

Figure 5. Tail length (l_t) presents an important design tradeoff between the induced pitchback-moment, maneuverability, and ability to transition.

2.3 *Controls*

Two 10 mm brushed DC motors provide approximately 0.2 Nm of torque, sufficient for the robot to climb at a weight of up to 400 g (the 300 g robot plus up to 100 g of payload). An ATmega328 microcontroller is used to control the motion of the robot, making it possible to accurately command the rotation of the rotary microspines depending on remote user or sensor input. A hybrid open-loop control architecture was used to create one set of gains at low throttle for controlled climbing, and another set at high throttle for high speed ground travel.

A 7.4 V, 180 mAh LiPo battery pack provides DROP with approximately 20 minutes of mission life; however, DROP can accommodate up to a 720 mAh battery pack to provide more than 1 hour of operation. This extended mission life is at the expense of increased weight and reduced climbing speed. The current prototype uses a 2.4 GHz off the shelf nanny-cam as the audio/visual payload. While this proof of concept payload successfully streams information to the operator, it is not intended to represent the state of the art in camera or microphone technology.

3. Results

The current DROP prototype has the ability to travel at a ground speed of 45 cm/s, climb concrete faces up to 85° at 25 cm/s, and survive impacts from 3 meters (without camera attached). Unlike most other microspine climbing robots, DROP also has the ability to transition from horizontal to 90° vertical surfaces with ease [Fig 6]. These horizontal to vertical surface transitions were possible at a variety of speeds, including full throttle, with a high degree of consistency. At this time, however, travel on fully vertical, 90° surfaces is sporadic and is not accomplished with a high degree of reliability. Transitions from vertical to horizontal (i.e. onto a rooftop) have been limited by the torque of the motors. The platform is highly maneuverable, with turn-in-place capability and near-vertical climbing at headings up to 20° from center. Impact testing, performed by driving the robot off rooftops of various heights, resulted in consistent failure of the hub connecting the microspine mechanism to the body at heights greater than 3 meters. During testing the microspine mechanism did not fail structurally and the elastomeric flexures prevented the microspine hooks from undergoing plastic deformation. Improved hub designs are expected to lead to increased durability of the robot. A video of DROP's overall performance has been accepted for publication and can be found here [16].

Figure 6. DROP uses the driving wheels to transition from flat ground to vertical walls. The platform was also able to transition partially onto rooftops, but in its current form, lacked the torque required to pull the tail over the lip. A subsequent prototype is currently being developed with improved torque capability and new iterations of the rotary microspine design.

4. Discussion

This paper introduced the first prototype of DROP, a small two-wheeled reconnaissance robot designed for fast traverse of both horizontal and vertical rough surfaces, and the corresponding transitions. DROP has several potential applications including use as a remote presence for disaster relief efforts, in combat scenarios for the military, and for commercial inspection of concrete structures like dams and buildings. The ability to be transported easily and deployed rapidly makes this robot ideal for many situations where time and space are at a premium. As impact resistance is improved, the potential to deploy DROP from UAVs opens the possibility for DROP to be deployed remotely and perform extended perch and stare missions.

The rotary microspine mechanisms used on DROP extend the state of the art in climbing robots by demonstrating rapid transitions from horizontal to vertical surfaces and improving climbing speeds on rough near-vertical inclines (85°). Achieving near vertical climbing using a wheeled design has been accomplished in other robots (magnetic, adhesive), however, none of the robots using these mechanisms are capable of climbing concrete, brick, or stone surfaces. When compared to existing microspine climbing platforms operating on fully vertical surfaces, DROP's climb speed on near-vertical surfaces is more than 5x that of RiSE (4 cm/s) [8] and more than 10x that of Spinybot II (2.3 cm/s) [9]. With continued development, DROP is expected to retain this rate of climb on fully vertical surfaces in future iterations.

The combination of size and durability exhibited by DROP fills an underserved niche in reconnaissance robots. This initial prototype displays the functionality of a rotary microspine climbing mechanism for real-world applications. A revised prototype is in development concentrated on achieving a 90° vertical climbing ability. Ensuing iterations of the platform will extend this functionality by exploring additional rotary microspine concepts, improving feedback control for more stable climbing, developing extended durability through design and material selection, full integration of the audio/video system, testing of UAV deployment, and further miniaturization of the platform.

Acknowledgments

The research was carried out at the Jet Propulsion Laboratory, California Institute of Technology, under a contract with the National Aeronautics and Space Administration.

References

1. http://www.qinetiq-na.com/products-talon.html (2011)
2. http://www.irobot.com/gi/ground/510 PackBot/ (2011)
3. P. Birkmeyer, K. Peterson, and R. Fearing, "Dash: A dynamic 16g hexapedal robot," *IEEE International Conference on Intelligent Robots and Systems* (2009)
4. M. Murphy, W. Tso, M. Tanzini, and M. Sitti, "Waalbot: An agile small-scale wall climbing robot utilizing pressure sensitive adhesives," *IEEE International Conference on Intelligent Robots and Systems* (2006)
5. N. Kohut, A. Hoover, K. Ma, S. Baek, and R. Fearing, "Medic: A legged millirobot utilizing novel obstacle traversal," *IEEE ICRA* (2011)

6. A. Asbeck, S. Kim, A. McClung, A. Parness, and M. Cutkosky, "Climbing walls with microspines," *IEEE ICRA* (2006)
7. A. T. Asbeck, S. Kim, M. R. Cutkosky, W. R. Provancher, and M. Lanzetta, "Scaling hard vertical surfaces with compliant microspine arrays," *Int J Robot Res*, vol. 25, no. 12, pp. 1165–1179 (2006)
8. M. Spenko, G. Haynes, J. Saunders, M. Cutkosky, and A. Rizzi, "Biologically inspired climbing with a hexapedal robot," *JOURNAL OF FIELD ROBOTICS* (2008)
9. S. Kim, A. Asbeck, M. Cutkosky, and W. Provancher, "Spinybotii: climbing hard walls with compliant microspines," Advanced Robotics, 2005. *ICAR '05. Proceedings., 12th International Conference on*, pp. 601 – 606 (2005)
10. A. L. Desbiens, A. Asbeck, and M. Cutkosky, "Landing, perching and taking off from vertical surfaces," *International Journal of Robotics Research* (2011)
11. A. L. Desbiens, A. Asbeck, and M. Cutkosky, "Hybrid aerial and scansorial robotics," *IEEE ICRA*, (2010)
12. A. Parness and D. Channel, "Gecko super suit," *Television Episode: Prototype This!* episode 11 (2009)
13. http://www.reconrobotics.com/products/recon-scout_XT.cfm (2011)
14. A. Saunders, D. Goldman, R. Full, and M. Buehler, "The rise climbing robot: body and leg design," *SPIE*, vol. 6230 (2006)
15. W. Provancher, J. Clark, W. Geisler, and M. Cutkosky, "Towards penetration-based clawed climbing," *CLAWAR* (2004)
16. C. McKenzie and A. Parness. "Video presentation of DROP: The durable reconnaissance and observation platform." *IEEE ICRA* (2012)

DEVELOPMENT OF A PERISTALTIC CRAWLING ROBOT WITH BELLOWS-TYPE ARTIFICIAL RUBBER MUSCLES FOR SEWER PIPE INSPECTION

TAKAICHI YANAGIDA, TARO NAKAMURA, MASATO YOKOJIMA AND KAZUNORI ADACHI†

Faculty of Science and Engineering, Department of Precision Mechanics
Chuo University, 1-13-27 Kasuga, Bunkyo-ku, Tokyo 112-8551, Japan

In this study, we evaluate the traveling performance of a robot in sewer pipes that imitates the peristaltic crawling locomotion of an earthworm. This robot can pass through complex networks of pipes and pipes coated with lubricating liquids. The actuator of the robot is composed of bellows-type artificial rubber muscles. The robot performs earthworm-like locomotion by supplying air pressure to the artificial muscles. We believe that the developed robot can be used to inspect pipe lines that could not be inspected previously.

1. Introduction

Regular inspection of sewer pipes is essential because serious accidents could occur if a sewage system breaks down. For example, sagging of roads or sewage water overflow in the event of bottlenecks of oil or other extraneous substances.

At present, pressure pipes are mostly employed as sewer pipes. There are two primary reasons for this. One of them even if several pipes are plumbed, it is possible to plumb the pipes freely. Second, they are highly durable and inexpensive and have a short installation time. However, there are problems associated with the inspection of pressure pipes. Large-diameter pipes can be inspected; however, entire lengths of pipes with diameters less than 150 mm cannot be inspected. For example, wheel-type robots [1]–[3] have been developed but they are not easily miniaturized because of their complex structures. The robots currently employed for this task are required to move inside sewage systems containing both bent and vertical pipes. In addition, because of the draining of domestic wastewater, the pipes have slimy surfaces that promote the growth of bacteria. However, there currently exist no methods that can be used to inspect all types of sewer pipes.

In this study, we develop a peristaltic crawling robot using bellows-type artificial rubber muscles in order to enable the inspection of previously unexplored pipes. In addition, we conduct driving tests in a bent pipe, a vertical pipe, and a complex pipeline.

2. Peristaltic Crawling Pattern of an Earthworm

Peristaltic crawling is the way that insects such as an earthworm and inchworm etc. move forward. Our robot imitates the earthworm.

The earthworm consists on numerous segments (about 110-200). Figure 1-(a) shows that a structure of the segment of the earthworm. The earthworm has two layers of muscle; the circular muscle that is arranged inside of a skin circumferentially, and longitudinal muscle that is arranged inside of the circular muscle in axial direction. The former elongates and the latter contracts the segments in the axial direction [4]. Figure 1-(b) shows the locomotion pattern of an earthworm by peristaltic crawling. The earthworm contracts and expands as following patterns:

(1)The earthworm contracts an anterior segment by longitudinal muscles.

(2)This contraction is transmitted to a rear segments and the anterior segments are elongated by the circular muscles simultaneously.

(3)The contraction segments generate the friction force between the segments and a surface. Because of friction force, reaction force to elongate contraction segments is generated.

Thus, the earthworm can move because of the repetition of "thick-short" and "thin-long".

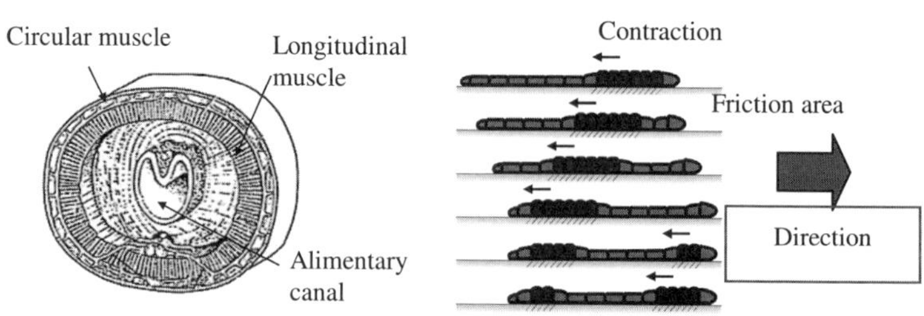

(a) Inner structure

(b) Peristaltic crawling

Figure 1. Earthworm locomotion.

3. Bellows-Type Artificial Rubber Muscles

In the developed robot, both the segments and actuator are represented by bellows-type artificial muscle. Figure 2 shows the mechanism of the artificial muscle. Figure 3 shows the section of the artificial muscle. The muscle is composed of low-ammonia natural rubber latex. This material has advantages of lightweight, low cost, and flexibility. It is molded into a tube, and a micro carbon sheet is inserted in the long axial direction. The range of this sheet can be extended by elongating the rubber [5].

The actuator is controlled by air pressure. When air is supplied to the actuator, the artificial muscle contracts in the axial direction and expands in the radial direction. The reason is because the rubber expands although the carbon sheet does not expand.

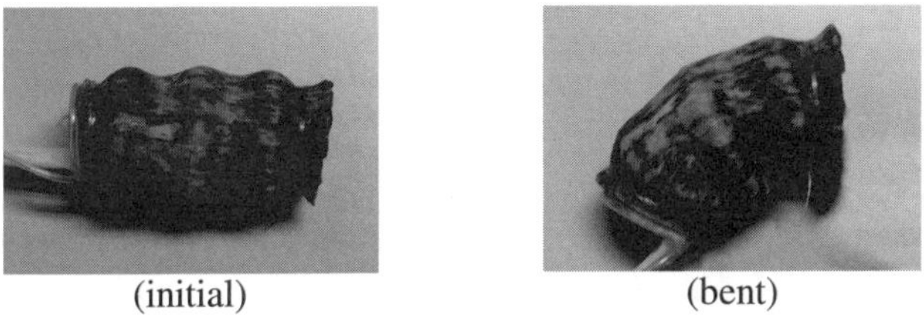

Figure 2. Bellows-type artificial rubber muscle

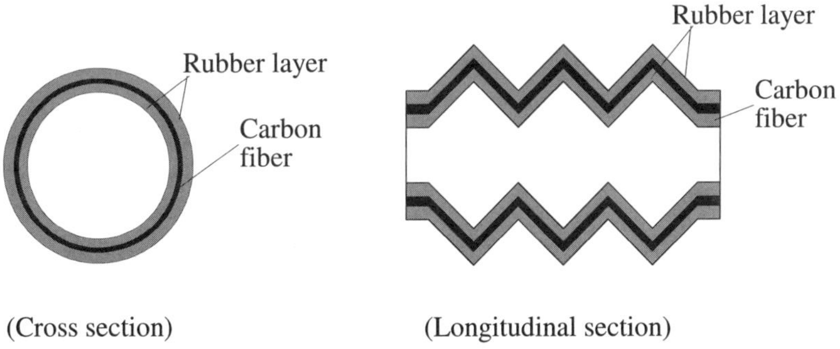

Figure 3. Sections of the bellows-type artificial rubber muscle.

4. Peristaltic crawling robot

Figure 4 demonstrates the mechanism of the robot. The robot's specifications are shown in Table 1. The robot consists of four units and three connects that have a rubber tube over bellows. The units have one artificial muscle over each bellows. Air tubes are contained in the bellows. When air is supplied to the chamber between the artificial rubber muscle and bellows, flanges slide on the air tube along the guide in the flange. Only a single point on the air tube is fixed in the flange. Thus the units are independently controllable.

Figure 4. Peristaltic crawling robot using bellows-type artificial rubber muscles.

Table 1. Specifications of the robot.

	Extension	Contraction
Robot length [mm]	270	238
Unit length [mm]	49	41
Outer diameter [mm]	30	51
Connect length [mm]	15	

5. Experimental results and discussion

With four segments, this robot can have the two movement patterns shown in Figure 5. Pattern 2-1-1 (contracted units" → "transmitted waves in the contracted units" → "number of waves") imitates the motion of an earthworm. Pattern 7SM represents a seven-step motion.

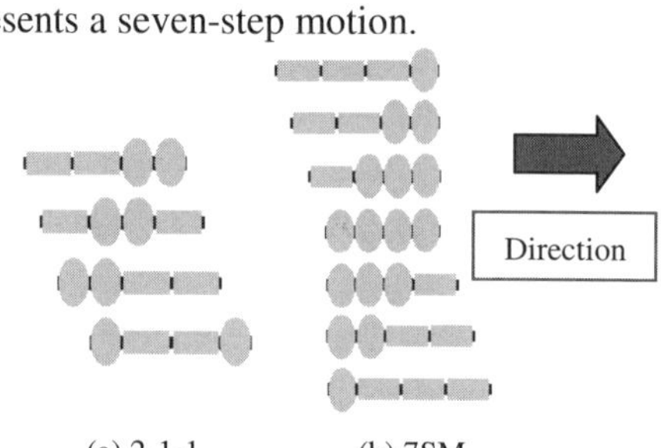

Figure 5. Movement patterns

5.1. *Driving test in a straight pipe*

We first performed an experiment in a straight pipe using the movement patterns. In addition, to examine the effect of slimy surfaces on the robot's motion, we constructed a pipe coated with a lubricant. The viscosity of the lubricant was 10000–15000 mPa·s. Moreover the inner diameter of the pipe was 50 mm. Figure 6 shows the speed of the robot for each movement pattern.

We then computed the average speeds as follows. For the 2-1-1 pattern, the speed was 3.87 mm/s in the pipe coated with a lubricant and 3.25 mm/s in the pipe with no lubricant. For the 7SM pattern, the speed was 4.47 mm/s in the pipe coated with a lubricant and 3.56 mm/s in the pipe with no lubricant. These results indicate that with the two patterns, the speed of the robot is greater in the pipe coated with a lubricant.

We believe that the reason for this is that friction between the robot and pipe wall is reduced. The friction occurs in order to allow sagging under the robot's own weight when the robot extends. In addition, the friction interrupts the robot's motion when the robot is extended. This friction is reduced with the use of a lubricant, and the speed is therefore faster for the pattern in the pipe coated with a lubricant.

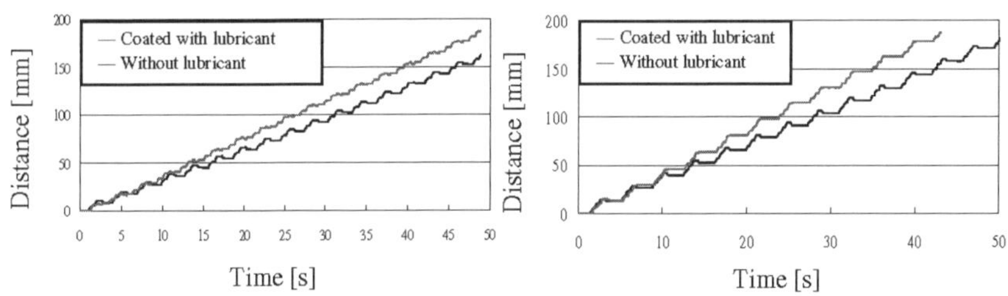

Figure 6. Speed of the robot in a straight pipe

5.2. *Driving test in an actual bent pipe*

We performed an experiment in a bent pipe with the movement patterns. In addition, to examine the effect of slimy surfaces on the robot's motion, we constructed a pipe coated with the lubricant. The interior diameter of the pipes was 50 mm, and the outer curvature radius of the bent pipe was 90 mm. The bent pipe is shown in Figure 7. Figure 8 shows the process of the experiment while Figure 9 shows the speed of the robot for each movement pattern. The arrow indicates the position of the tip of the robot.

We then computed the average speeds as follows. For the 2-1-1 pattern, the speed was 2.58 mm/s in the pipe coated with a lubricant and 2.71 mm/s in the pipe with no lubricant. For the 7SM pattern, the speed was 3.90 mm/s in the pipe coated with a lubricant and 3.00 mm/s in the pipe with no lubricant. These results indicate that the speed of the robot in an actual bent pipe is slower than that in a straight pipe.

We believe that the speed of the robot in an actual bent pipe is reduced in order to increase friction in the bent pipe because the robot straightens itself in the bent pipe. Thus, the friction increases and the speed decreases.

In addition, for the 7SM pattern, the speed of the robot is faster in the pipe coated with a lubricant. We believe that the speed after entering the bent pipe are factors because the speed is considerably more with the 7SM pattern in the straight pipe coated with a lubricant.

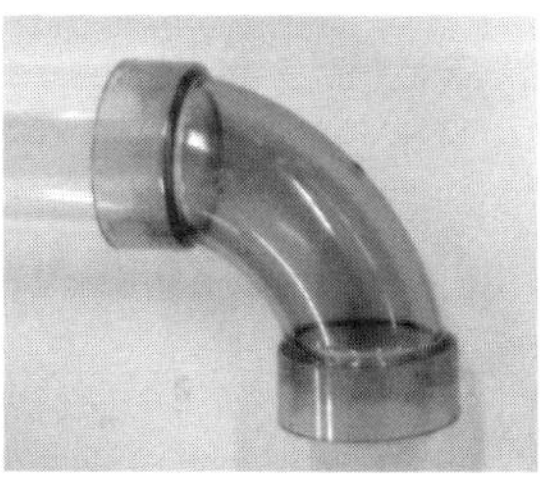

Figure 7. A bent pipe

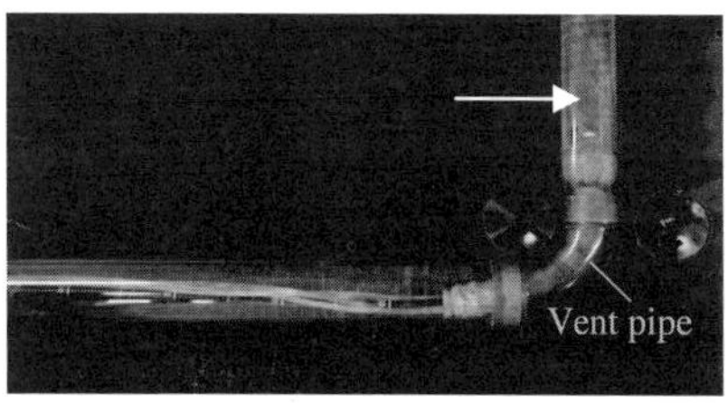

Figure 8. Experiment conducted in a bent pipe.

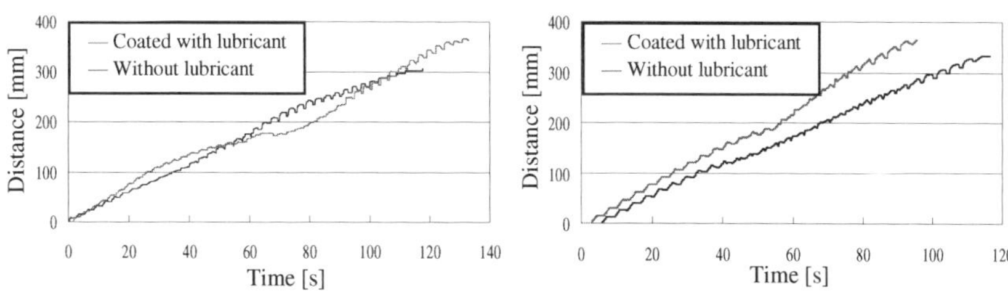

(a) 2-1-1

(b) 7SM

Figure 9. Speed of the robot in a bent pipe

5.3. *Driving test in a vertical pipe*

We performed an experiment where the robot moved through a vertical pipe using movement pattern 2-1-1. In addition, to examine the effect of slimy surfaces on the robot's motion, we constructed a pipe coated with the lubricant. Figure 10 shows the speed of the robot with movement pattern 2-1-1. The average speeds were computed as 1.94 mm/s in the pipe coated with a lubricant and 2.30 mm/s in the pipe with no lubricant.

From these results, the speed of the robot is seen to be faster in the pipe without a lubricant. We believe that in this case, it is because of slippage due to the lubricant. The robot's outer diameter was 51 mm and the inner diameter of the pipe was 50 mm. It is believed that the robot did not achieve a sufficient gripping force because the outer diameter of the robot was insufficient for the slippery surface. To solve this problem, we believe that the robot's outer diameter should be made larger.

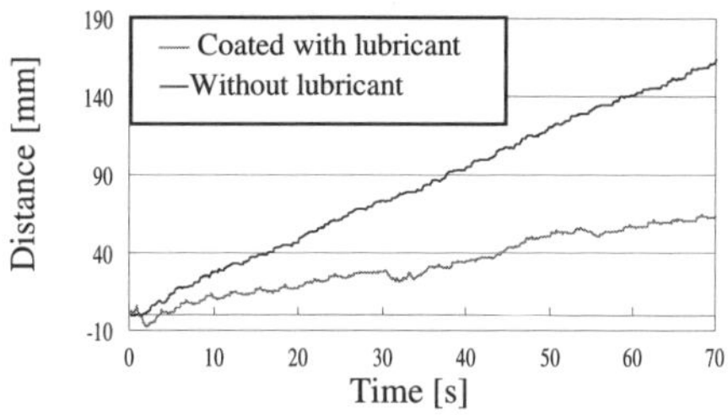

Figure 10. Speed of the robot in a vertical pipe

5.4. *Driving test in a complex pipeline*

Next, we performed an experiment where the robot moved through a complex pipeline coated with the lubricant using movement pattern 2-1-1. Figure 11 shows the apparatus for the experiment in the complex pipeline. Images of this experiment were taken every 60 s and are shown in Fig. 12. The arrow indicates the position of the tip of the robot. In this experiment, to prevent the robot from slipping in the pipe line, the robot was fitted with materials that generate friction.

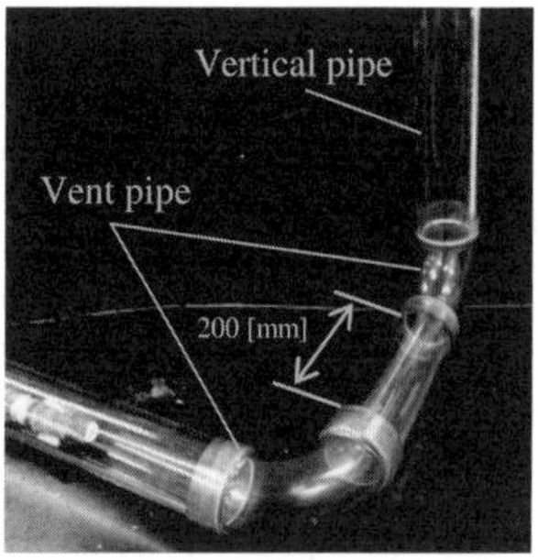

Figure 11. A complex pipeline

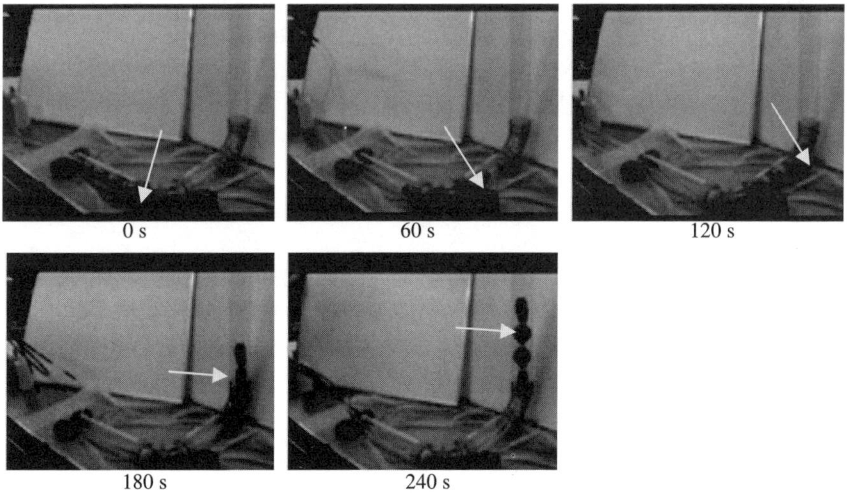

Figure 12. Experiment of the robot traveling a complex pipeline

We verified that the robot can pass through the entire length of the complex pipe line. Thus we think the robot have availability.

6. Conclusion

A peristaltic crawling robot was developed for the inspection of sewer pipes. We performed an experiment with a pipe, which has an inner diameter of 50 mm. In addition, we examined the effect of slimy surfaces on the robot's motion. Our results were as follows.

(1) We confirmed the robot's capability in a straight pipe using the movement patterns.
(2) The robot can pass through both bent and vertical pipes.
(3) The robot can pass through the entire length of a complex pipeline.

From these results, we believe that this robot is suitable for the inspection of sewer pipes.

In future, for practical realization, we will perform experiments with water flowing through the pipes. Moreover, we will model the movement of the robot and analyze its theoretical speed.

References

1. Peng Li, Shugen Ma, Bin Li, and Yuechao Wang, "Development of an Adaptive Mobile Robot for In-pipe Inspection Task," Proc. IEEE International Conference on Mechatronics and Automation, pp. 3622-3627, 2007
2. Tokuji Okada and Tsuyoshi Sanemori, "MOGER: A Vehicle Study and Realization for In-pipe Inspection Tasks," IEEE JOURNAL OF ROBOTICS AND AUTOMATION, VOL. RA-3, NO. 6. DECEMBER 1987
3. Amir H. Heidari, M. Mehrandezh, R. Paranjape and H. Najjaran, "Dynamic Analysis and Human Analogous Control of a Pipe Crawling Robot," Proc. IEEE/RSJ International Conference on Intelligent Robots and Systems, pp. 733-740, 2009
4. H. Sugi, "Evolusion of muscle motion," The University of Tokyo Press, p. 72, 1977 (in Japanese).
5. T. Nakamura, N. Saga, and K. Yaegashi, "Development of Pneumatic Artificial Muscle Based on Biomechanical Characteristics," Proceedings of IEEE International Conference on Industrial Technology (ICIT 2003), pp. 729-734.

DESIGN OF MULTI-TERRAIN CLIMBING ROBOT FOR PETROCHEMICAL APPLICATIONS

BRAISA KAMAGALUH, JOHN SUJITH KUMAR
Department of Electrical Engineering, South Westphalia University of Applied Sciences, Luebecker Ring 2, Soest, 59494, Germany

GURVINDER S. VIRK
Faculty of Engineering and Sustainable Development, University of Gävle, SE-801 76 Gävle, Sweden

A multi-terrain climbing robot provides more flexibility while doing inspection in scenarios where the environments comprise several types of landscape such as walls, steps, ceiling, and sloped ferromagnetic surfaces. This paper presents the development of a walking and climbing robot named Winspecbot which uses legs with magnetic padded feet to climb on steel pipes and is planned to have a hand gripper to perform inspection tasks. The robot is designed to carry about 1 kg inspection tools while it is climbing. With the use of Wi-Fi communications, the user is able to monitor the operational conditions of the robot when it is working in complex environments.

1. Introduction

The current state of the art in climbing robot has shown that the climbing robot can be used in applications such as window cleaning [1], petrochemical tank inspection [2], remote maintenance [3], and search and rescue applications [4]. Attachment methods also have been developed using various methods: Yukawa *et al.* [5] have developed a wall climbing robot using permanent magnets on the robot's wheels; H. Kim *et al.* [6] have developed a caterpillar robot using vacuum suction cups; S. Kim *et al.* [7] have developed a gecko-like robot which uses dry adhesive/Van der Wall's forces on the robot's legs; Inoue *et al.* [8] have developed an omni-directional hexapod robot which uses a special mechanical gait to hang on a grid structure; and Prahlad *et al.* [9] have developed a climbing robot which uses electro-adhesive forces. All these attachment methods have their own advantages and disadvantages. For industrial inspection, especially on steel pipes or vessel inspection tasks, robots with magnetic pads offer one of the most feasible and practical attachment methods. Therefore, in this paper, the development of a climbing robot with magnetic pads for steel pipe inspection is adopted; the robot is called Winspecbot

(Westphalia Inspection Robot). The mechanical design, control schematic, and the robot locomotion are presented here and form the main part of this paper. Several tests related to the performance of the climbing robot are also presented to provide a demonstration of the proof of concept of the overall design.

2. Mechanical design

2.1. *Body and legs design*

The mechanical design of the robot's body and legs is based on a 4-legged configuration where each leg has 4 degrees of freedom (dof) and a magnetic pad attached to the end effector of each leg. The robot's body consists of two layers where the bottom layer is designed to accommodate the hip of each leg and heavy items such as the batteries and the top layer is designed as a base for the microcontroller and the sensors. The motor on each joint is selected to achieve sufficient force so that the robot can climb with a maximum weight of about 5.4 kg. This maximum weight is based on the mass distribution of the elements presented in Table 1.

Table 1: Mass design calculation of Winspecbot.

Parts	Consisting of	Unit (gr)	Qty	Total (gr)
Leg	Servos, brackets, magnetic pad	760	4	2,800
Bottom Body	2 x Battery 6v, 2 x Aluminum plate	830	1	830
Top Body	Sensors, Microcontroller boards, Aluminum cap	570	1	500
	Spare for inspection tools			1,000
			Total :	5,370

To calculate the servo toque on each joint, we consider several manoeuvres such as the following:

- walking on non-ferromagnetic flat ground
- for ferromagnetic surfaces: climbing on sloped and vertical surfaces as well as moving upside down on ceilings.

The calculations are based on the trotting gait locomotion which represents the worse-case scenario for the robot requiring maximum torques. For this, two legs support the robot and the other legs are free to move in air. However, to reduce the complexity of the calculations, the following assumptions are made:

- The friction force on the bearings is ignored.
- The cross product between F and L is taken at its worse-case position, that is F and r are perpendicular ($\sin \theta x = 1$). Each leg is independent and works on its load as shown in Figures 1(a) and (b). The torques at the higher and

lower knees, and ankle (τ_A, τ_B, τ_C) are calculated when the robot is walking on a horizontal flat steel surface. The torque at the hip (τ_O) is calculated when the robot is climbing on a vertical flat steel surface.

- The friction force between the foot and the travel surface is strong enough to avoid the pad slipping from its standing position.
- The ankle is controlled to be perpendicular to the travel surface. The calculation of torque when the leg is stretching forward or upward is not considered because the robot locomotion does not move in this manner.

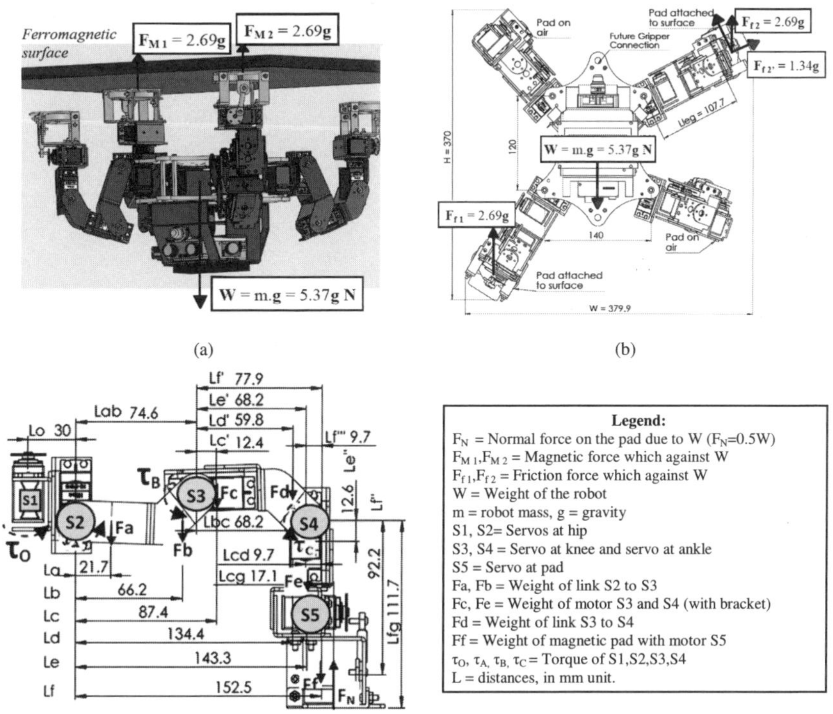

Figure 1: Torque analysis for worst case conditions (a) Upside down walking on ceilings, (b) Straight up/down climbing on walls, (c) One leg analysis when the leg is stretched to maximum, (d) Legend: All F are in Newton, g = gravity, L are in mm.

Using standard static and dynamic analysis, the required torques can be calculated and the results are shown in Table 2. We calculate the torque requirement to move a leg dynamically and to hold the load statically when the leg does not move. The servos installed on the robot are chosen to have ≈20% more torque than that is required.

Table 2: Torque calculations for the Winspecbot climbing robot.

Torque		Equation	Result (kg.cm)	Min. Servo Torque (kg.cm)
τ_A	Dynamic :	Fa.La + Fb.Lb + Fc.Lc + Fd.Ld + Fe.Le + Ff.Lf	6.98	**15.94**
	Static :	-Fa.La -Fb.Lb + (FN-Fc-Fd-Fe-Ff).Lab	15.94	
τ_B	Dynamic :	Fc.Lc' + Fd.Ld'+ Fe.Le' + Ff.Lf'	2.94	**14.44**
	Static :	-Fc.Lc'-Fd.Ld' +(FN-Fe-Ff).Lbc	14.44	
τ_C	Dynamic :	Fe.Le'' + Ff.Lf''	2.20	**15.01**
	Static :	-Fe.Le'''-Ff.Lf'''+FN.Lcd (horizontal walk)	0.28	
		or $F_{f2'}$.Lfg (vertical climb)	15.01	
τ_O	Dynamic :	Fa.(La+Lo) + Fb.(Lb+Lo) + Fc.(Lc+Lo) + Fd.(Ld+Lo) + Fe.(Le+Lo) + Ff.(Lf+Lo) (vertical climb)	8.72	**23.68**
	Static :	F_{f2}.Lleg - Fa.(La+Lo) - Fb.(Lb+Lo) - Fc.(Lc+Lo) - Fd.(Ld+Lo) - Fe.(Le+Lo) - Ff.(Lf+Lo) (vertical climb)	23.68	

2.2. *Design of magnetic feet*

The magnetic pads installed on Winspecbot are based on permanent magnets which can attach/detach to/from steel surfaces according to the locomotion mode of the robot. Permanent magnets have been chosen because they provide reliable steel attachment and do not need high current to hold such a heavy robot. Figure 2 shows the construction of the magnetic foot module which has two magnets made from nickel-plated neodymium.

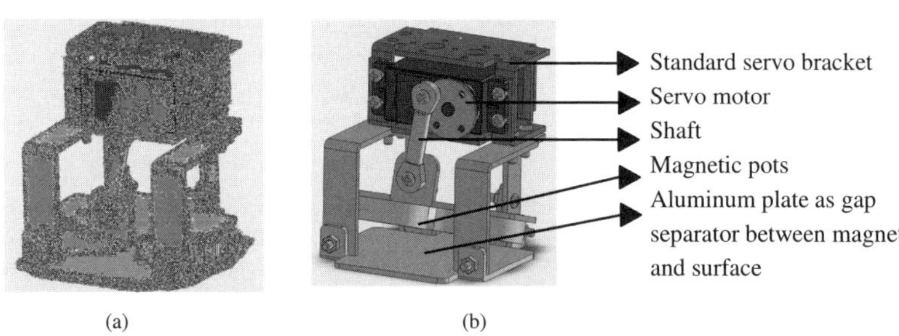

Figure 2: Design of magnetic foot module: (a) attaching position and (b) detaching position.

The diameter of the pot magnet is 32 mm and has a holding strength of 34 kg making then quite strong but the magnetic force will reduce when an air gap is introduced. In order to get a suitable attachment force via the magnetic feet, spacers have been added to adjust the gap between the magnets and the travel surface so that the actual magnetic force is sufficient to allow the robot to climb on the steel surface as required. With this configuration, high torque servos are not needed to do the detachment process but the robot still fulfils the requirement of providing 2.69 kg magnetic force as discussed in Figure 1(a).

3. Control system schematic

The control system used by Winspecbot is based on a centralized microcontroller model where the user has access from a PC host to control the movements of the robot and to monitor the operational environment of the robot. The control system diagram is shown in Figure 3. The central microcontroller is an AVR AT-Mega 2560 with 16 MHz clock speed. By using a polling method and sensor scheduling, the microcontroller manages all the data going in and out including controlling the 20 servos on the robot's legs. A Wi-Fi network is utilized to communicate with the robot and to stream video from an IP camera attached on the robot.

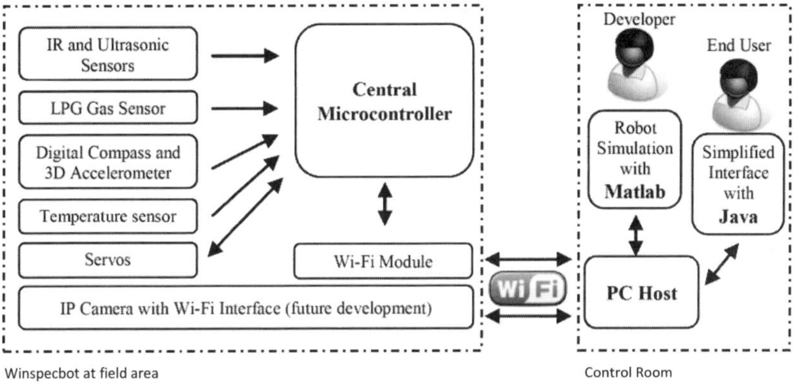

Figure 3 : Control system schematic of the Winspecbot robot

The system interface is developed using Matlab (for control strategy development) and Java (for the user interface). We develop the robot locomotion and simulate the robot model in Matlab as shown in Figure 4(a). This process is useful to know the robot behaviour when trajectory-following control strategies using inverse kinematics are implemented. Sometimes the inverse kinematic calculations yield unexpected results due to mechanical limitations of the servos. The robot user needs only to work on high level robot control tasks such as starting and stopping the robot; moreover he can focus on the monitoring aspects of the inspection application.

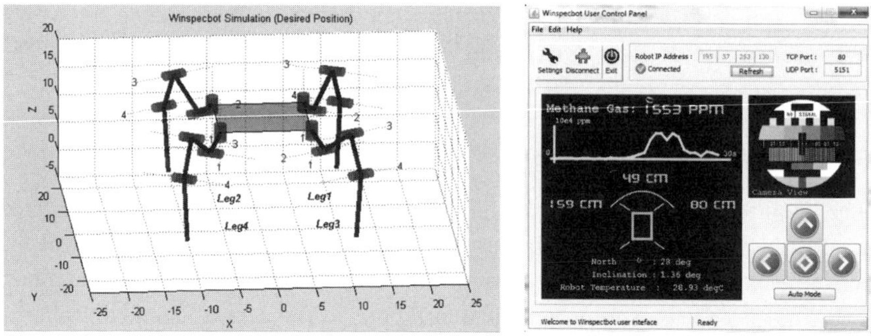

(a) (b)

Figure 4: (a) Matlab based user interface for simulating and designing advanced controllers and (b) Java based user interface.

4. Robot locomotion strategy

The robot locomotion is developed based on developing a suitable locomotion gait. Several gaits have been installed from which the user can select. The robot is also designed to accept new patterns from the user while it is operating. A specific communication protocol for this online control has been developed on the robot so that it is possible that all the servos can be controlled directly from Matlab in real-time.

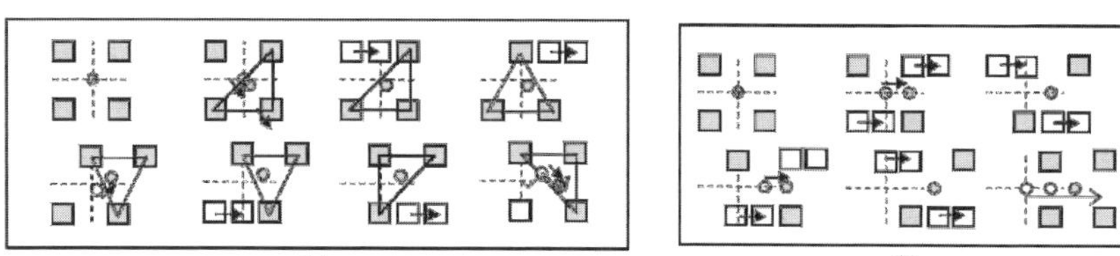

(a) (b)

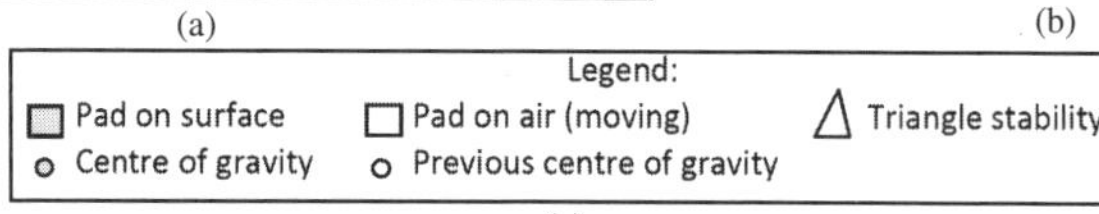

(c)

Figure 5: Winspecbot locomotion strategy: (a) Walk gait locomotion mode (b) Trot gait locomotion mode and (c) legend.

The robot movement depends on the surface on which the robot is travelling. On a smooth and non-ferromagnetic surface, Winspecbot will use its walking gait mode: three legs support the robot and one leg is free to perform its transfer phase (see Figure 5(a)). The three legs shall be placed so that the centre of gravity is inside the triangle-shape formed by the ground support area; the body is also needed to be moved during parts of the walking cycle to maintain the stability. On ferromagnetic surfaces, this triangle support area stability is not required as the magnetic forces can maintain sufficient stability. The robot can also use its trotting gait mode where 2 legs support the robot and the other two legs are free to move (see Figure 5(b)).

(a)

(b)

(c)

(d)

Figure 6: Winspecbot in several positions: (a) Walking gait, (b) Trotting gait (c) Two legs hanging on steel surface, and (d) Four legs hanging on a pipe surface.

5. Winspecbot Performance Evaluation

In this section, we present a short performance review on Winspecbot. We found that the concept of using 4 dofs in each leg to be sufficient for supporting the robot in its various locomotion modes. Three dofs are needed for the climbing: to locate the leg on the planned x, y, and z position and the last dof is to make the magnetic foot align to be parallel with the travel surface. The servo selection is critical to making an effective 4 dofs legged climbing robot. If the installed servos have insufficient torques then the robot will not be able climb; on the other hand if they are too high then the robot will be expensive, heavy, noisy and consume a lot of power. The Winspecbot robot consumes around 570 mA for the controller and 3.25 A for the servos when the robot in the standby position. The current for servos will increase up to 8.7 A when the robot is walking. We have implemented the walking gait and trotting gait locomotion modes. Both are providing a good stability for Winspecbot and pictures of Winspecbot in several positions can be seen in Figure 6. The actual tests for locomotion in ferromagnetic surfaces are still under development but we provide the feasibility of these tests by showing the attachment capability as shown in Figures 6 (c) and (d).

6. Conclusions

This paper has presented a systematic design and implementation of a 4-legged climbing robot named Winspecbot. Early results have shown that by using the proposed robot design and its magnetic feet modules, the robot can climb with sufficient force on vertical ferrous walls as well as being able to climb upside down on ceilings. Although the robot is successfully able to perform according to the design, the power consumed while climbing is high. Therefore the plans are to improve the energy efficiency for the climbing by improving its locomotion strategy using inverse kinematics. Future research on this robot will involve detailed modelling and simulation of the Winspecbot via Matlab so that more efficient joint movements may be performed to optimize the energy consumption in performing its various locomotion modes.

References

[1] Sun, Dong, Zhu, Jian and Tso, Shiu Kit., *A Climbing Robot for Cleaning Glass Surface with Motion Planning and Visual Sensing.* Vienna : s.n., 2007.

[2] Kalra, L.P., Guf, J. and Meng, Max., "A Wall Climbing Robot for Oil Tank Inspection." Kunming : IEEE/ROBIO, 2006. pp. 1523 - 1528.

[3] Saltarén, R., et al., "Parallel Climbing Robots for Construction, Inspection and Maintenance." Madrid : Automation and Robotics in Construction XVI, 1999.

[4] Eich, M., Grimminger, F. and Kirchner, F., "A Versatile Stair-Climbing Robot for Search and Rescue Applications." Sendai : s.n., 2008. IEEE SSRR. pp. 35 - 40.

[5] Yukawa, T., et al., "Design of Magnetic Wheels in Pipe Inspection Robot." Taipei : s.n., 2006. IEEE International Conference on Systems, Man, and Cybernetics.

[6] Kim, H., Kim, D. and Yang, H., "A Wall Climbing Robot with Vacuum Caterpillar Wheel System Operated by Mechanical Valve." Brussels : Proceedings of the 9th Int. Conf. on Climbing and Walking Robots, 2006. pp. 28-33.

[7] Kim, S., Spenko, M. and Trujillo, S., "Whole Body Adhesion: Hierarchical, Directional and Distributed Control of Adhesive Forces for a Climbing Robot." Rome : IEEE/ICRA, 2007. pp. 1268 - 1273.

[8] Inoue, K., et al., "Omni-directional Gait of Limb Mechanism Robot Hanging from Grid-like Structure." Beijing : IEEE/RSJ, 2006. pp. 1732-1737.

[9] Prahlad, H., et al., "Electroadhesive Robots - Wall Climbing Robots Enabled by a Novel, Robust, and Electrically Controllable Adhesion Technology." Pasadena : IEEE/ICRA, 2008. pp. 3028-3033.

MOBILITY PREDICTION OF ROVERS ON SOFT TERRAIN: EFFECTS OF WHEEL- AND TOOL-INDUCED TERRAIN DEFORMATIONS

DANIEL HOLZ[†], ALI AZIMI[‡], MAREK TEICHMANN[†] and JOZSEF KÖVECSES[‡]

[†]*CM-Labs Simulations Inc.*, [‡]*McGill University*

Planetary rovers play a key role in modern space exploration. Remotely controlled and operated on unknown terrain, these vehicles run the risk of being damaged, immobilized or simply used in inefficient ways. Their performance depends not only on their design and motion characteristics but also on the performed operation being adequate for the situation and environment they are momentarily in. Consequently, accurate prediction of a specific rover's performance in different situations and environments is a pre-requisite for their efficient and successful deployment. In this work, we combine physical models from terramechanics and soil mechanics to model a rover in highly complex operations, such as excavation tasks with a mounted bulldozer blade. We consider the impact of both wheel–ground and tool–ground interactions on the rover motion system by creating a fully coupled simulation of the dynamics of the rover and its environment. In simulations we show that the rover's mobility and its trajectory are impacted when modifying the way the rover is operated – a direct result of changes in rover–ground interaction.

1. Introduction

In the future of planetary exploration, reliable rover mobility prediction and accurate mission planning will become increasingly important. Planetary rovers, today a tool for remote assessment and investigation of unknown terrain, may in the future perform advanced support tasks such as assisting in the erection of outposts. Consequently, rover operations may likely become far more complex than they are today. However, today even simple operations hold great challenges for modern rovers. A rover, operated on unknown, unstructured terrain, e.g. being driven into highly soft ground, continuously creates changes to its immediate environment, which, if not reacted to appropriately, may drastically reduce its operation efficiency or ultimately even render the rover entirely immobile. Prediction of such events, like hardening and compaction of terrain, slip-sinkage of wheels, and tool–ground interaction, and their imminent impacts on rover mobility, can help in both rover design and mission planning. Consequently, the problem of rover mobility prediction is directly related to

- the dynamics of the rover motion system
- stress distributions in wheel–ground contact area
- traction and rolling resistance on hard and soft ground
- wheel-induced ground compaction and hardening, which give rise to the multi-pass effect
- horizontal ground deformations and the bulldozing effect

In order to have a representative rover simulation tool, as many aspects of the above as possible should be considered. This is especially important when trying to capture the dynamics of certain advanced, yet plausible rover operations. Such operations are for example a rover driving on partially soft and partially hard ground or a rover employed for excavation operations. This also includes failures, such as a defective wheel which has to be dragged along for the rover to continue its mission. In order to assess the usability and effectiveness of a specific rover design in one of the above situations, the interplay of all related effects and problems have to be considered at the same time in simulation. Here, we propose a simulation approach which considers all the above listed phenomena, and which is hence able to successfully represent complex rover operation scenarios. We model wheel–ground as well as tool–ground interactions and show that, due to the consideration of both, plausible rover mobility estimations can be achieved for rovers pushing soil.

In the following sections we outline the proposed method and provide details on wheel–ground and tool–ground interaction modeling. We present and discuss simulation results and close with a conclusion.

2. Method

In this work we combine physics models from both terramechanics and soil mechanics with a fully simulated and articulated rover chassis. Terrain deformations caused by wheel–ground and tool–ground interactions are persisted throughout the simulation. This is achieved through the usage of a heightfield-based terrain representation which captures the terrain surface as well as its current state of compaction at any location. We create a fully coupled dynamics simulation of both the rover and its environment through the usage and extension of CMLabs' multi-body dynamics toolkit Vortex[a]. Here, we model the rover articulation system and its geometry using Vortex constraints and rigid bodies. Also attachments such as a mounted bulldozing blade are modeled as such. Terrain reaction forces due to wheel–ground and tool–ground interactions are applied to the rover articulation system by adding additional, dynamically created constraints to the system and parametrizing them at runtime. Details are provided in the following sections.

2.1. *Wheel–Ground Interaction*

For the modeling of wheel–ground interactions we generate contact constraints at relevant locations of the wheel and parametrize them according to a suitable interaction model. Two fundamentally different representations for wheel and soft ground interaction have been developed in Vortex. They are (*i*) semi-empirical terramechanics models, like the works of Bekker[1] and Wong,[2] with necessary modifications as discussed by Azimi et. al,[3] and (*ii*) a higher-fidelity detailed model developed recently by Azimi et al.[4] Here, we select an implementation of the semi-empirical Bekker model[2] presented by Azimi et. al.[5] In this implementation vertical terrain compaction and hardening effects are considered, which allows for the simulation of the multi-pass effect.

2.2. *Tool–Ground Interaction*

For tool–ground interaction, we base our work on the hybrid particle-/mesh-based soil simulation method presented by Holz et. al.[6] It combines a mesh-based representation of soil at equilibrium – the *soil grid* – with a particle simulation based on the Discrete Element Method (DEM). When horizontal motions of the tool in contact with the grid are detected, such as a blade carving through the ground, DEM particles are introduced adaptively, replacing portions of the soil grid in front of the tool. The particles are simulated in Vortex using spherical rigid bodies and hence apply reaction forces to the tool. However, the soil grid itself does not cause any reaction forces. Here, the approach from Holz et. al.[6] is augmented by considering these missing forces. In order to model appropriate cutting forces of a blade penetrating the soil grid, the Fundamental Earth Moving Equation (FEE) of Reece,[7] and the method of trial wedges presented by McKyes[8]

---

[a]http://www.vxsim.com/

are employed. The forces acting on the soil wedge are shown in Fig. 1. Similar to Singh et. al.[9] we obtain a tool–ground interaction formulation which accounts for the slope of the terrain. The cutting force F per tool width required by the tool to fail the soil is computed as

$$F = \gamma g d^2 N_\gamma + c d N_c + Q N_Q + c_a d N_a \tag{1}$$

where

$$N_\gamma = \frac{(\cot\rho + \cot\beta)\sin(\alpha+\phi+\beta)}{2\sin(\delta+\rho+\phi+\beta)}, \quad N_Q = \frac{\sin(\phi+\beta)}{\sin(\delta+\rho+\phi+\beta)} \tag{2a}$$

$$N_c = \frac{\cos\phi}{\sin\beta\sin(\delta+\rho+\phi+\beta)}, \quad N_a = \frac{-\cos(\rho+\phi+\beta)}{\sin\rho\sin(\delta+\rho+\phi+\beta)} \tag{2b}$$

with gravity g, soil slope inclination angle α, tool/soil angle ρ, tool penetration depth d, soil failure angle β, soil internal friction angle ϕ, soil cohesion c, specific weight of the soil γ, tool/soil friction angle δ, tool/soil adhesion c_a and surcharge force per tool width Q.

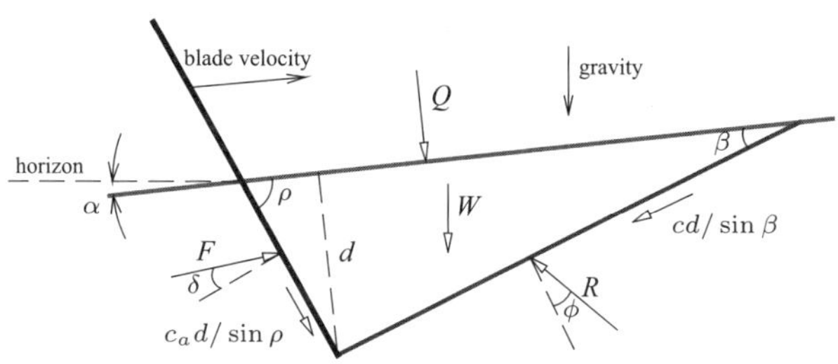

Fig. 1. Forces acting on the soil wedge.

As Holz et. al. we perform a detailed analysis of the contact region between tool and terrain. From this analysis we extract geometric information such as terrain surface inclination α, tool penetration depth d, tool width w, and tool/terrain surface angle ρ, all of which are relevant inputs to the FEE. In our case, the soil grid corresponds to a heightfield (HF). The positions of the HF vertices directly in front of the moving tool as well as the penetration points of the tool can, therefore, be easily identified as explained by Holz et. al.[6] In each simulation step, we compute the mean of these positions, p, and sample the linear velocity, v, of the tool at this point. The linear velocity vector projected on the horizontal $x-y$ plane of the HF, v_{xy}, serves as a look-ahead direction for the soil slope approximation. The line formed by the projected point p_{xy} and vector v_{xy} serves as

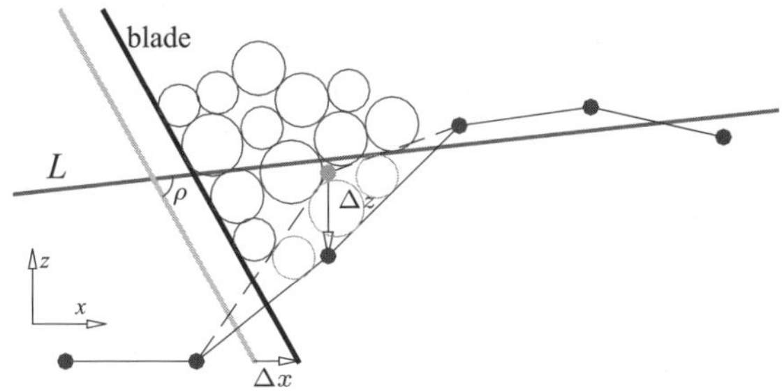

Fig. 2. Blade cutting through soil slope. The heightfield vertices (blue/green dots) serve as input for the soil slope approximation (L). The blade motion Δx causes heightfield deformation Δz and particle generation (green circles). Particles from previous simulation steps (disturbed soil) are depicted as blue circles.

input to Bresenham's line algorithm[10] used to identify a selected number of vertices ahead of the tool. As shown in Fig. 2, we compute the least-squares line of the corresponding vertex positions. The computed line, L, approximates the inclination angle of the terrain surface in front of the moving tool, α, which will be used as input to the FEE model. In our implementation we approximate the tool with a thin box geometry, representing a bulldozing blade. We can therefore readily compute the tool/terrain surface angle, ρ, by comparing the angle of L with the orientation of the box (cf. Fig. 2). The remaining geometric parameters such as tool width, w, and tool penetration depth, d, can be computed in a similar way. Finally, a 6-DOF constraint between tool and ground is added to the simulation and parametrized with the cutting force F obtained from Eq. (1) in order to model the soil resistance.

The combination of the FEE model with the DEM-based soil simulation provides a two-layered representation of tool–ground interaction. The undisturbed soil – the soil grid – forms the lower layer. Here, the soil reaction force computed by the FEE has to be overcome by the tool. Once this happens, a portion of the soil enters the disturbed state, represented by particles. While the tool continues to move forward, the previously generated particles accumulate in front of the tool, i.e., on top of the undisturbed soil further ahead. The particles hence form the upper layer. The FEE model represents this particle accumulation as surcharge force per tool width Q acting on top of the soil wedge (cf. Fig. 1). As a result the force required to fail the soil increases. The particles can therefore cause soil reaction force, on one hand, due to their direct contact with the tool and, on the other hand, due to the surcharge in the FEE model. However, we do not yet consider this effect in our implementation.

3. Results and Discussion

We show simulation results of a fully articulated rover – Neptec's Juno Rover[b] – driving on soft terrain. A bulldozing blade, mounted to the rover chassis, is used to highlight our tool–ground interaction method by adding bulldozing capabilities to the rover. Figure 3 shows a simulation of the modified Juno Rover performing a bulldozing operation on a circular mound of soil. Wheel–ground and tool–ground interaction parameters are listed in Table 1. In two simulation studies, we show correlations between soil

Table 1. Soil parameters used in the simulation

ϕ	δ	c(kPa)	c_a	$k_\phi(\mathrm{kN/m}^{(n+2)})$	$k_c(\mathrm{kN/m}^{(n+1)})$	n	K_d(m)	$\gamma(\mathrm{kg/m}^3)$
30°	10°	1.0	0	1814	1.37	1.0	0.025	1500

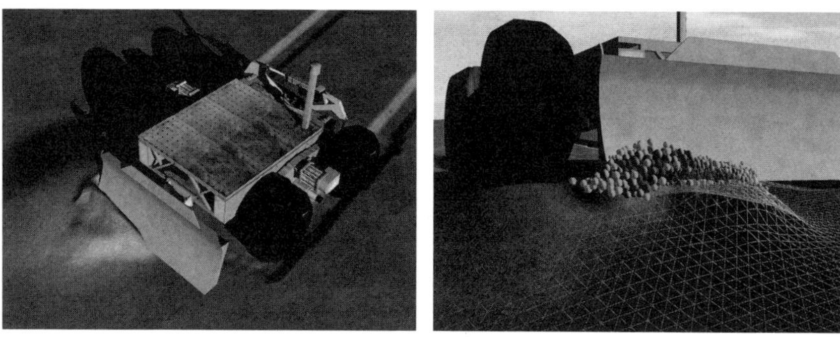

Fig. 3. Simulation of the Juno Rover performing bulldozing operation. The terrain is deformed both due to wheel–ground and tool–ground interaction modelling. Left: Particles are wrapped with a heightfield for visualization purposes. Right: Particles are depicted in green and blue representing high and low void ratio of the soil respectively.

reaction forces acting on the rover's bulldozing blade and the average wheel slip ratio. In the first simulation study, the rover cuts three trenches with different depths by modifying the blade height and inclination angle. The results are shown in Fig. 4. We see that the soil reaction force simulated with the FEE method increases with different blade postures. As a result the average wheel slip ratio increases. This shows that the bulldozing operation has a direct impact on the rover's mobility. In the second simulation study, the rover drives straight towards a mound of soil with 20cm height and a

[b]http://www.neptec.com/

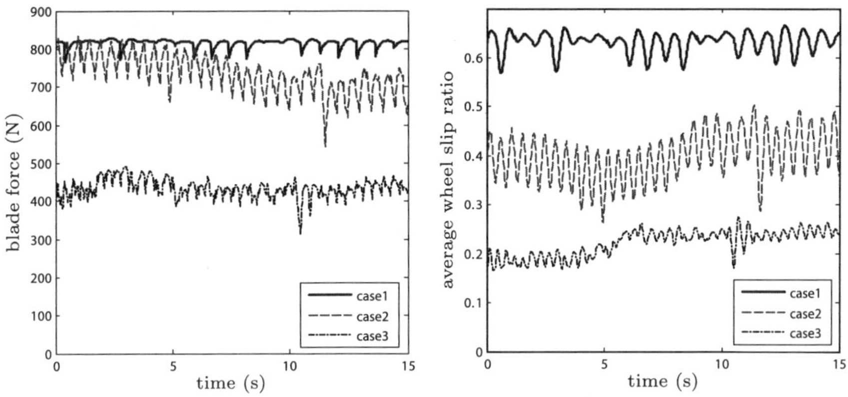

(a)

(b)

Fig. 4. (a) Blade force (FEE only) and (b) the resulting average slip ratio of all wheels. Cases 1, 2, and 3 correspond to average blade depth (d) of 0.13m, 0.14m, 0.11m, and average tool-soil angle (ρ) of 89°, 77°, and 71° respectively.

diameter of 2m. In a first run the bulldozing blade is up, not touching the ground through the entire simulation. In a second run, the bulldozing blade is at a height of 8cm and an inclination of about 60° to the horizon, causing strong soil deformations once the rover reaches the mound. Figure 5 shows the rover trajectories on the horizontal $x - y$ plane for both runs and the soil reaction forces for the second run. As can be clearly seen, the rover trajectory deviates from the original straight path when bulldozing forces are involved, acting only on one side of the blade. Also the average wheel slip ratio increases and the progress of the rover is slowed. Furthermore, we see that, as expected, the soil forces rise and fall steadily while the rover passes over the mound. The results also show that the DEM particle force contribution is not negligible.

4. Conclusion

In this work we presented a rover simulation method which combines terramechanics and soil mechanics models to capture the effects of both wheel–ground and tool–ground interactions. Through the usage of Vortex, all soil reaction forces are directly coupled with the rover articulation system – a fact which allows for the simulation of rovers in transient motion states, interacting with rapidly changing terrain surfaces. In various simulation runs, we showed that our method yields plausible rover motion estimations in bulldozing scenarios.

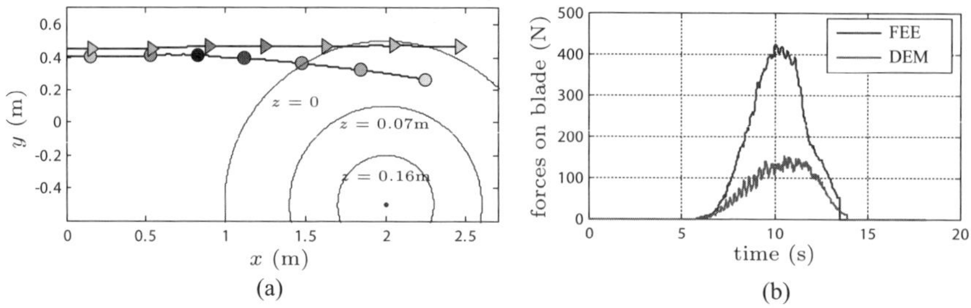

Fig. 5. (a) Rover trajectories while driving toward and over a mound of soil (blue circles) with a height of 20cm: triangles corresponds to a simulation run with high blade position, and circles to a run with low blade position causing mound deformations. The markers are placed every 2 seconds of the simulation and their color corresponds to the average wheel slip ratio ranging from a minimum of 0.05 (light color) to a maximum of 0.25 (dark color). Triangle markers are slightly offset in y direction for clarity. (b) Soil reaction forces (FEE and DEM particles) on the blade corresponding to the trajectory with circles.

References

1. M. G. Bekker, *Introduction to Terrain-Vehicle Systems* (Ann Arbor, The University of Michigan Press, 1969).
2. J. Y. Wong, *Terramechanics and Off-Road Vehicle Engineering: Terrain Behaviour, Off-Road Vehicle Performance and Design* (Elsevier, 2010).
3. A. Azimi, M. Hirschkorn, B. Ghotbi, J. Kövecses, J. Angeles, P. Radziszewski, M. Teichmann, M. Courchesne and Y. Gonthier, *Canadian Aeronautics and Space Journal* **57**, 24 (2011).
4. A. Azimi, J. Kövecses and J. Angeles, Wheel-soil interaction model for rover simulations based on plasticity theory, in *IEEE/RSJ International Conference on Intelligent Robots and Systems*, (San Francisco, USA, 2011).
5. A. Azimi, D. Holz, J. Kövecses, J. Angeles and M. Teichmann, Efficient dynamics modeling for rover simulation on soft terrain, in *50th AIAA Aerospace Sciences Meeting*, (Nashville, USA, 2012).
6. D. Holz, T. Beer and T. Kuhlen, Soil Deformation Models for Real-Time Simulation: A Hybrid Approach, in *6th Workshop in Virtual Reality Interactions and Physical Simulations*, VRIPHYS '09 (Germany, 2009).
7. A. R. Reece, *Institution of Mechanical Engineers* **179**, 16 (1964).
8. E. McKyes, *Soil cutting and tillage* (Elsevier, 1985).
9. O. Luengo, S. Singh and H. Cannon, Modeling and identification of soil-tool interaction in automated excavation, in *IEEE/RSJ International Conference on Intelligent Robotic Systems*, October 1998.
10. J. Bresenham, *IBM Systems Journal* **4**, 25 (1965).

DIDACTIC HUMAN ANTHROPOMORPHIC GRIPPER FOR AUTOMATION TEACHING

ALVARO J. URIBE, EDGAR DIAZ, HERNANDO LEON-RODRIGUEZ

Industrial Engineering Department, Nueva Granada Military University
Karrera 11 # 101-80 Bogota, Colombia

This paper presents the design, development and implementation of a 14 DOF human anthropomorphic gripper as a didactic tool for teaching automation basics using pneumatics and PLCs. The anthropomorphic hand's fingers behave like serial robots and their motion allows performing several types of power and precision grasps just as the human hand, this collaborative motion can be seen as robotics workcell where each finger is an actuated mechanism with a specific task. Through the programming of each mechanism, the objective of the pneumatic hand is to enhance the student's skills for executing collaborative actions through pneumatics and PLC basic concepts over the device. Robotics has been widely used as a teaching assistive tool in several areas such as in training surgical procedures, manipulation of elements in hazardous environments for us, kinematics, dynamics and path planning simulations for various industrial processes, or as didactic tools for inspiring school students for working with robotics. The system's architecture is composed of an anthropomorphic gripper controlled by a PLC along with the pneumatic actuators for simulating the muscles, and the user's inputs for executing the chosen tasks. For overcoming the physical limitations of having just one device, an offline virtual environment is also developed, so several users may program and perform practices for later execution on the real device.

1. Introduction

You Innovative devices tend to find inspiration in natural designs (dams, submarines, airplanes, humanoid, quadruped and insect-like robots), the human physiology is no exception as it allows the development of better and more efficient robotic devices due to our bodies inherent dexterity. One essential device in a robotics workcell is the gripper, which depending on the application; it may perform several tasks such as, welding, assembling, painting, assisting surgery and manipulation of delicate or hazardous materials. Common activities when programming robots involve teleoperation [1], offline programming [2], hazardous environments or prehensile tasks [3], [4], as well as grasp planning [5], design validation [6], and assistive rehabilitation [7].

Some other solutions resembling the human hand based also on serial mechanism are jaw liked and three fingered grippes used in several industrial applications and training tools, as presented in [8], [9], [10], [11], [12], [13].

Over the years the dexterity of the human hand has given birth to the development of devices mimics its functionality, this has resulted in the design and construction of experimental prototypes such as the Stanford/JPL, the Utah/MIT, the TUAT/Karlsruhe, the DLR [14], and the Nasa Robonaut [15] among several others [16].

Common educational tools available for teaching automation rely on laboratories pneumatic, hydraulic, mechanical or electric actuators, programmable logic controllers (PLC) [17], and automated conveyors with robotic arms. Additional to this hardware, certain software allow simulating offline queues for training and learning in cases where hardware limitations are present or for validating processes for guaranteeing that the hardware will not be damage during practices. For solving the need of simulation remote or local tools using interactive or virtual reality (VR) solutions provide access for assisting remotely the learning or training process independently of the physical workcell.

Considering the importance and impact of hand-based research and VR applications, this work proposes the development and implementation of an anthropomorphic robotic human hand for automation learning purposes focusing on the programming of various grasping scenarios through a CLP and the pneumatic muscles. An offline tool will also be used when several users may need to practice with the device so they can later run the programed solutions thus, overcoming the hardware limitations because of having just one hand.

This paper is organized as follows; in section II the human hand basics and the kinematics of the human finger are studied using the homogenous transformation matrix from forward kinematics analysis. In section III the anthropomorphic hand mechanical and pneumatic design is presented. In section IV the offline programming tool is described as well as the parameters for the virtual environment implementation, and finally in section V the results, the conclusions and future works are discussed.

2. Human Hand Analysis

The human hand has twenty DOF and it can perform various movements and geometry grasps [17], these DOF allow each finger to perform the flexion/extension and adduction/abduction movements. In [17] Cutosky, studied various grasping techniques and classified them in two main groups, precision and power types being the most common grasped objects based on prismatic and circular shapes. Main differences between these grasps, are the number of fingers used and the amount of contact that is applied to the object; therefore precision grasps may involve fewer fingers and solely their fingertips whereas power ones will need more fingers and greater contact with the object.

Similarly to how we reach objects from known shapes, where we rotate our fingers to prepare for a specific grasp, for calculating joint positions given known flexion/extension rotations, the forward kinematics supply sufficient information for executing both power and precision grasp tasks with the anthropomorphic proposed hand. Each finger can be analyzed as an open kinematic chain represented by a serial mechanism modelled as three serial connected bars and its behaviour studied with forward kinematics. The rotational and translational information for each link and its final effector can be obtained through the homogenous transform matrix analysis. This matrix can be obtained by using the Denavit-Hartenberg notation; this method was preferred over the geometric analysis as it allows increasing the number of DOF without recalculating the kinematics models, another reason is that, as the maximum number of DOF are 19 when multiplying the transformation matrixes the amount of zeros multiplied within each link do not affect the computing speed and processing, making it a suitable solution.

3. Anthropomorphic Hand Design

The mechanism design was based upon the skeletal human hand system as it gives the support for the fingers, in the same manner; the actuation system was inspired on muscles and tendons that were configured and implemented using pneumatic actuators. In Figure 1, both, the human inspiration and the projected device are presented; it is worth noting the physical resemblance between them, which guarantees a similar dexterity according to the considered DOF

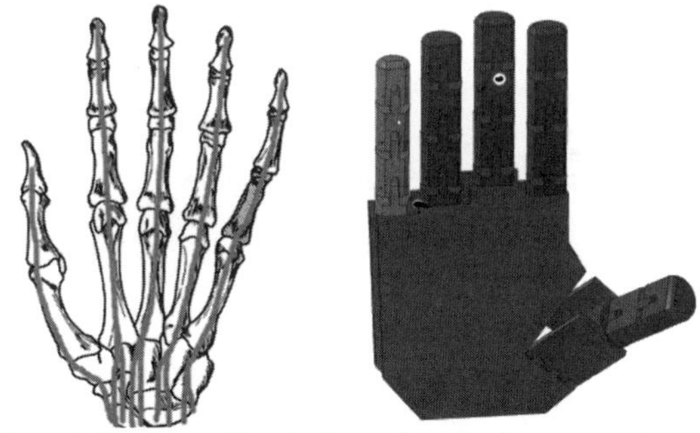

Figure 1: Hand-based inspiration and mechanism proposal.

In order to correctly design the gripper, we started by modeling each of the fingers, this allowed us to focus on each of the joints motion and actuation system through nylon strings attached to the pneumatic actuators located at the forearm of the robotics hand. The hand prototype is presented in Figure 2, from it, can be seen where the phalanx rotates and where the tendon passes through. Given that the most significant motions in grasping are the flexion and extension rotations, each finger abduction/adduction were not considered for simplifying the mechanism and actuation system. The CAD models serve as base for the rapid prototyping of the pieces so the physical model for validation is built faster than made manually

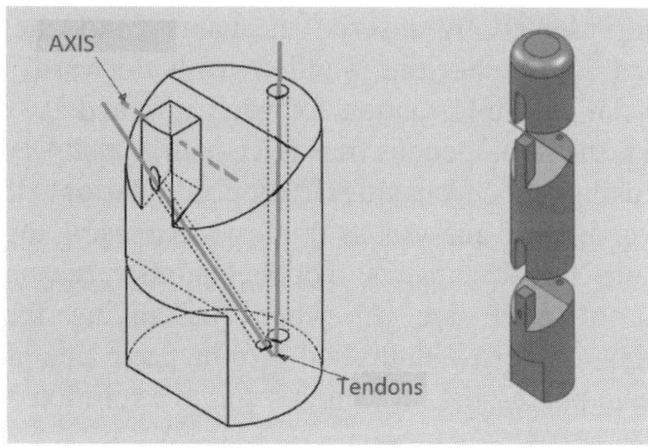

Figure 2: Phalanx and finger prototype.

For the actuation system, the pneumatic muscles were implemented using Festo's FC660PLC, these were preferred over direct current (DC) motors given the goal of pneumatics teaching and programming. The chosen actuators allow performing and controlling each of the finger's flexion and extension rotations through its contraction and expansion, and their main components are presented in Figure 3.

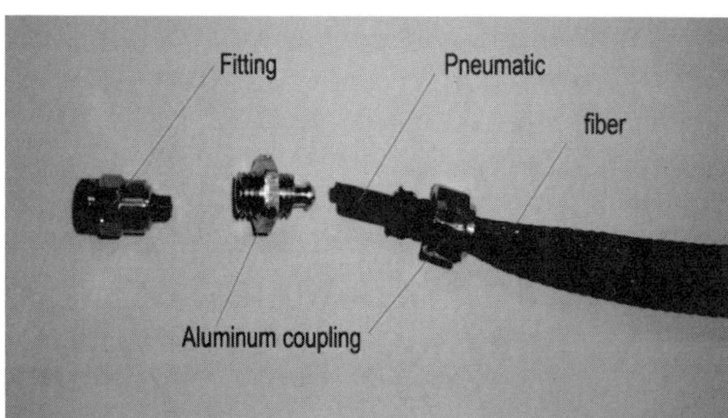

Figure 3: Pneumatic muscle.

Once each of the muscles was connected to the nylon strings, theses were connected to two sets of valves in communication with the corresponding outputs of the PLC. The activation and control every single muscle is done by controlling a logic sequence of instructions from the PLC. The hand is powered at 24V and it can be controlled manually or by a continuous sequence via Ethernet protocols. The setup for the muscles and PLC is presented in Figure 4, where the connection schematic is presented.

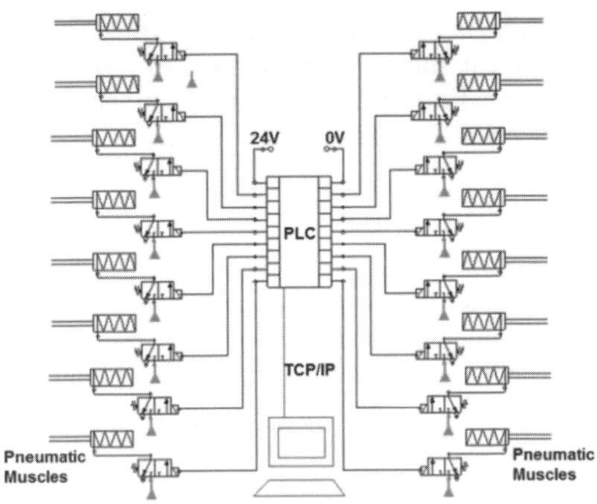

Figure 4: Connection schematics.

4. Offline Programming

The proposed offline programming architecture is presented in Figure 5; it covers input/output devices and information, kinematics processing, virtual instrumentation and task execution.

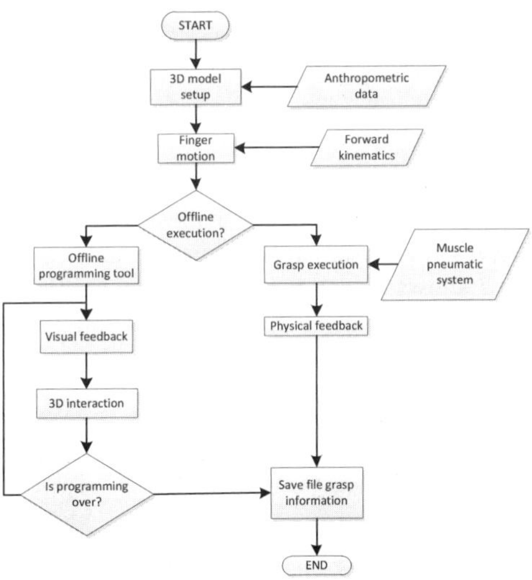

Figure 5: Proposed System architecture.

Java with its Java3D API (Application Program Interface), along with the Virtual Reality Modeling Language (VRML) standard are chosen given their portability and operational system independence. The implementation was based upon the work presented by Uribe in [18]. In Java3D the virtual assembly is organized following a hierarchical order, considering as base the palm of the anthropomorphic device, from it each finger is a branch, and each phalanx is located relative to the branch's origin point on the palm.

5. Experimental Results

Following the CAD design, the anthropomorphic hand was assembled and connected following the presented system's architecture. The modularity of the parts allow its fast and simple assembly and maintenance when necessary, each of the joints supports continuous motion so the devices is physically adequate for training and teaching automation basics. For testing the developed anthropomorphic hand, various grasps where programmed using different valve configurations resulting in the finger positions presented in Figure 6.

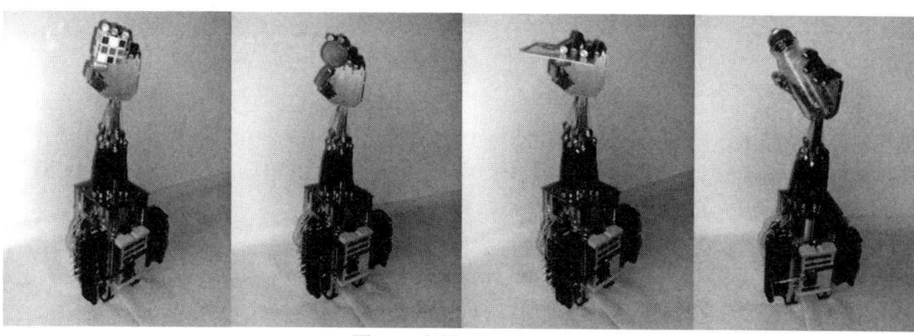

Figure 6: Executed grasps.

Virtual tests proved to be realistic when mimicking the finger's movement and the way it grasps the three predefined shapes. The navigation, lightning and button location gives the user a comfortable environment for using the application, Figure 7 shows a screen capture of the application, presenting from left to right the predefined shape buttons, the 3D canvas and finally the single finger movement buttons.

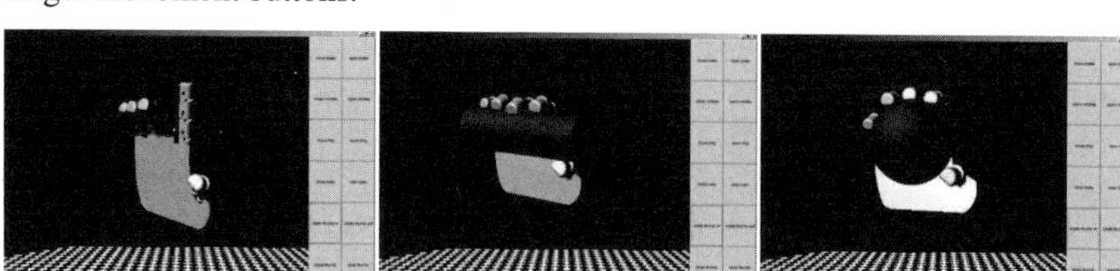

Figure 7: Virtual reality test performing a grasp motion.

After following the proposed methodology from this CAD assembly, the model was successfully exported through VRML and the files successfully edited for working properly in Java 3D for programming the device.

Finally the following conclusions can be obtained from the previous work. First, the proposed methodology allows the use of various geometries regardless of the 3D creation software, as VRML can be obtained from them. This characteristic offers the user portability as no specific modelling software is required; it also achieves low cost and OS independency when combined with the Java Application Program Interface, proving the viability of VR as a tool for complying with the needs of new devices.

Future work

Substantial work needs to be done in other to improve the performance of the hand in the areas of grasping and manipulation control. The system requires to uses different and more effective control model is the next major stage, and also, improve the mechanical joins of the fingers. Further developments can be made by using 3D user interfaces such as the Wiimote, Kinect® o haptic gloves for increasing user immersion and interaction. The virtual environment will be enhanced by adding more objects for simulating interactions between them and the hand.

Acknowledgments

The authors would like to thank the support of the Nueva Granada Military University of Colombia.

References

1. Afshari, K. E. & Payandeh, S. , Toward Implementation of Java/VRML Environment for Planning Training and Tele-Operation of Robotic Systems, *Americas Conference on Information Systems-AMCIS*, 1999
2. Belousov, I. R.; Chellali, R. & Clapworthy, G. J. , Virtual Reality Tools for Internet Robotics, *Eighth IEEE International Symposium on Distributed Simulation and Real-Time Applications*, 2004.
3. Miller and P. Allen, Graspit! a versatile simulator for robotic grasping, *Robotics & Automation Magazine*, IEEE, Vol. 11, Issue 4, pp. 110-122, 2004.
4. Miller, P. Allen, V. Santos, and F. Valero-Cuevas, From robotic hands to human hands: a visualization and simulation engine for grasping research, *Industrial Robot: An International Journal*, Vol. 32, pp. 55-63, 2005.
5. C. Goldfeder and P. Allen and C. Lackner and R. el Pelossof, *Grasp Planning via Decomposition Trees*, ICRA 07, 2007.
6. Y. Tsepkovskiy and L. Antonov and Cv. Kocev and F. Palis and N. Shoylev, Development Of A 3D And VRML Virtual Hand Models For Different Mechanical Gripper, *Journal of the University of Chemical Technology and Metallurgy*, 2007.

7. A. Kollreider, D. Ram, R. Scherer and P. Grieshofer, *Robotic hand/finger rehabilitation for apoplexy patients*, European Symposium Technical Aids for Rehabilitation – TAR 2007, pp. 41-42. Berlin – Germany.
8. H. Hirukawa, T. Matsui, S. Hirai, K. Konaka, and S. Kawamura, *A prototype of standard teleoperation systems on an enhanced vrml*, Proceedings of the 1997 IEEE/RSJ International Conference on Intelligent Robots and Systems, IROS '97, Vol. 3, Grenoble - France pp. 1801-1806.
9. M. Haage and K. Nilsson, *On the scalability of visualization in manufacturing*, in Proceedings. ETFA '99. 1999 7th IEEE International Conference on Emerging Technologies and Factory Automation, Vol. 1, pp. 43-51.
10. E. Freund and J. Rossmann, *Multimedia and virtual reality techniques for the control of ERA*, Proceedings of the 2001 IEEE international Conference on Robotics & Automation, Vol. 2, pp. 1921-1926.
11. R. Safaric, R. M. Parkin, C. A. Czarnecki, and D. W. Calkin, Virtual environment for telerobotics, *Integrated Computer-Aided Engineering*, vol. 8, pp. 95–104, 2001.
12. X. Yang, Q. Chen, D. Petriu, and E. Petriu, *Internet-based teleoperation of a robot manipulator for education*, Proceedings. The 3rd IEEE International Workshop on Haptic, Audio and Visual Environments and Their Applications, pp. 7-11, 2004.
13. U. Kartoun, H. Stern, and Y. Edan, *Virtual reality telerobotic system*, e-ENGDET 2004 4th international conference on e-engineering & digital enterprise technology, p. 10.
14. J. Butterfass, M. Grebenstein, H. Liu, and G. Hirzinger, *Dlr-hand II: next generation of a dexterous robot hand*, Proceedings 2001 ICRA. IEEE International Conference on Robotics and Automation, Vol. 1, pp. 109-114.
15. C. Lovchik and M. Diftler, *The robonaut hand: a dexterous robot hand for space*, Proceedings. 1999 IEEE International Conference on Robotics and Automation, Vol. 2, Detroit - MI - USA, pp. 907-912.
16. A. Bicchi, M. Gabiccini, and M. Santello. Modeling Natural and Artificial Hands with Synergies. *Philosophical Transactions of the Royal Society B*, 2011.
17. Jun-ya NAGASE, Norihiko SAGA, Toshiyuki SATOH, Koichi SUZUMORI, "Development and Control of a Multi-Fingered Robotic Hand Using a Pneumatic Tendon-driven Actuator", Journal of Intelligent Material Systems and Structures, 2011.
18. Cutkosky, M.R., On grasp choice, grasp models, and the design of hands formanufacturing tasks, *IEEE Transactions on Robotics and Automation*, vol. 5. pp. 269-279, 1989.
19. Alvaro J. Uribe, João M. Rosário, Oscar F. Avilés, "Anthropomorphic Gripper Virtual Environment for Automation Grasping Task", International Review of Mechanical Engineering IREME, ISSN: 1970 - 8734 Vol. 3. n. 5, pp. 547-552, 2009.

A TREE-CLIMBING ROBOT FOR INVASIVE INSECT DETECTION

BENZUN PIOUS WISELY BABU,
ERICK T. READ, JUSTIN A. GOSTANIAN, MICHAEL A. GENNERT
Robotics Engineering Program, Worcester Polytechnic Institute, Worcester, MA 01609, USA

This paper reviews progress in the development of a scansorial robot for invasive insect detection. It discusses the motivation for our approach, provides design considerations and implementation details, and presents progress to date. One notable feature of the robot is its use of vSLAM to map the tree under study. The robot is currently under development at WPI and this paper provides a summary of its status and future plans.

1. Introduction

Asian Longhorn Beetle (ALB) is an invasive insect species that poses a serious threat to the North American hardwood industry. In order to eradicate ALB, extensive surveys must be conducted around known infestation areas, infested trees destroyed, and neighboring uninfested host trees treated with pesticide. The beetle is difficult to observe directly; rather, it is generally detected indirectly through surveys that look for frass, oviposition wounds, and insect exit holes on tree branches and trunks (Figure 1).

Figure 1 (left): Adult Asian Longhorn Beetle and characteristic 3/8 inch exit hole. Photograph courtesy Dr. Clint McFarland, USDA. Figure 2 (right): Tree climbing robot chassis with 4 legs and clear acrylic body. Two extra legs can be fit in the middle at the provided slot. Battery not shown

The United States Department of Agriculture (USDA) is the lead agency for the ALB eradication program. Currently, the USDA conducts surveys of trees using human climbers: a slow, error-prone, hazardous, and expensive process. A medium-size maple tree can take an hour to survey by a team; with almost 2 million trees surveyed to date around the Worcester, MA infestation area this becomes an expensive proposition. Estimated damage of $3.5 billion annually [11] is caused by wood-boring insects such as ALB. We are developing a tree-climbing robot to efficiently conduct surveys and to support other scansorial activities. The robot is equipped with a single camera to generate a 3D visual representation of the tree surface.

A longer version of this paper appears in [13].

2. Related Work

Rise [1] and Dynoclimber [2] have demonstrated robots capable of climbing trees. These scansorial robots could be used in search and rescue, surveying, and monitoring applications. Much work has been carried out on the electromechanical design of the robots and controllers, but there has been less development of computer vision algorithms on scansorial robots for surveying and mapping. [3] describes a technique for 3D reconstruction of archeological artifacts, which shares some problem characteristics with mapping trees. Photogrammetry has developed solutions to 3D reconstruction based on Structure-From-Motion, but these techniques use offline processing and require extensive camera setup and calibration, hence are not immediately. [4] describes a reconstruction technique that relies on a stationary camera with different views of the object for reconstruction. This is similar to our problem, except that they use a stationary camera to map a moving object whereas we use a moving camera to map a stationary object. This problem is amenable to attack using Visual Simultaneous Localization and Mapping (vSLAM), a map building technique that uses visual odometry and photogrammetry to generate and refine world models as sensed by a mobile robot. [6-8] describe the vSLAM algorithm and various improvements. [9] compares the two most common approaches in vSLAM.

3. Detecting Asian Long Horn Beetle

Most of the ALB's life cycle is spent inside tree trunks as larvae, which makes it difficult to detect during the initial stages of infestation. There are five signs to detect the Asian Longhorn beetle.

1. **Oviposition holes** - These are scars left on the bark by the female beetle when she chews tree bark in preparation for laying an egg. The egg hatches and the larva enter the hardwood. Once in the hardwood, the larva is difficult to detect. Acoustic detection of ALB larvae [12] is not yet suitable in noisy urban environments.

2. **Exit Holes** - After the larva pupates the adult beetle emerges and chews out of the tree, creating an exit hole. Exit holes are 3/8 inch uniform size circular holes, which are easy to detect, remaining visible indefinitely.
3. **Frass** - The frass and boring dust produced by the feeding larvae are expelled out through the oviposition hole. Frass is also left behind when the exit holes are created. Old frass may be difficult to detect, as it can be washed away by rain or blown away by wind.
4. **Beetles** - Adult beetles are visible briefly during the summer when the beetles feed on the leaves and mate.
5. **Leaves** - The ALB has a specific feeding pattern, chewing the veins of the leaves leaving the reminder of the leaves intact.

Currently ground based surveys are carried out to detect the infestation using bucket trucks and tree climber.

4. Approach

To explore research issues involved in scansorial robots, we have developed a prototype tree climbing robot as shown in Figure 2. The robot is a legged system with 3 Degrees of freedom per leg. The robot has a wireless camera mounted on the front pointed downward at an angle of approximately 45 degrees, which gives a view for navigation as well as imaging the tree surface. The robot was designed to carry a single camera, rather than a stereo pair, in order to minimize weight and conserve power. Images captured by the robot are sent to a workstation where image processing, map generation, and display operations will be carried out. The workstation also serves as the operator's console (Figure 3). The robot is designed primarily to travel up and down; however, it can also turn to circumnavigate the tree and acquire images from multiple viewpoints. To provide detailed information for surveying, dense tree surface information needs to be generated. The representation of the tree in 3D must also be robust and should easily handle changes in the tree environment. vSLAM is a family of computer vision algorithms used for mapping and generating 3D information from image and distance/depth data. It has been applied to generating a dense map of world objects [10] Also vSLAM has shown good performance in outdoor environment [7]. In these applications, the robot is looking outward, through freespace. In our application, the robot is looking inward, toward the tree, essentially requiring an "inside-out" approach to vSLAM.

There are additional challenges to implementing vSLAM on tree surfaces. As we detect many feature points for matching, performing accurate correspondences becomes more difficult. A tree's curvature also provides an interesting challenge as it provides views of the same point from different angles and the common assumption of planar surfaces does not hold. We see that Visual SLAM is a suitable algorithm for mapping a tree surface with a moving camera.

5. Design

5.1. *Mechanical*

The robot consists of a modular design with 4 to 6 legs. The total weight of the robot with 4 legs is around 685g. Each leg features a spike, consisting of a sharp screw end, to firmly grip the outer bark of the tree (Figure 4), similar to how insects or clawed creatures climb trees. Each leg is a 3-DOF kinematic chain with three actuators. Six legs provide more varied gait options as only 3 points of contact on the tree are needed for stability. The 4-legged design only allows simple gaits in which only one leg moves at a time. The 4-legged design allows each leg to take a longer stride as the middle legs will not obstruct movement of the front and rear legs. The legs are the heaviest part of the robot; thus, by reducing the number of legs the weight and in turn power is reduced.

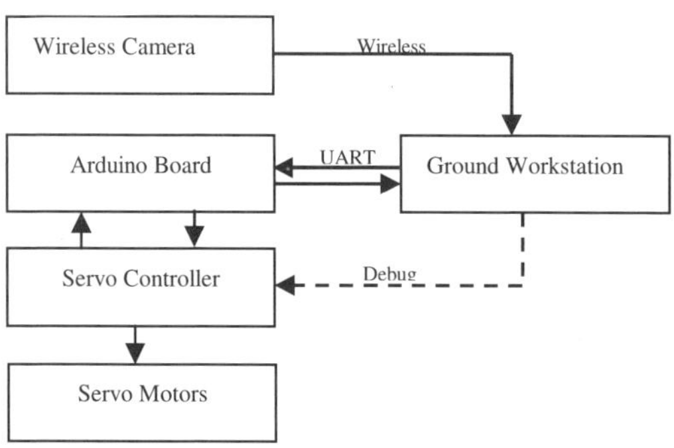

Figure 3 (left): Overall architecture of the system. Figure 4 (right): The robot gripping a tree log using the spike type legs.

5.2. *Electronics*

The robot was developed with the ultimate aim of being used for commercial purposes hence most of the electronic equipment is designed using off-the-shelf components. The robot has an onboard AVR microcontroller (Arduino Uno Board) and a servo controller (Polulu Mini Maestro Servo controller) for gait control and communication. There is currently no feedback from the servo motors for closed loop control. The Arduino Uno Board was selected due to its availability, low cost, and technical support. Metal gear micro servos were selected because of their high torque to weight ratio, which ensured low weight. To reduce weight further, the robot is powered by an external power supply link rather than an onboard battery.

5.3. *User Interface*

It was designed keeping in mind the need to accommodate users inexperienced with robot control. The user interface consists of simple buttons that allow the operator to navigate up, down, left or right (Figure 5). The controls are also mapped to the keyboard for further flexibility and ease of use. The UI also displays useful information, including the status of the robot.

Figure 5. Prototype user interface, including simulated image of tree as seen by robot.

6. Processing Steps

A filter-based approach is selected to implement the vSLAM algorithm. In this approach, the state of the system consisting of the robot pose and location along with the observed map features is estimated and updated iteratively as new observations are made.

1. **Image acquisition** - The images from the monocular camera onboard the robot are transmitted to the workstation computer over a wireless connection.
2. **Preprocessing** - The images from the camera are preprocessed by removing background details irrelevant to the tree by image masking. The remaining tree image is converted to a luminosity-invariant Lab color coordinate system to remove the effect of lighting variations that occur in outdoor environments.
3. **Feature Detection** - A reliable and fast feature detection algorithm, SURF [9] is used to determine the location of feature points (Mk). SURF uses a robust feature detector and compact feature descriptor that makes it suitable for real time operation. The detected feature points are added to a feature database upon which the model is built.
4. **Location and Pose Estimation** - An estimate of the camera location is made based on the system model and the estimates of the features in the map. The noise in the estimates will decrease as pose estimates converge with more observations. The current state of the camera X_k is obtained in

this step. With the camera state approximately known, the distance estimates Z_k to the features M_k are also calculated and added to the database. Also details from the current estimate of the map/model are used to further refine the location and pose estimate.

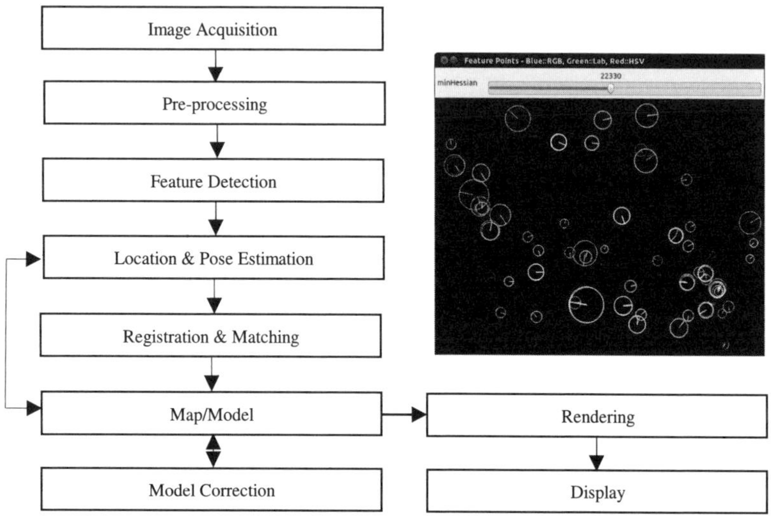

Figure 6(left). Flow chart describing the processing steps. Figure 7(top right). SURF algorithm run on a tree image. Results using the same algorithm on different color spaces are shown in Blue: RGB, Green: Lab, Red: HSV. Color combinations show features detected in more than one color space as Aqua: RGB and Lab, Purple: RGB and HSV, Yellow: Lab and HSV. White features indicate where all 3 color spaces find the same feature. Lab color space finds features (Green) that other color spaces miss

5. **Registration and Matching** - The next step consists of vSLAM / 3D registration where the collected feature points M_k and distance estimates Z_k from consecutive image sequences in the camera coordinate frame are mapped to points in the 3D world coordinate space of the tree. An Extended Kalman filter is used to update the estimates of the positions of the features in the world coordinate space and the camera state X_k. The state model consists of the robot's position (x, y, z) and pose (yaw, pitch, roll) along with the position of the features. In the current model the feature points are assumed to be static.
6. **Map/Model** - The map consists of all the feature points observed so far registered to the global coordinate frame of the tree and is used in the odometry estimation.
7. **Rendering** - The representation of the model in a form suitable for visualization by human observers is done at this step. Based on user-specified viewpoints, different views of the tree are generated. In order to

produce dense information, a triangulation can be generated using the feature points as mesh points.

8. **Model Correction** - The model is updated as new images are acquired. Loop closure occurs when the camera observes the same feature point again from a different viewpoint. Model correction is performed at longer intervals compared to model updating to reduce the computation time. The model correction further improves the accuracy of the tree model generated.

7. Results

A novel robot has been designed for climbing trees and performing visual inspection. It is currently being built and tested. Additional work is in progress to further develop the robot. We have collected a data set containing tree images from various viewpoints simulating the imagery that will be acquired by the tree climbing robot. The SURF algorithm has been implemented for feature detection. We compared the performance of the SURF feature detector using Luminance-Alpha-Beta (Lab), HSV, and RGB color spaces in order to evaluate their performance under different lighting conditions, and found that Lab detects more features than the other color spaces (Figure 7). The Extended Kalman Filter based vSLAM is under implementation.

8. Future Work

The mechanical design of the leg needs to be studied in detail to gain a better understanding of scansorial robots. Other mechanisms for gripping the tree such as micro also claws need to be studied. The robot is planned to be tele-operated with fixed gaits. Automatic gait evolution is another aspect of the robot that requires further work. Better control and communication algorithms on the microcontrollers are expected in future designs. Autonomous exit-hole detection will also be incorporated in the processing step. Also the vSLAM algorithm can be further adopted to incorporate the time changes that occur in the trees over a long period, thus providing more details to survey teams.

The completed robot is expected to be a valuable tool for the USDA to survey trees reliably, efficiently, and at low cost.

Acknowledgments

The authors would like to thank Dr. Clint McFarland, Federal Program Manager, Asian Longhorn Beetle Eradication Program-Massachusetts, Ms. Donna Fernandes, USDA-APHIS Operations Supervisor, and Dr. Alfred Rizzi, Chief Robotics Scientist, Boston Dynamics, for helpful and insightful discussions.

References

1. Kellar Autumn, Martin Buehler, Mark Cutkosky, Ronald Fearing, Robert J. Full, Daniel Goldman, Richard Groff, William Provancher, Alfred A. Rizzi, Uluc Saranli, Aaron Saunders and Daniel E. Koditschek, "Robotics in scansorial environments", Proc. SPIE 5804, 291 (2005).
2. J. Clark, et.al., "Design of a Bio-inspired Dynamical Vertical Climbing Robot", in Robotics: Science and Systems III, Atlanta, Georgia, 2007.
3. Bernardini, F.; Rushmeier, H.; Martin, I.M.; Mittleman, J.; Taubin, G.; , "Building a digital model of Michelangelo's Florentine Pieta", IEEE Computer Graphics and Applications, vol. 22, no.1, pp. 59-67, 2002.
4. Reitmayr, Gerhard, and Tom Drummond. "ProFORMA: Probabilistic Feature-based On-line Rapid Model Acquisition", Proc 20th British Machine Vision Conference (BMVC), pp. 1-11, 2009.
5. J. Civera, A. J. Davison, and J. M. M. Montiel. "Inverse depth parametrization for monocular SLAM", IEEE Transactions on Robotics (T-RO), 24(5):932–945, 2008.
6. K. Konolige and M. Agrawal. "FrameSLAM: From bundle adjustment to real-time visual mapping", IEEE Transactions on Robotics (T-RO), 24:1066–1077, 2008.
7. Silveira, G., E. Malis, and P. Rives. "An Efficient Direct Approach to Visual SLAM", IEEE Trans. Robotics (T-RO), 24.5, pp. 969-79, 2008.
8. Strasdat, H., J. M. M. Montiel, and A. J. Davison. "Real-Time Monocular SLAM: Why Filter?", IEEE International Conference on Robotics and Automation (ICRA), 2010.
9. H. Bay, T. Tuytelaars, and L. Van Gool. "SURF: Speeded up robust features", European Conference on Computer Vision (ECCV), 2006.
10. P. Paalanen, V. Kyrki, and J. Kamarainen. "Towards monocular on-line 3d reconstruction", European Conference on Computer Vision (ECCV) Workshop on Vision in Action, 2008.
11. Aukema, J. E., B. B. Leung, K. Kovacs, C. Chivers, K. O. Britton, J. Englin, S. J. Franke, R. G. Haight, T. P. Holmes, A. M. Liebhold, D. G. McCullough, and B. Von Holle. "Economic impacts of non-native forest insects in the continental United States". PLoS ONE 6(9): e24587. (doi: 10.1.371/journal.pone.0024587), 2011.
12. Mankin, R.W., Smith, M.T. 2011. Eavesdropping on coconut rhinoceros beetles, red palm weevils, Asian longhorned beetles, and other invasive travelers. Proceeding of Meetings on Acoustics. 14:010001:1-8, 2011
13. Babu, B.P.W., Read, E.T., Gostanian, J.A., Gennert, M.A., "A Tree-Climbing Robot for Invasive Insect Detection", Technical Report WPI-CS-TR-12-XX, Worcester Polytechnic Institute, Worcester, MA, 2012. http://www.wpi.edu/academics/cs/technical-reports.html.

SECTION–8

MODELLING AND SIMULATION OF CLAWAR

ROBUST PD-PID CONTROL OF A NEW CONFIGURATION OF TWO-WHEELED MACHINES UNDER VARIOUS OPERATING CONDITIONS

A M Almeshal[1], K M Goher[2], M O Tokhi[1], S A Agouri[1] and O Sayidmarie[1],
[1]Department of Automatic Control and Systems Engineering
The University of Sheffield, United Kingdom, e-mail: cop10ama@sheffield.ac.uk
[2]Department of Mechanical and Industrial Engineering
Sultan Qaboos University, Oman, e-mail: kgoher@squ.edu.om

This paper focuses on the design and implementation of a control algorithm for balancing a novel two-wheeled vehicle. The proposed design provides the vehicle with more flexibility in terms of increased degrees of freedom, which enable the vehicle to enlarge its work space. The model of the system mimics a double inverted pendulum on a moving base. Lagrangian dynamic formulation is used, due to its relative simplicity, to derive the system dynamics. The new developed configuration is of great importance in various applications including self balance robots, wheelchairs on two wheels, stability analysis of multi segment gaits and multi link cranes. The balancing of the machine in the upright position is achieved using a hybrid PD-PID control mechanism. Different external disturbance forces are applied to the intermediate body over different periods in order to test the robustness of the developed control mechanism.

1. Introduction

Inverted pendula are currently used as teaching aids and research experiments (Quanser, 2004). Many researchers have built their own inverted pendulum systems (Åström and Furuta, 1996; Brockett & Hongyi, 2003; Rubi, 2002) to suit their investigations.

Various control approaches have been developed and applied to invered pendula. Nundrakwang et al. (2006) developed a hybrid controller to swing up and stabilize the inverted pendulum on cart system from natural position. Benjanarasuth and Nundrakwang (2008) applied a similar hybrid control approach on rotational inverted pendulum system. They have used an energy-based controller to swing up the pendulum and state feedback with minimum-order observer controller to stabilize the pendulum. Similarly, Asa et al. (2008) applied the same approach on an inverted pendulum on cart system. Panya et al. (2008) developed a hybrid controller to swing-up and stabilize an inverted pendulum on cart system.

2. System description

The vehicle considered in this work comprises a rod on an axle incorporating two wheels as described in Figures 1 and 2. The intermediate body (IB) of the vehicle encompasses two segments; link-1 and link-2, where link-2 includes a set of two coaxial-parts connected by a linear drive to actuate the upper part and the attached payload. The vehicle is driven by four direct current (DC) motors; two motors drive the vehicle wheels and in turn the entire system, a motor drives link-2 and a linear actuator between the two parts of link-2.

The angular positions of link-1 and link-2; θ_1 and θ_2, are measured from the positive vertical Z axis. The linear displacement of the payload; Q, is measured, from position O_2, along link-2. In order to achieve a certain angular position from the upright vertical axis, the vehicle is linearly moved in the XY plane undergoing a planar motion in both directions.

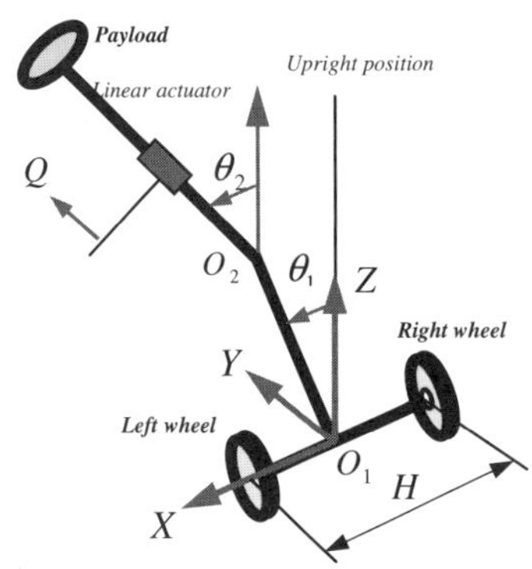

Fig. 1 Schematic of the vehicle

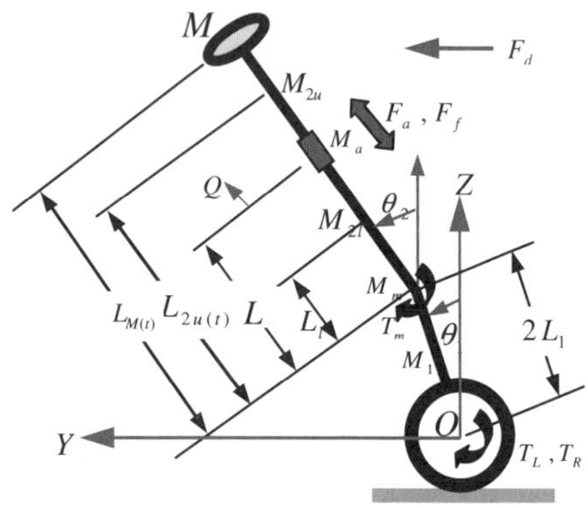

Fig. 2 Positions of vehicle main parts and Com

3. System modelling

Using Lagrange dynamic formulation for the system dynamics, the following dynamic equation can be expressed

$$\frac{dq}{dt}\left(\frac{\partial L}{\partial \dot{q}_i}\right)-\frac{\partial L}{\partial \dot{q}_i}=Q_i \tag{1}$$

The generalized coordinates of the system are chosen as $q_i=[\delta_L \quad \delta_R \quad \theta_1 \quad \theta_2 \quad Q]^T$ and the generalized force is expressed as $Q_i=\left[T_{LT} \quad T_{RT} \quad 0.5(T_{LT}+T_{RT}) \quad T_{MT} \quad F_{aT}\right]^T$, where $L=T_v-V_v$, is the Lagrangian function, Q_i is a generalized force, and q_i is generalized coordinate, δ_L and δ_R are the angular displacements of the left and right wheels around the X axis respectively, θ is the angular displacement of the IB, Q is the linear displacement of the payload along the IB, T_{LT} and T_{RT} are the resultant driving torques of the left and right wheels respectively, T_{MT} is the resultant torque of the actuator driving link-2, and F_{aT} is the resultant linear actuator force.The Lagrangian function of the system; L can be expressed as the difference between the system kinetic and potential energy as:

$$L = C_{21}(\dot{\delta}_L^2 + \dot{\delta}_R^2) + C_{22}\,\dot{\delta}_L\,\dot{\delta}_R + \frac{1}{2}C_8\,\dot{Q}^2 + \frac{1}{2}C_{16}\,\dot{\phi}_2 + C_{18}\,\dot{\theta}_1^2 + \frac{1}{2}(C_{20} + C_{12}Q + C_8Q^2)\dot{\theta}_2^2$$
$$+ C_9\frac{R_w}{2}L_1\,\dot{\theta}_1(\dot{\delta}_L + \dot{\delta}_R)\cos\theta_1 + \frac{R_w}{2}(C_{10} + C_8Q)(\dot{\delta}_L + \dot{\delta}_R)\dot{\theta}_2\cos\theta_2 +$$
$$2L_1(C_{10} + C_8Q)\dot{\theta}_1\,\dot{\theta}_2\cos(\theta_1 - \theta_2) - C_3 g\cos\theta_1 + (C_{15} + C_8Q)g\cos\theta_2 \quad (2)$$

Manipulating the above expressions yields the following five highly non-linear differential equations describing the vehicle dynamics alongside the driving moments and an external disturbance force as:

$$2C_{21}\,\ddot{\delta}_L + C_{22}\,\ddot{\delta}_R + C_9\frac{R_w}{2}L_1\,\ddot{\theta}_1\cos\theta_1 - C_9\frac{R_w}{2}L_1\,\dot{\theta}_1^2\sin\theta_1 + \frac{R_w}{2}(C_{10} + C_8Q)\ddot{\theta}_2\cos\theta_2$$
$$- \frac{R_w}{2}(C_{10} + C_8Q)\dot{\theta}_2^2\sin\theta_2 + \frac{R_w}{2}C_8\,\dot{Q}\,\dot{\theta}_2\cos\theta_2 = T_L - T_{fL} \quad (3)$$

$$2C_{21}\,\ddot{\delta}_R + C_{22}\,\ddot{\delta}_L + C_9\frac{R_w}{2}L_1\,\ddot{\theta}_1\cos\theta_1 - C_9\frac{R_w}{2}L_1\,\dot{\theta}_1^2\sin\theta_1$$
$$+ \frac{R_w}{2}(C_{10} + C_8Q)\ddot{\theta}_2\cos\theta_2 - \frac{R_w}{2}(C_{10} + C_8Q)\dot{\theta}_2^2\sin\theta_2 + \frac{R_w}{2}C_8\,\dot{Q}\,\dot{\theta}_2\cos\theta_2 = T_R - T_{fR} \quad (4)$$

$$2C_{18}\,\ddot{\theta}_1 + C_9\frac{R_w}{2}L_1(\ddot{\delta}_L + \ddot{\delta}_R)\cos\theta_1 - C_9\frac{R_w}{2}L_1(\dot{\delta}_L + \dot{\delta}_R)\dot{\theta}_1\sin\theta_1 + 2L_1(C_{10} + C_8Q)\ddot{\theta}_2\cos(\theta_1 - \theta_2)$$
$$-2L_1(C_{10} + C_8Q)\dot{\theta}_1\,\dot{\theta}_2\sin(\theta_1 - \theta_2) + 2L_1(C_{10} + C_8Q)\dot{\theta}_2^2\sin(\theta_1 - \theta_2) + 2L_1C_8\,\dot{Q}\,\dot{\theta}_2\cos(\theta_1 - \theta_2)$$
$$+C_9\frac{R_w}{2}L_1\,\dot{\theta}_1^2(\dot{\delta}_L + \dot{\delta}_R)\sin\theta_1 + 2L_1(C_{10} + C_8Q)\dot{\theta}_1^2\,\dot{\theta}_2\sin(\theta_1 - \theta_2) - C_3 g\,\dot{\theta}_1\sin\theta_1 = 0.5(T_{LT} + T_{RT}) \quad (5)$$

$$C_{20}\,\ddot{\theta}_2 + (C_{12}\,\dot{Q} + 2C_8Q)\dot{\theta}_2 + (C_{12}\,Q + C_8Q^2)\ddot{\theta}_2 + \frac{R_w}{2}(C_{10} + C_8Q)(\ddot{\delta}_L + \ddot{\delta}_R)\cos\theta_2$$
$$- \frac{R_w}{2}(C_{10} + C_8Q)(\dot{\delta}_L + \dot{\delta}_R)\dot{\theta}_2\sin(\theta_2) + C_8\frac{R_w}{2}\dot{Q}(\dot{\delta}_L + \dot{\delta}_R)\cos\theta_2 + 2L_1(C_{10} + C_8Q)\ddot{\theta}_1\cos(\theta_1 - \theta_2)$$
$$- 2L_1(C_{10} + C_8Q)\dot{\theta}_1^2\sin(\theta_1 - \theta_2) + 2L_1(C_{10} + C_8Q)\dot{\theta}_1\,\dot{\theta}_2\sin(\theta_1 - \theta_2) + 2C_8L_1\,\dot{\theta}_1\,\dot{\theta}_2\cos(\theta_1 - \theta_2)$$
$$+ \frac{R_w}{2}(C_{10} + C_8Q)(\dot{\delta}_L + \dot{\delta}_R)\dot{\theta}_2^2\sin\theta_2 - (C_{15} + C_8Q)g\,\dot{\theta}_2\sin\theta_2 - 2L_1(C_{10} + C_8Q)\dot{\theta}_1\,\dot{\theta}_2^2\sin(\theta_1 - \theta_2)$$
$$= T_M - T_{FM} - L_d F_d \quad (6)$$

$$C_8\,\ddot{Q} - \frac{1}{2}(C_{12} + 2C_8Q)\dot{\theta}_2^2 - C_8\frac{R_w}{2}\dot{\theta}_2(\dot{\delta}_L + \dot{\delta}_R)\cos\theta_2 - 2L_1C_8\,\dot{\theta}_1\,\dot{\theta}_2\cos(\theta_1 - \theta_2) + C_8 g\cos\theta_2 = F_a - F_{fa} \quad (7)$$

4. Control strategy

The strategy to control the system depends on developing a feedback control strategy of five control loops as shown in Figure 3. In order to drive the vehicle to undergo a specific planar motion in the XY plane, five feedback loops are developed. The angular position of the intermediate body is controlled by the measurement of the error in the position of link-1 and link-2. In order to control the position of the attached payload, a feedback control loop is developed with the error in the payload position as an input and the actuation force as the output of the control loop.

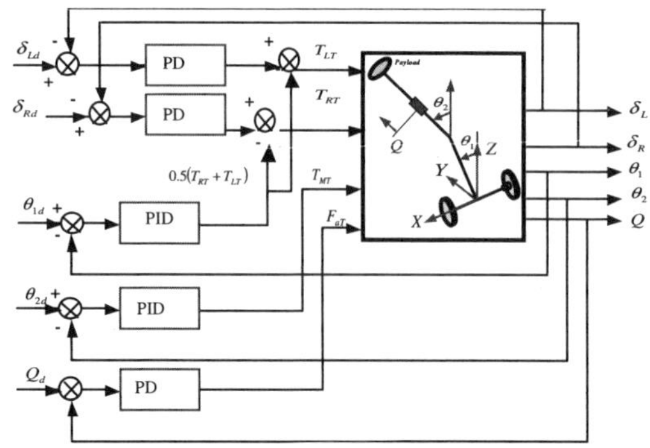

Fig. 3 Schematic description of the control strategy

The system considered in this paper is a multi input multi output (MIMO) system characterized by its high nonlinearity and highly coupled dynamics. The inputs to the system are the driving torques of the main motors; T_{LT} and T_{RT}, torque driving the motor activating link-2; T_{MT}, the linear actuator force; F_{aT} and the average of the motors torques to stabilize the link-1. The system possesses five outputs; the angular positions of the left and right wheels; δ_L and δ_R respectively, the angular positions of the IB segments; θ_1 and θ_2 and the displacement of the payload; Q.

Five control loops are used in which two of them for the wheels angular displacement, two for the pendulum links and one for the payload linear actuator displacement. As a first step to develop a controller for the system, a combination of Proportional-Derivative (PD) and Proportional-Integral-Derivative (PID) controllers were used to get the system response and as an initial controller design step. The cart is limited to travel with a maximum angular displacement of the wheels of 0.8 m in order to control the system within the linear region. The PD controllers are used to control the two wheels angular displacement and the displacement of the payload actuator. While PID controllers are used to control the tilt angles of link-1 and link-2. A residual on the response over the required steady state level is obtained when implementing a PD controller on link-2. In order to compensate for this phenomenon, an integral part has been added in order to remove the steady-state error that appears in the response.

5. Numerical simulation

System Simulink© model was built using the derived equations describing system dynamics. For the current stage of the research, the parameters of system controllers are tuned heuristically with the gains described in Table 1.

Table 1. Heuristically adjusted controllers gains

Feedback loop	***PID controller Gains***		
	K_P	K_I	K_D
Left wheel	50	*NA*	1
Right wheel	50	*NA*	1
First pendulum	10	*0.00001*	2
Second pendulum	30	15	10
Linear actuator	5	NA	60

5.1 Variation of disturbance force amplitude

Different levels of the applied disturbance force; 0 N, 40 N, 80 N, 160 N, 300 N, were used to test the developed control algorithms and examine their robustness.

Referring to Figure 4, which represents the response of the system for disturbances applied on the link-1; the disturbance resulted in some overshoots in the tilt angle of link-1. Increasing disturbance force increases the peak values by an average of 50%. The controller was able to stabilise link-1 to the set point within an average of 5 seconds. The high percentage of increase in peak value explains the fact that increasing the applied disturbance to a larger amplitude results in causing the vehicle to collapse. The settling time increased by an average of 6%. The right and left wheel angular displacements were affected by this disturbance and negative peaks are noted. This is explained by the fact that the cart is moving in the opposite direction of the applied distrubance in order to preserve the upright position of the pendulum. The titl angle of link-2 and the payload actuator displacement were unaffected by this disturbance.

The system performance due to applying disturbances on the link-2 is described in Figure 5. It can be noted from the figure that the disturbance caused oscillations at the tilt angle of link-2. The controller was able to balance the link-2 with an average of 6 seconds depending on the amplitude of the force applied. The larger the amplitude of the applied disturbance the larger the time the controller took to converge the system to the set point. With every increment of disturbance force amplitude, the peak value increased by an average of 60%. The settling time remained unaffected while the rise-time increased significantly after the first disturbance amplitude increment. The left and right wheels, the tilt angle of link-1 and the payload actuator displacement were not affected. This is due to the high damping effect caused by the different joints of the vehicle.

5.2 Variation of disturbance force duration

Further levels of the applied disturbance force were used to test the developed control algorithms and examine their robustness. Forces were applied onto the system at different durations; 1.25, 2.5, 7.5 and 12.5 *sec*. The corresponding system responses are shown in Figures 6 and 7.

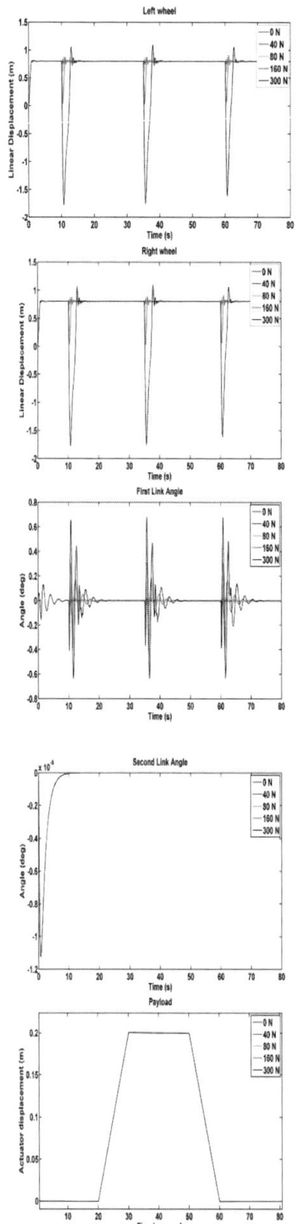

Fig. 4 System performance
Disturbance applied at the centre of link-1

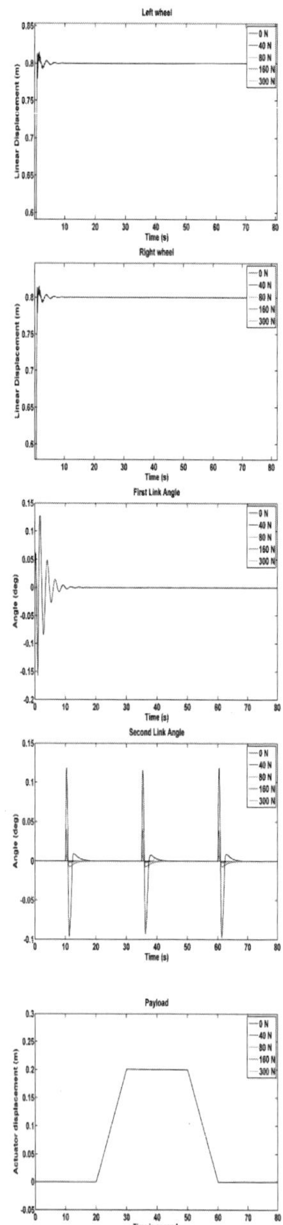

Fig. 5 System performance
Disturbance applied at the centre of link-2

Referring to Figure 6, which represents the response of the system for disturbances applied on link-1; an increase of the overshoot peaks can be observed. Increasing the disturbance duration increased the peak value by an average value of 40%. The settling-time increased by an average of 4% for each incremental increase in the disturbance duration. Furthermore, the rise-time fluctuated by increasing and decreasing depending on the duration of the applied disturbance. The right and left wheel angular displacements were affected by this disturbance and fluctuations can be noted. The tilt angle of link-2 and the payload actuator displacement were not affected by this disturbance.

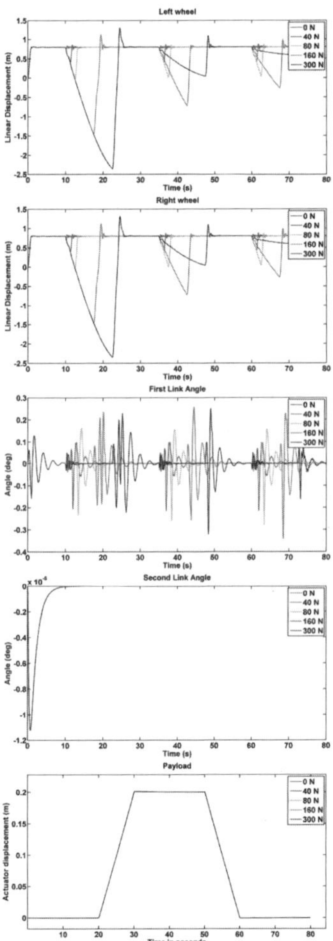

Fig. 6 System performance
Disturbance applied at the centre of link-1

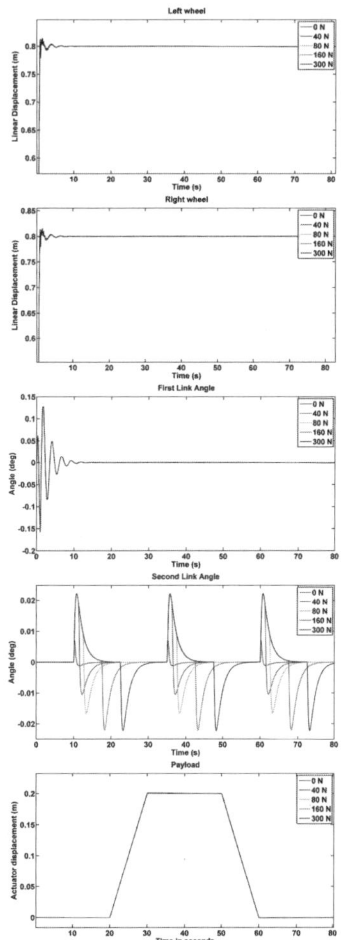

Fig. 7 System performance
Disturbance applied at the centre of link-2

Figure 7 represents the response of the system for disturbances applied on the link-2. The disturbances resulted in an oscillatory response; at disturbance durations up to 7.5 seconds, the controller was able to control the tilt angle of the second pendulum and stabilise the system within an approximate average of 9 seconds. On the other hand, the controller failed to cope with disturbances applied for a longer duration within the defined interval between two consecutive disturbance forces; assumed to be 10 seconds at this stage. Overshoots are noted in the system response with a maximum increase in peak value of 29%. The settling-time increased by an average of 3.5% at a time while the rise-time had significant average increase of 60%. Furthermore, the right and left wheel angular displacements, tilt angle of link-1 and the payload actuator displacement were not affected by this disturbance.

6. Conclusions

A PD-PID control approach has been designed and implemented on the developed model of a two-wheeled vehicle. The control technique used has been able to achieve the target orientation of the whole system with high degree of robustness while achieving system balance in the upright position. The control strategy showed the capability to overcome various levels of external impact applied over different durations at various positions of the vehicle.

References

[1] Asa, E.; Benjanarasuth, T.; Ngamwiwit, J.; Komine, N."Hybrid controller for swinging up and stabilizing the inverted pendulum on cart," Control, Automation and Systems, 2008. ICCAS 2008. International Conference on , vol., no., pp.2504-2507, 14-17 Oct. 2008

[2] Brokett, R.W. and Hongyi LI. (2003). A Light Weight Double Pendulum: Maximising The Domain Of Attraction, Proceedings of the IEEE Conference on Decision and Control, v 4, 2003, p 3299-3304.

[3] Goher K. M. Ahmad S. and Tokhi M. O., (2010), A new configuration of two wheeled vehicles: Towards a more workspace and motion flexibility, in the Proceedings of the 4th IEEE International Systems Conference, San Diego, CA, USA, April 5- 8. pp. 524 – 528.

[4] Åström K. J. and Furuta K., " Swinging Up a Pendulum by Energy Control", Proceedings of the 13th IFAC World Congress, San Francisco, CA, 1996.

[5] Nundrakwang, S.; Benjanarasuth, T.; Ngamwiwit, J.; Komine, N.; , "Hybrid Controller for Swinging up Inverted Pendulum System," Information, Communications and Signal Processing, 2005 Fifth International Conference on , vol., no., pp.488-492, 2006

[6] Benjanarasuth, T., Nundrakwang, S. 2008, Hybrid controller for rotational inverted pendulum systems. SICE Annual Conference. Pages 1889-1894. Doi: 10.1109/SICE.2008.4654969.

[7] Quanser Consulting Inc., www.quanser.com accessed (28/01/2012)

[8] Rubi, J.; Rubio, A.; Avello, A.; Swing-up control problem for a self-erecting double inverted pendulum, , IEE Proceedings Control Theory and Applications, Vol: 149 Issue:2, pp. 169 – 175, 2002.

HYBRID FUZZY LOGIC CONTROL APPROACH OF A TWO WHEELED DOUBLE INVERTED PENDULUM LIKE ROBOTIC VEHICLE

A M Almeshal[1], K M Goher[2], M O Tokhi[1], O Sayidmarie[1] and S A Agouri[1]

[1]Department of Automatic Control and Systems Engineering
The University of Sheffield, United Kingdom, e-mail: cop10ama@sheffield.ac.uk
[2]Department of Mechanical and Industrial Engineering
Sultan Qaboos University, Oman, e-mail: kgoher@squ.edu.om

This paper considers designing a hybrid fuzzy logic controller (FLC) for a two wheeled double inverted pendulum robotic vehicle. The design and implementation of the hybrid FLC controller is demonstrated. Fuzzy logic controllers have been proven to be powerful in controlling non-linear systems. In addition, FLC comprises model-free controlling approach to control complex systems that are difficult to be described and modelled mathematically. Implementation on the system model and simulations are illustrated to study the system behaviour. Simulations and results showing successful control of the systems are presented in this paper.

1. Introduction

The inverted pendulum system has always been of interest to control system engineers and researchers due to its nonlinearity and under actuated nature [1]. There are many variations of the inverted pendulum system. These include the linear inverted pendulum system, mobile inverted pendulum on cart, and rotational inverted pendulum. These have enabled researchers to study various applications including gait of humanoid robots, personal transporters, and self-balancing wheelchairs [3]. In recent years, Inverted pendulum systems have been extended to have multiple links, such as double inverted pendulum and triple inverted pendulum systems [4] [5]. This extension has increased the degrees of freedom and complexity of the system providing models for more complex applications.

A novel double inverted pendulum like two-wheeled robotic vehicle has been presented in [2]. Equations of motion describing system dynamics were derived and presented. In this paper, the model derived by Al-Meshal et al. (2011), will be controlled using a Hybrid Fuzzy Logic Controller. The developed model will be used as a foundation block for ongoing research of self-balancing wheelchairs with ability to maneuver on irregular and uneven surfaces with moving payload.

2. System model description

A schematic diagram of the designed vehicle is shown in Figure 1. The vehicle is designed based on the double inverted pendulum system model with novel modifications [1]. The vehicle consists of two links and a cart driven by two DC motors that in turn drive the entire system. In addition, the vehicle has a third DC motor to drive the second link. These motors will help stabilizing the system in the upright position by applying an appropriate control signal. The second link consists of two co-axial rods connected by a linear actuator to enable lifting up the payload to a demanded height. Therefore, the system has five degrees of freedom; translational motion with the right and left wheels, first and second links and linear actuator on the second link. The tilt angles of the first and second links are θ_1 and θ_2 respectively. The linear displacement of the payload is defined as *Q*, while the angular displacements of the left and right wheels are defined as δ_L and δ_R respectively.

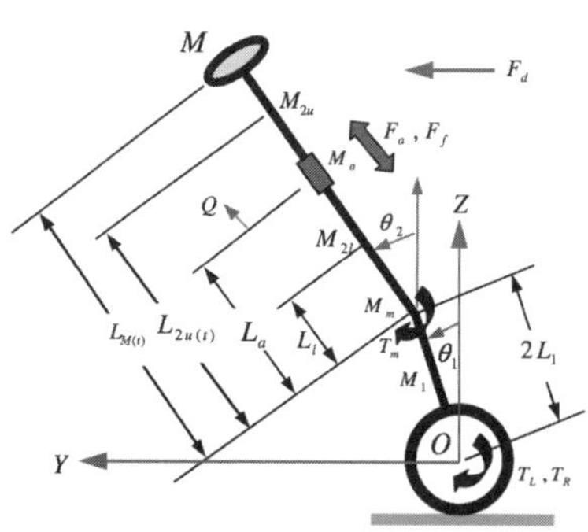

Figure 1. Schematic description of the vehicle

The system mathematical model consists of five differential equations. The system was modeled using Euler-Lagrange approach because of the system complexity and coupled nature. System equations of motion were derived in details in [2] and presented here by equation (1) to (5)

$$
\begin{aligned}
&2C_{21}\ddot{\delta}_L + C_{22}\ddot{\delta}_R + C_9\frac{R_w}{2}L_1\ddot{\theta}_1\cos\theta_1 - C_9\frac{R_w}{2}L_1\dot{\theta}_1^2\sin\theta_1 \\
&+\frac{R_w}{2}(C_{10}+C_8Q)\ddot{\theta}_2\cos\theta_2 - \frac{R_w}{2}(C_{10}+C_8Q)\dot{\theta}_2^2\sin\theta_2 \\
&+\frac{R_w}{2}C_8\dot{Q}\dot{\theta}_2\cos\theta_2 = T_L - T_{fL}
\end{aligned}
\tag{1}
$$

$$
\begin{aligned}
&2C_{21}\ddot{\delta}_R + C_{22}\ddot{\delta}_L + C_9\frac{R_w}{2}L_1\ddot{\theta}_1\cos\theta_1 - C_9\frac{R_w}{2}L_1\dot{\theta}_1^2\sin\theta_1 \\
&+\frac{R_w}{2}(C_{10}+C_8Q)\ddot{\theta}_2\cos\theta_2 - \frac{R_w}{2}(C_{10}+C_8Q)\dot{\theta}_2^2\sin\theta_2 \\
&+\frac{R_w}{2}C_8\dot{Q}\dot{\theta}_2\cos\theta_2 = T_R - T_{fR}
\end{aligned}
\tag{2}
$$

$$
\begin{aligned}
&2C_{18}\ddot{\theta}_1 + C_9\frac{R_w}{2}L_1(\ddot{\delta}_L+\ddot{\delta}_R)\cos\theta_1 - C_9\frac{R_w}{2}L_1(\dot{\delta}_L+\dot{\delta}_R)\dot{\theta}_1\sin\theta_1 \\
&+2L_1(C_{10}+C_8Q)\ddot{\theta}_2\cos(\theta_1-\theta_2) - 2L_1(C_{10}+C_8Q)\dot{\theta}_1\dot{\theta}_2\sin(\theta_1-\theta_2) \\
&+2L_1(C_{10}+C_8Q)\dot{\theta}_2^2\sin(\theta_1-\theta_2) + 2L_1C_8\dot{Q}\dot{\theta}_2\cos(\theta_1-\theta_2) \\
&+C_9\frac{R_w}{2}L_1\dot{\theta}_1^2(\dot{\delta}_L+\dot{\delta}_R)\sin\theta_1 + 2L_1(C_{10}+C_8Q)\dot{\theta}_1^2\dot{\theta}_2\sin(\theta_1-\theta_2) \\
&-C_3g\dot{\theta}_1\sin\theta_1 = \frac{1}{2}(T_{LT}+T_{RT})
\end{aligned}
\tag{3}
$$

$$
\begin{aligned}
&C_{20}\ddot{\theta}_2 + (C_{12}\dot{Q}+2C_8Q)\dot{\theta}_2 + (C_{12}Q+C_8Q^2)\ddot{\theta}_2 \\
&+\frac{R_w}{2}(C_{10}+C_8Q)(\ddot{\delta}_L+\ddot{\delta}_R)\cos\theta_2 - \frac{R_w}{2}(C_{10}+C_8Q)(\dot{\delta}_L+\dot{\delta}_R)\dot{\theta}_2\sin(\theta_2) \\
&+C_8\frac{R_w}{2}\dot{Q}(\dot{\delta}_L+\dot{\delta}_R)\cos\theta_2 + 2L_1(C_{10}+C_8Q)\ddot{\theta}_1\cos(\theta_1-\theta_2) \\
&-2L_1(C_{10}+C_8Q)\dot{\theta}_1^2\sin(\theta_1-\theta_2) + 2L_1(C_{10}+C_8Q)\dot{\theta}_1\dot{\theta}_2\sin(\theta_1-\theta_2) \\
&+2C_8L_1\dot{\theta}_1\dot{\theta}_2\cos(\theta_1-\theta_2) + \frac{R_w}{2}(C_{10}+C_8Q)(\dot{\delta}_L+\dot{\delta}_R)\dot{\theta}_2^2\sin\theta_2 \\
&-(C_{15}+C_8Q)g\dot{\theta}_2\sin\theta_2 - 2L_1(C_{10}+C_8Q)\dot{\theta}_1\dot{\theta}_2^2\sin(\theta_1-\theta_2) = T_M - T_{FM} - L_dF_d
\end{aligned}
\tag{4}
$$

$$
\begin{aligned}
&C_8\ddot{Q} - \frac{1}{2}(C_{12}+2C_8Q)\dot{\theta}_2^2 - C_8\frac{R_w}{2}\dot{\theta}_2(\dot{\delta}_L+\dot{\delta}_R)\cos\theta_2 \\
&-2L_1C_8\dot{\theta}_1\dot{\theta}_2\cos(\theta_1-\theta_2) + C_8g\cos\theta_2 = F_a - F_{fa}
\end{aligned}
\tag{5}
$$

3. Hybrid Fuzzy Logic Controller design

Two types of Hybrid FLC are designed and implemented to control the non-linear model of the two-wheeled vehicle. Proportional-Derivative-like fuzzy logic controller (PD-like FLC) is used to control the angular displacement of the wheels, the first link tilt angle and the payload linear actuator displacement. Where a PD plus integral fuzzy logic controllers (PD+I FLC) are used to control the links tilt angles to overcome the steady state error.

The inputs to the PD-like FLC are the error signal and the change of error. While the inputs for the PD+I FLC are the error signal, change of error and the sum of previous errors. Figure 2 and Figure 3 illustrate PD-like FLC and PD+I FLC respectively. Figure 4 presents the Simulink block diagram of the controlled system.

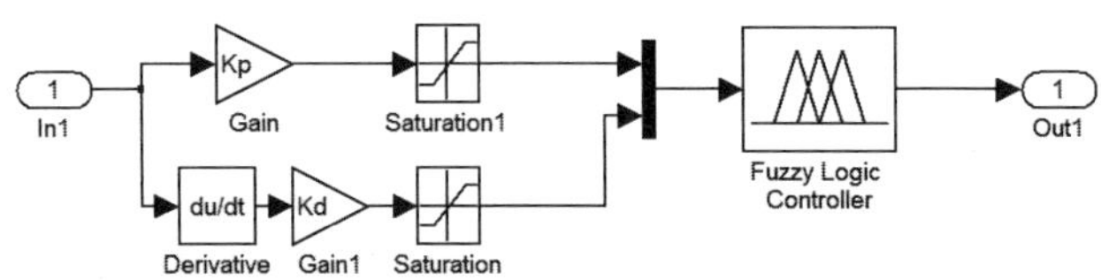

Figure 2: PD-Like Fuzzy logic controller

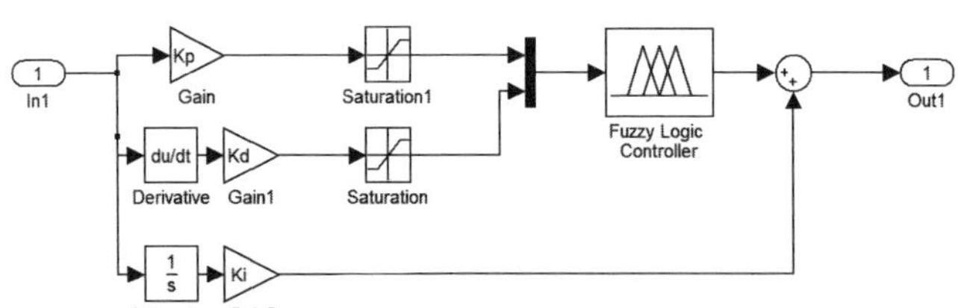

Figure 3: PD+I Fuzzy logic controller

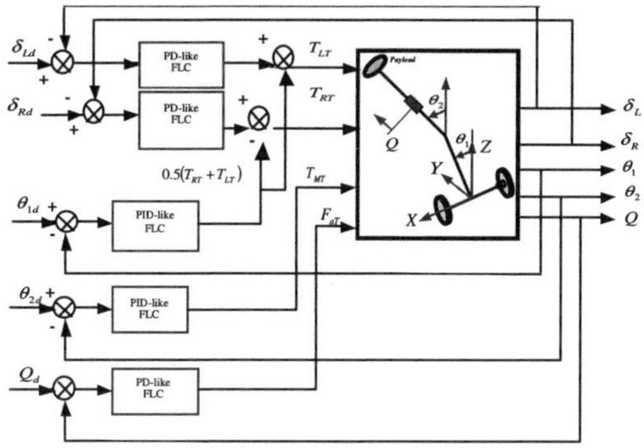

Figure 4: System simulink block diagram

The design of the FLC involves choosing a suitable fuzzy inference engine, defining the fuzzy rules and choosing the membership function type. At this stage of the research, the FLC presented will be based on Mamdani-type of fuzzy inference engine of the form:

If (error) is x and (change of error) is y then the (output) is z

where: x, y and z are linguistic variables describing the input and output levels. The selection of a Mamdani-type fuzzy inference engine is because of its sufficient description of input and output values and simplicity in formulation. Gaussian membership functions for the input and outputs are chosen and illustrated in Figure 5. The linguistic variables describing the inputs and outputs are chosen as Positive Big (PB), Positive Small (PS), Zero (Z), Negative Big (NB) and Negative Small (NS).

4. Simulation and Results

The generation of the fuzzy rules-base is based on the required system performance to minimize the system error between the output signal and the desired signal. Table 1 describes 25 fuzzy rules-base for the FLC output. The system was simulated with the controller gains of Table 2. Figure 6 illustrates the controlled-system response showing a stable balancing response.

Table 1. Fuzzy rules base

e \ ê	**NB**	**NS**	**Z**	**PS**	**PB**
NB	NB	NB	NB	NS	Z
NS	NB	NB	NS	Z	PS
Z	NB	NS	Z	PS	PB
PS	NS	Z	PS	PB	PB
PB	Z	PS	PB	PB	PB

Table 2. PD and PID controllers gains

Loop	***PID controller Gains***		
	K_P	K_I	K_D
Left wheel	2	-	2
Right wheel	2	-	2
First pendulum link	10	-	10
Second pendulum link	1500	2000	1000
Payload linear actuator	0.1	-	100

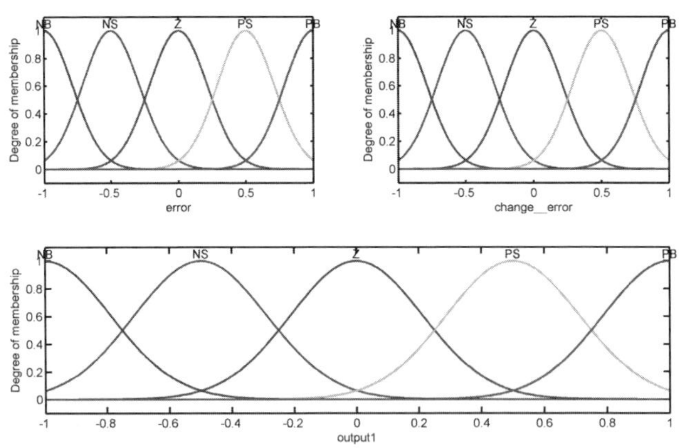

Figure 5: Gaussian fuzzy membership functions

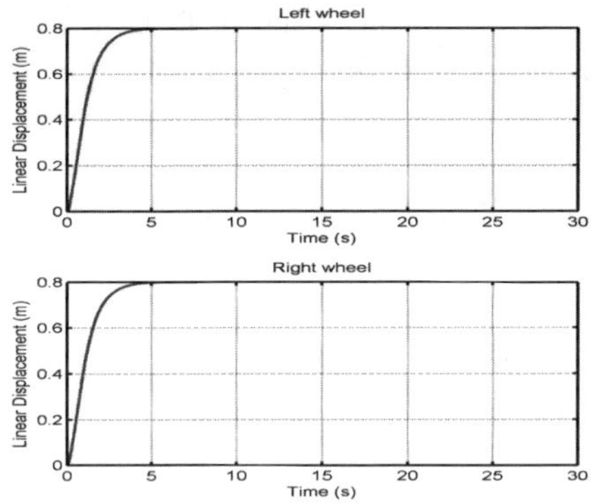

(a) Linear displacement of left and right wheels

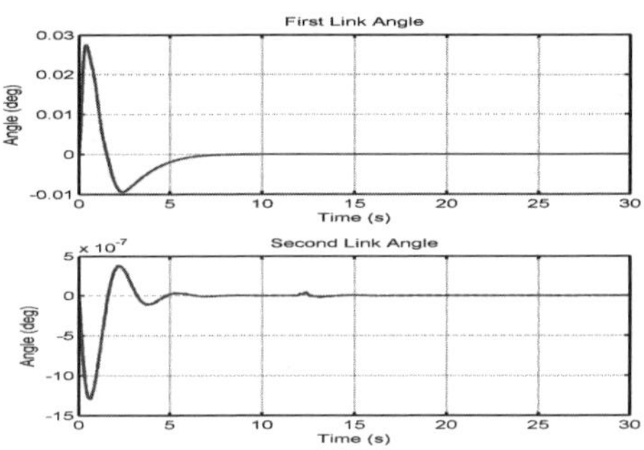

(b) Tilt angles of first and second links

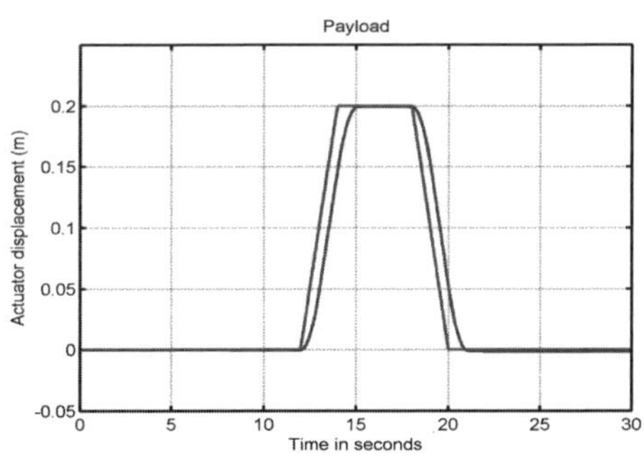

(c) Displacement of payload linear actuator

Figure 6: Controlled system response

The wheel's linear displacement is shown to have a very stable response without an overshoot, while the first and second link tilt angles have an improved overshoot when compared to the PID controller response presented in [2]. On the payload actuator response, it is clear that the controller forces the system to follow the desired input signal but with a slight delay in the rise time. The overall response of the system has been improved with the implemented hybrid FLC.

5. Conclusion

This paper presented the design of a hybrid fuzzy logic controller for the system model. The presented hybrid controller consists of PD-like FLC and PD+I FLC controllers to stabilise the system. The FLC is of a Mamdani-type inference engine with 25 fuzzy rules-base and Gaussian membership functions. Simulations have been presented showing a successful control approach of the system. With the promising results of the controller, further work will be carried out to improve the overall system response by optimising the FLC rules and testing the robustness of the controller.

References

1. K. Goher, S. Ahmad, M. O. Tokhi, "A new configuration of two wheeled vehicles: Towards a more workspace and motion flexibility", Proceedings of 4th Annual IEEE conference on Intelligent Systems, San Diego, CA, USA, 5-8 April 2010.
2. A. M. Al-Meshal, K. M. Goher, M.O. Tokhi, (2011, 8-10 September), Modelling of two-wheeled robotic wheelchair with moving payload. Proceedings of the 14th International Conference on Climbing and Walking Robots and the Support Technologies for Mobile Machines (CLAWAR 2011), Paris, France.
3. S. Ahmad S., M.O. Tokhi, "Modeling and Control of a Wheelchair on Two Wheels" Second Asia International Conference on Modeling & Simulation, 2008. AICMS 08. ,pp.579-584, 13-15 May 2008.
4. K.G. Eltohamy, and C.Y. Kuo, "Real time stabilisation of a triple link inverted pendulum using single control input" , IEE Proceedings on Control Theory and Applications , vol.144, no.5, pp.498-504, Sep 1997.
5. Bogdanov A., Optimal control of a double inverted pendulum on a cart [online], Oregon Health and Science University Technical Report CSE-04-006.

POWER MODELING OF THE XRL HEXAPEDAL ROBOT AND ITS APPLICATION TO ENERGY EFFICIENT MOTION PLANNING

Camilo Ordonez*, Nikhil Gupta, Emmanuel G. Collins, Jr., Jonathan E. Clark

Mechanical Engineering, Florida A&M University-Florida State University, Tallahassee, Fl 32310, USA
**E-mail: camilor@eng.fsu.edu*

Aaron M. Johnson

Electrical and Systems Engineering, University of Pennsylvania, Philadelphia, PA 19104, USA

Analysis of the power consumption for walking and running robots is particularly important for trajectory planning tasks as it enables motion plans that minimize energy consumption and do not violate power limitations of the robot actuators. This paper builds upon previous work on wheeled skid-steered robots, and for curvilinear motion of the XRL hexapedal robot, presents models of the inner and outer side torques and power requirements. In addition, the applicability of the power model to energy efficient motion planning is illustrated for a walking gait on a vinyl surface.

Keywords: skid-steering; turning for legged robots.

1. Introduction

The ability to predict power demands is an important capability for energetically constrained robot motion planning. In particular, the development of robot power models aids in four important areas of robot motion planning:

(1) Motion planning with actuator power limitations,
(2) Energy efficient path/trajectory planning,
(3) Task completion prediction based on available energy,
(4) Planning for refueling or recharging.

Recent work on skid-steered vehicles has led to the formulation of analytic dynamic and power models that rely on the terramechanics of wheel-terrain interactions, in particular, models based on the exponential relationship between shear stress and shear displacement at the contact patch. These models have been verified on flat and rigid surfaces for steady state turning maneuvers.[1] Reduced order models that relax some of the assumptions of the terramechanics models have also been derived for skid-steered vehicles.[2]

While the ground contact patterns for legged robots and in particular the RHex-like robots considered here,[3] are characterized by discrete rather than continuous contact, some work has been done to experimentally develop kinematic models of these systems.[3] There has been some related work focused on reducing power (or specific resistance) of these systems, but not specifically on modeling the power.[4]

The present work is motivated by the hypothesis that for certain regimes of operation (i.e., certain gait parameters), legged robots from the RHex family behave in a similar fashion to skid-steered robots while in general curvilinear motion. The remainder of the paper is organized as follows: Section 2 presents a short description of the experimental platform and its approximated kinematics. Section 3 details the developed torque and power models. Section 4 presents an illustrative example of the application of the developed model on energy efficient motion planning. Finally, Section 5 provides concluding remarks and directions for future work.

2. Robotic Platform and Approximated Kinematics

The XRL (X-RHex Light) experimental platform[5,6] employed in this research and shown in Fig. 1 is a very versatile, slightly lighter version of the X-RHex robot, capable of performing diverse gaits, including walking, jogging, running and pronking. The robot has 6 compliant C-shaped legs independently actuated by 50 Watt Brushless Maxon motors with a gear ratio of 18:1.

Fig. 1. XRL Robot Utilized as the Experimental Platform for this Work.

For this paper we will consider an alternating tripod gait for the XRL locomotion. Referring to Fig. 2, the alternaiting tripods correspond to a left tripod formed by legs 1,3, and 5 and a right tripod formed by legs 2,4, and 6.

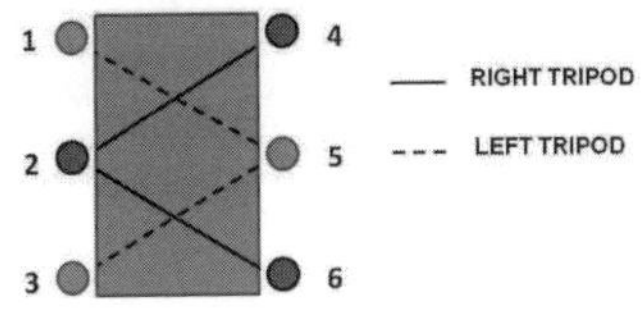

Fig. 2. Left and Right Tripods for XRL.

Each of the tripods follows at the low level a desired trajectory generated by a Bueheler clock,[3] which characterizes the gait by four parameters: the frequency $f\,[Hz]$, nominal duty factor $\overline{d_f}$, nominal leg offset $\overline{\phi_0}\,[rad]$ and the nominal leg sweep angle $\overline{\phi_s}\,[rad]$. These parameters are grouped into the control vector $u = \left[f, \overline{d_f}, \overline{\phi_0}, \overline{\phi_s}\right]^T$, where the frequency and duty factor are expressed in terms of the cycle t_c and stance t_s times by ($f = \frac{1}{t_c}$ and $\overline{d_f} = \frac{t_s}{t_c}$).

In order to turn while in motion, perturbations on the gait parameters for the inner (1,2,3) and outer (4,5,6) legs are introduced (hereafter we assume left hand turns). These perturbations are functions of a turn gain t_g and some extra constants (α, β, γ). The perturbed gait parameters for the inner legs are expressed by:

$$\phi_o^i = \overline{\phi_o} - t_g\alpha, \tag{1}$$

$$\phi_s^i = \overline{\phi_s} - t_g\beta, \tag{2}$$

$$d_f^i = \overline{d_f} - t_g\gamma. \tag{3}$$

The outer legs parameters ϕ_o^o, ϕ_s^o, and d_f^o are computed similarly but with opposite signs for the perturbations. For all experiments performed in this work, $\alpha = 0.05$, $\beta = 0.7$, and $\gamma = 0.0$. The robot kinematics were derived by conducting experiments for turning gains in $\{0.0, 0.04, 0.1, 0.2, 0.4, 0.5, 0.6\}$ and frequencies in $\{0.5, 0.6, 0.7, 0.8, 0.9, 1.0, 1.1\}Hz$ using a walking gait parameterized by $u = [f, t_g, 0.65, -0.314, 0.785]^T$. Robot forward velocity and turn radius were measured by tracking with a high-speed digital camera (Casio Exilim EX-F1) a set of LEDs mounted on the robot. Figs. 3 and 4 summarize the approximated kinematics for the walking gait on a vinyl surface. In particular, the robot forward velocity $v[m/s]$ and the robot turn radius $\rho[m]$ are approximated by:

$$v = 0.16f, \tag{4}$$

$$\rho = 13.92e^{-5.78t_g} + 0.764. \tag{5}$$

It is important to note that, as can be seen from Figs. 3 and 4, no clear dependence of the velocity on the turn gain and of the turn radius on the frequency was observed and that is the reason why Eqs. 4, and 5 depend only on one variable. To generate these models, the experimental data set was split using 80% of the data for model fitting and 20% for validation, which yielded average prediction errors of 4.1% for the forward velocity and 17.7% for the turn radius.

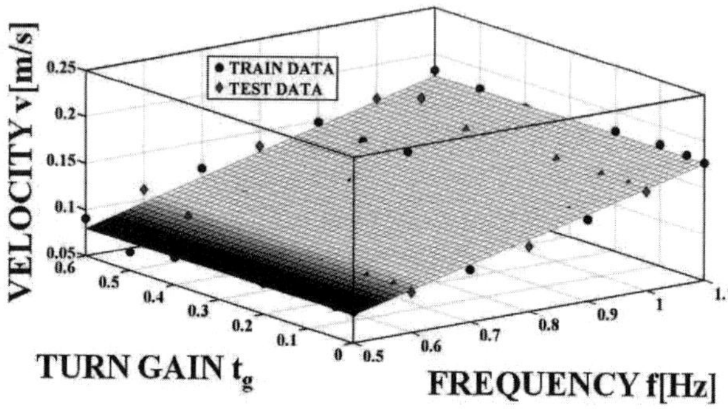

Fig. 3. XRL Kinematics (forward velocity).

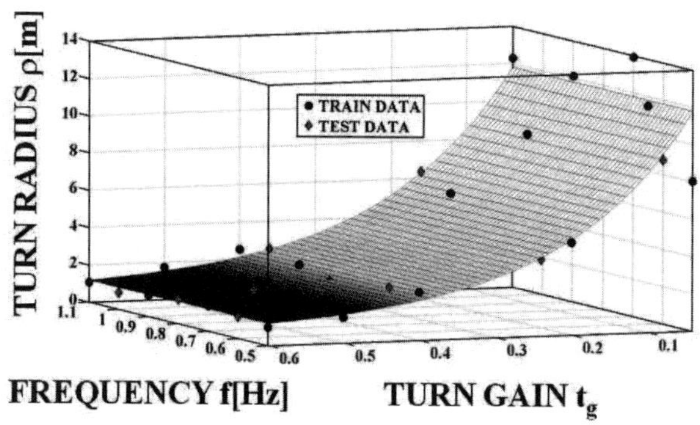

Fig. 4. XRL Kinematics (turn radius).

3. Torque and Power Model

During the experiments described in Section 2, motor torques τ for each leg were monitored. Then, the average torques per cycle for the inner and outer legs were estimated using:

$$\overline{\tau_i} = \overline{\tau_1(t) + \tau_2(t) + \tau_3(t)}, \quad t \in [0, tc], \tag{6}$$

$$\overline{\tau_o} = \overline{\tau_4(t) + \tau_5(t) + \tau_6(t)}, \quad t \in [0, tc], \tag{7}$$

where $\overline{x}$ is the average of x. Fig. 5 shows that the torques follow clear exponential trends as a function of turn radius, which is qualitatively similar to the torque vs turn radius curves observed for skid-steered robots.[1,2] Notice that as the turn radius increases, the torques tend to a small value

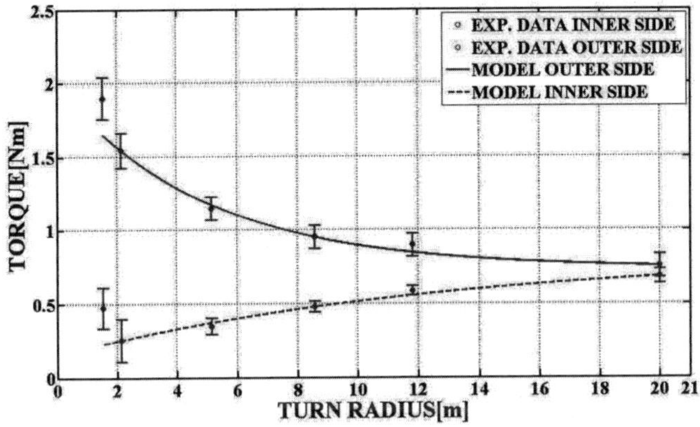

Fig. 5. XRL: Inner and Outer Side Torques as a Function of Turn Radius.

which is associated with the rolling resistance (error bars represent torque standard deviation for the different speeds).

The validation turn radii ρ are in the set $\{5.14, 11.81\}m$. For these ρ, the fitted models yield average prediction errors of 3.81% for the outer side and 6.04% for the inner side. For turn radii of less than $2.14m$, the model does not capture the behavior of the inner side torque, perhaps due to the fact that for these sharper turns roll motion is increased, making the gait less stable and modifying the steering dynamics significantly.

Power for the inner and outer sides was computed by adding the mechanical power and the electrical losses as follows:

$$P_i = \overline{\Sigma_{j=1}^{3}(|\tau_j(t)w_j(t)| + I_j^2(t)R)}, \quad t \in [0, tc], \tag{8}$$

$$P_o = \overline{\Sigma_{j=4}^{6}(|\tau_j(t)w_j(t)| + I_j^2(t)R)}, \quad t \in [0, tc], \tag{9}$$

where w_j is the angular velocity of leg j, I_j is the current through the motor corresponding to leg j and R is the phase to phase electrical resistance of each of the six motors. Motivated by insight gained from power modeling of skid-steered robots,[2] we look for a model of similar shape where power is related to velocity and turn radius by

$$P = a(\rho)v + b(\rho), \tag{10}$$

where a and b are coefficients dependent on the turn radius. Experimental curves shown in Figs. 6 and 7 relating power and speed for different turn radii were obtained for both the inner and outer robot sides. Straight lines were then fitted to data corresponding to each turn radius and their intercepts and slopes were used to estimate the power model parameters a and b for both robot sides. These parameters are summarized in Table 1. The power model was then evaluated at the test speeds v in the set$\{0.096, 0.128, 0.16\}m/s$, yielding average errors over all turn radii of 8.69% and 9.19% for the outer and inner side respectively.

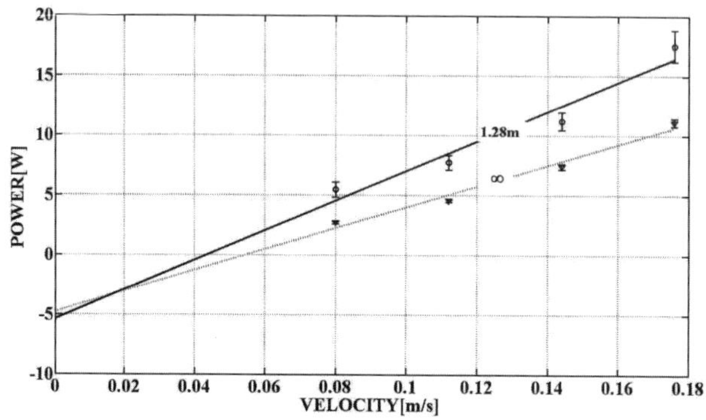

Fig. 6. Inner Side Power Curves (only two turn radii ($1.28m$ and ∞) are shown for ease of visualization).

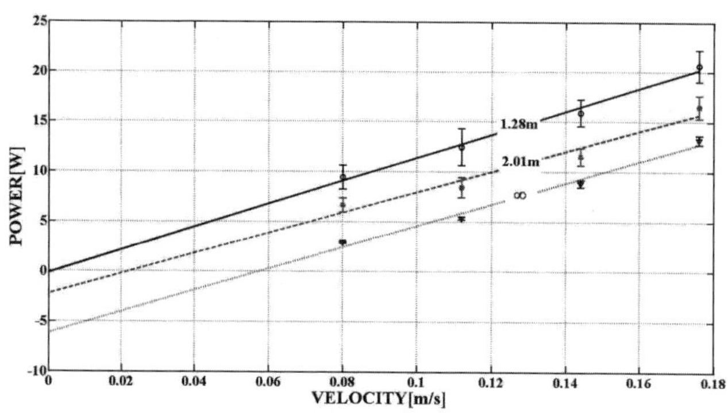

Fig. 7. Outer Side Power Curves (only three turn radii are shown for ease of visualization).

Table 1. Power Model Parameters.

INNER SIDE		OUTER SIDE	
$a_i(\rho)$	$b_i(\rho)$	$a_o(\rho)$	$b_o(\rho)$
$240.6e^{-1.26\rho} + 88.82$	$-1.40e^{-0.07\rho} - 4.35$	106.9	$9.21e^{-0.34\rho} - 6.12$

4. Application of the Proposed Model To Energy Efficient Motion Planning

Using the power models developed in Section 3, it is possible to generate the power surface illustrated in Fig. 8, which combines the power of both robot sides. This power surface was then integrated with Sampling-Based Model

Predictive Optimization (SBMPO), which is a motion planning algorithm capable of planning efficiently with dynamic models.[7] The outcome of this integration is the simulation result shown in Fig. 9, which compares paths obtained by optimizing a distance cost function and an energy cost function. Notice that energy efficient motion planning saved 33.5% in energy with only a 13.1% increase in distance traveled.

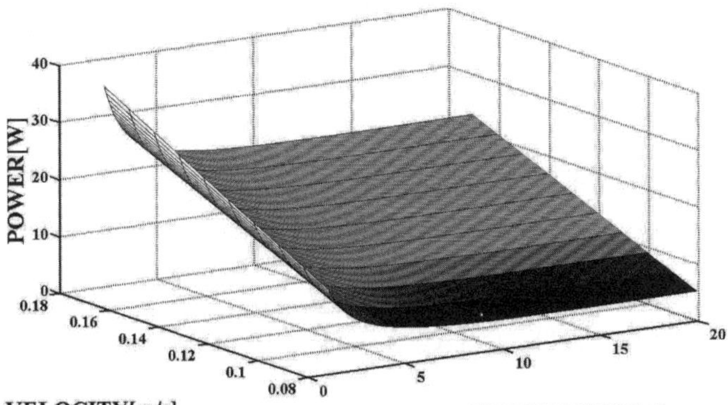

Fig. 8. XRL Power Surface.

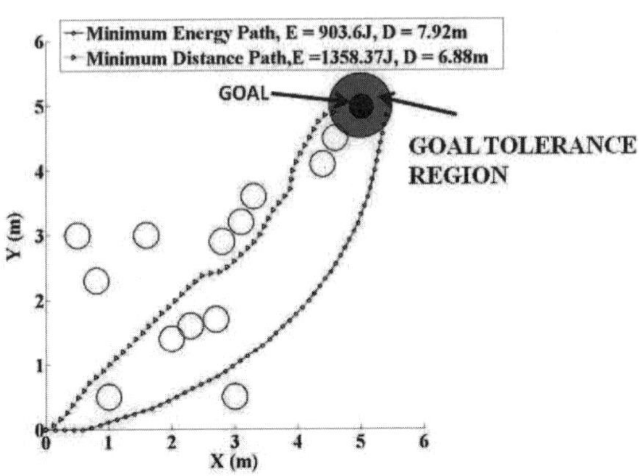

Fig. 9. Minimum Energy Vs Minimum Distance Paths.

5. Discussion and Future Work

Preliminary experimental results for a walking gait show significant similarities in the torque and power vs turn radius trends for the XRL and skid-steered wheeled vehicles. The observed similarities, suggest that at least to some extent the frictional forces involved in the curvilinear motion of both platforms share some commonalities and should be exploited in the future to obtain an analytical model of the dynamics of turning for RHex-like robots.

In addition, initial path planning results using the obtained power model for a walking gait show that paths obtained using energy optimization tend to be smoother than the paths obtained using distance optimization since the power curves show that more power is needed for turns with small radius. Future work will involve the verification of these observations for different gaits and the experimental validation of the energy estimates obtained using SBMPO.

Acknowledgments

This work was supported by the collaborative participation in the Robotics Consortium sponsored by the U.S. Army Research Laboratory under the Collaborative Technology Alliance Program, Cooperative Agreement DAAD 19-01-2-0012. The U.S. Government is authorized to reproduce and distribute reprints for Government purposes not withstanding any copyright notation thereon.

References

1. W. Yu, O. Chuy, E. Collins and P. Hollis, *IEEE Transactions on Robotics* **26**, 340 (2010).
2. O. Chuy, E. Collins, W. Yu and C. Ordonez, Power modeling of a skid steered, wheeled robotic ground vehicle, in *Proceedings of the IEEE Conference on Robotics and Automation*, (Kobe, Japan, 2009).
3. U. Saranli, M. Buehler and D. Koditschek, *The International Journal of Robotics Research* **20**, 616 (2001).
4. J. Weingarten, G. Lopes, M. Buehler, R. Groff and D. Koditschek, Automated gait adaptation for legged robots, in *Proceedings of the IEEE Conference on Robotics and Automation*, (New Orleans, USA, 2004).
5. G. C. Haynes, J. Pusey, R. Knopf, A. M. Johnson and D. E. Koditschek, Laboratory on legs: An architecture for adjustable morphology with legged robots, in *SPIE Unmanned Systems Technology XIV (Conference 8387)*, 2012. To Appear.
6. K. C. Galloway, G. C. Haynes, B. D. Ilhan, A. M. Johnson, R. Knopf, G. Lynch, B. Plotnick, M. White and D. E. Koditschek, *X-RHex: A Highly Mobile Hexapedal Robot for Sensorimotor Tasks*, tech. rep., University of Pennsylvania (2010).
7. D. D. Dunlap, C. V. Caldwell, J. E. G. Collins and O. Chuy, *Motion Planning for Mobile Robots Via Sampling-Based Model Predictive Optimization, Recent Advances in Mobile Robotics* (InTech, 2011). http://www.intechopen.com/books.

TRAJECTORY FOLLOWING CONTROL FOR OMNI-DIRECTIONAL WALKING OF A SIX-LEGGED ROBOT

HIROAKI UCHIDA, NORIYUKI SHIINA, HIROYUKI KONDOH

Department of Mechanical Engineering, Kisarazu National College of Technology, 2-11-1, Kiyomidai-higashi, Kisarazu, Chiba, 292-0041, Japan

This paper dealt with a control method for an achieving walking not to follow the desired angle of the rotating links between supporting phases. In this study, we verified the validity of LQI control method to control the position of the center of gravity and the yawing angle of the body. We examined the effectiveness of the proposed design method by 3D simulations where the reference orbit was semicircle and S-turns.

1. Introduction

The activity of a six-legged robot is expected under the extreme environments. As a walking control method of a six-legged robot, generally the orbit of the toe is designed, the target orbit of each leg joint is beforehand calculated from Inverse Kinematics, and the feedback(FB) control method(PD control) to follow the reference signal of each leg joint is used. However, since it is need to recognize walking environment beforehand and a walking orbit is designed after that, it is hard to say that it is practical control method under the unknown environments. The studies[1,2] of a quadruped robot have been reported where the design of the target orbit of each leg link for a circular orbit and a rotational orbit was treated in the multi-legged robot. However, in a circular walk of a quadruped robot, it is difficult to seek for a movement of a supporting leg and a swing leg easily. As far as we know, there was no report that treated a design method of a circular orbit and a rotational orbit for a six-legged robot. That's why there are many walking patterns for a six-legged robot compared with a quadruped robot and it becomes difficult to formulate qualitatively. In a walk of a multi-legged robot, if the projected point of the centre of gravity of the body to the walking surface exists in the polygon where support legs make, static stability is maintained. The control method that the position of the centre of gravity and the posture of the body become desired values is needed without designing the target orbit of a support legs in advance. From such viewpoints, we've studied the walking control method of a six-legged robot using a six-legged robot which has a leg mechanism, Hirose et.al proposed[3], divided into the link supporting the body and the link controlling the movement of a walking direction. Uchida et al.[4] reported the control method to control three axes of the height, pitching and rolling angles of the body by controlling the link to support the body. And, Uchida et al.[5] proposed the control method to control the body position in the walking plane and the yawing angle of the body in three support phase. By the proposed control method, the equations of motion around x, y direction and yawing angle of the body were derived, and optimal servo system (LQI) was applied. The validity of the proposed control method was verified only on the target orbit in the local coordinate of the robot, and it was not verified in a world coordinate. In this study, the target orbit was designed by the world coordinate, and the following performance to the target orbit transformed into a local coordinate of the robot was examined.

2. Six-legged Robot

Figure 1 shows the 3D CAD model of a six-legged robot. 3DCAD model was built in Virtual.Lab Motion (LMS Coop.). One leg of the six-legged robot consists of three joints that are a rotating part, a thigh part and a shank part. The advantages of the leg mechanism of the six-legged robot are that the pitching angle, the rolling angle and the height of the body are controlled by controlling the thigh links, the walking direction (x, y) and yawing angle of the body are controlled by controlling the rotating links.

3. Walking Planning

The effectiveness of the proposed method is verified by the walking pattern of three support legs and three swinging legs. Walking period is 9 sec.

4. Omni Directional Control Method

4.1. *Transformation from world coordinate into local coordinate*

Because the target orbit of a six-legged robot is given by a world coordinate, it is transformed into a local coordinate of the robot. The origin of a local coordinate of the robot is set at the center-of-gravity of the robot body. The relation between the error (e_{xr}, e_{yr}) in a local coordinate of the robot between the target orbit and a center-of-gravity position, and the yawing angle (θ_{yr}) of a body are shown in Figure 2. Error (e_{xr}, e_{yr}) are obtained in a world coordinate and calculated by using the coordinate transform matrix where rotating angle is yawing angle of the body.

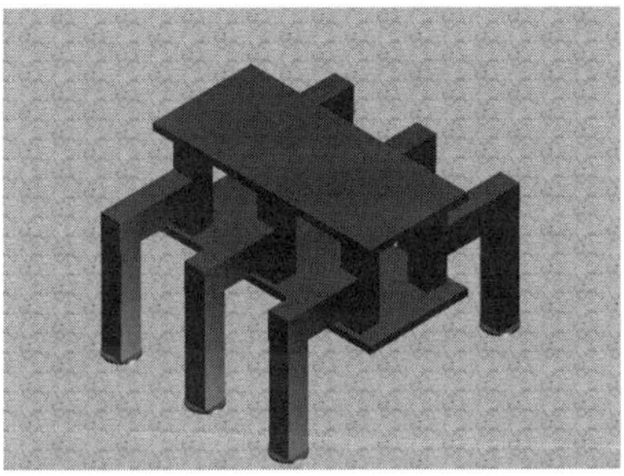

Fig.1 3D CAD model of a 6-legged robot

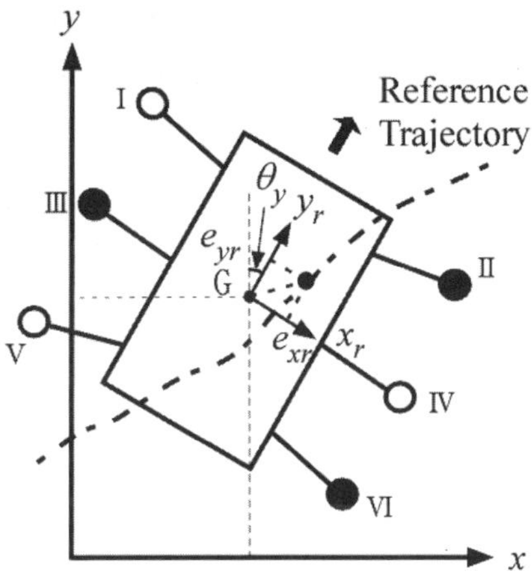

Fig.2 Relation between local coordinate of a robot and world coordinate

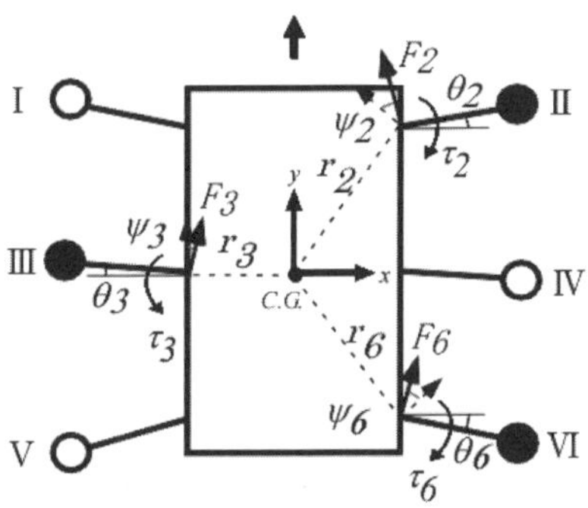

Fig.3 Mathematical model of walking direction in the 3 support legs

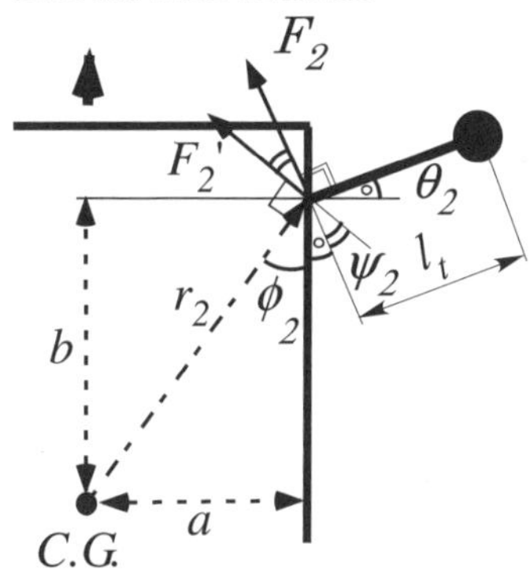

Fig.4 The relationship of ϕ , θ and ψ .

4.2. *Mathematical Model of the Body in the Walking Direction*

Figure 3 shows the relation of the force working at the rotational part in the 3 support legs. Here, support legs are Leg II, III and VI, and swing legs are Leg I, IV, and V. Moreover, the support legs are assumed to support the body firmly, and not to slip. y is the walking direction, x is the vertical direction to y, θ_z is the angle around z axis. Moreover, the force $F_i\left(i=1,\cdots,6\right)$ from the support leg acts on the rotational axis in the support leg. M is the weight of the body, I_z is the inertia around z axis, $\theta_i\left(i=1,\cdots,6\right)$ is the angle of the rotating link,

$r_i\,(i=1,\cdots,6)$ is the distance between the center of gravity of the body and to the center of the rotation in the rotating link. The motion equation of x-direction, y-direction and around θ_z are shown as follows:

$$\begin{cases} M\ddot{x} = F_2\sin(-\theta_2)+F_3\sin\theta_3+F_6\sin(-\theta_6) \\ M\ddot{y} = F_2\cos(-\theta_2)+F_3\cos\theta_3+F_6\cos(-\theta_6) \\ I_z\ddot{\theta}_z = F_2 r_2\cos\psi_2 - F_3 r_3\cos\psi_3 + F_6 r_6\cos\psi_6 \end{cases} \tag{1}$$

By defining state vector $x=\left[x, y, \theta_z, \dot{x}, \dot{y}, \dot{\theta}_z\right]^T$, input vector $u=\left[\tau_2, \tau_3, \tau_6\right]^T$, following state equation and output equation are obtained. Here, input matrix $B(t)$ becomes time variant matrix.

$$\dot{x}(t)=Ax(t)+B(t)u(t) \tag{2-a}$$

$$y(t)=Cx(t) \tag{2-b}$$

where,

$$C=\begin{bmatrix} 1 & 0 & 0 & 0 & 0 & 0 \\ 0 & 1 & 0 & 0 & 0 & 0 \\ 0 & 0 & 1 & 0 & 0 & 0 \end{bmatrix},\quad \tau_i = l_t F_i\,(i=1,\cdots,6),$$

$$A=\begin{bmatrix} 0 & 0 & 0 & 1 & 0 & 0 \\ 0 & 0 & 0 & 0 & 1 & 0 \\ 0 & 0 & 0 & 0 & 0 & 1 \\ 0 & 0 & 0 & 0 & 0 & 0 \\ 0 & 0 & 0 & 0 & 0 & 0 \\ 0 & 0 & 0 & 0 & 0 & 0 \end{bmatrix},\quad B(t)=\begin{bmatrix} 0 & 0 & 0 \\ 0 & 0 & 0 \\ 0 & 0 & 0 \\ \dfrac{\sin(-\theta_2)}{l_t M} & \dfrac{\sin\theta_3}{l_t M} & \dfrac{\sin(-\theta_6)}{l_t M} \\ \dfrac{\cos(-\theta_2)}{l_t M} & \dfrac{\cos\theta_3}{l_t M} & \dfrac{\cos(-\theta_6)}{l_t M} \\ \dfrac{r_2\cos\psi_2}{l_t I_z} & -\dfrac{r_3\cos\psi_3}{l_t I_z} & \dfrac{r_6\cos\psi_6}{l_t I_z} \end{bmatrix}$$

4.3. *Optimal Servo System*

The servo system that the system shown by Eq. (2) follows to the desired vector $r=\left[r_x, r_y, r_{yaw}\right]^T$ is designed.

$$\begin{cases} \dot{z}(t)=r(t)-Cx(t) \\ \dot{x}(t)=Ax(t)+B(t)u(t) \end{cases} \tag{3}$$

Equation (3) is described in equation form as follows:

$$\dot{x}_g(t) = A_g x_g(t) + B_g(t)u + f_g r \tag{4}$$

The control input to the actuator driving the rotating link is obtained to minimize the following equitation.

$$J = \int_0^\infty \left[x_g(t)^T Q x_g(t) + u(t)^T R u(t) \right] dt \tag{5}$$

Here, $Q(n \times n)$ and $R(m \times m)$ are the weighting matrix given by the design specification, and $Q \geq 0, R > 0$. The control input to minimize Eq.(5) is shown as following:

$$u_b^o(t) = -R^{-1} B_g^T(t) P(t) x_g \tag{6}$$

Here, $P(t)(n \times n)$ is the solution of the following Ricatti equation.

$$P(t)A_g + A_g^T P(t) - P(t)B_g(t)R^{-1}B_g^T(t)P(t) + Q = 0 \tag{7}$$

5. 3D Simulations

The walking simulations were done in Case I and Case II. Case I was that the reference orbit was semicircle of 0.5m radius and 1 period of walking is 9 sec. Total period of walking is 18 cycles. MATLAB/Simulink was used as a solver of simulation. Figure 5 shows the simulation results. Fig.5(a)~(c) show the time response of *x*, *y* and yawing angle of the body, respectively. And, Fig.5(d) shows

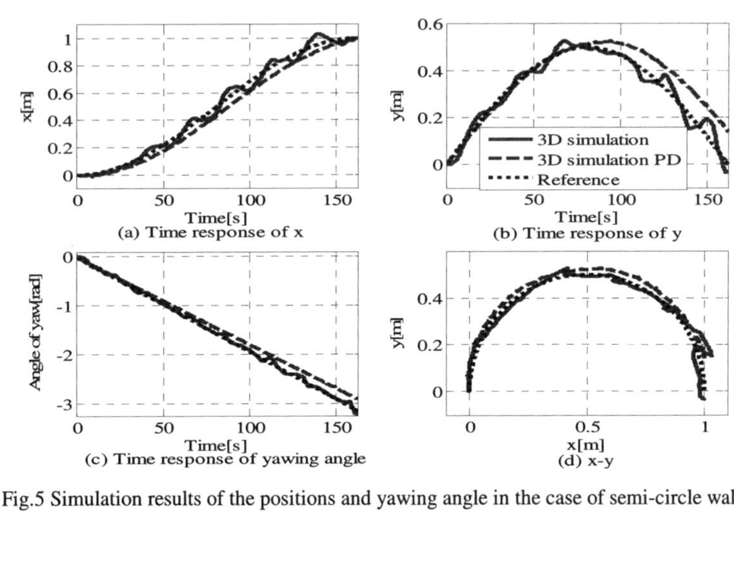

Fig.5 Simulation results of the positions and yawing angle in the case of semi-circle walking

walking trajectory in *x-y* plane. In Figures, the solid line shows LQI control, the dashed line shows the PD control to follow the target orbit of each leg joint obtained by Inverse Kinematics in the case that a six-legged robot can realize a semicircle walking, and the dotted line shows a target orbit. The weighting matrix of Eq.(5) was applied as $Q = diag(50,100,100,0,\cdots,0)$, $R = diag(1,1,1)$. In Fig.3(b), the rotating link was controlled by the proposed control method in the supporting phase and the position control to follow the desired orbit in the swing phase, respectively. Then, the ranges of the movements in the rotating links were $-20° \le \theta_{1,2} \le 60°$, $-30° \le \theta_{3,4} \le 30°$, $-60° \le \theta_{5,6} \le 20°$. And, when the rotating link exceeded it, the control input to the rotating link was zero. In Eq.(6), feedback system becomes time-varying system. This becomes a problem for an experimental set up. So, we designed the feedback system by the fixed values as follows. θ_1, θ_2 were set by 30[deg], in the case that θ_3, θ_4 were positive value, θ_3, θ_4 were set by 5[deg], in the case that θ_3, θ_4 were negative value, θ_3, θ_4 were set by -5[deg], θ_5, θ_6 were set by -30[deg]. In this case, there was a possibility that slipping was caused at the foot. However, in this study, even if slipping was caused, the simulations were continued without change because it has aimed to confirm the effectiveness of the proposed control method. The oscillatory response appeared in *x* and *y* directions between 14～18 period.

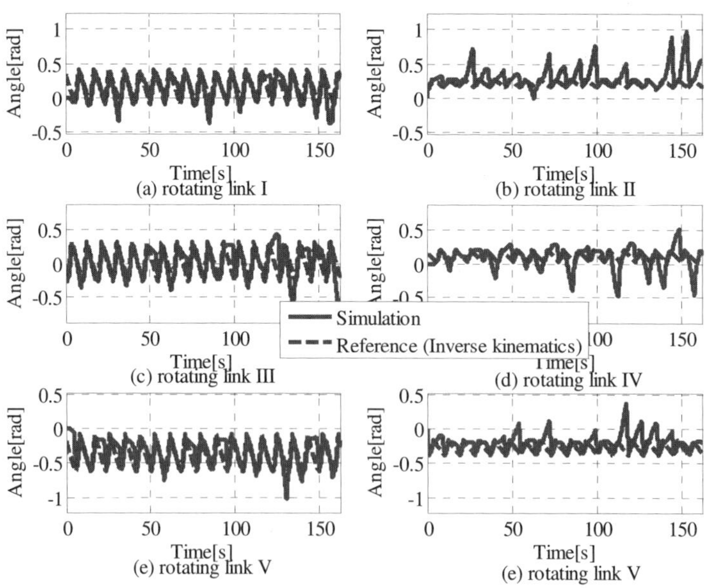

Fig.6 Simulation results of rotating angles in the case of semi-circle walking

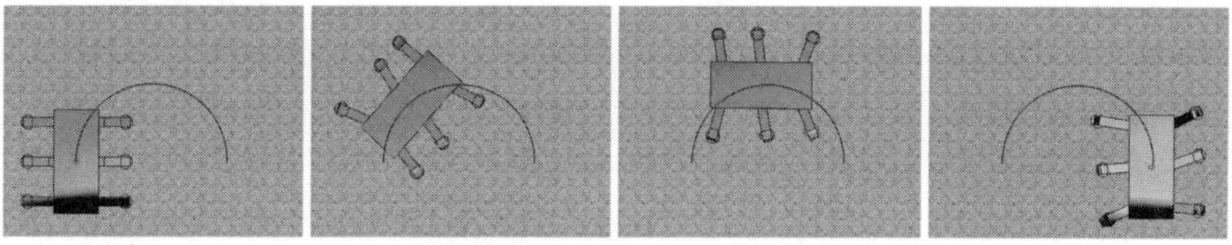

(a) 0sec (b) 40.5sec (c) 81sec (d) 162sec

Fig.7 Animations of semi-circle walking

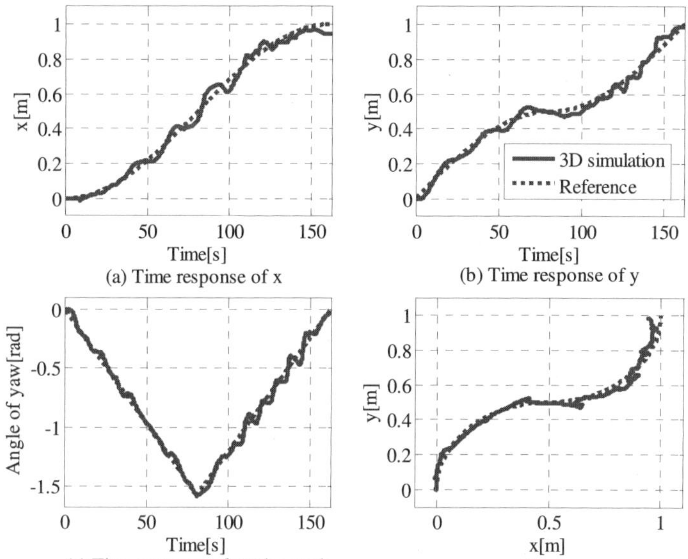

(a) Time response of x (b) Time response of y (c) Time response of yawing angle (d) x-y

Fig.8 Simulation results of S-turns walking

This phenomenon related to the movement to back shown in Fig.6(d). Moreover, the error of the yawing angle at the terminal point becomes -0.069[rad] in LQI control, the error of PD control becomes 0.240[rad]. Figure 6 shows the time responses of rotating links. The dashed line shows a target orbit obtained by Inverse Kinematics in the case that a six-legged robot can realize a semicircle walking. PD control shown as Fig.5 only controlled the each leg link to follow this desired orbit. Because it is a circular orbit, the amount of change of inside leg, Leg II, IV and VI for a swing phase becomes smaller than outside leg, Leg I, III and V. The solid line shows the time response of LQI control. The oscillatory response appeared in x and y directions in LQI control as shown in Fig.5. We can see that this oscillatory response was caused by the large movement of the rotating links. Figure 7 shows the animations. In Fig.7, the red solid line shows the target orbit of the trajectory of the center of gravity of the body. We can see that the six-legged robot was controlled so that the position of the center of

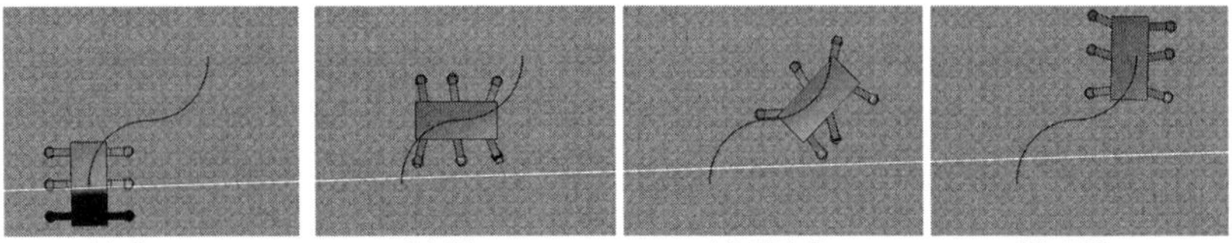

(a) 0sec (b) 81sec (c) 121.5sec (d) 162sec

Fig.9 Animations of S-turns walking

gravity and the yawing angle of the body follow the target orbit. Case II was that the reference orbit was S-turns of 0.5m radius and 1 period of walking is 9 sec. Total period of walking is 18 cycles. Figure 8 and 9 show the 3D simulation results and animations for this reference. In this case we can see that the six-legged robot was controlled to follow the target orbit. The validity of the proposed control method using LQI control system to follow the target orbit of the position of the center of gravity and the yawing angle of the body which was designed in a world coordinate system was verified by 3D simulations.

6. Conclusions

As a trajectory following control for omni-directional walking of a six-legged robot, LQI control system to follow the target orbit of the position of the center of gravity and the yawing angle of the body which was designed in a world coordinate system was designed. By using 3D model of the six-legged robot, the effectiveness of the proposed control method was examined in the case that the reference orbit was semicircle of 0.5m radius and S-turns of 0.5m radius. As a result, the validity of the proposed control method was verified.

References

1. S. Hirose, H. Kikuchi and Y. Umetani, The standard circular gait of a quadruped walking vehicle, *Advanced Robotics*, Vol. 1, No.2 (1986), pp143-164.
2. S. BAI, K. H. Low and T. ZIELINSKA, Quadruped free gait generation for straight-line and circular trajectories, *Advanced Robotics*, Vol. 13, No.5 (1999), pp513-538.
3. Hirose S. and Arikawa K., Development of Quadruped Walking Robot TITAN–VIII for Commercially Available Research Platform, Journal of Robotic Society of Japan, Vol. 17, No.8 (1999), pp1191-1197. (Japanese)
4. H. Uchida and K. Nonami, Attitude Control of a Six-legged Robot in Consideration of Actuator Dynamics by Optimal Servo Control System, *Climbing & Walking Robots Towards New Application, I-Tech,* ***pp*** 299-312 (2007).
5. H. Uchida and T. Kameyama Walking Direction Control of a Six-legged Robot by Optimal Servo System, *Proc. 14th Int. Conf. Climbing and Walking Robots and the Support Tech. for Mobile Machines,* ***pp*** 758-765 (2011).

GEOMETRIC MECHANICS FOR SAND-SWIMMING

Ross L. Hatton[1*], Yang Ding[2], Daniel I. Goldman,[2] and Howie Choset[1]

[1] *Robotics Institute, Carnegie Mellon University*
[2] *School of Physics, Georgia Institute of Technology*
* *rlhatton@cmu.edu*

Sand-swimming, a form of desert burrowing, offers interesting potential as a locomotion mode for robots operating in loose, granular environments. Unfortunately, the computational cost of modeling the relevant physics raises obstacles to a thorough exploration of the system dynamics. Geometric mechanics offers techniques for reducing the complexity of evaluating gaits, thereby offering the potential for exploring a gait design space; unfortunately, these tools have historically been restricted to systems with linear, analytical dynamics. In this paper, we present a framework for combining empirical data from nonlinear models with geometric gait evaluation methods. The resulting tools both reduce the computational costs of describing sand-swimming and reveal fundamental aspects of the motion.

Keywords: geometric mechanics, sand-swimming, gait

1. Introduction

Sand-swimming, a burrowing mode employed by some desert-dwelling animals, offers interesting possibilities for robotic locomotion in similar environments, especially where concealment or harsh surface conditions are a concern. To make use of this mode of locomotion, however, it is necessary to move beyond simple biological inspiration and identify its key physical principles. In our past work,[1,2] we have developed models of sand-swimming that capture the granular flows and forces acting on the locomoting system. These models are, however, computationally expensive, making it difficult to experiment with different body morphologies or gait patterns. In this paper, we develop computational procedures that allow us to both better understand the fundamental physics of locomotion in granular flow and to expediently evaluate gaits so that we can more readily explore the design space of mechanism and controller.

The tools developed in this paper have a solid foundation in geometric mechanics. This field offers several methods for reducing computational

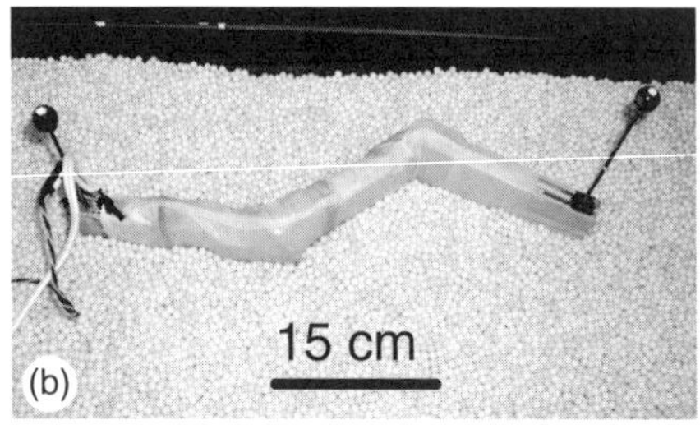

Fig. 1. (a) Sandfish lizard. (b) Three-link robot on bed of plastic particles. LED masts at the end of the robot allow for tracking while robot is submerged.

complexity in locomotion problems. First, it introduces notions of *symmetry* that present opportunities for efficient re-use of expensive calculations. Second, these symmetries identify effective gait cycles with geometric features on a system's configuration space,[3] allowing them to be directly identified with minimal trial-and-error iteration. Unfortunately, these methods have historically been restricted to systems with analytically-describable linear dynamics, and have not found wider use in considering systems in non-linear or non-analytic regimes.

To move past these limitations on geometric analysis, and bring their benefits to complex locomotion modes like sand-swimming, we have developed a hybrid approach that uses a geometric framework to organize empirical information gleaned from nonlinear physics models. This approach gives us considerable insight into sand-swimming, highlighting how gait amplitudes and patterns contribute to the displacements they induce. Additionally, the underlying framework is inherently generalizable to other physical domains, suggesting further avenues of exploration.

2. Sand-swimming

Sand-swimming is a form of locomotion exhibited by a variety of desert animals, such as the sandfish lizard in Fig. 1(a). These organisms move below the surface of the terrain, protecting themselves from heat and predators while stalking their own prey. An interesting feature of this motion is that the sand around the animal can act as a granular frictional fluid,[2] making the progress of the organism much more akin to swimming than it is to burrowing or walking. We have investigated sand-swimming at several levels, including computational modeling and both biological and robotic experiments. The key products of this research are a detailed (and computationally expensive) *discrete element model* (DEM) of the granular flow

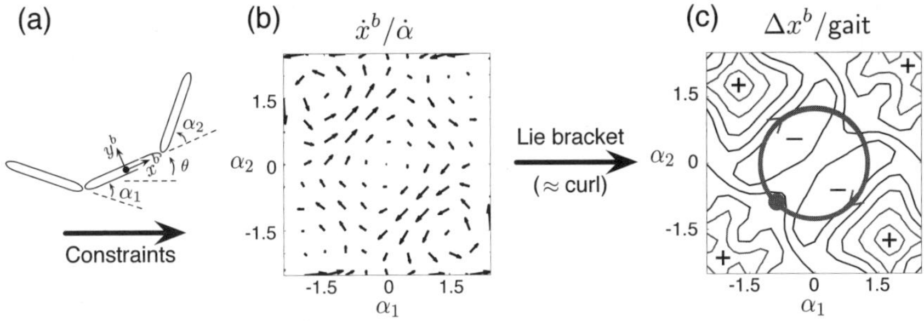

Fig. 2. (a) An example three-link system with joint angles α. (b) Many systems' dynamics can be reduced to linear constraints between a system's body velocity and its shape velocity $\dot{\alpha}$, represented by vector fields. (c) The net displacement from a gait corresponds to the areas it encloses on the components of the *Lie bracket* of the fields (approximately their curls). These plots are for a three-link sandswimmer, and were derived via the method described in §4.

around a sand-swimmer, capturing the motions of individual grains; a simpler (but still nonlilnear) *resistive force theory* model for the forces on a body moving through a granular medium, fit to empirical data from the DEM; and a robotic test platform, a three-link version of which appears in Fig. 1(b). These tools allow us to evaluate the motion of a sand-swimmer, but the time taken for a given evaluation (on the order of days for a DEM simulation of a given gait) makes extensive trials prohibitively costly.

3. Geometric Mechanics

Geometric mechanics applies principles from differential geometry to problems in classical mechanics.[4,5] A powerful application of these principles to locomotion is the reduction of many mobile systems' equations of motion to a set of shape-dependent linear constraints between its body velocity and its rate of shape change.[3,6–12] These constraints can be visualized as a set of vector fields, like that shown in Fig. 2(b). Applying a function called the *Lie bracket* (conceptually similar to curl) to these fields measures how the constraints vary over different shapes. By Stokes's theorem, the net displacement over any gait corresponds to the area it encloses on these Lie bracket functions.[3] For example, the gait in Fig. 2(c) negatively (counterclockwise) encloses a negative region of the function, and so produces a positive x^b displacement of the system.

Lie bracket approaches are based on an identity between the *exponential coordinates** $z(\phi)$ of the net displacement over a gait ϕ (a closed trajectory in the shape space) and a series whose first two terms correspond to the integral of the abstract *curvature* of the constraints over a region of the shape space bounded by ϕ.[13] The Lie bracket measures the *curvature* of the constraints encoded by the local connection, which also corresponds to the net translation induced by a differential oscillation in the system's shape. In two dimensions, the identity appears as

$$z(\phi) = \iint_{\phi} \overbrace{\underbrace{-\text{curl}\mathbf{A}}_{\text{nonconservativity}} + \underbrace{\left[\mathbf{A}_1, \mathbf{A}_2\right]}^{\text{curvature (Lie bracket)}} d\alpha + \text{higher-order terms}}_{\text{noncommutativity}}, \tag{1}$$

where the curl operator is applied individually to each row of $\mathbf{A}$, and $[\mathbf{A}_1, \mathbf{A}_2]$ is the *local* Lie bracket of the columns of $\mathbf{A}$ (taken as if $\mathbf{A}$ did not depend on the shape), which evaluates on $SE(2)$ as

$$\left[\mathbf{A}_1, \mathbf{A}_2\right] = \begin{bmatrix} a_{y1}a_{\theta 2} - a_{y2}a_{\theta 1} \\ a_{x2}a_{\theta 1} - a_{x1}a_{\theta 2} \\ 0 \end{bmatrix}. \tag{2}$$

In (1), the curl term measures the *nonconservativity* of the local connection, or how the constraints change over the shape space, preventing antipodal segments of a gait from pushing or pulling the system equally. The local Lie bracket and higher order terms correspond to the *noncommutativity* of the system's position space, *i.e.*, the extent to which translations with intermediate rotations do not commute, as in parallel parking maneuvers.

4. Hybrid Emprical-Geometric Mechanics

Geometric constraint models like those shown in Fig. 2 both improve the speed of computation—they are typically faster to evaluate than the raw forcing equations on a given system—and allow for simple characterization of gait effects via the Lie bracket. In past work, however, it has been

*The exponential coordinates of a position are the components of the constant body velocity required to reach that position in unit time, starting from the origin. A mapping between exponential coordinates and displacements is provided in various sources,[11] but for the purposes of this paper, it is sufficient to note that this mapping is an identity mapping for pure translation or pure rotation; i.e. $\exp\left([z_x, z_y, 0]^T\right) = (z_x, z_y, 0)$ and $\exp\left([0, 0, z_\theta]^T\right) = (0, 0, z_\theta)$.[11]

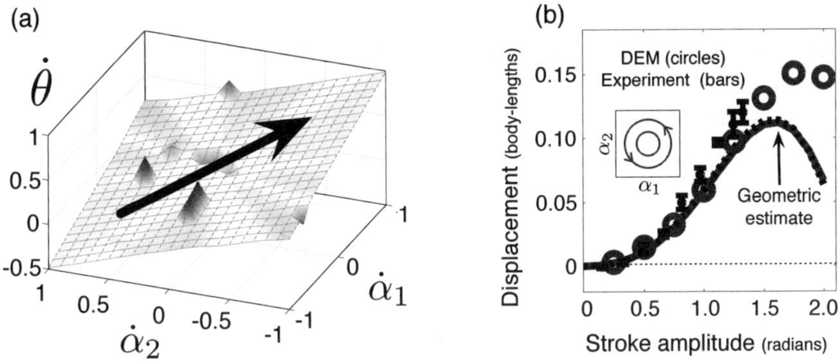

Fig. 3. (a) The body velocity of the sand-swimmer is empirically a linear function of the shape velocity $\dot{\alpha}$. (b) Geometric analysis of an empirically-fit sandswimmer model gives good accuracy, at a fraction of the computational cost.

assumed that these constraints can only be generated from analytically-describable linear physics models, such as the conservation-of-momentum constraints acting on a satellite or the low-Reynolds number flows around a micro-organism. Granular flow produces nonlinear forces that must often be evaluated through simulation, which would initially seem to disqualify it from such analysis.

In our work with sand-swimmers, however, we have observed that these nonlinear forces combine to produce nearly-linear system velocities, as shown in Fig. 3(a). Fitting linear functions to this velocity relationship (at a reasonably-dense sampling of shapes) produces a geometric constraint model that approximates the behavior of the sand-swimmer. This constraint model is orders of magnitude faster to evaluate than the DEM (seconds verses days), and, as shown in Fig. 3(b), gives an accurate approximation of the system motion for gaits with all but the largest joint motions. Further, by observing patterns in the Lie bracket function, we can see *why* the displacement grows and falls off with amplitude as it does: small amplitude gaits enclose sign-definite areas, so increasing the amplitude increases the net displacement, while very large gaits enclose both positive and negative areas, and so do not experience the same growth in displacement.

5. Optimal Coordinates

A fundamental problem with integrating the motion of the system via Lie brackets is that they do not completely measure the net displacement; instead, they measure the "forwards minus backwards" motion in each body direction, together with a first-order approximation of the non-commutative effects of rotation coupled with translation. Although planar rotations always commute with each other and are unaffected by translations, translations specified in body coordinates only commute when there are no intervening rotations. For example, moving a system forward and backward by one unit each returns it to the origin, but interposing a rotation between the two steps leaves the system with a net translation. This noncommutativity can be captured

This lack of commutativity was long viewed as inherent to the locomoting systems.[11] In our work on nonholonomic systems, however, we observed that in many instances the noncommutativity can be alleviated by an appropriate choice of body frame for the system.[3] In these choices of coordinates, the reference line on the system used to represent its orientation θ rotates very little in response to changes in shape, allowing the translation motions produced by those shape changes to "almost commute."

The optimal choice of body frame for a system are those for which the vector fields encoding the constraint are "smallest" – i.e. those for which a given change in shape produces the least translation and rotation of the system. We can directly solve for these optimal coordinates by first making an arbitrary choice of body frame (the center link is often a good choice, as it allows for convenient representation of the system kinematics, and finding the constraint vector fields for that frame. We can then find the body frame (out of the set of all body frames) whose motion relative to the first link during shape changes most cancels out the motion induced by the constraints. This operation takes the form of finding (conservative) gradient fields that are as close as possible to the constraint vector fields, as illustrated for orientation in Fig. 4. More details on this process are given in our prior work,[3] including an implementation of body-frame optimization based on a *Hodge-Helmholtz decomposition*.[14]

6. Conclusions

By combining our geometric and granular modeling tools, we have both increased the speed with which sand-swimmer gaits can be evaluated and gained insight into the forms taken by solutions to the gait equations. As we further explore sand-swimming and related motions, such as rubble-crawling, these tools will allow for rapid iteration of swimmer morphology and gait pattern, directing expensive simulations with the full suite of granular tools to the inputs where they will be most informative. We are also very interested in finding other applications of emperical-geometric models to provide insight into complex problems.

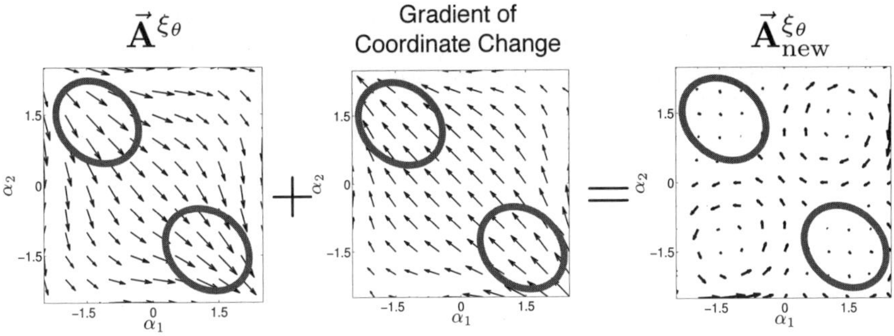

Fig. 4. Calculating the optimal body frame starts with finding the constraints in a convenient body frame, such as the middle link (left). The gradient of the optimal body frame's orientation relative to the middle link is the conservative vector field that is most opposite to the θ constraint field (center). The residual values after summing these two terms (right) are the new minimized constraint field, equivalent to finding the constraints in the optimal coordinates.

References

1. Maladen R, Ding Y, Li C, Goldman D (2009) Undulatory Swimming in Sand: Subsurface Locomotion of the Sandfish Lizard. *Science* 325:314.
2. Maladen RD, Ding Y, Umbanhowar PB, Kamor A, Goldman DI (2011) Mechanical models of sandfish locomotion reveal principles of high performance subsurface sand-swimming. *J Roy Soc Interface* 8:1332–1345.
3. Hatton RL, Choset H (2011) Geometric motion planning: The local connection, Stokes's theorem, and the importance of coordinate choice. *International Journal of Robotics Research* 30:988–1014.
4. A. M. Bloch *et al.*, *Nonholonomic Mechanics and Control*, Springer, 2003.
5. R. Abraham and J. E. Marsden, *Foundation of Mechanics*, Addison Wesley, 1985.
6. Shapere A, Wilczek F (1989) Geometry of self-propulsion at low Reynolds number. *Journal of Fluid Mechanics* 198:557–585.
7. Murray R, Sastry S (1993) Nonholonomic Motion Planning: Steering Using Sinusoids. *IEEE Transactions on Automatic Control* 38:700–716.
8. G. Walsh and S. Sastry, *On reorienting linked rigid bodies using internal motions*, Robotics and Automation, IEEE Transactions on, 11 (1995), pp. 139–146.
9. S Kelly and Richard M Murray, *Geometric Phases and Robotic Locomotion*, J. Robotic Systems, 12 (1995), pp. 417–431.
10. Ostrowski J, Burdick J (1998) The geometric mechanics of undulatory robotic locomotion. *International Journal of Robotics Research* 17:683–702.
11. Melli JB, Rowley CW, Rufat DS (2006) Motion Planning for an Articulated Body in a Perfect Planar Fluid. *SIAM Journal of Applied Dynamical Systems* 5:650–669.
12. Kanso E (2009) Swimming Due to Transverse Shape Deformations. *Journal of Fluid Mechanics* 631:127–148.

13. Radford JE, Burdick JW (1998) Local motion planning for nonholonomic control systems evolving on principal bundles. In *Proceedings of the International Symposium on Mathematical Theory of Networks and Systems* (Padova, Italy).
14. George B. Arfken, *Mathematical Methods for Physicists*, Elsevier, 6th ed., 2005.

PIPE INSPECTION PARALLEL-LINK ROBOT WITH FLEXIBLE STRUCTURE

SERGEY JATSUN

Theoretical mechanics and mechatronics department, South-West State University, Kursk, Russia

ANDREY YATSUN

Theoretical mechanics and mechatronics department, South-West State University, Kursk, Russia

SERGEY SAVIN

Theoretical mechanics and mechatronics department, South-West State University, Kursk, Russia

The present paper deal with problems related to movement along a pipe. These problems, deal with passing L- and T-shaped pipe junction, with variable diameter of the pipe, with a difference of pipe material characteristic, and with the problem, related with moving along a narrow pipe. In this paper the last problem is discussed. In the present article the in-pipe inspection robot consisted of two units is discussed. There are the penetrate unit and the driver unit there. Driver unit contains electric drives and that unit is placed outside of the pipe, while penetrate unit is placed inside.

Introduction

We are going to discuss some certain problems which are devoted to the movement through the pipes and have been mentioned in many papers on automation of pipe inspection. These problems deal with the following issues: passing L- and T-shaped pipe junction, variable diameter of the pipe, a difference of pipe material characteristic, moving along a narrow pipe [1-3]. The latter problem is investigated in the given paper. It is the problem of finding tiny drives with a considerable power which appears during designing such robot for movement inside a narrow pipe. The possibilities of using bimorph piezoelectric, magnetostrictive materials, pneumatic and electromagnetic actuators are described in several papers [4, 5]. The problem of adaptation to variable diameter of the pipe arises during the designing of that kind of robot. The efficiency of using the pneumatic actuator is showed in the paper [1]. This kind of structure requires providing the division of the device into two parts. One of them is the penetrate unit (the bubble), which is placed within the pipe. Another one is the driver unit - the source of forced air for the bubble. This is one of the ways of creating a small-sized robot for moving inside a narrow pipe with a variable diameter. Such mechanism resembles endoscope tube which is used for double-balloon enteroscopy [6].

In the given paper the in-pipe inspection robot consisted of two units: the penetrate unit and the driver unit. The driver unit is placed outside the pipe and contains electric drives, while the penetrate unit is placed inside the pipe. Both units are connected by flexible ropes. It is the way when we have all the advantages of electric drive, such as high quality of precision and controllability without power and size restriction. At the same time we can keep the size of the penetrate unit at minimum. Moreover, the robot is able to move inside a pipe with variable diameter thanks to the parallel-link structure of the penetration unit.

Description of the design

Outlined scheme of the robot's penetration unit and the parallel-link structure of the unit can be seen on the Figure 1.

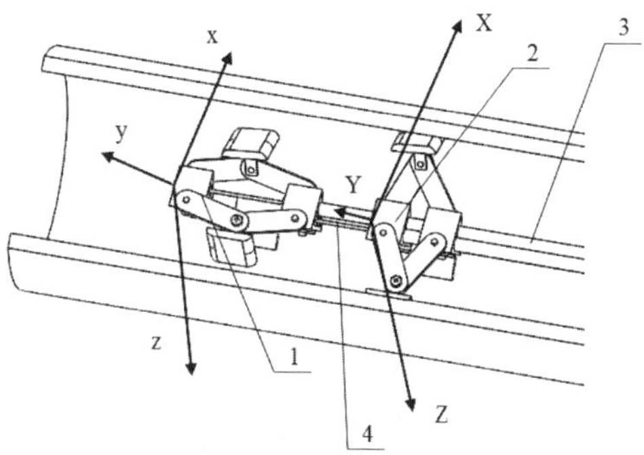

Figure 1. Scheme of the penetration unit; 1 – first platform; 2 – second platform; 3 – cable housings; 4 – actuators' inner cables

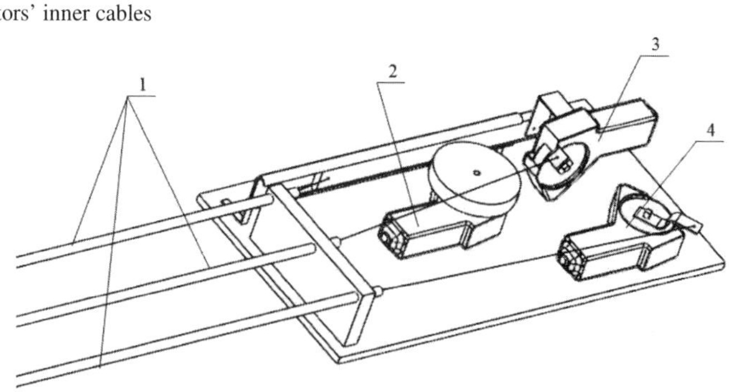

Figure 2. The scheme of the driver unit; 1 – cable housings; 2 – drive gear, operating the ropes, which connect the sections of the robot; 3,4 – drive gears, operating the ropes of separated sections;

As have been mentioned above, the robot consists of the penetration unit (the part of the robot which moves along the pipe) and the driver unit. The penetration unit consists of two sections which are the parallel-link mechanisms. These mechanisms consist of four elements, forming a pentahedron by connection with one another. The ends of the inner cable and the cable housing are connected to the two opposite apexes of the pentahedron. Movement of the inner cable, which is relative to the cable housing, makes the pentahedron change his form. Movement of one section relative to another is doing the same way. The robot moves by fixing one of the sections and moving another one. The movement of an inner cable, relative to the cable housing, is provided by an electric drive, which is placed inside the driver unit. The driver unit contains three electric drives as well as some electronic boards which being used for controlling drives and communicating with a computer. The penetration unit contains neither electronic nor electro mechanics component.

Experimental prototype

An experimental prototype was created for the purpose of studying robot's behavior in a real environmental conditions. The prototype consists of the robot itself, electronic control system connected with a computer, a computer and a tube. On the figure 3 the penetration unit is shown. On the figure 4 the driver unit is shown.

Size of the each penetration unit section is 200x130x50 mm. The driver unit size is 800x200x200 mm. Electrical motors installed in the unit have 40 watt output power. The supply voltage is 12 V. An operator controls the robot using special software. The computer connected with the robot by Arduino board device, which is connected with driver board devices. The driver board devices are switch the electrical motors.

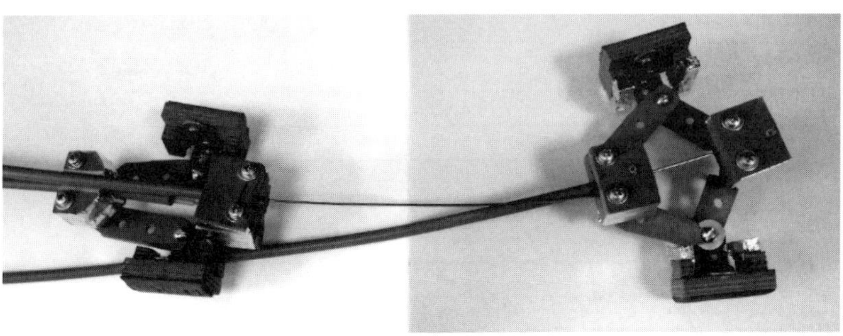

Figure 3 The penetration unit (a part of the experimental prototype)

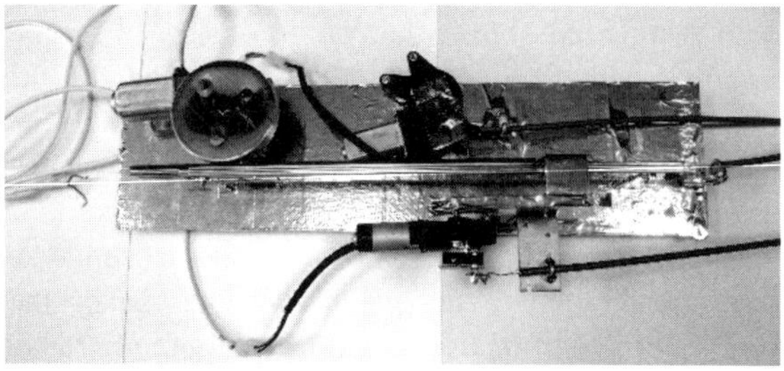

Figure 4. Driver unit (a part of the experimental prototype)

Experimental data

The diagrams of the first section of the penetration unit motion are shown on figures below.

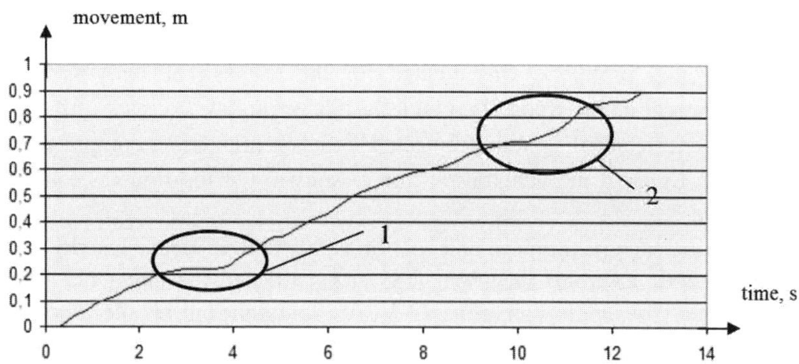

Figure 5. The motion of the robot inside a tube with obstacles; 1, 2 – zones were the robot passed an obstacle

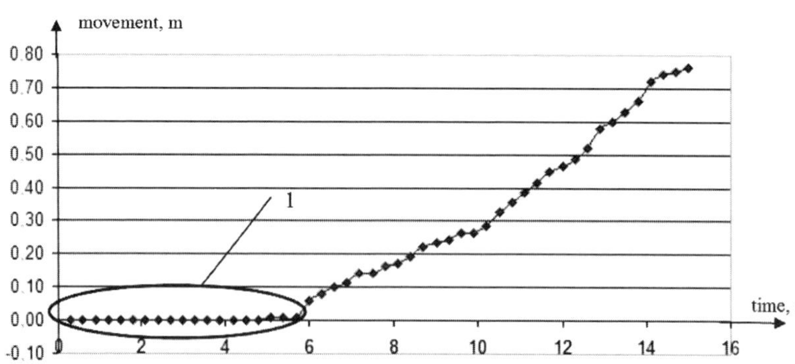

Figure 6. Motion of the robot inside a tube witch is bend (the angle is 10°); 1 – a zone, where a level of the force generated by driver unit was not sufficient.

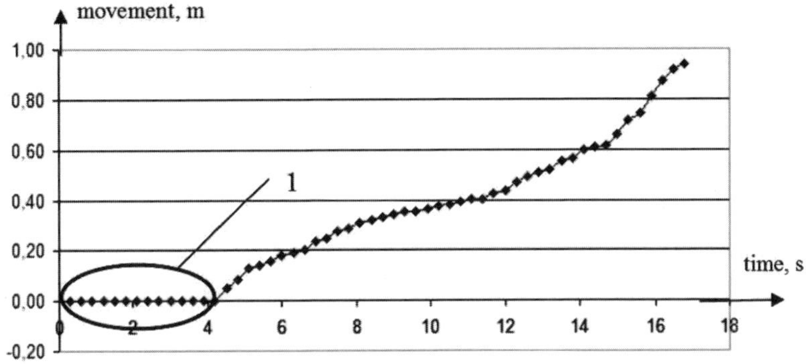

Figure 7. Motion of the robot inside a tube witch is bend (the angle is 20°); 1 – a zone, where a level of the force generated by driver unit was not sufficient.

Mathematical model

Let's consider the motion of the first platform in the situation when the second platform is fixed. The angular momentum of the first platform can be described as:

$$H_p = (I_{pxx}\omega_{px})i_p + (I_{pyy}\omega_{py})j_p + (I_{pzz}\omega_{pz})k_p \tag{1}$$

where I_{pxx}, I_{pyy} and I_{pzz} are, respectively, the mass moments of inertia of the platform about the X, Y and Z axes of the first platform coordinate frame [x y z]. The rate change of angular momentum is

$$\dot{H}_p = (\dot{H}_p)_{xyzp} + \omega_p \times H_p \tag{2}$$

The total torque acting on the first platform mass center P is given by:

$$\sum M_p = \sum_{i=1}^{3} {}^pP_i \times {}^pF_{pi}, \tag{3}$$

The ${}^pP_i$ is the vector from P to B_i, i = 1, 2, 3. By equating Eq. (2) with (3), we obtain:

$$\begin{aligned}
&\sum_{i=1}^{3} {}^pp_{iy}\,{}^pF_{piz} = I_{pxx}\dot{\omega}_{px} + (I_{pzz} - I_{pyy})\omega_{py}\omega_{pz},\\
&\sum_{i=1}^{3} -{}^pp_{ix}\,{}^pF_{piz} = I_{pyy}\dot{\omega}_{py} + (I_{pxx} - I_{pzz})\omega_{px}\omega_{py},\\
&\sum_{i=1}^{3} ({}^pP_{ix}\,{}^pF_{piy} - {}^pP_{iy}\,{}^pF_{pix}) = I_{pzz}\dot{\omega}_{pz} + (I_{pyy} - I_{pxx})\omega_{py}\omega_{px},
\end{aligned} \tag{4}$$

The rate change of angular momentum of the first platform is equal to the torque acting on P. With help of some manipulations we can observe that it is not sufficient to solve for the required force ${}^{i}F_{pix}$, and it is necessary to combine the following auxiliary force to determine this term.

$$-W_p + \sum_{i=1}^{3} R_i^i F_{pi} = m_p a_p, \tag{5}$$

where $W_p = \begin{bmatrix} 0 & 0 & W_p \end{bmatrix}^T$

W_p and m_p are, respectively, the weight and mass of the upper platform;

a_p denotes its acceleration. By expanding $R_i^i F_{pi}$, Eq. (5) can be expressed as follows:

$$\sum_{i=1}^{3}(R_{i11}{}^{i}F_{pix} + R_{i12}{}^{i}F_{piy} + R_{i13}{}^{i}F_{piz}) = m_p a_{px}$$

$$\sum_{i=1}^{3}(R_{i21}{}^{i}F_{pix} + R_{i22}{}^{i}F_{piy} + R_{i23}{}^{i}F_{piz}) = m_p a_{py}$$

$$\sum_{i=1}^{3}(R_{i31}{}^{i}F_{pix} + R_{i32}{}^{i}F_{piy} + R_{i33}{}^{i}F_{piz}) - W_p = m_p a_{pz} \tag{6}$$

The presented equations can be used for the description of carrying out of the constructions under various conditions. It is important to describe the dynamic properties of the construction so that it will help to define the exact location of the robot at every instant, which can be necessary during the certain kinds of work.

Several experiments took place within the frame of the work. The abilities of the robot's movement inside the pipes under various conditions and different variants of robot gate have been explored. The results of one of the experiments can be seen at the Figure 8.

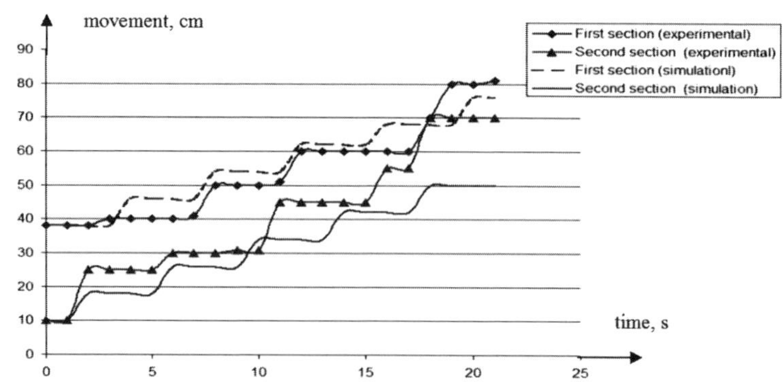

Figure 8. Results of one of the experiments compare with results of the simulation

We can see the graphics of movements of the first and the second sections of the robot relatively to the beginning of the pipe. The movement was carried out in the open-loop mode of the automatic control system in the frame of the experiment. As we can see, the sections are immobile most of the time, and then make a short movement, where the magnitude of the movement may vary due to the effect of friction force. This fact emphasizes the importance of creating the adequate mathematic model of the device, which is necessary for assessing the level of external factors influence on the work of the system.

Further work

The development of mathematical model, which will describe all the process of robot's movement in detail, is considered to be the next step of the given research. Simulation of the process of passing though obstacles like L- and T-shaped pipe junction and simulation of the moving process inside inclined pipe will be discussed. Experimental prototype has already been created and is going to be tested. It should be noted that singularity and stability analysis will be done in the future research as well as numerical simulations and experiments.

References list

1. Fukuda, T., Hosokai, H., Uemura, M. *Rubber gas actuator driven by hydrogen storage alloy for in-pipe inspection mobile robot with flexible structure.* Scottsdale, AZ, USA: Robotics and Automation, 1989. Proceedings., 1989 IEEE International Conference on, 1989. Print.
2. Yukawa, T.; Suzuki, M.; Satoh, Y.; Okano, H.; *Design of Magnetic Wheels in Pipe Inspection Robot.* Taipei: Systems, Man and Cybernetics, 2006. SMC '06. IEEE International Conference on, 2006. Print.
3. Sakamoto, S.; Hara, F.; Hosokai, H.; Kinoshita, H.; Abe, Y.; *Parallel-link robot for pipe inspection.* Industrial Electronics Society, 2005. IECON 2005. 31st Annual Conference of IEEE, 2005. Print.
4. Y. Sabzehmeidani, M. Mailah, M. Hussein, A. R. Tvakolpour, M. R. Safizadeh *Intelligent Control and Modelling of a Micro Robot for In-pipe Application.* Advances in Control, Chemical Engineering, Civil Engineering and Mechanical Engineering. Print.
5. PinKuan Liu, ZhiJie Wen and LiNing Sun *An in-pipe micro robot actuated by piezoelectric bimorphs.* Chinese Science Bulletin volume 54, number 12. Print.
6. Matsumoto T., Esaki M., Moriyama T., Nakamura S., Iida M. *Comparison of Capsule Endoscopy and Enteroscopy with the Double-Balloon Method in Patients with Obscure Bleeding and Polyposis* Endoscopy 2005, New York 2005. Print.

SECTION–9

PERCEPTION AND SENSOR FUSION

DESIGN AND EXPERIMENTATION OF A TRACKING SYSTEM FOR HUMANITARAN DEMINING APPLICATIONS*

ROEMI FERNÁNDEZ[1], CARLOTA SALINAS[1], HÉCTOR MONTES[1,2], PABLO GONZÁLEZ DE SANTOS[1], MANUEL ARMADA[1]

[1]Centre for Automation and Robotics CSIC-UPM
Ctra. Campo Real, Km. 0,200, La Poveda, Arganda del Rey, 28500, Madrid, Spain
[2]Faculty of Electrical Engineering, Technological University of Panama, Panamá
roemi.fernandez@car.upm-csic.es

This paper presents a tracking system that enables a careful observation of the deminers' scanning technique during the training tasks for landmine detection with hand-held detectors. The proposed system consists of two decoupled high resolution catadioptric cameras separated by a distance D, and with their focal points displaced a distance d along the z-axis. This configuration allows an easily image rectification, by means of an estimation of parallel epipolar lines for real-time stereo processing. With the acquired information, the movements of the hand-held detectors will be computed, and trainee performance feedback could be provided to the deminers in order to enhance their competencies.

1. Introduction

In 2010, a total of 4191 new landmine casualties were reported, 5% more than in 2009, and a total of 72 states, as well as seven disputed areas, were confirmed or suspected to be mine-affected [1]. Unfortunately, the consequences of the use of antipersonnel mines go far beyond the military field, since the activity of these weapons does not stop with the end of the hostilities. The landmines, cluster munitions and explosive remnants of war (ERW) can remain active for decades, they are not aware of negotiations or peace treaties, and do not distinguish between soldiers and civilians.

The presence of landmines also produces negative economic effects, as it denies access to the affected areas and their resources, causing deprivation and social problems among the affected populations. In addition, harmful effects on the environment are evident. They alter the characteristics of the soil, disrupting the ecological balance, and the soil degradation increases with the explosions. They also pose some threat to the local fauna. Therefore, the medical, social,

* This work is supported by TIRAMISU European Project – Grant Agreement N° 284747.

economic and environmental consequences are immense, which is why the elimination of antipersonnel mines is a vital requirement for many regions to recover their impact.

In military demining the objective is to clear a minefield as fast as possible using brute force, and usually a clearance rate of 80% to 90% is accepted [2]. On the other hand, humanitarian demining is more difficult and dangerous, as it requires the complete removal of all mines and ERW and the return of the cleared minefield to normal use [3].

Today, most humanitarian demining is done using hand-held detectors or sniffer dogs, attempting to carefully locate each explosive item and then either blow it up or burn it in situ, or render it safe and remove it for dismantling or disposal elsewhere.

Unfortunately, the proficiency of land mine detection equipment operators is highly variable. Expert operators of the most commonly-used hand-held detectors exhibit mine detection probabilities of 0.9 [4], but a simple mistake can jeopardize their lives. Thus, training is a very important aspect in order to improve the safety and effectiveness of the mine detection activities performed by the human operators.

This paper proposes a tracking system for improving the deminers' sweeping technique during close-in detection tasks with hand-held detectors. With this sensory system, the expert's skills could be studied by quantifying some critical performance variables, so that they can be used later as reference values for the training tasks. The rest of the paper is organized as follows. Section 2 describes the proposed sensory tracking system. Section 3 explains the projected methodology for representing the knowledge acquired by the sensory tracking system. Section 4 presents the preliminary experimental results and finally, Section 5 summarises major conclusions.

2. Tracking system

The sensory system approach is based on an omnidirectional catadioptric stereo system and a hand-held detector which is mounted with several LEDs of high intensity utilised as landmarks for the motion tracking. The proposed vision system provides 360º field of view (FOV) and stereo capability without sacrificing too much the image resolution. This is a great advantage in comparison with conventional imaging systems that are severely limited in their FOV. Moreover, acquisition of redundant information provides knowledge of the environment with greater clarity, precision and reliability.

A catadioptric system is the result of combining one refracting (dioptric) surface with other reflecting (catoptrics) in a carefully arranged configuration to capture a much wider field of view. Thus, a catadioptric system is normally compound of a conventional high-resolution camera acting as the refracting part and a curved mirror as the reflecting one. In order to acquire a single image containing the information of the whole scene, the camera and the mirror must be arranged in a configuration such that the entire system has a single effective viewpoint [5]. Therefore, a single viewpoint is a requirement for the generation of pure perspective images from the sensed images. To generate omnidirectional images, perfect quadratic surfaces are considered as the only candidates for mirror shapes. In this way, every incident ray of light that strikes the surface towards the mirror focus is reflected to the second focus. Since the geometry of the system is known, it is possible to compute the ray direction for each pixel and its irradiance value.

A stereo rectified configuration is accomplished with two decoupled high resolution omnidirectional catadioptric systems vertically aligned. Using this type of design, there exist no occlusions between the two systems, and it ensures that there exists parallax on 360º FOV. Thus, the rectification process provides epipolar radial lines onto the image plane and when they are projected onto the panoramic perspective they become parallel lines to the vertical axis. As a result, the depth is isotropic in all directions.

The metal detector that has been selected as hand-held detector for the experiments in this work is the AN-19/2 manufactured by Schiebel [6]. However, the training tool can be easily adapted to be utilised with any kind of hand-held detector. The LEDs of high intensity are placed along the hand-held structure: three on a light plastic plate that is located above the search head and four on the wand. As they are attached by means of plastic clamps, they can be easily installed and removed. The separation of the plastic plate from the sensor head was calculated for eliminating any chance of interference on the metal detector. With the utilisation of the LEDs as landmarks, the omnivision stereo system is able to detect and track the position and orientation of both the wand and the search head during the sweeping motions. Fig. 1 shows an overview of the proposed sensory tracking system.

3. Methodology for representing the acquired knowledge

The sensory tracking system described above will be utilized for acquiring information in two different scenarios: when the expert's skills are studied in order to quantify some critical performance variables and when the efficiency of the deminers is evaluated during the training tasks.

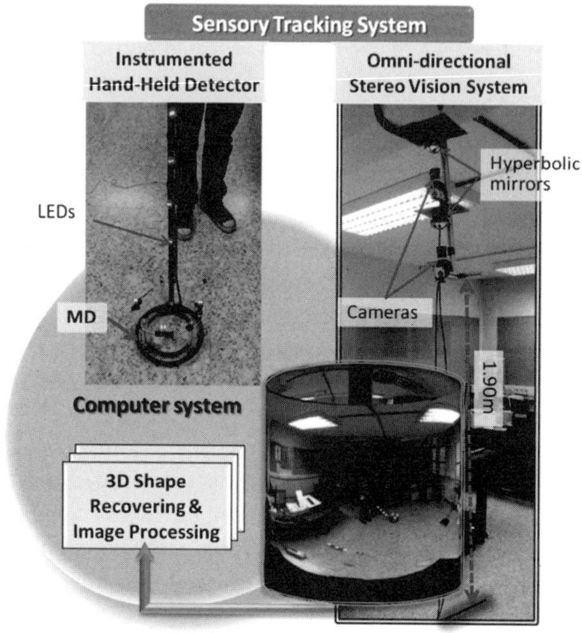

Figure 1. Overview of the sensory tracking system.

Previous analysis of the expert's hand-held detection activities shows that their techniques and strategies differed from conventionally taught operating procedures. Whereas other operators perform detection on the basis of auditory outputs pointing out the existence of conductive materials, experts use the onset and offset of outputs that occurred during sweeping motions of the hand-held detector to create spatial patterns that they compare to learned models [7]. They also modify their detection techniques to adapt to environmental variations (deserts, hilly rocky terrain, lands with a wide variety of vegetation). For this reason, the projected methodology proposes the utilization of the omnidirectional stereo tracking system to study the experts' skills in different environments and with different kinds of hand-held detectors. With the compiled information, some critical performance variables will be extracted, assessed, and quantified, so that they can be used after as reference values for the training task. The preliminary chosen critical variables that will be analysed are the safety distance to advance the detector search-head on each sweep, the sweep velocity, the scan height, and the inclination of the hand-held detector head with respect to the ground.

For the analysis and assessment of the trainee skills, the same set of critical variables proposed to parameterise the experts' performance is acquired and compared in real time with the reference training goals. In this way, the training tool is able to evaluate on-line the deminers' efficiency during the scanning tasks and feedback them important information for improving their competences. If the experts' skills have been studied previously, the reference training goals required for the assessment will be obtained from the results of the experts' performance. In case experts are not available for these experiences, trainee operators will be evaluated according to the reference values established in the standard procedures.

Since Fuzzy Set Theory has proved to be a worthwhile means for handling and representing experience-based heuristic knowledge, it will be used to representing the training goals. Given a universe of discourse U, a fuzzy value A is defined by means of a possibility distribution π^A defined over U [8]. Given a precise value $x \in U$, $\pi^A(x) \in [0,1]$ represents the possibility of A being precisely x. The possibility distribution π^A could be represented by means of a trapezoidal function specified by four parameters. In this way, $A = (\alpha, \beta, \gamma, \delta)$, $\alpha \leq \beta \leq \gamma \leq \delta$, where $[\beta, \gamma]$ represents the core, $\text{core}(A) = \{x | \pi^A(x) = 1\}$, and $]\alpha, \delta[$ represents the support, $\text{support}(A) = \{x | \pi^A(x) > 0\}$. The minimum and maximum values acceptable for the chosen critical variables are the beginning and the ending of the distribution support, respectively. In this way, for instance, if the deminer advances the hand-held detector with a lower velocity than the value corresponding with the beginning of the support, the scanning task is too time-consuming. If the deminer advances the hand-held detector with a higher velocity than the one corresponding with the ending of the support, the sweep is being carried out too fast for efficient and safe landmine detection. The beginning and ending of the core are the limits of the interval within the ideal values lie. If v_i is the sweep velocity output in m/s, the degree to which the training objectives are being achieved is $\pi^A(v_i)$. If $\pi^A(v_i) = 0$ then the acquired velocity is outside the range of ideal values. Conversely, if $\pi^A(v_i) = 1$, the acquired velocity is within the range of ideal values. The closer $\pi^A(v_i)$ is to 1, the closer the sweep velocity output is to the ideal values.

4. Preliminary results

The main goal of the experimentation phase was to evaluate the effectiveness of the sensory system to track the hand-held detector during the training sessions. Fig. 2 shows a representative pair of images selected from a large image sequence acquired with the omnidirectional catadioptric stereo image system and the detection results of the high intensity LEDs that are present in the scene. Note that not only the LEDs that are placed in the hand-held detector structure are detected but also the three LEDs that form the reference pattern utilised for recovering 3D information from the environment. Several steps are required to accomplish this detection. First, objects segmentation based on the shape and the colour of the LEDs is carried out. Then the detected LEDs that form the hand-held detector wand are clustered if they fit a lineal model. Finally, the detected LEDs that belong to the search head are clustered if they fit a triangle shape model.

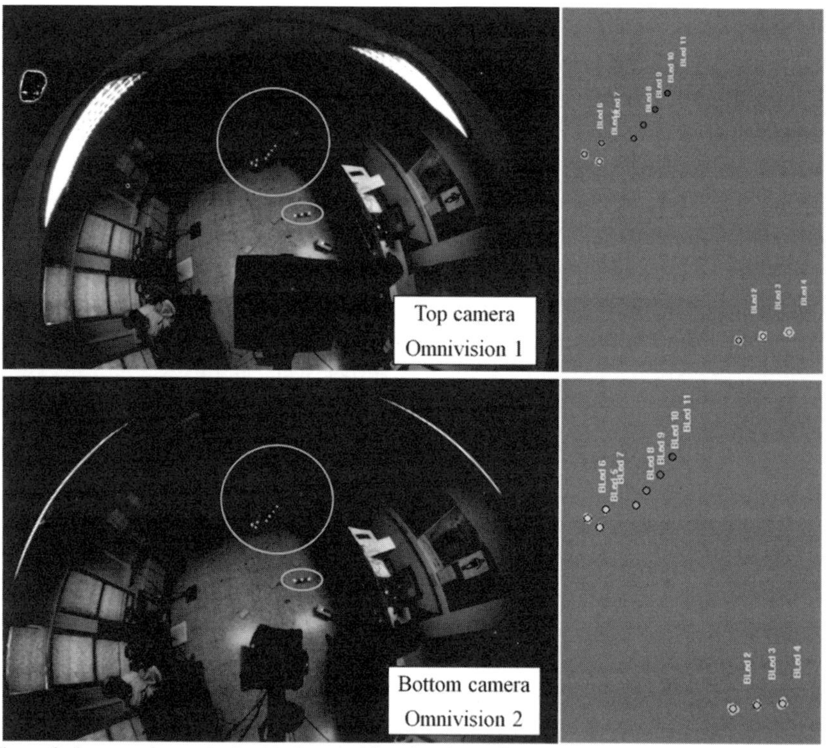

Figure 2. Image pair acquired with the omnidirectional catadioptric stereo system and the detection results of the high intensity LEDs that are present in the scene.

Fig. 3 displays the tracking results of the hand-held detector structure in the image plane with the omnivision system during the scanning motions carried out by a novice operator. Triangles represent the search head of the metal detector during the scanning motions, and are obtained from the detection of the three

LEDs placed in the search head. The lines correspond to the wand of the detector, obtained also thanks to the detection of the four LEDs installed on the same. The tracking of the wand has been painted just for the first scan in order to show more clearly the tracking of the search head. In this graph it is possible to observe that uneven sweep spacing is a major problem in novice operators. To ensure the safety of the operator during the detection task, each sweep across the lane must overlap the previous one by about one-half the width of the detector head. This requirement is difficult to be achieved by novice operators that should keep in mind a spatial image of the last sweep motion carried out. The feedback provided by the training tool could be especially useful in this case, helping the novice operator to create the required mental patterns.

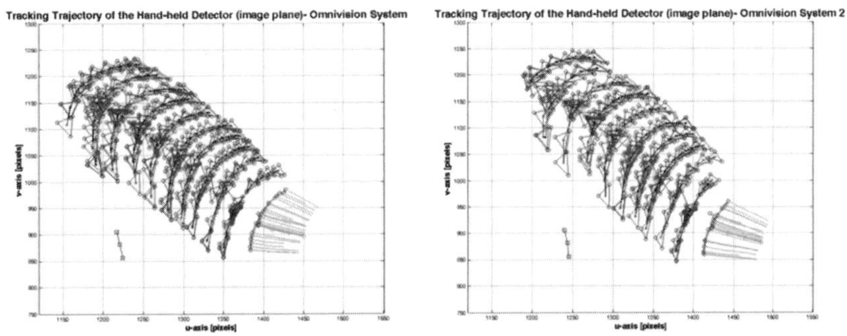

Figure 3. Tracking of the hand-held detector with the omnidirectional catadioptric stereo system.

These results confirm that with the omnivision system is feasible to detect and track a dynamic object without the necessity of moving the cameras. Moreover the omnidirectional system enables the training tool to be aware of its complete surrounding at all times. Omnivision system has no problem to detect simultaneously several hand-held detectors in distinctly different parts of its environment. In this way several operators could be evaluated simultaneously.

5. Conclusions

In this work a tracking system for improving the deminers' skills during close-in detection tasks with hand-held detectors was introduced. The approach is based on an omnidirectional catadioptric stereo sensory system and LEDs of high intensity utilized as landmarks for tracking the hand-held detector movements. With this sensory system, the experts' skills could be studied by quantifying some critical performance variables, so that they can be used later as reference values for the training task. The experience-based acquired knowledge is represented by means of Fuzzy Set Theory. Using the critical variables, the omnidirectional catadioptric stereo tracking system, and a friendly human interface, the proposed system will be able to evaluate the efficiency of the deminers during the scanning tasks and to feedback them important information for improving their competencies.

Acknowledgments

The authors acknowledge funding from the European Community's Seventh Framework Programme (FP7/2007-2013 TIRAMISU) under Grant Agreement Nº 284747 and partial funding under Robocity2030 S-0505/DPI-0176 and FORTUNA A1/039883/11 (Agencia Española de Cooperación Internacional para el Desarrollo - AECID). Dr. Roemi Fernández acknowledges support from CSIC under grant JAE-DOC. Dr. Héctor Montes acknowledges support from Universidad Tecnológica de Panamá and from CSIC under grant JAE-DOC.

References

1. Landmine Monitor Report 2011, Landmine and Cluster Munition Monitor. http://www.the-monitor.org/index.php/publications/dsplay?url=lm/2011/
2. U. Rosengard, T. Dolan, D. Miklush and M. Samiei. *IAEA Bulletin*, February (2001).
3. E. M. A. Hussein, E. J. Waller. Landmine detection: the problem and the challenge. *Applied Radition and Isotopes* **53**, 557 (2000).
4. J. J. Staszewski. *In Detection and Remediation Technologies for Mines and Minelike Targets IV*, Proceedings of the Society for Photo-Optical Instrumentation Engineers 13th Annual Meeting, **3710**, 766 (1999).
5. S. Baker, S. K. Nayar. A theory of single-viewpoint catadioptric image formation. Int. J. of Computer Vision, 35 (2), 177, (1999).
6. D. Guelle, A. Smith, A. Lewis and T. Bloodworth. Metal detector handbook for humanitarian demining. Luxembourg: Office for official publications of the European Communities, (2003).
7. J. Staszewski. *Proceedings of the 48th Annual Meeting of the Human Factors and Ergonomics Society*, **48**, 458 (2004).
8. L. Zadeh. *Inform. Sci.*, **8**, 199, (1975).

ROBOTIC MULTISENSORY SYSTEM FOR PRECISION AGRICULTURE APPLICATIONS

H. MONTES[1, 2], R. FERNANDEZ[1], C. SALINAS[1] and M. ARMADA[1]

[1]*Field and Service Robotics Group, Centre for Automation and Robotics-CAR (CSIC-UPM), Spain;* [2]*Faculty of Electrical Engineering, Technological University of Panama, Panama,* hector.montes@ csic.es

This paper presents the description of a robotic test bed which has four degrees of freedom, two prismatic and two rotational ones. The two prismatic joints provide the positioning in the Cartesian plane orthogonal to the ground and, besides, support to the pan-tilt unit, which provide pitch and yaw movements. The pan-tilt unit holds the perception system which consists of a multispectral camera, a Time-Of-Flight 3D camera and a hyperspectral system. This robotic multisensory system is intended to detect, locate and characterize different fruit on trees (sweet-peppers, apples and grapes). To perform these tasks, we present a strategy of active perception which takes into consideration issues like occlusion of the fruit, lighting, and others.

1. Introduction

The visual perception activity is exploratory, investigative, searching. For example, human beings use visual information feedback widely to plan and perform actions. Humans do not just see, but rather we look. And while we observe something thoroughly, our pupils adjust to the level of illumination, our eyes bring the world into sharp focus, our eyes converge and diverge, and we move our heads and/or change our position and posture to get a better perspective of what we want to see. This adaptability has been vital for survival in a dynamically changing world. Therefore, human beings, in a natural way, have been carrying out active perception of their environment efficiently.

Hence the active perception can be stated as a problem of controlling strategies applied to the data acquisition process which will depend on the current state of the data interpretation and the goal or the mission of the process [1]. Therefore, researchers must take into account the main requirements for developing an artificial active perception. It should be account the sensory system and therefore the development of algorithms for sensory fusion. Furthermore, it must take into account the strategies to observe the region of interest. This last consideration directly involves the development of a robotic platform that adapts to the environment in which it must operate.

On the other hand, some researchers have undertaken several works for the location of fruits on trees for their subsequent harvesting process. In [2] a survey of methods for locating fruit on trees was presented, emphasizing the kind of perception system used, the strategy applied to detect the fruit, and the performance in terms of correct and false detections as well as its sensibility to changes in color.

In other researches, the perception system (colour cameras) was placed in one fixed position relative to the tree [3-5]. In other studies, the perception system was placed in other positions, displacing horizontally the perception system to specific positions [6], and at different positions around the crop [7-8]. In some cases, placement of the perception system is performed manually; due to the fact that it improves the visibility percentage of fruit in the tree. In other cases, the cameras were placed on the harvesting robot.

In this paper an automatic test bed for locating fruits, which will provide location data of fruits in the tree to control system of a harvesting robot is presented. The test bed has four degrees of freedom, and supports the perception system which consists of a hyperspectral camera, Time-Of-Flight (TOF) camera, and a multispectral camera. Preliminary results with the use of this perception system have been developed at CAR labs and could be seen in [9].

2. Description of robotic testbed

For locating fruits (sweet-peppers, apples, grapes) on trees a robotic test bed has been designed. This first robotic prototype has four degrees of freedom, two prismatic and two rotational ones.

The two prismatic joints provide the movements in the horizontal and the vertical axes, or in other words, provide the positioning in the Cartesian plane. The two rotational joints constitute a pan-tilt unit where the perception system is installed for its orientation.

Figure 1 presents a sketch of the first prototype of the experimental platform made in Pro-Engineer. In this sketch it is possible to notice that the platform uses two perpendicular linear guides with rack and pinion drive to construct the X and Z axes. By controlling the movement of these two linear guides relative to each other, the platform is capable of operating with high accuracy and reliability within a rectangular plane.

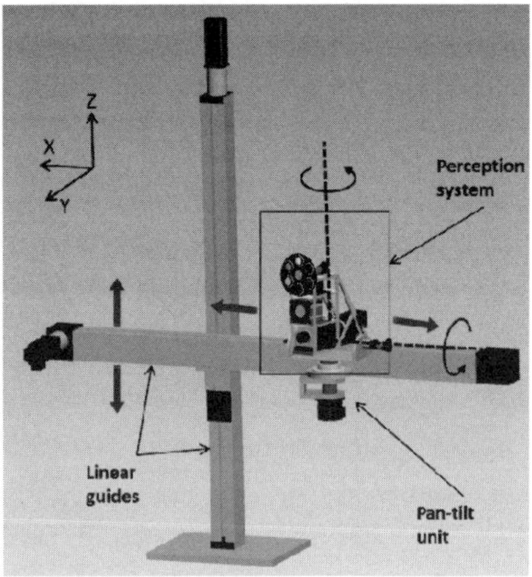

Figure 1. First prototype of the experimental platform that holds the perception system.

The linear guides can be moved in a range of speeds from 0.5mm/s up to 500mm/s. The linear motion system provides a position accuracy of ±0.1mm and can support loads up to 1500N. With these motion characteristics, the perception system is capable of "observing" fruits in its field of view and, subsequently, to move to another position quickly for the next observation process.

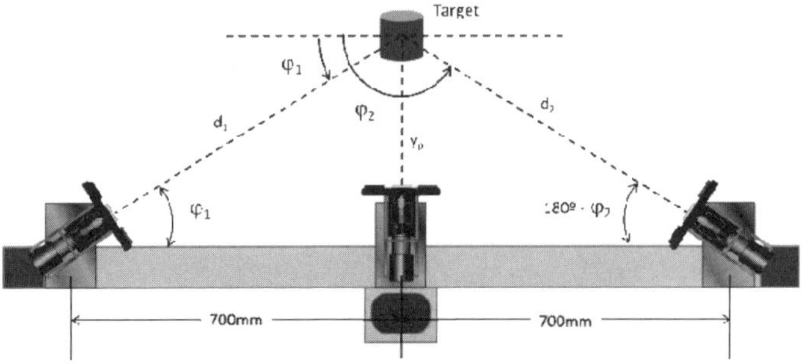

Figure 2. Azimuthal angles achieved by the pan-tilt unit.

The pant-tilt unit has the pitch movement limited to angular displacement of ±30° relative to the X-axis *(60°≤ β ≤120°)* due to mechanical constraints. The yaw movement has no mechanical constraint, so it could rotate 360° around the Z-axis. However, in the middle position of the X-axis, the rotational movement is limited by the vertical linear guide of the Z-axis. Therefore, the automatic yaw movement is restricted for azimuth angles out the range given by $0° \leq \varphi \leq 180°$. Figure 2 displays an outline where the azimuth angles that can be achieved by the unit are represented. Furthermore, with the pitch movement of the pant-tilt unit the zenith angles are obtained (see Figure 3).

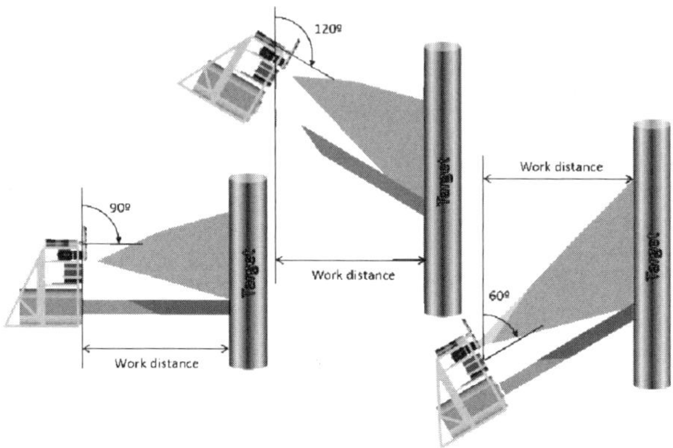

Figure 3. Zenith angles achieved by the pan-tilt unit.

3. Description of the perception system

The perception system must detect and locate fruits for three different crops: sweet-peppers, apples and grapes. Two different environments are considered, one of them in greenhouses for sweet-peppers and another in outdoors for apples and grapes. Therefore, the sensory system must be able to provide appropriate data for identification of the fruit in different environments, so that fruits are detected, located and characterized. Thus the perception system consists of a multispectral camera (a progressive color camera + filter wheel), a hyperspectral system and TOF 3D camera. Figure 4 shows the perception system.

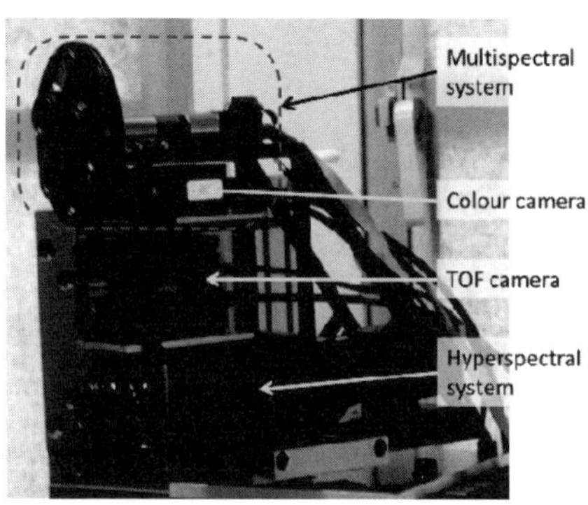

Figure 4. Close up view of the perception system.

4. Strategy of active perception

Detecting and locating fruits on trees is not a trivial task, because many fruits are occluded by the leaves, branches, other fruits, etc. In addition, it must consider the lighting in the environment in which the tasks of fruits detection must be performed. Thus, an active perception cycle that attempts to solve these issues is presented in the Figure 5 flow chart.

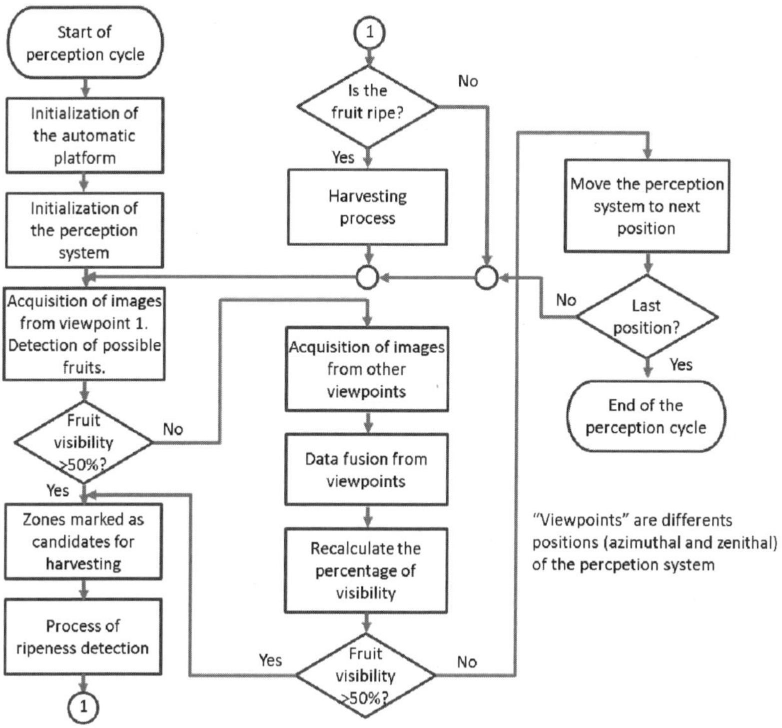

Figure 5. Flow chart of the active perception cycle.

In this paper, we present a part of the active perception that consists of the initialization of the automatic platform and the perception system, and the images acquisition from different "Viewpoints", taking into account the percentage of visibility of the fruit. Detection of ripeness is not presented in this manuscript.

5. Experimental results

Several experimental tests were carried out using the robotic multisensory system proposed in this article. Figure 6 shows the proposed system while an experimental test was performed. Figure 6(a) shows a photographic sequence of the end positions and central position of the robotic platform for obtaining fruit images (apples).

a)

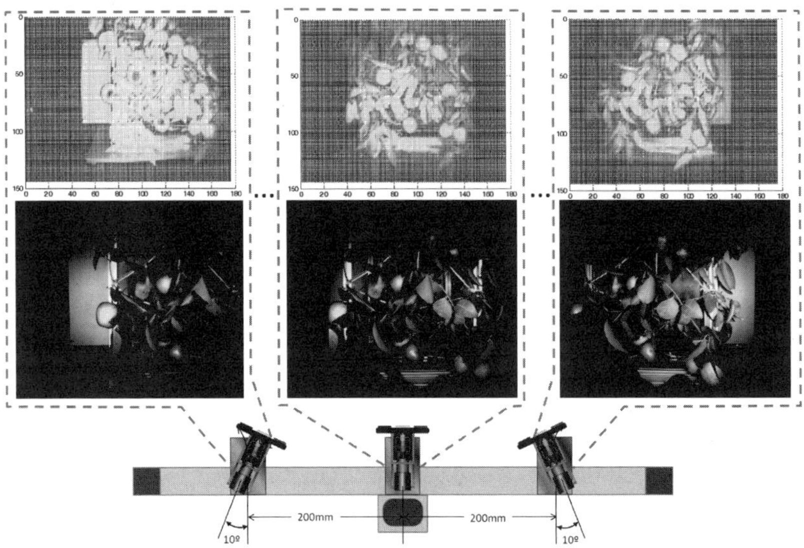

b)

Figure 6. (a) Robotic multisensory system in an experimental test. (b) Images of fruits obtained of the multisensory system with angles of azimuth (10°, 90°, and 170°), and zenithal of 90°, from TOF camera, and multispectral camera.

Figure 6(b) shows three pairs of pictures that represent the images obtained from the TOF camera (up) and images obtained from the multispectral camera (bottom).

The images acquired with multispectral system were obtained with a wavelength of 590nm, with a bandwidth of 20nm. For it was used a xenon light source which comprises bands of visible light and near infrared. The main idea is to distinguish the fruits of the leaves, branches, etc., in order to detect them, locate them, and characterize them, for later collection.

TOF camera is used to provide other data related to the separation distance between the multisensory system and fruits, and fruits size. Furthermore, performing a sensor fusion with both cameras we obtain sufficient information for the detection of the fruits.

The images obtained from the TOF camera, shown in Figure 6(b), are images filtered by using the confidence map and reflectance images supplied by the own camera.

6. Conclusions

In this paper a robotic test bed of four degree of freedom and the perception system have been presented. One strategy of active perception was presented, considering several "viewpoints", created by the azimuth and zenithal angles that the testbed provides.

The perception system used in the experiments described in this article consists of a multispectral camera (were presented images of 590nm) and a camera Time-Of-Flight 3D. The lighting used was a xenon light source that includes a bandwidth that we use in our experiments (visible and near infrared bands).

Performing a sensor fusion with data from the multi-sensory system can be differentiated the fruits of the leaves, branches, stems and of possible occlusions.

7. Acknowledgements

The authors acknowledge funding of this work under Crops Project - Grant Agreement N° 246252 of the 7FP, and partial funding under Robocity2030 S-0505/DPI-0176; FORTUNA A1/039883/11 (Agencia Española de Cooperación Internacional para el Desarrollo - AECID). Dr. Héctor Montes acknowledges support from Universidad Tecnológica de Panamá and from CSIC under grant JAE-Doc. Dr. Roemi Fernández acknowledges support from CSIC under grant JAE-Doc.

References

1. R. Bajcsy, Proc. *IEEE Special Issue on Computer Vision*, 76(8), 966 (1976).
2. A.R. Jiménez, R. Ceres, and J.L. Pons, *Trans. of the ASAE*, 43(6), 1911 (2000).
3. A. Plebe and G. Grasso, *Machine Vision and Applications*, 13(2), 70 (2001).

4. S. Kitamura and K. Oka, SICE-ICASE Int. Joint Conf., 353 (2006).
5. R. Chinchuluun and W.S. Lee, in *ASABE Paper No. 063010* (2006).
6. E.J. van Henten, J. Hemming, B.A.J. van Tuijl, J.G. Kornet, J. Meuleman, J. Bontsema, and E.A. van Os, *Autonomous Robots,* 13(3), 241 (2002).
7. K. Tanigaki, T. Fujiura, A. Akase, and J. Imagawa, *Computers and Electronics in Agriculture*, 63(1), 65 (2008).
8. D.M. Bulanon, T.F. Burks, and V. Alchanatis, *Environ. Control Biol.* 48(2), 81 (2010).
9. C. Salinas, R. Fernández, H. Montes, and M. Armada, in *Proc. Field Robotics CLAWAR2011*, 801 (2011).

CHARACTERIZATION OF THE MEASUREMENT ERRORS IN A MINIATURE LASER SCANNER FOR WALKING ROBOTS

Przemysław ŁABĘCKI, Michał NOWICKI, Piotr SKRZYPCZYŃSKI*

Institute of Control and Information Engineering,
Poznan University of Technology, ul. Piotrowo 3A, PL-60-965 Poznań
**E-mail: piotr.skrzypczynski@put.poznan.pl*

This paper analyses selected operational characteristics and measurement errors in the 2D laser scanner URG-04LX, which is a sensor suitable for walking robots due to its small size and weight. Quantitative errors in range measurement are analysed considering the intensity output available from the sensor. The obtained intensity values help to correct the range measurements with regard to various optical characteristics of the observed surfaces. Moreover, the most important qualitative errors – mixed measurements are characterized and compared to mixed measurements in the popular LMS 200 laser scanner. Finally, it is shown how taking into account the error characteristics of the URG-04LX scanner improves its performance as the main exteroceptive sensor of an autonomous walking robot.

Keywords: Laser scanner, Range measurement error, Walking robot

1. Introduction

Nowadays laser range finders and 2D/3D laser scanners are the most used exteroceptive sensors in mobile robotics. However, walking robots require compact, lightweight, and energy efficient sensors. Therefore, our six-legged walking robot Messor uses the Hokuyo URG-04LX miniature 2D laser scanners. On Messor one URG-04LX is tilted down, to scan the terrain ahead of the vehicle with the fanning laser beam, enabling terrain mapping.[7] The second URG-04LX is level-mounted in order to acquire profiles of the surrounding walls for self-localization with scan matching.[9] Both applications revealed limitations of this miniature scanner. Compared to the well-known Sick LMS 200 sensor, the Hokuyo URG-04LX not only has shorter range of measurements, but the measured distances to a much bigger extent depend on the optical properties of the surfaces being observed. Moreover, significant qualitative errors occur when the laser beam hits simultaneously two objects at different distances. Such errors are known in the literature as mixed measurements or discontinuous points.[1]

The Hokuyo URG-04LX was already characterized by others,[6,8] but in this paper we focus on the errors in range measurements of this sensor, and propose how to identify and (to some extent) remove these errors, particularly in the context of walking robot applications. To our best knowledge this paper is also the first that analyses the URG-04LX errors taking into account the intensity output, which is not well-documented by the manufacturer.

2. The URG-04LX laser scanner

The Hokuyo URG-04LX laser scanner exploits the phase-shift-based ranging technique,[4] in which the distance to a target is proportional to the difference of phase between the amplitude-modulated continuous wave (AMCW) of laser light, and a fraction of this wave reflected by the target surface back to the sensor. The URG-04LX uses two different modulation frequencies in order to avoid the distance ambiguity, which allows the sensor to detect distances between 0.02 m and 5.6 m. However, the actual measurement range is smaller – up to 4 m for most of the surface types. A near-infrared (785 nm) laser diode is used as the beam source.[4] The laser beam is deflected by a rotating mirror in the default range of 240° with the angular resolution of 0.352°. In this mode a single scan contains 683 range measurements. However, both the angular field of view and the angular resolution can be set by software. The default control protocol (SCIP 1.1) allows only for range measurements, while the extended one (SCIP 2.0) makes it possible to obtain also the received intensity values, and the AGC (Automatic Gain Control) voltage, depending on the mode being selected. Unfortunately, the modes allowing for additional measurements have smaller angular resolution: 0.704°for the intensity mode, and 1.056°for the intensity/AGC mode. According to the manufacturer the URG-04LX scanner measures ranges up to 1 m with the accuracy of 10 mm, and ranges between 1 and 4 m with the accuracy of 1% of the measured distance. However, this is specified only for a white sheet of paper, while the scanner suffers from a strong dependency between the accuracy of the measured range and the received light intensity because of the applied measurement principle.

3. Measured range errors

The methodology of the range accuracy characterization was largely inspired by the approach used in previous research on the sensor.[6,8] However, we experimented with the URG-04LX scanner using both the standard mode and the intensity modes available through our custom-made software. The simple mechanical setup we use allows to manually change the distance and incidence angle of the target surface, which is a sheet of paper of A4 size. Because the distance between the sensor and the target surface is measured by a 4 m long ordinary meter, we stacked the URG-04LX with the LMS 200 laser scanner with a common origin of the coordinate system in the xy plane, to obtain additional ground truth measurements.

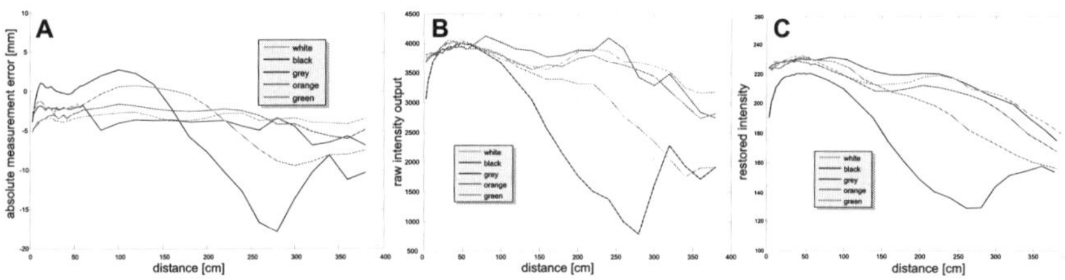

Fig. 1. Effects of surface type on range measurement accuracy (A), intensity (B), and restored intensity (C)

Results of our experiments are similar to what was already shown in Refs. 6 and 8 with regard to the basic characteristics of the URG-04LX sensor. Therefore, because of the tight page limit, we focus here on selected results obtained using the intensity and intensity/AGC modes, which so far were described in Ref. 5 only for the purpose of estimating lightness of the observed objects. The intensity I obtained from the URG-04LX scanner is not the true received intensity, but it is equivalent to the amplified output voltage of the APD (Avalanche Photodiode) used as the receiver in the device. Because in URG-04LX the amplification is controlled by the AGC unit, the intensity available to the user is a non-linear function of the true intensity. By reading the AGC voltage values from the sensor it is possible to eliminate the AGC influence, and to obtain I_{res} – the so-called restored intensity,[5] which is given by:

$$I_{\text{res}} = \frac{\sqrt{I} \cdot 1023}{v_{\text{AGC}}}, \tag{1}$$

where v_{AGC} is the AGC voltage, and 1023 is the range of the 10-bit A/D converter used on the voltage output. Results of some experiments with paper surfaces of different colours performed in the intensity/AGC mode are shown in Fig. 1. For the surfaces under study the absolute measurement error Δr_m fits within the limits specified by the manufacturer (Fig. 1A), however, the dependency between the error and the measured distance r_m is highly non-linear, particularly for the black paper.

Results of experiments in the intensity/AGC mode (Fig. 1A) suggest that there is a dependency between the received intensity and the range measurement error. This hypothesis is also supported by the literature. For example, Ref. 1 shows how this dependency develops in the phase measuring circuitry of an AMCW range finder. However, unlike Adams,[1] we do not have access to the internal structure of the sensor under study. Therefore, we try to approximate this dependency from the available experimental data. To this end we apply the approximation technique based on multi-dimensional Gaussians, which we already developed for approximation of the decision surface in the foothold selection problem of a walking robot.[2] The structure of the approximation polynomial is suggested by the generalized form of the Kolmogorov theorem:

$$\Delta r_m(r_m, I_{\text{res}}) = \sum_{q=0}^{m} c_q \cdot \exp\big(\lambda_{1,q}(r_m - a_{1,q})^2 + \lambda_{2,q}(I_{\text{res}} - a_{2,q})^2\big). \quad (2)$$

In (2) two-dimensional Gaussian functions are used, which are shaped to follow the available experimental data by the parameters $\lambda_{i,q}$ and $a_{i,q}$, with $i = 1, 2$ and $q = 0, \ldots, m$, where m represents complexity of the calibration model and is a design parameter. The vector of c_q coefficients can be efficiently computed from Least-Squares Fitting using the Gram matrix.[2] However, finding optimal values of $\lambda_{i,q}$ and $a_{i,q}$ is an optimization problem. Here the Particle Swarm Optimization (PSO)[3] is used to find optimal values of $\lambda_{i,q}$ and $a_{i,q}$. The approximation procedure results in a closed-form function that can be later used in real-time to correct the range measurements.

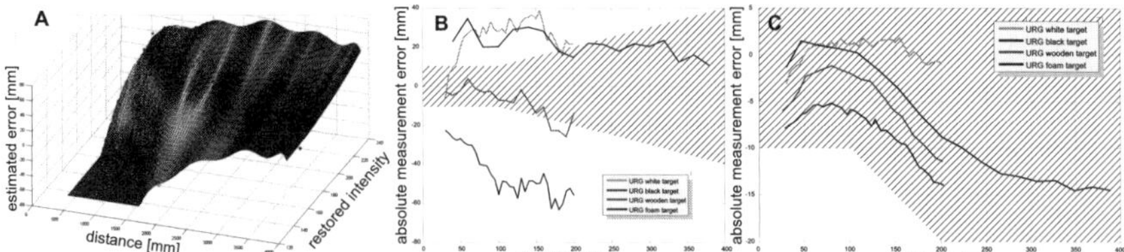

Fig. 2. Obtained calibration surface (A), measurement errors prior to calibration (B), and after calibration (C)

Figure 2A shows the surface that represents (2) for the collected calibration data. To verify the calibration we used range data collected on real target surfaces: white painted wall, matt black painted furniture, wooden plate, and white styrofoam plate (Fig. 2B). The light-grey background on the plot denotes the boundaries of the measurement errors given by the manufacturer. As one can see, the actual errors are far beyond these boundaries. However, when calibration according to (2) is applied (Fig. 2C), the range errors are kept smaller than 1% of the distance, for all the targets under study.

4. Qualitative errors

The calibration does not eliminate corrupted range measurements, which are not suitable for further processing. Some of these errors are caused by the mixed measurements, and they can be removed by clustering the range measurements and analyzing discontinuities in the scanned sequence.[2]

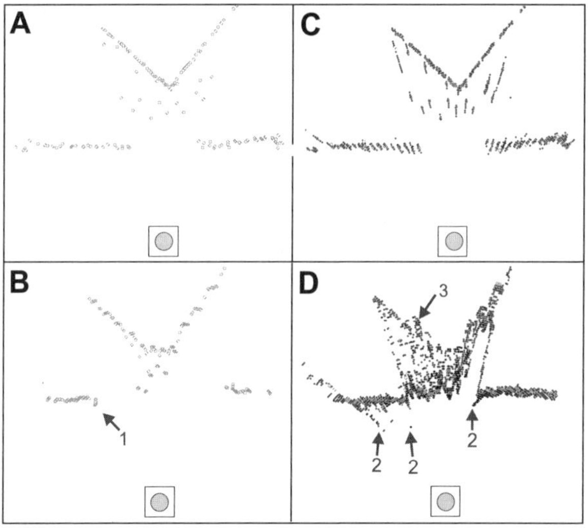

Fig. 3. Experimental comparison of mixed measurements in the LMS 200 and URG-04LX scanners

Taking advantage of our experimental setup, we made a comparison of the mixed measurement characteristics between the Hokuyo URG-04LX and Sick LMS 200 scanners. The most important differences in the spatial characteristics of mixed measurements are illustrated in Fig. 3. The number of mixed measurements (generated by a plastic net) in the LMS scan (Fig.

3A) is much smaller than in the scan of the URG sensor (Fig. 3B). Note that in the URG scan there are mixed measurements located in front of the closer obstacle (denoted as '1'). More differences appear if the two scanners are moved towards the net, making a pair of scans in one-centimeter increments. From Fig. 3C one can see that the mixed measurements of the LMS scanner are spatially isolated, but in the URG-04LX scan (Fig. 3D) there are mixed measurements in front of the closer object ('2') and some behind the farther object ('3').

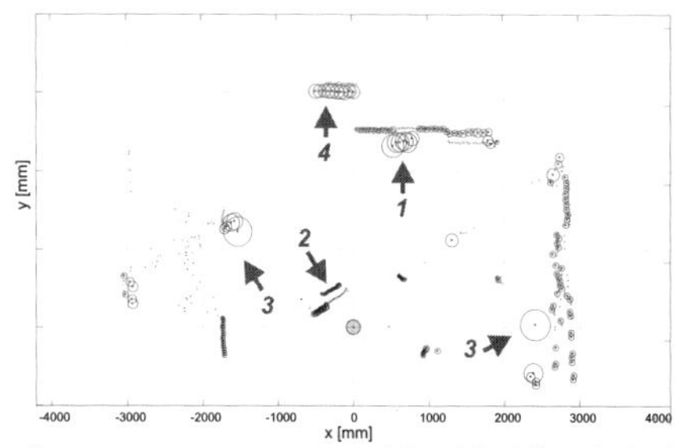

Fig. 4. Example range measurements identified by their intensity values

Erroneous range readings are also caused by black, highly reflective or semi-translucent surfaces. However, they can be removed by taking into account the obtained intensity values. Figure 4 shows an example laser scan from the URG-04LX working in the intensity mode (no AGC available). The intensity values of range measurements are visualized by the size of the black circles. A bigger circle means weaker intensity output. The green points indicate the measurements of the co-located LMS 200. Matt black objects ('1' in Fig. 4) are perceived at ranges shorter than the real distances, with low intensity. Shiny surfaces located close to the sensor can appear farther than in reality ('2'), with high intensity values. Small-size surfaces (e.g. table legs) often yield very weak intensity output, with much shortened distance measurements ('3'). However, lower intensity values are also registered for surfaces of good reflective properties, if they are located at bigger distances ('4'). Therefore, simple intensity tresholding is insufficient, and we experimentally identified the range-intensity characteristics of the most common erroneous readouts. We use thresholding in the range-intensity space to remove them, by discarding measurements of very low intensity values and high intensity values with short measured range.

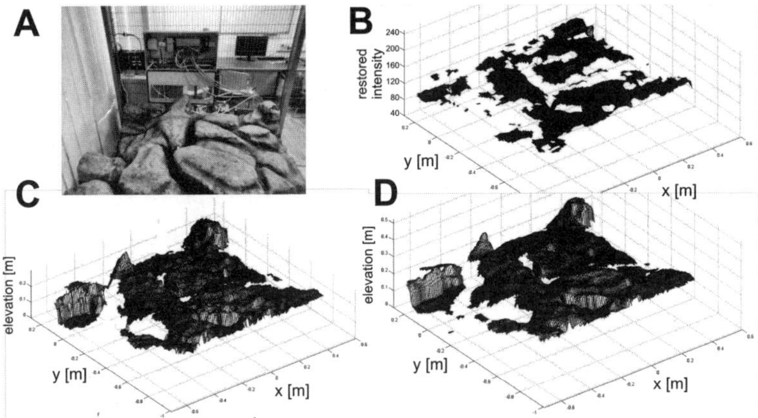

Fig. 5. Example terrain maps: experimental setup with terrain mockup (A), restored intensity (B), elevation with SCIP 1.1 (C), elevation with SCIP 2.0 (D)

5. Applications in the walking robot Messor

The self-localization system of the Messor robot that uses scan matching with a level-mounted URG-04LX scanner takes advantage of our characterization of the range data errors (cf. Fig. 4) by eliminating the vast majority of corrupted range measurements through range/intensity thresholding.[9] In this application we use the intensity mode because the number of 227 scanned points over the 240°field of view in the intensity/AGC mode is too small for the point-to-point scan matching method we use. The received intensity output alone gives no reliable characteristics of the scanned surfaces, but provides useful hints that some range readings may be in error.

Because of the differences in the spatial characteristics of the range measurement errors described in Sec. 4 we had to augment our terrain mapping system for the Messor robot with new procedures that eliminate the erroneous range measurements at the pre-processing stage, before these measurements are integrated into the elevation map. These procedures are described in details in Ref. 2. Using a realistic rocky terrain mockup (Fig. 5A) and an industrial robot arm to move the URG-04LX scanner (to have "perfect" localization) we experimented also with the intensity/AGC mode for terrain mapping. Although the range measurements in the intensity/AGC mode have much lower angular resolution, it is enough for short-range terrain mapping, as shown in Fig. 5, where an example map built in the default (range only) mode (C) is compared to a map built using the intensity/AGC mode (D). The restored intensity values can augment the usual elevation grid map by providing information about the optical characteristics of the terrain in front of the robot (Fig. 5B).

6. Final remarks

This paper characterizes the miniature URG-04LX laser scanner, which is suitable for walking robots. The contributions include (i) evaluation of the uncertainty of range measurements taking into account the restored intensity values; (ii) adaptation of our own Gaussian-based surface approximation method to correct systematic errors in the range measurements; (iii) analysis of the mixed pixel phenomenon and other quantitative errors due to varying optical properties of the observed surfaces. Moreover, we show that in applications of such a low-cost, miniature sensor, its error characteristics have to be carefully taken into account. This helps to obtain reliable range data for navigation of a walking robot that cannot use a more reliable full-size laser scanner.

References

1. Adams M. D. (1999) Sensor modelling, design and data processing for autonomous navigation, World Scientific.
2. Belter D., Skrzypczyński P. (2011) Rough terrain mapping and classification for foothold selection in a walking robot, J. of Field Robotics, 28(4), 497–528.
3. Eberhart R. C., Kennedy J. (1995) A new optimizer using particle swarm theory, Proc. Int. Symp. on Micromachine and Human Science, Nagoya, 39–43.
4. Kawata H., Ohya A., Yuta S., Santosh W., Mori T. (2005) Development of ultra-small lightweight optical range sensor system, Proc. IEEE/RSJ Int. Conf. on Intelligent Robots and Systems, Edmonton, 1078–1083.
5. Kawata H., Miyachi K., Hara Y., Ohya A., Yuta S. (2008) A method for estimation of lightness of objects with intensity data from SOKUIKI sensor, Proc. IEEE Int. Conf. on Multisensor Fusion and Integration for Intelligent Systems, Seoul, 661–664.
6. Kneip L., Tache F., Caprari G., Siegwart R. (2009) Characterization of the compact Hokuyo URG-04LX 2D laser range scanner, Proc. IEEE Int. Conf. on Robotics & Automation, Kobe, 1447–1454.
7. Łabęcki P., Rosiński D., Skrzypczyński P. (2010) Terrain perception and mapping in a walking robot with a compact 2D laser scanner, Emerging Trends in Mobile Robotics (H. Fujimoto *et al.*, eds.), Singapore, World Scientific, 981–988.
8. Okubo Y., Ye C., Borenstein J. (2009) Characterization of the Hokuyo URG-04LX laser rangefinder for mobile robot obstacle negotiation, Unmanned Systems Technology XI, Proc. SPIE 7332.
9. Skrzypczyński P. (2012) Laser scan matching for self-localization of a walking robot in man-made environments, Industrial Robot: An International Journal, 39(3).

GLOVE-BASED GESTURE RECOGNITION SYSTEM

MARIA EUGENIA CABRERA*, JUAN MANUEL BOGADO, LEONARDO FERMIN, RAUL ACUÑA, DIMITAR RALEV

Mechatronics Research Group, Simon Bolivar University, ELE Bldg 3rd Floor, Lab 302. Valle de Sartenejas
Caracas, Miranda 1080-A, Venezuela

Among the research topics developing within the Mechatronics Group at Simon Bolivar University, there's a field corresponding to Human-Computer Interaction (HCI), which is the center of this paper, regarding the design and implementation of a hand gesture recognition system, based on the information gathered with a data glove. The data glove sensors convey the degree of flexion in each finger of the hand. Additionally, an accelerometer sensor was placed on top of the glove to acquire the orientation of the hand in three axes. With this information, the system would begin a decision making process based on artificial intelligence tools, like neural networks, to recognize the American Sign Language (ASL) signs for fingerspelling. The whole system is executed by interfacing with a user, to collect the data and display the result of its classification.

1. Introduction

The use of gestures among humans, both in the form of pantomime or by using sign language, is closely linked to speech and represents an effective way of communication, used even prior to talking.

The formality of the set of rules chosen in each case is related to the validity of the performed gestures, which means that a pantomime gesture could be complimenting speech in a spontaneously and colloquial manner, or the same gesture be considered a part of a sign language and posses a particular meaning, whether it is to perform an action or convey some type of information (1).

There's an approach within HCI, relating to the communication structures between man and machines, and studying the physiological, psychophysical and psychological aspects associated to the brain and its abilities to interpret information. HCI generally focuses on the design, implementation and evaluation of computer systems, through which a human operator performs some kind of information exchange (2). Considering that the hands are used as a means of communication, one could think of a system where their movement could be interpreted as the input, to be analyzed and classified, using a determined set of rules that would pass along the information.

In order to decide which input device to acquire the user's hand movements, there are a few items to consider, regarding effectiveness, comfort of the user, and the selection also needs to be suitable to the working environment to fulfill the tasks set for the device within the system(3). In the topic of hand gestures recognition, the input device must capture the position of each finger on the hand, as well as the hand's orientation in its rotation axis. The most commonly used devices reported in the literature are data gloves for motion capture and cameras for artificial vision. A more detailed revision of the state of the art will appear in the following subsection.

The general idea of this particular project is based on the design of a system that allows the recognition of a set of established gestures, such as the alphabet in American Sign Language, performed by a user with a single hand, using a data glove as an input device. In that system, artificial intelligence tools will be used, like artificial neural networks, to achieve an interaction between a computer and the user, where the first classifies and recognizes the gestures executed by the latter. Within the alphabet of ASL, there are static and dynamic gestures: static gestures can be accomplished by maintaining a pose of the hand, while dynamic gestures join a specific pose of the hand with a movement.

1.1. *State of the Art*

Every research work starts by reviewing the literature on the subject, in order to get a sense of what people are doing all around the world and to know the possible applications and solutions presented for each topic specifically.

On the state of the art regarding the subject of gesture recognition, there are many teams in many countries approaching the matter with two particular goals; at least those two represent the most common among the reviewed literature. One goal, seeks for gesture recognition to aid hearing impaired people, giving them a way to interact with people who aren't fluid in such language, through the computer. The other goal relates to controlling robots, by the performed gestures of its operator.

Another feature reviewed on the literature, was about the tools used to achieve such said goals and the proposed solutions in each case. The work published by Parton (4), reviews a number of approaches within the field of Artificial Intelligence (AI) to effectively recognize and translate sign language: from Artificial Neural Networks (ANN), Hidden Markov Models (HMM), Artificial Vision, Virtual Reality, and motion capture using data gloves or electromyographic (EMG) sensors. Which leads to believe that there are two main approaches to recognizing hand gestures: one is based on motion capture of the user's hand, while the other bases on artificial vision to determine the gestures made in a specific time. Both approaches have their pros and cons.

Vision-based gesture recognition relies on image capture by one or more cameras and applying algorithms to extract visual features, in order to recognize the gesture made. The fact that the person making the gesture is totally detached from the system is one of the most important benefits in working with visual recognition. On the other hand, the system can be very susceptible to changes in the environment, like lighting or background figures and colors that could affect the process.

The work by Ricco & Tomasi (5), looks for recognizing features not only in the images where there is a gesture but they also consider in their algorithm the frames of transition to recognize gestures; while Gui *et.al* (6) focus on intrinsic characteristics of the gestures to recognize them. Another method reported by Malima *et.al* is to obtain the image centroid to recognize positions of the hand to command a robot to move. (7)

The approach based on capturing the user's motions, physically links the user to the system but in return, it detaches it from the changes in the environment giving the opportunity to use the system in different places and obtaining the same results. The problem with incorporating the user to the acquisition process is that it causes a discomfort on the user, since their hand is no longer as free as it would be in a normal scenario. The most common device used in the literature is the data glove, to acquire the flexion degree of each finger in the hand. Another system (8) bases its input device on EMG sensors placed on the forearm of an operator to control a robot with grasp recognition.

On most of the research work reviewed where a data glove was used the problem to be solved was the translation of sign language to written or spoken. As mentioned previously, most are based on a pre-established sign language to recognize a subset of letters on the alphabet in each case, commonly the static gesture letters. The work of Parvini *et.al* compares the result of three configurations for ANN classifying the static gestures of ASL acquired by a dataglove, one with neural networks alone, one with multi-layer networks working with template matching, and finally a gesture recognition system using bio-mechanical characteristics. The best results were obtained from the first and last configuration (9). In other papers, the system architecture is a bit more elaborate, like the case of Fels & Hinton, where they use two data gloves and a voice synthesizer with intonation control using a pedal. (10)

2. Architecture System Description

The proposed system for recognizing hand gestures begins and ends with the user. In between there are five stages: input, acquisition, preprocessing, classification and results presentation.

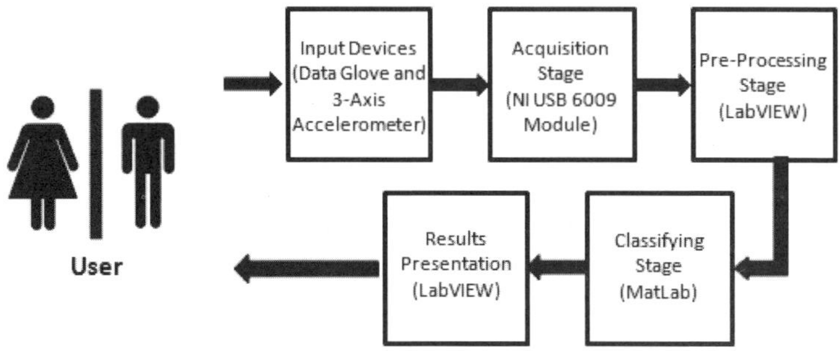

Figure 1. Diagram for the System Architecture.

As presented in Figure 1, the user places the input devices on his right hand; the input devices consist on the 5DT Data Glove 5 Ultra to gather information of the flexion degree of each finger of the hand and a 3-axis accelerometer place on the back of the hand to know its orientation. The data from the accelerometer goes through the acquisition stage, where an ADC converts the values of each axis to 14-bit numbers and passes them along to the computer for the preprocessing stage. In that stage, the information of the data glove, that communicates directly to the computer and the acquired data of the accelerometer go through a calibration and standardization process and then it is stored to use it as inputs in the classifying stage. Said stage is about using an artificial intelligence tool, such as artificial neural networks, to recognize gestures based on the experience of the previously collected data. With a trained network, an interface is developed to present the results of the gesture classification back to the users.

The proposed system innovates comparing to previous works by adding the accelerometer to gather information of wrist rotation, which allows identifying some ASL gestures for letters such as H and U that hold an identical position for all fingers in the hand and differ on the wrist rotation, another examples are K and P, also G and Q. Previous works, such as (9), select only one of the similar letters to input the system and be recognized by it. On figure 2, the similarities between the previously named letters are shown.

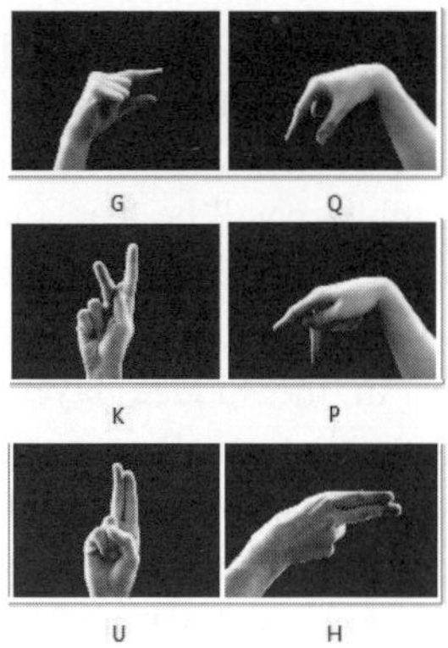

Figure 2. Similarities between ASL gestures for specific letters.

The final interface is made in National Instruments Labview® Software, as shown on Figure 3; the software is also used to gather the sensors information and store the data to be used in Matlab® for the ANN configuration and training.

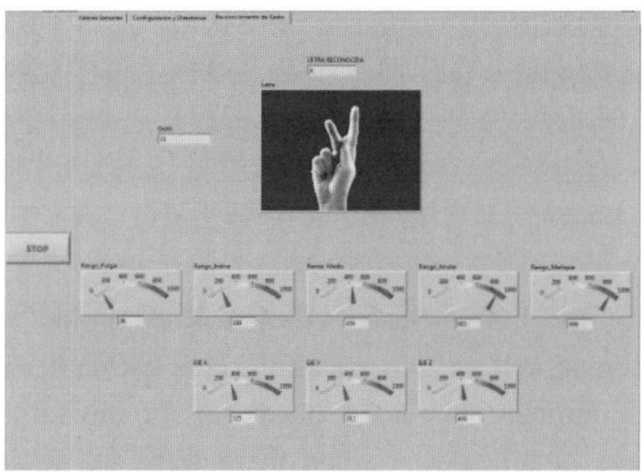

Figure 3. Interface in Labview® to present the final results of the Hand Gesture Recognition System

3. Classification Results

The Artificial Intelligence tool selected for this particular project was multilayer perceptron feed-forward neural networks with backpropagation as a training algorithm.

Several ANN were tried in order to obtain the best results in correctly classifying the data acquired and stored from the input devices. The total data used as input is 5 for each flexion sensor on the hand's fingers and 3 for each axis of the accelerometer. So the input layer of the system's network is set to 8. About the output layer, there are as many neurons as letters to classify and an extra one for gestures that aren't letters, such as transition gestures or gestures not included in the ASL alphabet; the letters to be included in the process are the ones based on static gesture, excluding J and Z. Which leaves the output layer set to 25 neurons. Each input cluster will be feed-forwarded through the hidden layer and the output layer to activate one of the neurons in the output layer. The values of the output layer may vary from 0 to 1 with a sigmoid function.

The number of neurons in the hidden layer (H.L) was the parameter changed throughout the selection of the best performing network. The other parameters were kept the same in each test: 5300 training patterns, with an average of 200 per letter and 500 for transition gestures. A set of 1200 patterns were left apart for validation purposes. Each training process was made with a learning rate of 0.015, a maximum of 100 iterations and a mean squared error target of 1.10^{-5}. The results are shown on Table 1.

Table 1. Training Results and Validation Percentages for different ANN configurations, changing the number of neurons in Hidden Layer.

H.L Neurons	Training MSE	Training Iterations	Training Time	Val. Classification %
40	$5{,}25.10^{-3}$	61	18'16"	92,02
50	$4{,}16.10^{-3}$	83	33'36"	92,7
60	$4{,}10.10^{-3}$	62	34'23"	94,07
70	$3{,}71.10^{-3}$	64	47'21"	93,39

4. Conclusions

The performance of the Glove-Based Gesture Recognition System is directly related to the effectiveness of the classification stage within it. Based on the results given by different number of hidden layers in the neural network, the best validation percentage obtained was when there were 60 neurons in the hidden layer. Among the many benefits of using neural networks for classification duties, there is the ability to adapt and generalize when properly trained, and they can be easily retrained to adapt its synaptic weights to variations in the system.

References

1. **A. Mulder.** *Hand Gestures for HCI.* Simon Fraser University : NSERC Hand Centered Studies of Human Movement Project, 1996.
2. **Hewett & others.** *Curricula for Human-Computer Interaction.* Carnegie Mellon University : The Association for Computing Machinery Special Interes Group on Computer Human Interaction, 1992.
3. **J. Preece & others.** *Human Computer Interaction.* Wokingham, England : Addison-Wesley Publishing Company, 1994.

4. **B. Parton.** *Sign Language Recognition and Translation: a multidisciplined approach from the field of Artificial Intelligence.* Journal of Deaf Studies and Deaf Education, Vol. 11, No. 11, págs. 94-101, 2006.
5. **S. Ricco, C. Tomasi.** *Fingerspelling Recognition through letter-to-letter transitions.*Ninth Asian Conference on Computer Vision, 2009
6. **L. Gui, J. Thiran, N. Paragios.** *Finger-spelling recognition within a collaborative segmentation/behavior inference framework.* 16th IEEE European Signal Processing Conference, 2008.
7. **A. Malima, E. Ozgur, M. Cetin.** *A fast algorithm for Vision-Based hand gesture recognition for robot control.* IEEE 14th Signal Processing and Communications Applications. págs. 1-4, 2006.
8. **S. Ferguson, G. Dunlop.** *Grasp Recognition from Myoelectric Signals.* Auckland : Proc. 2002 Australasian Conference on Robotic and Automation. págs. 83-87, 2002.
9. **F. Parvini & others.** *An approach to Glove-Based Gesture Recognition.* Lecture Notes in Computer Science. Proceedings of the 13th International Conference on Human-Computer Science Vol. 5611. págs. 236-245, 2009
10. **S. Fels, G. Hinton.** *Glove-Talk II - A neural-network interface which maps gestures to parallel formant speech synthesizer control.*IEEE Transactions on Neural Networks, Vol. 8, Issue 5, págs. 977-984, 1997.

REAL-TIME VISUAL PERCEPTION FOR TERRAIN MAPPING IN A WALKING ROBOT

Przemysław ŁABĘCKI, Piotr SKRZYPCZYŃSKI

Institute of Control and Information Engineering,
Poznan University of Technology, ul. Piotrowo 3A, PL-60-965 Poznań
E-mail: pl@cie.put.poznan.pl, piotr.skrzypczynski@put.poznan.pl

This paper presents a new method for terrain map building with hardware-embedded stereovision. A real-time range data post-processing method is proposed to remove artefacts, and a method for building and updating the grid elevation map from point cloud data is presented. The relatively low computational load of the embedded stereovision and the presented post-processing method allows for real-time application. Experimental verification of the map building algorithm is provided. Also integration of the stereovision-based map with the more precise map constructed from laser data is described.

1. Introduction

In our previous work[5] we presented a mapping system for the Messor hexapod robot, which used an active 2D range sensor, either a structured light vision system or the compact laser scanner Hokuyo URG-04LX. The obtained elevation map was then used for planning of the footholds.[1] However, the perception range of the tilted 2D sensor is too small for on-line path planning on the basis of the obtained terrain map.

A passive stereo vision system seems to be a good choice for long-range perception, due to its low power consumption and small size. Typically, calculating a dense depth map using a stereo system has high computational requirements, but advanced cameras, like the Videre Design STereo-On-a-Chip (STOC), provide such measurements in real time, thanks to data processing embedded in FPGA hardware. The stereo camera perceives the environment up to several metres from the robot, but the range data are sparse and corrupted by various artefacts. On the other hand, the laser scanner, already used on Messor, is much more precise in acquiring the terrain profile immediately in front of the robot. Therefore, we conceived a dual terrain representation, which employs a coarse "strategic" elevation map describing the more distant areas perceived by means of the camera, and a smaller, fine-grained "tactical" elevation map that surrounds the robot, which is updated from the URG-04LX range data.

This paper addresses the issues of visual perception for terrain mapping. We focus on the STOC camera for rough terrain mapping, considering limitations of this sensor, real-time performance, and describing integration of the stereo and laser scanner data.

Several papers already addressed the usage of sterovision for terrain mapping. In Ref. 6 usage of a STOC camera as a sensor for mobile robots in search and rescue tasks was considered. However, the artefacts produced by the camera were not taken into account. In Ref. 8, stereovision was used for terrain mapping with a walking robot, but no foothold planning was necessary due to the compliant legs of the RHex robot. Terrain mapping using stereovision was also performed with the LittleDog robot,[3] however none of the data processing was done onboard the robot and no stereo data filtering was employed.

2. STOC camera data post-processing

The STOC camera implements in hardware the Small Vision System (SVS) stereo matching algorithm.[4] The uncertainty of range measurements rises with the distance from the camera and the stereo processing often produces incorrect matches. Such errors occur as a result of insufficient image texture and around depth boundaries. False measurements are also produced when objects are located outside the horopter (i.e. too close to the camera) and on reflective surfaces. The STOC camera uses some built-in filters to reduce incorrect matches. However, these filters cannot handle errors produced by out-of-horopter objects or reflective surfaces, and they are generally too restrictive, removing useful information. Thus, there is a need for measurement post-processing. The artefacts that are most resilient to the built-in filters are those on the borders of objects with disparity values much different from the background. They manifest themselves as halos stretched from the foreground to the background object (Fig. 1B). The other group of artefacts is formed by out-of-horopter measurements, and artefacts appearing on reflective surfaces and texture-less areas. These artefacts manifest themselves as groups of points near the camera (Fig. 1C).

We propose to remove the artefacts belonging to both groups by analyzing the depth image, which is easily obtained from the disparity image. In a window based at every nonzero pixel of the depth image, two coefficients are calculated: (i) the point density and (ii) the depth span of the neighbor-

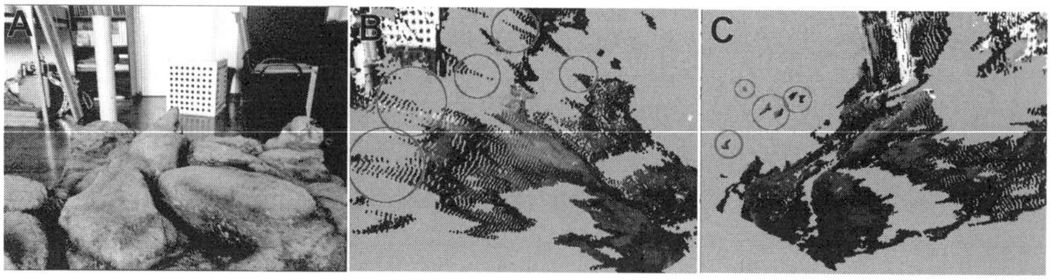

Fig. 1. Rough terrain mockup (A), and typical artefacts belonging to the first (B), and second (C) group

ing pixels. The depth span is simply the difference (in metres) between the highest and lowest point depth in the window. To calculate the point density, a cube is defined at the investigated point in 3D space. The dimensions of the cube are chosen so that the front face of the cube projected onto the image corresponds in size to the image window. The point density is then calculated as the ratio of points fitting into the cube to the total number of pixels in the window. Pixels of low density are likely to be a part of the halos while pixels with large depth span in the neighborhood are likely to be either a part of the halos or a part of an artefact from the second group. The values of the calculated coefficients are used to discard artefacts. Pixels with the following coefficient values are treated as faulty measurements: (i) density lower than 0.3, (ii) depth span greater than 0.8 m, (iii) density lower than 0.7 and depth span greater than 0.5 m. To avoid removing pixels on actual edges of the objects (which would also score high depth span and low density coefficient values), each time the window is limited to the boundaries of the object in which the center pixel is located. Object detection is performed on the gray image using watershed segmentation.

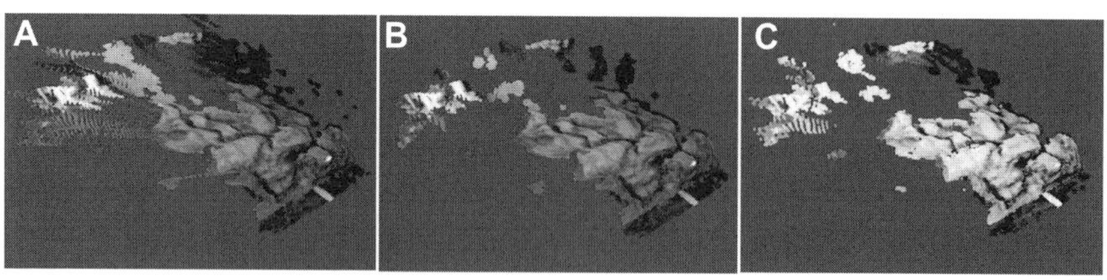

Fig. 2. Comparison of the filtration methods: raw point cloud (A), SOR filter (B), and the proposed method (C)

Figure 2 shows a comparison between the proposed post-processing method and the Statistical Outlier Removal (SOR) filter implemented in the well-known Point Cloud Library (PCL).[9] The SOR method is based on the concept described in Ref. 7. For each point of the cloud, it calculates the mean μ and standard deviation σ of distances, based on k nearest neighbours. The points which are outside $\mu \pm \alpha \cdot \sigma$ are removed (in the considered example, a value of $\alpha = 1$ was used). This algorithm doesn't need a fixed viewpoint to filter the outlieres and is more general than the proposed method. While the SOR method was slightly more effective in removing artefacts, it was about 30 times slower than our procedure. On a 1.6 GHz PC filtering of the point cloud shown in Fig. 2A took about 21 s using the SOR filter (Fig. 2B), whereas the proposed post-processing (Fig. 2C) needed only 0.65 s. Such performance allows real-time filtering of the STOC camera range data.

3. Terrain mapping

For the purpose of walking robot navigation and foothold selection, a grid-based elevation map is used. Two co-located grids of the same size are maintained: an elevation grid and a certainty grid. Each cell in the certainty grid holds a value that describes the certainty in the corresponding cell's estimate of the elevation.

Long-range perception with the STOC camera allows to create a strategic elevation map, reaching several metres from the robot. Because of the artefacts and empty areas, the precision of this map is quite low. Thus, it allows to roughly plan the path of the robot, but doesn't allow foothold selection, as this procedure requires an elevation grid, which reflects geometric properties of the ground at scale of the robot foot size.[1] Therefore the strategic elevation map is updated only from stereo data, while the smaller tactical map updating procedure uses mostly the URG-04LX range measurements. However, as the robot walks forward, the precise map gradually overlaps the area that was already mapped with the STOC camera. To re-use the already acquired elevation data in the precise map, we implemented a simple fusion mechanism. As the laser data are much more reliable, these data override the stereo data in all cells that have the certainty value above a given threshold (i.e. they were successfully measured several times). However, if there are low-certainty cells in the tactical map, they are filled with the elevation data from the corresponding cells of the map from stereo. Results of this fusion procedure are shown in Fig. 3 (stereo and laser data mapped to different colours to make them distinguishable). To update both the strategic and tactical maps, an estimate of the motion of the robot is needed. Because the STOC camera provides grayscale images as well as the disparity image, a visual SLAM algorithm[10] can be run paralell to the mapping algorithm to provide motion estimates.

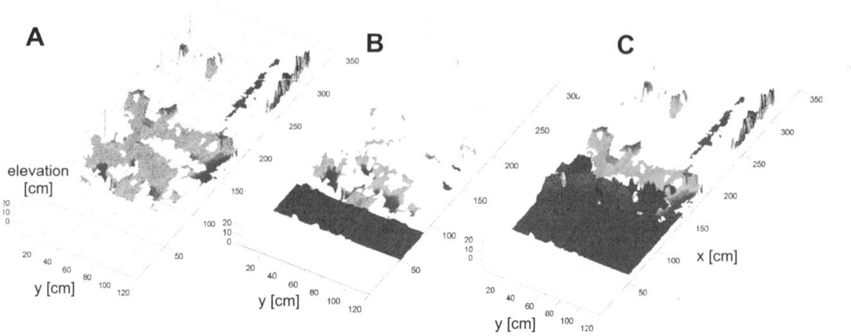

Fig. 3. Example map created from stereo data (A), and maps iteratively updated by using the laser data (B,C)

Each time new measurements are available, the measured points should be merged with the existing map. Because the procedure of updating the elevation map with the laser range data is identical to the one described in detail in Refs. 1 and 5, we focus here on the stereo vision data processing. Every elevation grid cell that contains new points can be updated with a new elevation value, if only the stereo range measurements are of better quality. To calculate the new elevation value, the space corresponding to the given cell is divided into vertically arranged bins of equal size. The bin that contains the most points from the new measurement is selected to calculate the new elevation and certainty value. The certainty value is calculated as the average certainty of all points from the selected bin, while the elevation is calculated as a weighted average of the vertical coordinates of the points accumulated in this bin, with certainty values of the points used as weights.

The certainty value of a single point is calculated as $c = \frac{1}{\Delta r} w$, where Δr is the depth resolution of the measurement, and w is a weighting factor introduced because the quality of the measurements degrades near the edges of the disparity image:

$$w = 1 - \frac{\sqrt{(u - C_x)^2 + (v - C_y)^2}}{\frac{1}{2} I_w}, \tag{1}$$

where (u, v) are the image coordinates of the given point, (C_x, C_y) the position of the camera's principal point, and I_w the width of the image.

The new values based on the points from the selected bin are used to update the map at the given cell, if the new certainty value is greater than the old one.

One last step in the map updating algorithm is the evaluation of the visibility constraint. Each newly updated cell is checked whether it is occluded by cells stemming from older measurements. If the certainty of the occluding cells is smaller than of the occluded cells, the old measurements are removed (Fig. 4).

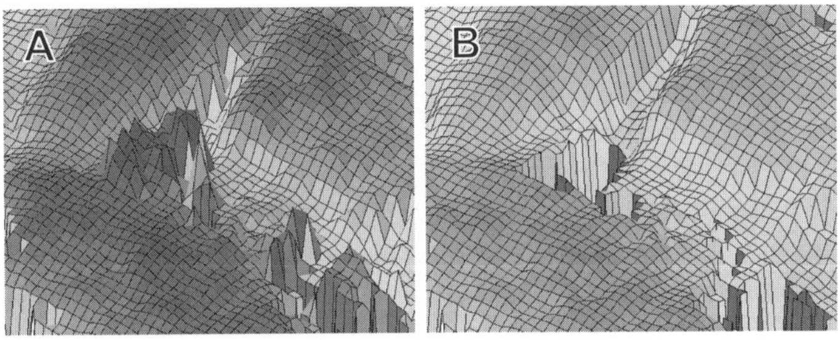

Fig. 4. Evaluation of the visibility constraint. Fragment of a map built without evaluation of visiblility (A),fragment of a map with the visibility constraint used (B)

4. Experimental results

Laboratory tests of the stereo-based mapping system were performed using a terrain mockup of size 150x150 cm. To compare the presented post-processing approach with the default filters in a quantitative way, a version of the Performance Index (PI), used also in Ref. 1 was applied:

$$PI = \frac{\sum_{i=1}^{n}\sum_{j=1}^{m} o_f(i,j)\cdot(h_f(i,j)-h_t(i,j))^2}{\sum_{i=1}^{n}\sum_{j=1}^{m} o_m(i,j)\cdot(h_m(i,j)-h_t(i,j))^2}\cdot\frac{N_{om}}{N_{of}}. \tag{2}$$

In the above equation, i and j are the coordinates of the given map cell, $h_f(i,j)$ and $h_m(i,j)$ are the values at the i,j coordinates of the elevation maps built from range data: filtered using the investigated approach or using minimal built-in filtering respectively, whereas $h_t(i,j)$ is the elevation value at i,j coordinates of the ground truth map. $o_f(i,j)$ and $o_m(i,j)$ can have zero or one values, according to whether the given cell in the map is empty or not, and N_{om} and N_{om} determine the total number of not empty cells in the respective elevation maps. The ground truth (Fig. 5A) needed to calculate the PI value was acquired by scanning the terrain mockup with a Hokuyo URG-04LX laser scanner moved by the KUKA KR-200 industrial robot.

The PI value tells how much the investigated approach reduces mapping errors with regard to a map build only with minimal filtering. Lower

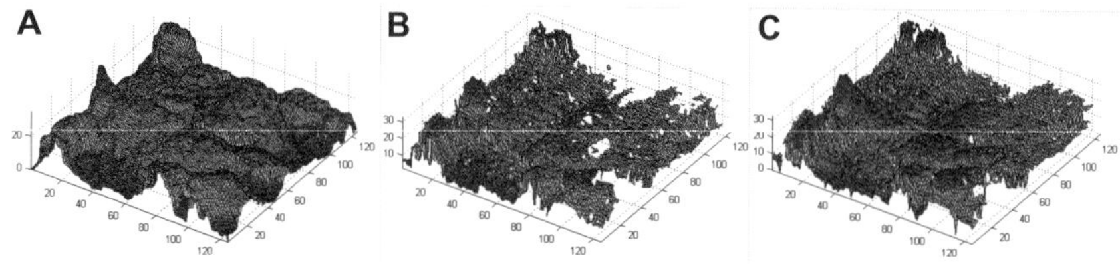

Fig. 5. Elevation maps: ground truth (A), data filtered with the built-in methods (B), and post-processed data (C)

PI values mean better reduction of the mapping errors. The PI value for the map created with built-in filters, set to default values (Fig. 5B) was 0.99, which means that most of the removed data was correct. The post-processing algorithm (Fig. 5C) scored a PI value of 0.91, while removing most of the artefacts.

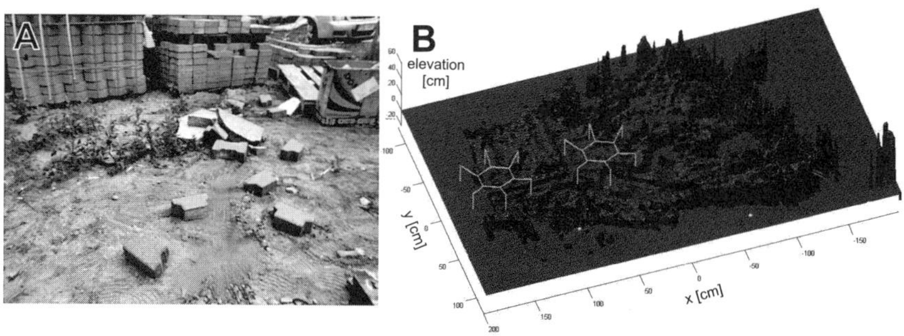

Fig. 6. Outdoor scene as seen by the STOC camera (A) and the resulting elevation map (B)

The described terrain mapping procedures were also verified in a small-scale outdoor experiment on a real rubble-covered terrain with some vegetation (Fig. 6A). Figure 6B visualizes the obtained elevation map, which was created using both the stereo and laser data, and then used to plan a path and footholds for the Messor robot. See Ref. 2 for details of the planning procedure.

5. Final remarks

This work presents a new method for terrain map building with hardware-embedded stereovision. A range data post-processing method is proposed to remove artefacts. While it isn't as much effective as other state-of-the-art methods, its main advantage is calculation speed, which allows for real-time applications. Integration of the map build with stereovision with the more precise map constructed from laser data is also described.

References

1. Belter D., Skrzypczyński P. (2011) Rough terrain mapping and classification for foothold selection in a walking robot, J. of Field Robotics, 28(4), 497–528.
2. Belter D., Łabęcki P., Skrzypczyński P. (2011) On-board perception and motion planning for legged locomotion over rough terrain, Proc. 5th European Conf. on Mobile Robots, Örebro, 195–200.
3. Kolter J. Z.,Kim Y., Ng A. Y.(2009) Stereo vision and terrain modeling for quadruped robots, Proc. IEEE Int. Conf. on Robotics and Automation, 1557–1564
4. Konolige K. (1997) Small vision systems: hardware and implementation, 8th Int. Symp. on Robotics Research, 111–116.
5. Łabęcki P., Rosiński D., Skrzypczyński P. (2011) Terrain map building for a walking robot equipped with an active 2D range sensor, J. of Automation, Mobile Robotics and Intelligent Systems, 5(3), 67–78.
6. Poppinga J., Birk A., Pathak K. (2009) A characterization of 3D sensors for response robots, RoboCup 2009: Robot Soccer World Cup XIII, J. Baltes et al., eds., LNAI Vol. 5949, Berlin, Springer, 264–275.
7. Rusu R. B., et al. (2008) Towards 3D point cloud based object maps for household environments, Robotics and Autonomous Systems, 56(11), 927–941.
8. Rusu R. B., et al. (2009) Leaving flatland: efficient real-time 3D perception and motion planning, J. of Field Robotics, 26(10), 841–862.
9. Rusu R. B., Cousins S. (2011) 3D is here: point cloud library (PCL), Proc. IEEE Int. Conf. on Robotics and Automation, 1–4
10. Schmidt A., Kasiński A. (2010) The visual SLAM system for a hexapod robot, Computer Vision and Graphics, L. Bolc et al., eds., LNCS Vol. 6375, Berlin, Springer, 260–267

A SENSOR DATA FUSION ALGORITHM FOR INDOOR SLAM USING A LOW-COST MOBILE ROBOT

LUCIANO BUONOCORE and AEROLINO DE ALMEIDA NETO
Universidade Federal do Maranhão, Department of Electrical Engineering
65085-581 - São Luis - MA, Brazil
E-mail: luciano@dee.ufma.br, areolino@dee.ufma.br

CAIRO LÚCIO NASCIMENTO JÚNIOR
Instituto Tecnológico de Aeronáutica, Division of Electronic Engineering
12228-900 - São José dos Campos - SP, Brazil
E-mail: cairo@ita.br

This article is concerned with the solution of the SLAM (Simultaneous Localization And Mapping) problem in an indoor environment using a low-cost mobile robot. The robot was constructed with a distance measurement sub-system composed of three types of sensors: a wireless webcam with a laser pointer (a visual sensor), two infrared sensors and an ultrasonic TOF (time-of-flight) sensor. An algorithm is proposed to fuse the noisy data provided by these sensors and a particle filter approach is employed to solve the SLAM problem. The proposed fusion algorithm was evaluated using two artificial indoor environments and the estimated maps were used to navigate the robot satisfactorily in these environments.

1. Introduction

Nowadays mobile robots are used in several sectors of the society and one of the main challenges for the researchers is how to make these machines intelligent and autonomous enough, such that they are able to perform their tasks with minimal human intervention. It is highly desirable that a mobile robot should be able to navigate autonomously in unknown environments. This means that the robot needs to construct a map of the environment and at the same time localize itself within this map, a problem commonly referred as SLAM (Simultaneous Localization And Mapping). There are many approaches to solve the SLAM problem such as fuzzy logic, artificial neural networks, and the Dempster-Shafer theory. However, the majority of the literature employs the so-called probabilistic SLAM solutions [1], using different combinations of sensors, such as laser scanner and CCD camera [2] or two sonars and six infrared sensors [3]. Many other sensor combinations may be found in the robotic literature.

The use of low-cost sensors in SLAM is more challenging than using high-cost sensors such as laser scanners, which are able to precisely and very rapidly measure distances up to 30 meters and cost typically around US$ 5,000. The authors of this article constructed a low-cost mobile robot, which was named SLAMVITA, to investigate the solutions to SLAM problem with a total cost around US$ 2,100.

In this article a particle filter approach known as Rao-Blackwellized or FastSLAM is used to solve the SLAM problem, where samples (particles) are employed to hold hypothesis of the robot poses along with the environment features estimated by an EKF (Extended Kalman Filter). The solution adopts the odometry motion model, where the outputs of the encoders are computed as control information [1].

This article is organized as follows. Section 2 introduces the SLAM problem and the approaches used to solve it. In Section 3 the proposed sensor data fusion algorithm is presented and shows the results of a simple experiment which was designed to measure its accuracy for environment mapping. Section 4 presents the experimental results obtained by the proposed solution to the SLAM problem in two small indoor environments.

2. SLAM Problem Overview

Figure 1 shows a graphic representation of the two main forms of probabilistic SLAM problems [1]: online SLAM and full SLAM. In the solution of the SLAM problem the robot poses and environment map are unknown estimated variables (***X*** and ***m***, respectively) that must be computed at every time step ***t***, and the accuracy of one variable is important to best estimate the other. This fact is referred in the literature as the "chicken and egg" problem.

This article uses a version of a particle filter known as Rao-Blackwellized or FastSLAM, where samples (particles) are employed to hold hypothesis of the

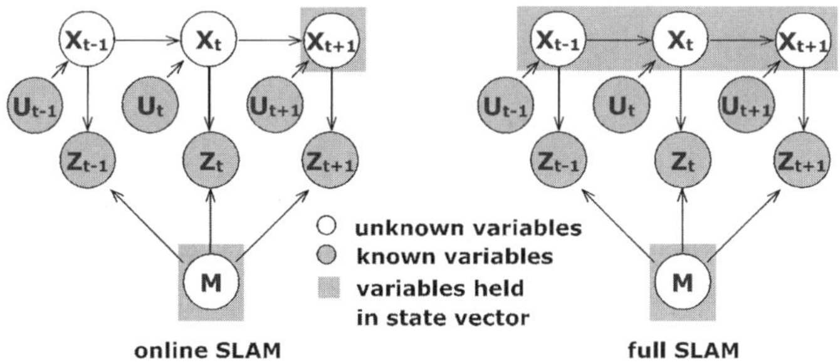

Figure 1. Forms of SLAM problems.

robot poses and environment features extracted by the proposed sensor data fusion algorithm. Moreover, the choice for FastSLAM version 1.0, instead of version 2.0, is due to the high quality odometry [1] shown by the SLAMVITA robot in the experiments.

3. Proposed sensor data fusion algorithm

A single sensor hardly ever can provide enough information about the environment because of its limited field of view. The solution to overcome such shortcoming requires the use of two or more sensors whose information must be fused. A three-level architecture for sensor data fusion is presented by Visser [4], in which three methods of data fusion may be applied: a) cooperative; b) competitive; and c) complementary. The sensor data fusion algorithm proposed in this article uses only competitive and complementary methods. In a competitive fusion each sensor produces its own estimated parameters, which are combined to generate a single set of parameters (in this case, distance and angle of the measured object). An example of competitive fusion occurs when the visual and infrared sensors measure the same object in the environment. Complementary fusion is applied when each sensor has only partial information of the environment. It aims to overcome the incompleteness of sensors being a classical example the array of several sonars attached around a mobile robot.

The SLAMVITA sensors were assembled on a panning turret on the top of the robot. The turret and the robot vertical axis are aligned and the turret can rotate around its axis to execute an 180° scan procedure. The parameters supplied by the sensors are the distance ($\boldsymbol{d}$) and angle ($\boldsymbol{\Theta}$) of the object (Figure 2). Due to their operating principles, the visual and infrared sensors provide reliable directional information (angle of the object), while the sonar is mainly a distance sensor [3]. This visual sensor has not been presented in the literature for mapping purpose and its operating principles and experimental results were presented in [5].

The proposed sensor data fusion algorithm, whose flowchart is shown in Figure 3, is informally described next and results from a simple mapping experiment are presented. It is based on the following considerations:

- It is assumed that the robot environment is composed of walls (which are always orthogonal to the floor but not necessarily to each other). The walls can have openings, such as doors. Therefore, the sensor data fusion algorithm aims to detect straight line segments instead of single points.

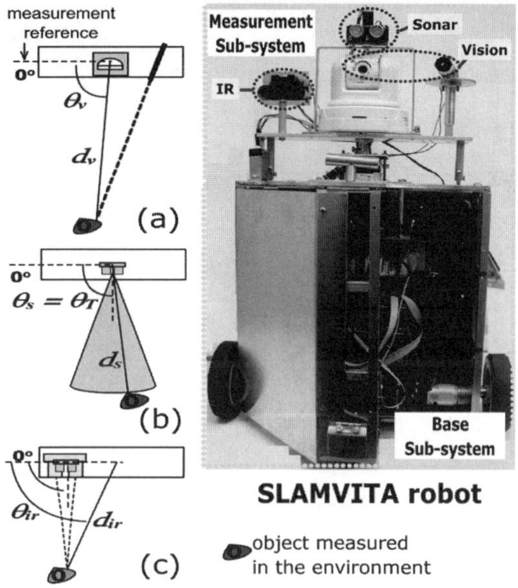

Figure 2. Measurement parameters (d, Θ) provided by the SLAMVITA sensors: (a) visual, (b) sonar and (c) infrared.

- The occupancy grid (OG) mapping technique is employed to represent the environment in two dimensions at the beginning of the fusion process. After that, the environment representation is changed to a feature-based map.
- The RANSAC algorithm [6] is employed to extract straight line segments from the directional (visual and IR) maps.
- The sonar measurements available in OG map (OGs) and in its raw data vector are used to eliminate noises in the directional OG maps (OGv and OGir) and to adjust the feature distances in the fused directional feature-based map, respectively.
- The distance parameters in the visual and IR sensors are considered to have the same weight, when the straight line segments are fused.

The input data (ID) shown in Figure 3 are the distance/angle contained in each raw data sensor acquired at each robot pose in an 180° scan. These sensor vectors are composed by 101 distance/angle parameters (1.8^{o} scan resolution). The algorithm generates the features extracted for each robot pose (output data, OD). The proposed fusion algorithm is composed by 4 specific algorithms and 3 fusion rules. The first 3 algorithms (A1, A2 and A3) process each sensor raw data to generate its respective OG map. The 3 rules (R1, R2 and R3) implement the data fusion methods mentioned above.

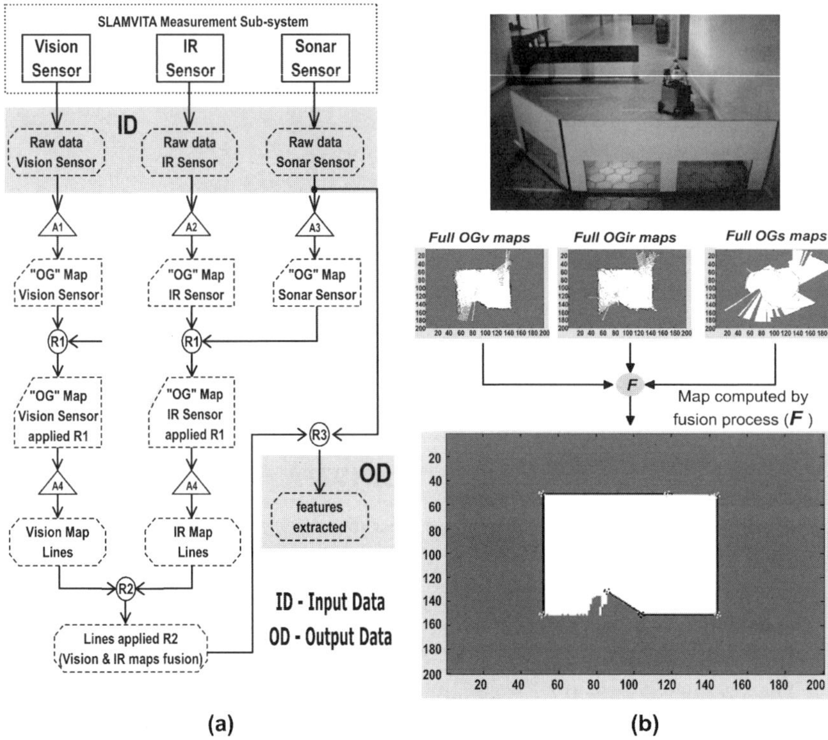

Figure 3. (a) flowchart of the proposed sensor data fusion algorithm and (b) comparison between maps generated by raw data sensors and the map outputted by the proposed fusion algorithm

The basic concept in R1 rule is as follows: no object detected by the sonar cone must be perceived by the other two directional sensors. The application of R1 rule justifies the use of the OG maps by the algorithm, where the inverse sonar model is employed [1]. Due to the surface reflection that is not normal to the sonar acoustic wave, the sonar raw data must be pre-processed before being employed to remove these wrong data, keeping only those which generate a true RCD (Region of Constant Depth) [7].

To extract line segments from directional maps, the proposed algorithm converts the OGv and OGir internal representation (after applying R1 rule) to x-y coordinates, which are held in separated vectors. Then, the RANSAC method is used in A4 algorithm to extract straight line segments (features) from those vectors, one map at a time. The line segments extracted from each map are combined in R2 rule, by applying a competitive or a complementary method. The competitive fusion is selected if both sensors have data at specific angular area. Otherwise, the complementary fusion is used, such as in the case when the visual sensor is unable to define the laser spot at the image.

The R3 rule adjusts, if possible, the feature distances of the fused directional map (after applying R2 rule), searching for RCD in the angular area that contains the normal straight line of each feature in the map. Whenever the RCD is found, the feature distance is adjusted using the sonar measurement.

Figure 3(b) shows the geometrical representation of the actual environment which, in comparison with each OG sensor map, is more consistent and noise free, with wall distance errors less than 1.5%. The map constructed by the proposed algorithm in this experiment is represented as full OG maps, instead of a feature-based one. However, in SLAM applications, the features for each robot pose must be processed and the environment representation generated by the algorithm is transformed to a feature-based map because of its reduced computational cost in the particle filter.

4. Experiments and Results

Two small indoor environments (8.8 m x 2.8 m or 24.64 m^2) were assembled for the SLAM experiments, and 700 particles were used by the FASTSLAM filter. The poses where the robot took measurements were manually chosen such that the robot firstly visited the inner loops of the environments. Figure 4 shows the real and estimated pose (black and green dots, respectively), the estimated maps (red lines) and the actual environments (black dotted lines). Some absolute differences between the real and estimated robot poses are shown in Table 1, where it can be seen that they were less than 1.5% for both experiments.

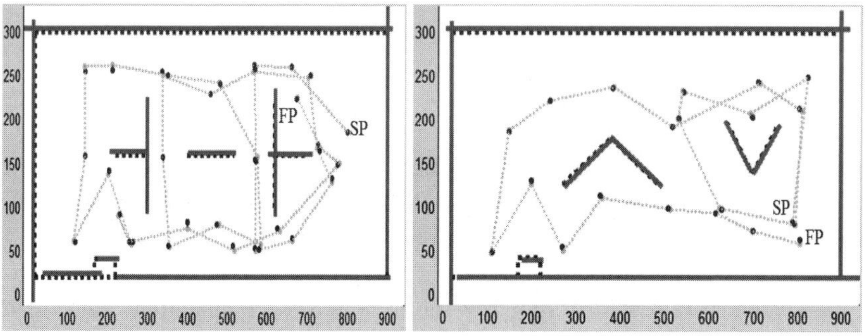

SP – Start Point; FP – Final Point; Robot estimated Path – green dot line

(a) (b)

Figure 4. Robot poses and estimated map for: (a) environment 1 and (b) environment 2.

Table1. Data information about some absolute differences between real and estimated robot poses in the SLAM experiments (ΔX and ΔY in centimeters and ΔΘ in degrees)

	No. Pose	ΔX (cm)	ΔY (cm)	Δθ (°)		No. Pose	ΔX (cm)	ΔY (cm)	Δθ (°)
Environment 1	1	0.00	0.00	0.00	**Environment 2**	1	0.00	0.00	0.00
	9	-2.01	-2.10	0.00		8	1.23	-8.96	-4.40
	12	0.01	-8.49	-2.26		10	2.80	-11.58	0.00
	18	-0.71	-1.13	0.00		18	8.97	-4.63	0.00
	23	-10.50	0.49	0.49		22	11.02	-3.60	-2.21
	28	-1.29	-13.19	0.00					
	33	0.72	0.38	0.00					

The absolute difference between the real and estimated robot poses is small in all of its components (X and Y coordinates and Θ angle). These results and the few number of robot poses to construct the map (33 and 22 for environments 1 and 2, respectively) are due to the precise robot odometry and to the efficient feature extraction obtained by the proposed algorithm. During the rotation and translation tests made to evaluate the robot odometry, the errors were less than 1° in rotation and 1% in translation. The largest absolute error was less than 4.7% (13.19cm/280cm, pose 28 in the Y axis, environment 1) and some others largest differences are highlighted in Table 1.

As shown in Figure 4, the features that compose the maps estimated by theirs associated EKF have shapes, lengths and orientations which are close to the ones in the real environment. However, some features have endpoints that exceed the real environment space. Also, a short feature (environment 2) appears in front of a large one, and two other short neighbors features are not mapped in the both experiments. A post-processing procedure should be able to remove these errors.

The maps obtained by the proposed sensor data fusion are consistent with the real environments, and allow the robot to safely navigate in these environments. In other experiments which made use of low-cost mobile platforms (such as presented by [3] that employs 2 sonars and 6 infrareds sensors), the solution to the SLAM problem has some restrictions due to the short measurement range of the infrared sensors.

Recently, the RGB-D camera became a low-cost option as a sensor for mobile robots [8], but it requires more complex computations and more expensive hardware.

5. Conclusions

Environment mapping is an important task performed by mobile robots for many applications, which usually requires sensor data fusion. A low-cost mobile robot was constructed with three sensor types: visual (camera and laser pointer), infrared and sonar. This paper presented the experimental results of the proposed sensor data fusion algorithm that generates the environment features. These features can then be used to solve the indoor SLAM problem.

To validate the proposed fusion algorithm, two experiments were carried out in different small indoor environments. The results showed that the estimated maps generated by the particle filter were similar to the actual environments and, therefore, they can be used for the robot navigation.

Further experiments including autonomous exploration and larger and more complicated environments are already in progress.

References

1. S. B. Thrun, W. Burgard, D. Fox. **Probabilistic Robotics**. *MIT Press* (2005).
2. J. A. Castellanos *et al.* Simultaneous Map Building and Localization for Mobile Robots: a Multisensor Fusion Approach. ***IEEE International Conference on Robotics & Automation***, *vol. 2*, Leuven, Belgium, pp. 1244 – 1249 (1998).
3. J. Vazquez, C. Malcolm. Fusion of triangulated sonar plus infrared sensing for localization and mapping. ***5th International Conference on Control and Automation,*** *vol. 2*, Budapest, Hungary, pp. 1097 – 1102 (2005).
4. A. Visser. Chapter 9 - Sensor Data Fusion. Unpublished. (*http://www.science.uva.nl/~arnoud/education/OOAS/fwi/Chap9.pdf*)
5. L. Buonocore, C. L. Nascimento Jr., A. Almeida Neto. Sistema de baixo custo para medição de distância em duas dimensões usando câmera IP sem fio. ***18º Congresso Brasileiro de Automática***, Bonito, MS, Brazil, pp. 2239 – 2246 (2010) (in portuguese).
6. M. Zuliani. RANSAC for Dummies-Draft. Unpublished. (*http://vision.ece.ucsb.edu/~zuliani/Research/RANSAC/docs/RANSAC4Dummies.pdf*)
7. Pandey, A. K., Krishna, K. M., Nath, M. Feature based grid maps for sonar based safe-mapping. ***XX International Joint Conferences on Artificial Intelligence***, Hyberdad, India, pp. 2172–2177 (2007).
8. P. Henry *et al.* RGB-D Mapping: Using Depth Cameras for Dense 3D Modeling of Indoor Environments. ***12th International Symposium on Experimental Robotics***, Delhi, India, (2010).

SECTION–10

PLANNING AND CONTROL

DYNAMIC POTENTIAL FIELD GENERATION USING MOVEMENT PREDICTION

R. ACUÑA, A. TERRONES, N. CERTAD-H, L. FERMÍN-LEÓN and G. FERNÁNDEZ-LÓPEZ

Simón Bolívar University, Mechatronics Group, ELE-328, Sartenejas 1080-A Miranda, Venezuela
**E-mail: racuna@usb.ve*

Path-planning in dynamic environments while meeting safety requirements for robots and humans is an open problem in robotics. The successful navigation in this kind of environments requires a certain level of anticipation to the future behavior of moving objects. In this paper we propose the use of movement prediction in the generation of dynamic artificial potential fields (DAPF), which will allow the robot to navigate in highly dynamic environments with a major degree of safety and effectiveness, especially in cases where the obstacles and objectives move at higher velocities than the controlled robots. Our approach is based on previous works on potential and velocity fields with additional considerations to avoid the well known local minimum problem and to anticipate the predicted path of objects, our solution tries to maintain the reactive characteristics of the artificial potential fields. The proposed method was tested in a holonomic simulation with 100% better results for the same scenarios than the original fields without prediction.

Keywords: Artificial Potential Field; Path-Planning; Local Minimum Problem; Velocity Field Generation; Movement Prediction; Autonomous Navigation

1. Introduction

The navigation using Artificial Potential Fields (APF) was presented for the first time by Khatib with the introduction of FIRAS, this function was designed to achieve real-time collision avoidance in unknown environments for manipulators and mobile robots.[1] This original idea has been in continuous development and adapted through the years in the field of robotics. However, there were many problems inherent to the APF approach which were summarized by Koren *et al.*[2] In an attempt to solve those problems, new methods emerged to modify the fields in order to cope with their limitations. The *Virtual Force Field Method* (VFF), the *Vector Field Histogram* (VFH)[3] and the enhanced VFH*[4] for local obstacle avoidance are some examples.

Despite its limitations, the APF are still widely used due to its real-time performance in many applications like planetary exploration rovers,[5,6] or in foraging tasks.[7] In other related work Li and Horowitz proposed the *Passive Velocity Field Control* (PVFC), in this method the motion task for manipulator control is encoded as a velocity field,[8] this method was improved in consecutive works by Li and Moreno.[8,9] However, this solution was inefficient in environments that change rapidly. Therefore new ways of dealing with the generation of references in an active fashion emerged, primary, in the works of Bruijnen,[10] ,Estévez[11] and more recently Pérez-D'Arpino,[12] but it is still considered as an open problem due to the limitless variations in dynamic settings. One of the many problems in dynamic environments is the likely presence of moving objects with greater or comparable velocities than the top speed of the robot. This creates a safety hazard, since in some cases the APF planner may take the robot to an area believed safe but that will be occupied by a fast moving object in the near future, in this scenario the robot may not be able to escape that area in the next iterations and a collision will occur. In this work a prediction scheme is introduced which modifies the fields in order to anticipate the future positions of moving objects on the workspace. This allows the controlled robots a mechanism to avoid the regions that will be occupied by moving obstacles, also this method is used to predict the position of goals and modify the fields to anticipate them.

2. Methodology

The solution proposed in this paper aims to solve general problems of navigation with motion prediction. In order to illustrate the most typical scenarios two main cases were selected: Predictive obstacle avoidance with and without defined objectives. In the predictive obstacle avoidance scenario without objetives, one robot lies on the path of a fast moving obstacle and it needs to avoid a collision. On the second case, the robot is traveling towards a defined objetive in a course that will pass through the path of a fast moving obstacle, once again the robot has to avoid the collision and also get to the goal. The proposed method is divided into two main parts: The predictor, responsible of the indexing and prediction of moving obstacles and goals; and the planner which defines the attractive and repulsive fields according to the position of the robot, the goals, the obstacles and their predicted paths. This method assumes the presence of a subsystem which provides the location information of all the objects in the workspace in each iteration. The primary experiments were implemented using a Python^{TM}

simulation with an holonomic platform. Additional tests were implemented in the robot development environment Webots^{TM} with physics using a simulated differential wheels robot Amigobot. The maximum speed and other physical properties of the simulated robot in both cases are similar to the those of the real Amigobot.

The proposed solution is organized as in Fig. 1.

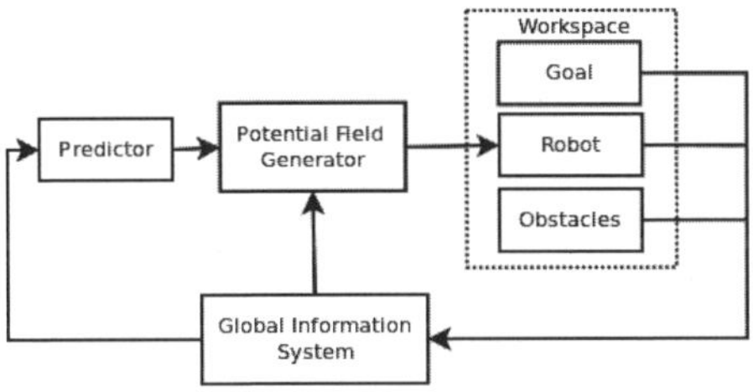

Fig. 1: Block Diagram of the predictor and potential field generation system.

3. Predictor and Potential Fields Definition

The system faces two constrains,[13] the response time available to plan a safe movement (which is a function of the system dynamics) and a temporal restriction about the certainty of the prediction in the future. In other words, it is imperative to plan rapidly but you can not plan far away in the future. In order to cope with this effect it is possible the use of a precise short term predictor, an inaccurate long term predictor or hybrid solutions. In any case, previous data is needed in order to predict whether a system model or historical data information. In our solution we assume that we do not know anything about the moving objects, therefore we can not make a prediction without studying their past actions. In this scenario the *Recursive Motion Function* (RMF)[14] is an ideal candidate, because it creates a model based on the historical positions of the indexed objects, this motion function can express a large number of simple and complex movement types like polynomials, ellipses, circles, sinusoids, etc. In the RMF algorithm, any polynomic function of degree D can be converted to a lineal recurrence after $D+1$ differentiations. Based on this information Tao proposed the following recursive function:[14] $o(t) = \mathbf{C}_1 \cdot o(t-1) + \mathbf{C}_2 \cdot o(t-2) + \cdots + \mathbf{C}_f \cdot o(t-f)$. Wich can be converted to matrix form, defined as the motion state $\mathbf{S}_o(t) = \mathbf{K}_o \cdot \mathbf{S}_o(t-1)$

Where $\mathbf{C}_i(1 \leq i \leq)$ is a $d \times d$ constant matrix (d is the dimension of the space), and f is an important parameter called the retrospect. The *motion state* $\mathbf{S}_o(t)$ is defined for an object o at time t as a vector representing its location at the f most recent timestamps. Where $\mathbf{K}_o$ is a constant $(d \cdot f) \times (d \cdot f)$ motion matrix for o.

In our solution we are using $d = 2$, and using the two most recent locations (retrospect $f = 2$) then equation 3 becomes:

$$\begin{bmatrix} o(t) \cdot \mathbf{x}_1 \\ o(t) \cdot \mathbf{x}_2 \\ o(t-1) \cdot \mathbf{x}_1 \\ o(t-1) \cdot \mathbf{x}_2 \end{bmatrix} = \begin{bmatrix} \mathbf{k}_{11} & \mathbf{k}_{12} & \mathbf{k}_{13} & \mathbf{k}_{14} \\ \mathbf{k}_{21} & \mathbf{k}_{22} & \mathbf{k}_{23} & \mathbf{k}_{24} \\ 1 & 0 & 0 & 0 \\ 0 & 1 & 0 & 0 \end{bmatrix} \cdot \begin{bmatrix} o(t-1) \cdot \mathbf{x}_1 \\ o(t-1) \cdot \mathbf{x}_2 \\ o(t-2) \cdot \mathbf{x}_1 \\ o(t-2) \cdot \mathbf{x}_2 \end{bmatrix} \tag{1}$$

The problem is finding the matrix K. We use the method of motion estimation proposed by Tao[14] using the actual location $\mathbf{l}_o(t)$ of the object o at time t and the h most recent timestamps to solve $S \cdot \mathbf{k}_l(t) = l$. In order to find the solution for this equation we need to find the inverse of S, Tao proposed *singular value decoposition* (SVD) but in our work we also tried other faster iterative methods like the gradient conjugate suitable for large retrospects f.

In our holonomic simulation we tested different types of movements like linear, polynomial and circular. In those cases a retrospect of $f = 5$ represented accurately the motion as indicated by Tao. However, when we add some noise to the input data the results change dramatically. In Fig. 2 the motion function of a sinusoid is predicted using $f = 5$ and the prediction is almost perfect. In Fig. 3 we add noise to the same sinusoid and in this case the predictor fails to estimate the exact motion behavior and the prediction is less accurate. If computational time is not an issue one way to solve this problem is to increase the value of the retrospect f, this mitigates the effect of the noise, in Fig. 4 we use for the same noisy data a retrospect $f = 40$, this prediction is less smooth but is more accurate than the case with a small retrospect.

Also, in our Webots simulation we implemented a Kalman filter that greatly reduce the effect of the noise in the prediction. In terms of computational time, an increment in the retrospect produces a larger S and the SVD computation time increases. For a retrospect $f = 100$ our predictor implementation in a Core 2 Duo CPU took $8ms$ while a large retrospect $f = 100$ took $30ms$. For single object tracking and prediction this may not be a problem, but with several objects this could be excessively high. An alternative is to solve the inverse of S using an iterative method like conjugate gradient and stop at n iterations, this produces an estimate of the inverse but the computational time is decreased with similar results,

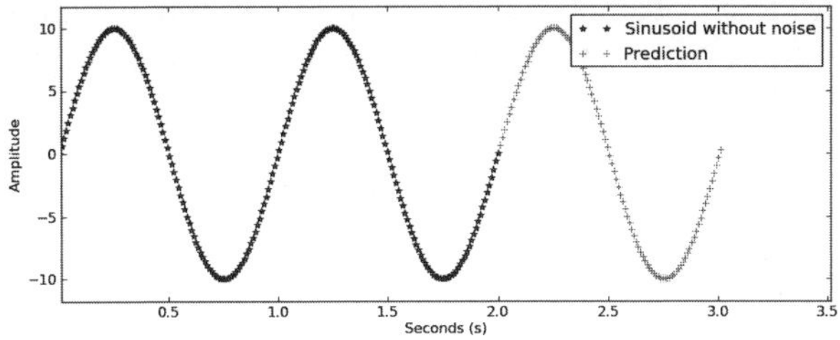

Fig. 2: Sinusoid prediction. Predicion Horizont $Ph = 100$, $f = 5$ and $h = 100$.

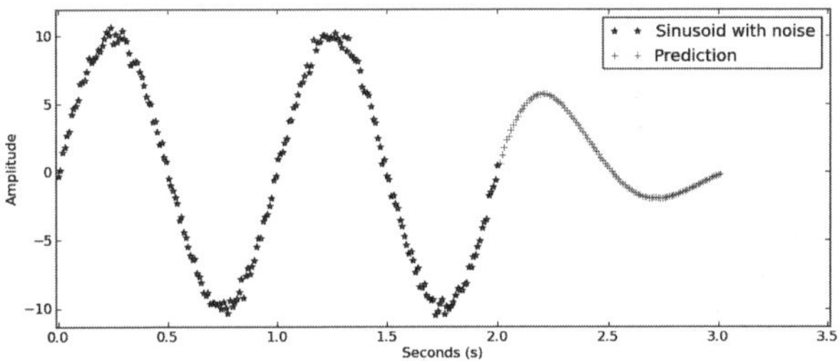

Fig. 3: Sinusoid with noise prediction. Predicion Horizont $Ph = 100$, $f = 5$ and $h = 100$.

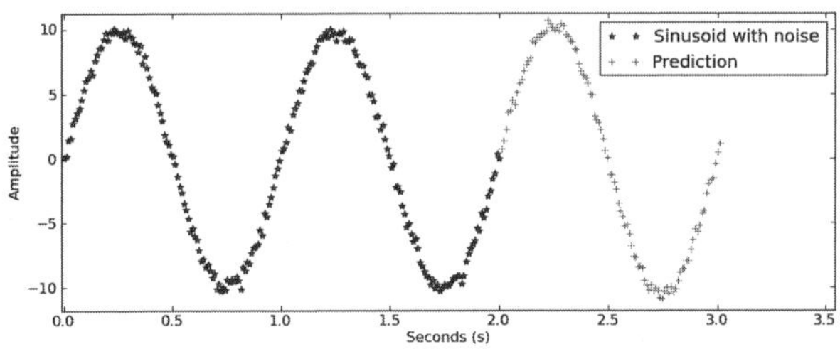

Fig. 4: Sinusoid with noise prediction. Predicion Horizont $Ph = 100$, $f = 40$ and $h = 100$.

for a retrospect of $f = 100$ the time to calculate the conjugate gradient was $8ms$.

The generation of vector references for the robot is based in potential surfaces. Basic potential functions for obstacles, goals and walls are described in the work of Estevez.[11] For the obstacles we use Gaussian Hills, which have limited range of action and continuous gradient. For the goals Conical Attractors are used that provide constant radial movement. The walls are considered obstacles and they have the same type of potential surface, with a maximum in the position of the wall and a Gaussian decay in a direction perpendicular to the wall. In our method the predicted path of the obstacles creates a repulsive potential field. Given a prediction $P(t)$ of dimension $d \times Ph$, for an object o, where d is the space dimension and Ph is the prediction horizon. A Gaussian potential field with σ_x and σ_y according to the size of the object o is created for each $P(t)\ \forall t \in [t, tPh]$. The individual contribution of each Gaussian Surface in each $P(t)$ is summed with the consecutive surfaces, this creates a Gaussian potential path along the predicted route of the object o. Also, a similar approach is used to avoid the local minima problem, a Gaussian path along recent historical positions of the robot favors the continuous movement of the robot towards the goal.

4. Results and Discussion

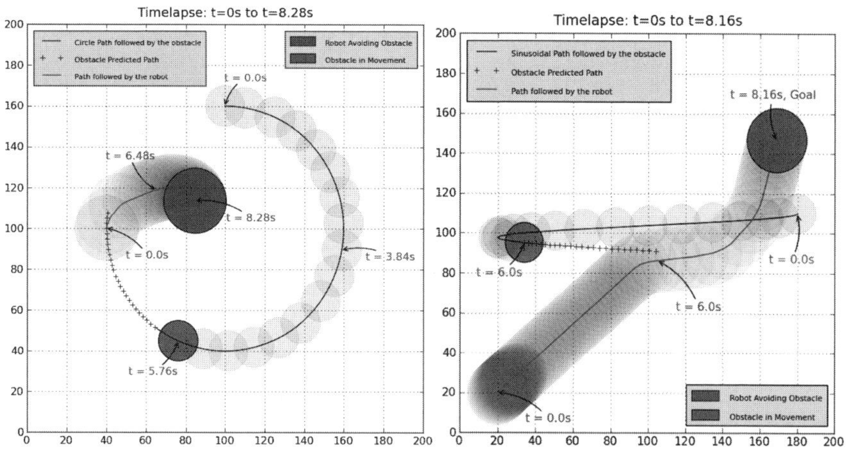

Fig. 5: Predictive obstacle avoidance. Left: without goals, Right: with goal at (170,150).

In Fig. 5 a fast moving obstacle (red) travels in a collision course to the slower Robot (blue), in Fig. 6 one obstacle travels in the path of the robot and another travels perpendicular. The predicted position of the obstacles and the generated fields allow the robot to prevent the collision, when this same tests where performed without prediction a collision always occurred. However the success of the proposed method depends on several factors, the training horizon H, the retrospect f, the prediction Horizon Ph and finally

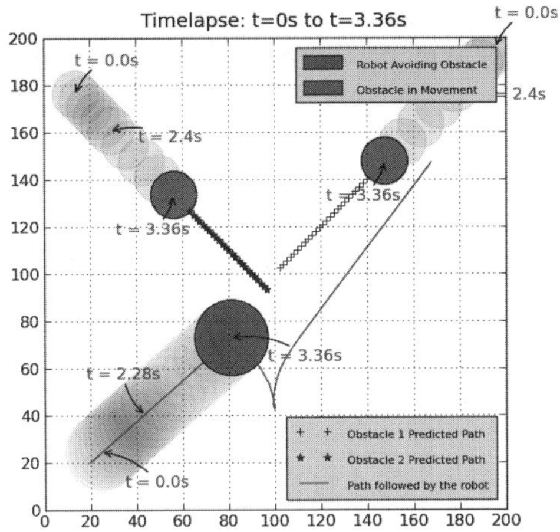

Fig. 6: Predictive obstacle avoidance. With Goal and Two Obstacles.

the weight and size of the potential path. The optimum Ph in this method is an open issue since it depends on both the relative speed between the robot and the obstacle and the relative direction of movement, which is not easy to determine since the movements are not always linear. A large value in Ph may increase the available time to successfully avoid the obstacle but a long predicted path will affect a large area of the workspace that may not be relevant and also increase the uncertainty of the prediction in the future. From our tests we conclude that it is safe to set Ph equal to the value of H. The size and weight of the Gaussian Potential Path PGP may change the outcome even if the prediction is accurate. If the Gaussian area and weight is too small the robot will not avoid the obstacle in time, in contrast a large area of effect will repel any robot in a vast area that in reality may be safe. A good value for the size of the PGP is the same width as the maximum of the obstacle. Finally if one obstacle suddenly changes direction, the prediction using RMF will fail and will be very unstable until H samples of the new type of movement have been indexed. In the meantime it is necessary to detect the moment of a sudden change in the movement pattern and disable the predictor.

Previous techniques for navigation using potential fields usually assume static obstacles or slow moving obstacles. Our paper presents a reactive solution using Gaussian repulsion zones based on the predicted positions of the obstacles to ensure collision-free trajectories in scenarios with fast moving obstacles. We believe that this method provides a safer reactive planner than other methods and open many doors for future research on the effect of the variations of the parameters to ensure safety and goal achievement.

References

1. O. Khatib, Real-time obstacle avoidance for manipulators and mobile robots, *The international journal of robotics research* **5**, p. 90 (1986).
2. Y. Koren, S. Member, J. Borenstein and A. Arbor, Potential Field Methods and Their Inherent Limitations for Mobile Robot Navigation, *Proceedings of the IEEE* , 1398 (1991).
3. J. Borenstein and Y. Koren, Real-time Obstacle Avoidance for Fast Mobile Robots in Cluttered Environments 1, *Robotics* , 572 (1990).
4. I. Ulrich and J. Borenstein, VFH*: Local Obstacle Avoidance with Look-Ahead Verification, *Robotics* , 2505 (2000).
5. M. Massari and G. Giardini, Autonomous navigation system for planetary exploration rover based on artificial potential fields, *Control of Systems* (2004).
6. J. Hagelbäck and S. J. Johansson, Using Multi-agent Potential Fields in Real-time Strategy Games, *Strategy* , 631 (2008).
7. O. Simonin, Construction of Numerical Potential Fields with Reactive Agents, *Science* , 1 (2005).
8. P. Y. Li and R. Horowitz, Passive velocity field control of mechanical manipulators, *IEEE Transactions on Robotics and Automation* **15**, 751 (1999).
9. J. Moreno and R. Kelly, On manipulator control via velocity fields, *Proceedings of the 15th IFAC World Congress,* **1**, 4 (2002).
10. D. Bruijnen, J. van Helvoort and R. van de Molengraft, Realtime motion path generation using subtargets in a changing environment, *2006 American Control Conference* , 4243 (2006).
11. P. Estévez, J. Cappelletto, R. Acuña, F. Alvarez and G. Fernandez-López, Coordinated Navigation Using Dynamically Varying Velocity Fields (2007).
12. C. Perez-D'Arpino, W. Medina-Melendez, L. Fermin, J. Guzman, G. Fernandez-Lopez and J. C. Grieco, Dynamic Velocity Field Angle Generation for Obstacle Avoidance in Mobile Robots Using Hydrodynamics, *Advances in Artificial Intelligence - IBERAMIA 2008* , 372 (2008).
13. C. Laugier, D. Vasquez, M. Yguel, T. Fraichard and O. Aycard, Geometric and Bayesian models for safe navigation in dynamic environments, *Intelligent Service Robotics* **1**, 51 (2008).
14. Y. Tao, C. Faloutsos, D. Papadias and B. Liu, Prediction and indexing of moving objects with unknown motion patterns, *Proceedings of the 2004 ACM SIGMOD international conference on Management of data - SIGMOD '04* , p. 611 (2004).

POSITION-BASED IMPEDANCE CONTROL OF A HYDRAULIC ACTUATOR FOR A WALKING ROBOT

SANG-RYU LEE

Intelligent Robot Engineering, University of Science & Technology, 176 Gajung-dong, 217 Gajungro Yuseong-gu, Daejeon, Korea

Department of Applied Robot Technology, Korea Institute of Industrial Technology, 1271-18 Sa3-dong, Sangrok-gu, Ansan, Korea

JUNG-SAN CHO, SANG-DEOK PARK

Department of Applied Robot Technology, Korea Institute of Industrial Technology, 1271-18 Sa3-dong, Sangrok-gu, Ansan, Korea

This paper presents position-based impedance control of a hydraulic actuator for a walking robot. Hydraulically actuated joints of the robot are modeled as a pendulum. After the derivation of a linearized actuator model, hydraulic parameters are acquired by an experiment. Because actual torque can be directly sensed from a load-cell attached to the actuator, the controller is simply designed despite lack of the knowledge of the environmental stiffness. To verify this control scheme, an experiment involved force interaction between the pendulum and the environment is carried out.

1. Introduction

Recently, a hydraulically actuated walking robot has become an active research field [1]-[3] because of their high power-to-weight ratio and fast response [8]. When the robot is walking, it requires the ability to handle interaction with its environment. So, in many cases, force control plays an important role. To date, the various robot force control algorithm has been developed [4], [5]. Unlike position control, however, it's a difficult problem to apply it to hydraulic actuators [6]. Indirect force control such as position-based impedance control [7], therefore, can be a suitable alternative. In this paper, we deal with joint torque control of a hydraulically actuated walking robot using position-based impedance control.

2. System model

2.1. *System configuration*

The hydraulically actuated leg and the actuator are shown in Fig. 1. In order to obtain force information, a load-cell is attached directly to each actuator.

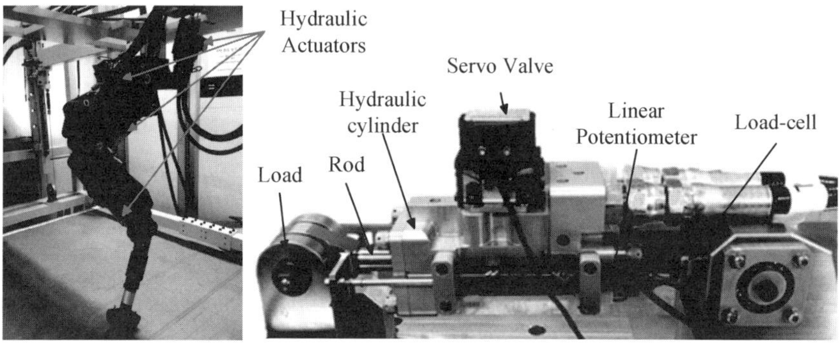

Fig. 1 Hydraulically actuated robot leg and actuator configuration

2.2. *Pendulum model*

Hydraulic actuated joints of the robot can be modeled as a pendulum with the same structure as in Fig. 2. To obtain the joint angle and torque from the linear potentiometer and load-cell, we use kinematic and dynamic equations.

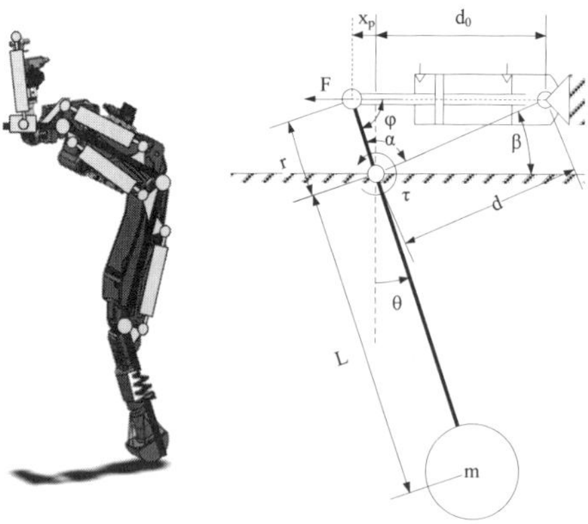

Fig. 2 Pendulum model

The kinematic equation of this pendulum, as shown in Fig. 2, is

$$\theta = \alpha + \beta - \frac{\pi}{2}, \tag{1}$$

where $\alpha = \cos^{-1}\left(\frac{d^2 + r^2 - (d_0 + x_p)^2}{2dr}\right)$,

and the dynamic equation is

$$\tau = r \times F = Fr\sin\varphi = I\ddot{\theta} + C\dot{\theta} + mgL\sin\theta\,, \tag{2}$$

where $\varphi = \cos^{-1}\left(\frac{(d_0 + x_p)^2 + r^2 - d^2}{2(d_0 + x_p)r}\right)$.

Parameters used for this pendulum system are given in Table 1.

Table 1. Pendulum parameters

Symbol	Description	Value	Unit
β	Fixed angle	0.249	rad
r	Moment arm	0.58	m
L	Link length	0.371	m
d_0	Initial Position of actuator	0.23	m
d	Fixed displacement	0.24	m
m	Mass	10	kg
C	Damping coefficient	-	$kg \cdot m^2/s$
I	Moment of inertia	2.33	$kg \cdot m^2$
g	Gravity	9.81	m/s^2

2.3. *System identification and hydraulic actuator model*

To obtain hydraulic actuator model, an experiment which is the frequency response of a current input versus a velocity is used considering the hydraulic actuator is a velocity source. The frequency response is shown in Fig. 3.

Consequently, we can model this system as the linearized second order system because the range of the system's motion is within 10 Hz at most. The hydraulic actuator model is

$$\frac{x_p}{u} = \frac{K}{s\left(\frac{s^2}{\omega_h^2} + \frac{2\zeta_h}{\omega_h}s + 1\right)}. \tag{3}$$

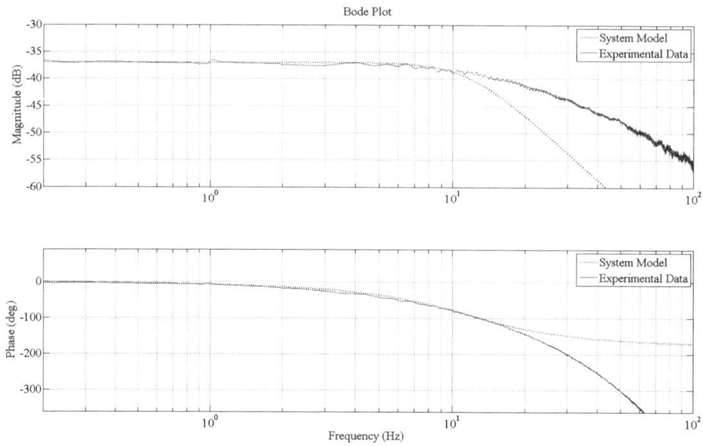

Fig. 3 Frequency response

Typical parameters for this hydraulic system are given in Table 2.

Table 2. Hydraulic actuator model parameters

Symbol	Description	Value	Unit
ζ_h	Damping ratio	0.68	-
ω_h	Natural frequency	72.25	rad/sec
K	Gain	14×10^{-3}	m/mA

3. Position-based impedance control

A force control method in this paper is that the contact force between the pendulum and its environment is controlled indirectly by an appropriate position trajectory. The reference position trajectory produces the required reference torque. If the pendulum contacts with its environment, the difference between the reference torque and the contact force modifies position trajectories based on mechanical impedance. In our system, the contact force can be sensed from the load-cell without knowledge of the environmental stiffness and location. The control block diagram is shown in Fig. 4.

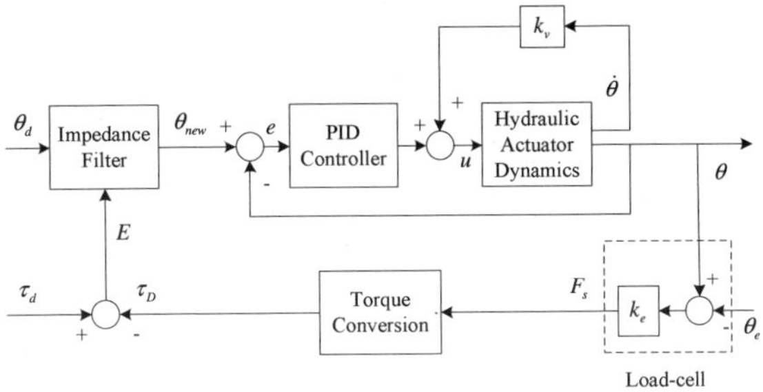

Fig. 4 Position-based impedance control

The reference torque is obtained as

$$\tau_d = I\ddot{\theta}_d + C\dot{\theta}_d + mgL\sin\theta_d \,. \tag{4}$$

The contact force can be easily calculated from geometry eq. (2) as

$$\tau_D = F_s r \sin\varphi \,. \tag{5}$$

We obtain the modified trajectory through the impedance filter as

$$\theta_i = \frac{1}{Is^2 + cs + k} E \tag{6}$$

and

$$\theta_{new} = \theta_d + \theta_i \,, \tag{7}$$

where $E = \tau_d - \tau_D$.

Finally, $k_v\dot{\theta}$ is incorporated in control input to compensate the damping characteristic of the hydraulic actuator [3].

$$u = \left(K_P + K_D s + \frac{K_I}{s} \right) e + k_v \dot{\theta} \tag{8}$$

4. Simulation result

Simulation is carried out based on the system model and controller derived in the previous two sections using MATLAB. Simulation result is shown in Fig. 5.

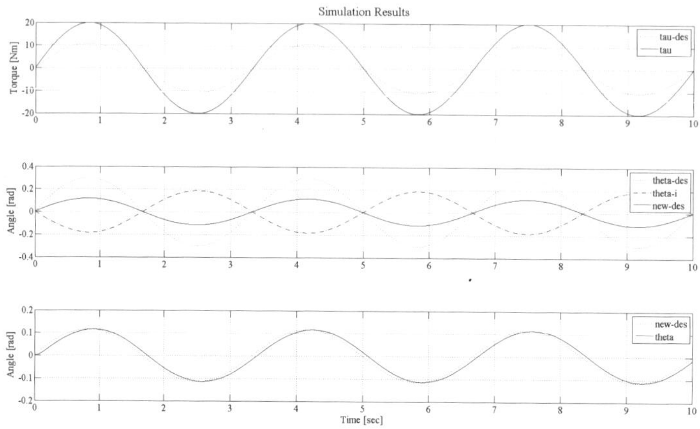

Fig. 5 Simulation result

It shows that the position is modified by the impedance filter.

5. Experimental results

To verify the effectiveness of position-based impedance control of the hydraulic actuator, the experiment is carried out.

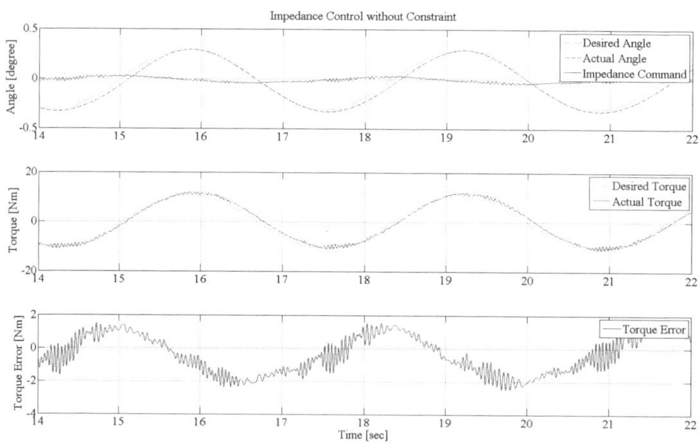

Fig. 6 Position-based impedance control without position constraint

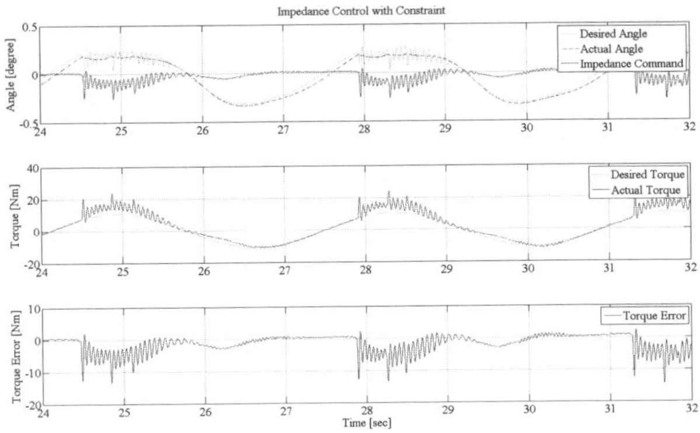

Fig. 7 Position-based impedance control with position constraint

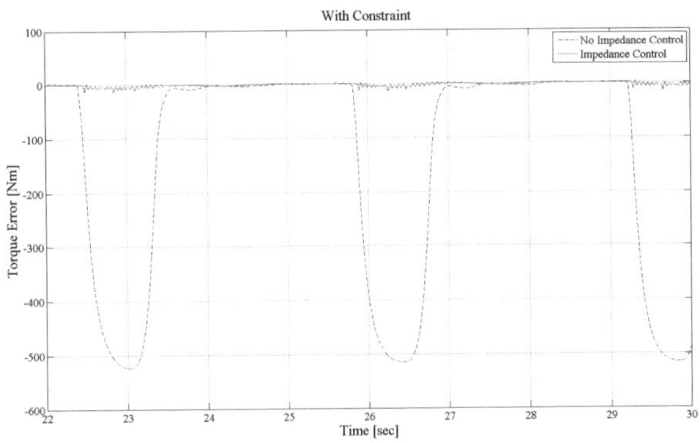

Fig. 8 Comparison between no impedance (position) control and impedance control

The parameters used in this experiment are given in Table 3.

Table 3. Parameters used in the experiment

Symbol	Description	Value	Unit
P_s	Supply pressure	190	bar
f	Frequency	0.3	Hz
K_p	Proportional gain	17	-
K_D	Derivative gain	0.2	-
K_I	Integral gain	0	-
k_v	Damping compensation	4.2	-
c	Impedance damping coefficient	0.1	-
k	Impedance spring coefficient	50	-

6. Conclusions

This paper has described the implementation of position-based impedance control of a hydraulic actuator. The experiment including force interaction between the pendulum and the environment was carried out to demonstrate the feasibility of the hydraulically actuated joint torque control. It is hereby expected to achieve compliant motion of a walking robot during the interaction.

References

1. Marc Raibert, Kevin Blankespoor, Gabriel Nelson, Rob Playter and the Bigdog Team, "BigDog, the Rough-Terrain Quadruped Robot," *Int. Federation of Automatic Control*, 10822-10825 (2008).
2. Michele Forcchi, Emanuele Guglielmino, Claudio Semini, Thiago Boaventura, Yousheng Yang and Darwin G. Caldwell, "Control of a Hydraulically-Actuated Quadruped Robot Leg," *proc. IEEE Int. Conf. on Robotics and Automation*, 4182-4188 (2010).
3. Darrin C. Bentivegna, Christopher G. Atkeson, and Jung-Yup Kim, "Compliant Control of a Hydraulic Humanoid Joint," *IEEE-RAS Int. Conf. on Humanoid Robots*, 483-489 (2007).
4. Ganwen Zeng and Ahmad Hemami, "An overview of robot force control," *Robotica*, vol. 15, 473-482 (1997).
5. D.E. Whitney, "Historical Perspective and State of the Art in Robot Force Control," *Int. J. of Robotics Res.* 6, 3-14 (1987).
6. A. Alleyne and R. Liu, "On the limitations of force tracking control for hydraulic servosystems," *ASME J. of Dync. Sys., Mesa. and Contr.*, vol. 121, no. 2, 184-190 (1999).
7. Homayoun Seraji, Richard Colbaugh, "Force Tracking in Impedance Control," *proc. IEEE Int. Conf. on Robotics and Automation*, vol. 2, 499-506 (1993).
8. Herbert E. Merritt, *Hydraulic Control Systems,* New York: John Wiley & Sons (1967).

MULTI-AXIS CONTROL SYSTEM FOR WALKING MACHINES

KRZYSZTOF KOZŁOWSKI[‡] and MATEUSZ MICHALSKI and PAWEŁ PARULSKI

Chair of Control and Systems Engineering, Poznan University of Technology, Ul. Piotrowo 3A, 61-138 Poznań
[‡]E-mail: krzysztof.kozlowski@put.poznan.pl

This paper gives a proposition for a multi-drive control system, which can be used as a testbed for prototyping walking robots of various kinematic structures, equipped with electric drives, particularly for tasks where synchronization of multiple axis is needed, as well as verification of different control algorithms. The example of the practical application of the proposed system was also given: it was used during prototyping process of a quadruped walking robot, powered by eight electric motors. The abilities and performance of the system were verified with the attempt to realize the dynamic gait of the robot.

1. Introduction

As walking machines are a field of intensive research with the wide spectrum of possible applications, the new prototypes of two-, four-, and more-legged robots are being constructed. Engineering of those new machines gives opportunity to optimize and study not only original mechanical solutions but also control algorithms devoted to walking in different environmental conditions with various tasks given to the robot. Walking robot's performance is based mainly on the continuous information about its state and surrounding obstacles.[1] The basic information about robot state includes the positions and velocities of particular legs links as well as forces exerted on it. Designing and building each new machine equipped with electric drives requires several similar steps, such as estimation of payloads, speed and acceleration requirements, and subsequently, selection of appropriate motors and their controllers. Mechanical solutions chosen by constructors are usually dedicated to a given prototype, while the demands for control part does not basically change from one project to another: it should enable fast enough measurements of robot's state and easy implementation of various control algorithms. Universal control system that would be suitable for a rapid prototyping of walking robots should have, first of all, modular and open structure, which supports a wide range of drives. As walking machines have usually a lot degrees of freedom, the control system should simultaneously support multiple driven axes, and have a moderate cost – the factor that comes into account when prototyped structures are changed frequently. This paper presents a proposition of solution for such multi-drive control system. As there are various implementations possible our proposition is meant as practical guide with suggestions useful for those interested in designing servo control systems.

The paper is organized as follows: Section 2 presents the proposed control system; Section 3 gives the presentation of system abilities and performance with exemplary case of quadruped walking robot WR-06. Section 4 describes experimental validation of a proposed control structure. The final remarks are given in the Summary, Section 5.

2. Description of the control system

The universal control system for multiple axes should primarily enable connection of a different types of motors (DC, brushless, stepper motors), be able to reliably realize a various control strategies, with possibility of easy change of the algorithm and its settings, and furthermore be scalable (adding or reducing the number of supported axes). Drive controllers are

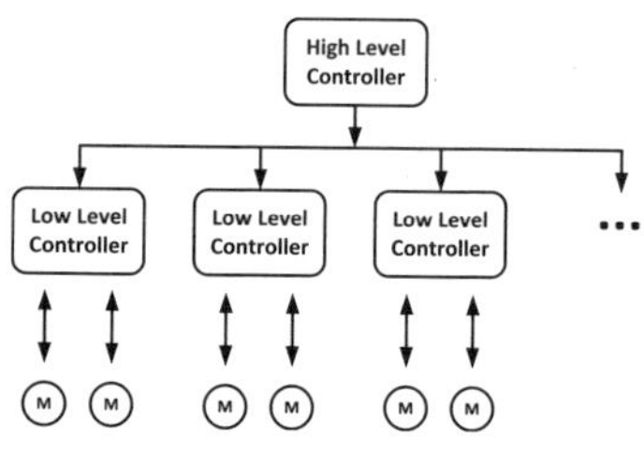

Fig. 1. Control system structure

often divided into two levels: the lower one, which handles the power bridge dedicated to the particular type of motor and works with high-frequency feedback loop, and the higher one, superior to the first one, which generates the reference values for the lower one. Following this well-known template, we propose similar, two-level division of the system, but in both manners: in the hardware and in the software. The structure of the control system is shown in Figure 1. The low-level control algorithms[2–6] will be realized by a low-level motor controllers based on DSP processors suited for drives control tasks,[7] and the high-level algorithms – by an easily programmable controller, which offers a wide range of communication capabilities.

2.1. *Low-level controllers*

The low-level axis controller has been based on commercially available DSP processors.

Such a choice allows for a quick exchange of the control algorithm, as well as guarantees the use of reliable hardware solutions. (With other possible choices, such as an FPGA, the designer must take care of correctness of every aspect of the system). A substantial outcome is also a modularity of the proposed solution.

To increase the versatility of a proposed solution, the low-level controller was further subdivided into two physical/logical parts. First one, the digital one, includes a DSP processor together with the accompanying interfaces. The second one is the power part, which is directly attached to the motor. It can be easily replaced to support different types of motors.

The controller combines the control and measuring functions. Fast data exchange with a higher level components is the one of the key objectives of this control level. The interface selected for this purpose must also be reliable and common industrial standard. The RS485 serial interface is proposed here.

The axis controller acts as a slave device. It was assumed that this controller will not initiate the data transmission but will only respond to queries sent from the master device.

2.2. *High level controller*

A PC computer with a real-time operating system can be used as a high-level controller. Such solution guarantees an universal and inexpensive platform, which enables an easy way to implement and verify various motor control strategies.

Tasks for each axis must be coordinated in a real time with sufficiently high frequency. This means the high-level controller must be governed by a real-time operating system. Here the QNX system was used, which meets the criteria for a hard real-time operating system.

The high-level controller's critical component is the efficient communication with axis controllers (and possibly any other sensory subsystems). Therefore it is equipped with two Moxa's multi-port communication cards

CP-168EL PCI-EXPRESS. Each of these cards support up to 8 serial ports transmitting data at a speed of 921600 b/s. To enable reliable data exchange with low-level controllers, each port was equipped with RS232-RS485 converter working in full-duplex mode.

3. Case study - walking robot WR-06

Application of the considered system will be presented on the example of four-legged walking robot designed and constructed at Chair of Control and Systems Engineering at Poznan University of Technology.[8]

The robot (Fig 2) is composed of four symmetrical legs. Each leg has two links and two driven joints. First link is connected to the body frame. The design enables easy addition of another degree of freedom to the leg by simple inserting of extra driven joint between the frame and the leg flange. Legs are driven by Maxon's 200W EC-powermax 30 brushless motors with

Fig. 2. Quadruped walking robot WR-06

planetary gearhead of reduction $n = 53$. The drives were chosen with regard for their advantageous power to mass ratio. Such drives configuration gives maximal static torque of approximately 20 Nm. Mass of the robot equals 22 kg while dimensions are $445 \times 420 \times 765$ mm (H×W×L) as of geometrical configuration presented in Fig. 2.

The universal control system described in the previous section was used to control the robot. The structure of the system was suited to the robot: at its simplest form it its equipped with 8 motors, and while each low-level controller can simultaneously handle two motors, at least four of such controllers were needed. The measurements of quantities that describe the state of each robot's leg are the basis for calculations of control signals for the robot as a whole – it was performed by the high-level controller. The multilevel control system used with WR-06 robot was given in Fig. 3. Structure of the control system could be easily upgraded with additional

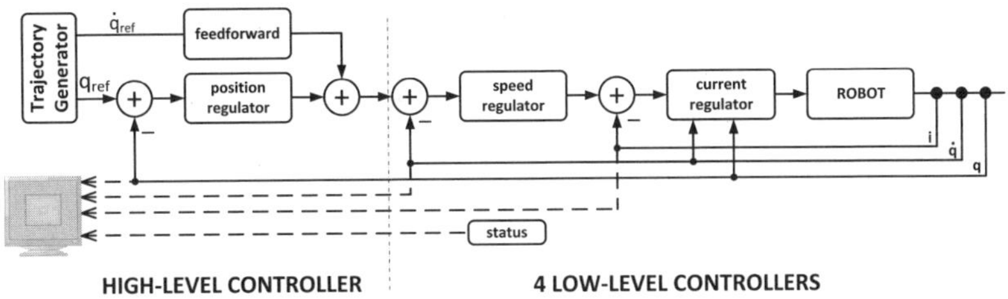

Fig. 3. Multilevel control system

measuring subsystems.

3.1. *Low-level controller*

An appropriate coordination of robot during walking demands a fast measurement of the following quantities: internal coordinates (leg joint angles), joint velocities and torques, detection of the leg contact with the ground as well as the orientation of the robot's body with respect to the ground. The configuration angles are determined with the measurements of drives shafts positions – impulse encoders of 500 impulses per rotation were used. The measurements of leg joints velocities are based on configuration angles changes. Robot joint torques are also delimited indirectly from the values of the measured motor currents. Hardware capabilities and computational performance of the Texas Instruments' F2812 processor[9] located in the digital part allowed for the simultaneous handling of two DC or two brushless motors. Each low-level controller used in this case was equipped with the power part that enables the control of the two Power Series Brushless motors: the International Rectifier IRAMX16UP60A power bridges were used for this purpose.

The firmware blocks that implement motor vector control are shown in (Fig. 4). Current control loop is updated with 18 kHz rate. Standard PID block is accompanied with two additional modules: gear friction compensation derived on the basis of rotor's speed and motor's characteristics and gravity decoupling, which determines the joint torque on the basis of the current leg configuration and its load. Outer speed control loop could be switched on or off and is updated with 2 kHz rate.

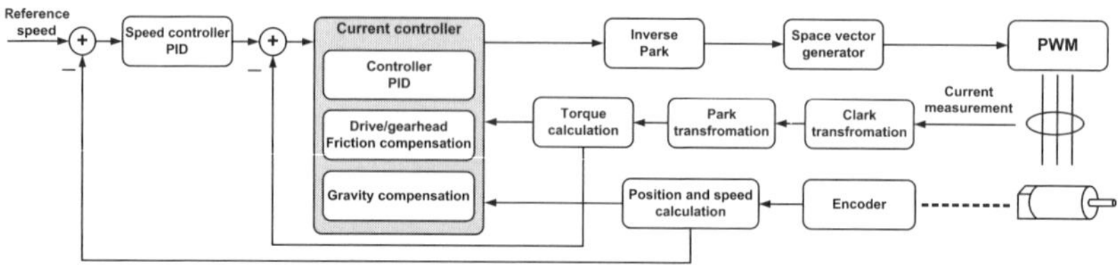

Fig. 4. Low level controller's software components

3.2. *High-level controller*

The main task of the high-level controller is legs coordination and generation of various gait patterns. The basic regulation type is leg's joints position regulation. This function is realized by a standard PID controller with feedforward. To this end robot's state is updated every 1 ms: master controller polls the slave controllers for the actual values of joint velocities and/or currents. The desired values of joint positions (or velocities or currents) that compose the robot's gait are generated with trajectory generator module. The generator procedure could be either hard-coded or desired values could be received from external source with ethernet link or additional serial port. On the operator's demand the controller can log the history of robot's state changes to a disk file. The graphical user interface allows to modify the low- and high-level controller's parameters, observation of the current state of each robot's leg (position, velocity and current of each drive) and the status of connected devices and measuring systems.

4. System tests

To verify the system's capabilities for the 4-legged robot a trot gait pattern was used.[10,11] For this type of gait one can distinguish four phases: *support phase* – two diagonal robot's legs are standing on ground and carrying the whole robot weight; *lift-off* of the supporting legs; *swing phase* – while the legs are not carrying the robot in this phase its links' masses are only significant motor loads; leg's *impact* with the ground.

The overall performance of the control system was good, as presented in Fig.5. A small errors of the angular position that occurred at every step stemmed from the discontinuity of velocity of reference trajectories, a force impulses in the leg's impact with the ground, and the dynamic changes of leg's load that resulted from it. Presented system may facilitate the selection of control algorithm's settings through an easy comparison of resulting trajectories. Based on this user can select the more appropriate

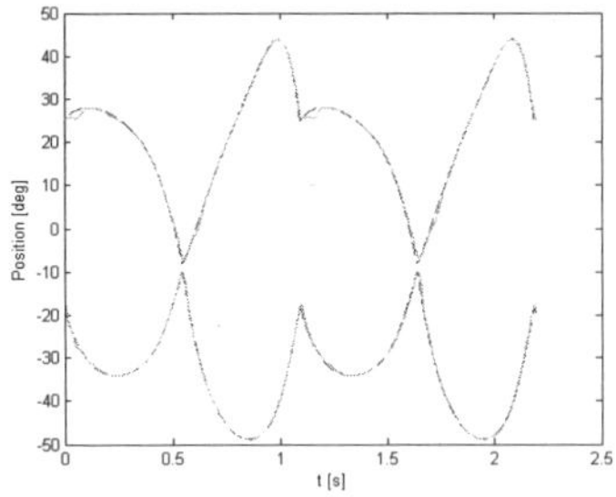

Fig. 5. Reference (red and blue lines) and measured (green line) angular positions for two trot steps

option or setting. For example, figure 6 presents the differences in joint's velocity outcomes obtained with a feedforward module switched off and on. It must be pointed out that the control system enables easy implementation

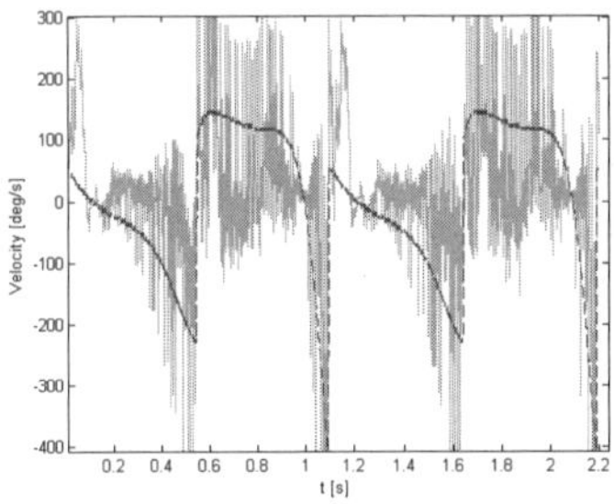

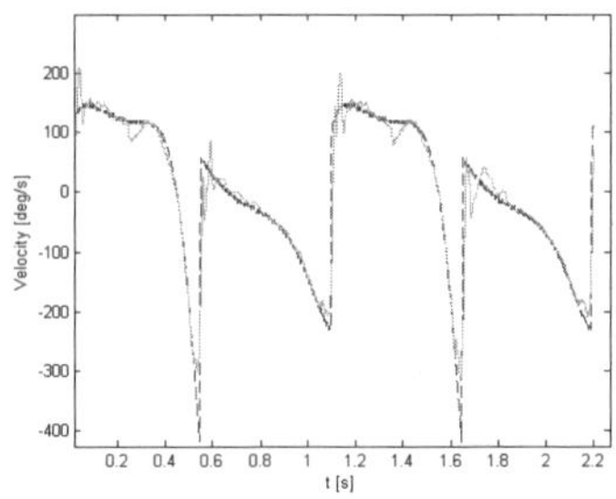

Fig. 6. The angular velocity reference values (red line) and measured (green line) from a trajectory generator for a different structures of the master controller a) PD controller, b) PD + feedforward controller

of different types of algorithms. Particular tasks could be bound to different levels of the system depending on the data required for the task, its feedback frequency, and the need of drives synchronization.

5. Conclusions

The presented multi-axis control system could be used with many different walking machines - the simultaneous handling of up to 16 motors covers the needs of wide range of current constructions. Its usefulness has been proved during prototyping process of a quadruped walking robot WR-06. The analysis of system's properties has shown that the construction is capable of performing dynamic gait while control system will manage to receive state information and transmit commands sufficiently fast.

The system allows the user to implement any control algorithm on both high- as well as low-level controller for different types of motors. Designed control system was tested both for one- and four-legged walking machines. Conducted experiments proved that the system meets its expectations and that the positional control loop running at 1 kHz enable realization of reference trajectories in configuration space with high enough precision.

Authors are grateful to Aaron Ames for his inspiring advice, suggestions and providing preliminary reference trajectories for our robot.

References

1. A. Morecki and J. Knapczyk, *Foundations of Robotics.* Warsaw: WNT, 1993, (in Polish).
2. Z. Li and H. Yang, "Speed control of brushless dc motor using human-simulated intelligent control," in *Industrial Mechatronics and Automation, 2009. ICIMA 2009. International Conference on*, may 2009, pp. 236–240.
3. A. Zentai and T. Daboczi, "Improving motor current control using decoupling technique," in *Computer as a Tool, 2005. EUROCON 2005.The International Conference on*, vol. 1, nov. 2005, pp. 354–357.
4. H. Grabner, W. Amrhein, S. Silber, and W. Gruber, "Nonlinear feedback control of a bearingless brushless dc motor," *Mechatronics, IEEE/ASME Transactions on*, vol. 15, no. 1, pp. 40–47, feb. 2010.
5. Z. Li, J. Chen, G. Zhang, and M. Gan, "Adaptive robust control for dc motors with input saturation," *Control Theory Applications, IET*, vol. 5, no. 16, pp. 1895–1905, 3 2011.
6. J. Gao and Y. Hu, "Direct self-control for bldc motor drives based on three-dimensional coordinate system," *Industrial Electronics, IEEE Transactions on*, vol. 57, no. 8, pp. 2836–2844, aug. 2010.
7. L. Dessaint, B. Hebert, H. Le-Huy, and G. Cavuoti, "A dsp-based adaptive controller for a smooth positioning system," *Industrial Electronics, IEEE Transactions on*, vol. 37, no. 5, pp. 372–377, oct 1990.
8. M. Michalski, M. Kowalski, and D. Pazderski, *Quadruped walking robot WR06 -Design, control and sensor subsystems*, ser. Lecture Notes in Control and Information Sciences, 2009, vol. 396.
9. *TMS32OF2810, TMS320F2811, TMS320F2812, TMS32OC2810, TMS320C2811, TMS3 -Signal Processors Data Manual*, Texas Instruments, Dallas, Dec 2003.
10. A. Muraro, C. Chevallereau, and Y. Aoustin, "Optimal trajectories for a quadruped robot with trot, amble and curvet gaits for two energetic criteria," *Multibody System Dynamics*, vol. 9, no. 1, pp. 39–62, 2003.
11. Y. Aoustin, C. Chevallereau, and A. Formal'sky, "Numerical and experimental study of the virtual quadrupedal walking robot-semiquad," *Multibody System Dynamics*, vol. 16, no. 1, pp. 1–20, 2006.

LOCAL DISTRIBUTED CONTROL FOR MULTI-ROBOT NAVIGATION

A. TERRONES*, R. ACUÑA, N. CERTAD, L. FERMÍN-LEÓN and G. FERNÁNDEZ-LÓPEZ

Simón Bolívar University, Mechatronics Group, ELE-328
Sartenejas 1080-A Miranda, Venezuela
**E-mail: aterrones@usb.ve*

In this work a distributed navigation control strategy for multi-robot systems based only in the local information provided by the robots is presented. Each robot calculates the vector forces required for navigation towards a goal avoiding possible obstacles on the workspace and traveling in formation with other members of the swarm. A control architecture was implemented for the generation of the vector force used for navigation, based on three basic behavior agents responsible of calculating individual force components needed for navigation to the target, obstacle avoidance and swarm navigation. Also, an event driven control system is in charge of assigning a priority level to each force component generated by the agents of basic behaviors. Results obtained in simulations show the effectiveness of the control system in the navigation of a system of multiple robots under different obstacle configuration.

Keywords: Swarm, Distributed Systems, Artificial Potential Fields, Vector Force Fields, Event Driven Control

1. Introduction

The problem of coordinated navigation arises when the task is considered too complex to be efficiently executed by a single robot. Tasks such as reconnaissance, rescue and cleaning are complex enough to be performed by one robot in an acceptable time limit.[1,2] However, the presence of several platforms working together may optimize runtime work,[3] allows the use of simpler robots and introduces system redundancy.[2] Raynolds[4] developed an algorithm for the simulation of birds and fish flocks by using only three rules responsible of collision avoidance and velocity matching between agents and centralization of the group. Arkin *et. al.*[5] address the problem of flock navigation using a creative approach based on potential fields. Olfati-Saber[6,7] attacks the problem of multi-robot navigation using a vector field control for each robot. Estévez *et. al.*[8] used dynamically varying potential fields with insistence mechanisms for the coordination of a group of robots. In order to avoid being trapped in local minima, random components were added to the fields. Gayle *et. al.*[1] proposed the use of global information system and vector force field control to achieve local coordination of the robots in the system. McLurkin[9] develops a set of individual behaviors that serve as a basis for emergent behaviors.

The control scheme proposed in this work is decentralized, each robot computes its own control action based on data acquired from the environment. For this purpose, the agents are in charge of generating force vectors associated with different basic behaviors and sum them according to a priority plan based in several events. To solve the local minima problems, we propose an auxiliary agent whose function is grouping the obstacles in clusters based on their proximity.

2. Control Architecture

The proposed control scheme (Fig. 1) is based on three main agents responsible for calculating the control actions necessary to navigate towards the target, avoid obstacles and travel together.

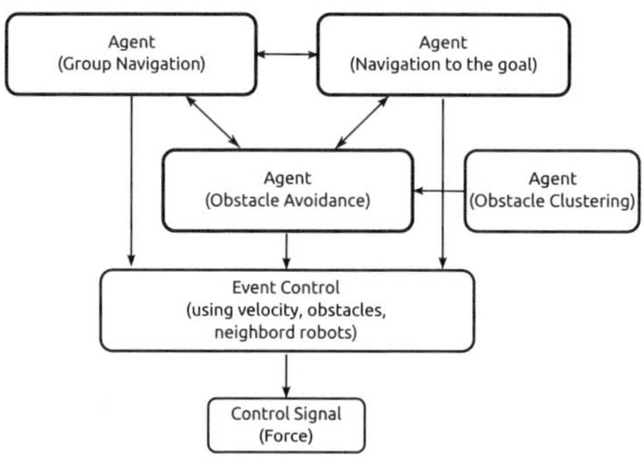

Fig. 1: General outline of a navigation controller based on agents and events.

Furthermore, there is one auxiliary agent in charge of the clustering of visible obstacles. Once calculated, the basic force vectors are added according to this equation

$$\mathbf{u}_i = \alpha_i \mathbf{u}_i^{target} + \beta_i \mathbf{u}_i^{group} + \gamma_i \mathbf{u}_i^{obstacle} - \delta_i \mathbf{u}_i^{damping} \tag{1}$$

were α_i, β_i, γ_i and δ_i are the coefficients of priority of each one of the components calculated in function of the states of each robot.

2.1. *Navigation to the goal*

The control function to the target is given by a simple potential well, which is described by the following equation

$$\mathbf{u}_i^{target} = max(\mathbf{q}_{target} - \mathbf{q}_i^{robot}, \mathbf{u}_{max}) \tag{2}$$

were $\mathbf{q}$ indicates a position vector. The Fig. 2 shows the graphical representation of the vector $\mathbf{u}_i^{target}$.

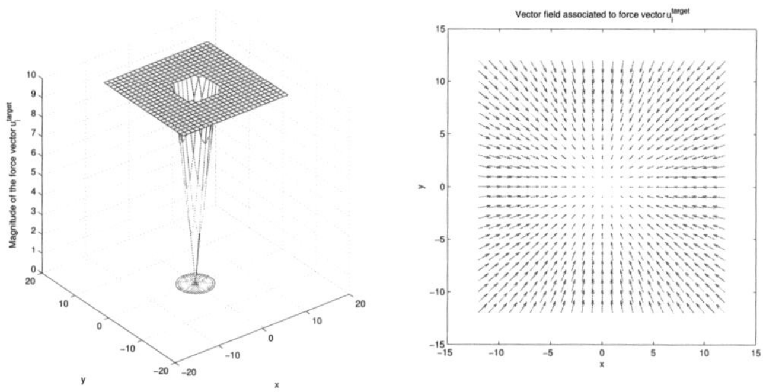

Fig. 2: Graphical representation of the force vector $\mathbf{u}_i^{target}$. The maximun value is set to 10.

2.2. *Group Navigation*

For the group navigation, a potential function with the following equation is used

$$\mathbf{u}_i^{group} = \sum_{j \in N_i} [\phi(\| \mathbf{q}_j^{robot} - \mathbf{q}_i^{robot} \| - d_{formation})\mathbf{n}_{ij} + c_1(\mathbf{p}_j^{robot} - \mathbf{p}_i^{robot})] \tag{3}$$

based on the work of Olfati-Saber.[6,7] However, a different function is used for $\phi(n)$ in order to avoid oscillations around $x = 0$. The following equation define $\phi(n)$

$$\phi(n) = \frac{a+b}{2}\left(\sqrt{1+(n+c)^2} - \sqrt{1+c^2}\right) + \frac{a-b}{2}n \tag{4}$$

where a, b and c define the shape of the function.

The $\mathbf{q}$ and $\mathbf{p}$ vectors denotes the position and velocity of the robots respectively, and N_i is the robot neighbourhood consisting of those robots within the range of vision of robot i. The vector $\mathbf{n}_{ij}$ is a unit vector parallel to the vector of minimum distance between the robots j and i. The coefficient c_1 represents the K coefficient in a PID controller, which is responsible of velocity matching between robots. The parameter $d_{formation}$ is the desired distance between a robot and another. In Fig. 3 a graphical representation of the $\phi(n)$ function is showed.

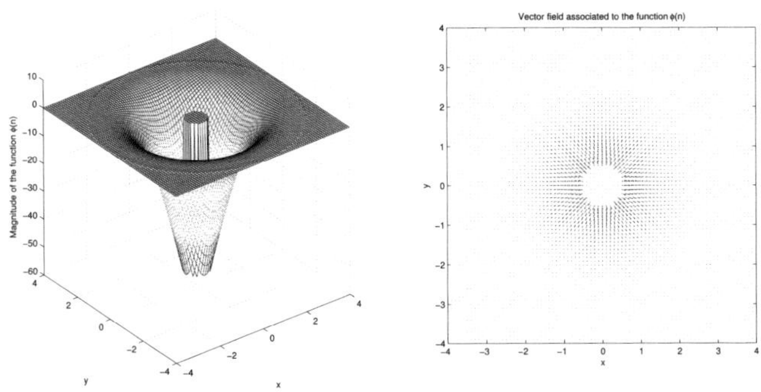

Fig. 3: Graphical representation of $\phi(\| \mathbf{q}_j^{robot} - \mathbf{q}_i^{robot} \| - d_{formation})$, with $a = 3$, $b = 45$, $c = 1.807$ and $d_{formation} = 4m$.

2.3. *Obstacle Avoidance*

For obstacle avoidance, the following control action is proposed

$$\mathbf{u}_i^{obstacle} = \mathbf{F}_g + \mathbf{F}_p + \mathbf{F}_{\mathbf{n}_{ij}}^{\perp} \tag{5}$$

were $\mathbf{F}_g$ corresponds to a gyroscopic force,[10,11] $\mathbf{F}_p$ is a potential field force defined by the function $\phi(n)$ defined previously and $\mathbf{F}_{\mathbf{n}_{ij}}^{\perp}$ is a force vector perpendicular to the distance vector between a robot and a obstacle.

The vector $\mathbf{F}_p$ is used to divert the robot from his original path, in order to avoid the incoming obstacle. $\mathbf{F}_p$ is responsible for keeping the robot at a certain distance from the obstacle. This approach allows the robot to be able to navigate through narrow areas without colliding with nearby obstacles. The vector $\mathbf{F}_{\mathbf{n}_{ij}}^{\perp}$ is used to allow the robot to surround the obstacles, which, in conjunction with the vector $\mathbf{F}_p$ and the obstacle clustering agent, allows the robot to escape from local minima.

2.4. *Auxiliary agents*

In addition to the basic behaviors agents discussed above, there are another agent whose function is to provide additional support for the obstacle avoidance behavior. This cluster agent is responsible for grouping the obstacles

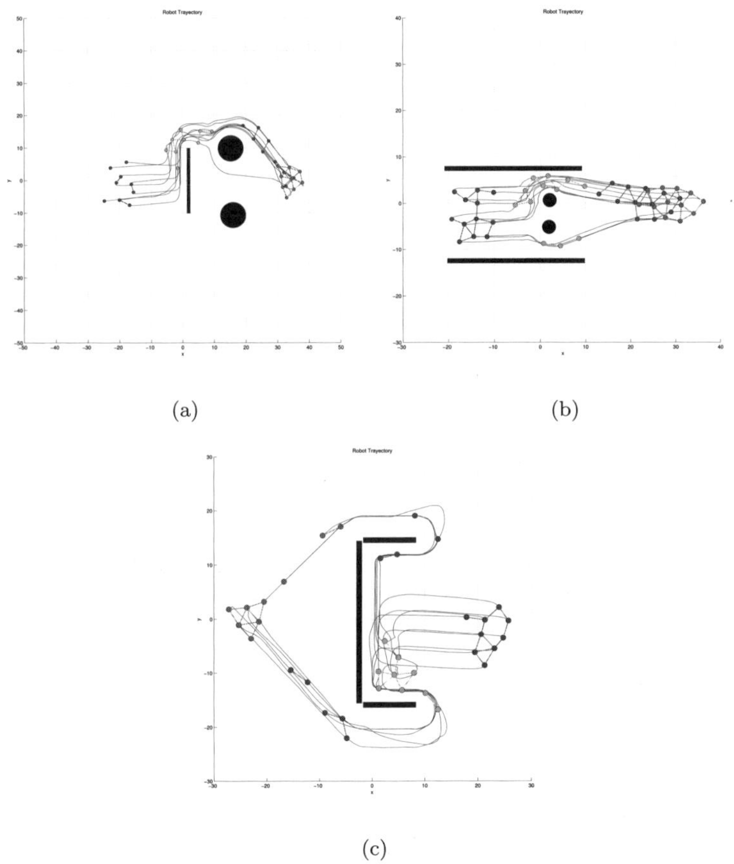

Fig. 4: Simulation results with different arrangements of obstacles: (a) Wall and 2 cylinders with target at (35, 0) and 9 robots, (b) Narrow corridor with target at (29, 0) and 12 robots, (a) C shaped obstacle with target at (-22, 0) and 9 robots. Different colors in each graph represents intermediate positions of each robot. Each colored line represents the line of sight between robots.

detected, so the robot can avoid being trapped in a local minima. To accomplish this, each distance vector to an obstacle detected is compared to the others to determine the angle formed between them. If this angle is less that a threshold, the vectors are then averaged to create a new virtual obstacle. This procedure is repeated until it is impossible to further reduce the number of obstacles.

2.5. *Priority System*

This system is responsible for modifying the values of the coefficients α_i, β_i, γ_i, δ_i as a function of several parameters such as: the presence of obstacles, number of robots belonging to N_i, robot velocity, relative location of neighboring robots and the target, so the robot can be adapted to the enviroment.

3. Results and Discussion

Simulation tests with a different number of robots and obstacles with different arrangements show the effectiveness of the implemented control architecture and its capability to achieve flock navigation under certain circumstances and avoid being trapped in local minima. The simulations were performed on a simulator developed by one of the authors, running on a PC with a Intel®Core^{TM}i7, with a time-step of $30ms$. All distances are in meters. The robots used in the simulations are of the holonomic type, with a maximun radius of vision of 4 meters, and a maximum speed of $2m/s$. The results of three performed simulations are showed in Fig. 4. The arrangement of obstacles for each simulation is such that it creates a potencial

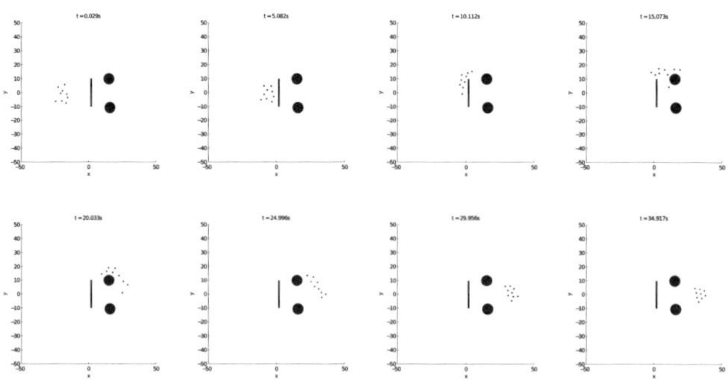

Fig. 5: Consecutive snapshots of the simulation results showed in Fig. 4a: Wall and 2 cylinders with target at (35, 0) and 9 robots.

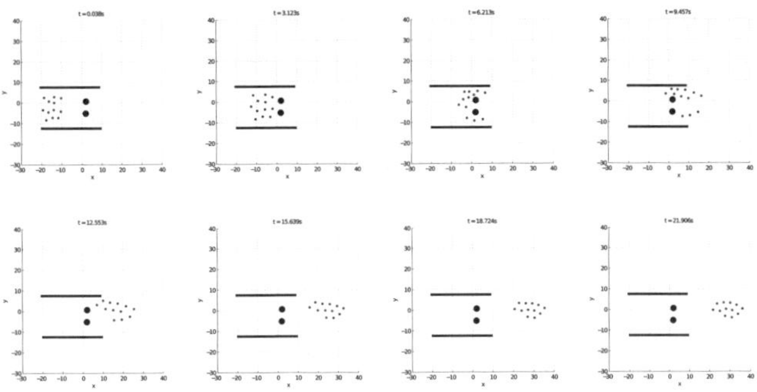

Fig. 6: Consecutive snapshots of the simulation results showed in Fig. 4b: Narrow corridor with target at (29, 0) and 12 robots

well around some obstacles. Traditional approaches to robot navigation and obstacle avoidance fail to prevent the robots of being trapped in a local minima. However, our approach succesfully demonstrates the ability to avoid being trapped in a potencial well, manifested in the fact that all robots, in each simulation, reach the target with no collision at all. The Fig. 4b shows the group's ability for navigation in narrow spaces while avoiding collision between them and the enviroment. Additionally, there is partial group fragmentation in each simulation, due to the fact that a total group cohesion is not absolutely requiered to accomplish their navigation objetive.

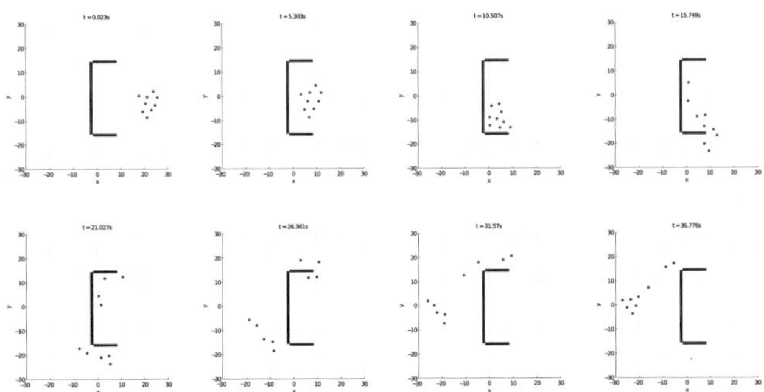

Fig. 7: Consecutive snapshots of the simulation results showed in Fig. 4c: C shaped obstacle with target at (-22, 0) and 9 robots

4. Conclusions

In this paper a distributed navigation control strategy for a multi-robot system was developed. This control system is based in three basics behaviors. In addition, a event driven control system was developed for managing execution priority levels of each basic behavior. To avoid potencial wells, an extra agent whose function is grouping obstacles in clusters was designed. Simulations performed for different workspaces shows the effectiveness of the algorithm allowing a group of robots to navigate in an unknown environment without colliding.

References

1. R. Gayle, W. Moss, M. Lin and D. Manocha, Multi-robot coordination using generalized social potential fields, *IEEE International Conference on Robotics and Automation* , 106 (2009).
2. W. Burgard, M. Moors, C. Stachniss and F. Schneider, Coordinated multi-robot exploration, *Robotics, IEEE Transactions on* **21**, 376 (2005).
3. N. Kwok, Q. Ha, V. Ngo and S. Hong, Particle Swarm Optimization-based coordination of a group of construction vehicles, in *Proc. Intl. Symposium on Automation & Robotics in Construction*, 2006.
4. C. W. Reynolds, Flocks, herds and schools: A distributed behavioral model, *ACM SIGGRAPH Computer Graphics* **21**, 25 (1987).
5. R. Arkin and T. Balch, Behavior-based formation control for multi-robot teams, *IEEE Transactions on Robotics* **XX**, 1 (1999).
6. R. Saber, W. Dunbar and R. Murray, Cooperative control of multi-vehicle systems using cost graphs and optimization, *Proceedings of the 2003 American Control Conference, 2003.* , 2217 (2003).
7. R. Olfati-Saber, Flocking for Multi-Agent Dynamic Systems: Algorithms and Theory, *IEEE Transactions on Automatic Control* **51**, 401(March 2006).
8. P. Estévez, J. Cappelletto, F. Alvarez, R. Acuña, G. Fernández, J. C. Grieco and M. Armana, *Navegacion coordinada de un sistema multirrobot usando campos de velocidades dinamicos*, in *5° Workshop, Cooperación en Robótica*, 2009, pp. Vol. 1. pp. 79 – 103.
9. J. McLurkin, Stupid robot tricks: A behavior-based distributed algorithm library for programming swarms of robots, PhD thesis, (Massachusetts Institute of Technology, 2004).
10. D. Chang and J. Marsden, Gyroscopic forces and collision avoidance with convex obstacles, *New Trends in Nonlinear Dynamics and Control and their Applications* , 145 (2004).
11. S. Shadden, J. Marsden and R. Olfati-Saber, Collision avoidance for multiple agent systems, *42nd IEEE International Conference on Decision and Control (IEEE Cat. No.03CH37475)* , 539 (2003).

SECTION–11
SERVICE ROBOTS

MULTI-FUNCTIONAL WALL CLIMBING ROBOT

V.G. GRADETSKY and M.M. KNYAZKOV.
The Institute for problems in Mechanics of Russian Academy of Science, Russian Federation, 119526 Moscow, prospect Vernadskogo 101-1, e-mail: gradet@mail.ru

The new type of multifunctional inspection robot is suggested intended for environment surveillance by means of on-board inspection equipment. The structure of inspection equipment is interchangeable depends on solving tasks, such as nondestructive testing, detection of flame or fume. The sensors and information treatment system combine into inspection module.

The multifunction inspection machine can move over horizontal, slope or vertical surface on the stepping mode operations. The structure of a robot is modular design. The paper presents description of main robot's modules, including inspection one, and technical parameters of the all machine.

The R&D were supported by Grants RFFI[1]

1. Introduction

One of important application area of robots is inspection of the objects and environmental surveillance by means of on-board sensory system that forming information – measuring system. Examples of hazardous environment infrastructure are distribution of explosives, oil platforms, cargo container ships, nuclear power plants, oil transportation pipes, etc. There are mobile robots intended for hazardous condition prevention [1]. In the frame of CLAWAR system were developed wall climbing inspection Robots for monitoring, mobility, measurement of inspection tasks [2]. The robots have successful applications in extreme conditions. As experience shown, it was necessary to equipped mobile robots with number of sensors to produce required task and control system with some elements of Artificial Intelligence to produce simple decision making problems [3]. As far as technological requirements were expanded, the climbing machine improved to the more convenient for various applications to fulfill the customer's demands. Following to the requirements

[1] Work partially supported by grant RFFI №11-01-92002ННС_a, №12-08-00374-a and №11-01-12110-офи-м-2011

depending on new tasks and potential applications the perfection of robot inspection machines accomplished to solving such problems as:
- multifunctional inspection task realization on the modular design;
- sensory and measurement – information systems evaluation (fig. 1);
- remote and intelligent control creation;
- inspection system modification and it adaptation to robot motion;
- concentration of the attention to reliability and special hardware and software modules [4-8].

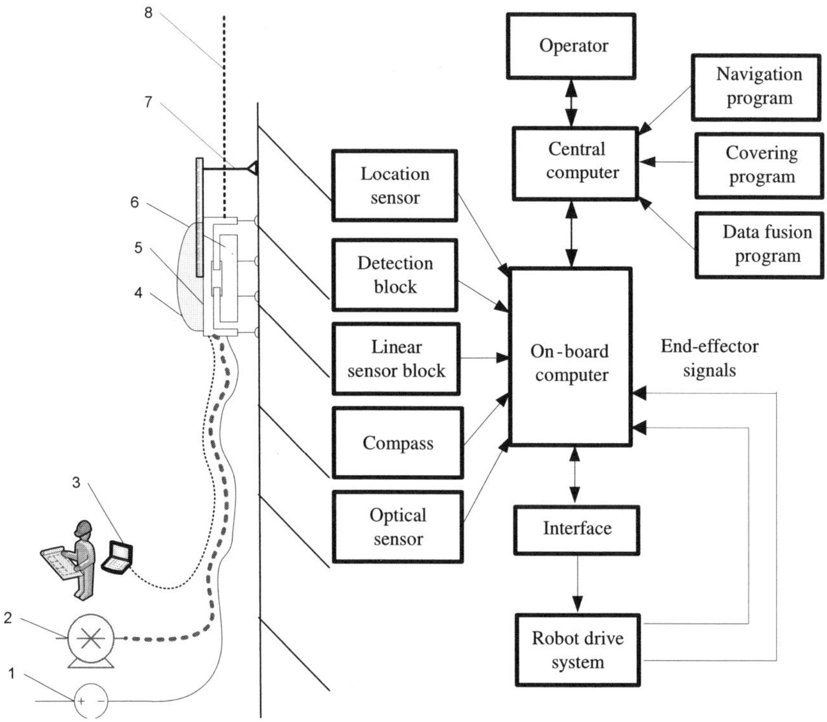

Fig. 1 Structure of the distributed control robotic system.
1- electrical suppply, 2- air suppply, 3- control console, 4- techniqiul equipment, 5- external platform, 6- internal platform, 7- detector-sensor, 8- pantograph.

2. Structure of multifunctional inspection machines

Multifunctional inspection machine is modular technological system and it permits to realize motion over horizontal, slope and vertical surfaces. The machine includes following modules–transport (Fig.2), technological inspection

(Fig.3), control, information-measuring, supply (Fig. 4-5). The open modular design satisfy the changing of the structure depends on inspection solving tasks.

Fig. 2 Internal and external platforms of transport module.

Transport module includes body, drive system and vacuum pedipulators. Body consists of two platforms-main-marching and internal-rotating. Two pneumo-drivers are installed on main platform and their satisfy translation motion of machine. Every pedipulators has 3 vacuum suckers.

The technological module offers inspection and measurement procedure, measuring and transport synchronization motion and trajectory of motion for measuring equipment relative to surfaces.

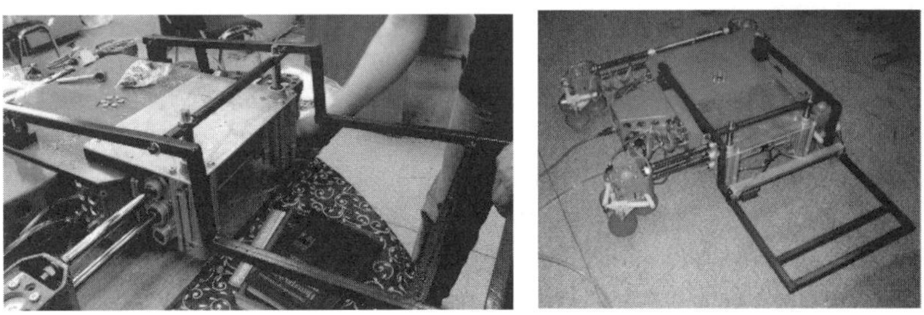

Fig. 3. Surveillance equipment disposition on technological platform.

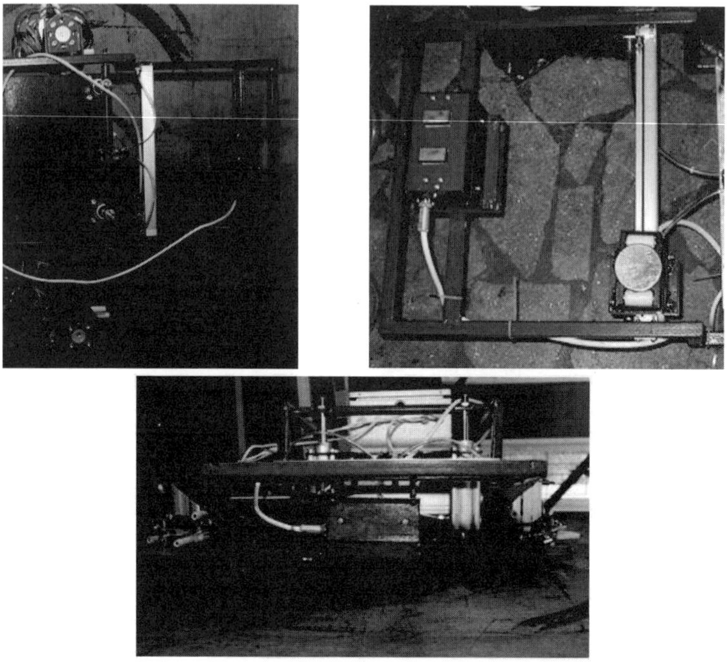

Fig. 4. Measuring sensors and equipment allocation on pantograph.

Technological module permits to install on the board-pantograph different types of measuring systems and sensors, such as scanning sensors for explosive detection or measuring sensor of the surface detection. It is possible realize continuous displacement for measuring purposes. For example, the nondestruction testing produces measuring of the material peculiarities, such as structure or thickness.

Programming consol unit permits to control from PC with typical characteristics and monitories. The interface RS232 intended for connection with transport module and it subsystems. Connection with measuring system realized by technology Fast Ethernet or wireless Wi-Fi 802.1a/b/g. Individual consol permits produce such operations as calibration, initial measurements, setting of elementary conditions, turning etc.

Information – measuring module consists of sensors for produce inspection procedures, such as non- distractive testing, mine detection, fume and fire detection. The system has also open modular design and number of the sensors depends on solving tasks. The information treatment system of the module produces generalization and representation of the measuring data.

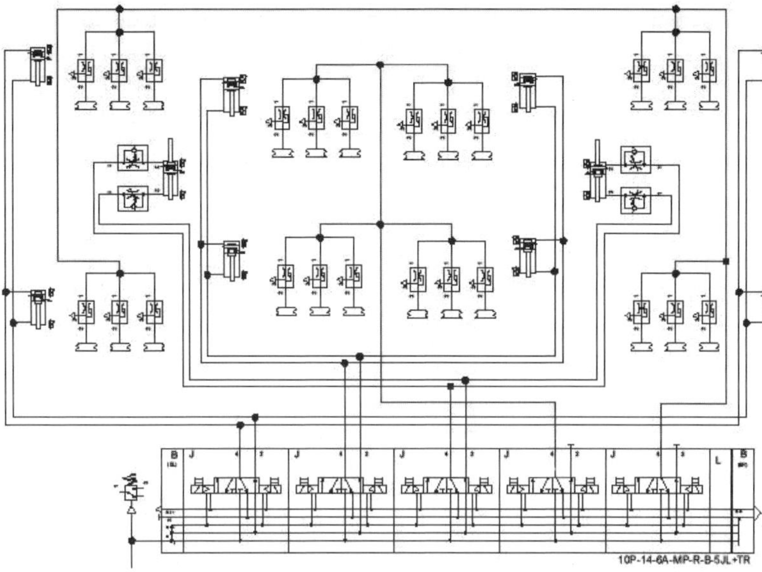

Fig. 5. Generalization and representation of measuring data.

3. Conclusion

In the paper multifunctional surveillance wallclimbing machine is delivered. The machine has interchangeable structure depending on solving tasks, for example flame, and explosive detection, nondestructive testing. As one of the main example, is the paper present application of the machine for nondestructive testing of a ship surfaces. Machine has a modular design that permits to use it for many applications depend on the solving problems, what is important from many view points. The authors tanks all colleagues took, place in the design manufacturing and testing this machine. Authors will present demonstration videos of mobile robot in action to explain some details.

References

1. Y. Baudoin, D. Doraftel, G. De Cubber, S. Berrabah, E. Colon, C. Pinzon, A. Maslowski, Y. Bedrowski "View-Finder: Robotic Assistance to Fire-Fighting Services". Proceed. of 12th International Conference on Climbing and Walking Robots. Sept. 2009, World Scientific Publ. Co. Pt., Ltd, pp. 397-406.
2. T. Sattar, B. Bridge, S. Chen, Z. Zhao "Development of CLAWAR systems that combine the tasks of monitoring, mobility, manipulation, and

measurement for industrial inspection tasks", Proceed. of 6th International Conference on climbing and Walking Robot. Sept. 2003, Professional Engineering Publishing Limited, UK, pp. 699-706.

3. Gradetsky V.G. "Wall climbing Robot: Evolution to Intelligent Autonomous Vehile". Proceed. of First CLAWAR Intern. Conference, Brussels, Belgium, Nov. 1998, pp.53-60.
4. M. Acheroy, Y. Baudoin, M. Piette "Belgian Project on Humanitarian Demining (HUDEM)", Procced. of First CLAWAR Intern. Conference, Brussels, Belgium, Nov. 1998, pp. 215-218.
5. http://www-robot.mes.titech.ac.jp/robot/walking/ninja/ninja_e.html
6. M. Armanda, I. Cobano, E. Garcia, P. Conzalez de Santos "Configuration of a legged robot for humanitarian demining arctivities", Procceed. of HUDEM 2005, Publ. JST- Japan Science and Technology Agency, Tokio, Japan, pp. 131-135.
7. A.A. Colle et al "Design and Performance of the Portsmouth Climbing Robot", Proceedings of 7th ISARC, Bristol, England, 1990.
8. D.S. Cooke, N.D. Newer, T.S. White, S. Galt, B.L. Luk "Implementation of Modularity in Robug IV – Preliminary Results", Proceed. of Second CLAWAR Intern. Conference, Portsmouth, UK, Sept. 1999, pp. 393-403.

ANALYSIS OF SLIDING SUCTION CUPS FOR NEGATIVE PRESSURE ADHESION OF A ROBOT CLIMBING ON CONCRETE WALLS

D. SCHMIDT* and K. BERNS

Robotics Research Lab, University of Kaiserslautern,
*Kaiserslautern, Germany, * E-mail: dschmidt@cs.uni-kl.de*
http://agrosy.cs.uni-kl.de/cromsci

J. OHR

Guilliard & Dörr GmbH,
Sachsenheim-Ochsenbach, Germany
http://www.guilliard-doerr.de

The success and efficiency of wall-climbing robots is not only a question of closed-loop control and electronics. Also materials have a large influence on the operability to make the systems light-weighted or more robust. This paper presents findings based on experiments to find an optimal material for inflatable adaptive sealings. Demands of such sliding sealings are robustness and good sliding characteristics but also fexibility for high sealing performances. The paper gives an overview on several materials and on their characteristics.

Keywords: Climbing Robot, Negative Pressure Adhesion, Adaptive Sealings

1. Introduction

Wall-climbing robots are a wide-spreaded research area all over the world using different locomotion and adhesion principles like magnets,[1] suction cups[2] and claws[3] or new approaches like electroadhesion[4] or van-der-Waal adhesion.[5] Nevertheless, no robot is available so far which can be used to inspect large concrete buildings. The wall-climbing robot CROMSCI[6] (figure 1, left) tries to close this gap by using active negative pressure adhesion in combination with an omnidirectional drive system. This setup has been chosen to make the system applicable for prevention maintenance which requires high velocities of up to 9.81 m/min for a sufficient fast navigation between inspection points and to carry a payload of 10 kg in terms of inspection sensors.

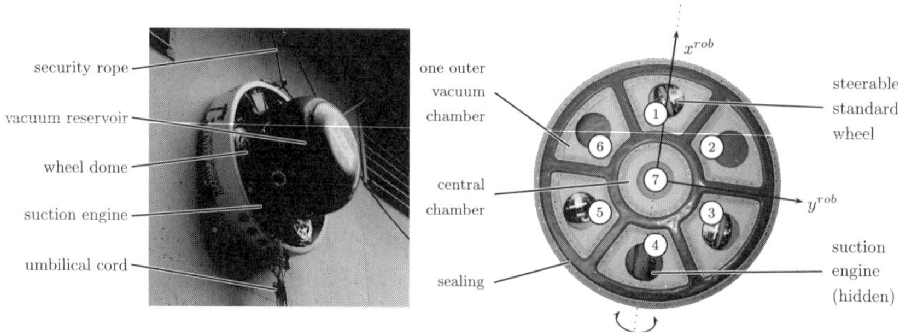

Fig. 1. CROMSCI (left) with its inflatable rubber sealing (bottom view, right)

This paper presents results to find the best materials for the inflatable sealings (figure 1, right) which are needed to seal the negative pressure adhesion system to the ambient air. In this special application the sealing has to deal with a very rough surface, temperature differences of up to 100° C and needs to be relative leak-tight. Literature mainly deals with static rubber sealings[7] or sliding suction cups[8] on glass surfaces, but the combination of rubber, rough concrete and negative pressure adhesion is nearly unexplored. So far, a silicone rubber of shore hardness (SH) 27 is used with a sliding material[9] which has been glued on the inflatable rubber. This combination works as a proof-of-concept, but is not robust enough for longer periods of inspection. Therefore, a couple of other rubber materials is taken into account and tested against demands which might be conflictive like robustness and flexibility. The main challenge is now to find the optimum which fulfills all requirements in a sufficient manner:

Low friction value: Sideward forces of the sealings must be reduced since they have to be overcome by the drive system.

Low wear: The abrasion should be low enough to be able to perform the desired task for a certain period of time.

Robustness: It has to be strong enough to avoid penetration by protrude obstacles or sharp edges as they are common on concrete surfaces.

High flexibility: To be able to overcome small steps the sealing has to be flexible enough to allow a working range of ± 1 cm.

Highly leak-proof: The sliding surface has to remain smooth enough for leak-tightness even on structured surfaces.

The next section will introduce the used test stand for determining key characteristics of the materials. The materials itself and the experimental results will be shown in section 3, the conclusion follows in final section 4.

2. Dynamic Test Stand

The dynamic test stand itself consists of different parts as shown in figure 2. The *cartesian robot* can move its tool center point (TCP) in a total range of $1.7 \times 1.5 \times 0.5\,\mathrm{m}$ with a velocity of up to $350\,\mathrm{mm/s}$. At the TCP the *experimental platform* with the adhesion system is mounted. Beside the adhesion system as shown in figure 3 this platform consists of the necessary electronic components, integrated pressure valves and sensors for controlling the sealing pressure. Beside the *vacuum cleaner* which is used as suction engine for the adhesion force, also a *pressure pump* for inflating the sealings and a *personal computer* are not integrated into the platform but connected to the system from outside.

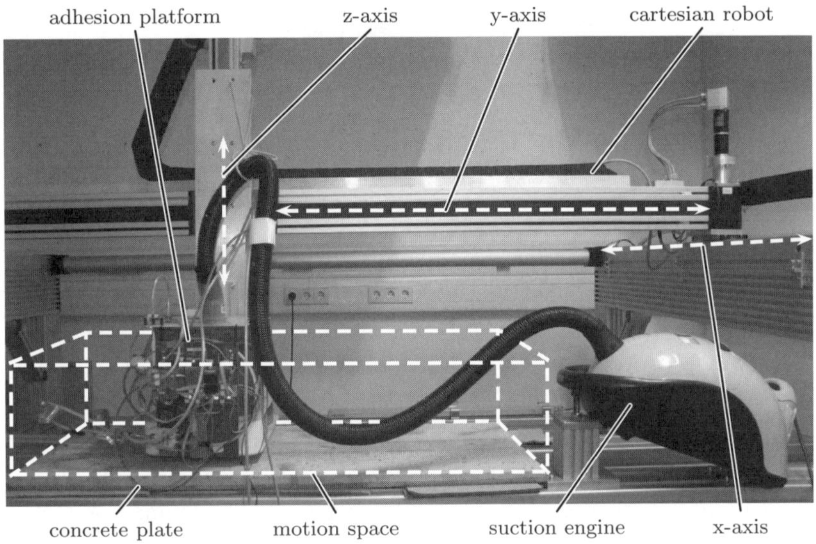

Fig. 2. The complete dynamic test stand with experimental platform (including the adhesion system), concrete surface, suction engine and cartesian robot with its three motion axes. The dashed box indicates the motion space of the platform.

The adhesion system (figure 3) is able to generate a desired adhesion force and to measure upcoming forces via a load cell connecting the system to the cartesian robot. The diameter of the round structures is $25\,\mathrm{cm}$, the volumes are $4\,\mathrm{l}$ of the adhesion chamber and $10\,\mathrm{l}$ of the reservoir. The drawing also shows the air flow: Via leakage areas at the sealings ambient air enters the adhesion system (1). Of course, these leakages have to be reduced but as one can imagine the system would not be able to move on a concrete surface if the sealings are 100 % leak-tight. The negative pressure inside of the chamber is controlled by the valve opening (2) connecting pressure chamber and vacuum reservoir. The outer valve (3) allows the raising of reservoir pressure whereas the suction engine (4) lowers it.

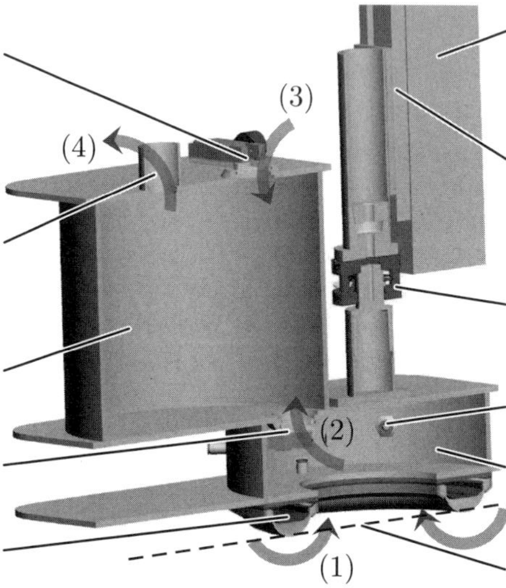

Fig. 3. Construction drawing of the adhesion system consisting of an adhesion chamber, a negative pressure reservoir and corresponding control elements.

3. Material Characteristics

The presented test stand is used for different static and dynamic experiments to determine a couple of material characteristics as presented in sections 3.1 and 3.2. Nevertheless, some additional trials on common industrial test stands become necessary to get more information. These experiments will be described in section 3.3.

3.1. *Static Experiments*

In the first experiments on the test stand some static values of the different sealing materials are determined. This includes leakage values and pressure force induced by the sealing depending on the distance between test stand and concrete surface. Table 1 sums up the examined rubbers. Sealing pressure (1.9 Bar) and negative pressure for adhesion (−50 mBar compared to ambient pressure) are set fix during these experiments. The graphs in figure 4 represent the leakages of the used materials at different heights (set via the z-axis of the cartesian robot) of the adhesion system while it is not in motion. Here one can test the static leakages (the lower the better) as well as the flexibility (the wider the range the better). It is obvious that the leakage increases with growing distance between adhesion system and ground surface. Furthermore, the softest sealing (MVQ-30) with SH 30 has the highest flexibility (± 1.2 mm) and is highly leak-proof, whereas sealing NBR-S-40 with a steal coating was leak-proof only in a very small range (± 0.4 mm). Very interesting are the results of neoprene (CR-40) which is of the same hardness as MVQ-40 and NBR-40 but much more leak-tight. Materials with a SH above 50 are not tested on this test stand since the needed pressure for inflation would be too high.

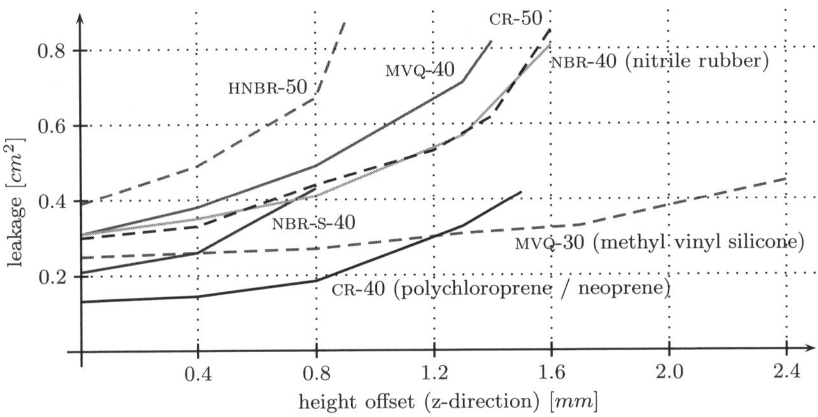

Fig. 4. Static leakages at varying distances between adhesion system and concrete plate with different sealing materials and shore hardnesses.

3.2. *Dynamic Experiments*

In a second test series the adhesion system is moved by the cartesian robot in y-direction (compare fig. 2) back and forth with a velocity of about 100 mm/s to retrieve the dynamic parameters. Figure 5 shows an excerpt of the logged data with materials of SH 40 on top and the others below. In this cutout one motion from left to right of the adhesion system is depicted – except MVQ-30 which had its turning point at about $t = 10$ s as it can be seen by the ratio value going down. This ratio of sideward forces $F_{xy} = \sqrt{F_x^2 + F_y^2}$ and pressure force F_p, which are measured by the integrated load cell, allows a conclusion on the friction value. In an optimal case it would be exactly the friction value, but so far it only allows a relative statement.

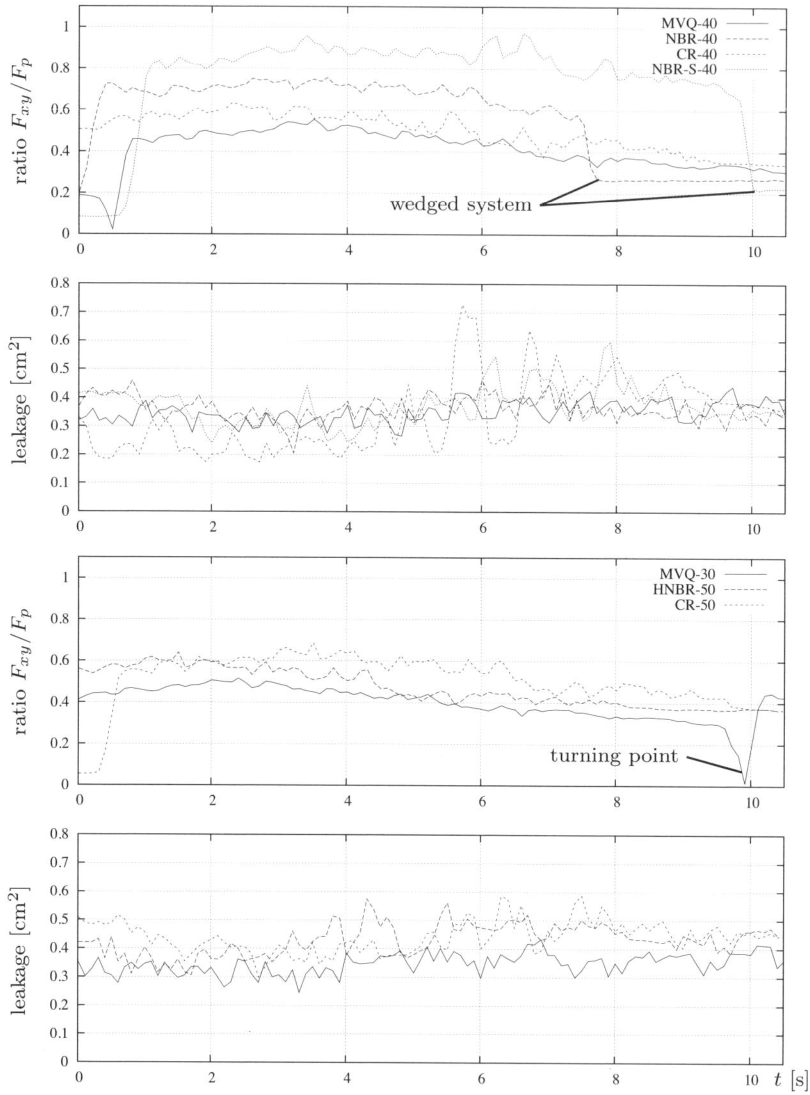

Fig. 5. Results of the dynamic experiments with adhered and moved platform using seven different materials.

In the graphs one can notice a surprisingly high difference in these ratio values although the shore hardnesses are (nearly) the same. Unfortunately, it is not possible to perform a large number of test cycles automatically because the adhesion system wedges itself if sideward forces are high. This problem occurs because of irregularities on the ground and results in an emergency stop of the motors which have to be reactivated manually. In the final application this problem does not exist since the distance between robot chassis and ground is set fix via the wheels.

3.3. *Additional Experiments*

In addition some more key data are examined on separate test stands: The abrasion has been analyzed on a test bench with a rotating barrel and abrasive paper. The values given in table 1 show the reduction of material weight in percent after 20 m. The elasticity is illustrated in needed air pressure to expand the sealing 1 cm on a separate test stand. The tensile strength as given in the last column has been determined on an universal testing machine as it is commonly used in industrial applications. Table 1 sums up the average values of all experiments, the best values are highlighted.

Table 1. Characteristics of the used materials: static flexibility, dynamic leakage and force ratio and finally common industrial key data.

material – abbreviation-SH	flexibility [± mm]	leakage [cm^2]	force ratio	abrasion [%]	elasticity [Bar]	tension [N/mm^2]
fluorocarbon rubber						
– FPM-50	-	-	-	21.4	-	5.8
methyl vinyl silicone						
– MVQ-30	**1.20**	0.34	**0.46**	**3.0**	**0.9**	5.7
– MVQ-40	0.70	0.34	0.48	4.4	1.3	6.7
– MVQ-50	-	-	-	10.8	1.6	9.9
natural rubber						
– NR-40	-	-	-	16.1	**0.9**	18.4
– NR-45	-	-	-	14.9	1.3	10.4
nitrile rubber						
– HNBR-50 (hydrogenated)	0.45	0.43	0.51	5.9	1.6	**22.6**
– NBR-40	0.80	0.36	0.70	12.9	1.0	5.8
– NBR-50	-	-	-	12.2	1.7	8.6
– NBR-S-40 (steel coated)	0.40	0.35	0.88	-	1.1	-
polychloroprene						
– CR-40	0.75	**0.30**	0.59	8.4	1.2	12.3
– CR-50	0.80	0.42	0.60	11.9	1.5	10.3

4. Conclusion

The sealing material is very essential for robust operability and economic application of the developed inspection system. A malfunction of the sealings caused by abrasion or penetration can lead to severe damages or a total loss of the system. So far, the results are very promising and give hints on the feasability of the different rubber materials for the sealings.

Nevertheless, a final decision can not be taken since a one-fits-all material does not exist. It seems as if a combination of materials like a soft and flexible sealing (e.g. MVQ-30) with a more robust sliding surface (e.g. HNBR-50) might be the best solution. The next steps are now to produce compound material probes and to perform additional test series to find the optimum in flexibility, leak-tightness and robustness with these combinations.

Acknowledgements

This research was funded by the german *Bundesministerium für Wirtschaft und Technologie* (BMWi) under the program *Zentrales Innovationsprogramm Mittelstand* (ZIM) and took place in cooperation with *Fraunhofer Institute for Industrial Mathematics* (ITWM), Kaiserslautern, Germany.

References

1. W. Shen, J. Gu and Y. Shen (2005), *Proposed Wall Climbing Robot with Permanent Magnetic Tracks for Inspecting Oil Tanks*, International Conference on Mechatronics and Automation (ICMA), 2072-2077
2. Y. Yoshida and S. Ma (2010), *Design of a Wall-Climbing Robot with Passive Suction Cups*, International Conference on Robotics and Biomimetics (ROBIO), 1513-1518
3. M.J. Spenko, G.C. Haynes, J.A. Saunders, M.R. Cutkosky, A.A. Rizzi, R.J. Full and D.E. Koditschek (2008), *Biologically inspired climbing with a hexapedal robot*, Journal of Field Robotics, Volume 25, Issue 4-5, 223-242
4. H. Prahlad, R. Pelrine, S. Stanford, J. Marlow and R. Kornbluh (2008), *Electroadhesive Robots - Wall Climbing Robots Enabled by a Novel, Robust and Electrically Controllable Adhesion Technology*, International Conference on Robotics and Automation (ICRA), 3028-3033
5. B. Aksak, M.P. Murphy and M. Sitti (2008), *Gecko Inspired Micro-Fibrillar Adhesives for Wall Climbing Robots on Micro/Nanoscale Rough Surfaces*, International Conference on Robotics and Automation (ICRA), 3058-3063
6. D. Schmidt, C. Hillenbrand and K. Berns (2011), *Omnidirectional locomotion and traction control of the wheel-driven, wall-climbing robot,* CROMSCI, Robotica Journal, Volume 29, Issue 07, Cambridge University Press, 991-1003
7. H. Ahmadian, J.E. Mottershead and M.I. Friswell (1997), *Parameterisation and Identification of a Rubber Seal*, International Modal Analysis Conference (IMAC), 142-146
8. Z. Qian, Y. Zhao, Z. Fu and Y. Wang (2006), *Fluid Model of Sliding Suction Cup of Wall-climbing Robots*, International Journal of Advanced Robotic Systems, Volume 3, Issue 3, 275-284
9. C. Hillenbrand, D. Schmidt, K. Berns, T. Leichner, T. Gastauer and B. Sauer (2007), *Development of a sealing system for a climbing robot with negative pressure adhesion*, 10th International Conference on Climbing and Walking Robots (CLAWAR), 115-124

AN IMITATION LEARNING APPROACH FOR TRUCK LOADING OPERATIONS IN BACKHOE MACHINES

C. MASTALLI, J. CAPPELLETTO, R. ACUÑA, A. TERRONES, and G. FERNÁNDEZ-LÓPEZ

Simón Bolívar University, Mechatronics Group, ELE-328, Sartenejas 1080-A Miranda, Venezuela
** E-mail: carlosmastalli@usb.ve*

This paper presents a motion planning and control system architecture development for autonomous earthmoving operations in excavating machines such as loading a dump truck. The motion planning system is imitation learning based, which is a general approach for learning motor skills from human demonstration. This scheme of supervised learning is based on a dynamical movement primitives (DMP) as control policies (CP). The DMP is a non-linear differential equation that encode movements, which are used to learn tasks in backhoe machines. A general architecture to achieve autonomous truck loading operations is described. Also, the effectiveness of our approach for truck loading task is demonstrated, where the machine can adapt to different operating scenarios.

Keywords: Imitation Learning, Dynamical Movement Primitives, Excavating Robots, Backhoe Machine.

1. Introduction

Everyday, earthmoving operations are performed at every time around the world. Moreover, the earthmoving industry is an important economic and productive activity. Furthermore, backhoe machines operations are repetitive and their workspace is complex and non-structured, where they have to interact with humans, trucks and obstacles. Finally, there are a few research works oriented to develop a fully autonomous excavating robot because of the complexity of the problem and their tasks.

On other hand, for service robotics, imitation learning based motion planning systems have been developed to solve objects grasping and manipulation,[1] or biped robot locomotion;[2] for which it has been raised the dynamical movement primitives approach that represent unidimensional discrete and rhythmic movement. In consequence, the autonomous reproduction of movement using learning by demonstration (LbD) could be feasible in any kind of task performed by backhoe machines. Therefore, it is necessary to demonstrate that movement can be generalized in other contexts, i.e. for several positions and orientations of the target in truck loading operations.

In contrast of Autonomous Loading System (ALS),[3] the motion planning system is implemeted through an imitation learning with DMPs based on truck position estimation. This method has the potential advantage to generate motion plans that solve obstacle avoidance problem in the same way as Park.[4]

Finally, Section 2 describes some of the current works for complete autonomy for earthmoving operations, and an overview of our approach. Section 3 expounds the details of imitation learning with DMPs for unidimensional discrete movements. In Section 4, it is shown the implementation of imitation learning with DMPs in a typical backhoe machine for truck loading operations. The inverse kinematics and proposed control system is presented, with the results for truck loading movement generation with our autonomous system. Finally, in Section 5 are shown the comments for the obtained results and futures works.

2. Motion Planning Systems for Excavator Robots

Currently, there are few works on complete autonomy for earthmoving operations, because of the environment complexity and the type of tasks for this kind of machines. The backhoe machines usually execute three main operations, which are digging a foundation or leveling a mound of soil and truck loading. In Singh[5,6] the task planning was formulated as constraint optimization in action space, and in Singh and Cannon[7] it was developed a multi-resolution planning systems that performs the digging operations autonomously. In other hand, in Stentz[3] have been raised a hardware and software architecture for autonomous truck loading operations. In this paper, it has been developed an architecture of perception systems similar to,[3] but with some improvements like: the scanning plane is parallel to the ground and the addition of a stereo camera. The perception systems have to determine and estimate the pose of the dump truck, as shown in Fig. 1.

In other hand, humans can execute complex task because of the abilty to learn movement primitives, i.e. dunk the ball to the basket in a basketball game. Therefore, imitation learning methods can be used for motion planning in complex backhoe machine operations. Thus, this approach could allow the execution of complex task.

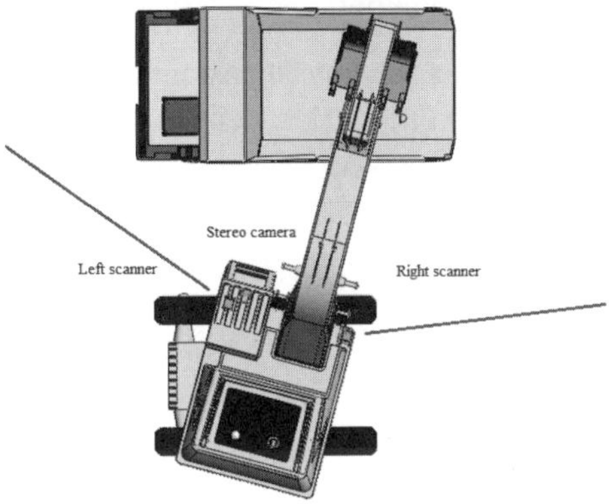

Fig. 1: Perception system architecture.

Finally, it is developed a motion planning system based in imitation learning with Dynamical Movement Primitives. DMPs are used to adequately encode movement primitives; and the learning process becomes a locally weighted regression problem.[8]

3. Imitation Learning with DMPs

This section briefly describe the imitation learning approach using dynamical movement primitives framework that is a potential way to simplify learning control policies.

3.1. *Dynamical Movement Primitives*

A Dynamical Movement Primitive is a non-linear dynamic equation that code the basic behavioral pattern (i.e. rhythmic or discrete movements),[9] and its formulation is based in attractor theory. Therefore, the purpose of this control policy is to reach the goal state with a particular trajectory shape, independent of its initial state. Besides, DMPs are a compact representation of high-dimensional planning policies and it must have the following properties:

- The convergence to the goal state must be guaranteed.
- The DMP formulation must generate any desired smooth trajectory.
- DMP have to be temporal and spatial invariant.

- The formulation must be robust against perturbations due to the inherent attractor dynamic.

A good DMP formulation for unidimensional discrete movements is developed for grasping and manipulation tasks by Pastor.[1] This representation is a set of differential equations referred to as a transformation system (1), and other as a canonical system(3). Note that the equation (1) can be interpreted as a linear spring-damper system perturbed by an external forcing term $f(s)$.

$$\begin{aligned} \tau\dot{v} &= K(g-x) - Dv - K(g-x_0)s + Kf(s) \\ \tau\dot{x} &= v \end{aligned} \tag{1}$$

where x and v are position and velocity system; x_0 and g are the initial and goal position; τ is a temporal scaling factor; K is a stiffness constant; D the damping constant; and $f(s)$ is a nonlinear function which can be learned for a determine smooth trajectory demonstrated. The K and D gains must be choosen such that the system is critically damped. The non-linear function $f(s)$ is defined as:

$$f(s) = \frac{\sum_i \omega_i \psi_i(s) s}{\sum_i \psi_i(s)} \tag{2}$$

where $\psi_i(s) = \exp(-h_i(s-c_i)^2)$ are Gaussian basis functions, with center c_i and width h_i, and ω_i are the weights that have to be learned.

$$\tau\dot{s} = -\alpha s \tag{3}$$

Note that with this formulation, the spatial and temporal invariance is satisfied, because $f(s)$ depends on a phase variable s. The phase variable is monotonically decreasing (from 1 to 0) according to (3).

Thus, the process of learning from demonstration consists in computing the set of weights ω_i for a given desired trajectory $(x(t), \dot{x}(t), \ddot{x}(t))$, where this requires computing the non-linear function by the equation (4). Therefore, to solve the learning problem it is necessary to minimize the error criterion $J = \sum_s (f_{target}(s) - f(s))^2$; this represent a locally weighted regression problem.[8]

$$f_{target}(s) = \frac{\tau\dot{v} + Dv}{K} - (g-x) + (g-x_0)s \tag{4}$$

4. Autonomous Backhoe Machine

To develop autonomous truck loading operations in backhoe machine is important to consider the following three aspects: a perception system that can recognize and estimate the position of the dump truck, a motion planning system that generate appropriate movements to reach the goal, and finally, a control system that adequately command the actuator of the machine. As a result, it was developed an imitation learning with DMP approach for the motion planning in truck loading operations.

4.1. *Imitation Learning with DMPs in Backhoe Machine*

Imitation learning using DMPs in backhoe machines is achieved allocating at least one transformation system per Degree Of Freedom (DOF). Thus, each DMP is set up with four transformation systems in order to encode the kinematics variables (x, y, z, q), where (x, y, z) is the cartesian position of end-effector and q is a joint position (pitch angle) of bucket. These state variables represent adequately the characteristics of loading truck operations, and also other kind of tasks performed by backhoe machines, because the orientation of the bucket is a relevant variable in any kind of operation.

The configuration of employed the dynamical movement primitives is illustrated in Fig. 2, where a single DMP generate discrete movements primitives in two coordinates frame concurrently. Moreover, we can see the non-linear target function to be learned (orange signal) and the generation of four states (red signal) in a typical truck loading operation (see Fig. 2).

Thus, the general architecture of the intelligent agent (IA) is illustrated in Fig. 3, which we can see the process of imitation learning, acquisition of environmental information, motion planning and control.

For the generation of autonomous truck loading tasks, the machine operator show a trajectory demonstrated $\left[\mathbf{x_d}(t)\ \dot{\mathbf{x}}_\mathbf{d}(t)\ \ddot{\mathbf{x}}_\mathbf{d}(t)\right]^T$ with the task parameters $\left[\mathbf{x_0}\ \mathbf{g}\right]$ (orange dashed lines) by the operation of the machine. Then, the imitation learning with DMPs system (blue system) find the weights ω_i based in the information of the tasks. And finally, the perception system send the particular task parameters in order to generate a motion plan and execution of the truck loading operation (red lines and boxes).

4.2. *Inverse Kinematics and Control System*

Because DMPs are encoded in cartesian (end-effector position) and joint (bucket orientation) positions, it is necessary to implement an inverse kine-

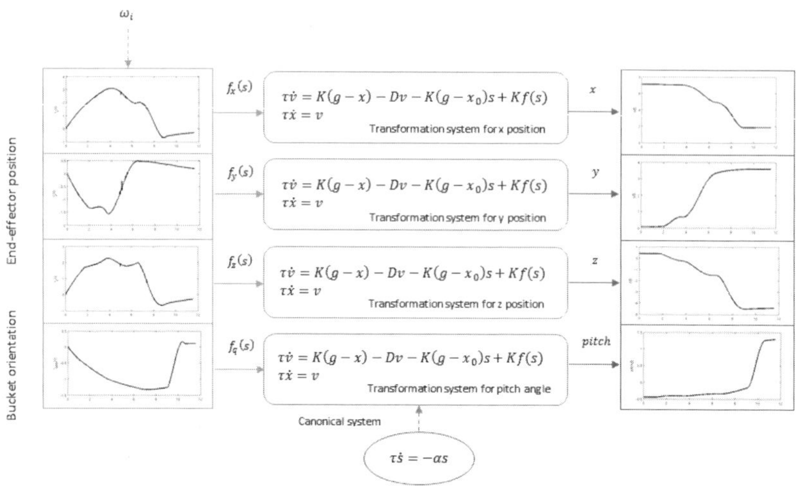

Fig. 2: The DMP implemented with four transformation systems, that is used to generate movements in backhoe machines.

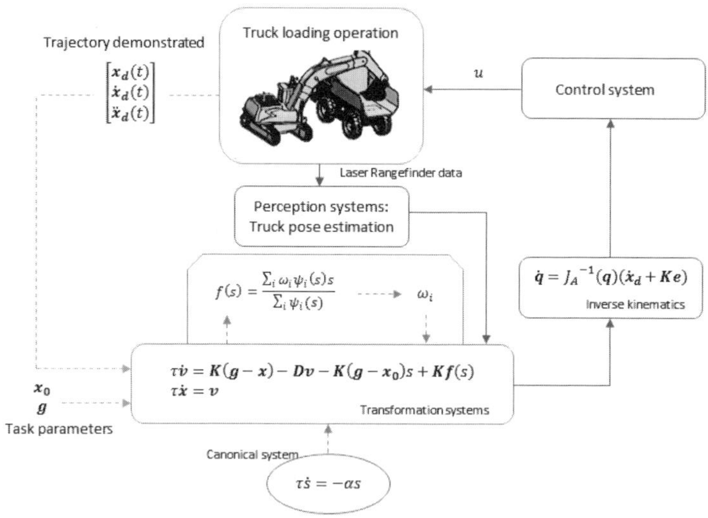

Fig. 3: The general architecture developed for autonomous truck loading operations.

matics algorithm that maps the cartesian coordinates to joint positions. For this purpose, it was formulated an inverse kinematics algorithm with Jacobian inverse as shown in Fig. 3. This computes the positions and velocities of turret, boom and stick joints.

The movement generation is transformed into the position and velocities references for the feedback controller (PID controller). Then, the controller performs appropriate torque commands $\mathbf{u}$ for the actuators.

In Fig. 4, we show the details about the implementation of inverse kinematics and control algorithm, where a feed-forward law control is implemented.

4.3. *Truck Loading Movement Generation*

To generate an autonomous truck loading operation, a DMP is selected and adjusted with the task information from perception systems: position to load the truck ($\mathbf{g}$) and the current state backhoe machine arm ($\mathbf{x_0}$) (see Fig. 3). In Fig. 5, it can be seen how the imitation learning with DMPs works in a typical left sided truck loading operation, where eight different cases of movement generation are shown. Thus, the green lines represent the movements generation for four different goal states based in the demonstrated trajectory (black line). The blue lines have the same four goals positions as green but with a new initial position. It can be observed that the motion planning follows the same shape trajectory desired while reaching any goal state.

4.4. *Backhoe Machine Simulation*

The simulations were implemented on an Open-Source Multi-Robot Simulator called Gazebo.[10] In the simulation of the autonomous truck loading operation, kinematic and dynamic properties for a typical backhoe machine were employed. Besides, it was implemented a set of range finder lasers with properties similar to those of Hokuyo lasers.

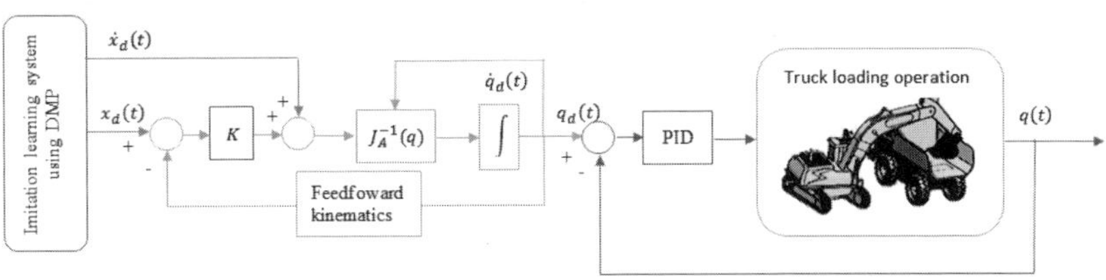

Fig. 4: The inverse kinematics and control algorithm implemented.

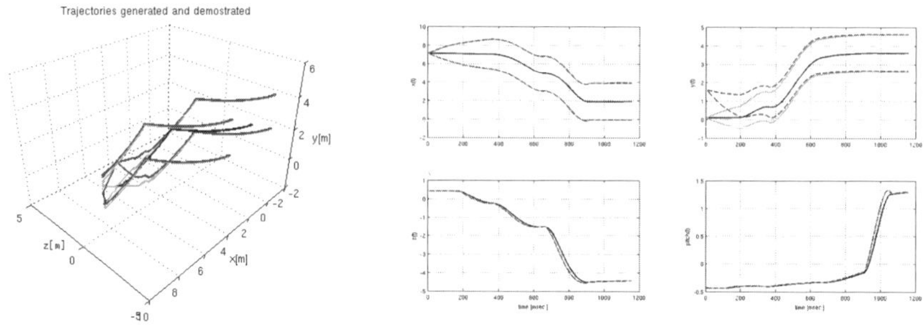

Fig. 5: The truck loading trajectory demonstrated (black whites) and generation movements for new goals (green lines), and news initial states (blue lines).

First, it was teached a movement for making a left truck loading operation. Second, the intelligent agent learned this movement. Third, the perception system recognize and estimate the position of the truck. Fourth, the DMP is selected and adjusted according to the position of truck. Fifth, the machine learning compute step by step the desired movement. Finally, this desired movement is mapped to joint position and velocity references for the generation of the torque commands for the actuators.

In Fig. 6 and 7, it can be seen how imitation learning with DMP approach works for autonomous truck loading operations; where in both cases

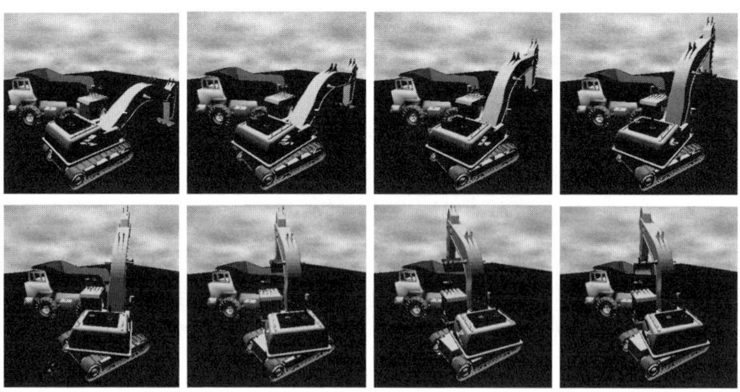

Fig. 6: The autonomous truck loading operation with $g = [1.87\ 3.60\ -4.44\ 1.3]$.

Fig. 7: The autonomous truck loading operation with $g = [2.50\ 4.00\ -4.44\ 1.3]$.

have the same initial state but with different goal states, g=[1.87 3.60 −4.44 1.3] and g=[2.50 4.00 −4.44 1.3], respectively.

In both cases, an appropiate performance of the truck loading task can be observed. Thus, the bucket position is adequate to avoid solid spilling. Therefore, the utility of the imitation learning with DMPs approach is demonstrated, where the system responded appropriately.

Besides, it is important to highlight that this method codes easily (from human demonstration) a high-dimensional control policy that performs in the same way as an expert operator. It must be noticed that this is not possible with non-learning methods.

5. Conclusion and Futures Works

This paper proposed a novel framework of motion planning system for autonomous truck loading operation. It was suggested an imitation learning approach as a motion planning scheme for backhoe machines. Numerical simulations demonstrated the reproduction and effectiveness of this architecture of supervised learning for the generation of motor skills in excavator intelligent agent, which is robust to change in target positions.

The approach could be implemented in other kind of operations as digging a foundation or leveling a mount of soil for the machine here modeled. The autonomous truck loading architecture presented could be extended to complete automation of earthmoving operations in backhoe machines. Also, this technology could be easily extended to other kinds of similar machines (i.e. bulldozers, cranes, etc).

Future works will address the generation of a extended DMP library for most common operations in a backhoe machine. It is important to develop a high-level task planning that make decisions of type of truck loading (i.e. right or left truck loading) based in the perception system; this would choose the most appropriate DMP for the task. In the future, we are interested in developing an efficient algorithm to build a map of occupancy that feeds a potential field generation system for obstacle avoidance. And finally, we want to test the complete architecture in a typical backhoe machine.

References

1. P. Pastor, H. Hoffmann, T. Asfour and S. Schaal, Learning and generalization of motor skills by learning from demonstration, *2009 IEEE International Conference on Robotics and Automation* , 763(May 2009).
2. J. Nakanishi, Learning from demonstration and adaptation of biped locomotion, *Robotics and Autonomous Systems* **47**, 79(June 2004).
3. A. Stentz, J. Bares, S. Singh and P. Rowe, A robotic excavator for autonomous truck loading, *Proceedings. 1998 IEEE/RSJ International Conference on Intelligent Robots and Systems. Innovations in Theory, Practice and Applications* , 1885 (1997).
4. D. Park, H. Hoffmann and S. Schaal, Combining dynamic movement primitives and potential fields for online obstacle avoidance, in *Fourth International Symposium on Adaptive Motion of Animals and Machines*, (1) (Case Western Reserve University, Cleveland, OH, 2008).
5. S. Singh and R. Simmons, Task planning for robotic excavation, *Proceedings of the IEEE/RSJ International Conference on Intelligent Robots and Systems* , 1284 (1992).
6. S. Singh, Developing plans for robotic excavators, in *Proc. of the ASCE Conf. on Robotics for Challenging Environments*, (Citeseer, 1994).
7. S. Singh and H. Cannon, Multi-resolution planning for earthmoving, *Proceedings. 1998 IEEE International Conference on Robotics and Automation (Cat. No.98CH36146)* , 121(May 1998).
8. C. Atkeson, A. Moore and S. Schaal, Locally weighted learning, *Artificial intelligence review* **11**, 11 (1997).
9. S. Schaal, J. Peters, J. Nakanishi and A. Ijspeert, Learning movement primitives, in *Robotics Research*, eds. P. Dario and R. Chatila, Springer Tracts in Advanced Robotics, Vol. 15 (Springer Berlin / Heidelberg, 2005) pp. 561–572.
10. N. Koenig and A. Howard, Design and use paradigms for gazebo, an open-source multi-robot simulator, in *IEEE/RSJ International Conference on Intelligent Robots and Systems*, (Sendai, Japan, 2004).

PARCELBOT: A TRACKED PARCEL TRANSPORTER WITH HIGH OBSTACLE NEGOTIATION CAPABILITIES

MARK H. HOEPFLINGER, DAVID BASCHUNG, C. DAVID REMY, MARCO HUTTER, and ROLAND SIEGWART

Autonomous Systems Lab, Institute of Robotics and Intelligent Systems, Swiss Federal Institute of Technology (ETHZ), Zurich, Switzerland, markho@ethz.ch

NICKLAS L. HOCH, and BERND WERTHER

Volkswagen AG, Wolfsburg, Germany

This paper introduces ParcelBot, a tracked robot designed as an assistance device for packet delivery services. Its application field requires high mobility in an artificial environment, including the ability to overcome fairly large obstacles such as steps or stairs. The platform is moving on tracks but consists of a mechanism to tilt the robot, so that it behaves like a wheeled system. Compared to pure tracked or wheeled robots, the combination leads to an increased obstacle climbing ability, superior maneuverability and higher energy efficiency due to less friction losses when navigating on flat ground. The paper focuses on the mechanical design of the prototype and highlights a selected number of experiments.

1. Introduction

The prototype, that is presented in this paper, was designed for the following application scenario: It is intended to serve as a human operated transporter for parcel delivery services, capable of carrying significant payload from a delivery car to the end customer. This scenario imposes that the robot has to be able to climb up artificial obstacles, typically stairs and has to be able to navigate on them robustly and with sufficient speed. The operation in a human defined environment on a very specific task clearly restricted the design choices. The size of the robot was influenced through specifications of the parcel delivery services (payload dimensions and weight), but also by the local building law and official recommendations from the council for accident prevention (stairs step height, width and inclination, [1]). While many existing robots partially full-filled the imposed requirements (such as [2],[3],[4],[5],[6],[7]), none of them seemed to be ideal for our specific scenario. Some of the existing platforms did not feature the required obstacle negotiation abilities, some were not able to handle sufficient payload and others were not applicable due to the need of elaborate perception to allow for robust locomotion. Hence we needed to develop a novel system that masters these specifications in a single device. This paper introduces the hardware prototype and mainly focuses on the mechanical design and its evaluation, while the electronic setup as well as the control software are just roughly en-lighted.

2. Obstacle climbing method

Due to the requirement to be able to navigate on stairs, the robotic platform has been designed as a tracked vehicle. Since the tracks alone (without a massive track height) did not provide the needed obstacle climbing abilities, an additional mechanism had to be incorporated. It consists of a lever arm with two passive castor wheels. While the arm is stowed in the main body, the robot behaves as a common tracked vehicle. With deployed arm (see Fig. 1, a), the ParcelBot acts like a purely wheeled system. This configuration is applied to move on flat ground and allows to navigate with less slippage which results in a higher turning rate, less friction losses and also reduced abrasion of the floor/tracks, that is important when performing on private stairways. The arm is as well used to lift the robot onto the obstacles. The process of stair climbing is illustrated in Fig. 1. While the robot approaches the obstacle, the lever arm is deployed. Once the tips of the tracks are on the obstacle, the lever arm is inserted into the main body. From then on, the robot takes advantage of the tracks to move on the stairs. At the end of the stairway, the robot stops its forward motion and deploys the lever arm again to quit the obstacle.

2.1. *Mechanical design*

The actuation unit for the tracks consists of a motor (Maxon EC 60, 400 W), a planetary gearbox (Maxon GP 81, reduction rate of 25:1) and a shaft extension and coupling system (Fig. 2). An in-track design has been chosen, which means that all parts required for the track actuation were embedded inside of the track unit. This holds for the mechanics, but also for the motor control electronics. Beside a simple mechanical mount, only electrical power and control signal connections have to be made to actuate the track. This largely simplifies the usage as well as the protection and

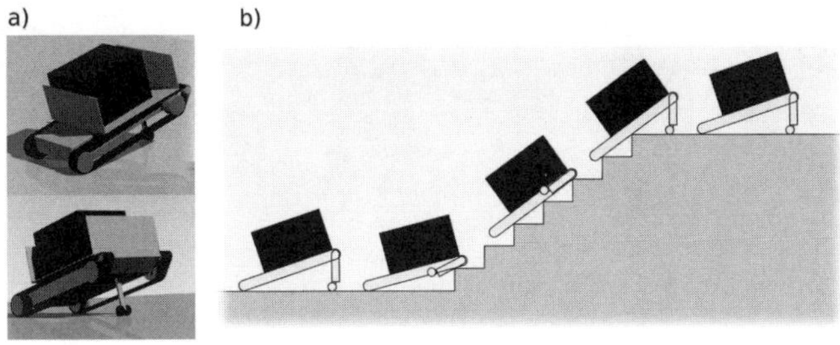

Fig. 1. a) Concept of the ParcelBot, a tracked platform that can be inclined by a wheeled lever arm. In the inclined configuration, the robot behaves on flat ground like a purely wheeled system. b) shows an illustration of the stair climbing process.

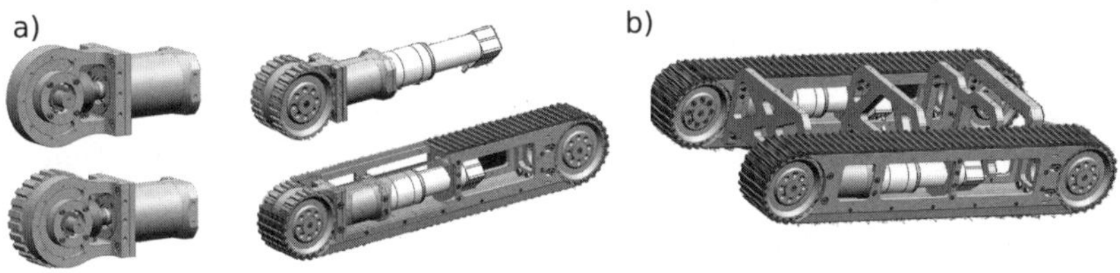

Fig. 2. CAD image of the actuation system of the tracks. a) shows drawings of the generic shaft extension and coupling unit and its integration into the track unit. b) illustrates the complete robot with the main body and the left and right track units.

sealing. The torque requirements and therefore the motor/gear combination for the track actuation has been defined based on two operating cases and on the payload and system specifications (Table 1): In the nominal case, both tracks of the robot are in contact with the stairs with inclinations up to 35 degrees. In the worst case scenario, only one track is in operation. The torque requirements were calculated to be 27 Nm (8 Nm on flat terrain) for the nominal case and 36 Nm (11 Nm on flat terrain) for the peak load. Based on that, the gear ratio has been determined to be 25:1 for the planetary gear and 2:1 for the bevel gear inside the coupling system. This resulted in a continuous torque of 26.6 Nm and peak torques of approx. 35 Nm with drive speeds of up to 0.5 m/s.

Another design focus was made on modular and reusable mechanical parts. Therefore the same actuation unit could be used for the lever arm mechanism by just mounting a different planetary gearbox. The lever arm was one of the most critical elements of the mechanical design of the robot (Fig. 3). Since it has to incline the fully loaded robot and to sustain significant impact collisions (e.g. during the training phase of the operator), it

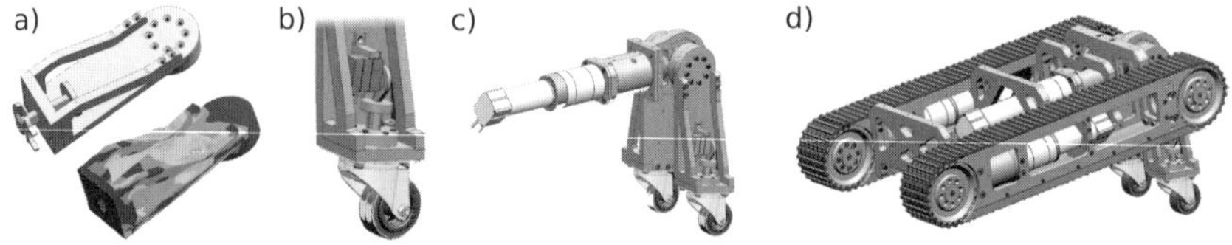

Fig. 3. a) Drawing of the main arm element of the castor wheel mechanism. b) The clutch to block the steering axis of the castor wheels. c) The assembled arm with the generic actuation unit and finally c) a drawing of the assembled robot with the castor arm.

was designed powerful and robust. A performed FEM analysis on selected parts with static forces (up to 1000 N) and impact forces of up to 3000 N confirmed the robustness of the lever arm (Fig. 3, a). Beside these considerations, the dimensions and mounting position, which strongly influence the stability on obstacles, were carefully defined. The high torque requirements of the lever arm mechanism were satisfied by choosing a reduction rate of 308:1 of the planetary gear, while the bevel gear ratio remained at 2:1. This resulted in a continuous torque of approx. 240 Nm and peak torques of up to about 310 Nm, with a turning rate of around 12 degrees/s. The tip of the lever arm was equipped with two castor wheels. During the deployment and the suspension of the arm, the steering axis of the castor wheels had to be blocked. A clutch has been developed that can actively be released by a servo motor.

The design of the main body of the robot was kept very simple. It consisted of a robust mechanical structure to connect the two track modules with the lever arm and a box to store the electronics including power converters, signal converters, and computers.

Table 1. Key data of ParcelBot

total length	960	mm	per track: Cont. torque:	26.6	Nm
total width	800	mm	per track: Max. torque:	35.0	Nm
total height	200	mm	lever arm: Cont. torque:	240	Nm
total mass	80	kg	lever arm: Max. torque	310	Nm
max. tested payload	20	kg	max. forward velocity:	0.5	m/s
max. slope angle	35	°			

3. Control overview

Due to the teleoperation of the robot, a simple control scheme could be selected. Fig. 4 shows a block-diagram of the different electrical components involved to control the robot and the data-flow from the operator to the robot. As a user interface, a gaming controller (Nintendo Wii remote) has been applied. This allows an operator to input commands, but also to receive feedback from the robot via LEDs and sound output. The remote control is used to set the reference velocities of the robot body (the translational velocity v_B and the rotational velocity ω_B), as well as the joint angle of the lever arm ϕ_C and a signal f_c to enable or disable the clutch to block the rotational movement of the two castor wheels. The commands are sent via Bluetooth to the on-board PC (ASUS EEE PC) of the ParcelBot. The on-board computer calculates the reference turning speed ω_R and ω_L of the right and left driving motors based on a differential drive kinematic model of the robot:

$$\omega_R = \frac{v_B + \frac{1}{2}\omega_B D}{R_R} \qquad \omega_L = \frac{v_B - \frac{1}{2}\omega_B D}{R_L}$$

D denotes the distance between the centers of the two tracks and ω_B the rotational velocity around the center of the robot respectively the center of the back axis, if the robot is in its inclined configuration. R_L and R_R stand for the radius of the left and right driving wheels (resp. the track height) that actuate the tracks. The reference motor signals are sent over CAN to the motor controllers (Maxon EPOS 70/10), that executed the velocity control loops for the track actuation and a position control loop for the lever arm actuation.

The clutches are operated by sending PWM signals to their actuators (Standard Servo Motors). To generate the signals, a custom made microprocessor board (with a Microchip dsPIC33) is used. The signal to enable/disable the clutch is sent by the on-board PC via serial connection.

4. Experiments

The manufactured prototype was tested in different experiments. To characterize the locomotion performance, turn-on-spot maneuvers in normal and inclined configuration, obstacle climbing and navigation experiments on stairs have been executed. Fig. 5 a, shows a sequence of pictures of the inclination process, while b depicts the stair climbing process. Fig. 5 c contains a sequence of picture of the robot while navigating on the stairs. The robot successfully climbed up and down step obstacles with heights of up

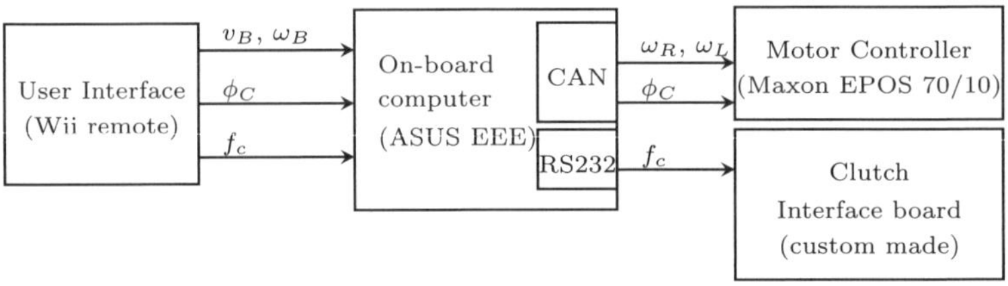

Fig. 4. Block diagram of the main hardware components involved in the control of the ParcelBot and the data flow from the operator to the robot.

Fig. 5. Pictures of the robot deploying the lever arm (a) and climbing up the first steps of a stair (b). c) illustrates the navigation capabilities on stairs (as overlay of images)

to 30 cm. For the stair climbing maneuvers, the geometrical properties of the first step were most critical. Stairs with a comparably low first step led to a low inclination angle of the robot and made it more difficult to climb on the second step (Fig. 5, b). This difficulty was already identified in the design process and did not affect the performance of the robot in the selected experiments. In contrast, the experiments illustrated that the tracks of the robot were not ideal for the navigation on steep stairways with rather slippery material properties. Conveyor belts have been used as tracks for the first prototype. Those belts where rather stiff with a low coefficient of friction and therefore increased the slippage of the robot. Using custom made tracks with appropriate material properties and grouser shapes would have reduced the slip effects. Nevertheless, navigation on common concrete stairs was easily possible (Fig. 5, c).

5. Discussion

While many existing platforms seemed to have been suited for the given application, every compared system showed certain drawbacks related to

Fig. 6. a) Sequence of pictures of the robot entering a stairway from the top. The pictures in b) show the inclined robot that performs a turn-on-spot maneuver around the center of the back axis of the robot

the task. Conventional tracked robots ([5]) usually require a track height in the range of the double step obstacle height to climb on and don't allow smooth transitions on the stairs. Tracked systems with flippers or other suspension mechanisms ([3],[2], [4]) would have been appropriate for smooth stair climbing but locomotion on flat ground would result in higher friction losses than for wheeled systems. Wheeled climbing robots, such as the Shrimp ([?]) or the Octopus ([?]) posses high obstacle climbing capabilities, but their performance on stairs strongly depends on the step properties (mainly on the step height and step length) and robust navigation on the stairs becomes very hard. Our platform combines the advantages of tracked and wheeled systems. In the nominal configuration, it behaves like a ordinary tracked robot, in inclined configuration rather like a wheeled system. This helps to reduce the friction losses for motion on flat ground and increases the robustness for navigation on stairs. The lever arm also support the obstacle climbing process by allowing to lift the robot onto the obstacles.

While the payload ratio of about 0.25 is rather low for such a system, there is still room for improvement (it has to be mentioned that the payload ratio of 0.25 does not reflect the effective payload capabilities of the robot but rather what has been specified and tested in the climbing experiments. Driving on flat ground was tested with a payload of 80 kg). The mechanical structure is build very rigid with high safety factors. Optimizing the ratio of payload was not part of the study and should be addressed in a next design iteration.

Heading towards a real application, certain additional features would be beneficial. First, the level of autonomy would have to be increased. The teleoperation of the robot was sufficient for a preliminary study, but of course, it would be desirable to disburden or even forgo an operator. Further, an additional mechanism to shift the position of the payload (and therefore the center of gravity or the robot) during run-time would help to increase the robustness of the system on steep obstacles and in our case, would lower the constraints that are posed on the payload location to allow for smooth obstacle climbing without tilting e.g. when leaving stairways.

6. Conclusions

Various specifications and requirements given by the concrete application largely restricted some of the design choices. The others as locomotion method or actuation principles were to be selected. The choice of using a tracked vehicle with a deployable castor wheel mechanism proved to be very qualified for the task. The design and assembly process could be simplified by constructing the track modules as independent units with a simple interface to the main body and by designing a generic motor module. This module was applied to actuate the tracks with torques of around 25 Nm and, with just simple changes, the lever arm requiring torques of up to 300 Nm. Experiments demonstrated that the robot successfully performed in simple application scenarios. The teleoperation of the robot by a roughly skilled operator lead to a smooth obstacle climbing maneuver.

References

1. Swiss council for accident prevention, Guidelines for stairs construction (www.bfu.ch).
2. F. Michaud, D. Letourneau, M. Arsenault, Y. Bergeron, R. Cadrin, F. Gagnon, M.-A. Legault, M. Millette, J.-F. Pare, M.-C. Tremblay, P. Lapage, Y. Morin, J. Bisson, S. Caron, "AZIMUT, a leg-track-wheel robot," *Intelligent Robots and Systems (IROS)*, (2003)
3. iRobot PackBot series, iRobot Corporation, (www.irobot.com)
4. RAPOSA, IdMind -Engenharia de Sistemas, Lda., (www.idmind.pt)
5. Foster-Miller Talon, Foster-Miller, Inc., (www.foster-miller.qinetiq-na.com)
6. Inuktun Services Ltd., (www.inuktun.com)
7. RoboProbe Technologies Inc., (www.roboprobe.com)
8. R. Siegwart, P. Lamon, T. Estier, M. Lauria, and R. Piguet, "Innovative design for wheeled locomotion in rough terrain", *Robotics and Autonomous Systems* 40, p. p. 151-162 (2002)
9. M. Lauria, Y. Piguet, and R. Siegwart, "Octopus -An Autonomous Wheeled Climbing Robot", *Proc. of The Fifth International Conference on Climbing and Walking Robots (CLAWAR)*, (2002)

SECTION–12

SERVICE ROBOT STANDARDS AND STANDARDIZATION

INTERNATIONAL ROBOT SAFETY STANDARDISATION

GURVINDER SINGH VIRK

Faculty of Engineering and Sustainable Development, University of Gävle, Gävle, Sweden, and CLAWAR Association Limited, Sheffield, United Kingdom.

The paper presents results of international standardization work items within ISO and IEC to develop safety requirements for new types of non-industrial robots, namely, personal care robots, and for medical robots. The key requirements for the three defined types of personal care robots focus on how the intended close robot-human interactions can be facilitated without causing harm. For medical robots, the key issue is how the degree of autonomy that is included in the new definition of a robot can be handled in medical electrical equipment to ensure basic safety and essential performance. The standardization work is continuing, and the paper describes the current thinking within the work groups and summarises the results that have been achieved.

1. Introduction

Although robot research covers a wide area, robot commercialisation comprises mainly the area of industrial robots, which developed during the 70s in the automation of car factories, and a stable market has developed since then. Today there are significant EU industrial robot companies comprising Kuka (Germany), Comau (Italy), and ABB (Sweden). However, this conventional industrial robotics market has become saturated, and has not shown significant growth for many years, causing the companies to look for fresh opportunities to expand into new potential markets; in addition, non-traditional robot companies are showing interest in adopting robots in emerging new sectors. Most of these new emerging applications require close robot-human interaction, which is in stark contrast to the classical uses of industrial robots where, in the main, robots and humans must be kept physically separated by real or virtual cages to ensure human safety while the robot is operational. This separation of robots and humans has been enshrined in the industrial robot safety standard; but due to the requirements from many manufacturing applications, there has been a little relaxation in the recent revision ISO 10218 [1] to allow a robot-human collaborative mode for industrial robots under defined and controlled situations. This industrial robot collaborative mode does not allow close robot-human interaction, and still no contact with humans is permitted.

The need to formulate ISO safety requirements for the new robot sectors where “close” and “contacting” situations are allowed to perform intended tasks, such as provide physical assistance, is urgently needed for facilitating innovative robot applications in both the industrial and non-industrial environments. In fact, this is exactly the role of standardization, namely to ensure products are safe, are of sufficient quality, and work as intended, also to define clear and unambiguous provisions in order to facilitate trade and communication; hence, international standards allow companies to sell their products anywhere they want because these standards contain all the information needed about differences between individual countries they want to export to. By using an international standard, a company loses less time (and money), and is able to design products that operate in a predictable way anywhere in the world. Safety standards are the most important types of standards as they state the requirements that must be met if a product is to be compliant and hence deemed to be safe. But, when a safety standard is missing, it can be extremely difficult (and expensive) for companies to carry out the vast testing needed to gather the data on their products to satisfy specific regulations of individual countries, as the burden of proof to demonstrate that the products are safe must still be met.

Safe robots, that can be deployed in close proximity to humans and be able to make contact with humans, will open up large new robot markets in personal care robots. However, we are at a classic “catch 22 situation”, since 1) we need the new robot safety standard to create the non-industrial markets, but 2) it is normally industry which formulates the requirements to go into a standard, and, as this new industry is not yet in existence, except for many SMEs working in a wide sector, there is insufficient critical-mass or know how to produce the detailed requirements needed for the standard. In this regard, new robots for some emerging service robotic areas are beginning to become available as small consumer products, where the safety issues are low but larger more powerful robots, although being talked about, cannot yet be commercialised. Because of such concerns, robot safety issues are starting to become a critical barrier to these new innovations since no company will risk taking the needed robot products to the market place as the fear of litigation if the robots cause serious harm to their human users is potentially very high.

One way to break this vicious circle is to ask researchers to play the key role in providing the knowhow in the new sectors because of their expertise in developing robot prototypes. It was with this thinking that CLAWAR Association Ltd. initiated the current major robot standardization work by using its academic partners to lobby many of the ISO members in the EU on the need for the work, and offering to provide experts to assist in the new working groups. This work started in around 2005 to formulate the new safety requirements for non-industrial robots involving personal care robots and medical robots. Because of this strong involvement of CLAWAR, annual updates of the standardization work items have been presented at CLAWAR conferences to keep the robotic community informed on progress, and also to align researchers to the global need. Some of the new international standards are almost ready to be published, representing a key stage in robotics which is likely to grow exponentially in the next few years.

Although there are many organisations involved in international standardization activities (DICOM, IEEE, OMG as well as national efforts), in fact only two bodies matter, namely International Organization for Standardization (ISO; www.iso.org), and International Electrotechnical Commission (IEC; www.iec.ch) who have been responsible for the majority of the international standards that are in current use. It is clear that, to maximise the impact of any new robot standards, they must be produced under ISO/IEC governance. IEC has, in the past, tended to focus on products which are electrotechnical in nature, and where the primary hazards are due to electricity, whereas ISO has been responsible for all other areas, including robotics where the hazards are mainly of a mechanical nature (such as crushing, trapping, cutting, impacting, etc. situations). However, as technology has developed and become multi-disciplinary, the boundary between ISO and IEC has become blurred, and there is now considerable overlap in their activities, although efforts are made to collaborate whenever possible.

The paper presents an update on two key ISO/IEC standardization projects in the robotic area, which are as follows:

- Personal care robot safety: the work aims to formulate the safety requirements for three types of personal care robots which have been defined, namely, mobile servant robots, physical assistant robots, and person carrier robots. The work is being carried out under ISO TC 184/SC 2/ WG7.
- Medical electrical equipment and systems using robotic technology (Medical robots): this work aims to formulate safety requirements for medical robots and how they may be regulated as medical devices. The work is being carried out under a joint working group (JWG9) between IEC SC 62A and ISO TC 184/SC2.

Both projects have the author as the Convenor, with CLAWAR Association providing the support; the work for both groups is well underway, and the latest results of each are presented in this paper. Section 2 will present the progress report for personal care robots, and section 3 describes the work of the medical robot group. The paper ends with some conclusions about the gaps in knowledge, and how these may only be successfully filled using good collaborations between the robot research and the robot standardization communities.

2. Personal care robot safety requirements

Virk [2], [3] and Virk & Moon [4] have described the background to the growth in international robot standardization from its initiation in 2005, when a study group was created to investigate the global robot standardization requirements and this led to three new ISO work groups being set up in 2006. One of these was the WG7 group whose task was to formulate the safety requirements for personal care robots, and the mission was to define the "rules and regulations" to allow close robot-human interaction for performing intended tasks and to consider cases when the human is the "work piece" of the robot. Brief summary details of this WG7 work group are as follows:

- Work group set up in Sept 2006 and 1st mtg on 20 Oct 2006 to explore the key issues and produce a committee draft document to start the new work item.
- A new work item on "personal care robot safety" was proposed and approved by international balloting. The work started in Feb 2008 with 8 participating countries (France, Germany, Hungary, Japan, Korea, Sweden, UK, and USA)
- ISO/DIS 13482 balloting document submitted in April 2011 with the number of participating countries in WG7 having grown to 12 (China, France, Germany, Hungary, Italy, Japan, Korea, The Netherlands, Sweden, Switzerland, UK, and USA).

The DIS (Draft International Standard) balloting has led to the document being approved by 14 "P" members voting "YES" (Canada, China, Czech Republic, Finland, France, Germany, Hungary, Italy, Japan, Korea, The Netherlands, Romania, Sweden, Switzerland, and UK) and 1 voting "NO" (USA), and a significant number of comments have been submitted.

The personal care robot safety standardization work is currently resolving the DIS comments received so that the FDIS (Final DIS) document can be obtained. In fact, the WG7 group has been already formulating its own comments to improve the DIS document, and much work has been carried out. One major issue has been to review the title and scope of the standard. The title has been changed to "ISO/DIS 13482: Robots and robotic devices – Safety requirements for service robots – Personal care robots" and the scope is the following:

> This international standard (IS) specifies requirements and guidelines for the inherent safe design, protective measures, and Information for use of personal care robots, and it has a special focus on three types of personal care robots, namely "mobile servant robot", "physical assistant robot", and "person carrier robot". These robots typically perform tasks as exemplified in Annex C for all users. The standard describes hazards associated with the use of these robots, and provides requirements to eliminate, or adequately reduce, the risks associated with these hazards. This IS covers human-robot physical contact applications.
>
> Significant hazards are presented in Annex A (List of significant hazards for personal care robots) and this standard describes how they should be

dealt with for each personal care robot type. Robotic devices used in personal care applications are also covered by this standard and should be treated as personal care robots. Personal care robots may need to adhere to additional standards and regulations where appropriate; for example motor vehicle regulations when person carrier robots are operating on public roads.

This scope is focussed on human care related hazards but where appropriate, includes domestic animals or property (including other robots).

Another significant result is that low risk and normal risk categories for each type of personal care robot have been defined so that two levels of normative performance levels (PLs) can be included in the standards which comply with internationally agreed risk assessment ([5]-[7]). In addition, the low risk category has been included to facilitate the development of markets for personal care robots. The issues agreed to be important in each level and category are given next.

1. Mobile servant robot

- Type 1.1 (Low risk): Small & light weight & slow & no manipulator
- Type 1.2 (General): Large OR fast OR with manipulator

2. Physical assistant robot

- Type 2.1 (Low risk): Low physical assistance (user can overpower personal care robot)
- Type 2.2 (General): High physical assistance (user cannot overpower personal care robot)

3. Person carrier robot

- Type 3.1 (Low risk): Standing, single passenger & indoor flat surfaces & is slow (less than 1.6m/second) & is lightweight & is semi-autonomous
- Type 3.2 (General): Multiple passengers OR non-standing passengers OR outdoor OR fast (possibly limited to 20kph via the scope) OR uneven surfaces OR autonomous.

Another important issue which has come up in the work program, and which cannot be resolved adequately at the present time, is the provision of normative numerical design data to allow personal care robot manufacturers to design the new types of robots within internationally agreed safety margins. This data relates to forces, pressures, and speeds which personal care robots should be allowed to possess, and such may vary depending on the type of users (adults, children, elderly person, disabled person, pregnant women, etc).

There have been some studies carried out on the pain tolerance limits of adults and robot-human collisions, on various parts of the human body, to study significant injury mechanisms [8]-[11], but the results are not widely accepted and cannot be used in international standards in a normative manner. It is therefore essential that there is major international collaboration between researchers and the standardization communities to plan the research activities to define what is needed. The main issue identified within WG7 is the open and internationally agreed experimental protocols that must be followed by a number of countries so that the results can be objectively compared and made acceptable as international normative design data for different categories of people (adults, children, elderly persons, pregnant women, etc) and personal care robot application (mobile servant, physical assistant, person carrier, etc). This requires the various funding agencies in the different regions to discuss and formulate collaborative R&D programmes.

3. Basic safety and essential performance of medical robots

The personal care robot safety standardization work has indicated that medical robot applications also need to be investigated, as has been described in Virk & Moon [3]; these studies have led to the creation of a joint working group (JWG9) under IEC lead, with over 50 experts from Brazil, Canada, China, France, Germany, Hungary, Italy, Japan, Korea, The Netherlands, Romania, Switzerland, UK, and USA. The main medical standardization committee within IEC is TC62 (Electrical equipment in medical practice) which has the following 4 sub-committees (SCs):

- 62A: Common aspects of electrical equipment used in medical practice
- 62B: Diagnostic imaging equipment
- 62C: Equipment for radiotherapy, nuclear medicine & radiation dosimetry
- 62D: Electromedical equipment

At this time it is not yet clear exactly where the new medical robot standards will belong, but SC62A and SC62D are the most likely candidates. From the early investigations within ISO, it has been clear that medical robot safety standards must fit the IEC 60601 family of standards for medical electrical equipment so that a medical robot will need to be regulated as a medical device and not as a machine. The view from IEC TC62 (who is responsible for producing the 60601 standards) has been that the possibility for a collateral standard should be explored, first focussing on the additional safety issues introduced by medical robots which are not already covered in the current 60601 family of documents.

Most of the work thus far for identifying the distinction has focussed on the implications from the latest definition for "robot", viz: *actuated mechanism programmable in two or more axes with a degree of autonomy, moving within its environment, to perform intended tasks*. The key term here appears to be "degree of autonomy" which has not been considered within IEC TC 62's work to date. In other words, the safety standardization of medical electrical equipment (MEE) has not considered the implications when MEE possesses "a degree of autonomy" so, hence, considering medical robots as MEE means that the hazards introduced by MEE having autonomous capabilities need to be investigated; this is being explored as being the basis of the new collateral standard.

To do the detailed work, JWG9 has formulated the meaning of degree of autonomy (DoA) for medical robots, and has used the definition of autonomy from WG1, viz: *ability to perform intended tasks based on current state and sensing, without human intervention.* This has led to the following definition for degree of autonomy of MEE: *extent to which operator intervention is incorporated in the MEE.* The range of DoA has been agreed to be from >0 to full autonomy. The next step will be to identify the potential hazards which are introduced, due to DoA, which need to be investigated, and then perform a risk assessment to ensure suitable safety requirements are formulated. The hazard identification is the next task to be carried out.

4. Conclusions

The paper has presented a summary update on the ISO/IEC international robot standardization work in personal care robots and in medical robots. In this respect, the personal care safety standard (ISO/DIS 13482) has progressed to the DIS balloting phase, and comments are being resolved to progress it to FDIS, and final publication of the international standard is expected in August 2013. The medical robot standardization work has formally started by the creation of JWG9, and the work is investigating what additional hazards are introduced into MEE if the equipment is designed possessing autonomous capabilities, and how the additional hazards that arise can be mitigated to design acceptable medical robots.

A key finding of the standardization work is that normative design data is needed to assist manufacturers to produce acceptable personal care robots, and obtaining this data requires major international collaboration to define the experimental protocols and duplication of the work in several countries to ensure widespread acceptance of the results. It is hoped that publicising these needs will create more dialogue between researchers, standardisation groups, and research funding organisations to produce safe and reliable robot products.

Acknowledgments

The author would like to thank the experts on the ISO TC 184/SC 2/WG 7 and IEC SC 62A & ISO TEC 184/SC 2 JWG9 groups who are doing excellent work to produce the safety standards needed to create and sustain the new robot market. The author is partially supported by the UK DTI under the AITS travel scheme for attending the international ISO/IEC meetings; this financial support is gratefully acknowledged.

References

1. ISO 10218 standard: ISO 10218-1:2011 Robots and robotics devices – Safety requirements for industrial robots, Part 1: Robots and ISO 10218-2:2011 Robots and robotics devices – Safety requirements for industrial robots, Part 2: Robot systems and integration, 2011.
2. Virk GS, Safety standard for personal care robots, Proc of the 12th Int Conf CLAWAR'2009, Istanbul, Turkey, pp 147-154, 2009.
3. Virk GS, Personal care robot safety, Proc of the 13th Int Conf CLAWAR'2010, Nagoya, Japan, pp 1333-1339, 2010.
4. Virk GS and Moon S, Safety for emerging service robots, Proc of the 14th Int Conf CLAWAR'2011, Paris, France, pp 956-963, 2011.
5. EC Machinery Directive, Directive 2006/42/EC of the European Parliament & Council, 17 May 2006 on Machinery, and amending Directive 95/16/EC.
6. ISO 12100:2010, Safety of machinery – General principles for design – Risk assessment and risk reduction.
7. ISO 13849-1:2006/Cor 1:2009, Safety of machinery – Safety-related parts of control systems – Part 1: General principles for design; and ISO 13849-2:2003 Safety of machinery – Safety-related parts of control systems – Part 2: Validation.
8. Yamada Y, Suita K, Imai K, Ikeda H and Sugimoto N, A failure-to-safety robot system for human-robot coexistence, Robotics & Auto. Systems, Vol 18, Issues 1-2, pp 283-291, 1996.
9. Yamada Y, Hirasawa Y, Huang SY, Umetani Y, Suita K, Human-robot contact in the safeguarding space, IEEE/ASME Trans. on Mechatronics, Vol.2, No.4, pp.230-236, 1997.
10. Haddadin S, Albu-Schäffer A, Frommberger M, Rossmann J, and Hirzinger G, The DLR Crash Report: Towards a standard crash-testing protocol, for robot safety - Part I: Results, 2009 IEEE Inte Conf on Robotics and Automation, Kobe Int Conf Centre, Kobe, Japan, May 12-17, 2009.
11. Haddadin S, Albu-Schäffer A, and Hirzinger G, Soft-tissue injury in robotics, ICRA 2010.
12. Haddadin S, Albu-Schäffer A, Haddadin F, Rossmann J, and Hirzinger G, Experimental safety study on soft-tissue injury in robotics, RAM, 2011.
13. Medical Device Directive; Directive 2007/47/EC of the European parliament and Council, 5 Sept 2007 amending Council Directive 90/385/EEC relating to active implantable medical devices, Council Directive 93/42/EEC concerning medical devices and Directive 98/8/EC concerning the placing of biocidal products on the market.
14. IEC 60601-1:2005; Medical electrical equipment Part 1: General requirements for basic safety and essential performance.
15. IEC 62061 ed1.0; Safety of machinery – Functional safety of safety-related electrical, electronic and programmable electronic control systems.
16. IEC 62061 ed1.0; Safety of machinery – Functional safety of safety-related electrical, electronic and programmable electronic control systems.

REVISION OF VOCABULARY STANDARD FOR ROBOTS AND ROBOTIC DEVICES IN ISO

SEUNGBIN MOON

Department of computer Eng., Sejong University, Gwangjin-gu, Seoul, 143-747, Korea

SOON-GEUL LEE

Department of Mechanical Eng., Kyung Hee University, Global Campus, Giheung-gu, Gyeonggi-do, 446-701, Korea

This paper summarizes recent revision of vocabulary standard in ISO TC 184/SC 2 (robots and robotic devices). The previous standard on vocabulary was published in 1994, and it focused mainly on industrial robots. A working group on vocabulary was formed to revise the existing standard and to expand it to cover the terminologies for service robots. The final version has been developed by experts and published in March 2012. Some new additions are summarized in this paper.

1. Introduction

As the service robot market is gradually emerging in recent years, standard development organizations, such as, ISO, IEC, OMG, and ASTM have initiated standardization activity in robotic field recently. [1, 2, 3].

ISO (International Organization for Standardization) TC 184/SC 2 (robots and robotic devices) has been active in developing robotic standards from its initiation in 1983[4]. It has developed 10 International Standards and 1 Technical Report, up to now. They include vocabularies, coordinate systems, safety requirements, performance measures, and mechanical interfaces. Since 2006, it started to develop standards for service robots in addition to the traditional industrial robots[5, 6, 7].

IEC (International Electrotechnical Commission) has been interested in the issues related to household vacuum cleaning robots and robotic lawn mower. Its main interest lies on the safety and performance issues of these robots[8].

OMG (Object Management Group) Robotics DTF(Domain Task Force) was established in 2005. Primary goal of Robotics DTF is to standardize robotic

This work was supported by Standard Technology Advancement Program from Ministry of Knowledge Economy, Korea.

middleware and its architecture. It has published three standards on middleware up to now[9].

ASTM (American Society for Testing and Materials) International has established committee E54.08(Operational Equipment) to develop standards on urban search and rescue robots. It has published standards on urban search and rescue robots, including vocabulary and performances.

ISO 8373(Vocabulary) was first published in 1994 and contained mainly terms for industrial robots. This second edition is expanded to include both industrial robots and service robots. The revision has been initiated from 2007 and the new work item proposal was accepted in 2010. In this paper, key vocabularies introduced in the current revision are described.

2. Basic vocabularies

As is well known, standards published by ISO are widely utilized throughout the world. Vocabularies defined in ISO 8373 are not only used in other ISO standards but also accepted as basic terms in the industry and the research communities of robotics field. For example, IFR (International Federation of Robotics) utilized these definitions in its annual World Robotics, a market report.

As expected, the vocabulary standard is essential before developing a series of standards such as interfaces, performance measures, and safety requirements, since it clarifies the meanings of terminologies used in these specific documents. The new edition contains 180 terms, including new terms for service robots.

The well respected definition of "robot" is now revised to include both industrial robots and service robots.

Robot

actuated mechanism programmable in two or more axes with a degree of autonomy, moving within its environment, to perform intended tasks.

NOTE 1 A robot includes the control system and interface of the control system.

NOTE 2 The classification of robot into industrial robot or service robot is done according to its intended application.

One of the important aspects of the robot is "autonomy" which is defined as below.

Autonomy

ability to to perform intended tasks based on current state and sensing, without human intervention.

New definition of robot is quite broad and might include some mechanisms which may not be considered as robot. One of discussions we had during the course of discussion was to add an additional phrase of "exhibiting typical locomotive motions of human or animal, such as walking, flying, and swimming". After extensive discussion, we decided not to add this phrase since it might narrow the definition so much that some innovative mechanisms could be excluded. We believe that current definition reflects the nature of both industrial and service robots.

The old definition of robot was the same as industrial robot, but now it is used only for industrial robots. The definitions of industrial robot and service robot are given below

Industrial robot

automatically controlled, reprogrammable, multipurpose manipulator, programmable in three or more axes, which may be either fixed in place or mobile for use in industrial automation applications.

Service robot

robot that performs useful tasks for humans, society or equipment excluding industrial automation applications.

NOTE 1 Industrial automation applications include, but are not limited to, manufacturing, inspection, packaging, and assembly.

NOTE 2 While articulated robots used in production lines are industrial robots, similar articulated robots used for serving food are service robots.

Basically, robot is classified into either industrial robot or service robot, and the classification is done by its application not by its mechanical structures. Service robot is further classified into personal service robot and professional service robot.

Personal service robot

service robot used for a non-commercial task, usually by a member of the general public.

Professional service robot

service robot used for a commercial task, usually by properly trained operator.

We also decided to introduce the term of "robotic device" to designate the mechanisms which might not be defined as robots according to the strict definitions of industrial robots or service robots, but utilized essential robotic technologies. The definition of "robotic device" is given as below.

Robotic device

actuated mechanism fulfilling the characteristics of industrial robot or service robot, but lacking either number of programmable axes or the degree of autonomy.

EXAMPLES Power assist device; teleoperated device; two-axis industrial manipulator.

3. Additional vocabularies

We also added some vocabularies to clarify the meanings so that it can be used in precise context in other standard documents. One example is the distinction between "collaboration" and "cooperation". While "collaboration" designates the situation where human and robot work together, "cooperation" is used for multiple robot cooperation. This distinction is essential in safety requirement documents, since the safety nature in "collaboration" mode is very critical.

Collaboration

work done by robot(s) and human(s) together to fulfil a task.

Robot cooperation

interaction between multiple robots to ensure that their motions are effective together for the task.

Definitions on terminologies, mainly intended for mobile service robots were also introduced as below.

HRI (human robot interaction)

information and action exchanges between human and robot to perform a task.

Mobile platform

assembly of all components of the mobile robot which enables locomotion.

Mobile platform coordinate system

coordinate system referenced to one of the components of a mobile platform.

Navigation

deciding on and controlling the direction of travel, derived from localization and environment map.

Dead reckoning

method of obtaining the pose of a mobile robot using only internal measurements from known initial pose.

Proprioceptive sensor

Internal state sensor

robot sensor intended to measure the internal state(s) of robot.

Exteroceptive sensor

External state sensor

robot sensor intended to measure the states of robot's environment or interaction of the robot with its environment.

4. Conclusion

In this paper, the recent revision of ISO 8373 at TC 184/SC 2/WG 1(Vocabulary and characteristics) was summarized. Experts from USA, UK, Germany, France, Sweden, Japan, and Korea has been actively engaged in the WG 1 activity during last few years. As the robot industry is widened to include both industrial robot and service robot, we needed to expand the vocabulary standard to harmonize the terminology used in both areas. We wish that the use of ISO 8373 will facilitate the development of robotic products in the market.

References

1. G. S. Virk, "New standards for new robots", in *Proc. 10th Int. Conf. on Climbing and Walking Robotics (CLAWAR 2007)*, pp 698-707, 2007.
2. G. S. Virk, S. Moon, and R. Gelin, "ISO standards for service robots," in *Proc. 11th Int. workshop on Climbing and Walking Robots and the Support Technologies for Mobile Machines*, pp. 133-138, 2008.
3. S. Moon, S. Rhim, Y.-J. Cho, G. S. Virk, "International standardization activities in robotic fields", in *Proc. 12th Int. workshop on Climbing and Walking Robots and Support Technologies for Mobile Machines*, pp. 140-146, 2009.
4. S. Moon and G. S. Virk, "Survey on ISO standards for industrial and service robots", in *Proc. ICROS-SICE Int. Joint Conf.,* pp. 1878-1881, 2009.
5. G. S. Virk, "Safety standard for personal care robot", in *Proc. 12th Int. Conf. on Climbing and Walking Robots and Support Technologies for Mobile Machines*, pp. 147-154, 2009.
6. C. Harper, S. Dogramadzi, and M. O. Tokhi, "Development in vocabulary standardization for robots and robotic devices", in *Proc. 12th Int. Conf. on Climbing and Walking Robots and Support Technologies for Mobile Machines*, pp. 155-162, 2009.
7. S. Moon, S.-G. Lee and K.-H. Park, "Recent progress of robotic vocabulary standardization efforts in ISO," in *Proc. SICE Annual Conf.*, pp. 266-268, Aug. 2010.
8. S. Rhim, J. Ryu, K. Park, and S. Lee, "Performance evaluation criteria for autonomous cleaning robots", in *Proc. IEEE Int. Symp. on Computational Intelligence in Robotics and Automation*, pp.167-172, 2007.
9. T. Kotoku and M. Mizukawa, "Robot middleware and its standardization in OMG", in *Proc. SICE-ICASE Int. Joint Conf.*, pp.2028-2031, 2006.

PERFORMANCE EVALUATION OF CLAWAR AND PERSONAL CARE SYSTEMS

FABIO BONSIGNORIO

RoboticsLab, University Carlos III de Madrid, Av. Universidad, 30
Leganes, 28911 Madrid, Spain

Heron Robots, V.R.Ceccardi, 1 18
Genova,16121 ,Italy

The difficulties in the comparison of functionalities and capabilities of robotics and cognition autonomous or partially autonomous agents slows the pace of advancement of research in this field and limits the exploitations of results by the industry. We will give a brief summary of activities in benchmarking and experimental methodologies, including the challenges for cognitive systems workshops organized by euCognition. As new ISO standards on personal care systems will soon be available, we should start to elaborate shared norms on performance evaluation of robotic partially or totally autonomous systems (in particular on risk evaluation).

1. Introduction

Although research practice is generally good in robotics, there are frequently not coped methodological issues, which usually make it difficult to replicate results, as mandatorily required by a proper scientific approach. As a consequence it is usually not easy to compare solutions and properly evaluate results. Reliability of robotics solutions is thus compromised by the dubious nature of reference models. This is probably at the origin of the lack of widely accepted benchmarking procedures in the field. Euron GEM Sig and IEEE RAS TC-PEBRAS have organized a number of workshops, benchmarking activities and competitions with the aim of facilitating the emergence of a strong benchmarking culture and improve scientific methodology in the robotics research.

We will give a survey of past, ongoing and future activities in the field, including the challenges for cognitive systems workshops organized by euCognition. We will elaborate on the utter importance of performance evaluation procedures to enable useful applications of already achieved research results. In particular risk assessment for partially or totally autonomous systems. Regulatory issues are seen as a serious bottleneck to the development of a significant scale market for robotic personal care systems.

These issues are now substantially overcome, as new ISO standards on personal care systems will soon be available. The next step should be to elaborate shared norms on performance evaluation of robotic partially or totally autonomous systems.

For some reasons we still have not seen a full bloom of new product-services exploiting the many results of robotics and cognition research. Although there might be still serious scientific problems to be solved, it seems apparent that the limitation in the quantitative methods to compare results might be a serious bottleneck. In particular these limitations make difficult to estimate the reliability of a CLAWAR robot in performing a task, which makes difficult to evaluate risks connected to autonomous systems and to properly define for example adequate insurance schemes, and define appropriate legal regulations.

Although there is a clear epistemological motivation in pursuing the replicability of research results and benchmarking, there are even stronger practical reasons to do that.

The research activities in the robotic fields are huge and it is huge the number of published papers. In many cases it is a common observation that it is not easy to compare the strength of the methods proposed for the same functionality, but in many if not most cases it is extremely time consuming even to simply replicate the published experiments.

In order to allow the practical exploitation of the many results obtained it should be at least necessary to be able to:

- validate the results by replicating them
- compare the results in term of the chosen performance criteria

It is not easy to that in practice, in a large majority of cases, even if we restrict our attention to the higher impact specialized journals, although in recent years many authors use to provide supplemental materials, like code, data etc., in order to allow the other researchers to replicate or evaluate their results. This happens in particular in the SLAM community.

It is natural to take inspiration from older more established research domains, like medicine and life science, where complexity of the studied objects is not smaller than in robotics.

For example in biology and medicine, there are well established experimental procedures, 'protocols', to deal with the behaviors of the quite complex systems they investigate, as if you look at an autonomous agent, like in biology or medicine, but also in robotics, you need a statistically significant samples of the behavior of the agent in a wide enough set of conditions while performing a wide enough set of tasks. In particular 'failures' of the systems are important, in the perspective of the *practical* utilization of results.

For example, the US NCI guidelines for drafting a clinical trial protocol, [16] emphasize signaling 'adverse events', the definition of 'criteria for response assessment', the necessity of defining clearly principal and secondary hypotheses to be validated.

As another example, a NCI acceptable trial specify the statistics of the experiments, with reference to the study objectives:

1. Method of randomization and stratification
2. Total sample size justified for adequate testing of primary and secondary hypotheses
3. Error levels (alpha and beta)
4. Differences to be detected for comparative studies
5. Size of the confidence interval of the estimates

These issues are now widely understood and shared by the research community and the EU Commission, as today project asking for funding in EU FP7 (7th Framework Program) are required to provide a sound methodology for benchmarking results.

In the following we will first give a short survey about current activities, then we will briefly discuss the open issues and the work still to be done.

2. A survey of current research in replication and benchmarking

2.1. *Conference Tracks*

There are some interesting conference tracks like NIST, IEEE and ACM Permis and the IARP, IEEE RAS series of workshop on Dependable Robots.

A significant number of workshops have been organized in the latest years by the Euron GEM (Good Experimental Methodologies) and Benchmarking Special Interest Group and by the IEEE RAS TC-Pebras (Performance

Evaluation and Benchmarking of Robotic and Automation Systems), Permis workshops aim to define metrics and methodologies for the evaluation of performances of intelligent systems, in particular quantitative metrics to be able to compare such things as: the level of autonomy, human-robot interaction, collaboration.

Other examples are Workshops on Technical Challenges for Dependable Robots in Human Environments co-sponsored by IARP and the IEEE Robotics and Automation Society. The 2005 workshop was co-sponsored by EURON, the EU network of excellence in robotics.

The concept of dependability requires that the capability of performing a set of tasks, in a set of changing environments, can be repeatedly shown, with given measured performances.

As mentioned, from 2006, at IROS/ICRA/RSS and Clawar conferences were organized a long series of workshops on performance metrics of robots and experimental methodologies in robotics, under the auspices or with the support of the Euron GEM Special interest group, [18], and the IEEE RAS TC-Pebras, established in 2009, [20]. A couple of related workshops, more focused on cognitive capabilities of artificial and natural systems, were organized in Rapperswil, Switzerland, and Oxford by the euCognition II network, [26, 27]. A joint workshop on these topics was organized by the Euron GEM Sig and euCognition II in Vasteras, Sweden, in 2010 at the European Robotics Forum.

2.2. *Benchmarking proposals*

Due to the huge diversity of robotic architectures and approaches and the intrinsic difficulty of measuring the performances of machines, which aim to be flexible and autonomous, so far, there are benchmarks defined only for a limited subset of the tasks which can be performed by a robot [5]. Here below follows a not exhaustive list of the most relevant initiatives.

Radish is a repository of standard data sets with a main focus on localization and mapping. At present it contains mostly logs of odometry, laser, sonar and other sensor data taken from real and simulated robots and environment maps created by robots or manually.

Universitat Jaume I developed a visual servoing simulation environment called JaViSS. It is written in Java with graphical rendering. Manipulator kinematics code is based on the Robotics Toolbox for Matlab by Peter Corke. It is intended as a tool to simplify the testing and comparison of different visual servoing approaches.

Another initiative at Universitat Jaume I offers a set of experiment data on visually-guided grasping of planar objects with a Barrett hand. This set of standard experiments allows the definition of benchmarks and associated performance metrics. This allows comparing different algorithms and implementations for visually-guided grasping.

The Karlsruhe Institute of Technologies has developed a set of benchmarks for home service related tasks, [24, 25], while the Technical University of Munich, M. Beetz's group, [21], has developed a query based language to classify service robotics tasks at home, in particular related to cooking.

The RAWSEEDS project is an SSA (Specific Support Action) in the EU 6th Frame Program, providing a comprehensive, benchmarking web accessible repository and toolkit for SLAM (Simultaneous Localization And Mapping). The NIST USAR (Urban Search And Rescue) 'after disaster' scenarios, ranked as yellow, orange and red, are used in RoboCup USAR. They provide a useful conventional reference scenario for USAR applications together with USARsim the open source simulation environment based on the Unreal Tournament gaming engine [17].

2.3. *Competitions*

RoboCup, [9] is probably the most famous competition in robotics. RoboCup was initially mostly focused on soccer game as a primary domain, and organizes the Robot World Cup Soccer Games and Conferences. Soccer is a very good testbed for multi (robot) agent technologies.

New competitions in search and rescue, based on NIST scenarios, and home assistance have been added. FIRA robot soccer league, a Korean KAIST initiative, is a similar one.

Eurobot is an international competition with chiefly educational purpose.

The DARPA Grand Challenge is a famous competition for outdoor robot race on an about 200km circuit in the desert.

In 2007 DARPA has organized the Urban Challenge where many robot vehicles with an urban traffic scenario.

An interesting cleaning robot competition was organized in 2002, in Lausanne, jointly with IEEE/RSJ Int. Conference on Intelligent Robots and Systems (IROS 2002). As an example task was to clean within 10 minutes 5x5 m room covered with sugar.

The European Land-Robot Trial (ELROB), organized by the German Federal Armed Forces (Bundeswehr), is an outdoor robot demonstration with no real competitions or prizes, but otherwise similar to the DARPA Grand Challenge.

It focuses on mobility and RSTA (Reconnaissance, Surveillance, and Target acquisition). It took place in 2006 for the first time, in 2007 a civilian version, ELROB-C, was organized for the first time, in Switzerland.

Today new competitions have been developed to foster research in service robotics, two interesting competitions are RoboCup@Home and RoboCup@Work.

A limitation of RoboCup when dealing with higher level cognitive functions is due to its 'Rulebook approach': a long list of rules and scoring procedure somehow limit the possibility of the team to get an advantage from flexibility and adaptivity and favors a design for the benchmark approach which is not particularly useful for research.

2.4. *Research coordination activities*

The RoSta project,[4], started in 2007, is a two-year coordination action, within the EU 6th Framework Program.

The objective of RoSta, which is linked also with standard related activities within IEEE, is to identify the action needed to start formal standards development and the establishment of de facto standards in service robotics. ALFUS (Autonomy Levels For Unmanned Systems) ,[8] , is a similar US federal agencies ad hoc working group focused on unmanned system autonomy metrics.

The Euron Good Experimental Methodology and Benchmrking Special Interest Group has the objective of addressing the need for replication of experiments and benchmarking in robotics. A first outcome are the GEM Review Guidelines which aim to help define the condition for experiment replication and broad result replication in robotics research [19], which can downloaded from [18].

As mentioned above, within euCognitionII and III, are organized meeting on the 'Challenges for cognitive systems'. In 2009 IEEE RAS constituted the TC-Pebras to address benchmarking issues in robotics and automation.

3. Novel approaches

In [12] and [13], see fig. 1, from [13], and in other experimental works 'entropy measures' on the 'sensory-motor' coordination of different 'robotics' equipment have shown that information metrics can be used to classify, at least, and to get an insight on (semi) autonomous robotics devices, which show an 'emergent behavior'.

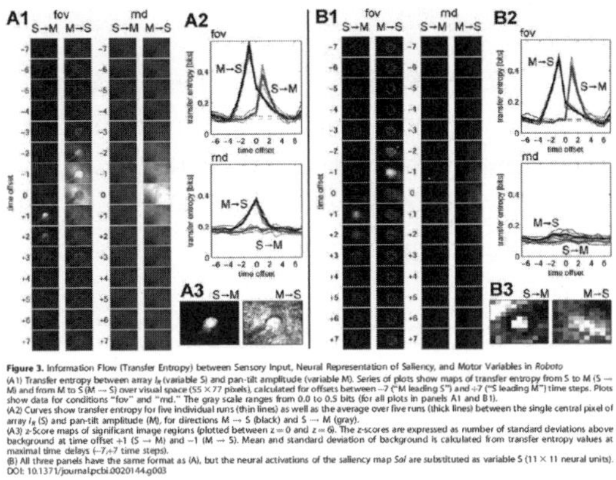

Fig.1. Sensory motor coordination information metrics measures (Lungarella and Sporns, 2006)

In [14], entropy measures are used to rank enviroment complexity, with reference to the navigation task. In [11] an approach integrating task and environment complexities is proposed.

HRI experimental research is sometime conducted by means of protocols deriving from psychology.

It might help to divide the robot functionalities into levels with an approach similar to the communication OSI level, starting, for instance, from the physical level, to the control, perception, planning and 'cognitive' levels.

4. Conclusions and Future Work

There is an increasing awareness of the importance, first of all practical, of having shared and repeatable benchmarking procedures. Not knowing the statistical repeatability of task execution and of the risk of failures for a partially or totally autonomous agent makes difficult the practical utilization of new research results. Saying that a robot is able to perform a given set of task in a given set of environments, without specifying with which probability distributions and how it will react to ‘adverse events’ make its utilization risky and *de facto* prevent it by making difficult the estimation of the liability, also in the mundane monetary sense, for the user and the supplier of the system.

It is thought that benchmarking procedures should be included in the regulations developed by bodies such as ISO.

References

1. James S. Albus, Metrics and Performance Measures for IntelligentUnmanned Ground Vehicles. In Proceeding of the performance Metrics for Intelligent System Workshop, (2002).
2. Angel P. del Pobil, Why do We Need Benchmarks in Robotics Research?, International Conference on Intelligent Robot and Systems, Beijing, China, (2006).
3. http://www.nist.gov/el/isd/permis2012.cfm
4. http://www.robot-standards.eu/
5. http://www.robot.uji.es/benchmarks
6. http://www.rawseeds.org/
7. http://www.euron.org/activities/SIGs.html
8. http://www.isd.mel.nist.gov/projects/autonomy_levels/
9. http://www.robocup.org
10. http://www.darpa.mil/grandchallenge/
11. J. Minguez, J. Osuna, and L. Montano. A Divide and Conquer Strategy based on Situations to achieve Reactive Collision Avoidance in Troublesome Scenarios. In ICRA2004, New Orleans, USA, (2004).
12. L. Olsson, C.L Nehaniv, D. Polani, Information Trade-Offs and the Evolution of Sensory Layouts, In Proc. Artificial Life IX, (2004).
13. M.Lungarella and O. Sporns, Mapping Information Flow in Sensorimotor Networks, *PLOS Computational Biology*, 2, 10, pp. 1301-1312 , (2006).
14. A.Lampe and R.Chatila, Performance measures for the evaluation of mobile robot autonomy, In ICRA2006, Orlando USA, (2006).
15. J.Hallam and G.Hayes, Benchmarks for mobile robotics?,In Towards Intelligent Mobile Robots: scientific methods in mobile robotics, Manchester University, School of Computer Science, (1997).
16. https://www.ctnbestpractices.org/
17. http://usarsim.sourceforge.net/
18. http://www.heronrobots.com/EuronGEMSig
19. F. Bonsignorio, J.Hallam and Angel P. Del Pobil (eds), GEM Guidelines, Euron GEM Sig Report, (2008)
20. http://tab.ieee-ras.org/committeeinfo.php?tcid=35
21. M. Beetz, F. Stulp, P. Esden-Tempski, A. Fedrizzi, U. Klank, I. Kresse, A. Maldonado, F. Ruiz, Generality and legibility in mobile manipulation, Autonomous Robots Journal (Special Issue on Mobile Manipulation), 28(1):21-44,(2010)

22. K. Matheus and A. M. Dollar, Benchmarking grasping and manipulation: Properties of the objects of daily living, in Proceedings of the 2010 IEEE/RSJ International Conference on Intelligent Robots and Systems (IROS 2010), (2010)
23. A. P. del Pobil, "Benchmarks in robotics research," in Lecture Notes for IROS 2006 Workshop II(WS- 2), (2006)
24. R. Dillmann, "Ka 1.10 benchmarks for robotics research," 2004.
25. B. Len, S. Ulbrich, R. Diankov, G. Puche, M. Przybylski, A. Morales, T. Asfour, S. Moisio, J. Bohg, J. Kuffner, and R. Dillmann, OpenGRASP: A toolkit for robot grasping simulation, in 2nd International Conference on Simulation, Modeling, and Programming for Autonomous Robots (SIMPAR), Darmstadt, Germany, (2010)
26. http://www.eucognition.org/index.php?page=challenges-ii-general-information
27. https://www.eucognition.org/index.php?page=eucogii-workshop-challenges

MEDICAL DEVICE STANDARDS – WHY THEY ARE SO IMPORTANT

CAROL HERMAN
Senior Vice President of Standards Policy and Programs
Association for the Advancement of Medical Instrumentation (AAMI)

The paper presents an overview of medical standards and the role they play in medical device safety. The standardization issues in the new and emerging areas of medical robots is discussed contrasting the differences for formulation the safety and essential performance requirements for them. How and why different requirements are needed is highlighted.

1. Introduction

"The medical device industry is different. Why? Because, next to perhaps nuclear power plants, medical devices are some of the most highly regulated products sold on the world market" [1]. Over the past 50 years, medical devices have become more complex right along with healthcare. A doctor's office today looks very different from that same office of decades ago, and with the recent shift to home healthcare, patient safety has become even more challenging. Most medical devices were not designed or intended to be used in the home environment yet they are sent home every day with patients who receive minimal training, if any, on the safe and effective use of that device. And the widespread adoption of medical devices or equipment that connects with the internet introduces more risk and another challenge to ensuring patient safety. The threats to patient and operator safety can now come from external sources, such as software errors, data transmission errors, and physical connection errors. It is a brave new world for healthcare. Voluntary consensus standards are the best tool to help meet the needs of the producers (medical device manufacturers) and the regulators to help ensure the safety and effectiveness of medical devices.

2. The Keys to the Process

Consensus standards are the workhorse of the medical device regulatory review process. While the data is not there to definitively support a reduced review time, it is possible that if the reviewers of premarket submissions did not have voluntary consensus standards to rely on, the U.S. Food and Drug Administration's (FDA) Center for Devices and Radiological Health (CDRH) would need more reviewers and longer review times to effectively meet their regulatory requirements.

The FDA's mission, simply stated is to promote and protect the public health by helping safe and effective products reach the market in a timely way, monitor products for continued safety after they are in use and help the public get the accurate, science-based information needed to improve health. In 1997, the U.S. Food and Drug Administration Modernization Act (FDAMA) envisioned that to best achieve this mission was to fully utilise voluntary consensus standards because, if used effectively, they could help reviewers of medical device premarket submissions to either shorten review times or re-target the review so that more time could be spent on those elements of the submission requiring more in-depth review. The latter could be accomplished because reviewers could rely on the appropriate use of standards they have recognized and have confidence in. That confidence comes from participating in the development and revision of the most necessary and relevant medical device standards. The appropriate use of standards is important to the success of the program, and industry plays a significant role in that success. If the wrong standard or the wrong criteria within a standard are used, this can slow down the process and ultimately require more review time. Consequently, industry is very important, and working closely with the agency and following the appropriate standards and guidance can play a significant role in an expeditious product review.

But not all regulators regulate in the same way. The countries within the European Union operate very differently and while they rely heavily on international consensus standards, they do not rely on the traditional product review process. And elsewhere in the world are other variations of these approaches. For more than 20 years, the medical device regulators have worked together with a goal of standardizing medical device regulations and processes. And while this goal has proven to be an impossibility, one reality is that international standards are the key - the one commonality. **Voluntary consensus standards are developed by all interested parties,** that is part of the definition of consensus standards - government regulators sit alongside those it regulates, those who use the devices, and those the regulator is sworn to protect. The standards development process is much like a three-legged stool: one leg is the regulators; one leg is the producers or industry; and the third leg is the users including clinicians, academicians, and patients or patient advocacy groups. The more all parties work together, the greater the product improvement, resulting in better patient safety.

3. The Near Future

Regulators continue to make significant improvements in processes, and collaborate with each other around the world. However, regulators do not have the resources needed to adequately keep up with the pressure of globalization. Manufacturers also face pressure to lower costs and improve products while trying to keep pace with emerging technologies and a fast paced global economy. Nanotechnology, robotics, interoperability, systems of systems – just to name a few – will continue to present new and exciting challenges. Global trends will have major implications for all regulators and the medical device industry in the next decade. Strong, global, safety standards will continue to be needed, as will faster progress towards developing these standards in order to meet the growing demand.

Medical robot standardization is a good example of the unique challenge facing the medical device industry and healthcare in general. Until two years ago, there were no medical robot standards but medical robots were and are still being developed; it is therefore clear standards are needed to ensure these new "medical devices" can be used in the medical environment without introducing harm to patients. It has been a struggle moving from the foundation of industrial robot standards to addressing the challenges of developing new standards specifically for the intended use of the medical robots in the healthcare environment and addressing all the essential principles and requirements that help mitigate risk and ensure patient safety. Today, medical robot standards are being developed in a joint working group (JWG9) of the International Organization of Standardization (ISO) and the International Electrotechnical Commission (IEC) ensuring that the needed consensus standards will eventually exist and provide a standardized approach to the visionaries bringing these products to market.

What is a medical robot? There is not yet a current standardized definition but the standardization experts working on this have come up with the following possible suggestions:

1. **robot** or **robotic device** intended to be used as **medical electrical equipment (MEE)** or as **medical electrical systems (MES)**
2. actuated mechanism with a **degree of autonomy** moving within its environment, to perform the intended tasks and used as **MEE** or **MES**.
3. **robot** or **robotic devic**e intended to be used as a **medical device**.

Note: the bolded words are formally defined terms.

How is a robot defined like this different from an industrial robot (which is defined in ISO 10218-1:2011 [2] as "automatically controlled, reprogrammable multipurpose manipulator, programmable in three or more axes, which can be either fixed in place or mobile for use in industrial automation applications") which have been used successfully for decades, Or to ask an even wider question: "How is a medical robot different from other new types of personal care robots whose safety requirements are also being formulated?" To give an idea of an answer and the key issues that have to be considered, it is necessary to step back a bit and look at some key robot definitions from ISO 8373 [3] and ISO/DIS 13482 [4]:

- **robot:** actuated mechanism programmable in two or more axes with a degree of autonomy, moving within its environment, to perform intended tasks
- **autonomy:** ability to perform intended tasks based on current state and sensing without human intervention
- **service robot:** robot that performs useful tasks for humans or equipment excluding industrial automation applications
- **personal care robot:** service robot that performs aiding actions and actions contributing directly towards improvement in the quality of life of humans, excluding medical applications. This may include physical contact with the human to perform the task.

The key between these other types of robots and medical robots is that the former are machines whereas the latter are medical devices. Both need to be safe and not cause harm but normally speaking, medical device safety in straightforward terms can be a matter of life and death. If an industrial robot malfunctions and does not install the dashboard of an automobile correctly, it is a serious problem - it effects the workflow of the assembly line, it may require a temporary shutdown of the line, at the very least that automobile needs to be pulled off the line and managed differently. Similarly, if a personal care robot malfunctions, while providing physical assistance to a user, the person will loos the support and will inconvenienced to have to use his/her own muscle-power for completing the physical task.

However, if a medical robot malfunctions, while performing a surgical procedure, the consequences could be catastrophic for the patient and the medical staff responsible for the safety of that patient. Just having an automatic shut-off if a malfunction or failure occurs is not enough in medical care applications because not completing the medical procedure could also be catastrophic. We need to know that these devices were designed and manufactured to do what they are intended to do without causing adding undo risk or harm to the patient. Consequently, using the same standard(s) from industry is not acceptable in medical care. But learning from other sectors can be very helpful. What parts of the industrial robot standards would be true whether the robot was used for industrial purposes or for medical purposes? Are there lessons learned that can and should be applied as medical robots emerge? There is much to be learned from other industries and standards developers know this. One sure tenant of standards development is to not duplicate effort or reinvent the wheel. Leverage the good work done in other standards, even if they are in other product sectors. If the engineering or principles are the same or similar, there is likely value in what has already been done and building on that or citing those as informative or normative references is a Best Practise in standards development.

There are no medical robots approved for use in the United States but they are getting closer. There are devices referred to as robots but are actually very sophisticated assist devices as they do not have the autonomy required to be a medical robot. But the technology is moving fast and designs and prototypes are on the horizon. All of which will need good standards to help ensure the greatest minds have had the opportunity to think through the essential requirements and principles to ensure patient safety.

IEC SC62A & ISO TC184/SC2 JWG9-Medical electrical equipment and systems using robotic technology (Medical robots) has been set up April 2011 after extensive work within ISO to develop the safety standard for medical robots. ISO has traditionally been responsible for developing robot standards and has expertise on robot safety focussing on industrial robots, whereas IEC has been involved in the standardization of medical equipment safety for many years and has produced a vast library of safety standards (collateral and particular types) for medical electrical equipment, which come under the IEC 60601 family. The medical robot standardization work is in its initial stages, and it is clear that medical robots will be classified and regulated as medical devices rather than as machines. This poses interesting boundary questions which are being pondered within JWG9; these include the following:

- Where is the boundary between medical robots and non-medical robots: many personal care robots can potentially have medical and non-medical applications, for example, assistive exoskeleton type robots can be used for rehabilitation of injured persons (which is clearly a medical application) as well as physically helping a tired person to carry a heavy load (clearly a non-medical application) but there are some applications that are not so clear, namely an assistive exoskeleton for helping the normal mobility of healthy elderly persons.
- Identifying where is the boundary between medical electrical equipment and medical robots: after detailed discussions within the ISO and IEC working groups, the key issue that has been identified here has been the fact that robots can possess a "degree of autonomy", whereas currently IEC standards on medical electrical equipment do not allow any autonomous capability. JWG9 is currently defining what "degree of autonomy" means in a medical

robot setting, how it may be measured and the consequences (in the risk assessment) from having autonomous functions. JWG9 is currently working on developing a collateral standard for medical robots where autonomy will be the main issue to be investigated to identify the additional hazards and the basic safety and essential performance requirements that will be needed.

4. Process Improvement

The medical device standards development process needs to evolve if it intends to keep pace with product development. Three key improvements can be made that will significantly impact the process and the future of standards development:

- **Virtual interactions in developing standards:** This not only reduces travel expenses, but can also allow for more frequent meetings and greater participation. Standards developers have seen the decline of regulators and users/clinicians participating in the standards development process, which is not helpful and creates imbalance. There are many committees developing standards of great importance where only one regulator is at the table. With the limited resources of both government and industry, it has become vitally important that we begin to think differently about how standards get developed. While nothing can replace the face-to-face interactions of expert scientists and policy makers, getting standards completed more quickly is a major concern for the entire community. We no longer have the luxury of taking five or more years to develop a standard. Good standards – with performance criteria wherever possible – are needed, and they are needed in a timely way. This is a challenge of utmost importance to the standards developers and participants for the coming decade.

- **More performance criteria:** In the early days of medical device standards development there was more on performance criteria in standards helping to answer the questions: "Does it do what it is supposed to do?"; and "Can it do it without introducing unexpected harm to the patient?". But industry has marched on, and products have become more complicated and sophisticated, making it more difficult to provide the coveted criteria and/or limits. There has also been concern that too much specificity in standards can contribute to a stifling of innovation and to trade issues. But the pendulum may have swung too far. Regulators need and want good performance criteria and important limits to be included in standards whenever possible. This not only helps the producer, but also can help ensure the consistency in the review process that is so desired by the producer.

- **New generation of participants:** More people need to participate in the process and they need to be under 50 years of age. When we look around the committee rooms, we tend not to see young, energetic, passionate engineers, scientists, clinicians; we tend to see older, but still passionate participants. The value of standards and the standards development processes need to be taught as part of the curriculum of our colleges and universities and those learned individuals need to get involved in the process if standards are going to continue to shape the design and development of new medical technology and ensure a safe environment for patients and the professionals who use the devices.

5. Conclusions

Voluntary consensus standards help make medical devices safer and more effective. Improvements can be made in medical device standards but they are still the best and surest way to universally keep medical devices as safe as possible. The value of the standards cannot be overstated, and if developed in an open, transparent process, will maintain support from regulators, greatly improve medical devices worldwide, and help contain the cost of healthcare. But the benefits of these improvements far exceeds just having regulatory support. The benefits of strong global standards include stronger global participation, better and safer products, and improvements to public health around the world.

References

[1] John G. Abbott, Ph.D., Director, Global Standards, Philips Healthcare.
[2] ISO 10218-1:2011, Robots and robotic devices – Safety requirements for industrial robots – Part 1: Robots
[3] ISO 8373:2012, Robots and robotic devices – Vocabulary.
[4] ISO/DIS 13482, Robots and robotic devices – Safety requirements for service robots – Personal care robots, 2012.

ROBOT ETHICS

M. O. TOKHI
Department of Automatic Control and Systems Engineering
The University of Sheffield, UK

Robots are developed to provide assistance and support to humans in various sectors of life. However, their development and deployment raises ethical questions that need to be addressed, robot ethics has emerged to encourage developments in robotics in such a way that are shared by different social groups and beliefs. From a socio-technical systems perspective three distinct ethical issues that may be considered in robotics: Ethical action of humans through, or with, robots, designing robots to act ethically, and ethical relationships between humans and robots. The former two involve human ethics of designers, manufacturers and users of robots, and the latter involves how humans treat robots and vice versa. While the intention is to increase the human quality of life with new technological developments, which is ethically good, the responsibility in complying with ethical issues is shared among the various groups involved. This paper discusses robot ethics from a moral perspective and how this may be translated into the design and deployment of a robot and further to distribution of responsibility among the parties concerned.

1. Introduction

Humans always face dilemmas of having to choose a certain option to act upon, where each available option will have certain level of moral obligation associated with it. Ethical consideration in practice is to seek means for resolving these issues. There are various schemes that may be adopted for containing these moral deliberations. However, the choice of the scheme, if any, to use is ultimately up to the individual. Robots are technological built systems, and for them to be able to resolve such dilemmas these ethical schemes must be built-in and chosen by designers. It is not clear that even those systems that could learn ethical rules or behaviour, would qualify to make moral decisions autonomously, and the designer of these learning methods would still be responsible for their effectiveness. It might be possible for a robot to develop to a point where its designers and programmers are no longer responsible for its actions, like the parent of a child is not generally held responsible for their actions once they become adults.

Different cultures and religions treat interventions in sensitive fields differently. Examples include reproduction, neural therapy, implantations, and privacy. In some cultures certain groups have fewer rights than others. Such cultural differences will amount to the range of ethical debates in relation to robots. These will lead to the range of ethical specifics within the cultural dimension. Cultural differences also impact on perception of natural versus artificial. An example is attitude of different people towards surgical or organ implantation. In case a robot is designed for the purpose of providing assistance or support to humans, there may be no ethical concern. However, if the robot is intended for use in applications which come in conflict with cultural perception of people then this raises a whole range of ethical issues.

Robots in close contact with humans will have the capacity of gathering data and information about the humans. This will raise questions about privacy and data protection of the humans as these data may be available to others. Moreover, issues related to human safety, health, education, legality of the technology and how to ensure the design is not misused raise ethical questions. Furthermore, technology drivers that may relate to financial gains will raise the question of whether responsibility should be at an individual level or organisational level, or should the responsibility be shared.

2. Moral agency

The newly emerging area of robot ethics raises questions that highlight how robots may be made to behave in morally acceptable ways, and how morality applies in the case of human beings. Since machine morality requires an engineering solution, abstract speculation about the nature of moral action must be translated into a design specification and then implemented in a real, working machine. This will require determining a concrete set of mechanisms that can regulate moral behaviour.

Kant (1981) considers a "moral agent" as a person suspended between inclination and duty, where inclination is derived from desire which, in turn, is always fixed to something outside the self and where duty is determined by pure practical reason as action in conformity with the "universality of a law as such". This suspension does not make the person moral, but rather able to be moral, and success as a moral agent is left to be decided by whether the person follows inclination or duty in a situation where the two diverge.

There are at least three distinct ethical issues that may be considered in robotics (Asaro, 2006):

(a) Ethical action of humans through, or with, robots, i.e. humans acting as ethical agents. This involves the ethics of people who design and use robots.
(b) Designing robots to act ethically, i.e. whether robots will be able to act truly as ethical agents. This involves the ethical systems that are built into robots.
(c) Ethical relationships between humans and robots. This involves the ethics of how people treat robots and how should robots treat humans.

where, the first two above involve the designers, manufacturers and users of the robot. Further ethical questions arising from relationship between humans and robots include, whether it is ethical to create artificial moral agents, or provide sophisticated robots with ethical reasoning capabilities, or create robotic soldiers, police officers, nurses, whether robots should have rights, and whether robots be treated as objects or as subjects.

Robots will need to be considered as socio-technical systems for the above ethical issues to be addressed properly (Asaro, 2006). In this sense one may consider an agency scale from amoral agent to fully autonomous moral agent, where a robot might move gradually along the scale as it acquires greater capabilities and ethical sophistication. Accordingly, while at the lower end of the scale the robot may be considered as "causal agent", where the robot may not be held responsible for its actions, and as it moves to the higher end of the scale it is considered as "moral agent", and hence held responsible for its actions. Sullins (2006) poses three requirements for a robot to be seen as a moral agent; (a) the robot is significantly autonomous from any programmer or operator of the machine, (b) The robot's behaviour can be analysed or explained only by ascribing to it some predisposition or intention to do good or harm, and (c) The robot moral agency requires the robot to behave in a way that shows an understanding of responsibility to other moral agents. It is thus argued that robots adhering to these requirements will have moral rights as well as responsibilities regardless of their status as persons.

Thus, considering robots in a moral agency framework, the distribution of responsibility may best be addressed in a socio-technical systems context through legal theory rather than moral theory. Such an approach will allow preventing robots from doing harm, as well as preventing humans from unjustly avoiding responsibility for their actions.

3. Distribution of responsibility

The primary aim of robot ethics in considering an individual robot should be to develop the means to prevent the robot from doing harm, i.e. harm to people, to itself, to property, to the environment, to people's feelings, etc. This in simple terms means designing robots that do not pose serious risks to people in the first place. As ability and complexity of robots increase, it will become necessary to develop more sophisticated safety control systems that prevent the most obvious dangers and potential harms. Further, as robots become more able to understand and interpret human actions, they will require greater social, emotional, and moral intelligence. For robots that are capable of engaging in human social activities, and thereby capable of interfering in them, it might be expected that they behave morally towards people, e.g. not to lie, cheat or steal, etc., even if people are not expected to act morally towards robots.

Moral responsibility is different from legal responsibility. However, legal responsibility forms a good basis for considering many of the issues in robot ethics. There is no single universally accepted moral theory, and there are only a few generally accepted moral norms (Allen et al., 2000), and while there are differing legal interpretations of cases, and differing legal opinions among judges, the legal system ultimately tends to settle questions of responsibility well in both criminal law and civil law.

Considering established policies for fair distributions and build them into robots, one can look at the construction of robots that follow certain policies as being essentially like the adoption and enforcement of policies in institutions, and can seek ways to challenge them, and hold institutions and robot makers accountable for their policies. The establishment of institutional policies is also a way of isolating individuals from the moral responsibility of taking certain decisions.

4. Service robots

There are numerous societal benefits to the development and deployment of service robots. They are useful in hazardous environments, and in performing domestic tasks in the home. They are used in the agriculture sector, assisting doctors and surgeons, demining tasks and for education and entertainment. However, the use of service robots poses potential risks and ethical issues. Among the various applications, two areas of growing interest, namely robots in personal care for children and the elderly, and robots in military applications, are considered in this section.

Robots in personal care for children and the elderly is a growing area of technology. Several companies have developed child-minding robots that have facilities for video-games, verbal quiz games, speech and face recognition, and conversation. Robotics technology in this sense is developed to contribute to enriching the childhood educational environment. The robots can be controlled by mobile phone or from a PC that allows input from camera and remote talking from caregivers.

Research has demonstrated close bonding and attachment to child-minding robots by children, and, in most cases, children prefer a robot to a toy (Sharkey, 2008). Although short-term exposure may be enjoyable and entertaining, humanly care and attention that is needed is not provided by robots. The children could be left without human contact for many hours a day or perhaps for several days, and this varying degrees of social isolation will have possible psychological impact on the children.

The concern with ageing population relative to the younger caregivers in many countries has motivated the development of sophisticated elder-care robots. These robots can help the elderly to maintain independence in their own homes. However, their presence and over-reliance on them could lead to the risk of leaving the elderly in the exclusive care of the robot, and in social isolation.

The development of robotic soldiers is an area of serious discussion and public debate. Projects on development of semi-autonomous and autonomous weapons systems are well-funded. There are military research projects developing mobile weapons platforms with sophisticated sensory systems, tracking and targeting facilities for selective use of lethal force. The ethical issues that arise in using multiple robots in combat scenarios include the problem of how to discriminate between combatants and innocents in a close-contact encounter. Many of the serious ethical questions posed by such systems have already been contained into the military command and control framework, where commanders are responsible for issuing orders and the soldiers are for carrying out those orders. In cases of war crimes, it is the high-ranking commanders who are usually held accountable, while the soldiers who actually carry out the orders are not held responsible. Accordingly, a conscious effort is being exercised to keep "humans-in-the-loop" of robotic and autonomous weapons systems. However, as with advances in technology the level of sophistication in these systems will increase and they become fully autonomous, so too will the complexity of regulating their actions, and so too will the pressure to design such systems to deal with that complexity automatically and autonomously.

5. Conclusion

Development of robots for application in various sectors of the society brings socio-economic benefits. However, ethical issues that are raised in relation to action of humans, robots and the inter-relation between humans and robots need to be addressed by the robot designer, manufacturer and user of the robot within a socio-technical systems framework. This is to take account of a balanced approach between risks posed and benefits gained from the technology.

References

Allen, C., Varner, G. and Zinser, J. (2000). Prolegomena to any future artificial moral agent, *Journal of Experimental and Theoretical Artificial Intelligence*, 12, pp. 251-261.

Asaro, P. M. (2006). What should we want from a robot ethic? *International Review of Information Ethics*, 6 (12), pp. 9 – 16.

Kant, I. (1981). *Grounding for the metaphysics of morals*. (J. W. Ellington, Trans.). Indianapolis: Hackett Publishing Company. (Original work published in 1785)

Sharkey, N. (2008). The ethical frontiers of robotics. *Science*, 322, pp. 1800 – 1801.

Sullins, J. P. (2006). When is a robot a moral agent? *International Review of Information Ethics*, 6 (12), pp. 23 – 30.

AUTHOR INDEX